Breeding and Biotechnology of Flowers

VOLUME – 1: Commercial Flowers

Anil K. Singh
Professor and Head
Department of Horticulture
Institute of Agricultural Sciences
Banaras Hindu University
Varanasi, Uttar Pradesh, India

NEW INDIA PUBLISHING AGENCY
New Delhi – 110 034

NEW INDIA PUBLISHING AGENCY
101, Vikas Surya Plaza, CU Block, LSC Market
Pitam Pura, New Delhi 110 034, India
Phone: + 91 (11)27 34 17 17 Fax: + 91(11) 27 34 16 16
Email: info@nipabooks.com
Web: www.nipabooks.com

Feedback at feedbacks@nipabooks.com

ISBN: 978-93-83305-35-3 (Volume -1)

ISBN: 978-93-83305-40-7 (Volume -2)

ISBN: 978-93-83305- 61-2 (Set)

DEDICATED
TO MY BELOVED FATHER

LATE SHRI L.P. SINGH

WHO TRIED HIS LEVEL BEST TO SHAPE ME.

फाउण्डेशन फॉर एडवॉन्समेण्ट ऑफ एग्रीकल्चर एण्ड रूरल डेवलपमेन्ट
FOUNDATION FOR ADVANCEMENT OF AGRICULTURE AND RURAL DEVELOPMENT

प्रो. पंजाब सिंह
Prof. Panjab Singh
FNAAS, FIAS, FNIE
President
Ex. Secretary DARE GOI & DG ICAR
Ex. Vice Chancellor, BHU

Foreword

Flowers symbolize beauty, purity, peace and love. These are intricately associated with social beliefs and no social function is complete without the use of flowers especially in a country like India. From a symbol of love, flowers have transformed into an industry, generating both income as well as employment. In present scenario floral industry is an energetic, global fast growing industry which has achieved eloquent growth rates during the past few decades. The globally important flower crops are Rose, Chrysanthemum, Tulip, Lilium, Gerbera, Carnation, Anthurium, Cyclamen, Freesia, Orchids (Cymbidium, Dendrobium, Cattleya), Gladiolus, whereas, crops like Alstroemeria, Lisianthus, Gypsophila, Heliconia, Cyclamen, Zantedeschia, Kangaroo Paws, etc. are new emerging important flower crops. Marigold, Jasmine, Tuberose, China aster and Crossandra are commercial flowers of India. Since, cultivation of flowers are more pragmatic endeavour than other crops provide environmental security for immediate living surroundings, thus intensive research in floriculture particularly on the crops were initiated earlier. Higher education in Floriculture and Landscaping has been initiated in various universities and institutes at M.Sc. (Ag.) and Ph.D. levels. Breeding programmes on flowers and ornamental plants have been taken up in many ICAR institutions and Agricultural Universities. The genetic base of a number of floricultural crops has been broadened and many improved cultivars have been developed despite the fact that breeding is relatively ponderous in many floriculture crops. Inspite of such development, varieties developed of commercially important flowers are under Plant Breeder's Right Act and parentages are not disclosed to save the intellectual property right. Good literature on breeding of commercial flowers is

Saraswati Kunj, Narainpur (Dafi) Varanasi - 221005 • Phone : 0542 - 2575678 (O)
E-mail : faardvaranasi@gmail.com • Website : www.faard.org

scanty and hardly less than half dozen books are available worldwide. Therefore, the need for a compilation of good book on this aspect is being felt by everyone.

This book "Breeding and Biotechnology of Flowers, Vol. I: Commercial Flowers" will vividly highlight genetical and breeding application in flower crops covering wide range of aspects. The present book contains information on breeding of globally important and new emerging commercial cut and loose flower crops, potted and dry flower crops. The author provides crop-specific origin and distribution, species, biology, genetics, breeding objectives, state-of-the-art breeding/genetics, classical breeding, interspecific hybridization, mutation, molecular breeding and genetic engineering. The present publication has been organized in such a way that it will be useful to everyone interested in the flowers and ornamental plants particularly to students, flower breeders, hobbyists, scientists and amateur gardeners.

I personally congratulate and appreciate the efforts of Dr. Anil K. Singh, Prof. & Head, Department of Horticulture, Institute of Agricultural Sciences, Banaras Hindu University, Varanasi to compile the valuable information in the form of book "Breeding and Biotechnology of Flowers, Vol. I: Commercial Flowers". I am sure that this publication will be very useful to all, who are concerned with floriculture.

Panjab Singh

Saraswati Kunj, Narainpur (Dafi) Varanasi - 221005 • Phone : 0542 - 2575678 (O)
E-mail : faardvaranasi@gmail.com • Website : www.faard.org

Preface

Floriculture is getting attention globally due to change in life style of people. Floriculture industry is expanding steadily and continuously. Commercial floriculture is based on cut flowers, loose flowers, saplings and hybrid seeds world-wide. In this book an effort is made to compile the fundamentals of flower breeding alongwith achievements made in the improvement of commercial flowers in India and abroad. The book "Breeding and Biotechnology of Flowers, Vol I: Commercial Flowers" dealt fundamentals chapter History of Flower Breeding, Conventional Breeding Methods, Mutation and Polyploidy Breeding, Molecular and Transgenic Breeding, F_1 Hybrid Seed Production in Ornamental Crops, Intellectual Property Rights alongwith breeding of 28 commercial crops i.e. Alstroemeria, Anthurium, Antirrhinum, Carnation, China aster, Chrysanthemum, Crossandra, Cyclamen, Freesia, Gerbera, Gladiolus, Gypsophila, Heliconia, Jasmine, Kangaroo Paws, Lilium, Lisianthus, Marigold, Rose, Stock, Tuberose, Tulip, Zantedeschia and Orchids (Cattleya, Cymbidium, Dendrobium, Paphopedilum, Vanda). Various breeding problems including pre and post fertilization barriers discussed in length with the techniques followed to overcome from it and evolved improved and outstanding varieties for global market. Efforts made by various workers/scientists have been compiled by citing about 2252 references. Achievements have also been depicted in 162 Tables. I trust this book will be helpful to the M.Sc. & Ph.D. students, scientists, teachers, breeders, hobbyists and gardeners.

I feel immense pleasure to express my heartfelt gratitude to Dr. Panjab Singh, Ex-V.C., BHU, Varanasi, JNKVV, Jabalpur and Secretary (DARE) & DG (ICAR) for his inspiring guidance, encouragement and blessings. Author is grateful to Dr. Lalji Singh, V.C., BHU, Varanasi for encouragement and support. I have great pleasure in acknowledging moral support, encouragement and valuable critical criticism received from Prof. Ravi P. Singh, Director, & Plant Breeder, Institute of Agricultural Sciences, BHU, Varanasi.

I must pay gratitude to my Guruji Dr. Chetan Singh and Dr. O.P. Chhonkar, who always inspired me to write papers and books. Support and help received from Late Prof. J.N. Singh, Ex- Head Horticulture, Prof. A.N. Maurya, Ex-Director (I. Ag. Sci.), Prof. B.P. Singh, Prof. M.M. Syamal, Prof. S.P. Singh, Prof. B.K. Singh, Dr. Anand K. Singh and Dr. A.K. Pal, Department of Horticulture, Prof. A.K. Singh, Head, Department of Extention Education, Prof. S.K. Singh, Department of Soil Science & Agril. Chemistry, Institute of Agricultural Sciences, BHU, Varanasi are duly acknowledged.

I wish to express my appreciation for my Ph.D. students Ms. Anjana Sisodia and Dr. Shiva Jauhari, Ex-Marigold Breeder, Syngenta India Ltd., Pune for their dedication in collection of literature, for compilation and some valuable suggestions regarding the manuscript.

The author is thankful to Dr. J.P. Tiwari, Ex-Dean & Registrar, Prof. Santosh Kumar, Prof. K.K. Misra, GBPUA&T, Pantnagar, Prof. V.B. Singh, Ex-Dean Agriculture, Nagaland University, Prof. D.B. Singh, Ex- Dean Agriculture, SHIATS (Allahabad Agril. Institute), Allahabad, Dr. R.L. Misra, Ex- Coordinator, AICRP on Floriculture, Dr. S.K. Singh, Dr. T. Janakiram, Dr. Kanwar P. Singh, Dr. Krishna P. Singh, Dr. K.V. Prasad, Dr. A.K. Tiwari, IARI, New Delhi, Dr. F.U. Khan, SKUAS&T, Srinagar for their encouragement and providing materials related to floriculture.

I also appreciate the efforts made by my Ph.D. students Dr. Anuj Kumar, Ms. Jaymala Singh and Mr. R.S. Reddy and M.Sc. (Ag.) students Ms. Raimani Hembrom, Ms. Hemlata and Ms. Asmita in collection of literature and compilation. I appreciate the efforts of the staff of M/s New India Publishing Agency, New Delhi for printing the book in the nice form.

I don't have words to express my emotional feeling towards my parents and relatives Late Prof. S.P. Singh (Canada), Mrs. Kamala Singh, Dr. D.K. Singh & Mrs. Kiran Singh, Dr. Manoj & Tanu (USA), Praveen & Atishi, Rahul & Nimesh for their encouragement and support. I must express my appreciation for my wife Leela, brother Ashok, daughter Suchi and son Praveen who have been affectionate, understanding and patient during the long hours I spent on the manuscript.

Anil K. Singh

Contents

Section III: Cut and Loose Flower Crops

Section V: Cut and Dry Flower Crops

Section VI: Loose Flower Crops

Section VII: Orchids

Section VIII: Colour Plates

Glossary

A

Abaxial: The side or surface located away from or on the opposite side of the axis, as lower surface of a leaf dorsal.

Acclimatization: Acclimatization is the ability of crop or variety to become adapted to new climate and soil conditions.

Achene: A dry one-seeded indehiscent fruit with the seed distinct from the fruit wall. It may be smooth, as in the buttercup, or feathery, as in Clematis.

Acropetal succession: It is floral arrangement to form bunch of flowers in which flowers are arranged so that new flowers are at top and old flowers are at bottom such system is found in racemose type of arrangement. e.g. Gladiolus, Gulmohar.

Actinomorphic: Capable of being divided into equal halves along any diameter, as the flowers of the rose or tulip.

Adaxial: The side or surface facing or nearest to the axis of an organ, such as the upper surface of a leaf; ventral.

Additive gene: A form of allelic interaction in which there is no dominance.

Adventitious bud: The bud occurring elsewhere on the plant (such as on trunks, roots or leaves) except from the leaf axils, stem tips and plant apices.

Alba: White form of a flower.

Alien gene: A gene transferred to the cultivated species from another related or unrelated species.

Alleles: It is one of a number of alternative forms of the same gene or same genetic locus (generally a group of genes).

Alloploid: A hybrid individual or cell with two or more sets of chromosomes derived from two different ancestral species; depending on the number of multiples of haploid sets, alloploid.s are referred to as allodiploids, allotriploids, allotetraploids, allopentaploids, allohexaploids, etc.

Amphidiploids: A plant having the sum of the chromosome numbers of two parental species, owing to the doubling of the chromosomes in a hybrid of two species.

Anaphase: It is the fourth phase of mitosis, the process that separates the duplicated genetic material carried in the nucleus of a parent cell into two identical daughter cells.

Androecium: The androecium is the male reproductive whorl of the flower. It is composed of stamens. Each stamen consists of filament, anther and connective.

Anther: The pollen producing organ of the flower.

Anther cap: In orchids, a cap of tissue that covers the clusters of pollen.

Anthesis: The period during which a flower is fully open and functional.

Anthocyanin: Class of water-soluble pigments

usually found in the vacuoles of plant cells; anthocyanins are often blue, pink, or yellow.

Apetalous: Having no petals.

Asynchronous: Asynchronous refers to things that do not occur at the same time. As asynchronous flowering refers to the phenomenon of consecutive flowering of all population of a given species in a large region.

Autopolyploid: An organism having more than two sets of chromosomes, all of which were derived from the same species.

Autotetraploid: An individual or strain whose chromosome complement consists of four copies of a single genome due to doubling of an ancestral chromosome complement.

Axil: The angle formed between a leaf stalk and the stem to which it is attached. In flowering plants, buds develop in the axils of leaves.

Axis: The longitudinal support on which organs or parts are arranged, the stem and root, the central line.

B

Back bulb: Pseudobulb that has completed its growth cycle and become dormant. It serves as a storage organ and may bear a dormant bud. Under favourable condition, this bud may develop into a new plant if the pseudobulb is severed from the plant.

Band: A thin strip of flexible material used to encircle and bind one object or to hold a number of objects together.

Bicarpellary: An ovary having two carpels.

Bigeneric: Derived from parents of two different genera.

Biotechnology: Manipulation of organisms or their components to make useful product.

Bisexual: Presence of both male and female parts in the same flower.

Bivalent: During the prophase of meiosis I, homologous chromosomes pair and form synapses. The paired chromosomes are called bivalents. The bivalent has two chromosomes and four chromatids, with one chromosome coming from each parent.

Bizarrers: When the under-colour is yellow and marking red to brown of tulip flower they called bizarres.

Blossom: The flower of a plant, especially of one producing and edible fruit or the state of flowering.

Bracteole: A leaflike organ subtending a flower in an inflorescence that is itself subtended by a bract.

Bracts: A modified leaf growing just below a flower or flower stalk.

Brochi: Any of the fine, thin-walled, tubular extensions.

Broken: Mixed colour of tulip flowers called broken.

Bud: A small protuberance on a stem or branch, sometimes enclosed in protective scales and containing an undeveloped shoot, leaf or flower.

Bybloems: When the under-colour is white and markings lilac or purple of tulip flowers they are called bybloems.

C

Calliclone: A colony derived from a single plant cell that has been removed from a suspended culture and plated out in a semisolid medium.

Callosity: A hardened or thickened part of a plant or a callous condition.

Calyx: The whorl of sepals of a flower collectively forming the outer floral envelope or layer of the perianth enclosing and supporting the developing bud; usually green.

Campanulate: Bell-shaped, as a corolla.

Capitula: Any globose or knoblike part, as a flower head.

Capsule: A dry dehiscent fruit that develops from two or more united carpels.

Carpel: One of the structural units of a pistil, representing a modified, ovule-bearing leaf or a simple pistil or a single member of a compound pistil.

Centripetal: Developing or progressing inward toward the centre or axis, as in the head of a sunflower, in which the oldest flowers are near the edge and the youngest flowers are in the centre.

Chiasma: A cross-shaped structure commonly observed between nonsister chromatids during meiosis; the site of crossing-over. X-shaped configuration seen in tetrads during the latter stages of prophase I of meiosis. They represent physical crossovers between DNA molecules.

Chimera: A single plant organism with genetically distinct cells from two different zygotes or from a mixed-species callus.

Chromogen: A compound that forms coloured compounds on oxidation.

Chromosome: A thread like strand of DNA in the cell nucleus that carries the genes in a linear order.

Clone: A group of organisms or cells of the same genetic constitution that are descended from a common ancestor by asexual reproduction, as by cuttings, grafting, etc., in plants.

Cohesion of stamens: Cohesion is defined as the fusion of the members amongst the same whorl, confined to different parts of stamens only.

Cohesion: Fusion among stamens in a flower.

Colchicine: A pale yellow crystalline alkaloid extracted from seeds or corms of the autumn crocus. It is used in the treatment of gout and to create polyploidy plants by inhibiting chromosome separation during meiosis. Formula: $C_{22}H_{25}NO_6$.

Coloumn: A long structure in a flower, such as that of an orchid, consisting of the united stamens and style.

Combining ability: It is the capacity of an individual to transmit superior performance to its offspring. It is the phenomenon with which inbred lines when crossed give rise to hybrid vigour.

Composite varieties: A composite variety is a plant population in which at least 70% of its progeny result from the crossing of the parent lines.

Concave: Curving inwards or having one or two surfaces curved or ground in the shape of a section of the interior of a sphere.

Connective: It is a patch of tissues connecting the parallel anther lobes.

Convex: Curving out or bulging outward, as opposed to concave.

Corolla: The petals of a flower considered as a group or unit and usually of a colour other than green; the inner whorl of the perianth.

Cosuppression: An alternative term for RNA silencing which probably evolved as part of plants. defence against viral attack, but which has become important in the context of plant transgenesis.

Cross: The progeny that result from transferring pollen from one plant to the flower of another; the act itself.

Cross pollination: Transfer of pollen from the anther of a flower to the stigma of a flower of another plant.

Cryopreservation: Preservation of seeds of pollen by immersing them in liquid nitrogen.

Cultivar: A race or variety of a plant that has been created or selected intentionally and maintained through cultivation.

Cuticle: In plant cuticles are a protective waxy covering produced only by the epidermal cells of leaves, young shoots and all other aerial plant organs without periderm.

Cybrid: A hybrid cell produced by introducing nuclear material from one organism into cell of another organism from which nucleus has been removed.

Cyme: An inflorescence in which the primary axis bears a single central or terminal flower that blooms first.

Cytogenetics: The branch of biology that deals with heredity and the cellular components, particularly chromosomes, associated with heredity.

Cytomixis: Extrusion of chromatin from one cell into the cytoplasm of an adjoining cell.

Cytoplasm: The protoplasm of a cell

contained within the cell membrane but excluding the nucleus: contains organelles, vesicles and other inclusions.

D

Deciduous: The trees or shrubs shedding all leaves annually at the end of the growing season and then having a dormant period without leaves.

Dehiscence: The spontaneous opening at maturity of a plant structure, such as a fruit, anther or sporangium to release its contents.

Determinate: Growth in which the axis, or central stem, being limited by the development of the floral reproductive structure, does not grow or lengthen indefinitely.

Diadelphous: The filaments are united into two bundles, anthers remaining free. e.g., Sweet pea Pisum.

Diallel cross: A diallel cross is a mating scheme used by plant breeders or geneticists to investigate the genetic underpinnings of quantitative traits.

Dichogamy: Having pistils and stamens that mature at different times thus promoting cross-pollination rather than self-pollination.

Dicotyledonous: Belonging or pertaining to the Dicotyledoneae and having two cotyledons.

Didynamous: Having four stamens in two pairs of unequal length.

Diploid: The state of having each chromosome in two copies per nucleus or cell. A cell or an organism having two chromosome sets or twice the haploid number, or an individual having two chromosome sets in each of its cells. This can be seen by the presence of 2 of each type of chromosome in a cell nucleus at interphase.

Diplontic selection: Competition among diploid cells within multicellular organisms. This can take place only when dominant mutation or other dominant-acting change(s) occur.

Disc florets: Florets in centre of chrysanthemum inflorescence, conspicuous in daisy or anemone-flowered types.

Distal: Situated farthest from the centre, median line, or point of attachment or origin.

Domestication: The process of bringing the wild species under cultivation is called domestication.

Dominance: Ability of an allele to express itself in heterozygous state.

Dorsal: Pertaining to the surface away from the axis, as of a leaf, abaxial.

Dorsifixed: Describing an anther that is joined to the filament for some distance along its dorsal edge.

Dysgenesis: The state of being unable to produce offspring.

E

Ecogeographic: Of or pertaining to ecogeography, the study of the geographical distribution of living things in terms of ecology.

Emasculation: Emasculation is the process of removing anthers from bisexual flowers without affecting the female reproductive part (pistil), which is used in various plant hybridization techniques.

Embryogenesis: Morphological and physiological development of the embryo from the embryoless seeds.

Embryoid: A mass of plant tissue that resembles an embryo.

Endosperm: Nutritive material in seeds of the Anthophyta, formed from the triple fusion nucleus.

Epicalyx: A calyx-like extra whorl of floral appendages, positioned below the calyx. The individual segments resemble sepals and are termed episepals.

Epipetalous: When stamens adhere to petals, it is called epipetalous condition.

Epiphyte: Plant which naturally grows upon another plant or other means of support above the earth.s soil but is not parasitic; deriving it.s needed moisture from the air.

Epistasis: An interaction between nonallelic genes, especially an interaction in which one gene suppresses the expression of another.

Ethological: The scientific study of animal behavior, especially as it occurs in a natural environment.

Explant: To remove (living tissue) from the natural site of growth and place in a medium for culture.

***Ex situ*:** Out of normal or natural habitat and away from the indigenous place of origin.

Eye: Undeveloped bud growth which will ultimately produce new growth.

F

Fertile: Bearing functional reproductive structures such as seeds or fruit or material such as spores or pollen.

Fertilization: The union of male and female gametes to form a zygote.

Filament: The filament is a slender stalk of the stamen which bears anthers.

Filiform: Having the form of a thread.

Flecks: Mark or dot with small patches of colour or particles of something.

Floriferous: Bearing or capable of bearing many flowers.

Florigen: Hypothesized hormone-like molecules responsible for controlling and/ or triggering flowering in plants. Florigen is produced in the leaves and acts in the shoot apical meristem of buds and growing tips.

Flower: The reproductive structure of some seed-bearing plants, characteristically having either specialized male or female organs or both male and female organs, such as stamens and a pistil, enclosed in an outer envelope of petals and sepals.

Fruit: The ripened ovary or ovaries of a seed-bearing plant, together with accessory parts, containing the seeds and occurring in a wide variety of forms.

G

Gamete: A reproductive cell having the haploid number of chromosomes, especially a mature sperm or egg capable of fusing with a gamete of the opposite sex to produce the fertilized egg.

Gametogenesis: The process in which cells undergo meiosis to form gametes (male or female). The development and production of the male and female germ cells required to form a new individual.

Gametophyte: The haploid stage of a plant life cycle that produce gametes (by mitosis).

Gamosepalous: Having the sepals united.

Geitonogamous: Stamens and pistil both do not mature simultaneously in the individual flowers. The stigma of the individual flower unfolds after the pollen is discharged from the flower.

Geitonogamy: Pollination of a flower with the pollen from another flower on the same flowering plant.

Gene: Basic unit of hereditary consisting of a sequence of DNA that occupies a specific location on a chromosome and determines a particular characteristic in an organism or a segment of DNA, occupying a specific place on a chromosome, that is the basic unit of heredity.

Genetic variability: The potential for a given characteristic or genotype to vary within a population when faced with a particular influence. As the genetic variability of a population increases, so does its resistance to environmental and genetic influences and ultimately extinction. Consequently, genetic variability is directly tied to biodiversity and evolution.

Genome: The complete genetic information (either DNA or, in some viruses, RNA) of an organism, typically expressed in number of basepairs.

Genomic: Of or pertaining to a genome.

Genus: A taxonomic category ranking below a family and above a species and generally consisting of a group of species exhibiting similar characteristics.

Germ cell: Reproductive cell which, when united with a germ cell of the opposite sex, develops into a new individual.

Germ tube: Short, juvenile hyphae arise from certain types of spores upon germination.

Germination: Resumption of active metabolism by a spore or seed, rupture of its outer wall and beginning of renewed growth.

Gigantism: The condition in which a plant is far larger than normal in size.

Gynoecium: The female reproductive organs of a flower; the pistil or pistils considered as a group.

Gynogenesis: A type of reproduction by parthenogenesis that requires stimulation by a sperm to activate the egg into development but occurs without fusion of sperm and egg nuclei.

Gynostemium: The united stamens and pistil of an orchid.

H

Haploid: Cell having only one complete set of chromosomes, ordinarily half the normal diploid number.

Haplotype: A group of alleles that are transmitted together.

Hermaphrodite: A flower having both male and female reproductive organs.

Heterobeltiosisis: Increased vigour in fitness traits of F1 hybrids of two genetically dissimilar parents over the better parent, while heterosis is over the mid parents (mean of both parents).

Heterochromatic: Having a pattern of mixed colours.

Heterogenesity: The quality or state of being heterogeneous.

Heteromorphic: A pairs of homologous chromosomes differing from each other in size or form.

Heterosis: The increase in growth, size, yield or other characters in hybrids over those of the parents.

Heterozygotes: A hybrid containing genes for two unlike forms of a characteristic, and therefore not breeding true to type.

Heterozygous: Having different alleles at one or more corresponding chromosomal loci.

Homeosis: It is the transformation of one body part into another, arising from mutation in or misexpression of specific developmentally critical genes.

Homoeologous: The term also spelled homeologous, is used to describe the relationship of similar chromosomes or parts of chromosomes brought together following interspecies hybridization and allopolyploidization, and whose relationship was completely homologous in an ancestral species.

Homogenesity: The quality or state of being homogeneous.

Homologous chromosomes: Also called homologs or homologous are chromosome pairs of approximately the same length, centromere position and staining pattern with genes of the same characteristics at corresponding loci.

Homology: A fundamental resemblance based on common descent.

Homozygotes: An organism with identical pairs of genes with respect to any given pair of hereditary characters, and therefore breeding true for that character.

Homozygous: Having the same alleles at a particular gene locus on homologous chromosomes.

Hybrid: Resulting progeny from the union of two different species (known as primary hybrid), or of a species and a hybrid, or of two hybrids (known as a complex hybrid).

Hypanthium: An enlargement of the floral receptacle bearing on its rim the stamens, petals, and sepals and often enlarging and surrounding the fruits as in the rose hip.

Hypocotyl: The part of a plant embryo or seedling that lies between the radicle and the cotyledons. Upon germination, the hypocotyl pushes the cotyledons above the ground to develop. It eventually becomes part of the plant stem.

Hypogynous: A flower having the floral parts such as sepals, petals and stamens, borne on the receptacle beneath the ovary, as in the buttercup.

I

Ideogram: Any graphic sign or symbol.

Impair: To cause to diminish, as in strength, value or quality.

Imperfect flowers: Flowers that only have one set of reproductive organs, so they are considered to be either male or female. Flowering plants can fall into two categories of imperfect flowers- monoecious or dioecious.

***In vitro*:** In an artificial environment.

***In vivo*:** Within a living organism.

Incompatibility: Sexual reproduction, it is the inability of the gametes to unite.

Incomplete flower: Flowers lacking any of the four whorls or parts (sepals, petals, stamens, or pistils).

Indehiscent: Indehiscent is a term used to describe fruits that do not open or split spontaneously to release their seeds, or anthers that do not open to discharge pollen.

Infructescences: The fruiting stage of an inflorescence. In other words, Infructescence is the ensemble of fruits derived from the ovaries of an inflorescence. It usually retains the size and structure of the inflorescence.

Indeterminate growth: Growth in which a plant axis is not limited by development of a terminal flower bud or other reproductive structure and so continues to elongate indefinitely (as in racemose inflorescence).

Indigenous: Native to a country or region.

Inflorescence: A characteristic arrangement of flowers on a stalk.

Infructescence: The fruiting stage of the inflorescence.

Inheritance: The process of genetic transmission of characteristics from parents to offspring.

Intergeneric: Occurring between different genera.

Interracial: Conducted, involving, or existing between different races or ethnic groups.

Interspecific: Arising or occurring between species.

Intervarietal: Hybridization between two varieties of the same kind or form.

Intrageneric: Within a genus.

Intrastylar pollination: It is the method used to overcome the self and cross incompatability in hybrids and species of lilies in which the pollens are placed at the stylar canal.

Introgression: Crossing of two species producing a fertile hybrid, which backcrosses for several generations with one of the parent species. In this way, genetic material and characteristics is transferred from one species to another.

Isolation: A state of separation between two plant varieties.

Isomorphic: Being of identical or similar form, shape, or structure or having sporophytic and gametophytic generations alike in size and shape.

K

Karyokinesis: Nuclear division in mitosis or meiosis.

Karyotype: The characterization of the chromosomal complement of an individual or a species, including number, form and size of the chromosomes.

Keiki.s: Occasionally an orchid plant will bear a little plantlet off of its flower stem or pseudobulb. These little baby plants are often called Keiki.s. Keiki are seen more frequently on Phalaenopsis, Epidendrum and Dendrobium than on other orchids.

L

Labellum: The often enlarged petal of an orchid flower.

Land races: An ancient or primitive cultivated variety of a crop plant.

Ligula: Pl. ligulae. A strap-shaped or tongue like structure, especially a mouth part in certain insects.

Ligulate flower: A species of compound flower, the florets of which have their corollets flat, spreading out toward the end, with the base only tubular.

Ligulate: Having the shape of a strap.

Limb: Any of the main branches arising from the trunk or a bough of a tree.

Line: A group of identical pure-breeding diploid or polyploid organisms, distinguished from other individuals of the same species by some unique phenotype and genotype.

Linkage: Genes to be found on the same chromosome are not randomly assorted but tend to be inherited together.

Lip: Orchid labellum.

Loci: The plural of locus.

Locular: Having, formed of, or divided into small cavities or divided into compartments by septa the locular ovary of a plant.

Locus: The position of a gene on a chromosome.

Luxuriance: Luxuriance is the increased vigour and size of specific hybrid.

M

Malformed flower: Faultily or anomalously formed flower.

Meiosis: Two cell divisions, occurring one after the other following only one round of DNA replication, shows chromosome pairing and chromosome separation in the first division and chromatid separation in second, leads to the formation of haploid daughter cells.

Mericlone: An artificially produced clone of an orchid which is deemed to be of significant quality to merit this kind of replication. They are produced by taking a small piece of tissue from the specimen plant, mincing the piece into tiny pieces and then growing the tissue in a laboratory into multiples of identical plants.

Meristem: The undifferentiated plant tissue from which new cells are formed as that at the tip of a stem or root.

Micropropagation: Making new plants by any of the laboratory techniques, including meristem tissue propagation and sterile seed culture.

Mitochondria: Subcellular double membrane organelles surrounding a matrix, with the inner membrane folded into finger-like projections called cristae, found in eukaryotes. Mitochondria are self replicating and contain DNA.

Mitosis: The process in cell division by which the nucleus divides, typically consisting of four stages, prophase, metaphase, anaphase, and telophase, and normally resulting in two new nuclei, each of which contains a complete copy of the parental chromosomes. Also called *karyokinesis.*

Monoadelphous: Stamens whose filaments are united into a single tube like group. The stamens of flowers of leguminous plants are often monoadelphous.

Monochasium: A cymose inflorescence in which each branch gives rise to one other branch only.

Monocot: Member of the division anthophyta, class liliopsida. This class is characterized by having one cotyledon.

Monoecious: Flowering plants having the male and female reproductive organs in separate flowers on the same plant.

Monogenic: Pertaining to or influenced by a single gene.

Monosomic: A diploid cell missing a single chromosome. A cell or individual that is basically diploid but that has only one copy of one particular chromosome type and thus has chromosome number 2n-1.

Morphogenesis: The development of form and structure in an organism during its growth from embryo to adult or the evolutionary development of form in an organism or part of an organism.

Morphology: The branch of biology that deals with the form and structure of organisms.

Multipetalled: Having numerous petals.

Multiple allele: An allele of a genetic locus having more than two allelic forms within a population.

Mutation: A sudden departure from the parent type in one or more heritable characteristics, caused by a change in a gene or a chromosome.

N

Native plant: Plants that occurs and grows naturally in a specific region or locality.

Natural Hybrid: Hybrid that occurs in the wild without the help of humans.

Nectar: It is a sweet liquid made in special glands called nectaries in flowering plants. Located deep within a flower, at the base of its petals.

Nectariferous: Plants having nectar glands in flower or other parts.

Non-additive gene: Of, relating to, or being a genetic effect that is not additive.

Non-recurent parent: In backcross method, the parent from which one or few genes are transferred to the recurrent parent are called donor parent.

Nucleolus: A small, typically round granular body composed of protein and RNA in the nucleus of a cell. It is usually associated with a specific chromosomal site and involved in ribosomal RNA synthesis and the formation of ribosomes.

O

Octaploid: Plant cell having eight haploid sets of chromosomes.

Outbreeding: The breeding of distantly related or unrelated individuals, often producing a hybrid of superior quality.

Ovary: It is the part of the pistil which holds the ovule(s) and is located above or below or at the point of connection with the base of the petals and sepals.

Ovule: Ovule literally means .small egg.. The portion of ovary that contains the embryo sac and egg cell, which after fertilization develop into a seed.

P

Pachytene: The third stage of the prophase of meiosis during which the homologous chromosomes become short and thick and divide into four distinct chromatids.

Palate: A projection in the throat of such flowers- e.g. the snapdragon.

Panicle: A branched indeterminate inflorescence in which the branches are racemes, so that each flower has its own stalk (called a pedicel) attached to the branch.

Pantaporate: Type of pollen.

Papilionoid: Butterfly-like.

Papillae: A minute projection on the surface of a stigma, petal or leaf.

Paralogous gene: Two genes or cluster of genes at different chromosomal location in the same organism that have structural similarities indicating that they derived from a common ancestral gene and have since diverged from the parent copy by mutation and selection or drift.

Parent: A plant that has produced one or more plant similar to itself.

Particle bombardment: Particle bombardment is a technique for introducing foreign DNA into a cell culture.

Pedicel: The stalk bearing a single flower of an inflorescence.

Pedicellate: Also called as peduncle, any short stalk bearing an organ or organism.

Pedigree: A record of ancestry.

Peduncle: A primary flower stalk, supporting either a cluster or a solitary flower.

Pentamerous: Flowers having five members in each whorl.

Perfect flower: A flower having both stamens and pistils.

Perianth: The floral structure comprised of the calyx and corolla especially when the two whorls are fused.

Petal: A division of the corolla; one of a circle of modified leaves immediately outside the reproductive organs, usually brightly coloured.

Petaloid: Having the form or appearance of a petal.

Petalous: Flower having petals.

Phenotype: The expressed characteristics of an organism. The phenotype is determined by an interaction between the environment and the genotype and between dominance and epistatic relationships within the genotype.

Phyllotaxis: It is the arrangement of leaves on a plant stem.

Phylogeny: Phylogeny, the history of the evolution of a species or group, especially in reference to lines of descent and relationships among broad groups of organisms.

Pistil: The female, ovule-bearing organ of a flower, including the stigma, style, and ovary.

Plastid: A double membrane organelle found in the cytoplasm of the plant cells. Many plastids, e.g. amyloplasts and elaioplasts, are storage organelles. Chloroplasts contains the chlorophylls, carotenoids, electron transfer compounds, and enzymes.

Plastome: It is the genetic material that is found in plastids in plant cells (for example in the chloroplast). It composes part of the entire genome of photosynthetic organisms.

Plastome: The part of the genome of a photosynthetic plant found in its plastid.

Ple A gene: Pleiotropic gene.

Pod: A superior one or many seeded fruit of two valves which split open.

Pollen parent: The plant that contributes male reproductive material, the pollen.

Pollen tube: Structure formed by germinating pollen that absorbs nutrients, and in some groups carries the sperm cells to the eggs.

Pollen: It is a fine to coarse powder containing the microgametophytes of seed plants, which produce the male gametes (sperm cells).

Pollination: Transfer of pollen from the stamen (male part of the flower) to the pistil(female part of the flower), which results in the formation of a seed.

Pollinator: Agent of pollination.

Pollinium: (Pl. pollinia) A mass or packet of pollen grains specialized for transfer to other flowers as a unit by pollinating insects. Orchids and milkweeds produce pollinia.

Pollinizer: Plant in fruit plantation which provides pollen for self-unfruitful varieties.

Polyadelphous: The filaments are united to form many bundles, anthers being free, e.g., Bombax, etc.

Polyandrous: The stamens are free from one another. Anthers and filaments remain completely free. e.g. papaver, ranunculus, etc.

Polymorphic: Population having more than two distinct morphological variants. **Polymorphism:** Two or more clearly different phenotypes exist in the same population of a species or in other words, the occurrence of more than one form or morph.

Polyploidy: The condition of having three, four or more sets of chromosomes instead of the two present in diploids.

Predominance: The state or condition of being greater in number or amount.

Primary introduction: Introduction of new crops is called primary introduction.

Primordium: The first recognizable, histologically differentiated stage in the development of an organ or a part.

Progeny: The immediate descendant or descendants of a plant.

Proginator: Someone who gives rise to a family line.

Protandrous: Flower in which the anthers release their pollen before the stigma of the same flower is receptive.

Protandry: The maturation of the male reproductive organs before those of the female. For example, in many members of the Compositae and Leguminosae the pollen is released from the anthers before the stigma in the same flower is receptive.

Protoclone: Regenerated plant derived from protoplast culture or a single colony derived from protoplasts in culture.

Protocol: The established code of procedure or behavior in any group, organization or situation.

Protocorm: The tuber-like swollen part of an orchid seed; the first stage of growth after germination.

Protogynous: A flower in which the stigma is receptive before the pollen is shed from the anthers of the same flower.

Protoplast: Cell without cell wall, cell wall completely or partially removed using either mechanical or enzymatic means.

Pseudobulb: An enlarged, aboveground portion of stem, present in many tropical orchids, in which moisture is stored.

Pubescent: Covered with hairs, especially if short, soft and down-like.

Pureline: Progeny of a single homozygous, self-pollinated plant.

Pustular: A small blister-like elevation of the epidermis, often opening to expose spores.

Putative: The definition of putative refers to something commonly accepted as true or thought by most to be true.

Q

Quadrigeneric: Pertaining to four genera, primarily used to describe hybrids which contain genetic material from four different genera.

Quadrivalent: Having a valence of four or tetravalent.

R

Ray floret: Any of the small strap-shaped flowers in the flower head of certain composite plants.

Receptacle: The expanded tip of a flower stalk or axis that bears the floral organs or the group of flowers in a head.

Reciprocal cross: A cross, with the phenotype of each sex reversed as compared with the original cross, to test the role of parental sex on inheritance pattern. A pair of crosses of the type genotype A (female) × genotype B (male) and genotype B (female) × genotype A (male).

Recombinant: An organism or cell in which genetic recombination has taken place.

Recurrent parent: In backcross breeding, the parent to which one or few genes from the donor parent are transferred or called recipient parent.

Regeneration: It is the act or process of coming back, growing a new or the re-growth of tissues and organs from differentiated tissues.

Reservoir: A place where a great stock of germplasm is accumulated.

Resistance: Ability of an organism to exclude or overcome, completely or in some degree, the effect of a pathogen or other damaging factor.

Rostellum: It is a projecting part of the column in Orchidaceae flowers, and separates the male stamen from the female gynoecium, commonly preventing self-fertilization.

Ruffle: A strip of frilled or closely pleated fabric used for trimming or decoration.

S

Secondary introduction: Introduction and exploitation of variability for genetic improvement of existing crop is called secondary introduction.

Seed parent: In hybrid seed production the parent that produces seed or male sterile parent is called seed parent.

Seed pod: Capsule bearing the seeds of an orchid.

Seed: Fertilized and ripened ovule, consisting of the plant embryo, varying amounts of stored food material, and a protective outer seed coat.

Segregation: The separation of paired alleles or homologous chromosomes, especially during meiosis, so that the members of each pair appear in different gametes.

Self fertile: Able to be pollinated by its own flowers.

Self incompatible: Incapability of fertile pollen to fertilize the ovule of same flower or genotype.

Selfing: Transfer of pollen grains from the stamens, the flower parts that produce them, to the ovule-bearing organs or to the ovules (seed precursors) themselves.

Self pollination: Transfer of pollen from anther of a flower to the stigma of the same flower or of another flower on the same plant.

Self sterile: Incapable of self fertilization.

Semi alba: White flower with a coloured lip.

Sepal: A division of a calyx; one of the outermost circles of modified leaves surrounding the reproductive organs.

Sepaloid: Resembling or characteristic of a sepal.

Solitary: Borne singly or alone; not in clusters.

Spadix: A fleshy spike of minute flowers, usually enclosed within a spathe, as in the calla lily and anthurium.

Spathe: A large, leaflike, often showy bract that encloses a flower cluster or spadix, as in the calla lily and anthurium.

Species: A fundamental category of taxonomic classification, ranking below a genus or subgenus and consisting of related organisms capable of interbreeding.

Spheroidal: A body that is shaped like a sphere but is not perfectly round, especially an ellipsoid that is generated by revolving an ellipse around one of its axes.

Spike: Technically an unbranched flower stem with short stalked or stalkless flowers long inflorescence having multiple florets.

Spinose: Plants bearing many spines.

Spiral: Winding in a continuous and gradually widening (or tightening) curve, either around a central point on a flat plane or about an axis so, a spiral pattern.

Splash: Petal of an orchid flower that modifies its petals by duplicating the colouring found on the lip; a type of peloric condition.

Spore: The reproductive organ in cryptogams which in function corresponds to a seed but possesses no embryo.

Sporophyte: The form of a plant in the alternation of generations that produces asexual spore.

Sporulation: The formation of spores; especially: division into many small spores (as after encystment).

Spur: Hallow, sac-like or tubular extension of the base of the lip, usually bearing nectar.

Stalk: The pedicel of flower.

Stamen: The pollen-producing reproductive organ of a flower, usually consisting of a filament and an anther.

Staminoid: A pollenless stamen.

Sterile: Not producing or incapable of producing offspring.

Stigma: Part of the pistil of a flower which is receptive to pollen.

Stout: Strong in structure or substance; solid or substantial.

Strain: A number of plants which have similar characteristics such as colour blends, different enough to be designated, but not sufficiently distinct to be considered a horticultural variety.

Sympodial: One of the two forms of orchid growth (the other is monopodial), where in each new growth arises from the rhizome of a previous growth, and each new growth is completely capable of bearing an inflorescence. (sympodial means .with conjoined feet. and in biology is often used to refer to the outward morphology or mode of growth of organism).

Synandrous: Here both filaments and anthers of the stamens are united throughout their length. e.g., Cucurbita, Lagenaria, etc.

Synchronous flowering: The phenomenon of coincident, simultaneous flowering of all population of a given species in a large region. It may involve many plants, but not necessarily all plants of that species.

Syngenesious: In this case, stamens are united by their anthers while filaments are free as in Asteraceae.

Syntopy: The position of an organ relative to other organs.

T

Tapetum: A nutritive tissue within the sporangium, particularly within an anther.

Taxon: (pl. taxa), A taxonomic category or group, such as a phylum, order, family, genus, or species.

Tepal: One of the divisions of a flower perianth, especially one that is not clearly differentiated into petals and sepals, as in gladiolus, lilies and tulips.

Tester: Plants or lines that are used for testing during hybridization.

Tetrageneric: Resulting from interbreeding members of four genera.

Tetraploid: A cell or organism having four times the haploid number of chromosomes in the cell nucleus.

Throat: The opening of a tubular corolla or calyx where the tube joins the limb.

Tracheids: Elongated cells in the xylem of vascular plants that serve in the transport of water and mineral salts.

Transformation: It is the genetic alteration of a cell resulting from the direct uptake, incorporation and expression of exogenous genetic material (exogenous DNA) from its surrounding and taken up through the cell membrane(s).

Transgene: It is a section of genetic material from one organism which appears in the DNA of another organism.

Transposable element (TE): It is a DNA sequence that can change its position within the genome, sometimes creating or reversing mutations and altering the cell.s genome size. Transposition often results in duplication of the TE.

Transposable element: Elements are DNA sequences that are capable of mediating their own movement (transposition) to new locations within the genome they inhabit.

Trigeneric: Resulting from interbreeding members of three genera.

Trilocular: Having three chamber like divisions or cavities, as the capsule of a plant.

Triploid: A cell or organism having chromosome number that is three times the basic or haploid number in cell nucleus.

Trisomy: An organism with an extra chromosome in addition to the normal complement, i.e. 2n + 1. Trisomy typically arises when a normal gamete (n) fuses with one containing an extra chromosome (n + 1). Trisomics often have abnormal phenotypes and are frequently less fertile than diploids.

Trivalent: Chromosome formed by three homologous chromosomes that lie close together or appear to join completely during meiosis.

V

Variability: The quality of being subject to variation.

Variegated: Leaves which are marked with multiple colours.

Variegation: Appearance of different coloured zones in the leaves and sometimes in the stems of the plants.

Variety: Subdivision of a species that groups plants with a distinct form that is passed along to the progeny.

Viable: Capable of living, developing or germinating under favourable conditions.

W

Whorl: A radial arrangement of three or more petals, stamens, leaves etc.

Z

Zygomorphic: A flower capable of being cut in only one plane so that the two halves are mirror images.

Zygotes: The cell produced by the union of two gametes, before it under go cleavage.

Section I

Fundamentals of Flower Breeding

Chapter – 1

History of Ornamental Breeding

WORLD HISTORY

Work on ornamental crops was first initiated in Krelage Nursery founded by Ernst Heinrich Krelage for bulbs, lilium and dahlias in 1811. During 1895, Taiwan Agricultural Research Institute was established for ornamental crops. The first yellow corn Golden Bnatam was listed in Burpee's Catalogue in 1902. In 1909 Primadonna F_1 hybrid variety of begonia was released by the Benary Seed Company in Germany and in the same year Atlee established seed farms at Floradale in California, Erfurt in Germany Anglia in UK and Kashmir valley in India for annual garden flowers. Luther Burbank did excellent breeding works on ornamental trees, flowers and fruits in 1926.

David developed the first successful hybrid in Nasturtium (1934) and Red and Gold hybrids of marigold (1937). In 1940, Burpee was successful in chromosome doubling and transformation of wild flowers and developed "Bright Scarlet" and "Super Tetra Series" in antirrhinum and "Jumbo Scarlet" in zinnia. He produced new strains of marigold, zinnia, nasturtium and snapdragon. William Houston in 1940 initiated the Ageratum breeding. Morden Research station was set up in Western Canada in 1945 for ornamental crop breeding. In 1968, Dr. Gustav A.L., Mehlquist was recognized as an international authority on breeding of ornamental plants. George J. Ball breeder of peas, asters and calendulas was a pioneer for the supply of dwarf impatiens and orange impatiens 'Super Elfin' in 1968. Israeli breeders during 1990 gave more attention towards the development of carnation, lily and anemone varieties. Longwood Gardens in 2008 got success in breeding of stock. The first record of chrysanthemum hybridization, germination and selection of superior seedlings occurred in 1827 by M. Bernet.

Roses are complex hybrids involving interspecific hybridization, polyploidy (triploid, tetraploid and aneuploid) having high female and male sterility, cytological abnormalities, and heterozygous genotypes perpetuated by vegetative propagation. Many Hybrid Tea and Floribunda are either tetraploid or triploid and a few may be aneuploid also. Most of the Polyantha roses are diploid. Besides poor seed set and low seed germination are some other disadvantages. Hence, they do not afford much success in achieving the desired objectives in breeding unlike sexually propagated species. It is a general experience of rose breeders that the chance of obtaining a new rose worth introducing in trade is only one in a thousand. In spite of these limitations, creating new roses by hybridization has been most fascinating and rewarding for both rose hobbyists and professional nurserymen.

Many of the famous rose breeders of the world have had several generations' experience in developing rose varieties, like Meilland (France), Harkness (U.K.), Wheatcroft (U.K.), McGredy (Ireland, now in New Zealand), Tantau (West Germany), Kordes (West Germany), Poulsen (Denmark) etc. The other important rose breeders are Anderson, Bennett, Boerner, Cocker, Conard-Pyle, Delbard, De Ruiter, Dickson, Dot, Eddie, Fryer, Gaujard, Gregory, Jolly, Lammert, Lapererriere, Lindquist, Mallerin, Moore, Norman, Robinson, Saville, Schwartz, Spies, Swim, Warriner, Weeks, Williams and a few others. In Japan rose breeding is carried out by only two nurserymen, Suzuki (Keisi Rose Nursery) and Teranishi (Itami Rose Nursery), but there are a number of amateur rose hybridists. The two other famous rose nurseries of the USA, Jackson and Perkins, and Armstrong, also evolve several new varieties every year. There are also many amateur rose hybridists in the world, particularly in the USA and Canada where they have a Rose Hybridizers' Association (U.S.A.). More than 20,000 rose varieties have been evolved in the world (Modern Rose: VIII) and about 200-300 new roses are developed every year. As a result of intensive breeding, there is almost no distinction between Hybrid Tea and Floribunda now. According to a recent classification adopted by the World Federation of Rose Societies, there are Large-flowered and Cluster-flowered rose types instead of Hybrid Tea and Floribunda but Polyantha is still recognized as a separate class.

Publications in orchids like "Orchids Growers Manual" by William in 1885, "The Manual of Orchidaceous Plant" by James Veitchand Sons in 1894 and Sanders orchid guide in 1901 provided early days information. The importance of establishing accurate record of the parentage of hybrids in orchid was first recognized by Mr. Fred K. The main work of Sanders was the "Sander's List of Orchid Hybrids" Volume I (1854-1945) and Volume II (1946-1960). He initially listed innumerable number of natural and artificial hybrids of orchids. Polyploidy hybrids boosted the Hawaii's fledging flower industry on *Dendrobium* orchids. In the 1970's double diploids of *Dendrobium* Jaquelyn Thomas (*D. phalaenopsis*× *D. gouldii*) was released by the University of Hawaii (De and Bhattacharjee, 2011).

In 1874, Max Leichtlin obtained the first gladiolus hybrid from the cross of *Gladiolus gandavensis* Van Houtte with *G. saundersii* Hook. The modern hybrids are designated as *G.* × *grandiflorus*, which is a complex of about 11 species, among them: *Gladiolus carneus, Gladiolus dalenii, Gladiolus oppositiflorus, Gladiolus gandavensis, Gladiolus primulinus*, *Gladiolus papilio* and *Gladiolus saundersii*. Nowadays the cultivars have a large diversity of colours, sizes, textures, numbers and types of flowers and particularities of forms of the inflorescence (Anderson and Park, 1989; Buch, 1978 and Ohri and Khoshoo, 1985b).

The significant breakthrough in Lilium breeding occurs during the past 50 years in The Netherlands, Korea and Japan. The Netherlands and France are the primary producers of tulip bulb and this crop was introduced from Turkey to Europe. In 1594 and around 1630 the tulip flower for the first time in The Netherlands. Primarily in *Tulipa gesneriana* the continued breeding and improvement work was carried out in The Netherlands. *Freesia* shifted from Europe to South Africa and several diploid, triploid, tetraploid, double-flowered and F_1 hybrids cultivars are introduced.

An important event in the history of bougainvillea took place with the discovery of a crimson bougainvillea in Cartagena, a Spanish port in the Mediterranean, by Mrs. R.V. Butt. Originally thought to be a distinct species, it was named *B. buttiana* in her honour. However, it was later discovered to be a natural hybrid of a variety of *B. glabra* and possibly *B. peruviana* - a "local pink bougainvillea" from Peru. Natural hybrids were soon found to be common occurrences all over the world. For instance, around the 1930s, when the three species were grown together, many hybrid crosses were created almost spontaneously in East Africa, India, the Canary Islands, Australia, North America, and the Philippines.

A.B. Stout started breeding work in *Hemerocallis* in the New York Botanical Garden, USA and produced first selection "Mikado" in 1929. More than 40000 cultivars *Hemerocallis* cultivars have been registered and more than 15000 of them are available commercially. Before 20th century controlled pollination were not widely used in roses and new cultivars were developed from open pollinated flowers of the existing cultivars. Modern rose breeders are engaged in the studies of inheritance of traits, molecular genetic techniques. In 1997, Weigel identified ornamental plants for cut flowers, foliage and cut plants.

The Ornamental Plants Germplasm Center (OPGC) was founded in 2004 by the U.S. Department of Agriculture as part of US National Plant Germplasm System to establish a new gene bank for herbaceous ornamental plants. The "Plant Patent Act" was enacted in 1930 in USA by "an act of congress" through considerable lobbying by fruits and ornamental plant breeders led by Luther Burbank. In 1970 the plant variety protection was enacted in USA to provide intellectual property rights (IPR) for sexually propagated plants and tuber multiplied plants. During 2004, Canadian food inspection agency raised among plant breeders and importers of plant of ornamental plants about the regulation for plant with novel traits. Now a days series of techniques and genetic markers have been developed to estimate genetic diversity. DNA markers have been used as a tool for a large number of applications ranging from utilization of a gene to plant improvement by marker assistance selection. In recent years various markers have been developed for large number of plants to assess the genetic diversity and population studies, varietal identification, construction of genetic linkage maps and marker assisted selection for traits of interest. DNA fingerprinting revolutionized identification of ornamental plant varieties during 2009. In 2007 recombinant DNA technology was started for development of transgenic ornamentals through genetic engineering manipulation of flower shape, plant height control, floral scent, vase life, intensification-reduction-elimination or synthesis of flower colour pigments.

INDIAN HISTORY

India with its rich ecogeographical diversity is ideally suited to grow ornamental crops. Several species can be grown successfully almost round the year in one or another part of the country. India abounds in its wealth of ornamentals indigenous to this country such as *Rhododendron* (43 species), Orchids (1234), *Primula* (43), *Lonicera* (23), *Aster* (14), *Begonia* (64), *Impatiens* (241), *Iris* (13), *Jasminum* (43), *Hedychium* (24), *Pandanus* (7), *Ixora* (37), *Gardenia* (14), *Crinum* (12), *Lilium* (11), *Barleria* (26), *Ipomoea* (57), *Tabernaemontana* (14), *Thunbergia* (10), *Bauhinia* (37) etc. (Khoshoo, 1968). Poor conservation of this rich and valuable resource largely due to the interference of man through clearing of forest and other areas, has depleted this wealth, which had considerable potential to generate much greater wealth in time to come. Under these circumstances, more attention needs to be paid to generate and create more variability through genetic means (Malik and Dadlani, 1986).

Genetic improvement of ornamental crops began in our country in the early sixties at a few institutions like the Indian Agricultural Research Institute (IARI), New Delhi; National Botanical Research Institute (NBRI), Lucknow and a few agricultural and traditional Universities. Later with the establishment of the Indian Institute of Horticultural Research (IIHR) at Bangalore in 1968, the work on these crops was strengthened. At IARI, it was largely due to the personal interest of the then Director of the Institute, Dr. B.P. Pal, that the work on flower crops was initiated. The research efforts at this institute have been generally concentrated on improvement of roses (Pal, 1972, 1982).

Enthusiasts like Dr. B.P. Pal and Dr. M.S. Randhawa had their personal involvement in the development of floriculture. In India the ornamentals arrived by the nurseryman in 1935 when the rose variety, Dr. S.D. Mukherjee was evolved by hybridization by B.K. Choudhary at Mihijam, Santhal Parganas (West Bengal) and the variety Ramakrishna Dev in 1941 by B.S. Bhattacharjees at Deoghar (Bihar).

In 1956 the maintenance and conservation of orchid germplasm at Shillong led to the beginning of scientific research in floriculture in India. The then Prime Minister Late Pandit Jawahar Lal Nehru visited Shillong and saw the excellent collection of Indian orchids made by an amateur, Mr. Mustafi and asked Dr. B.C. Roy, the then Chief Minister of West Bengal to provide adequate facilities for the same. The orchid germplasm which was maintained by Mr. Mustafi was brought and maintained in the glasshouse at the Lloyd Botanic Garden, Darjeeling. During 1957, Indian Council of Agricultural Research (ICAR), New Delhi, set up a sub-committee for considering the pattern of research to be adopted in ornamental crops. The committee expressed the view that a distinction has to be made between ornamental flowers and plants which were imported from other countries every year, and those which are indigenous to India or have undergone long acclimatization under Indian conditions. It was recognized that in former group of plants and flowers, no immediate research was necessary, the main problem with them being proper maintenance and multiplication of imported stocks. In second group, however, there was need for undertaking organized research with a view to improving them to the maximum extent possible (Chadha and Bhattacharjee, 1995).

The research work on floriculture was started during 1950 at the

- Indian Botanic Garden now National Botanical Research Institute (NBRI), Lucknow.
- The botany departments of Central Universities had also started research programmes on cytology of ornamental plants.
- In 1960, a scheme on "Cytological and Physiological studies in relation to Floriculture" was implemented by the Indian Council of Agricultural Research (ICAR), New Delhi, for two years.
- Subsequently, a Coordinated Scheme on Floriculture" was started at nine centres for undertaking research on different ornamentals during 1960-65 (Table 1).

Table 1: Coordinated centre on floriculture during 1960-65.

S. No.	State	Centres	Crop allotted
1.	Andhra Pradesh	Hyderabad	Bougainvillea
2.	Assam	Shillong	Azalea, Camellia, Orchid and Hibiscus
3.	Delhi	IARI	Rose
4.	Karnataka	Bengaluru	Bougainvillea, Croton and Hibiscus
5.	Maharashtra	Pune	Canna
6.	Tamil Nadu	Coimbatore	Crossandra, Chyrsanthemum and Jasmine
7.	Tamil Nadu	Ootacamund	Azalea, Camellia, Cacti and Succulents
8.	Uttar Pradesh	Saharanpur	Chrysanthemum, Jasmine, Marigold and Rose
9.	West Bengal	Darjeeling	Orchids

The ICAR, New Delhi for the period 1960-65 also sanctioned a Co-ordinated scheme for the maintenance and propagation of seeds and bulbs of imported ornamental plants at Shimla (H.P.), Mussorie (U.P.) and Ootacamund (T.N.). Two more schemes were initiated by ICAR at Coimbatore namely, Investigation on Ideal Storage Conditions for Seeds of ornamental plants during 1963-66 and Research on Commercial Flower Crops: Jasmine, crossandra and chrysanthemum from 1962-67.

The research on floriculture was intensified at the Indian Agricultural Research Institute, New Delhi from 1962 and at the Indian Institute of Horticultural Research, Bengaluru from 1968 onwards with the establishment of the Institute. Further strengthening of the research on floriculture was initiated in 1970-71 with the starting of a multidisciplinary All India Coordinated Floriculture Improvement Project by the Indian Council of Agricultural Research (ICAR), in which several State Agricultural Universities participated actively (Chadha and Bhattacharjee, 1995). The division of Floriculture and Landscaping at IARI, New Delhi was established in 1983.

ALL INDIA COORDINATED RESEARCH PROJECT ON FLORICULTURE

All India Coordinated Research Project on Floriculture was started in 1970-71 during Fourth Five Year Plan. In the beginning the objective was to undertake research on export oriented problems with a view to establishing trade in seeds, cut flowers and ornamental plants and thus the project was entitled as "The Coordinated Export Oriented Project on Floriculture". Later on the project was reviewed in 1972 and its objectives broadened to cover all aspects of research on ornamental plants and its title was changed to "All India Coordinated Floriculture Improvement Project (AICFIP)". Under AICFIP initially investigation was carried out on 25 ornamental crops under 60 different research projects. The project work was reviewed in 1978 and it was decided to limit the work only on 15 important crops like rose, gladiolus, carnation, chrysanthemum, tuberose, orchid, jasmine, hibiscus, bougainvillea, gerbera, pansy, marigold, aster, nymphaea and cyclamen. In 1982 it was further decided to concentrate and intensify research efforts on 5 crop only, which were identified for their export potential namely, rose, gladiolus, orchid, carnation and chrysanthemum. During Seventh Five Year Plan in view of the changed priorities work was also taken up on five seasonal flowers like pansy, petunia, marigold, aster and antirrhiunum and five bulbous plants namely begonia, tulip, narcissus, iris and hyacinth (Anon., 2013) and various centres identified to carry out the research work on these crops (Table 2, 3 & 4).

Table 2: First 5 centres sanctioned for the AICRP on Floriculture.

S. No.	Name of the centres	Place
1.	Indian Agricultural Research Institute (IARI)	New Delhi
2.	IARI Regional Station	Katrain, H.P.
3.	Indian Institute of Horticultural Research	Bengaluru, Karnataka
4.	Indian Botanic Garden, B.S.I.	Kolkata, West Bengal
5.	Botanical Survey of India	Shillong

Table 3: Four centres added in the AICRP on Floriculture during V Five Year Plan

S. No.	Name of the centres	Place
1.	Punjab Agricultural University	Ludhiana, Punjab
2.	Y.S. Parmar University of Horticulture& Forestry	Nauni, Solan
3.	Mahatma Phule Krishi Vidyapeeth, C.R.F.R.S.	Ganeshkhind, Pune
4.	Kerala Agricultural University	Vellanikkara

In 1977 the two centres of B.S.I. located at Kolkata and Shillong were shifted to Bidhan Chandra Krishi Viswavidyalaya, Kalyani, West Bengal and NEH Research Complex of ICAR, Shillong. During Sixth Five Year Plan 4 new centers were added to this project (Table 4).

Table 4: Four new centres added in the AICRP on Floriculture during VI Five Year Plan

S. No.	Name of the centres	Place
1.	Tamil Nadu Agricultural University	Coimbatore
2.	Rajasthan College of Agriculture	Udaipur
3.	Horticulture Experiments and Training Centre	Chaubattia, U.P.
4.	Orchid Sanctuary, Merong Forest	Kalimpong, West Bengal

A part of the research programme allotted to B.C.K.V., Kalyani centres has been shifted to the B.C.K.V. Farm at Pedong, Kalimpong similarly a part of the work allotted to the centre at T.N.A.U, Coimbatore, has been shifted to T.N.A.U. Regional Station at Yercaud and Kodaikanal during Seventh Plan. Two new centres were also sanctioned and started functioning from 1987-88 during seventh plan in the following universities (Table 5).

Table 5: Two centres added in the AICRP on Floriculture during VII Five Year Plan.

S. No.	Name of the centres	Place
1.	Andhra Pradesh Agricultural University	Hyderabad
2.	S.K. University of Agricultural Science and Technology	Shalimar, Srinagar

In Eight Five Year Plan, another centre has been proposed under this project at Bhubaneswar in order to cover the Coastal Region of Eastern India. Two voluntary centers located at following places (Table 6).

Table 6: Name of the one voluntary centres

S. No.	Name of the centres	Place
1.	University of Agricultural Sciences	Bengaluru
2.	National Botanical Research Institute	Lucknow

OBJECTIVES

The All India Coordinated Research Project on Floriculture started operating with 60 projects on 25 crops during the VI plan. The utility and efficiency of these programmes were reviewed critically during the end of each Five Year Plan and the number of crops was subsequently reduced to 15 and 5 during VI and VII Plan periods, and the number of research projects was reduced to 43 and 26 respectively, during the above plans (Kumar *et al.*, 2011). Following are the major mandates.

- Collection, maintenance and assessment under different agro-climatic conditions of important floricultural plants germplasm including the wild species from different parts of the country and abroad.
- Development of new and improved varieties and types by hybridization, mutation etc. and F_1 hybrid seed production.
- Improvement of agro-techniques for the production of cut flowers, bulbs and plants for domestic and export markets.
- Standardization of propagation techniques including embryo/tissue/meristem culture for quicker multiplication.
- Developemnt of suitable pre and postharvest technology for cut flowers.
- Identification and control of major diseases and pests of major commercial flowers.

I.C.A.R. INSTITUTES INVOLVED IN FLORICULTURE RESEARCH

IARI, New Delhi

During the year 1959 the systematic research on ornamental crops started at IARI, New Delhi in the Division of Botany under an ICAR coordinated scheme entitled 'Cytogenetical and Physiological Studies in Relation to Floriculture' and it formed the basis for strengthening of floriculture research work during Third Five Year Plan. IARI and its regional station Katrain, Himachal Pradesh carried out the research work on floriculture. The floriculture work at IARI was under the Division of Vegetables Crops and Floriculture. However in 1983, Division of Floriculture and Landscaping separated from the Division of Vegetable Crops and Floriculture (Chadha, 1993).

IIHR, Bengaluru

Establishment of IIHR at Hessarghatta, Bengaluru expanded the research work on floriculture. IIHR is the first institute where a Division of Floriculture and Landscape Gardening was established in 1969. In 1982 the Division was renamed as Division of Ornamental Crops.

ICAR Research Complex for NEH Region, Shillong

In 1971 work on ornamental crops was started at this Institute and presently concentrated on the *in vitro* propagation of orchids.

National Research Centre (NRC) for Orchids, Gangtok (Sikkim)

NRC on orchids has been approved to be set up in Sikkim during the VIII plan. Mission oriented research on all aspects of orchid improvement and their production technology was under taken by this centre.

Directorate of Floricultural Research (DFR), New Delhi

In India, there was no full-fledged institute on floriculture. Keeping in view of the importance of the sector the Department of Agricultural Research and Education (DARE) with approval of the Planning Commission, Govt. of India has decided to establish Directorate of Floricultural Research (DFR) at New Delhi during XI Plan. Formally DFR was launched on 10th December 2009 at IARI, New Delhi. DFR is coordinating research on floriculture on genetic resource utilization, development of cultivars, production technology, etc.

CSIR INSTITUTES

National Botanical Research Institute, Lucknow

NBRI engaged in the collection, introduction and hybridization of ornamental crops during 1953-1966 in order to broaden the base of genetic variability. Later on researches were aimed to achieve the twin national objectives of export promotion and import substitution.

CSIR Complex, Palampur

During 1986 research on floriculture was started at this institute with more emphasis on greenhouse technology.

OTHER INSTITUTES

Bhabha Atomic Research Centre (BARC), Mumbai

This Institute is engaged in research programme on the development of new cultivars and micropropagation of ornamentals.

Botanical Survey of India (BSI)

During 1973-76 investigations were carried out on the standardization of agrotechniques of several ornamental plants like Hippeastrum, Gerbera, Tuberose, Dahlia, Orchids and Nymphaea at Indian Botanic Garden, Howrah of BSI. It was established in 1787 and further may be mentioned that the Indian Botanic Garden is the largest of its kind in Asia. Inside the garden there are Central National Herbarium, up-to-date Botanical library and a series of laboratories.

AGRICULTURAL AND TRADITIONAL UNIVERSITIES

Agricultural Universities

Almost in all Agricultural Universities, floriculture has been given importance as a subject for teaching or research or both (Table 7).

Table 7: Universities and aspects on which research work is going on.

S. No.	Universities	Crops/Aspects on which working
1.	T.N.A.U., Coimbatore	Jasmine, chrysanthemum, crossandra, tuberose, hibiscus, barleria, marigold, rose, dahlia, hollyhock, nerium, carnation, gladiolus and orchids
2.	B.C.K.V., Kalyani	Gladiolus, marigold, orchid, anthurium, rose and jasmine
3.	Y.S.P.U.H.F., Solan	Carnation, chrysanthemum, gladiolus, rose, begonia, hyacinth and tulip
4.	U.A.S., Bangaluru	Jasmine, marigold, tuberose, chrysanthemum and aster
5.	K.A.U., Vellanikkara	Anthurium, hibiscus, gladiolus and orchid
6.	M.P.K.V., Rahuri	Chrysanthemum, gladiolus, carnation, rose, petunia and aster
7.	P.A.U., Ludhiana	Gladiolus, carnation and chrysanthemum
8.	M.P.U.A.T., Udaipur	Gladiolus, chrysanthemum, rose and marigold
9.	S.K.U.A.S.T., Srinagar	Begonia, chrysanthemum, gladiolus, hyacinth, iris, narcissus and rose

Traditional Universities

S. No.	Name of University	Crops/Aspects on which working
1.	Department of Botany, University of Delhi	Flower bud development, opening and physiology of senescence in ornamental, cytogenetical aspects of vast range of ornamental plants
2.	College of Agriculture, University of Kolkata	Pioneering work has been carried out on plant propagation, use of growth regulating chemicals, standardization of agro-techinque and related physiological and biochemical aspects

OTHER ORGANIZATIONS

Lalbagh, Bengaluru: This garden got collection of vast range of ornamental plants, and is equipped with a good library and laboratories to undertake research work on floriculture.

The Agri-Horticultural Society of India, Alipore, Kolkata: This society has a very rich and vast collection of various kinds of ornamental plants. It has a library, laboratory and plant houses for conducting research work on floriculture.

Regional Plant Resource Centre (RPRC), Bhubaneshwar: The major activity of this centre is plant introduction. It provides research opportunities in the field of tissue culture, cytogenetics, physiology and biochemistry. The centre has the richest collection of cactus in India. At this centre the major aspects of research on cacti include standardization of rapid propagation techniques, hybridization, application of physical and chemical mutagens to induce variability, study of biochemical basis of stock-scion relationship, monstrosity and fasciation.

Ad-hoc Schemes

Some ad-hoc schemes funded out of the Agriculture Produce Cess Fund of ICAR are also in operation as listed below:

- Standardization of micropropagation of some important indigenous flowers of Eastern Region (Assam Agricultural University, Jorhat).
- Genetic amelioration of tuberose for high yield of essential oil (IIHR, Bengaluru).
- Collection, evaluation, cataloguing and genetic improvement for better commercial varieties with export potential in *Anthurium* spp. (Kerala Agricultural University, Vellanikkara).
- Exploration and domestication of some wild ornamentals from Western Ghats Forests of Kerala (Environmental Resource Research Centre, Trivandrum).
- Standardization of agro-techniques in foliage plants (Kerala Agricultural University, Vellanikkara).

Research achievements of AICRP on floriculture and varietal improvement

India has a very rich germplasm collection of various ornamental crops, which consists of both indigenous and exotic plants. All India Coordinated Research Project (AICRP) on Floriculture was established during IV Five-Year Plan in the year 1971-72 to carry out nation-wide research by linking ICAR institutes with State Agricultural Universities (SAUs). At present the Coordinated Project has 22 centres (Anon., 2013) which includes 16 budgetary, 4 institutional and 2 voluntary centres (Table 8).

Table 8: AICRP on Floriculture centres with mandate crops.

S. No.	Centre	Year of start	Mandate crops
Budgetary Centres			
1.	BCKV, Mohanpur, WB	1972	Carnation, orchid, anthurium, tuberose, gerbera
2.	YSPUH&F, Solan, HP	1975	Gladiolus, carnation, chrysanthemum, tulip, daffodils, lilium, alstroemeria
3.	KAU, Vellanikkara, Kerala	1975	Orchid, anthurium, gerbera
4.	MPKV, Pune, Maharashtra	1975	Rose, gladiolus, carnation, chrysanthemum, tuberose, gerbera
5.	PAU, Ludhiana, Punjab	–	Rose, gladiolus, chrysanthemum, tuberose, gerbera, lilium
6.	MPUAT, Udaipur, Rajasthan	1980	Rose, gladiolus, chrysanthemum
7.	HRS, TNAU, Yarcaud, TN	1982	Carnation, gladiolus, chrysanthemum, orchid, anthurium, tuberose, gerbera
8.	HCRI, TNAU, Coimbatore	1982	Carnation, gladiolus, chrysanthemum, orchid, anthurium, tuberose, gerbera
9.	UBKV, Kalimpong, WB	1985	Gladiolus, carnation, orchid, anthurium, alstroemeria
10.	ANGRAU, Hyderabad	1987	Gladiolus, chrysanthemum, tuberose
11.	SKUAS&T, Srinagar, J&K	1987	Gladiolus, orchid, chrysanthemum, tulip, daffodils, lilium, alstroemeria
12.	HRS, AAU, Kahikuchi, Azara, Guwahati, Assam	2001	Gladiolus, orchid, chrysanthemum, anthurium, tuberose, gerbera
13.	OUA&T, Chiplima, Odisha	2001	Rose, chrysanthemum, orchid, anthurium
14.	GBPUA&T, Pantnagar	2001	Rose, gladiolus, chrysanthemum, tuberose, gerbera
15.	BAU, Ranchi, Jharkhand	2001	Gladiolus, chrysanthemum, gerbera
16.	RAU, Samastipur, Bihar	2010	Rose, tuberose, gladiolus, marigold
Institutional Centres			
17.	IARI, New Delhi	1971	Rose, gladiolus, chrysanthemum, tuberose
18.	IARI Regional Station, Katrain, HP	1971	Gladiolus, carnation, gerbera, tulip, daffodils, lilium, alstroemeria
19.	IIHR, Bangalore	1971	Rose, gladiolus, carnation, chrysanthemum, orchid, anthurium, tuberose, gerbera
20.	ICAR Research Complex for NEH Region, Barapani	1971	Orchid, anthurium, gerbera
Voluntary Centres			
21.	UAS, Bangalore	1977	Carnation, anthurium
22.	HCRI, TNAU, Periyakulam	2010	Marigold, tuberose, chrysanthemum

Genetic improvement

Genetic improvement in the ornamentals has been accomplished through induced mutations, polyploidy and hybridization, which play an important role in creating larger variability. Hybridization is associated, at all stages, with rigorous selection. Another approach used with much success for improvement of ornamentals has been the induced polyploidy (Malik and Dadlani, 1986). Some of the important results achieved with these approaches in the improvement of some of the ornamental crops in India are summarised below.

Heterosis breeding

There is a strong preference in the flower trade for F_1 hybrids for the obvious advantages of uniformity and increased vigour. Many ornamentals, in which F_1 hybrids have proven their success commercially in other countries and in which some useful results have been obtained in this country, include antirrhinum, aster, calendula, carnation, marigold, pansy, petunia, zinnia etc. These attempts have, however, not yet yielded any commercially acceptable F_1 hybrids, except, perhaps, in marigold, where, as mentioned earlier, the F_1 triploid developed at NBRI. Lucknow, has been found very promising. Nevertheless, various studies conducted in this regard have given us some useful new information, as well as good basic material for further work, or in some cases a few hybrids, which should be tested on a larger scale.

Malik (1979) in his genetic studies with diploid and tetraploid progenies on snapdragon (*Antirrhinum majus*) observed appreciable heterosis in the F_1 hybrids for all the characters studied. Overdominance was noticed in the inheritance of all the characters. Mahal (1972) also reported that the F_1 hybrids between diploid and tetraploid types were better than the parents and had longer blooming period. Swarup *et al.* (1975) observed heterosis for all the characters studied in balsam (*Impatiens balsamina*) and reported better performance of F_1 hybrids than the parents.

Khoshoo and Arora (1971) developed four free flowering heat resistant hybrids in verbena by crossing *Verbena tenuisecta* and *V. hybrida.* These hybrids have reasonably large and highly showy flowers. Since the hybrids have genic male sterility (Arora and Khoshoo, 1967) and normally do not set seeds, it makes them highly floriferous and free flowering. Swarup *et al.* (1973b) observed high heterosis in the intervarietal crosses in hollyhocks (*Althea rosaea*) for most of the characters studied. Singh (1971) developed cytoplasmic male sterility in petunia. Incorporation of such lines in breeding would result in reducing the production cost of hybrid seeds. In petunia, since flower, and not the seed, is the commercial product, maintenance and utilization of such a line does not require restorer lines. He also reported positional sterility in petunia. Its practical implication will be that such individuals could be pollinated directly for raising hybrids without emasculation.

Induced mutations

Mutation breeding is an established method for plant improvement, particularly in the vegetatively propagated ornamentals, which are highly heterozygous and frequently polyploid. This method of crop improvement is particularly advantageous when due to some barriers, genetic or otherwise, the conventional hybridization method becomes difficult to use. Another advantage is that one could go for improvement in one or a few characters of an otherwise outstanding cultivar.

Misra (1977, 1978) reported 2-5 kR dose of gamma rays to be optimum for inducing mutants in gladiolus and obtained some promising colour mutants in the varieties Oscar. Jawagenaar and Picardy. Gupta and Shukla (1974) found bougainvilleas, in general, to be sensitive to radiation. Abraham and Desai (1977) suggested administering recurring low doses of radiation to increase surviving populations and induce mutations. The bougainvillea varieties developed in India through induced mutations are Arjuna from Partha. Jaya and Jayalaxmi Variegata from Jayalaxmi. Lady Hudson of Ceylon Variegata from Lady Hudson of Ceylon and Silver Top from Versi Colour (Gupta, 1979; Gupta and Nath, 1977 and Nath *et al.*, 1983).

For the improvement of large flowered chrysanthemums, which do not set seeds, mutation breeding resulted in the selection of mutants in M_1 and M_2 generations. In a few cases, mutations were also selected after recurrent irradiation and fractionation of dose. The dose for obtaining mutants ranged from 1 to 3 kR (Gupta, 1966; Gupta and Shukla, 1971b; Gupta and Jugran, 1978). The mutants isolated in pure form and released as new cultivars/from these studies are Anamika, Himani, Kapisb, Lohita, Gairik, Tamara, Asha, Shukla, Basant, Pingal, Kansya, Aruna, Ashankit, Jhalar, Kanak, Kunchita, Nirbhaya, Nirbhik, Shefali, Svarnim and Shveta (Gupta, 1979). Das and Ghosh (1972) observed three distinct changes (yellow, orange, brown) from the original pink colour in variety Pride of Madford. In annual chrysanthemum Jain and Mazumdar (1957) and Jain *et al.* (1961) observed changes in flower's from single to double of various types and from nontubular to tubular corolla.

Induced polyploidy

Induced polyploidy has been used with much success in the improvement of ornamental plants all over the world. There are many reports of commercial exploitation of autopolyploidy in antirrhinum, petunia, pansy, marigold etc. Autoploidy confers some useful properties to the ornamentals, *e.g.* increased vigour, later and longer blooming periods, larger flowers and intensification of flower colour and fragrance.

In India, induction of autoploids in ornamentals like *Alyssium, Ibeds* and *Linaria* was started by Tandon and his coworkers (Bali and Tandon, 1959), though their objective was not commercialisation of the resultant autoploids. The role played by induced polyploidy and its use in hybridization, living us new perspectives in bougainvillea breeding, has already been mentioned. Pal and Khoshoo (1977) induced an improved tetraploid in ornamental amaranth (*Amaranthus tricolor* var. *salicifolius*) which was short, sturdy, and had thicker ornamental leaves, intense and more lasting colour. Mahal (1972) and Mahal *et al*. (1968) reported autotetraploids in antirrhinum (*Antirrhinum majus*) which were hardier, slower in growth, shorter, and with large flowers they had a longer flowering period and individual flowers also remained fresh for a longer period. In verbena, Arora (1972) observed that induced triploids had vigorous growth habit, broader leaves, larger flowers and sterile pollen. Jalil *et al*. (1974) induced autotetraploids in *Impatiens balsamina* and reported increase in height, size of leaves. petals and flowers, and later and longer blooming. The autotetraploid of *Dilflorphotheca aurantica* was gigas with longer lasting, and deeper coloured flowers (Rizvi, 1973).

VARIETAL IMPROVEMENT

Rose

The rose breeding in India was initially done by a few private nurserymen, largely through making selections in the progeny raised from open-pollinated seeds. However, systematic work in rose breeding was initiated by Dr. B. P. Pal at IARI in 1960 (Pal, 1969 and Malik and Dadlani, 1986).

- Rose breeding has received highest importance in India, and as a result about 780 accessions of varieties belonging to various classes of roses *i.e.* Hybrid Tea, Floribunda, Miniature and Climbing groups were collected and maintained at Delhi, Bhubaneswar, Ludhiana, Hessaraghatta, Ranchi, Pantnagar and Udaipur centers (Swarup *et al*., 1973). The main objective of rose improvement has been to evolve varieties with attractive flower colour, form and fragrance, floriferousness, disease and pest resistance, and suitable for growing under tropical and subtropical conditions.

- Studies at IARI to identify varieties with high male and female fertility helped in making successful crosses and evolving a large number of improved varieties. The studies on inheritance of the characters indicated that crosses among white or yellow flowered varieties produced hybrids resembling one of the parents only. A greenish white flowered variety crossed with a pure white one showed greenish white colour as dominant. Similarly, deep yellow was found to be recessive to light yellow, and pink shades dominant over dark red colour (Swarup *et al*., 1973).

- According to the Department of Agricultural Research and Education, ICAR, Government of India, in 1999-2000, rose varieties Dr. B. P. Pal, Blue Delight, Anurag, Ashirwad, Ico, Ambassador, Blue Ocean and Aditya were promising at the national level; Ashirwad, Anurag and Siddhartha for exhibition purpose; Black Delight, Blue Ocean and Aditya for garden display and Ico, Ambassador and Kanchi for cut and loose flowers at Hyderabad.

- The first rose variety named as Dr. S.D. Mukherjee was developed in India by B.K. Roy Chaudhary, a nursery man of Mihijam in Santhalpragna during 1935. Dr. B.P. Pal contributed significantly to rose breeding evolving 105 new varieties. Promising hybrid-tea roses bred by Dr. B.P. Pal include Dr. Homie Bhaba, Sir C.V. Rama, Dr. M.S. Randhawan, Shantipal, Mrs. K.B. Sharma and Kamala Devi Chattopdhaya. Promising floribundas roses developed by Dr. B.P. Pal are Delhi Princess, Banjaran, Chittchor, Madhura, Suryakiran, Tarang and Manmatha. At IARI New Delhi improved cultivars of roses in different categories namely Hybrid Rose, Floribundas, Polyanthes, Miniatures and Climbers include Raktgandha, Pusa Bahar, Mrinalilni, Raktima, Pusa Gaurav, Pusa Pitamber, Mother Teresa, Pusa Priya and Dr. B.P. Pal.
- At IARI New Delhi Division of Floriculture and Landscaping newly released variety include Pusa Mansij, Pusa Ajay, Pusa Arun, Pusa Sathabandh and Pusa Komal and also released thornless rose varieties e.g. Pusa Mohit and Pusa Komal. Outstanding Floribundas developed at IARI include Himangini, Prema, Sadabahar, Lahar, Shabna, Deepshika and Mohini. Pusa Christina, Striped Chirstiandior, Abhisarika and Madhosh are the mutants developed at IARI New Delhi. Unique chocolate brown bloom colour hybrid variety Mohini attracted the attention of rose breeders across the globe.
- The rose variety Rose Sherbet bred by Dr. Pal is highly fragrant, with the rose oil content of 0.03%. Some other notable fragrant Indian varieties are Sugandha, Dulhan, Diva Swapna, Rajhans, Anuraag, Charugandha, Jawahar, Haseena, Indian Princess, Madhura, Mridula, Nurjehan, Kanakangi, President Radhakrishnan etc. Mechak, Mrinalini, Dr. Homi Bhabha, Haseena, Raktagandha, Raja Surendra Singh of Nalagarh, Nishada and Surkhab can be cited as examples of varieties with shapely flowers. Some of the important floribunda varieties bred in India are Delhi Princess, Banjaran, Prema, Sadabahar, Neelambari, Chandrama, Surya Kiran etc. These have become very popular with the rose growers and compete well with the exotic varieties in their class at various flower/rose shows (Malik and Dadlani, 1980, 1984).
- Hybrid Tea roses developed at NBRI Lucknow include Light Pink Prize, Mrinalini Stripe, Pink Contempo, Pink Montizuma, Summer Holiday Mutant and Winter Holiday Mutant. Floribundas developed at NBRI Lucknow Angara, Curir, Pink Imperator, Salmon, Beauty Mutant, Sharda, Sukumari, Twinkle, Yellow Contempo and Zorina Park. While Windy City Mutant is an improved miniature, climbing Cri Cri type climbing type.
- The inheritance studies on fragrance in roses have shown that the hybrid progenies invariably have fragrance in their flowers if one of the parents has fragrant flowers. However, sometimes crosses involving nonfragrant parents also give hybrid seedlings having fragrant flowers (Swarup *et al.*, 1973a).

Chrysanthemum

In chrysanthemums, hybridization at NBRI, Lucknow, led to the development of better plant types and also varieties blooming out of season, thus enabling development of varieties with compact growth habit, requiring no pinching and staking, suitable for pot culture, e.g. Hemant Singar, Suhag Sringar etc.

- Work on evolving new chrysanthemum varieties has been under way mainly at the NBRI, Lucknow, and the IIHR, Bengaluru. In chrysanthemum, a total of 1115 accessions of varieties are maintained at Hessaraghatta, Ludhiana, Pune, Solan, Coimbatore, Bhubaneswar, Hyderabad, Udaipur, Lucknow, Srinagar, UAS, Bengaluru, Ranchi and Pantnagar centres. At the IIHR, Bengaluru emphasis had been on breeding high-yielding cultivars for garland and pot-culture and 5 new cultivars have been developed. The PAU, Ludhiana, and TNAU, Coimbatore, each has evolved 4 cultivars.

- Some important varieties developed at this Institute include Birbal Sahani, Jyoti, Kundan, Rim Jhim, Sbarada, Sharad Mala, Varsha and Vasantika. The other good chrysanthemum varieties developed in our country include Indira, Red Gold and Rakhee at IIHR, Bengaluru, and Shanti, Vasanti, Gul-e-Shahir and Baggi at Punjab Agricultural University, Ludhiana (Malik *et al.*, 1984).

- According to DARE in 1999-2000, at Lucknow, a new small-flowered, red, Anemone type, mini chrysanthemum variety, Diana, was released. A desirable natural mutant was isolated and multiplied in Nilima and a promising hybrid 87-17-1 was identified for future release at Hessaraghatta.

- Introduction of early and late blooming cultivars from abroad, particularly from Japan, and hybridizing them with the local varieties resulted in offseason cultivars (Kher and Bhutani, 1979). This has enabled development of varieties which bloom in different seasons of the year and it has now become possible to get chrysanthemums almost all the year round.

- New varieties, Mutant No. 9 and Hybrid No. 10, Hybrid No. 11, Hybrid No. 12, Hybrid No. 13, Sunil, Indira, Sonaly, Tara, Selection 5 and Shyamal were suitable for commercial flower production. Shyamal, Baggi, IIHR Sel. 5 and Indira were promising at Pune, whereas Basanti, Sharadmala, Baggi, Red Gold, CO-1, IIHR 13 and Vasantika at Hyderabad. Indira, IIHR Sel 6, Meera and Red Gold performed well in respect of flower yield at Udaipur.

- Newly developed varieties of chrysanthemum *viz.*, Arka Ravi (at Hessaraghatta, Solan, Udaipur and UAS Bengaluru), Punjab Gold (at Hessaraghatta and Pune), Mother Teresa (at Hessaraghatta), CO-2 and Indira (at Coimbatore), Yellow Gold (at Solan), Ratlam Selection and Basanti Local (at Hyderabad), Sonali Tara, Sunil, Baggi and IIHR Hybrid 11 (at Pune) were found to be very promising. Baggi,

Gul-E-Sahir, Pujab Gold and Shanti are developed at PAU Ludhiana. Baggi in chrysanthemum was recorded outstanding.

- Important varieties developed through open pollinated seedling selection at NBRI Lucknow are Birbal Sahni, Apsara, Jayanti, Jubli, Kundan, Manjari, Ratna, Sunnel, Lalima, Jaya, Neelima, Basantika, Sunayna, Sharadh Sringar, Haldighati and Jyoti. Important mutants developed at NBRI Lucknow include Sonali, Navneet, Purnima, Kumkum, Shabnam, Batik and Himani.
- At IARI, New Delhi, Pusa Anmol and Pusa Centenary through mutation breeding cvs. Ajay and Thaichen Queen, respectively at IIHR Bengaluru varieties Chandrakant, Chandrika, Indira, Kirti, Neelima, Pankaj, Rakhi, Ravikiran, Red Gold and Yellow Star. Mutant developed at IIHR Bengaluru named Yellow Gold by inducing irradiation by gamma radiation. CO-2 and CO-2 are developed at TNAU Coimbatore.
- Pusa Arunodaya is a gamma ray induced pink coloured mutant of cv. Thai Chen Queen, which is orange in colour. Pink coloured ray florets (65D) appeared as chimers which were used for in vitro regeneration to establish as a new variety (Anon., 2013).
- Pusa Kesari is also a gamma ray induced red coloured mutant of cv. Thai Chen Queen. Red coloured ray florets (171A) appeared as chimers which were used for *in vitro* regeneration to establish as a new variety (Anon., 2013).
- Pusa Aditya is an open pollinated seedling of cv. Jaya and Pusa Chitraksha is an open pollinated seedling of cv. Lal Pari and Pusa Sona is an open pollinated seedling of cv. Sadbhawana developed by IARI, New Delhi (Anon., 2013).

Gladiolus

In recent years, gladiolus has become a very popular cut flower among the florists in India. The emphasis in this crop has been on breeding varieties with attractive flower colour, long spike with large number of florets, mainly for cut flowers, and high rate of corm multiplication. The important varieties evolved in our country include Apsara (IIHR), Agni Rekha (IARI), Mayur (IARI), Meera (IIHR), Nazrana (IIHR), Poonam (IIHR), Sapna (IIHR), and Suchitra (IARI) (Malik *et al.*, 1984).

- In gladiolus, 637 accessions of varieties were maintained at Ludhiana, Yercaud, Srinagar, Hessaraghatta, Udaipur, Solan, Katrain, Hyderabad, Delhi, Kahikuchi, Lucknow, Ranchi, Pantnagar, Shillong and Kalimpong.
- The success of *G. psittacinus* hybrid at the NBRI inspired research in evolving different coloured varieties suitable for growing in the plains and resulted in the release of 11 cultivars. Notable among these are Jwala, Priyadarshini and Gazal, all of which are open-pollinated selections. Further, through systematic

hybridization involving gladiolus Friendship (2n = 60) with *G. tristis* (2n = 30), 8 new triploid (2n = 45) cultivars were evolved.

- According to DARE, on Gladiolus Shagun and Shringarika were released by Delhi centre. Hybrids Amal and Tambri performed well at Vellanikkara. Gladiolus Snow White, Suchitra, Punjab Morning, Jester, Aldrion, Golden, Wave and Thrombolina were promising at Hyderabad.
- In gladiolus, Arka Kesar, 87-22-1, hybrid 84-7-11, 86-3-4, 84-4-8, Swarnima and Shagun have been found performing better.
- Varieties developed at IIHR Bengaluru through hybridization include are the Apsara, Darshan, Dheeraj, Kumkum, Meera, Najrana, Poonam, Sagar, Sapna, Shakti, Arka Kesar, Arka Suwerna and Sindhur.
- Varieties developed at IARI New Delhi include Agnirekha, Anjali, Archana, Bindiya, Chandini, Chirag, Dhanwantri, Mayor, Neelam, Pusa Suhag and Sarang, Shweta, Suchitra, Sunaiyna, Shagun, Pusa Kamini , Pusa Urmil, Pusa Swapnil, Pusa Mohini, Pusa Sukanya, Pusa Ranglal, Pusa Gunjan, Pusa Swarnima, Sringarika and Suryakiran. The mutant variety 'Shobha' was developed by gamma rays irradiation of corms of cv. Wild Rose at IIHR Bengaluru (Negi and Raghava, 1986).
- Hybrids identified for release. Two hybrids Berlew × Heady Wine (Tall) and Berlew × Heady Wine were identified for release as Pusa Unnati and Pusa Srijana respectively (Anon., 2013).

Bougainvillea

Bougainvillea is an important and popular flowering plant for landscaping. Progress in bougainivillea breeding is limited by the extensive pollen and/or seed sterility (Ninan *et al.*, 1959 and Zadoo and Khoshoo, 1968). This restricts the choice of female and male parents only to the few relatively fertile types, which more often are not good cultivars. After detailed studies on the causes of sterility, fertility was restored by colchiploidy (Khoshoo and Zadoo, 1969). This enabled immediate widening of germplasm by including such cultivars in breeding programmes that were otherwise very useful but so far beyond the reach of bougainvillea breeders. As a result, now seeded analogues of several sterile but very popular varieties have been created. These have been used at NBRI, Lucknow, to develop a number of very promising, colourful and floriferous, often bicoloured hybrid varieties at triploid, tetraploid and aneuploid levels (Khoshoo, 1979).

About 150 varieties have been developed in India through hybridization, sports and mutation breeding. The Division of Floriculture and Landscaping, IARI, New Delhi, has been designated as the International Registration Authority for Bougainvillea

Cultivars, and the first checklist described more than 300 cultivars of bougainvillea. A number of very promising, colourful and floriferous, often bicoloured, hybrid bougainvillea (triploid, tetraploid and aneuploid) ideal for pot-culture have become handy. Bloched-bract varieties like Begum Sikander, Wajid Ali Shah and Mary Palmer Special provide a visual feast in pots.

- Five varieties of bougainvillea developed by Dr B P Pal in 1959, *viz.* Dr R.R. Pal, Sonnet, Stanza, Spring Festival and Summer Time, are popular even today.
- At Bhaba Atomic Research Centre (BARC), Mumbai, several varieties were developed through mutation breeding with the use of gamma rays.
- The IIHR, Bengaluru, has also released 6 cultivars, *viz.* Chitravati, Dr. H.B. Singh, Jawahar Lal Nehru, Purple Wonder, Sholay and Usha. A colchicine-induced mutant with bright magenta bracts has been isolated from cv. Zakariana.
- Bhabha, Lalbagh, Trinidad, Thimma, Yellow Queen were developed at Lalbagh, Bengaluru (Bose, 1972).
- Promising varieties developed in India are Beauty, Dream, Gagarin, Ganga Swami, Jayalaxmi, Krumbiegel, Lalbagh, Mahatma Gandhi, Mrs. H.C. Buck, Mrs. Lanchaster, Partha, Raman, Sonnet and Usha. Begamsikander, Wajid Ali Shah and Chitra have been developed from planned inter specific crosses. Dr. R.R. Pal, Summer Time, Spring Festival, Maharaja of Mysore, Pink Beauty, Pixie, and Rose Queen have been developed at natural interspecific hybrid. Four varieties Dr. H.B. Singh, Purple Wonder, Chitrawati, and Mary Palmer, special have been developed through intra-specific hybrid.
- About 50 cvs. with leaf Variegation have arised from bud sports with species *Bougainvillea* × *buteana* giving rise to maximum number (25) followed by 13 in *B. peruviana*, 8 in *B. spectadilous* and 3 in *B. spectroglabara*. The varieties developed through induced mutation are Arjun, Jaya, Lady Hudson of Shilong, Variegata and Silver Top.
- Tetraploidy was induced in number of cvs. Namely Shipra, Mrs. M.C. Clean, President Roosevelt, Lady Marybaring, Zakariana and Theema with a restore fertility at NBRI Lucknow. 6 promising varieties Dr. B.P. Pal, Tetra Mrs. M.C. Clean, Begum Sikander, Wajid Ali Shah, Mary Palmer Special and Chitra have been released using tetraploidy.

Carnation

- In carnation, 121 accessions were collected and maintained at Solan, Yercaud, Srinagar, Ludhiana, Katrain, Kalimpong and UAS, Bengaluru.
- At the IARI, New Delhi, mutation-breeding experiments of different lines of carnation resulted in some interesting mutants with variegated leaf (Kaicher, 1988).

Gerbera

- A total of 286 accessions of gerbera were collected, maintained and evaluated at Shillong, Pune, Yercaud, Hessaraghatta, Kahikuchi, Ranchi, Srinagar, Kalyani, Ludhiana and Vellanikkara.

Daffodils, Lilium, Tulip and Alstroemeria

- During 2004-05, daffodils, lilium, tulip and alstroemeria are the new crops that were included in AICFIP.
- About five cultivars of tulip were collected, maintained and evaluated at Srinagar, Solan and Katrain.
- In daffodils, five cultivars were maintained at Katrain, Srinagar and Solan.
- 17 varieties of Lilium were collected, maintained and evaluated at Solan, Srinagar and Katrain.
- In alstroemeria, 15 varieties were maintained at four centres *viz.*, Kalimpong, Srinagar, Katrain and Solan.

Hibiscus

- In *Hibiscus,* hybridization and selection work at IIHR, Bengaluru and Tamil Nadu Agricultural University, Coimbatore, have led to release of a large number of promising varieties, like Aikta, Anuradha, Basant, Geetanjali, Dilruba, Priya, Smt. Indira Gandhi, Smt. Kamla Nehru, Pilnnagai and the intergeneric hybrid *(Hibiscus rosa-sinensis* X *Hibiscus arboreus)* Thilagam (Malik *et al.*, 1984).
- In hibiscus, hybridization and selection work at the IIHR, Bengaluru, has led to the release of 25 varieties which are floriferous and bear attractive flowers.
- The TNAU, Coimbatore, has developed an inter-generic hybrid (*Hibiscus rosa-sinensis* × *Malvaiscus arboreus*) named Thilagum (Carmine red) in addition to a seedling selection Punnagai. Another hybrid named CO-1 hibiscus has also been released by the TNAU. It is floriferous and produces apricot-yellow flowers with signal-red throat. Lalbagh, Bengaluru, has also developed 22 attractive varieties.
- Twenty five varieties have been developed at IIHR, Bengaluru either as a result of planned crosses or open pollinated seedling selection which include Aikta, Arunodaya, Ashirwad, Basant, Benazeer, Bharatsundari, Chitralekha, Dilruba, Geetanjali, Jogan, Nartki, Nazneen, Queen of Hessaraghatta, Ratna, Red Gold, Red Saturn, Shanti, Smt. Indira Gandhi, Smt. Kamala Nehru and Tribal Queen.
- The Lalbagh Botanic Gardens, Bengaluru has developed 22 varieties which includes Aldrin, Apollo-II, Armstrong, Bangladesh, Collins, Dr. B.P. Pal, FLT, Lt. R. Massey, Gnat, Indira Gandhi, Mother, Nijlinpgappa, President, Rachaiah, Rastrapati, Soyuz-11, Veerendrapatil, Victor, Valdislav and Zakirhussain.

- At TNAU, Coimbature, 2 varities *i.e.* Punnagai (CO-1) and Thilagan (CO-2) have been bred.

Jasmine

- The TNAU, Coimbatore, has released 3 varieties of *J. auriculatum* and 2 varieties in *J. grandiforum.* Initially, work carried out in *J. auriculatum* led to the identification of a high-yielding, gall-mite-resistant clone Pari Mullai. The clone is characterized by long flowering season of 9 months with flower bud yield of 7,800 kg/ha and concrete recovery of 0.34%. Subsequently, the university released CO-1 Mullai and CO-2 Mullai. The yield of flower buds in CO-2 Mullai is reported to be 8,800 kg/ha with 0.34% concrete recovery. CO-2 Mullai is a cross between a mutant (S1) × Long Point (JA 1). It yields 1, 100 kg flower buds/ha accounting for an increase of 26.88% over CO-1 Mullai. In *J. grandiflorum*, CO-1 Pitchi is a clonal selection from Lucknow type which yields 10,000 kg flower buds/ha with concrete recovery of 0.29%. CO-2 Pitchi is an induced mutant from CO-1 (Muthukrishanan and Pappiab, 1980; Muthuswamy and Madhava Rao, 1972; Veluswamy *et al*., 1973 and Malik *et al*., 1984).

- Surabhi a high-yielding variety from *J. grandiflorum* has been developed by clonal selection at the IIHR, Bengaluru. The potential flower yield is 10,000 kg/ha with high concrete recovery.

- Commercially important species that is *J. sambac*, *J. auriculatum* and *J. grandiflorum* have been developed either through open pollinated seedlings selection or through clonal variation.

- Varieties developed at TNAU, Coimbatore are In *J. sambac*, varieties such Sujimallige, Gundumalli, Motia, Bela, Verupakshi, Iruvatchi, Madanbam, Ramabanam, Elusuttumallige, Adukkumalli and Mohara. In *J.auriculatum*, CO-1 Mulliai, CO-2 Mullai, Pari Mullai, Large Round, Long Point and Short Point and CO-1 Pitchi in *J. grandiflorum.*

- A highly fragrant and high yielding variety Surabhi has been released by IIHR, Bengaluru.

Orchids

- In India, breeding work on orchids was started at the IIHR, Bengaluru, during mid-seventies. The Institute has developed 2 hybrids, *viz*. IIHR 164 and IIHR 38, in Vanda and *Dendrobium* respectively.

- The centres namely Vellanikkara, Yercaud, Kalyani, Bhubaneswar, Kahikuchi, Shillong, Kalimpongand UAS Bengaluru collected and maintained 165 accessions of various genera, species and hybrids of orchids.

- The All India Coordinated Project on Floriculture in orchid carried out at Kalimpong, Shillong, Kalyani and Vellanikkara (Chadha, 1986).

Dahlia

- In dahlia, several varieties have been developed under giant decorative (11) medium decorative (6), small decorative (6) and cactus group (1), all by amateur growers.

Tuberose

- In tuberose, more than 50 accessions were collected, maintained and evaluated at Hessaraghatta, Coimbatore, Kahikuchi, Kalyani, Hyderabad, Lucknow, Ludhiana, Pune, Delhi and Vellanikkara.
- Two mutants, *viz.* Rajat Rekha and Swarna Rekha in single flowered and a double flowered tuberose respectively, were obtained through gamma irradiation (1 to 5 kr) at the NBRI. In Rajat Rekha there are silvery white streaks along the middle of the blade, whereas in Swarna Rekha golden-yellow streaks are present along the margins of the blade (Gupta, 1979).
- Prajwal and Vaibhav had been recorded outstanding.
- Many varieties have been developed through planned crosses and open pollinated seedling selection in India which includes Bikkus Mother, Bikkus Vivek, Jyotsana, Lord Budhha, Probodh, Sarda Devi, Swami Gourisvarananada, Swami Lokeshwarananda and Swami Madhawananda.
- Other varieties developed in India through hybridization include Basudev, Blue Monarch, Chitchor, Disco, Jayanti and Sribhabani.
- One variety Manali which has magenta streaked flower has been developed from white pearl variety at the IARI, Regional station, Katrain (Kullu-valley).
- Kenya White, Kenya Blue, Kenya Yellow and Manjushri have been evolved as result of bud sports.

Amaryllis and Hippeasterum

- Several varieties have been released by the NBRI, Lucknow. They include cv. Kiran a triploid, cv. Samrat a tetraploid and Jyoti a semi-double. The IARI, New Delhi, has released variety Suryakiran. A variety named Anjana has been developed from an inter-varietal cross between Bouquet and White Favourite (Jana, 1991).
- Hybrids developed at NBRI, Lucknow, namely Gorgeous, Apollo, Diara, Sydney, Percy Lancaster, Aries, Cares, Charon, Coquette, Imperor, Nizam, Phoenix, Saturn, Snow White, Samrat, Kiran, Jyoti, Poonam, Deepali, Apurab and Chitwan.

- Suryakiran has been developed at IARI, New Delhi by crossing *Amaryllis stylossum* red with *A. stylossum* white.

Croton

- Most of the breeding work on varietal development of this crop has been carried out at Lalbagh, Bengaluru, and at the IIHR, Bengaluru. At the IIHR, Bengaluru, 15 hybrids of crotons have been developed (Anon., 1983).
- Improved cultivars developed at IIHR, Bengaluru popular among growers include Akali, Phoola Singh, Arkavati, Gen. Kumarmangalam, Gnat, Lt. Gen. Yadavindra Singh of Patiala, Pinkelegans, Pokhran, Sanjay, S.B. Lal, Saheed Bhagat Singh, Sonar Bangala, Valmiki, Vijavantha and Vikrant.

Perennial portulaca

- At NBRI, Lucknow six varieties developed through mutagenesis using gamma radiation include Jhumka, Karanapali, Lalita, Mukta, Vibhuti and Ratnam (Sane, 1986).

Marigold

Marigold is one of the most important cut flowers of India, being widely used both for religious and decorative purposes. Singh and Swarup (1971, 1973a, 1973b) studied the genetics of African marigold and reported appreciable heterosis for all the characters studied and also found dominance and epistasis playing a major role in the inheritance of days to flowering and flower weight. The double flower character was found to be monogenic dominant and the type and colour of flowers governed by two genes each. Transgressive segregation for various characters was observed in the F_2 generation, indicating its importance in the improvement of these characters.

- In marigold, a F_1 hybrid triploid has been developed at the NBRI, Lucknow, by using male-sterile African diploid marigold (*Tagetes erecta*) and male-fertile French tetraploid (*T. patula*). The F_1 hybrids are dwarf, highly floriferous and free flowering with uniform golden-yellow heads that bear nearly one-and-a half times more flowers by weight per unit area (Jalil *et al.*, 1974 and Jalil, 1976).
- At the IARI, New Delhi, a F_1 hybrid of French marigold obtained from crosses between Petite Orange × Rusty Red and Naughty Marrieta × Pygmy showed increase in the weight and number of flowers respectively than that of parents. In African marigold, F_1 hybrids between Hawaii × Cupid Orange and Alaska × Spun Gold were superior to parents (Bose, 1972).
- Pusa Sankar-1 is F_1 hybrid has been developed by using apetalous type of male sterility.

China aster

- From the germplasm collections at the IIHR, Bengaluru, 25 pure lines were developed by a single plant selection. Of these AST 1 and AST 2 were found very promising.
- Varieties developed at IIHR, Bengaluru are Kamini, Poornima, Sashank and Violet Cushion. Hybridization followed by repeated selection resulted in 6 promising pure lines, which were numbered IIHR 35, IIHR 18, IIHR 19, IIHR 26, IIHR 31 and IIHR 35 (Anon., 1988).
- According to DARE in 1999-2000 China aster Kamini, Poornima, Phule Ganesh Pink and Phule Ganesh Violet were recommended for release at the national level.
- Phule Ganesh Purple, Phule Ganesh Pink, Phule Ganesh White and Phule Ganesh Violet are developed at MPKVP, Pune.

Barleria

- The TNAU has released a variety CO-1 Barleria. It is a clonal selection from the local type, and bears attractive pink flowers.

Amaranthus

- Systematic hybridization and selection programme in amaranthus has resulted in the development of 8 cvs., *viz*. Amar Kiran, Amar Poet, Amar Prithu, Amar Parvati, Amar suikiran, Amar Tarang, Amar Raktab and Amar Mosaic.
- A tetraploid cultivar Amar Tetra was evolved through colchiploidy.
- Amar Shola, a hybrid amaranth, is a selection from a cross within *Amaranthus caudatus* complex involving a grain type and an ornamental type (Pal, 1970).
- Amar Shola, Amar Kiran, Amar Mosiac, Amar Poet, Amar Parvati, Amar Tetra, Amar Prithu, Amar Raktabh and Amar Tarang are developed at NBRI, Lucknow.

Antirrhinum

- Tetra giant snapdragons were raised at the NBRI, Lucknow, from some F_1 hybrids. The resultant tetraploids are hardier, sturdier, stockier and shorter than corresponding diploids. The Tetra giant have more flowering stems large deep coloured flowers and long lasting than diploid (Khoshoo, 1990).

Hollyhock

- At the IARI, 4 F_1 hybrids have been developed from intervarietal crosses in hollyhock. These are Pusa Apricot Supreme, Pusa Pastel Pink Supreme, Pusa Pink Beauty and Pusa Yellow Beauty.

- Deepika, Dulhan, Gouri, Pusa Sweta, Pusa Krishna, Pusa Lalima, Pusa Gulabi are seven open pollinated varieties developed at IARI, New Delhi.

Coreopsis

- Pusa Tara is a spontaneous mutant developed at IARI, Regional Station, Katrain (Anon., 1991).

Zinnia

- By recurrent selection from the irradiated seeds of *Zinnia elegans*, a mixed coloured variety resistant to leaf-curl virus, has been evolved at the IARI, New Delhi (Swarup and Raghava, 1974).

Anthurium

- In anthurium, 206 varieties were collected, maintained and evaluated at Hessaraghatta Coimbatore, Yercaud, Kalimpong, Kahikuchi, Shillong, Kalyani and Vellanikkara.

REFERENCES

Abraham, V. and Desai, B.M. 1977. Radiation induced variegated mutant in bougainvillea. *Current Sci.,* **46**:351.

Anderson W. and Park, R. 1989. Growing Gladioli. Timber Press. Portland, Oregon. pp. 166.

Anonymous. 1983. Research on Ornamental Crops at IIHR. IIHR News Highlights, **4**(2): 14.

Anonymous. 1988. Research Highlights (1971-1985). All India Coordinated Floriculture Project (ICAR). AICFIP Tech Bull., **4**: 21.

Anonymous. 1991. Progress Report (1990-1991). All India Coordinated Floriculture Project (ICAR), New Delhi. pp. 177.

Anonymous. 2013. Annual Report (2012-13). Directrate of Floricultural Research, New Delhi.

Arora, O.P. 1972. Cytogenetical studies of some ornamental verbenas. Ph.D. Thesis. Kanpur University, Kanpur.

Arora, O.P. and Khoshoo, T.N. 1967. Breakdown and restoration of fertility in *Verbena* hybrids. *Iridian J. Genet.,* **21**: 275.

Bali, P.N. and Tandon, S.L. 1959. Morphological and cytological studies of the induced polyploids in *Alyssium maritimum. Genetica,* **30**: 129.

Bose, T.K. 1972. Floriculture in India in the last two decades. *Indian Farming*, **22**: 13-18.

Buch, P.O. 1978. The species. *In*: N. Kaeing and K. Crowley (eds.). The world of the *Gladiolus*. Edgewood Press, Edgewood, Maryland. pp. 2-7.

Chadha, K.L. 1986. Research on Ornamental Crops at IIHR. *In*: Chadha, K.L. and Choudhary, B. (eds.). Ornamental Horticulture in India. Publication and Information Division, ICAR, New Delhi. pp 1-6.

Chadha, K.L. 1993. Floriculture Research in India. *J. Ornamental Hort.*, **1**(1): 1-17.

Chadha, K.L. and Bhattacharjee, S.K. 1995. Ornamental plants research in India- History, Infrastructure and Achievements *In:* K.L. Chadha and S.K. Bhattacharjee (eds.). Advances in Horticulture: Ornamental Plants, Vol. 12-II. Malhotra Publishing House, New Delhi. pp. 1-41.

Das, P.K. and Ghosh, P. 1972. Improvement of some vegetatively propagated ornamentals. *Abstr.* Third Int. Symp. on Subtropical and Tropical Hort., Bangalore.

De, L.C. and Bhattacharjee, S.K. 2011. Ornamental Crop Breeding. Aavishkar Publishers, Jaipur. pp 4-8.

Gupta, M.N. 1966. Induction of somatic mutations in some ornamental plants. *Proc.* All· India Symp. Hort., Calcutta: 107.

Gupta, M.N. 1979. Mutation breeding of some vegetatively propagated ornamentals. *In*: Progress in Plant Research. Silver Jubilee Publication, NBRI. Today and Tomorrow's Printers & Publishers, New Delhi, **II**: 75.

Gupta, M.N. and Jugran, H.M. 1978. Mutation breeding of chrysanthemum. II. Detection of gamma-ray induced somatic mutations in M_1. *J. Nuclear Agric. Biol.*, **7**: 50.

Gupta, M.N. and Nath, P. 1977. Mutation breeding in bougainvillea II. Further experiments with cvs. Partha and President Roosevelt. *J. Nuclear Agric. Biol.*, **6**: 122.

Gupta, M.N. and Shukla, R. 1971a. Effects of gamma irradiation on some scented roses. *Japanese J. Breeding,* **21**: 129.

Gupta, M.N. and Shukla, R. 1971b. Mutation breeding of chrysanthemum. I. Production of new cultivars by gamma·ray induced somatic mutation inVM_1. *Proc.* Int. Symp. on Use of Isotopes and Radiation in Agriculture and Animal Husbandry Research, New Delhi. 164.

Gupta, M.N. and Shukla, R. 1974. Mutation breeding in bougainvillea. *Indian J. Genet.*, **34A**: 1296.

Jain, H.K. and Majumdar, P.K. 1957. Some cytological and phenotypic effects of X-irradiation in annual chrysanthemum. *Indian J. Hort.*, **14**: 1.

Jain, H.K., Bose, A.K., Satpathy, D. and Sur, S.C. 1961. Mutation studies in annual chrysanthemum. I. Radiation induced variation in Dower form. *Indian J. Genet.*, **21**: 68.

Jalil, R. I976. Induction and assessment of some colchiploids in some ornamentals. M. Phil. thesis. Aligarh Muslim University, Aligarh.

Jalil, R., Khoshoo, T.N. and Pal, M. 1974. Origin, nature and limit of polyploidy in marigolds. *Curr. Sci.*, **43**: 717.

Jalil, R., Zadoo, S.N. and Khoshoo, T.N. 1974. Colchiploid balsams. *Nucleus,* **17**: 118.

Jana, B.K. 1991. Anjana- A new hybrid amaryllis, *Hort. J.*, **3**: 6-8.

Kaicker, U.S. 1988. Breeding of carnation (*Dianthus caryophyllus* L.)- A review. *Haryana J. Hort. Sci.*, **17**(3-4): 166-76.

Kher, M.A. and Bhutani, J.C. 1979. Floriculture at the National Botanical Research Institute. *In*: Khoshoo, T.N. and Nair, P.R. (eds.), Progress in Plant Research. Today and Tomorrow's Pub., New Delhi, **2**: 127-36.

Khoshoo, T.N. 1968. Genetic improvement of ornamentals in India-prospects, experimental approaches and suggestions. *Indian J. Genet.*, **28A**: 187-98.

Khoshoo, T.N. 1979. Genetics in relation to plant evolution and improvement. *In*: Khoshoo, T.N. and Nair, P.K. (eds.). Progress in Plant Research. Silver Jubillee Pub., NBRI. Today and Tomorrow's Pub., New Delhi, **2**: 7-14.

Khoshoo, T.N. 1990. Creativity in ornamentals. *Indian Hort.*, **34**(4): 25-30.

Khoshoo, T.N. and Zadoo, S.N. 1969. New perspective in *Bougainvillea* breeding. *J. Hered.*, **60**: 357-60.

Khoshoo. T.N. 1979. Cytogenetics in relation to plant evolution and improvement. *In*: Progress in Plant Research. Silver Jubilee Publication, NBRI, Today and tomorrow's Printers and Publishers, New Delhi, II.

Khoshoo. T.N. and Arora, O.P. 1971. Free flowering hybrid verbenas. *Amer. Hort. Mag.*, **50**: 16.

Khoshoo. T.N. and Zadoo, S.N. 1969. New perspectives in *Bougainvillea* breeding. *J. Hered.*, **60**: 357.

Kumar, R., Saha, T.K., Kumar, P.N. and Kumae, G. 2011. Quinquennial Review (2006-2011). AICRP on Floriculture. Directorate of Floricultural Research. New Delhi.

Mahal, C. 1972. Cytogentics of *Antirrhinum* and *Lantana.* Ph.D. Thesis. Kanpur University, Kanpur.

Mahal. C., Pal, M. and Khoshoo, T.N. 1968. Artificial and natural polyploids in *Antirrhinum. Curr. Sci.*, **37**: 5.

Malik, R.S. 1979. Study of heterosis in antirrhinum (*Antirrhinum majus* L.). Ph.D. Thesis. Agra University. Agra.

Malik, R.S. and Dadlani, N.K. 1984. Rose cultivation in India. *Ind. Hort.*, **29**(2): 27.

Malik, R.S. and Dadlani, N.K.1980. Improvement of roses. *IARI Res. Bull*. No. 5. New Delhi.

Malik, R.S. Dadlani, N.K. 1986. Dr. B.P. Pal and rose improvement in India. *Delhi Gdn. Mag.*, **25**(1): 1.

Malik, R.S. Dadlani, N.K. 1986. Genetic improvement of ornamental plant. *Indian J. Genet. Plant Breeding*, **46**(1): 253-66.

Malik, R.S., Singh, B. and Dadlani, N.K. 1984. Three decades of research on floriculture in India. *Ind. Hon.,* **29**(2): 9.

Misra, R.L. 1978. Induced mutations in gladioli through gamma radiation. *Gladiolus Annual,* 71.

Misra. R.L. 1977. Induction of mutations in gladiolus. *Plant Sci.*, **9**: 99.

Muthukrishanan, C.R. and Pappiab, C.M. 1980. Jasmine breeding-a resume. *Proc.* National Seminar on Prodn. Technol. for Common Flower Crops. TNAU, Coimbatore.

Muthuswami, S. and Rao, V.N.M. 1972. Tamil Nadu's fragrant new Parimullai. *Indian Hort.*, **17**(3): 28.

Nath, P., Banerji, B.K. and Gupta, M.N. 1983. Spontaneous and induced mutations in bougainvillea. Newsletter, Bougainvillea Society of India, III(1).

Negi, S.S. and Raghava, S.P.S 1986. Gladiolus. *In*: Chadha, K.L. and Choudhary, B. (eds.). Ornamental Horticulture in India. ICAR Pub., New Delhi. pp. 86-103.

Ninan, T., Singh, M.P. and Swaminathan, M.S. 1959. Meiotic behaviour and pollen fertility in some variaties of bougainvilleas. *J. Indian Bot. Soc.*, **38**: 140.

Ohri, D. and Koshoo, T.N. 1985b. Cytogenetical evolution of garden *Gladiolus. The Nucleus*. **28**: 216-21.

Pal, B.P. 1969. A note on rose breeding. *Indian J. Hort.*, **26**: 35.

Pal, B.P. 1972. The Rose in India (second edition). I.C.A.R., New Delhi.

Pal, B.P. 1982. Rose breeding-some thoughts and experiences. *The Indian Rose Annual.* **2**: 10.

Pal, M. 1970. Cyto-anatomical studies in some Indian Amaranthaceae. Ph.D. Thesis, Agra University, Agra.

Pal, M. and Khoshoo, T.N. 1977. Evolution and improvement of cultivated amaranths. VIII. Induced autotetraploidy in grain types. Z. *Pflanzenzucht.*, **78**: 135.

Rizvi, S. 1973. Colchiploidy in *Dimorphotheca aurantica. Plant Sci.,* **5**: 39.

Sane, P.V. 1986. Floriculture Research at the NBRI. *In*: Chadha, K.L. and Choudhary, B. (eds.). Ornamental Horticulture in India. ICAR Pub., New Delhi. pp 20-37.

Singh, B. and Swarup, V. 1973b. Inheritance of qualitative characters in African marigold. *Indian J. Genet.*, **33**: 373.

Singh, B. and Swarup, V. 1971. Heterosis and combining ability in African marigold. *Indian J. Genet.*, **31**: 407.

Singh, B.P., Gupta, P.N. and Rana, R.S. 1993. New introductions in ornamentals. *J. Ornamental Hort.*, **1**(1): 68-72.

Singh. B. and Swarup, V. 1973a. Inheritance of flowering, flower weight and flower number in African marigold. *Indian J. Genet.*, **33**: 172.

Singh. F. 1971. Cytogenetics of some ornamentals with particular reference to garden *Hibiscus*. Ph. D. Thesis. Karnataka University, Dharwar.

Swamp. V., Raghava, S.P.S. and Balakrishanan, K.A. 1975. Heterosis in balsam. *Indian J. Genet.*, **35**: 69.

Swarup, V. and Raghava, S.P.S. 1974. Induced mutations for resistance to leaf curl virus and its inheritance in garden zinnia. *Indian J. Gene. Plant Breed.*, **34**(1): 17-21.

Swarup, V., Kaicker, U.S., Malik, R.S. and Singh, A.P. 1973. Rose breeding in India. *Indian J. Gene. Plant Breed.*, **33**(1): 43-51.

Swarup, V., Ragbava, S.P.S. and Balakrishanan, K.A. 1973b. Genetical studies in hollyhock. *Indian J. Genet.*, **35**: 69.

Swarup. V., Kaicker, U.S., Malik, R.S. and Singh, A.P. 1973a. Rose breeding in India. *Indian J. Genet.*, **33**(1): 43.

Vetuswamy, P., Thangaraj, T. and Muthuswami, S. 1973. Morphological variation in *J. grandldiflorm* L. *Madras Agric. J.*, **60**: 1695.

Zadoo, S.N. and Khoshoo, T.N. 1968. Cytogenetical studies on *Bougainvillea.* A case of interchange heterozygosity. *Genetica,* **39**: 353.

Chapter – 2

Conventional Breeding

INTRODUCTION

Various procedures which are used for genetic improvement of crop plants are known as breeding methods. The breeding approaches that are more commonly used for genetic improvement of crop plants are called conventional breeding methods. Following conventional breeding methods were followed in improvement of crops.

(A) Plant introduction

(B) Pureline selection

(C) Mass selection

(D) Progeny selection

(E) Synthetic and composite breeding

(F) Pedigree method

(G) Bulk method

(H) Backcross method

(I) Heterosis breeding

(J) Clonal selection

(A) PLANT INTRODUCTION

The acquisition of superior varieties by importing them from other areas accomplishes the same purpose as developing superior varieties in deliberate breeding programmes (Allard, 1966). Primary introduction and secondary introduction are two types of plant introduction. Introduction of new crops is called primary introduction. Introduction and exploitation of variability for the genetic improvement of existing crops is called secondary introduction. Plant introduction accomplishes two jobs (Sharma, 2002) as given below.

1. It provides for isolation of superior types for direct use.
2. It builds the base for future breeding.

Domestication and acclimatization are two important processes of plant introduction. The process of bringing the wild species under cultivation is called Domestication. Acclimatization is the ability of a crop or crop variety to become adapted to a new climatic and soil conditions. Crop or crop species are also introduced to serve the aesthetic interest or fancy for uncommon and bizarre forms, e.g., ornamental plants and shrubs, like bougainvillea, *Jacaranda mimosaefolia*, *Spathodea campanulata*, phlox, aster, snapdragon, etc. Thus, through plant introduction, the wild relatives of cultivars/local populations- land races and sometimes, completely new crop species can be obtained.

SOURCES OF GERMPLASM

The germplasm is obtained through plant introduction from three main sources

1. Wild species and primitive forms of crops in primary centres of diversity/origin.
2. Plant migrants to secondary centres of cultivars where their diversity may be augmented by human interference.
3. Unique products of plant breeding.

Procedure for plant introduction

1. Correspondance
2. Exchange
3. Purchases
4. Gifts
5. Travellers
6. Foreign embassies
7. Deliberate intent for specific types for plant breeding

8. Friendly government exchanges
9. Plant exploration in centres of diversity

Organizations concerned with plant introduction in India are

i. Botanical Survey of India, Calcutta, for medicinal and botanical plants.
ii. Forest Research Institute, Dehradun, for trees and forest plants.
iii. Central Research Institutes and Agricultural Universities for horticultural and field crops.
iv. National Bureau of Plant Genetic Resources (NBPGR), New Delhi is the main organization dealing with plant introduction from foreign countries, along with the maintenance of collection of both exotic and indigenous origin. It is the main channel in India of in and out movements of plant materials after fulfilling the quarantine requirements.

At International level, crop plants or germplasm are collected by International Plant Genetic Resources Institute (IPGRI), Rome, Italy.

(B) PURELINE SELECTION

The term pureline was first used by Johannsen in 1903. The progeny of a single, homozygous, self pollinated plant is called pureline. The concept of pureline theory was developed by Johannsen in 1926 while working with Princess variety of common bean (*Phaseolus vulgaris*). Main features of purelines are

1. All the plants within a pureline have the same genotype as the plant from which the pureline was derived.
2. Purelines are homozygous and homogeneous.
3. Purelines are highly uniform.
4. The variation within a pureline is environmental and non-heritable.
5. Purelines have narrow genetic base.
6. Purelines have narrow adaptation.

Highly self pollinated seed propagated herbaceous annual flowers include China aster, balsam, sweet pea, lupin, clianthus, gypsophila, bells of Ireland, salvia, saponaria, sweet william, carnation and dianthus. In these flowers, there is negligible cross pollination in nature. In flowers like statice, linaria, lobelia, nemasia, dahlia and salpiglossis, cross pollination is also quite frequent. In crops like larkspur, godetia, nicotiana, nigella, phlox and verbena, cross pollination occurs to a considerable extent (Watts, 1980 and Swarup, 1997).

General procedure for pureline selection

The pureline selection has following 3 steps (Singh, 2009).

1. Selection of individual plants from a local variety or some other genetically variable and homozygous population. In the first year 200-300 plants are selected on the basis of their phenotypes.
2. Visual evaluation of individual plant progenies: In second year individual plant progenies are grown and undesirable progenies are rejected. Here, number of plant progenies are carried forward are drastically reduced.
3. Yield trials: In the third year remaining progenies are planted in a preliminary yield trial and inferior progenies are rejected. From fourth year to sixth year replicated yield trials are conducted at several locations, inferior progenies are rejected, disease resistance and quality tests are done. In seventh year best progeny is released as new variety and seed or plant propagules multiplication begins for distribution.

Advantages of pureline selection

1. Pureline varieties are extremely uniform i.e., homozygous and homogeneous.
2. Maximum possible improvement over the original variety is achieved.
3. Helps in isolating best genotypes for disease resistance, insect resistance, yield, quality, earliness, etc. from a mixed population of an old variety.
4. This is an easy and cheap method of crop improvement.
5. Pureline variety can be easily identified in seed certification programmes.

Disadvantages of pureline selection

1. Pureline varieties have narrow adaptation.
2. Due to narrow genetic base the pureline varieties cannot be improved beyond certain limit.
3. Takes more time compared to mass selection.

(C) MASS SELECTION

Mass selection is the method of crop improvement in which individual desirable plants are selected on the basis of phenotype from a mixed population, their seeds are bulked and used to grow next generation is called mass selection (Singh, 2010).

Main features of mass selection

1. It is used in both self and cross pollinated crops.
2. The varieties developed by this method have wider adaptability and broad genetic base.
3. The mass selected variety is homozygous but heterogeneous as it is a mixture of several purelines.
4. Compared to the varieties developed by pureline selection the varieties developed by mass selection are less uniform.
5. Mass selected variety have heritable variation.
6. Mass selected variety is less prone to pest and disease attack due to broad genetic base.

Advantages of mass selection

1. Mass selection is the simplest, easiest and quickest method of developing cross pollinated crops due to the absence of progeny testing and no control over pollination.
2. It is the simplest and effective method to improve the adaptability of an open pollinated variety.
3. For developing still more productive hybrids mass selection is very effective in increasing the yield base of parents.
4. New variations can be developed quickly.
5. Revamping of 'Run out' varieties can be done quickly.
6. As the generation interval is minimized gain/year is more in mass selection.
7. Better utilization of germplasm pool in mass selection.

General mass selection scheme

The process of mass selection involves selection or retention of a large number of individual plants on the basis of superior phenotype and composting the harvested seeds without progeny testing. En masse sowing of mixture of seeds is done to raise the next generation. Till the desired level of uniformity is achieved the process is continued. Mass selection is employed most often for purification of existing improved varieties. Mass selection is female oriented as there is no control over the male parent. Progeny testing is not done in mass selection. The aim of mass selection is (Sharma, 2002) as under.

i. To produce new varieties from heterogeneous local land races, or from racial blends.

ii. To purify mixed varieties.

iii. To maintain the purity of the variety.

Table 1: Differences between pureline selection and mass selection.

	Pureline selection		Mass selection
1.	Pureline selection is used in self pollinated and often cross-pollinated crop species.	1.	Mass selection is used in both self and cross pollinated crop species.
2.	The new variety is a pureline.	2.	The new variety is a mixture of purelines.
3.	A pureline is homozygous and homogeneous.	3.	The selected plants are homozygous and heterogeneous.
4.	Selection within a pureline is not effective.	4.	Selection within mass selected variety is effective.
5.	The variety has narrow adaptability.	5.	The variety has wide adaptability.
6.	The developed variety is highly uniform.	6.	The developed variety is less uniform.
7.	No genetic variation within the pureline.	7.	Genetic variation is present in the mass selected variety.
8.	Careful progeny testing is necessary.	8.	No need of progeny testing.
9.	New variety release takes 6-7 years.	9.	New variety release takes 7-8 years.

(D) PROGENY SELECTION

Progeny test is also called Vilmorin Principle as this was developed by Louis de vilmorin. The test of the genotype value of an individual based on the performance of its progeny is called progeny test. Progeny selection refers to selection of plants from a diverse population on the basis of their progeny test. Progeny selection refers to selection of plants from a diverse population on the basis of their progeny test. In this selection procedure superior plants are selected from a heterogeneous population on the basis of their progeny performance. The progeny test is useful in two ways, *viz.* (1) in understanding whether a plant is true breeding (homozygous) or segregating (heterozygous) for a particular character and (2) in the assessment of the breeding value of a plant. If the selected plant is really superior, its progeny will also exhibit superior performance and vice versa.

Main features of progeny selection

The main features of progeny selection described by Singh (2010) are given below.

1. Selection of superior plants is based on the genotype, *i.e.* progeny performance.

2. Progeny selection is commonly used in cross pollinated and often cross pollinated species.

3. In cross pollinated species, three types of materials *viz.* (i) open pollinated seeds, (ii) Self seeds and (iii) top cross or test cross seeds of selected plants can be used in producing the progenies for testing. Generally, 10-50 seeds of each selected plants are grown for progeny testing.

4. Progeny test should be conducted in replicated trial to get more reliable results. The multilocational or multiseasonal test is considered ideal for progeny testing.

5. Those plants whose progeny performance is superior for the character under consideration are bulked together to produce the next generation.

Progeny selection is an efficient method of selecting superior genotypes on the basis of progeny performance of selected plants. Finally, plants whose progeny performance is good are selected and rests are discarded. This method is very simple and convenient. The main demerit of progeny selection is that there is no control over pollination. The plants of superior progeny are also pollinated by the plants of inferior progenies.

(E) SYNTHETIC AND COMPOSITE BREEDING

Synthetic variety: A variety which is developed by intermating in all possible combinations a number of inbred lines with good general combining ability and mixing the seeds of F_1 crosses in equal quantity is referred to as synthetic variety (Singh, 2010).

Main features of synthetics

1. Inbreds, clones and open pollinated varieties can be used as lines for developing synthetic variety.

2. Constituent genotypes are selected on the basis of GCA.

3. Generally, 5-8 lines with good general combining ability are used to constitute a synthetic variety.

4. Synthetics exploit more of additive gene action compared to hybrids which exploit more of non-additive gene action.

5. Synthetics have high stable yield and are polymorphic.

6. Reconstitution of synthetics should be done after every 4-5 years.

7. Due to heterogeneous nature and broad genetic base synthetic varieties have better disease resistance.

8. Yield of synthetic varieties more than open pollinated varieties but less than single and double crosses.

Steps for the derivation of synthetics from inbreds: It is described by Chopra (2008) as given below.

a. Isolation of one generation selfed lines (S_1).
b. Testing these lines in top crosses for high GCA for yield and other characters.
c. Intercrossing the superior lines to produce the F_1 generation of synthetic.
d. Advancing the F_1 synthetic in isolation under open pollination.

Advantages of synthetics

1. In areas where cost of hybrid seed is very high or availability is poor, farmers can keep their own seeds by growing synthetics.
2. Due to broad genetic base and greater variability more flexible to environmental changes.
3. Appreciable amount of heterosis can be exploited by using synthetic varieties in species where hybrid seed production is difficult.
4. Compared to single cross and double cross hybrids, seed of synthetic varieties is cheaper.

Disadvantages of synthetics

1. Synthetics exploit only GCA whereas, the single and double cross hybrids exploit both GCA and SCA. So, the performance of synthetics is less than single and double cross hybrids.
2. Synthetics are less uniform compared to hybrids due heterogeneity.
3. Performance of synthetics reduced by using lines with less GCA.
4. Synthetics can be developed in cross pollinated crops only whereas the hybrids can be developed in both self and cross pollinated crops.

Composite variety: Mixing the seeds of several phenotypically outstanding lines and encouraging open pollination to produce crosses in all combinations among the mixed lines constitute composite variety.

Main features of composites

1. Composites are seed mixture of advanced generation material of intervarietal and interracial cross.
2. Mixing the seed of various genotypes which are similar in maturity height, seed size, colour, etc. is the base material of composite variety.

3. To make use of heterosis in cross pollinated crops composite varieties are developed.
4. Farmers use their own seed for 3-4 years. The variety is maintained by open pollination.
5. Composites are more stable and highly adaptable to environmental fluctuations.
6. In composites, it is not possible to reconstitute the variety because in segregating populations gene frequency changes with time.
7. Heterogeneous populations of several homozygotes and heterozygotes is the genetic constitution of composite varieties.

Steps in development of composite variety

1. Selection of base material.
2. Intermating of selected genotypes.
3. Evaluation of crosses.
4. Mixing of parental material of superior crosses in equal quantity.

Several open pollinated varieties with similar maturity duration, height, seed size and seed colour are selected and intermated in all possible combinations. Crosses exhibiting little or no inbreeding depression in F_2, F_3 and F_4 generations are identified and selected for the development of composite cultivars. Seed multiplication is carried out by open pollination in isolation for one or two generations. Composite cultivars also exploit more of additive gene action (Singh, 2010).

(F) PEDIGREE METHOD

Pedigree is the record of ancestry of an individual selected plant for its various generations. The selection procedure which is used in segregating population of self pollinated species and keeps proper records of plants and progeny selected in each generation is known as pedigree breeding.

Main features of pedigree method

1. It is used for developing new varieties in self pollinated species.
2. It is used for developing inbreds in cross pollinated species.
3. For each selected plant pedigree records are maintained over generations.
4. Selection is based on genotypic value of progeny test.
5. The variety developed by pedigree method is homozygous and homogenous because it is a progeny of a single homozygote.

Pedigree procedure for handling segregating generations of crosses in self pollinated crops

The selected parents are crossed by planting in a crossing block. 10-30 F_1 seeds are space planted and the seed is harvested in bulk. In F_2 2000-10000 plants are space planted and 100-500 superior plants are selected and their seed is harvested separately. In F_3 and F_4 individual plant progenies are space planted and superior plants are selected. In F_5 and F_6 individual plant progenies planted in multi-row plots and superior plants are selected from superior progenies. Preliminary yield trial may be conducted in F_6. In F_7 generation, preliminary yield trials with 3 or more replications are conducted to identify few superior lines and quality tests also conducted. From F_8-F_{10} generations coordinated yield trials, along with disease and quality tests are conducted. In F_{11} generation, seed is multiplied and breeder seed distribution begins.

Advantages of pedigree method

1. It offers chances of obtaining transgressive segregants.
2. Pedigree method provides information about mode of inheritance of various qualitative characters.
3. The selection is based on genotypic value of progeny test rather than phenotypic value.

Demerits of pedigree method

1. Pedigree records for all selected plants over generations are to be maintained by the breeder which takes lot of time.
2. Huge breeding material has to be handled by the breeder which sometimes become difficult.

(G) BULK METHOD

The bulk method of handling segregating generations of a cross or composite cross population refers to a procedure, where segregating generations are advanced as a pool of genotypes without any artificial selection (Chopra, 2008). The method was first used by Nilsson-Ehle in wheat crop. As indicated by Allard (1960) Nilsson-Ehle had well appreciated and applied the important features of the bulk population method as under.

1. Natural selection eliminates poorly adapted low yielding types from the population.
2. Growing of a large population was essential for increasing the chances of recovering high yielding types.
3. The bulk method was the easiest for handling a large population as no record keeping and harvesting of individual plants was required.

4. Advancing generations by bulking will increase the homozygosity of individual plants in self pollinated crops. Selections made after a few generations of bulking are likely to breed true.

Harlan *et al.* (1940) proposed the bulk method to handle a large number of composite crosses involving several parents. In this method yield testing of F_2 bulks from a number of crosses and discarding the poor yielding crosses.

The selected crosses are then advanced by growing a large F_2 population (5000 to 10000 plants) at normal plant density and harvesting the material in bulk at maturity for planting the harvested seeds in a similar size plot in the following season. This process is repeated as many times as desired by the breeder. Normally by F_6 generation, most of the plants in the bulk population will reach a high level of homozygosity. In F_7 and F_8 single plant progenies are selected for yield tests.

Since, selection of individual plants in F_2 and F_3 for yield is ineffective (Sakai, 1954, McGinnis and Shebeski, 1968, Knott, 1972, Depauw and Shebeski, 1973, and Knott and Kumar, 1975), bulk handling of segregating generations seems to be the easiest procedure for propagating large populations of self pollinating plants from crosses to establish homozygous lines.

Mass-pedigree method proposed by Harrington (1937) as a means of overcoming the limitations of pedigree selection and single seed descent method suggested by Goulden (1939) for advancing segregating generations of self pollinated crops are the modifications of bulk method.

(H) BACKCROSS METHOD

Crossing of F_1 with either of its parents is termed as backcross. In the backcross method of breeding, the hybrid and the progenies in the subsequent generations are repeatedly backcrossed to one of the parents of the F_1. The main objective of the back cross method is to improve one or two specific defects of a high yielding variety (Singh, 2009). The parent which donated the desirable character is called donor plant. It is also called non-recurrent parent. The parent which receives the desirable character is called recipient parent also called recurrent parent as it is repeatedly used in backcrossing.

Backcrossing is a well-known and long established breeding scheme where a characteristic is introgressed from a donor parent into the genomic background of a recurrent parent. The characteristic could be a trait, a gene or even an anonymous locus or chromosome segment. In successive generations, progeny are selected for the characteristic of interest and then backcrossed to the recurrent parent. This ensures that the proportion of genome from the donor parent tends to zero as generations accumulate, except for the part hosting the characteristic of interest (Hospital, 2005). The objective is to reduce the latter to the smallest size necessary. If selection is applied for the desired characteristic only, then the proportion of donor genome is expected to be reduced by one-half (50%) at each generation, except on the chromosome holding the characteristic. On this chromosome, the rate of decrease is slower (Hanson, 1959 and Stam and Zeven, 1981 and Naveira and Barbadilla, 1992) resulting in linkage drag. Obviously, if selection can also be applied against the donor genome proportion, then its rate of decrease can become faster. Selection on phenotypic resemblance to the recurrent parent (or against the donor) has long been used by breeders. Selection can also be based on molecular marker alleles typical of either parent. Historically, this was among the first suggested uses of molecular markers to assist breeding programmes (Tanksley and Rick, 1980, Beckmann and Soller 1983 and Burr *et al.*, 1983). Reduction of linkage drag is the most difficult goal to achieve because of the selection for the target. Hence, this is where use of marker-assisted selection (MAS) is most rewarding (Young and Tanksley 1989 and Hospital 2001).

Backcross and introgression are useful for genetic improvement in breeding programmes. Backcrossing is also useful to dissect the genetic architecture of quantitative traits because it isolates a gene, or chromosomal region, in a different genetic background (the genetic background of the recurrent parent). In fact, it is one of the few reliable methods to validate the additive effect of a quantitative trait locus (QTL) or a candidate gene. In addition, backcrossing could be used prior to, or in conjunction with, QTL detection to increase the precision of QTL mapping (Hospital, 2005).

In addition to conventional breeding methods, various aspects of the use of molecular markers (for controlling the target genes, accelerating the recovery of recurrent genome or reducing linkage drag) to improve the efficiency of introgression in backcross breeding programmes have been investigated from a theoretical standpoint in recent years. These were reviewed (Visscher *et al.*, 1996, Whittaker, 2001, Dekkers and Hospital, 2002 and Hospital, 2003). Whether it is called marker-assisted introgression or marker-assisted backcross, use of markers in backcross breeding programmes is efficient. As far as theory is concerned, relatively simple calculations based on classical Mendelian genetics (probability of recombination) and simulations show that the use of markers may improve backcross breeding at all levels. Several studies have shown that selection on markers for the recovery of genetic background provides a gain in time equivalent to about two backcross generations. Selection against genetic drag can save tens of generations (Young and Tanksley, 1989), not necessarily at high cost (Hospital 2001). This is true even if the target gene is in fact a QTL located with a given error on the genetic map (Visscher *et al.*, 1996 and Hospital and Charcosset 1997). However, the number of targets is then limited. It is generally not possible to introgress more than four or five QTL, even with the largest population sizes. However, this assumes that the QTL is a 'true' QTL, that is, that it has an effect on the trait of interest (not a false positive) and that this effect is sustained over the breeding programme; specifically, the effect will be unmodified by changes of genetic background, environment or epistatic relationships with other genes. This assumption is not always true for all QTL, as will be discussed below when reviewing real-world QTL introgression experiments (Hospital, 2005).

Because backcrossing isolates a gene or chromosomal region in a different genetic background (the genetic background of the recurrent parent), it helps to dissect the genetic architecture of quantitative traits. In fact, it is one of the few reliable methods to validate the additive effect of a QTL or candidate gene after it is putatively detected. In addition, backcrossing could be used prior to detection to increase the precision of QTL mapping or exploit wild genetic resources. Introgressing one gene into a different genetic background removes or modifies the possible epistatic interactions between that gene and the rest of the genome; it is then useful to study the additive (non-epistatic) effect of the gene. Combinations of introgression lines with different genes and different backgrounds can thus be used to study epistatic interactions.

The advanced backcross QTL analysis proposed by Tanksley and Nelson (1996) is similar to the RSB scheme in that a QTL from an exotic resource is introgressed into 'elite' genetic background by selection. However, selection here is not necessarily for the trait of interest, but rather against the unwanted phenotypic characteristics of the donor. It could also be 'natural' selection (i.e. no conscious selection). This approach has been used with some success (Tanksley and Nelson 1996 and Bernacchi *et al.*, 1998).

Main features of backcross method: It is described by Singh (2010) as given below:

1. It is used in both self and cross pollinated species.
2. It is used to improve specific character of a well adapted variety.
3. It includes donor and recipient parents in crossing.
4. New variety is similar to parent except for the character under transfer.

Requirements for backcross programme

1. Suitable recurrent parent such as well adapted variety deficient in one character.
2. Donor or non recurrent parent for desirable character.
3. High heritability of the character under transfer.
4. Suitable number of back crosses are to be made. Generally, 6-7 backcrosses are sufficient for recovering the genotype of original variety with new character.

Advantages of backcross method

1. It is used for transfer of oligogenic traits such as disease resistance into cultivated varieties.
2. Intervariety transfer of simply inherited characters and quantitative characters.

3. It is used for developing nearly isogenic lines and multilines.
4. Interspecific gene transfer of simply inherited characters.
5. It is used for transfer of male sterility and restorer genes to various agronomic bases.

Drawbacks of backcross method

1. If there is tight linkage between some desirable and undesirable characters then the undesirable characters are also transferred along with desirable characters.
2. It is more successful for transfer of monogenic and oligogenic traits than polygenic traits.

It is likely that the old backcross-breeding scheme still has a lot to offer the genetics community. Improved selection methods based on marker information are still sought to make it more efficient, leading to improved agricultural populations of plants, finer description of the genetic architecture of quantitative traits, better understanding of epistasis between QTL and wider knowledge of genotype by environment interactions. Finally, it may help us tap into the genetic resources present in the wild relatives of most crops (Hospital, 2005).

(I) HETEROSIS BREEDING

The superiority of F_1 hybrid over its parents in one or more characters is called heterosis. The term heterosis was coined by Shull in 1914.

Main features of heterosis

1. F_1 hybrid exhibits superiority over both parents in different characters.
2. The heterosis has association with specific combining ability.
3. The superiority is confined to F_1 generation and is under genetic control.
4. Deleterious recessive genes are concealed in heterozygous condition.

Estimation of heterosis

Heterosis is estimated in three ways, *viz.* heterosis over the mean value of both the parents, heterosis over the better parent and heterosis over the standard cultivar/hybrid.

Advantages of hybrids

1. Hybrids have high productivity.
2. They have high uniformity.

3. They are attractive than synthetics, composites and open pollinated varieties.

Demerits of hybrids

1. Hybrid seed is costly.
2. Seed replacement should be done every year.
3. Fresh seed is to be purchased every year.

(J) CLONAL SELECTION

A clone is a group of plants produced from a single plant through asexual reproduction. Improvement of asexually propagated crops by selecting superior clones is known as clonal selection. For clonal selection a wide range of clones from various sources are collected. Superior clones can be isolated from three types of materials, *viz.* local variety, introduced variety and intercrossed populations.

1. In the first year collected clones are grown, from which superior clones are selected.
2. In the second year the selected clones grown separately and superior clones are selected on visual observation of morphological traits.
3. The selected clones evaluated in multilocation trials for 3-4 years with standard checks.
4. The best clone superior to the check variety is selected based on the yield, disease resistance and quality.

The numbers of female and male strobili assessed for 99 clones in a *Pinus densiflora* Sieb. et Zucc. seed orchard for four consecutive years showed that differences for the production of both female and male strobili among clones were very large and statistically significant for all years examined. The most abundant quartile of clones produced 48.1 and 37.3% of the total female and male strobili, respectively, averaged over the 4-year period. Differences in female strobilus production among clones were more variable than in male. Pearson's correlations between years for flower production were positive for all combinations of years. Spearman's rank correlation coefficients between years with moderate female flowers were highly significant. Both correlations between female and male flowers were moderate and positive for all 4 years. Broad-sense heritabilities for flower production on an individual ramet basis varied from 0.19 to 0.41, while those on a clone-mean basis were much higher, from 0.57 to 0.80. Maleness index showed that sexual symmetry in good and poor flower years deviated more from the ideal panmictic situation than that in intermediate years. Parental balance curves showed that orchard clones contributed more equally to the seed crop when flower production was moderate (Kang, 2000).

Merits of clonal selection

1. Clonally propagated varieties are highly stable and uniform.
2. Clonal selection helps in maintaining the purity of the clones.
3. It helps in isolating the best genotype from a mixed population of asexually propagated crops.
4. Besides avoiding the inbreeding depression it also preserves the heterosis present in clones.

Demerits of clonal selection

1. Clonal selection only conserves the variability already present in the population but cannot create new variability.
2. Clones are prone to new races of pathogen. So, hybridization is prerequisite by which not only resistance to pathogen is inducted into clones but also new variability can be created in them.

Table 2: Differences between pureline and clone.

	Pureline		Clone
1)	Progeny of a single homozygous self fertilized plant is called pureline.	1)	Progeny of a single asexually propagated plant obtained by asexual means.
2)	It is related to self pollinated species.	2)	It is related to asexually propagated species.
3)	All the genotypes are identical, homozygous and homogeneous.	3)	All the genotypes are identical, heterozygous and homogeneous.
4)	It has narrow adaptability.	4)	It has wide adaptability.
5)	Genetic variation may occur due to natural outcrossing, spontaneous mutations and mechanical mixtures.	5)	Genetic variation may occur due to bud mutations, mechanical mixtures and occasional sexual reproduction.
6)	Maintained by natural self-pollination.	6)	Maintained by asexual propagation.

SUMMARY

Methods of procedure in breeding self-fertilized species of crop plants have been fairly well standardized. The same cannot be said with reference to cross-fertilized crops. These vary greatly in their mode of reproduction and breeding behavior (Kirk, 1933). Improvement of these species presents a variety of problems, depending on how they are to be cultured and utilized. Methods of breeding should be used, which are appropriate to the species concerned and the objective to be attained. Breeding methods commonly employed can be grouped under (A) Plant introduction (B) Pureline selection (C) Mass selection (D) Progeny selection (E) Synthetic and composite breeding (F) Pedigree method (G) Bulk method (H) Backcross method (I) Heterosis breeding

(J) Clonal selection. These are discussed and compared with respect to their applicability to different species and specific problems.

Plant introduction provides for isolation of superior types for direct use and builds the base for future breeding. Primary introduction and secondary introduction are two types of plant introduction. Domestication and acclimatization are two important processes of plant introduction. Introduction accomplishes the same purpose as developing superior varieties in deliberate breeding programmes. Highly self pollinated seed propagated herbaceous annual flowers include China aster, balsam, sweet pea, lupin, clianthus, gypsophila, bells of Ireland, salvia, saponaria, sweet william, carnation and dianthus. In these flowers, there is negligible cross pollination in nature. In flowers like statice, linaria, lobelia, nemasia, dahlia and salpiglossis, cross pollination is also quite frequent. In crops like larkspur, godetia, nicotiana, nigella, phlox and verbena, cross pollination occurs to a considerable extent (Watts, 1980 and Swarup, 1997).

The progeny of a single, homozygous, self pollinated plant is called pureline. Pureline varieties are extremely uniform i.e., homozygous and homogeneous. Maximum possible improvement over the original variety is achieved. This is an easy and cheap method of crop improvement. In nature there exists within the species definite hereditary habitat types which represent the genotypical response of the species-population to a particular habitat. "Mass selection" provides a ready means of capitalizing on the work which nature has accomplished. The progeny test provides the only reliable method of evaluating the breeding potentialities of parent plants. The information to be gained from such tests differs greatly in significance and precision depending on whether the progeny is grown from selfed seed or from hybrid seed of crosses between homozygous or heterozygous individuals. It is of supreme importance that progenies shall be subjected to the incidence of those environmental factors which presumably may influence the success of the new production (Kirk, 1933).

A variety which is developed by intermating in all possible combinations a number of inbred lines with good general combining ability and mixing the seed of F_1 crosses in equal quantity is referred to as synthetic variety. In areas where cost of hybrid seed is very high or availability is poor, farmers can keep their own seeds by growing synthetics. Mixing the seeds of several phenotypically outstanding lines and encouraging open pollination to produce crosses in all combinations among the mixed lines constitute composite variety. To make use of heterosis in cross pollinated crops composite varieties are developed. The selection procedure which is used in segregating population of self pollinated species and keeps proper records of plants and progeny selected in each generation is known as pedigree breeding. In bulk method segregating generations are advanced as a pool of genotypes without any artificial selection. Bulk method was the easiest for handling a large population as no record keeping and harvesting of individual plants was required. In heterosis breeding has association with specific combining ability. F_1 hybrid exhibits superiority over both parents in different characters.

Improvement of asexually propagated crops by selecting superior clones is known as clonal selection. A clone is a group of plants produced from a single plant through asexual reproduction.

REFERENCES

Allard, R.W. 1960. Principles of Plant Breeding. John Wiley & Sons, New York.

Beckmann, J.S. and Soller, M. 1983. Restriction fragment length polymorphisms in genetic improvement:methodologies, mapping and costs. *Theor. Appl. Genet.*, **67**: 35-43.

Bernacchi, D., Beck-Bunn, T., Eshed, Y., Lopez, J., Petiard, V., Uhlig, J., Zamir, D. and Tanksley, S. 1998. Advanced backcross QTL analysis in tomato. I. Identification of QTLs for traits of agronomic importance from *Lycopersicon hirsutum. Theor. Appl. Genet.*, **97**: 381-397.

Burr, B., Evola, S.V., Burr, F.A. and Beckmann, J.S. 1983. Application of restriction fragment length polymorphism to plant breeding. *Genet. Eng.*, **5**: 45-59.

Chopra, V.L. 2008. Plant Breeding Theory and Practice. Oxford & IBH Publishing Co. Pvt. Ltd. New Delhi.

De Pauw, R.M. and Shebeski, L.H. 1973. An evaluation of an early generation yield testing procedure in *Triticum aestivum. Canadian J. Pl. Sci.*, **53**: 465-70.

Dekkers, J.C.M. and Hospital, F. 2002 The use of molecular genetics in the improvement of agricultural populations. *Nat. Rev. Genetics.*, **3**: 22-32.

Goulden, C.H. 1939. Problems in plant selection. *In*: R.C. Burnett (ed.). *Int. Genet. Congr. Proc.* 7th (Edin burgh) Cambridge University Press. pp. 132-33.

Hanson, W.D. 1959. Early generation analysis of lengths of heterozygous chromosome segments around a locus held heterozygous with backcrossing or selfing. *Genetics*, **44**: 833-7.

Harlan, H.V., Martini, M.I. and Stevans, H. 1940. A study of methods in barley breeding. USDA *Technical Bulletin* 720, Washington, D.C. pp. 720-940.

Harrington, J.B. 1937. The mass-pedigree method in the hybridization improvement of cereals. *J. American. Soc. Agron.*, **29**: 379-84.

Hayes, H.K. and Garber, R.J. 1919. Synthetic production of high protein corn in relation to breeding. *J. American. Soc. Agron.*, **11**: 309-19.

Hospital, F. 2003. Marker-assisted breeding. *In*: H.J. Newbury (Ed.), Plant Molecular Breeding, Blackwell Scientific, London. pp. 30-56.

Hospital, F. 2005. Selection in backcross programmes. *Philosophical Transactions: Biological Sciences*, **360**(1459): 1503-11.

Hospital, F. and Charcosset, A. 1997. Marker-assisted introgression of quantitative trait loci. *Genetics*, **147**: 1469-85.

Kang, K.S. 2000. Clonal and annual variation of flower production and composition of gamete gene pool in a clonal seed orchard of *Pinus densiflora. Canadian J. Forest Res.*, **30**(8): 1275-1280.

Kirk, L.E. 1933. The progeny test and methods of breeding appropriate to certain species of crop plants. *The American Naturalist*, **67**(713): 515-31.

Knott, D.R. 1972. Effects of selection for F_2 plant yield on subsequent generations in wheat. *Canadian J. Pl. Sci.*, **52**: 721-6.

Knott, D.R. and Kumar, J. 1975. Comparision of early generation yield testing and a single seed descent procedure in wheat breeding. *Crop Sci.*, **15**: 295-9.

Mc Ginnis, R.C. and Shebeski, L.H. 1968. The reliability of single plant selection yield in F_2. *In:* Wheat Genetic, Symposium Proceedings 3rd. pp. 109-14.

Naveira, H. and Barbadilla, A. 1992. The theoretical distribution of lengths of intact chromosome segments around a locus held heterozygous with backcrossing in a diploid species. *Genetics*, **130**: 205-9.

Sakai, K.I. 1954. Theoretical studies on the technique on plant breeding I: value of heritability in bulks and plant progenies of successive hybrid generations, in autogamus plants. *Japanese J. Breed.*, **4**: 145-148.

Sharma, J.R. 2002. Principles and Practice of Plant Breeding, Tata McGraw-Hill Publishing Company Limited, New Delhi.

Singh, P. 2010. Essentials of Plant Breeding. Kalyani Publishers, Ludhiana, 467 p.

Stam, P. and Zeven, A.C. 1981. The theoretical proportion of the donor genome in near-isogenic lines of self-fertilizers bred by backcrossing. *Euphytica*, **30**: 227-38.

Swarup, V. 1997. Ornamental Horticulture. Macmilan India Limited, 462 p.

Tanksley, S.D. and Nelson, J.C. 1996. Advanced backcross QTL analysis: a method for the simultaneous discovery and transfer of valuable QTLs from unadapted germplasm into elite breeding lines. *Theor. Appl. Genet.*, **92**: 191-203.

Tanksley, S.D. and Rick, C.M. 1980. Isozymic gene linkage map of the tomato: applications in genetics and breeding. *Theor. Appl. Genet.*, **57**: 161-70.

Visscher, P.M., Haley, C.S. and Thompson, R. 1996. Marker assisted introgression in backcross breeding programmes. *Genetics*, **144**: 1923-32.

Watts, L. 1980. Flower and Vegetable Breeding, Grower Books, London.

Whittaker, J.C. 2001. Marker assisted selection and introgression. *In*: D.J. Balding, M. Bishop and C. Cannings (eds.). Handbook of Statistical Genetics. Wiley, New York. pp. 673-695.

Young, N.D. and Tanksley, S.D. 1989. RFLP analysis of the size of chromosomal segments retained around the *tm-2* locus of tomato during backcross breeding. *Theor. Appl. Genet.*, **77**: 353-59.

Chapter – 3

Mutation and Polyploidy Breeding

MUTATION BREEDING

Mutation refers to sudden heritable change in phenotype of an organism or permanent change in the number, kind and sequence of nucleotides in genetic material, while individuals showing these changes are known as mutants. Mutation was first discovered by Wright in 1791 in male lamb which had short legs. Hugo de vries coined the term mutation. In crop improvement the breeders are always look for wide range of variability and stabilization of desirable traits. Mutation breeding is very useful in inducing new variability, which is an essential requirement of any plant breeding programme. The modern day cultivars are the resultants of various breeding techniques like selection, hybridization, tissue culture and mutagenesis including some biotechnological tools. Novelty in commercial traits like colour, size, shape, length, post harvest life of flowers etc. are always valued and preferred by consumers. Mutation breeding by exposing planting material with irradiation has play vital role in inducing novelty in a number of flower crops.

TYPES OF MUTATION

Based on occurrence

(1) Spontaneous mutation: Occur spontaneously in nature and not present in sufficiently high frequency and spectrum.

(2) Induced mutation: Artificially produced by the treatment of various mutagenic agents and present in sufficiently high frequency and spectrum.

Depending upon visibility

1. Macro mutation: Causes distinct morphological changes in phenotype and found in qualitative characters. It is also called as oligogenic mutation.
2. Micro mutation: Causes invisible phenotypic changes and observed in quantitative characters. It is also called as polygenic mutation.

Depending upon survival of an individual

1. Lethal
2. Sub-lethal
3. Sub-vital
4. Vital

Mutagens and their mode of action

Mutagens refer to various physical or chemical agents, which greatly enhance the frequency of mutations. Mutagens are of two types as given below.

1. Physical mutagens
2. Chemical mutagens

1. Physical mutagens: Includes various types of radiations i.e.

i. X-rays
ii. Gamma rays
iii. Alpha particles
iv. Beta particles
v. Fast and thermal (slow) neutrons
vi. Ultra violet rays.

2. Chemical mutagens: Includes various types of chemicals i.e.

i. Colchicine
ii. Sodium azide
iii. P-amino benzoic acid
iv. Ethylene imine-copper sulfate complex (ECC)
v. Alpha-trichloro methyl-N-ethylene imine carbinol (TCEC)
vi. N-methyl-N-nitroso urea (MNU) - super mutagen

vii. Ethylene imine (EI)

viii. Dimethyl sulfate

ix. Ethyl methane Sulfonate (EMS)

x. N-nitroso-N-methyl urethane (NMU)

Features of mutation breeding

Most of the induced mutations are harmful to the organism or have deleterious effects. Only about 0.1% of induced mutations are useful in crop improvement.

1. Induced mutagenesis is more effective for the improvement of qualitative characters. It is limited value for the improvement of polygenic characters.
2. Mutations are generally recessive but dominant may occur.
3. Mutation occur at random, however, some genes show higher mutations rate than the other.
4. Mutations are recurrent i.e. some mutations may occur again and again.

Process of mutation breeding

The mutation breeding process consists of four important steps these are:

1. **Choice of material:** The best adapted variety of crop should be chosen for mutagenesis, because such variety has all acceptable agronomic features.
2. **Choice of mutagen:** It depends upon the plant parts to be treated.
3. **Mutagenic treatment:** Mutagenic treatment take three things into account (1) plant species, (2) dose of mutation and (3) duration of treatment. LD_{50} is considered optimum by many workers.
4. **Handling of treated material:** The handling of mutagen treated material differs in seed propagated species and vegetatively propagated species.

Application in crop improvement

Induced mutations play an important role in crop improvement. The main uses of induced mutation are:

1. Development of improved varieties.
2. Induction of males sterility.
3. Production of haploids.
4. Creation of genetic variability.

5. Overcoming of self incompatibility.
6. Improvement in adoption.

Advantages of mutation breeding

1. Mutation breeding is a cheap and rapid method of developing new varieties as compared to backcross, pedigree and bulk breeding.
2. Mutation breeding is more effective for the improvement of oligogenic characters such as disease resistance than the polygenic traits.
3. Mutation breeding is simple, quick and best way when a new character is to be induced in vegetatively propagated crops.

Limitation of mutation breeding

1. Most of the mutations are deleterious and undesirable.
2. Identification of micro-mutations, which are more useful to a plant breeder is usually very difficult.
3. Since mutations are produced at a very low frequency, a very large plant population has to be screened to identify and isolate desirable mutants.

Mutation breeding has limited scope for the genetic improvement of quantitative or polygenic characters.

MUTATION BREEDING IN FLOWER CROPS

Mutation breeding is an established method for crop improvement and has played a major role in the development of many new flower colour/shape mutant cultivars in ornamentals. In rose 45 days after irradiation, percentage of survived explants was determined to assess LD_{50}. The LD_{50} was determined on survival rate and calculated at 67 and 66 Gy in cv. Mourossia and Apollo respectively. As related to the other recorded parameters, the doses that caused 50% reduction in height in cv. Apollo and cv. Mourossia were 70 and 68 Gy respectively. Thus, the most appropriate treatment dose for these cultivars can be indicated between 60 to 70 Gy. This dose range provided sufficient survival rates and plant growth, basic requirements for performing a successful mutant selection and is valuable in planning a breeding programme based on mutation induction in roses (Kahrizi *et al*., 2011).

The effect of chronic (gamma field) and acute (gamma room) radiation and *in vitro* culture on mutation induction of flower colour in chrysanthemum had been studied. The combination of both methods yielded a mutation rate 10 times higher than the conventional chronic cutting method, and also produced non-chimeric mutants. Flower colour mutation could be more readily induced in plants regenerated from petals and buds, than from leaves. The combined method of chronic irradiation with floral organ cultures proved to be of particularly great practical use in mutation breeding, not only of flower species but of other species as well (Nagatomi and Degi, 2009).

Irradiated corms of cultivars Blue Isle, Dixieland, Frisky and White Frosting produced some interesting mutations. The new forms showed crimson and light-red flower colouration. The colour changes were evidently associated with chromosome reorganization in the apical L_2 layer of the main shoot of the corm, resulting in quantitative and qualitative changes in anthocyanin synthesis (Sedelnikova, 1988).

A new active *hAT* type transposable element designated *Tdic101*, the movement of which caused a bud mutation in carnation that led to a change of flower colour from purple to deep pink. In carnation, the movement of transposable elements plays an important role in the emergence of a bud mutation (Momose *et al.*, 2013). In carnation, chronic gamma irradiation gave a wider mutation spectrum than that produced by acute gamma irradiation. Chronic irradiation may be useful for minimizing radiation damage (Okamura, 2006).

Details of mutation breeding of various flower crops given in their chapters.

In several ornamental crops spontaneous mutant have been found, some bud sport are also reported. Several improved varieties have been evolved by using mutagens. These are given in Table 1 & 2.

Table 1: Spontaneous mutant or Bud sport varieties of flower crops.

S.No.	Crop	Varieties
1.	Bougainvillea	Alick Lancaster, Anindita, Archana, Bhabha, Daya, Fantasy, Meera Sport, Gangamma, Gangaswamy, Jane Stansfeld, Jawaharlal Nehru, Jayalakshmi Variegata, Kuvempo, Laxminarayana, L.N. Birla, Louise Wathen, Louise Wathen Medipicta, Louise Wathen Variegata, Magenta Queen,Mahatma Gandhi Variegated, Marigowda, Mary Palmer, Mataji Agnihotri, Mrs. Butt Magenta, Mrs. Butt Scarlet, Mrs. G.S. Randhawa, Mrs. Mc Clean Nirmal, Munivenkatappa, Odisee, Parthasarthy, Preeti, Purple King, Purple Prince, Purple Queen, Rao, Red Glory Improved, Scarlet Glory, Scarlet Queen Variegated, Shweta, Shubhra, Soundarya, Surekha, Suvarna, Thimma, Vellayani, Yellow Queen, Zakeriana, Zakir Hussain, Godrej Cherry Blossom.
2.	Chrysanthemum	Kasturba Gandhi, Sonar Bangla, White Cloud, Sharda, Queen of Tamluk, R. Venkatraman, William Turner, J.S. Lloyd, White Ball, NBRI-Pushpangadan.
3.	Dahlia	Gallery Art Nouveau, Kenya White, Kenya Yellow, Kenya White.
4.	Gladiolus	Swarnima.
5.	Pansy	Lydia's Legacy, Poussieres d'Etoiles, Zampieri, Dudine.
6	Petunia	Revolution Pastel Pink No. 2, Tiger Light, Revolution White, Sun Inferno.

Table 2: Varieties of flower crops developed through induced mutation

S.No.	Crop	Varieties
1.	Alstroemeria	Canaria Stagula, Yellow Tiger Stavero, White Wings Staretto, Harmony Stabroza, Rosita Stareza.
2.	Begonia	Saanred, Silver Peridot, Glennis Crouch, Jose, Lightening.
3.	Bougainvillea	Arjuna, Jaya, Poultoni Variegata, Silver Top, Los Banos Variegata, Mahara Variegata, Pallavi.
4.	Carnation	Pink Sim, White Sim.
5.	Chrysanthemum	Anamika, Aruna, Asha, Ashankit, Basant, Basanti, Himani, Kumkum, Man Bhawan, Navneet, Navneet Yellow, Nirbhaya, Purnima, Sharad Har, Sonali, Navneet Yellow, Nirbhaya , Tulika, Yellow Gold, Colchi Bahar, Pusa Anmol, Pusa Centenary, Pusa Arunodya, Pusa Kesari.
6.	Crossandra	Lakshmi, Dr. APJ Abdul Kalam, Maruvur Arasi, Kanakadhara, Vijaya Kanakambaram, Raj Kankambaram, Subasu Neelambari.
7.	Geranium	Evka (*P. peltatum*), Orapin (*P. zonale*).
8.	Gladiolus	Shobha, Subhangini.
9.	Hibiscus	Suminokura, Hongdansim 2, Yongkwang, Gyewolhyang.
10.	*J. grandiflorum*	CO-2 Pitchi.
11.	Lilium	Mies Bouwman, TX 68-1.
12.	*Phlox paniculata*	David's Lavender, Twister, Shockwave, Becky Towe, Fancy Feelings, Empty Feelings, Elisabeth, Silvermine, Pleasant Feelings, Party Punch, Pure Feelings, Blushing Shortwood, Goldmine, Rubymine.
13.	*Phlox subulata*	Bubble Gum, Cotton Candy.
14.	*P. glaberrima*	Triple Play.
15.	Hybrid Tea rose	Abhisarika, Tata Centenary, Pusa Christina, Madhosh, Light Pink Prize, Stripped Christian Dior, Gippeumi, Honami, Candy Avalanche.
16.	Floribunda rose	Angara, Saroda, Curio, Twinkle, Sharda, Sukumari.
17.	Tuberose	Rajat Rekha, Swarna Rekha.
18.	Tulip	Bartigon, William Copland, Murillo, Apeldoorn.

POLYPLOIDY BREEDING

Polyploid cells and organisms are those containing more than two paired (homologous) sets of chromosomes. Most eukaryotic species are diploid, meaning they have two sets of chromosomes -one set inherited from each parent. However polyploidy is found in some organisms and is especially common in plants.

Polyploidy is pervasive in plants and some estimates suggest that 30–80% of living plant species are polyploid, and many lineages show evidence of ancient polyploidy (paleopolyploidy) in their genomes. Huge explosions in angiosperm species diversity appear to have coincided with the timing of ancient genome duplications shared by many species. It has been established that 15% of angiosperm and 31% of fern speciation events are accompanied by ploidy increase. Polyploid plants can arise spontaneously in nature by several mechanisms, including meiotic or mitotic failures, and fusion of unreduced (2n) gametes.

Two types of polyploids were recognized based on the chromosome constitution of the individuals, *viz.*, auto (polysomic) and allo (diosomic) polyploids.

(1) **Autopolyploids:** Individuals have three or more homologous chromosome sets (genomes) derived from a single species.

(2) **Allopolyploids:** Homologous chromosome sets derived from two or more distinct species. The frequencies of polyploids in angiosperms have been estimated to be anywhere in between 47% to 70%, depending on the criteria used by various workers (Grant, 1981), and regarded as an important mechanism of speciation and adaptation in plants.

ORIGIN AND PRODUCTION OF DOUBLED CHROMOSOME NUMBERS

Cells/individuals having doubled chromosome numbers may originate in one of the following several ways as given below.

(1) **Spontaneous:** Chromosome doubling occurs occasionally in somatic tissues and unreduced gametes are also produce in low frequencies.

(2) **Physical agents:** Heat or cold treatments, and X–ray or gamma-ray irradiation may produce polyploids in low frequencies.

(3) **Regeneration *in vitro*:** Polyploidy is a common feature of the cells cultured *in vitro*. Some of the plants regenerated from callus and suspension cultures may be polyploids.

(4) **Colchicine treatment:** Colchicine treatment is the most effective and the most widely used treatment for chromosome doubling. It has been used with great success in a large number of crop species belonging to both dicot and monocot groups. Pure colchicine is $C_{22}H_{25}O_6N$. It blocks spindle formation and thus inhibits the movement of sister chromatids to the opposite poles. The resulting restitution nucleus includes all the chromatids; as a result, the chromosome number of the cell is doubled.

(5) **Other chemical agents:** Several other chemical have polyploidizying effect. Notable among them are, 8-hydroxyquinoline and nitrous oxide. These chemicals are much less effective than colchicine and are not commonly used.

Levin (1983) concluded that “chromosome doubling may propel a population into a new adaptive sphere” and “bring about abrupt, transgressive, and conspicuous changes in the adaptive gestalt of populations within micro evolutionary time.

POLYPLOIDY BREEDING IN FLOWER CROPS

Besides classical cross breeding and selection, more modern breeding techniques offer an extended potential to introduce new genetic variation in ornamentals plants. Artificial polyploidization can by itself be used to create new variation. Polyploid plants in general are more robust and sturdier than diploids. Frequently, larger flowers, bigger leaves and more vigorous growth can be observed. The induction of polyploidy is also a common technique in interspecific crosses to overcome the sterility in F_1 progenies or to generate genotypes with equal ploidy level when interploidy crosses are problematic. Interspecific and even intersubgeneric hybridizations are highly interesting to combine desired traits from genetically distinct genotypes (Huylenbroeck, 2010).

Some of the investigations on allopolyploids, derived from *Festuca-Lolium*, *Alstroemeria* and *Lilium* species hybrids, have revealed that 2n gametes can be most useful for the introgression of alien genes and chromosomes into cultivars. An important feature of using sexual polyploidization in the case of allopolyploids is that introgression can be achieved through recombination due to genetic crossing-over between alien chromosomes as well as addition of alien chromosomes, which is extremely difficult or impossible to achieve in the case of colchicine induced allopolyploids (Ramanna and Jacobsen, 2003).

A plant regeneration system investigated in *Gerbera priorii*, *G. tristis* and *G. virescens* from callus culture and chromosome doubling showed that, the highest efficiency of chromosome doubling was observed with 0.05% colchicine treatment for 72 h. The regenerants of the wild species with doubled chromosomes will be powerful materials for gladiolus breeding (Suzuki *et al.*, 2005).

The scales of bulbs treated with 1 g kg^{-1} colchicine have resulted in the production of lily polyploid forms. The polyploid plants regenerated after pre-treatment by colchicine may be selected in early stage by phenotype of young bulblets. Broad, fleshy bulb scales setting right angle to growth axis were characteristic for regenerated tetraploids. The produced polyploid plants, in comparison with diploids, had a larger diameter of the flower (+2 up to 2.5 cm), more extended plant height (+10 up to 20 cm), increased flower count (+2) and was coming into bloom 4-10 days later than diploids of the same variety (Balode, 2008).

Tetraploids were successfully induced from *in vitro* plantlets of diploid *Gerbera jamesonii* Bolus cvs. Cabana and Dalma (2n=2x=50) by dipping shoot tips with colchicine. Most efficient condition for inducing tetraploids was with 0.10% colchicine for 48 h. The chromosome number of part mutants have doubled to 100 (2n=4x=100). As compared with diploid, tetraploid of *Gerbera jamesonii* Bolus was characterized by larger stoma area, larger flower head, thicker leaves colour, leaves tend to round, thicker leaves, larger stem diameter and anthocyanin colouration of peduncle bases were obviously strong (Li *et al.*, 2009).

Rhododendron L. 'Fragrantissimum Improved' is an attractive cultivar with showy, fragrant flowers but has limited potential for breeding because it is a sterile wide hybrid. Protocols for *in vitro* regeneration and polyploid induction were developed for this cultivar as a means to potentially restore fertility and enhance ornamental traits. A percentage of homogenous, tetraploid shoots was recovered from treatments of 30 micro M oryzalin for 1 (13%) or 3 (13%) days and 7.5 micro M oryzalin for 7 (20%) or 14 (7%) days (Hebert *et al*., 2010).

A protocol for *in vitro* polyploid induction using oryzalin was developed for *Dendrobium*, *Epidendrum*, *Odontioda*, and *Phalaenopsis* orchids (Orchidaceae). Protocorms and protocorm-like bodies (PLBs) in liquid nutrient media were subjected to oryzalin treatments. The optimal treatments were 14.4 micro M for 6 days in *Dendrobium* and *Odontioda*, 57.7 micro M for 6 days in *Epidendrum* and 14.4 micro M for 3 days in *Phalaenopsis* (Migvell and Leonhardet, 2011).

Apart from this polyploidy breeding of various crops have been discussed in their respective chapters.

There are number of species of various ornamental plants are polyploidy (Table 3). Artificial polyploidization created new variation and number of varieties evolved through it (Table 4).

Table 3: Polyploidy levels in various ornamental plants.

S. No.	Flower crop	Ploidy level	Ploidy species
1.	Amaryllis	Hexaploid	*Amaryllis belladonna*
		Heptaploid	*A.blumenvia*
2.	Anthurium	Diploid	*Anthurium andreanum, A. Hookerii, A. magnificent*
		Triploid	*A. scandens*
		Tetraploid	*A. digitatum, A. wallisii*
3.	Carnation	Tetraploid	*Dianthus chinensis*
4.	Dahlia	Tertraploid	*Dahlia imperialis*
		Octaploid	*D. veriavilis, D. coccinea, D. rosea*
5.	Gladiolous	Pentaploid	*Gladiolus psittacianus*
6.	Jasmine	Triploid	*Jasminum primulinum, J. sambac*
		Tetraploid	*J. flexile, J. angustifolium*
7.	Lily	Triploid	*Lilium tigrinum*
8.	Marigold	Diploid	*Tagetes erecta, T. tenuifolia*
		Tetraploid	*T. patula, T. minuta, T. biflora, T. remotiflora*
9.	Narcissus	Diploid	*Narcissus pseudonarcissus, N. poeticus*
		Triploid	*N. pseudonarcissus hispanicus*
		Hexaploid	*N. bulbocodium*
10.	Primula	Diploid	*Primula frondosa*
		Tetraploid	*P. farinose*
		Hexaploid	*P. scotica*
		Octaploid	*P. scandenavica*

(*Contd.*)

(*Contd. Table 3*)

S. No.	Flower crop	Ploidy level	Ploidy species
11.	Rose	Diploid	*Rosa gigantean, R. mutliflora, R. chinesis, R.moschata, R. wichuriana*
		Triploid	*R. bourborniana*
		Tetraploid	*R. gallica, R. damascena, R. foetida*
12.	Tulip	Triploid	*Tulip alanata, T. stellata*
		Tetraploid	*T. clusiana,*
		Pentaploid	*T. cluciana*

Table 4: Varieties developed through polyploidy breeding in ornamental plants.

S. No.	Flower crops	Ploidy level	Varieties
1.	Amaranthus	Tetraploid	Amar Tetra.
2.	Amaryllis	Triploid	Kiran.
		Tetraploid	Samrat, Tetra Apricot, Tetra Starzynski.
3.	Antirrhinum	Tetraploid	Tetra Giant, Tetra Guilt, Velvet Beauty, Red Shades.
		Triploid	Velvet Beauty.
4.	Bougainvillea	Triploid	Cyphiri, Temple Fire, Lateritia, Perfection, Poultoni Special.
		Tetraploid	Crimson King, Princess, Mahara, Mangifica, Shubra, Mrs. McClean, President Roosevelt, Lady Mary Baring, Thimma, Zakariana.
5.	Cacti	Triploid	Gold Charm.
		Tetraploid (Hatiara)	Amherst, Louise, Mira.
		Tetraploid (Schlumbergera)	Peach Parfait.
		Hexaploid	Cambridge, Santacruz Sanibel.
6.	Crossandra	Triploid	Delhi.
		Tetraploid	Orange Crossandra, Luteạ Yellow, Sebaculis Red.
7.	Day Lily	Diploid	Barbara Michael Ruffled, Master Piece, Ruffled Perfection.
		Tetraploid	Tetra Apricot, Tetra Peach, Crest Wood Series, Wedding Band, Winter in Eden, Bill Norris.
8.	Gladiolus	Triploid	Manmohan, Monohar, Manhar, Mukta, Manisha, Mohini, Triloki, Sanyukta.
		Tetraploid	Major cultivars
		Aneuploid	Archana, Arun.
9.	Lilium	Triploid	Pari Sienne, Tetra Connecticut King, Hornback's Gold, Eros, Rosepoint Lace, Red Velvet, Jasper, Embarrassment, Pink Giant, Tiger Babies.
		Tetraploid	Avignon, Bold Knight, Apricot Supreme.
10.	Marigold	Triploid	Seven Star, Show Boat, Nugget.
11.	Orchids	Amphidiploid (Dendrobium)	Jacqueline Thomas Y166, Walter Oumae 4N.
		Tetraplolid (Phalinopsis)	River Band.
		Tetraploid (Oncidium)	Popcorn.
		Tetraploid (Spathoglottis)	Lion.

(*Contd.*)

(*Contd. Table 4*)

S.No.	Flower crops	Ploidy level	Varieties
12.	Petunia	Double Haploid	Michael.
		Auto Tetraploid	State Fair, Old Mexico.
13.	Rose	Triploid	Prema, Surekha, Surya.
		Tetraploid	Major Garden Cutivars.
		Aneuploid	Mohini.
14.	Stock	Aneuploid	Snow Flake.
15.	Tulip	Tetraploid	Judith Leyster, World's Favourite.

REFERENCES

Balode, A. 2008. Investigation of polyploidy in the breeding of lilies research for rural development. *Proce*: Annual 14th International Scientific Conference, Jelgava, Latina. pp. 347-52.

Grant, V. 1981. Plant speciation. Columbia Univ. Press.2nd ed. New York.

Hebert, C.J., Touchell, D.H, Ranney, T.G., Lebude, A.V.2010. *In vitro* shoot regeneration and polyploidy induction of Rhododendron 'Fragrantissimum Improved'. *Horti Sci.*, **45**(5): 801-4.

Huylenbroeck, Laere, J.V. and Van, K. 2010. Breeding strategies for woody ornamentals. *Acta Horticulture*, **885**: 391-401.

Kahrizi, Z.A., Kermani, M.J., Amiri, M.E. and Vedadi, S. 2011. Identifying the correct dose of gamma-rays for *in vitro* mutation of rose cultivars. *Acta Horticulturae*, **923:** 121-7.

Levin, O.A. 1983. Polyploidy and novelty in flowering plants. *American Naturalist*, **122**: 1-25.

Li, H., Yan, B., Zhang, T., Jiang, Y., Zhang, H., Yu, L., and Li, S. 2009. Preliminary studies on polyploidy mutation of cut flower *Gerbera jameson***ii** Bolus**.** *Acta Horticulturae Sinica*, **36**(4): 605-10.

Migvell, T.P. and Leonhardet, K.W. 2011. *In vitro* polyploidy induction of orchids using oryzalin. *Scientia Horticulturae*, **130**(1): 314-19.

Momose, M., Nakayama, M., Itch, Y., Umemoto, N., Toguri, T. and Ozeki, Y. 2013. An active hAT transposable element causing bud mutation of carnation by insertion into the flavonoid 3-hydrolase gene. *Mol. Gene. Genomics*, **288**(3/4): 175-84.

Nagatomi, S. and Degi, K. 2009. Mutation breeding of chrysanthemum by gamma field irradiation and *in vitro* culture. *Proce*: Induced Pl. Mutations in the Genomics Era. Symposium on International Joint FAO/ IAEA, pp. 258-61.

Okamura, M. 2006. Flower breeding by quantum beam technology and its commercialization. *Gamma Field Symposia*, **45**: 77-89.

Ramanna, M.S. and Jacobsen, E. 2003. Relavance of several polyploidization for crop improvement- A review. *Euphytica*, **133**: 3-18.

Sedelnikova, L.L. 1988. Mutations in flower colour of gladioli from gamma irradiated corms. *Byulleyen' Glavnogo Botanicheskogo Sada*, **148**: 55-60.

Suzuki, K., Jakatsu., Gonai, T. and Kasumi, M. 2005. Plant regeneration and chromosome doubling of wild gladiolus species, *Acta Horticulturae*, **673**(1): 175-81.

Chapter – 4

Molecular and Transgenic Breeding

MOLECULAR BREEDING

Molecular breeding is the application of molecular biology tools, often in plant breeding.

Genetic markers

Genetic marker is defined as a gene whose phenotypic expression is usually easily discerned, used to identify an individual or a cell that carries it, or as a probe to mark a nucleus, chromosomes or locus (King and Stansfield, 1990). Three broad classes of genetic markers are morphological markers, biochemical markers and DNA or molecular markers.

(1) Morphological markers

Morphological markers are based on the visually assessable traits *viz.,* morphological and agronomic traits. However, they are limited in number.

Features of morphological markers

i. They are easy to identify, there is no need of any complex analysis. Trait of interest can be identified visually.

ii. Morphological markers are inherited qualitatively or quantitatively.

iii. Frequency of availability of morphological markers is less in number i.e., morphological markers may not be available for all the traits of interest.

iv. They are influenced by environment.

v. Since they are influenced by environment, they are unstable in inheritance.

vi. They are stage specific *i.e.*, a particular marker will be useful to identify in a particular stage of plant and not any stage of the plant.

vii. They are influenced by genotype and developmental stages of the plant.

Besides drawbacks, the morphological markers are useful to the plant breeders for the selection purpose in certain situation.

(2) Biochemical/protein markers

Genetic markers based on the gene product are called biochemical markers. Here the identification of trait/genotype of interest is based on the banding pattern of the proteins.

Isozymes: These are the enzymes, which have varying molecular weight and perform the same function. They are influenced by environment, genotype and ontogeny.

These are products of gene expression. Just based on the banding pattern of the proteins, one cannot identify a particular character/trait of interest. During gel running there is a chance of degradation of proteins and which will lead to multiple banding instead of single banding. The isozyme markers can be useful, provided if the genetics of these markers are known.

Genetic causes of isozyme variation

1. Gene duplication with subsequent mutation at gene loci.
2. More than one gene contributes to the structure of protein.
3. Other possible variations are: Binding of high polypeptides to varying number of coenzymes or other prosthetic groups.

It is always advisable to use genetically well defined isozymes as a marker. To overcome these problems DNA markers are used by the breeders (Scandalios, 1974).

(3) DNA markers (molecular markers)

Genetic markers relaying on DNA assay are called molecular markers. Molecular markers should be considered as normal genes, as they usually do not have any biological effect, and instead can be thought of as constant land marks in the genome (Semagn *et al.*, 2006). While cutting the DNA into segments several million beads/fragments can be obtained. Therefore, millions of individual fragments of DNA can be used, as the markers, so called DNA markers or molecular markers.

The DNA markers are classified based on

(a) Mode of transmission

i. Biparental nuclear inheritance

ii. Maternal nuclear inheritance

iii. Maternal organelle inheritance or paternal organelle inheritance.

(b) Mode of gene action

i. Dominant Markers.

If the bands of any one of the parent in the F_1 progeny is obtained then they are called dominant markers.

ii. Codominant markers.

If the bands of both the parents appear in the progeny, then the marker is said to be codominant marker.

(c) Method of analysis

i) Hybridization based markers e.g. RFLP.

ii) PCR based markers e.g. RAPD, SSR, AFLP.

Desirable properties of molecular markers

It is described by Semagn *et al.* (2006). In brief it is given below.

(1) High polymorphism.

(2) Codominant inheritance.

(3) Frequent occurrence and even distribution throughout the genome.

(4) Selectively neutral behaviour.

(5) Easy access.

(6) Easy and fast assay.

(7) Low cost and high throughput.

(8) High reproducibility.

(9) Transferability between laboratories, populations and/or species.

No molecular markers are available yet that fulfill all these requirements. However, according to the kind of study to be undertaken, one can choose among the varieties of molecular marker systems, each of which combines at least some of these desirable properties.

Indirect selection for a desired plant phenotype based on the banding pattern of linked molecular markers is called marker assisted selection. Also called as marker aided selection or marker assisted breeding or gene assisted selection.

Steps in marker assisted breeding

(1) Selection of parents: Parents with divergent origin or contrasting characters are selected.

(2) Development of breeding populations: Generally, F_2 progeny required to study.

(3) Isolation of DNA: Done by using detergents, chelating agents and centrifugation. DNA can be isolated even from seedlings and there is no need to wait for flowering or seed development stage.

(4) Scoring for markers: Polymorphism between parents and F_2 progeny is determined.

(5) Correlation with morphological traits.

Various genes responsible for different types of action, regulating plant morphology, flavonoid synthesis, etc. are given in Table 1, 2 and 3.

Table1: Gene responsible for ornamental crop improvement (Awasth and Hanur, 2009).

Gene for vase life improvement

Gene	Action
ACC synthase	Inhibition with reduced ethylene production
ACC oxidase	Inhibition with reduced ethylene production
ACC deaminase	Over expression with reduced ethylene production
Etr 1	Expression of a defective gene with reduced ethylene production
ERS	Expression of a mutated gene with reduced ethylene production
Gene for regulating plant morphology and development	
Lazy, TAC1	Branching angles of tillers
GA insensitive (gai)	Stem elongation and plant height
Brassinosteroid	Plant height
Phytochrome	Harvest index and shading response
Rol C	Plant branching and architecture
Clavata, Wuschel	Establishment and maintenance of shoot apical meristem
MAX	More axillary branching

Table 2: Various gene responsible for flavonoid synthesis isolated from ornamental crops (Benmeir *et al.*, 2002).

Gene responsible for flavonoid synthesis

Gene isolated	Crops
Chs	Antirrhinum, Chrysanthemum, Callistephus, Dianthus, Lisianthus, Mathiola, Rosa, Torenia, Ranunculus
Gchs1, Gchs3	Gerbera
ChsA-E	Ipomea
Ochs3,4,8	Orchid
ChsA, ChsJ	Petunia
Chi	Antirrhinum, Dianthus, Lisianthus
ChiA, ChiB	Petunia
F3h	Dianthus, Chrysanthemum, Antirrhinum, Callistephus, Ipomea, Mathiola, Orchid, Petunia
F3'h[a]	Antirrhinum, Dianthus
F3'h	Petunia
F3'5'h	Catharanthus, Lisianthus, Gentian, Petunia, Torenia
Fls	Petunia, Rosa
AfnsII	Antirrhinum,
FnsII	Gerbera
Dfr	Antirrhinum, Gentian, Gerbera, Dianthus, Lisianthus, Forsythia, Ipomea, Orchid, Perilla, Petunia, Rosa, Torenia
Ans	Antirrhinum, Forsythia, Perilla, Petunia, Torenia
Ufgt	Antirrhinum
3Gt	Gentiana, Forsythia
Gst	Petunia

Table 3: Different gene incorporated in ornamental crops

S.No.	Gene incorporated	Method	Crops
1.	Chs	Sense/ Anti-sense orientation	Chrysanthemum
2.	Gchs, I/gdfr	Sense orientation, Anti-sense orientation	Gerbera
3.	Chs	Anti-sense orientation	Lisianthus
4.	Chs	Anti-sense/Sense orientation	Rosa cv. Royalty
5.	Anti-f3h	Anti-sense orientation	Carnation

MARKER ASSISTED BREEDING IN FLOWER CROPS

A high-density genetic map with a number of anchor markers has been created to be used as a tool to dissect genetic variation in rose. Linkage maps for the diploid 94/1 population consisting of 88 individuals were constructed using a total of 520 molecular markers including AFLP, SSR, PK, RGA, RFLP, SCAR and morphological markers. Seven linkage groups, putatively corresponding to the seven haploid rose chromosomes, were identified for each parent, spanning 487 cM and 490 cM, respectively. The average length of 70 cm may cover more than 90% of the rose genome. An integrated map was constructed by incorporating the homologous parental linkage groups, resulting in seven linkage groups with a total length of 545 cm. The present linkage map is currently the most advanced map in rose with regard to marker density, genome coverage and with robust markers, giving good perspectives for QTL mapping and marker-assisted breeding in rose. The SSR markers, together with RFLP markers, provide good anchor points for future map alignment studies in rose and related species. Codominantly scored AFLP markers were helpful in the integration of the parental maps (Yan *et al.*, 2005). SCAK16583 was a species-specific marker of *R. multiflora* that could be used in a marker-assisted breeding programme for hybrid identification of *R. multiflora* (De Feng *et al.*, 2013).

SSR markers acquired by transcriptome sequencing are potentially useful for marker-assisted breeding and genetic analysis in the genus *Chrysanthemum* and its related genera (Hai Bin *et al.*, 2013). It is feasible to estimate and differentiate chrysanthemum population's structure effectively using SRAP markers, and the markers obtained in this study are promising in molecular assisted breeding (Wang *et al.*, 2012).

Both the characterization and the genetic purity certification of cultivars of *Gladiolus* spp. by means of morphological markers were made. The characteristic of flower colour is the safest morphological descriptor for distinguishing gladiolus cultivars. Depending on the characteristics of the contaminant cultivars and on the ones under analyses, it is possible to use morphological descriptors for the genetic purity certification in gladiolus cultivars (Ferreira *et al.*, 2011). Random amplified polymorphic DNA (RAPD) was used for genetic fingerprinting of *Limonium* spp., *Alstroemeria* spp., carnation and *Prunus* spp. (Burchi, 2002).

Bacterial wilt caused by *Burkholderia caryophylli* is one of the most damaging disease of carnations. World's first wilt-resistant carnation cultivar, Karen Rouge selected from a cross between carnation cultivar Miracle Rouge and resistant line 4AZ31-5 is a deep yellowish-red standard-type cultivar. Karen Rouge has a high degree of resistance to bacterial wilt and its productivity and other traits desirable for cut flowers are similar to those of modern commercial cultivars. Marker-assisted selection with the STS-WG44 marker, which is linked to a QTL with a large effect for bacterial wilt resistance derived from *Dianthus capitatus* ssp. *andrzejowskianus* reduced the labour and time required for testing the resistance of breeding populations. This is also first cultivar produced by marker-assisted selection in carnation (Yagi *et al.*, 2012).

RAPD analysis was used for the identification of molecular markers that could be associated with flower vase life. RAPD analysis revealed several bands that allow discriminating significantly the individuals with extended longevity (Benedetti *et al.*, 2003).

Magnetic capture-hybridization and reverse-transcription polymerase chain reaction (MCH/RT-PCR) approach was investigated for the specific detection of *Tobacco rattle virus* and for the detection of viruses within the *Potexvirus* group in leaves, dormant bulbs and corms of flower bulbs of different species. Dilution series of TRV-infected tulip leaf sap showed that MCH/RT-PCR was 70,588 times more sensitive than enzyme-linked immunosorbent assay (ELISA) and was similar to that of RT-PCR. This method is amenable for routine indexing of stock plants as part of a management plan to reduce the propagation of virus-infected plants (Miglino *et al.*, 2007). Important molecular markers used in several prominent ornamental plants are given in Table 4.

Table 4: Use of molecular markers in ornamental plants

S. No.	Flower crops	Markers	Uses	References
1.	*Alstroemeria* spp.	RAPD	Genetic variation and Genotypic identification	Beyermann *et al.*, 1992; Anatassopoulos and Keil, 1996; Han *et al.*, 1999
		AFLP	Genetic diversity and Construction of linkage map	Han *et al.*, 2000
2.	*Anthurium* spp.	RAPD	Genetical relationships of cultivars	Ranamukhaarachchi *et al.*, 2001
3.	*Caladium*	AFLP	Genetic relationship	Loh *et al.*, 1999
4.	*Camellia* spp.	Microsatellite	Development of population genetic characterization	Kaundun and Matsumoto, 2002
5.	*Cephalotaxus*	AFLP	Analysis of genetic diversity	Zhang *et al.*, 2000
7.	*Cymbidium*	RAPD	Identification of cultivars and analysis of genetic diversity	Obara-Okeyo and Kako, 1998
8.	*Cynodon*	DAF	Genetical relationship	Caetano-Anolles *et al.*, 1995
9.	*Dahlia*	AFLP	Detection of genetic diversity and analysis cultivars	Wegner and Debener, 2008
10.	*Dendrathema*	RAPD/DAF	Genetic variability	Wolff and Peters-Van Rijn, 1993; Wolff *et al.*, 1995; Scott *et al.*, 1996; Wolff, 1996
11.	*Dianthus caryophyllus*	RAPD	Selection for resistance to *Fusarium oxysporium,* Genetic relationship among cultivars	Scovel *et al.*, 2001 Tejaswani *et al.*, 2008
	Dianthus spp.	RFLP	Selection for disease resistance	Scovel *et al.*, 2001
		Microsattelite	Identification of *Dianthus* varieties	Vainstein *et al.*, 1991
12.	*Euphorbia* (Poinsettia)	RAPD / DAF	Identification and genetic relationship	Ling *et al.*, 1997; Starman *et al.*, 1999

(*Contd.*)

(*Contd. Table 4*)

S No.	Flower crops	Markers	Uses	References
13.	*Gladiolus* spp.	RAPD	Genetic relationship	Takatsu *et al.*, 2001
14.	*Geranium*	DAF	Genetic relationship	Starman and Abbitt, 1997.
15.	*Gerbera*	RFLPs	Genetic fidelity Microsattelite	Tzuri *et al.*, 1991;
16.	*Heliconia*	RAPD	Genetic relationship	Marouelli *et al.*, 2010
17.	*Hemerocallis fulva*	AFLP	Genetic similarity	Tomkins *et al.*, 2001
18.	*Hibiscus*	AFLP	Genetic relationship	Cheng *et al.*, 2004
		RAPD	Genetic variation	Barik *et al.*, 2006
19.	*Iris setosa, I. laevigata, I. uniflora*	RAPD	Genetic identification	Arnold *et al.*, 1991
20.	*Ixora coccinea, I. javanica*	RAPD	Genetic diversity among cultivars	Rajaseger *et al.*, 1997
21.	*Juniperus*	RAPD	Finger printing	Adams and Demeke, 1993
22.	*Lilium* spp.	RAPD	Genetic relationship, genetic variation, phylogenetic relationship, identification of species	Yamashigi, 1995
		RFLP	Identification of parentage	Haruki *et al.*, 1998
23	*Lupinus* spp.	RAPD/AFLP	Phylogenetic relationship, linkage map	Talhinhas *et al.*, 2003
24	*Magnolia*	AFLP	Variation among species and cultivars	
25	*Ophrys aranaeola*	Microsatellite	Characterization of microsatellite and cross species amplification	Soliva *et al.*, 2000
26	*Opuntia*	RAPD	Identification and genetical analysis	
27	*Orchids*	RFLP	Analysis of maternal linkage	Xue *et al.*, 2010
28	*Osteospermum*	RAPD/AFLP	Genetic variation	Faccioli *et al.*, 2000
29	*Ozothamnus*	RAPD	Genetic diversity	Ko *et al.*, 1996
30	*Paeonia*	RAPD	Classification	Suo *et al.*, 2005
31	*Pelargonium*	RAPD/DAF	Identification of cultivars and genetic variation, Genetic relationship, Cultivars identification	Renou *et al.*, 1997; Starman and Abbit, 1997, Barcaccia *et al.*, 1999; Beecher *et al.*, 2000; Lesur *et al.*, 2000
		ISSR	Identification of variety	De Wet *et al.*, 2008
32.	*Petunia*	RAPD/ DAF	Linkage map, genetic relationship Molecular phylogenetic analysis	Cerny *et al.*, 1996; Jianhua *et al.*, 1997

(*Contd.*)

(*Contd. Table 4*)

S No.	Flower crops	Markers	Uses	References
		AFLP	Identification of clones containing Rf gene, Gene mapping	Bentolila and Hanson, 2001
		RFLP	Genetic analysis and mapping	Strommer *et al.,* 2000
		FISH	Genetic map	Ten Hoppen *et al*., 1996
33.	*Phyllostachys*	RFLP	Species identification Genetic variation Evolution	Friar and Kochert, 1994
		AFLP	Phylogenetic relationship	Hodkinson *et al*., 2000
		ISSR	Identification of hybrids	Lin *et al.,* 2010
34.	*Rhododendron*	RAPD	Clonal stability, genetical variability, cultivar identification	Dunemann and Kahnau 1998; De Riek *et al*., 1999; Naito *et al*., 1999
		AFLP	Genetic similarity	Escaravage *et al*., 1998
		Microsatellite	Analysis of genetic structure	Kameyama *et al*., 2001
		RFLP	Plastid DNA inheritance	
35.	*Rosa*	RADP	Linkage map, percentage analysis, genetical analysis, genetic variability, classification of genus, germplasm analysis	Tzuri *et al*., 1991; Hubbard *et al*., 1992; Rajapakse *et al*., 1992; Torres *et al*., 1993; Vainstein and Ben Meii, 1994; Debener *et al*., 1996; Gallego and Martinez, 1996; Matsumoto and Fukuri, 1996; Millan *et al*., 1996; Debener *et al*., 1996
		AFLP	Identification of varieties	Zhang *et al*., 2001
		RFLP	Cultivar identification	Hubbard *et al*., 1992
	R. multiflora	Microsatellite	Construction of genetic linkage	Hibrand-Saint Oyant *et al*., 2008
36.	*Scaevola*	RAPD	Genetic variability	Swoboda and Balla, 1997
37.	*Syringa*	RAPD	Genetic relationship	Marsolais *et al*., 1993
38.	*Vanda*	RAPD	Genetic relationship	
39.	*Viola*	RAPD	Genetic relationship	
40.	*Zantedeschia*	RAPD	Cultivar identification	Hamada and Hagimori, 1996

TRANSGENIC BREEDING

A transgenic plant is one that contains a gene or genes which have been introduced artificially into the plant's genetic makeup using a set of several biotechnology techniques collectively known as recombinant DNA (rDNA) technology. The inserted genes called transgenes may come from another same or a different plant species, or a completely unrelated kind of organism like bacteria or animals. The process of moving genes from one species to another is called transformation.

Once a transgenic plant is created, the transgenes can be inherited along with the rest of the plant's genes through normal mating by pollination. The offsprings are also transgenic when they acquire the transgenes this way. Transgenic plants are often referred to as genetically engineered (GE) plants or genetically modified (GM) plants.

Genetic improvement of crop plants, domestic animals and useful micro-organisms, through genetic engineering, in relation to their economic use for mankind is referred to as transgenic breeding.

Advantages of transgenic crop breeding

(1) It is a rapid method of crop improvement.

(2) It overcomes barriers of cross incompatibility.

(3) It is useful in solving those problems which cannot be solved by conventional methods.

(4) It permits the gene transfer from any source including animals.

Methods of foreign gene transfer

1. Plasmid method.
2. Particle bombardment method.
3. Direct DNA uptake method.
4. Micro-injection method.

1. Plasmid method

In general, the *Agrobacterium* method is considered preferable to the gene gun, because of a greater frequency of single-site insertions of the foreign DNA, which allows for easier monitoring. *Agrobacterium tumefaciens* is a soil bacteria that has the ability to infect plant cells with a piece of its DNA. When the bacterial DNA is integrated into a plant chromosome, it effectively hijacks the plant's cellular machinery and uses it to ensure the proliferation of the bacterial population. *A. tumefaciens*, causes crown gall disease in many ornamental plants. The DNA in *A. tumefaciens* cell is contained in the bacterial chromosome as well as in another structure known as a Ti (tumor-inducing) plasmid. The Ti plasmid contains a stretch of DNA termed T-DNA (~20 kb long) that is transferred to the plant cell in the infection process and a series of *vir* (virulence) genes that direct the infection process. *A. tumefaciens* can only infect a plant through wounds. When a plant root or stem is wounded it gives off certain chemical signals. In response to those signals, the *vir* genes of *A. tumefaciens* become activated and direct a series of events necessary for the transfer of the T-DNA from the Ti plasmid to the plant's chromosome.

In this method, the tumor inducing (Ti) region is removed from the T-DNA (transfer DNA) and replaced with the desired gene and a marker, which is then inserted into the organism. The steps taken include:

i. isolate the genes of interest from the source organism.

ii. develop a functional transgenic construct including the gene of interest; promoters to drive expression; codon modification, if needed to increase successful protein production; and marker genes to facilitate tracking of the introduced genes in the host plant.

iii. incorporate into a useful plasmid.

iv. introduce the transgenes into plant cells.

v. regenerate the plants cells.

vi. test trait performance or gene expression at lab, greenhouse and field level.

The T-DNA then enters the plant cell through the wound. T-DNA waits until the plant DNA is being replicated or transcribed, then inserts itself into the exposed plant DNA (Galun and Breiman, 1997). To harness *A. tumefaciens* as a transgene vector, scientists have removed the tumor-inducing section of T-DNA, while retaining the T-DNA border regions and the *vir* genes. The transgene is inserted between the T-DNA border regions, where it is transferred to the plant cell and becomes integrated into the plant's chromosomes (Wong, 1997).

2. Particle bombardment method

The particle bombardment device, also known as the gene gun, was developed to enable penetration of the cell wall so that genetic material containing a gene of interest can be transferred into the cell. Today the gene gun is used for genetic transformation of many organisms to introduce a diverse range of desirable traits. Plant transformation using particle bombardment follows the same outline as *Agrobacterium*-mediated method.The particle bombardment method starts with coating tungsten or gold particles (microprojectiles) with plasmid DNA. The coated particles are coated on a macro-projectile, which is accelerated with air pressure and shot into plant tissue on a petri plate (Gan, 1989). A perforated plate is used to stop the macro-projectile, while allowing the microprojectiles to pass through to the cells on the other side. As the microprojectiles enter the cells, the transgenes are released from the particle surface and may incorporate into the chromosomal DNA of the cells. Selectable markers are used to identify the cells that take up the transgene. The transformed plant cells are then regenerated into whole plants using tissue culture. Particle bombardment also plays an important role in the transformation of organelles such as chloroplasts, which enables engineering of organelle-encoded herbicide or pesticide resistance in crop plants and to study photosynthetic processes. Limitations to the particle bombardment method relative to *Agrobacterium*-mediated transformation include frequent integration of multiple copies of the transgene at a single insertion site, rearrangement of the inserted genes, and incorporation of the transgene at multiple insertion sites. These multiple copies can be linked to silencing of the transgene in subsequent progeny (Yao *et al.*, 2006). This method was employed for the transformation of several monocotyledonous ornamental crops such as tulip (Wilminks *et al.*, 1992), gladiolus (Kamo *et al.*, 1995a), lily (Langeveld *et al.*, 1995 and Van der Leede-Plegt *et al.*, 1997; Watad *et al.*, 1998 and Lipsky *et al.*, 2002), *Ornithogalum* (De Villiers *et al.*, 2000) and is widely used for many other crops.

3. Direct DNA uptake method

In electrophoration by passing brief intense electrical pulse through a suspension of (wall-less) cells in a solution of DNA results in the introduction of significant quantities of DNA into the cell mechanism. It can be used for the transfer of foreign genes into protoplasts made from monocots or dicots. Thus, the range of crop plants which can be used for gene transfer is not dependent on pathogen–host recognition, as in the case of *A. tumefaciens.* Furthermore, the genes to be transferred do not have to be recloned into T-DNA, bypassing an elaborate and time-consuming step. Electrophoration can transfer genes efficiently into a large population of plant protoplasts, and is not as tedious and time-consuming as is microinjection. Thus, large numbers of plant protoplasts can be transformed at one time with confidence that at least some of these protoplasts treated will be transformants.

Electrophoration is done with electrophorators, appliances that create an electro-magnetic field in the cell solution. The cell suspension is pipetted into a glass or plastic cuvette which has two aluminium electrodes on its sides. For bacterial electrophoration, typically a suspension of around 50 microlitres is used. Prior to electroporation it is mixed with the plasmid to be transformed. The mixture is pipetted into the cuvette, the voltage and capacitance are set, and the cuvette is inserted into the electrophorator. Immediately after electroporation, one millilitre of liquid medium is added to the bacteria (in the cuvette or in an eppendorf tube), and the tube is incubated at the bacteria's optimal temperature for an hour or more to allow recovery of the cells and expression of antibiotic resistance, followed by spreading on agar plates. The success of the elecrophoration depends greatly on the purity of the plasmid solution, especially on its salt content. Solutions with high salt concentrations might cause an electrical discharge (known as arcing), which often reduces the viability of the bacteria.

Advantages

i. Specialized vectors are not needed.

ii. Useful for high-efficiency transient expression of foreign genes in plants and for this, no selectable marker is needed.

Disadvantages

i. It can be only used with protoplasts - frequency of stable transformation is low (0.001 or less).

ii. In stable transformants, the integrated DNA is extensively rearranged this can lead to some mitotic and meiotic instability.

4. Microinjection method

In microinjection DNA can be introduced into cells or protoplast with the help of very fine needles or glass micropipettes having the diameter of 0.5 to 10 micro metre. Some of the DNA injected may be taken up by the nucleus. Computerized control of holding pipette, needle, microscope stage and video technology has improved the efficiency of this technique. The advantages, limitations and applications are listed below:

Advantages of microinjection

1. Frequency of stable integration of DNA is far better as compare to other methods.
2. Method is effective in transforming primary cells as well as cells in established cultures.
3. The DNA injected in this process is subjected to less extensive modifications.
4. Mere precise integration of recombinant gene in limited copy number can be obtained.

Limitations of microinjection

1. Costly.
2. Skilled personnels required.
3. More useful for animal cells.
4. Embryonic cells preferred for manipulation.

5. Knowledge of mating period, oocyte recovery is essential.
6. Method is useful for protoplasts and not for the walled cells.
7. Process causes random integration.
8. Rearrangement or deletion of host DNA adjacent to site of integration are common.

Applications of microinjection

1. Process is applicable for plant cell as well as animal cell but more common for animal cells.
2. Technique is ideally useful for producing transgenic animal quickly.
3. Procedure is important for gene transfer to embryonic cells.
4. Applied to inject DNA into plant nuclei.
5. Method has been successfully used with cells and protoplast of tobacco, alfalfa, etc.

TRANSGENIC BREEDING IN FLOWER CROPS

The global flower industry thrives on novelty. Genetic engineering is providing a valuable means of expanding the floriculture gene pool so promoting the generation of new commercial varieties. There has been extensive research on the genetic transformation of different flowering plant species and many ornamental species have now been successfully transformed, including those which are most important commercially. To date, more than 30 ornamental species have been transformed, including anthurium, begonia, carnation, chrysanthemum, cyclamen, datura, daylily, gentian, gerbera, gladiolus, hyacinth, iris, lily, lisianthus, orchid, pelargonium, petunia, poinsettia, rose, snapdragon and torenia (Deroles *et al.*, 2002). New ornamental plant varieties are being created by breeders in response to consumer demand for new products. In general terms engineered traits are valuable to either the consumer or the producer. At present only consumer traits appear able to provide a return capable of supporting what is still a relatively expensive molecular breeding tool. Commercialization of genetically engineered flowers is currently confined to novel coloured carnations. The production of novel flower colour has been the first success story in floriculture genetic engineering. However, further products are expected given the level of activity in the field. Other traits that have received attention include floral scent, floral and plant morphology, senescence of flowers both on the plant and post-harvest and disease resistance. To date, there are only a few ornamental genetically modified (GM) products in development and only one, a carnation genetically modified for flower colour, in the marketplace. There are approximately 8 ha of transgenic carnation in production worldwide, largely in South America. The other breeding programs on colour modification or alteration of plant architecture and height remain focussed on rose, gerbera and various pot plant species (Yalcin-mendi *et al.*, 2006).

Antisense *ETR1* cDNA from rose cv. Texas is constructed and transferred into petunias creating transgenic plants that are less sensitive to ethylene, which opens the possibility to obtain ethylene insensitive cut rose cultivars through molecular breeding (Mei *et al.*, 2007). In addition to the application of molecular markers, a regeneration and transformation protocol for roses has been established and a number of transgenic lines were generated expressing different antifungal proteins (Debener *et al.*, 2004).

There are no commercial cultivars of gladiolus that are resistant to virus for breeding so an alternative strategy of transgenic gladiolus plants containing the *uidA* gene under control of the *GUBQ1* promoter have been found to show GUS expression in the leaves, roots and callus derived from the transgenic plant (Kamo and Young, 2007). Transgenic *Gladiolus* plants transformed with the bean yellow mosaic virus (BYMV) coat-protein (CP) gene in either sense or antisense (AS) orientation were developed using biolistics. It appeared that BYMV infection was delayed in the CP- and AS-transgenic lines but that the transgenes did not prevent eventual infection of BYMV. This is the first report of developing a floral bulb crop with antiviral genes to BYMV (Kamo *et al.*, 2005). *Fusarium* rot caused by *Fusarium oxysporum* f.sp. *gladioli* (*Fog*) is one of the most serious diseases of *Gladiolus*, both in the field and in bulbs in storage. In order to study the mechanisms of pathogenesis of this fungus, *Fog* is transformed with *Agrobacterium tumefaciens* binary vectors containing the hygromycin B phosphotransferase (*hph*) gene and fluorescence reporter genes EGFP (green), EYFP (yellow) or ECFP (cyan) using the AGL-1 strain of *A. tumefaciens*. Hygromycin B (100 micro g/ml) resistant colonies were observed only when acetosyringone was added to the co-cultivation medium. Transformed lines will be useful for identifying pathogenicity related genes, screening transgenic resistance and in studies of host-pathogen interactions (Lakshman *et al.*, 2012).

Ground-cover chrysanthemum is an important landscape plant but it cannot be resistant to water logging. The traditional breeding method cannot solve this problem. Therefore, the advantages about targeted improvement and short breeding cycle of transgenic technology are particularly important in chrysanthemum breeding. Furthermore, the safety problem of molecular breeding is the focus in society. This concern could be settled if a non-antibiotic mannose selection system using the phosphomannose-isomerase (*Pmi*) gene is used. The *Vgb* gene is isolated from the bacterium *Vitreoscilla*. The strict aerobic bacterium *Vitreoscilla stercoraria* expresses hemoglobin-like protein, VHb, which facilitates oxygen transportation and efficiently dissociates the oxygen for aerobiosis. Only transformed cells are capable of utilizing mannose as a carbon source. A vector, pCAMBIA1301-*Pmi-Vgb*, with *Vgb* gene and *Pmi* gene was constructed. With this vector the genes *Pmi* and *Vgb* have been transformed into ground-cover chrysanthemum 'Fenditan' by *Agrobacterium*-mediated method. The antibiotics-resistant seedlings were respectively detected by PCR with primers depending on *Vgb* (Wang *et al.*, 2012).

Apart from above information transgenic breeding of different flower crops i.e. anthurium, carnation, orchids, etc. have also been discussed in their respective chapters. Transgenics obtained in several ornamental plants. Some of them are given in Table 5.

Table 5: Transgenics obtained in various ornamental plant species.

S.No.	Ornamental plants	Transformation method	References
1.	*Alstromeria* sp.	*Agrobacterium tumefaciens*	Kim *et al.*, 2007
		Microprojectile	Lin *et al.*, 2000
2.	*Anthurium andraeanum*	*Agrobacterium tumefaciens*	Chen and Kuehnle, 1996
3.	*Antirrhinum majus*	*Agrobacterium tumefaciens*	Cui *et al.*, 2001
		Agrobacterium rhizogenes	Hoshino *et al.*, 1998
4.	*Begonia* sp.	*Agrobacterium tumefaciens*	Kiyokawa *et al.*, 1996
			Kishimoto *et al.*, 2000
5.	*Cyclamen persicum*	*Agrobacterium tumefaciens*	Aida *et al.*,1999
			Sironi *et al.*, 2000
			Boase *et al.*, 2000
6.	*Dahlia*	*Agrobacterium tumefaciens*	Otani *et al.*, 2013
7.	*Datura* sp.	*Agrobacterium tumefaciens*	Ducrocq *et al.*, 1994
		Agrobacterium rhizogenes	
8.	*Delphinium* sp.	*Agrobacterium tumefaciens*	Hirose *et al.*, 1998
9.	*Dendranthema* sp.	*Agrobacterium tumefaciens*	Yepes *et al.*, 1999
			Takatsu *et al.*, 1999
		Microprojectile	Yepes *et al.*,1995
10.	*Dianthus caryophyllus*	*Agrobacterium tumefaciens*	Miroshinchenko and Dolgov, 2000, Yoshimura *et al.*, 2007
		Agrobacterium tumefaciens/ Agrobacterium rhizogenes	Nakano *et al.*, 1999
		Microprojectile	Zuker *et al.*, 1995
			Hasegawa *et al.*, 2003
11.	*Eustoma grandiflorum*	*Agrobacterium tumefaciens*	Hecht *et al.*, 1997
		Agrobacterium tumefaciens	Giovannini *et al.*, 1996
		Microprojectile	Takahashi *et al.*, 1998
12.	*Euphorbia pulcherrima*	Electrophoration	Vik *et al.*, 2000
13.	*Forsythia* × *intermedia*	*Agrobacterium tumefaciens*	Rosati *et al.*,1996
14.	*Gentiana* sp.	*Agrobacterium rhizogenes*	Vinterhalter *et al.*, 1999
		Microprojectile	Hosokawa *et al.*, 2000
15.	*Gerbera hybrida*	*Agrobacterium tumefaciens*	Nagaraju *et al.*, 1998
			Reynoird *et al.*, 2000
16.	*Gladiolus grandiflorus*	*Agrobacterium tumefaciens*	Bablu and Chawala, 2000
		Microprojectile	Kamo *et al.*, 1995a
			Kamo *et al.*, 1995b

(Contd.)

(*Contd. Table 5*)

S.No.	Ornamental plants	Transformation method	References
17.	*Hemerocallis* sp.	Microprojectile	Ling *et al*., 1998
	Hibiscus syriacus	*Agrobacterium tumefaciens*	Yang *et al*., 1996
18.	*Hyacinth*	Microprojectile	Langeveld *et al*., 2000
19.	*Iris germanica*	*Agrobacterium tumefaciens*	Jeknic *et al*., 1999
20.	*Linum* sp.	*Agrobacterium tumefaciens*	Ling and Binding, 1997
21.	*Lilium longiflorum*	Microprojectile	
22.	*Orchid* sp.	*Agrobacterium tumefaciens*	Belarmino and Mii, 2000
		Microprojectile	Yang *et al*., 1999
			Knapp *et al*., 2000
23.	*Ornithogalum* sp.	Microprojectile	De Villiers *et al*., 2000
24.	*Osteospermum* sp.	*Agrobacterium tumefaciens*	Allavena *et al*., 2000
25.	*Pelargonium* sp.	*Agrobacterium tumefaciens*	Krishnaraj *et al*.,1997
			Boase *et al*., 1998b
		Agrobacterium rhizogenes	Pellegrineschi and Davoliomariani, 1996
26.	*Petunia hybrida*	*Agrobacterium tumefaciens*	Van der Meer, 1999
		Vaccum Infltn	Tjokrokusumo *et al*., 2000
27.	*Rhododendron* sp.	*Agrobacterium tumefaciens*	Pavigerova *et al*., 1997
		Microprojectile	Hsia and Korban, 1998
28.	*Rosa* sp.	*Agrobacterium tumefaciens*	Souq *et al*., 1996
			Van der Salm *et al*., 1997

REFERENCES

Adams, R.P. and Demeke, T. 1993. Systematic relationships in *Juniperus* based on random amplified polymorphic DNAs (RAPDs), *Taxon*., **42**: 553-71.

Aida, R. and Shibata, M. 1995. *Agrobacterium*-mediated transformation of Torenia (*Torenia fournieri*). *Breed. Sci.,* **45**: 71-74.

Aida, R., Hirose, Y., Kishimoto, S. and Shibata, M. 1999. *Agrobacterium tumefaciens*-mediated transformation of *Cyclamen persicum* Mill. *Plant Sci*., **148**: 1-7.

Allavena, A., Giovannini, A., Spena, A., Zottini, M., Accotto, G.P., Vaira, A.M. and Berio, T. 2000. Genetic engineering of *Osteospermum* spp: A case story. *Acta Hort*., **508**: 129-33.

Anatassopoulos, E. and Keil, M. 1996. Assessment of natural and induced genetic variation in *Alstroemeria* using random amplified polymorphic DNA (RAPD) markers. *Euphytica*, **90**: 235-44.

Arnold, M.L., Buckner, C.M. and Robinson, J.J. 1991. Pollen-mediated introgression and hybrid speciation in Louisiana irises. *PNAS,* **88**(4): 1398-402.

Awasth, C. and Hanur, V.S. 2009. Genes to fulfil consumers demand. *Indian Horticulture*, **54** (10): 30-33.

Bablu, P. and Chawla, H.S. 2000. *Agrobacterium*-mediated transformation in gladiolus. *J. Hort. Sci. Biotech*., **75**: 400-04.

Barcaccia, G., Albertini, E. and Falcinelli, M. 1999. AFLP Fingerprinting in *Pelargonium peltatum*: Its development and potential in cultivar identification. *J. Hort. Sci. Biotech*., **74:** 243-250.

Barik, S., Senapati, S.K., Aparajita, A., Mohapatra, A. and Rout, G.R. 2006. Identification and genetic variation among *Hibiscus* species (Malvaceae) using RAPD markers. *Z. Naturforsch*., **61**: 123-28.

Beecher, S.A., Steinmetz, K.,Weising, K., Boury, S., Peltier, D., Renou, J.P., Kahl, G. and Wolff, K. 2000. Microsatellite for cultivar identification in *Pelargonium. Theor. Appl. Genet*., **101**: 643-51.

Belarmino, M.M. and Mii, M. 2000. *Agrobacterium*-mediated genetic transformation of a *Phalaenopsis* orchid. *Plant Cell Rep*., **19**: 435-42.

Benedetti, L., De Burchi, G., Bruna, S., Mercuri, A. and Schiva, T. 2003. Use of molecular markers to improve cut flowers longevity in carnation. *Acta Horticulturae*, **624**: 343-45.

Benmier, H., Zuker, A., Weiss, D. and Vainstein, A. 2002. Molecular control of floral pigmentation: anthocyanin. *In*: Breeding for ornamental: Classical and Molecular Approaches (ed.) A. Vainstein. Kluwer Academic Publishers, Netherlands. pp. 256-57.

Bentolila, S. and Hanson, M. 2001. Identification of a BIBAC clone that co-segregates with the petunia Restorer of fertility (*Rf*) gene. *Molecular Gene. Genom.*, **266**(2): 223-30.

Berio, T., Morreaale, G., Giovannini, A., Allavena, L., Arru, L., Faccioli, P., Terzi, V. and Pecchioni, N. 2001. RAPD and AFLP genetic markers for the characterisation of *Osteospermum* germplasm. *Acta Hort.*, **546**:171-76.

Beyermann, B., Numberg, P., Weihe, A., Meixner, M., Epolen, J.T. and Bomer, T.1992. Fingerprinting genomes with oligonucleotide probes specific for simple repetitive DNA sequences. *Theor. App. Genet.*, **83**: 691-94.

Bhatia, R., Singh, K.P., Sharma, T.R., and Jhang, T. 2010. Evaluation of the genetic fidelity of *in vitro*-propagated gerbera (*Gerbera jamesonii* Bolus) using DNA-based markers. *Plant Cell Tiss. Organ Cult.*, pp.1-5.

Boase, M.R., Bradley, J.M. and Borst, N.K. 1998b. An improved method for transformation of regal pelargonium (*Pelargonium* × *domesticum* Dubonnet) by *Agrobacterium tumefaciens. Plant Sci.*, **139**: 59-69.

Boase, M.R., Spiller, G.B. and Peters, T.A. 2000.Transgenic *Cyclamen persicum* Mill. produced from etiolated hypocotyls, stably express the *gus*A reporter gene in petals, scapes, leaf laminae, petioles and corms 27 months after infection with *Agrobacterium tumefaciens. SIVB Congress on In Vitro Biology*, **36**: 1035.

Burchi, G.A. 2002. Genetic characterization of ornamental germplasm through biotechnological approaches. *Itaus Hortus*, **9**(5): 28-33.

Caetano-Anollés, G., Bassam, B.J. and Gresshoff, P.M. 1991. DNA amplification fingerprinting: A strategy for genome analysis. *Plant Molecular Biology Reporter*, **9**(4): 294-07.

Caetano-Anollés, G., Callahan, L.M., Williams, P.E., Weaver, K.R. and Gresshoff, P. M. 1995. DNA amplification fingerprinting analysis of bermudagrass (*Cynodon*): genetic relationships between species and interspecific crosses. *Theor. Applied Gene*, **91**(2): 228-35.

Cerny, T.A., Caetnotano- Anolles, G., Trigiano, R.N. and Starman, T.W. 1996. Molecular phylogeny and DNA amplification fingerprinting of *Petunia* taxa, *Theor. Appl. Genet.*, **92**: 1009-16.

Chen, F.C. and Kuehnle, A.R. 1996. Obtaining transgenic anthurium through *Agrobacterium*-mediated transformation of etiolated internodes. *J. Amer. Soc. Hort. Sci.*, **121**: 47-51.

Cheng, Z., Lu, B.R., Sameshima, K., Fu, D.X. and Chen, J.K. 2004. Identification and genetic relationships of kenaf (*Hibiscus cannabinus* L.) germplasm revealed by AFLP analysis. *Genetic Resources Crop Evolution*, **51**(4): 393-01.

Christey, M.C. 2001. Use of ri-mediated transformation for production of transgenic plants. *In Vitro Cellular and Developmental Biology - Plant*, **37**(6): 687-700.

Cui, M.L., Takayanagi, K., Kamada, H., Nishimura, S. and Handa, T. 2001. Efficient shoot regeneration from hairy roots of *Antirrhinum majus* L. transforemed by *rol* type MAT vector system. *Plant Cell Rep.*, **20**: 55-59.

De Feng, Z., Aoki, Y., Kageyana, K. and Fukui, H. 2013. Development of species specific SCAR markers for identification of rose species, *Rosa multiflora. J. Japanese Soc. Horti. Sci.*, **82**(1): 78-82.

De Villiers, S.M., Kamo, K., Thomson, J.A., Bornman, C.H. and Berger, D.K. 2000. Biolistic transformation of Chincherinchee (*Ornithogalum*) and regeneration of transgenic plants. *Physiol. Plant*, **109**: 450-55.

De Wet, L. A.R., Barker, N.P., and Peter, C.I. 2008. The long and the short of gene flow and reproductive isolation: Inter-Simple Sequence Repeat (ISSR) markers support the recognition of two floral forms in *Pelargonium reniforme* (Geraniaceae). *Biochemical Systematics Ecology.*, **36**(9): 684-90.

Debener, T. 2002. Molecular markers as a tool for analysis of genetic relatedness and selection on ornamentals, *In*: A. Vainstein (ed.). Breeding for Ornamentals: Classical and Molecular Approaches. pp. 329-45.

Debener, T., Bartels, C. and Mattiesch, L. 1996. RAPD analysis of genetic variation between a group of rose cultivars and selected wild rose species. *Mol. Breed.,* **2**: 321-27.

Debener, T., Linde, M., Dohn, A. and Kaufmann, H. 2004. The utilization of molecular tools for rose breeding and genetics. *Acta Horticulturae*, **630:** 29-42.

De Riek, J. Dendauw, J., Mertens, M., Heursel, G. and Van Blockstaele, E. 1999. Validation of criteria for the selection of AFLP markers to assess the genetic variation of a breeders collection of evergreen azaleas *Theor. Appl. Genet.*, **99**: 1155-65.

Deroles, S.C., Boase, M.R., Lee, C.E., Peters, T.A. 2002. Gene transfer to plants. *In*: Vainstein, A. (ed). Breeding for Ornamentals: Classical and Molecular Approaches. Kluwer Academic Publishers, Dordrecht. pp.155–196.

Ducrocq, C., Sangwan, R.S. and Sangwannorreel, B.S. 1994. Production of *Agrobacterium*-mediated transgenic fertile plants by direct somatic embryogenesis from immature zygotic embryos of *Datura innoxia*. *Plant Mol. Biol.*, **25**: 995-1009.

Dunemann, F. and Kahnau, R. 1998. Genetische Fingerabdrucke zur molekularen identifzierung und verwandtschaftsanalyse bei *Rhododendron* und immergrune Laubgeholze, Jahrbuch, Deutsche Rhododendron gesellschaft, Bremen, pp. 8-20.

Escaravage, N., Questiau, S., Pornon, A., Doche, B. and Taberlet, P. 1998. Clonal diversity in a *Rhododendron ferrugineum* L. (Ericaceae) population inferred from AFLP markers. *Mol. Eco.*, **7**(8): 975–82.

Esmaiel, N.M., Al- Doss, A.A. and Barakat, M.N. 2012. *In vitro* selection for resistance to *Fusarium oxysporum* f.sp. dianthi and detection of genetic polymorphism via RAPD analysis in carnation. *J. Medicinal Plants Res.*, **6**(23): 3997-004.

Faccioli, P., Terzi, V., Pecchionoi, L., Berio, T., Giovanni, A. and Allavena, A. 2000. Genetic diversity in cultivated *Osteospermum* as revealed by random amplified polymorphic DNA analysis. *Plant Breeding*, **119**: 351-55.

Ferreira, C.A., Pinho, E.V., Paiva, P.D., Pereira, G.S. and Diniz, R.P. 2011. Identification of cultivars and certification of genetic purity of gladiolus through morphological markers. *Ciencia C Agrotecnologia*, **35**(4): 692-700.

Friar, E. and Kochert, G. 1994. A study of genetic variation and evolution of *Phyllostachys* (Bambusoideae: Poaceae) using nuclear restriction fragment length polymorphisms. *Theor. Appl. Genet.*, **89**(2-3): 265-70.

Galun, E. and Bruman, A. 1997. Transgenic Plants Imperial College Press, London, 376 pp.

Gallego, F.J. and Martinez, I. 1996. Molecular typing of rose cultivars using RAPDs. *J. Hort. Sci.*, **71**: 901-08.

Gan, C. 1989. Gene Gun Accelerates DNA-Coated Particles To Transform Intact Cells. *The Scientist*. **3**(18):25.

Giovanni, A., Pecchioni, N. and Allavena, A. 1996. Genetic transformation of Lisianthus (*Eustoma grandiflorum* Griseb) by *Agrobacterium rhizogenes. J. Genet. Breed.*, **50**: 35-39.

Giovanni, A., Zottini, M., Morreale, G., Spena, A. and Allavena, A. 1999. Characterization of ornamental Datura plants transformed by *Agrobacterium rhizogenes*. *In Vitro Cell Dev. Biol., - Plant*, **33**: 101-06.

Goh, C.J., Kumar, P.P. and Yau, J.C.K. 1995. Genetic variation detected with RAPD markers in *Heliconia*. *Acta Hort.*, **420**: 72-4.

Hai Bin, W., Jiang JiaFu Chen, SuMei Qi, Xiangyu Peng, Hui Li Pi Rui, Song Ai Ping, Guan Zhiyang Fang, Weitsin Liao and Yuan Chen Fa Di. 2013. Next generation sequencing of the *Chrysanthemum nankingense* (Asteraceae) transcriptome permits large scale unigene assembly and SSR markers discovery. *Plosone*, **8**(4): 62293.

Hamada, K. and Hagimori, M. 1996. RAPD-based method for cultivar-identification of calla lily (*Zantedeschia* spp.). *Scientia Horticulturae*, **65**(2/3): 215-218.

Han, T.H., De Jeu, M., Van Eck, H. and Jacobsen, E. 2000. Genetic diversity of Chilean and Brazilian *Alstroemeria* species assessed by AFLP analysis. *Heredity*, **84**(5): 564–69.

Han, T.H., van Eck, H.J., de Jeu, M.J. and Jacobsen, E. 2001. Use of AFLP marker technique for classification of *Alstroemeria* spp. and construction of a linkage map in *Alstroemeria aurea. Acta Hort.*, **546**:151-58.

Han, T.H., van Eck, H.J., De Jeu, M.J., and Jacobsen, E. 1999. Optimisation of AFLP fingerprinting of organisms with a large-sized genome: a study on *Alstroemeria* sp. *Theor. Appl. Genet.*, **98**: 465-71.

Haruki, K., Hosoki,T. and Nako, Y. 1998. Tracing the Parentages of Some Oriental Hybrid Lily Cultivars by PCR-RFLP Analysis. *Journal of the Japanese Society for Horticultural Science*, **67**(3): 352-59.

Hasegawa, T., Manago, N. and Kato, Y. 2003. Production of transgenic carnation introduced sarcotoxin gene through particle bombardment. *Research Bulletin of the Aichi-ken Agricultural Research Center*, **35**: 143-47.

Hecht, M., Ecker, R., Stav, R., Watad, A.A., Hirchberg, J., Mann, V., Altman, A. and Ziv, M. 1997. *Agrobacterium*-mediated transformation of Lisianthus (*Eustoma grandiflorum*). *Proce*. 3rd International ISHS Symposium on *In Vitro* Culture and Horticultural Breeding, Jerusalem, Israel. *Acta Hort*, **447**: 339-40.

Hibrand-Saint Oyant, L., Crespel, L., Rajapakse, S., Zhang, L. and Foucher. 2008. Genetic linkage maps of rose constructed with new microsatellite markers and locating QTL controlling flowering traits. *Tree Genetics Genomes*, **4**(1): 11-23.

Hirose, Y., Aida, R. and Shibata, M. 1998. *Agrobacterium tumefaciens*-mediated transformation of Delphinium (*Delphinium* spp.) *Breed. Sci.*, **48**: 117.

Hodkinson, T.R., Renvoize, S.A., Chonghaile, G.N., Stapleton, C.M.A. and Chase, M.W. 2000. A Comparison of ITS Nuclear rDNA Sequence Data and AFLP Markers for Phylogenetic Studies in *Phyllostachys* (Bambusoideae, Poaceae). *J. Plant Res.*, **113**(3): 259-69.

Hoshino, Y. and Mii, M. 1998. Bialaphos stimulates shoot regeneration from hairy roots Snapdragon (*Antirrhinum majus* L.) transformation by *Agrobacterium rhizogenes*. *Plant Cell Rep.*, **17**: 256-61.

Hoshino, Y., Turkan, I. and Mii, M. 1998. Transgenic bialaphos-resistant Snapdragon (*Antirrhinum majus* L.) produced by *Agrobacterium rhizogenes* transformation. *Sci. Hort.*, **76**:37-57.

Hosokawa, K., Matsuki, R., Oikawa, Y. and Yammura, S. 2000. Production of transgenic gentian plants by particle bombardment of suspension-culture cells. *Plant Cell Rep.*, **19**:454-58.

Hsia, C.N. and Korban, S.S. 1998. Micro-projectile-mediated genetic transformation of Rhododendron hybrids. *American Rhododendron Society Journal*, **52**: 187-191.

Hubbard, M., Kelly, J., Rajapakse, S., Abbott, A. and Ballard, R. 1992. Restriction fragment length polymorphisms in Rose and their use for cultivar identification. *HortScience*, **27**(2): 172-73.

Jeknic, Z., Lee, S.P., Davis, J., Ernst, R.C. and Chen, T.H.H. 1999. Genetic transformation of *Iris germanica* mediated by *Agrobacterium tumefaciens*. *J. America Soc. Hort. Sci.*, **124**: 575-80.

Jianhua, Z., McDonald, M.B. and Sweeny, P.M. 1997. Testing for genetic purity in *Petunia* and *Cyclamenet* seed using random amplified polymorphic DNA, *HortScience*, **32**: 246-47.

Kameyama, Y., Isagi, Y. and Nakagosh, N. 2001. Patterns and levels of gene flow in *Rhododendron metternichii* var. *hondoense* revealed by microsatellite analysis. *Mol. Eco.*, **10**(1): 205-16.

Kamo, K. and Young, H.J. 2007. A polyubiquitin promoter isolated from gladiolus and its expression in gladiolus plants. *Acta Horticulturae*, **763**: 251-7.

Kamo, K., Blowers, A., Smith, F. and Vaneck, J. 1995a. Stable transformation of Gladiolus by particle gun bombardment of cormels. *Plant Sci.*, **110:**105-11.

Kamo, K., Blowers, A., Smith, F., Vaneck, J. and Lawson, R. 1995b. Stable transformation of Gladiolus using suspension cells and callus. *J. Amer. Soc. Hort. Sci.*, **120**: 347-52.

Kamo, K., Gera, A., Cohen, J., Hammond, J., Blowers, A., Smith, F. and Eck, J.V. 2005. Transgenic gladiolus plants transferred with the bean yellow mosaic virus coat-protein gene in either sense or antisense orientation. *Pl. Cell Reports*, **23**(9): 654-63.

Kaundun, S.S. and Matsumoto, S. 2002. Heterologous nuclear and chloroplast microsatellite amplification and variation in tea, *Camellia sinensis. Genome*, **45**(6): 1041-48.

Kim, J.B., Raemakers, C.J.J.M., Jacobsen, E. and Visser, R.G.F. 2007. Efficient production of transgenic *Alstroemeria* plants by using *Agrobacterium tumefaciens*. *Annals of Applied Biology*, **151**(3): 401–12.

King, R.C. and Stansfield, W.D. 1990. A dictionary of genetics. 4th ed. Oxford Univ. Press, New York-Oxford. pp. 188.

Kishimoto, S., Aida, R. and Shibata, M. 2000. *Agrobacterium tumefaciens*- mediated transformation of *Begonia × hiemalis*. *J. Japanese Soc. Hort. Sci*., **69**: 424.

Kiyokawa, S., Kikuchi, Y., Kamda, H. and Harada, H. 1996. Genetic transformation of *Begonia tuber hybrida* by *Rirolgenes*. *Plant Cell Rep*., **15**: 606-09.

Knapp, J.E., Kausch, A.P. and Chandlee, J.M. 2000.Transformation of three genera of orchid using the *bar* gene as a selectable marker. *Plant Cell Rep*., **19**: 893-98.

Ko, H.L., Henry, R.J., Beal, P.R., Moisander, J.A. and Fisher, K.A. 1996. Distinction of *Ozothamnus diosmifolius* (Vent.) DC genotypes using RAPD. *Hort.Science,* **31**: 858-61.

Krishnaraj, S., Bi, Y.M. and Saxena, P.K. 1997. Somatic embryogenesis and *Agrobacterium*-mediated transformation system for scented geraniums (*Pelargonium* sp. Frensham). *Planta*, **201**: 434-40.

Kumar, P.P., Yau, J.C.K. and Goh, C.J. 1998. Genetic analysis of *Heliconia*spp and cultivars with random amplified polymorphic DNA (RAPD) markers. *J. American Soc. Hort. Sci.*, **103**: 91-97.

Labra, M., Grassi, F., Bardini, M., Imazio, S., Guiggi, A., Citterio, S., Banfi, E. and Sgorbati, S. 2003. Genetic relationships in *Opuntia* Mill. genus (Cactaceae) detected by molecular marker. *Plant Sci*., **165**(5): 1129–136.

Lakshman, D.K., Pandey, R., Kamo, K., Bauchan, G. and Mitra, A. 2012. Genetic transformation of *Fusarium oxysporium* f. sp. *gladioli* with *Agrobacterium* to study pathogenesis in gladiolus. *European J. Pl. Patho*., **133**(3): 729-38.

Langeveld, S.A., Gerrits, M.M., Derks, A.F.L.M., Boonekamp, P.M. and Bol, J.F. 1995. Transformation of lily by *Agrobacterium*. *Euphytica*, **85**: 97-100.

Langeveld, S.A., Pham, K., van Schadewijk-Nieuwstad, M., Burma, D., Langens -Gerrits, M., Bach, A. and Derks, T.F.L.M. 2000. Genetic transformation of bulbous crops.VIIIth Symposium on flower Bulbs. *Acta Hort*., **62**.

Lee, Y.H. 1991. Genomic constitutions and flower characteristics of selected Aranda orchid cultivars. *Euphytica*, **54**(3): 251-54.

Lesur, C. Baury, S., Weising, K., Kahl, G., Wolff, K., Becher, S.A. and Peltier, D. 2000. Comparision of seven molecular techniques for *pelargonium* cultivar identification. *Acta Hort.*, **508**: 297-301.

Lim, S.H., Teng, P.C.P., Lee, Y.H. and Goh, C.J. 1999. RAPD analysis of some species in the genus *Vanda* (Orchidaceae). *Ann. Bot.*, **83**: 193-96.

Lin, H.S., van der Toom, C.J.G., Raemakers, K.J., Visser, R.G.F., De Jue, M.J. and Jacobsen, E. 2000. Genetic transformation of *Alstroemeria* using particle bombardment. *Mol. Breed*., **6**: 369-77.

Lin, X.C., Lou, Y.F., Liu, J., Peng, J.S., Liao, G.L. and Fang, W. 2010. Crossbreeding of *Phyllostachys* species (Poaceae) and identification of their hybrids using ISSR markers. *Genet. Mol. Res.*, **9**(3): 1398-04.

Ling, H.Q. and Binding, H. 1997. Transformation in protoplast cultures of *Linum usitatissimum* and *L. suffruticosum* mediated with peg and with *Agrobacterium tumefaciens*. *J. Plant Physiol*., **151**: 479-88.

Ling, J.T., Huang, J. and Sauve, R. 1998. Genetic transformation and regeneration of transgenic 'Stella de Oro'daylily. *In Vitro Cell. Dev. Biol.-Plant*, **34**: 1053.

Ling, J.T., Sauve, R. and Gawel, N. 1997. Identification of *Poinsettia* cultivars using RFLP marker, *HortScience,* **32**: 122-24.

Lipsky, A., Cohen, A., Gaba, V., Kamo, K., Gera, A. and Watad, A. 2002. Transformation of *Lilium longiflorum* plants for Cucumber Mosaic Virus resistance by particle bombardment. *Acta Hort.,* **568**: 209-14.

Loh, J.P., Kiew, R., Keet, A., Gan, L.H. and Gan, Y.Y. 1999. Amplified fragment length polymorphism (AFLP) provides molecular markers for the identification of *Caladium bicolor* cultivars. *Ann. Bot.*, **84**: 155-61.

Marchant, R., Power, J.B., Lucas, J.A. and Davey, M.R. 1998b. Biolistic transformation of rose (*Rosa Hybrida* L.). *Annals of Botany*, **81**: 109-14.

Marouelli, L.P., Inglis, P.W. Ferreira, M.A. and Buso, G.S.C. 2010. Genetic relationships among *Heliconia* (Heliconiaceae) species based on RAPD markers. *Genet. Mol. Res.*, **9**(3): 1377-87.

Marsolais, J.V., Pringle, J.S. and White, B.N. 1993. Assessment of random amplified polymorphic DNA (RAPD) as genetic markers for determining the origin of interspecific lilac hybrids. *Taxon,* **42**:531-37.

Matsumoto, S. and Fukuri, H. 1996. Identification of *Rosa* cultivars and clonal plants by random amplified polymorphic DNA. *Sci. Hort.*, **67**: 49-54.

Mei, Z.Y., Ma, Y.L., Qing L.C. and Wen, Qi. 2007. An antisense ETR1 cDNA from rose can reduce the ethylene sensitivity of petunias. *Acta Horticulturae*, **751:** 473-9.

Miglino, R., Jodlowska, A., Pappu, H.R. and Schadewijik, T.R.V. 2007. A semi automated and highly sensitive streptavidin magnetic capture-hybridization RT-PCR assay: Appliction to genus wide or species-specific detection of several viruses of ornamental bulb crops. *J. Virology Method*, **146:** 155-64.

Millan, F., Osuma, F., Cobos, S., Torres, A. and Cubero, J.I. 1996. Using RAPDs to study phylogenetic relationships in *Rosa. Theor. Appl. Genet.,* **92**: 273-77.

Mitiouchkina, T.Y. and Dolgov, S.V. 2000. Modification of Chrysanthemum flower and plant architecture by *Rol*C gene from *Agrobacterium rhizogenes* introduction.The 19th International Symposium on Improvement of Ornamental Plants. *Acta Hort.*, **508**: 163-14.

Nagaraju, V., Srinivas, G.S.L. and Sita, G.L. 1998. *Agrobacterium*-mediated genetic transformation in Gebera hybrid. *Curr. Sci.*, **74**: 630-34.

Naito, K., Isagi, Y., Kameya, Y. and Nakagoshi, N. 1999. Population structure in *Rhododendron metternichi* var. *hondoense* assessed with micro-satellites and their amplication for conservation. *J. Plant Res.,* **112**: 405-12.

Nakano, M., Koike, Y., Watanabe, Y. and Suzuki, S. 1999. Production of transgenic plantlets of carnation via *Agrobacterium*-mediated transformation of petal explants. *Bulletin of the Faculty of Agriculture, Niigata University*. **51**(2): 105-14.

Obara-Okeyo, P. and Kako, S. 1998. Genetic diversity and identification of *Cymbidium* cultivars as measured by random amplified polymorphic DNA (RAPD) markers. *Euphytica,* **99**: 95-101.

Otani, Y., Chin, D.P. and Mii, Masahiro. 2013. Establishment of *Agrobacterium*-mediated genetic transformation system in *Dahlia. Plant Biotech.,* **30**: 135–39.

Pathania, N.S. and Misra, R.L. 2001. Characterization of gladiolus mutants using RAPD markers. *J. Orna. Hort.*, **4**(2): 65-8.

Pavingevora, D., Briza, J., Koytek, K. and Nicedermeierova, H. 1997. Transformation of *Rhododendron* spp. using *Agrobacterium tumefaciens* with a GUS-intron chimeric gene., *Plant Sci.*, **122**: 165-71.

Pellegrineschi, A. and Davoliomariani, O. 1996. *Agrobacterium rhizogenes*-mediated transformation of scented Geranium. *Plant Cell Tiss. Org. Cult.*, **47**: 79-86.

Rajapakse, S., Hubbard, M., Kelly, J.W., Abbotr, A.G. and Ballard, R.E. 1992. Identification of rose cultivars by restriction fragments length polymorphism. *Sci.Horti.*, **52**: 237-45.

Rajaseger, G., Tan, H. T. W., Turner, I. M. and Kumar, P. P. 1997. Analysis of Genetic Diversity among *Ixora* cultivars (Rubiaceae) using Random Amplified Polymorphic DNA. *Ann. Bot.*, **80**(3): 355-61.

Ranamukhaarachchi, D.G., Henny, R.J., Guy, C.L. and Li, Q.B. 2001. DNA Fingerprinting to Identify Nine *Anthurium* Pot Plant Cultivars and Examine Their Genetic Relationship. *HortScience*, **36**(4):758–60.

Renou, J.P., Aubry, C., Serveau, M. and Jalouzot, P. 1997. Evaluation of genetic variability in the genus *pelargonium* using RAPD markers. *J. Hort. Sci.*, **72:** 229-37.

Reynoird, J.P., Dewitte, W., Prinsen, E., Noin, M., Chriqui, D. and Van Onckelen, H. 2000. Shooty tumours induced on Gerbera hybrid leaf explants by *A. tumefaciens* strain 82319 are characterized by high endogenous cytokinin levels. The 19th International Symposium on Improvement or Ornamental Plants. *Acta Hort.*, **508**: 45-48.

Rosati, C., Cadic, A., Renou, C. and Duron, M. 1996. Regeneration and *Agrobacterium*-madiated transformation of *Forsythia × intermedia* Spring Glory. *Plant Cell Rep.,* **16**: 114-17.

Ryutaro, H., Sanae, K., Michio, S. and Yukio, H. 1999. Production of transgenic cyclamen plants by *Agrobacterium tumefaciens* method. *Yasai Chagyo Kenkyu Seika Joho.*, 1998: 75-76.

Scandalios, J.G. 1974. Isozymes in development and differentiation. *Annual Rev. Pl. Phy.*, **25:** 225-258.

Schum, A. 2003. Mutation breeding in ornamentals: an efficient breeding method? *Acta Horticulture*, **612**: 47-60.

Scott, M.C., Caetan-Anolles, G. and Trigiano, R.N. 1996. DNA amplification fingerprinting identifies closely related chrysanthemum cultivars. *J. Amer. Soc. Hor. Sci.,* **121**: 1043-48.

Scovel, G., Ovadis, M., Vainstein, A., Reuven, M. and Ben-Yephet, Y. 2001. Marker assisted selection for resistance to *Fusarium oxysporum* in the greenhouse carnation. *Acta Hort.*, **552**: 151-56.

Semegn, K., Bjornstad, A. and Ndjiondjop, M.N. 2006. An overview of molecular marker methods for plants. *Af. J. Biotech*, **5**(25): 2540-68.

Sironi. F., Rufforli, B., Giovannini, A. and Allavena, A. 2000. Morphogenesis an genetic transformation of *Cyclamen persicum* Mill, 4th International Symposium on *In Vitro* Horticultural breeding, pp.80.

Soliva, M., Gautschi, B., Salzmann, C., Tenzer, I. and Widmer, A. 2000. Isolation and characterization of microsatellite loci in the orchid *Ophrys araneola* (Orchidaceae) and a test of cross-species amplification. *Molecular Ecology*, **9**(12): 2178–179.

Souq, F., Coutus Thevenot, P., Yean, H., Delbard, G., Maizere, Y., Barbe, J.P., Boulay, M., Morisot, A. and Ricci, P. 1996. Genetic transformation of roses, 2 examples: one on morphogenesis, the other on anthocyanin biosynthetic pathway, Second International Symposium on Roses, Antibes, France. *Acta Hort.,* **424**: 381-38.

Starman, T.W. and Abbit, S. 1997. Evaluating genetic relationships of *geranium* using arbitrary signatures from amplification profiles, *HortScience,* **31**: 729-41.

Starman, T.W., Duan, X. and Abbit, S. 1999. Nucleic acid scanning technique distinguished closely related cultivars of *Poinsettia*. *HortScience,* **34**: 1119-22

Strommer, J., Gerats, A.G.M., Sanago, M. and Molnar, S.J. 2000. A gene-based RFLP map of petunia. *Theor. Appl. Genet.*, **100**: 899-05.

Stummer, B.E., Smith, S.E. and Langridge, P. 1995. Genetic transformation of *Verticordia grandis* (Myrtaceae) using wild-type *Agrobacterium rhizogenes* and binary *Agrobacterium* vectors. *Plant Sci.,* **111**: 51-62.

Suo, Z., Li, W., Yao, J., Zhang, H., Zhang, Z. and Zhao, D. 2005. Applicability of leaf morphology and intersimple sequence repeat markers in classification of tree peony (Paeoniaceae) cultivars. *HortScience*, **40**(2): 329-34.

Swoboda, I. and Bhalla, P.L. 1997. RAPD analysis of genetic variation in the Australian fan flower, Scaevola. *Genome*, **40**(5): 600-06.

Takahashi, M., Nishihara, M., Yamamura, S., Nishizawa, S., Irifune, K. and Morikawa, H. 1998. Stable transformation of *Eustoma grandiflorum* by particle bombardment. *Plant Cell Rep.,* **17**: 504-07.

Takatsu, Y., Miyamoto, M., Inoue, E., Yamada, T., Manabe, T., Kasumi, M., Hayashi, M., Sakuma, F., Marubashi, W. and Niwa, M. 2001. Interspecific hybridization among wild Gladiolus species of southern Africa based on randomly amplified polymorphic DNA markers. *Scientia Horticulturae*, **91**(3-4): 339-48.

Takatsu, Y., Nishizawa, Y., Hibi, T. and Akutsu, K. 1999. Transgenic Chrysanthemum (*Dendranthema grandiflorum* (Ramat.) Kitamura) expressing a rice chitinase gene shows enhanced resistance to gray mold (*Botrytis cinerea). Sci. Hort.,* **82**: 113-23.

Talhinhas, P., Neves-Martins, J. and Leitao, J. 2003. AFLP, ISSR and RAPD markers reveal high levels of genetic diversity among *Lupinus* spp. *Plant Breeding*, **122**(6): 507-10.

Tejaswini, Paramesh, H., Sreedhara, S. A. and Lalitha, A. 2008. Molecular markers for working out genetic relationship among genotypes of carnation (*Dianthus caryophyllus* L). *Indian J. Genet.*, **68**(1): 93-5.

Ten Hoopen, R., Robbins, T.P., Fransz, P.F., Montijn, B.M., Oud, O., Gerats, A.G.M. and Nanninga, N. 1996. Localization of T-DNA insertions in petunia by fluorescence in-situ hybridization: Physical evidence for suppression of recombination. *The Plant Cell.,* **8**(5): 823-30.

Tjokrokusumo, D., Heinrich, T., Wylie, S., Potter, R. and McComb, J. 2000. Vacuum infiltration of *Petunia hybrid* pollen with *Agrobacterium tumefaciens* to achieve plant transformation. *Plant Cell Rep.,* **19**: 792-97.

Tomkins, J.P., Wood, T.C., Barnes, L.S., Westman, A. and Wing, A.R.A. 2001. Evaluation of genetic variation in the daylily (*Hemerocallis* spp.) using AFLP markers. *Theor. Appl. Genet.*, **102**(4): 489-96.

Torres, A.M., Millan, T., and Cubero, J.D. 1993. Identifying rose cultivars using random amplified polymorphic DNA markers, *HortScience,* **28**: 333-34.

Tzuri, G., Hillel, J., Lavi, V., Haberfeld, A. and Vainstein, A. 1991. DNA fingerprint analysis of ornamental plants. *Plant Sci.,* **76:** 91-7.

Vainstein, A. and Ben Meii, H. 1994. DNA fingerprint analysis of roses, *J. Amer. Soc. Hort. Sci.,* **119**: 1099-103.

Vainstein, A., Hillel, J., Lavi, U. and Tzuri, G. 1991. Assessment of genetic relatedness in *Carnation* by DNA fingerprint analysis. *Euphytica*, **56**: 225-29.

Van der Leede-Plegt, L.M., van Kornberg, B.C.E., Franken, J., van Tuyl, J.M., van Tunen, A.J. and Dons, H.J.M. 1997. Transgenic lilies via pollen-mediated transformation. *Acta Hort.* **430:** 529-30.

Van der Salm, T.P.M., van der Toorn, C.J.G., Bouwer, R., Tencate, C.H.H., van der Kreiken, W.M. and Dons, H.J.M. 1997. Production of *rol* genes transformed plants of *Rosa hybrid* L. and characterization of their rooting ability. *Mol. Breed.,* **3**: 39-47.

Van der Meer, I.M. 1999. *Agrobacterium*-mediated transformation of *Petunia* leaf disks.*Plant Cell Culture Protocols*, **111**: 327-334.

Vanessa, E.T.M., Ashworth, Bart C., O'Brien and Elizabeth, A., Friar. 1999. Fingerprinting *Juniperus communis* L. cultivars using RAPD markers. *Madrono*, **46**(3): 134-41.

Vik, N., Gjerde, H., Bakke, K. and Hvoslef-Eide, A.K. 2000. Stable transformation of *Poinsettia* through DNA-electrophoresis, 4th International Symposium on *In Vitro* Culture and Horticultural Breeding, 9.

Vinterhalter, B., Orbovic, V. and Vinterhalter, D. 1999. Transgenic root culture of *Gentiana punctate* I. *Acta Societatis Botanicorum Poloniae,* **68**: 275-80.

Wang, J. and Bao, M.Z. 2005. Characterisation of genetic relationships in pansy (*Viola wittrockiana*) inbred lines using morphological traits and RAPD markers. *J. Hortil. Sci. Biotech*., **8**(5): 537-42.

Wang, L.R.W., Chen Dai Silan, Luo Xinyan, Li Yan, Li Bao Quin, Zhu Jan Lu, Jie and Liu Quin Quian. 2012. The association analysis of phenotypic traits with SRAP markers in Chrysanthemum. *Scientia Agricultura Sinica*, **45**(7): 1355-64.

Wang, Y., Shi, S., Li, X., Gao, Y. and Zhang, Q. 2012. Mannose selection system to screen *Vgb* transgenic chrysanthemum plant. *Acta Horticulturae*, **953:** 223-30.

Watad, A.A., Yun, D.J., Matsumoto, T., Niu, X., Wu, Y., Kononomicz, A.K., Bressan, R.A. and Hasegawa, P.M. 1998 .Microprojectile bombardment mediated transformation of *Lilium longiflorum. Plant Sci.,* **110**: 155-64.

Watad, A.A., Yun, D.J., Matsumoto, T., Niu, X., Wu, Y., Konowicz, A.K., Bressan, R.A. and Hasegawa, P.M. 1998. Microprojectile bombardment-mediated transformation of *Lilium longiflorum. Plant Cell Rep.*, **17:** 262-67.

Wegner, H. and Debener, T. 2008. Novel breeding strategies for ornamental dahlias. II: Molecular analyses of genetic distances between dahlia cultivars and wild species. *Eur. J. Horti. Sci.*, **73**(3): 97-103.

Wilmink, A., Vande Ven, B.C.E. and Dons, J.J.M.1992. Expression of the GUS gene in the monocot tulip after introduction by particle bombardment and *Agrobacterium*. *Plant Cell Rep*., **11** :76-80.

Wolff, K. 1996. RAPD analysis of sporting and cherism in chrysanthemum. *Euphytica,* **89**: 159-64.

Wolff, K. and Peters-Van Rijn, J. 1993. Rapid detection of genetic variability in Chrysanthemum (*Dendranthema grandiflora* tzvelio) using random primers. *Heredity*, **71**: 335-41.

Wolff, K., Zietkiewitz and Hofstra, H. 1995. Identification of Chrysanthemum cultivars and stability of DNA fingerprints, DNA fingerprint pattern. *Theor. Appl. Genet.*, **91**: 439-47.

Wong, D.W.S. 1997. The ABCs of gene cloning. Chapman & Hall. New York.

Xue, D., Feng, S., Zhao, H., Jiang, H., Shen, B., Shi, N., Lu, J., Liu, J. and Wang, H. 2010. The linkage maps of *Dendrobium* species based on RAPD and SRAP markers. *J. Genet. Genom.*, **37**(3): 197–04.

Yagi, M., Tani Kawa, N.M., Shibata, M. and Onozaki, T. 2012. Breeding carnations (*Dianthus caryophyllus* L.) for resistance to bacterial wilt (*Burkholderia caryophylli*). *Acta Horticulturae*, **937:** 427-431.

Yalcin-Mendi, Y., Buzkan, N. and Dolekoglu, C. 2006. Application and commercialization of transgenic ornamental plants. *In*: Jaime A. Teixeira da Silva (eds.), Floriculture, Ornamental and Plant Biotechnology, pp. 133-39.

Yamashigi, M. 1995. Detection of section specific random amplified polymorphic DNA (RAPD) markers in *Lilium. Theor. Appl. Genet.,* **91**: 830-35.

Yan, Z., Denneboan, C., Hollendorf, A., Dolstra, O., Debener, T., Stam, P. and Visser, P. B. 2005. Construction of an integrated map of rose with AFLP, SSR, PK, RGA, RFLP, SCAR and morphological markers. *Theor. Appl. Genet.,* **110**(4): 766-77.

Yang, J., Lee, H.J., Shin, D.H., Oh, S.K., Seon, J.H., Paek, K.Y. and Han, K.H. 1999. Genetic transformation of *Cymbidium* orchid by particle bombardment. *Plant Cell Rep.,* **18**: 978-984.

Yang, L.Y., Vazquwz-Tello, A., Hidaka, M., Masaki, H. and Uozumi, T. 1996. Agrobacterium-mediated transformation of *Hibiscus syriacus* and regeneration of transgenic plants. *Plant Tissue Culture Letters*, **13**(2): 161-67.

Yao. Q., Cong, L., Chang, J. L., Li, K. X., Yang, G. X. and He, G. Y. 2006. Low copy number gene transfer and stable expression in a commercial wheat cultivar via particle bombardment. *J. Experimental Botany*, **57**(14): 3737-746.

Yepes, L.M., Mittak, V. and Slightom, J.L. 1999. *Agrobacterium tumefaciens* versus biolistic mediated transformation of Chrysanthemum cvs. Polaris and Golden Polaris with nucleocapsid protein genes of three Tospovirus. International Symposium on Cut Flowers of Tropic Species. *Acta Hort.,* **482**: 209-18.

Yepes, L.M., Mittak, V., Pang, S.Z., Gonsalves, C., Slightom, J.L. and Gonsalves, D. 1995. Biolistic transformation of Chrysanthemum with the nucleocapsid gene of tomato spotted wilt virus. *Plant Cell Rep.,* **14**: 694-98.

Yoshimura, Y. Ohishi, K. Narita, R. Ohya, T. 2007. Production of transgenic carnation introduced sarcotoxin gene. *Research Bulletin of the Aichi-ken Agricultural Research Center*, **39**: 1-6.

Zhang, D. 2000. Discrimination and genetic diversity of *Cephalotaxus* accessions using AFLP markers. *J. American Soc. Hort. Sci.*, **125**(4): 404-12.

Zhang, D., Besse, C., Cao, M.Q. and Gandelin, M.H. 2001. Evaluation of AFLPs for variety identification in modern rose (*Rosa hybrida* L.). *Acta Hort.*, **546**: 351-57.

Zhang, D., Dirr, M.A. and Prince, R.A. 2000. Discrimination and genetic diversity of *Cephalotaxus* accessions using ALFP markers. *J. Amer. Soc. Hor. Sci.*, **125**: 404-12.

Zuker, A., Chang, P.L.F., Ahroni, A., Cheah, K., Woodson, W.R., Bressan, R.A., Watad, A.A., Hasegawa, P.M. and Vainstein, A. 1995. Transformation of carnation by microprojectile bombardment. *Sci. Hort.,* **64**: 177-85.

Chapter – 5

F_1 Hybrid Seed Production in Ornamental Crops

INTRODUCTION

Hybrid crops are derived from an established and well-proven breeding method used in ornamental sectors. The development of hybrid crops has been one of the major factors behind a dramatic increase in global crop yields. Hybrid is produced by crossing between two genetically dissimilar parents. Pollen from male parent called pollen parent will pollinate, fertilize and set seeds in female called seed parent to produce F_1 hybrid seeds. For production of a hybrid, crossing between two parents is important, the crossing process will results in heterosis or hybrid vigour. Heterosis is defined as the increase in size or vigour of a hybrid over its parent. This boost in performance, combining the best yield, quality and agronomic characters from each parent, is the result of hybrid breeding.

F_1 hybrid seed offers to the growers a significant benefits in terms of yield improvement, agronomic performance and consistency of end-use quality. But attempting to farm-save hybrid varieties will result in segregating populations producing highly variable offspring, which can lead to reduced yield, disease resistance and quality. As this factsheet explains, it doesn't make economic or agronomic sense to save seed of an F_1 hybrid. In self pollinated crops it is difficult to cross but in cross pollinated crops it is easier. In nature to create genetic variability and for its wider adaptation in different environmental conditions, flowering plants has adopted many mechanisms for cross pollination. Cross pollination results in genetic heterogeneity and show wider adaptations. F_1 hybrids are in immense usage throughout the world with the rapid increasing interest in flowers, however the total number of the hybrids developed in flowers is still less. In India hybrid seed production is taken up in Bengaluru, Pune, Delhi, Punjab, Haryana, Jammu and Kashmir and Himachal Pradesh for export.

F_1 hybrids in ornamental crops were known even before Shull 1911 when he propounded the classical theory of hybrid vigour in plant breeding for the first time. In 1909 F_1 hybrid variety Prima Donna in begonia was released by Benary Seed Company in Germany. The first commercial F_1 hybrid in petunia was produced by Japan during World War II, in 1942. From 1950 onwards, F_1 hybrids were available in ornamental flowers from The Netherlands, Denmark, Germany, USA and UK besides Japan and produced in flower crops like ageratum, anemone, gerbera, primula, petunia, cyclamen, marigold, pansies, begonia, geranium, portulaca, dianthus, balsam, stock, wall flower, ornamental sunflower, gazania, hollyhock, calceolaria and zinnia. Flower breeders became familiar with the F_1 hybrid production of those crops also which were formerly propagated by vegetative parts such as pelargonium and gerbera. A few important hybrids were first released in petunia in 1940-50, geranium single in 1960, antirrhinum, marigold, pansy and zinnia in 1965, ageratum in 1966, geranium double in 1970, dianthus, impatiens, portulaca in 1976-77, gerbera in 1980 and carnation in 1981.

ADVANTAGES OF F_1 HYBRIDS

i. In petunia, pelargonium, *Begonia semperflorens*, etc., the superiority of F_1 hybrids over conventional varieties is overwhelmingly different in producing an effect like an intensively coloured carpet, while the conventional varieties often produce a shimmer, which disappears with distance, particularly in some cut annuals like aster, calendula, solidago and chrysanthemum and potted plants such as *Pelargonium zonale*, *Anthurium scherzerianum*, etc.

ii. Duration of the flush that is the blooming *Ageratum houstonianum*, *Begonia semperflorens*, etc. exceeds that of non-hybrids considerably.

iii. Uniformity is the most essential components of F_1 hybrid e.g. begonia, petunia, geranium, etc. There are individual plants of outstanding performance in each conventional variety, but they occur as minority. To raise their frequency to majority and to achieve uniformity, more or less all the individuals should be of outstanding performance. This has always been the fascination for breeders which only F_1 hybrids are able to exhibit.

iv. Dwarf and compact plants, abundant basal branching, profuse free flowering, earliness, enhanced blooming season, doubleness, large-sized flowers, novelty in flower colour and shapes, tolerance/resistance to biotic and abiotic stresses e.g. petunia, dianthus and balsam.

v. Doubleness in petunia can only be derived through hybridization. It has also been possible to grow hundred per cent double flowers in stocks (Hansen Series) by discarding dark green foliage seedlings and selecting pale green-leafed seedlings growing under cool conditions.

vi. Flowers of triploid hybrids like Nugget in marigold from cross of *Tagetes erecta* (diploid) × *Tagetes patula* (tetraploid) being sterile with no fertilization, seed sets remain fresh on the plant for longer duration.

vii. Unusual shape of the flower has been derived in Butterfly hybrids of antirrhinum as compared to the standard varieties. In such hybrids, each floret is like a colourful butterfly exposing more colour.

viii. In carnation florist double is an unfixed hybrid from the single crossed with super double, doubleness being incompletely dominant. For example, the perpetual flowering carnation seems to have been developed in Europe even before 1750 and is reported to be a hybrid between *Dianthus caryophyllus* (2n = 30) and *Dianthus chinensis* (2n = 30, 60).

ix. Hybrid vigour can be preserved where the plants can be propagated by vegetative means. In F_1 hybrids Carefree geranium are more vigorous and uniform when grown from cutting than F_2 plants obtained from their seeds.

x. Greater vigour expressed as yield and targeted desirable traits.

xi. Greater adaptability to varying environmental conditions because all the genes are present in heterozygous state.

xii. Expression of the advantageous character as governed by dominant genes.

xiii. Exclusivity of control by the breeder, since he alone retains the parents which together produce the commercial product.

xiv. Stability of the variety over a period of years.

xv. Proprietary control of the variety by the originator.

DISADVANTAGES OF F_1 HYBRID

i. Seeds saved from F_1 hybrid plants will not produce plants that are true to the parent type. Therefore, fresh seed need to purchase for each crop.

ii. F_1 hybrid seed is expensive as it has to be recreated by crossing the parent inbred lines again. Self pollination of the parent inbred lines lead to poor quality plants called 'selfs'. Preventing selfs is complicated and costly which also contributes to the expense of F_1 hybrid seed.

iii. Due to the cost of maintaining the inbred lines, quite a lot of seed has to be sold for a hybrid to be commercially viable. For this reason hybrids are often only offered for a few years before coming off the market leaving gardeners to seek a replacement for favourite cultivars.

iv. Because of narrow genetic base, F_1 hybrids have less adaptability.

v. Vulnerability to epidemics.

Factors responsible for hybrid seed production

Various factors affecting hybrid seed production are as under

- Mechanism of pollination
- Favourable environment
- Flower types/sex forms
  - Cleistogamous
  - Protoandrous
  - Protogynous
  - Male sterility
  - Self incompatibility

Requisites of hybrid seed production

1. Breeders responsibilities
   - To develop inbred lines
   - Identification of specific parental lines
   - To develop system for pollen control
2. Major problems for breeders and producers
   - Maintenance of parental lines
   - Separation of male and female reproductive organs
   - Pollination

Characteristics of parental lines

- High seed yield
- Good pollen production
- Good seed characteristics
- Long shedding period
- Male sterility
- Plant height
- Lodging resistant
- Fertility restoration

TECHNIQUES FOR F_1 HYBRID SEED PRODUCTION

Flowering plants have evolved a number of devices to encourage cross pollination.

1. **Dicliny:** Flowers are unisexual. In monoecious plants male and female flowers are borne on the same plant while in dioecious plants male flowers are borne on different plants.
2. **Dichogamy:** Time of anther dehiscence and stigma receptivity are different forcing times for cross pollination. The time gap between the two may vary from one day to many days. In protandry anthers dehisce earlier than the stigma receptivity. In protogyny stigma become receptive earlier than the anther dehisce.
3. **Self-incompatibility:** Self fertilization is avoided by recognizing the self pollen by the stigma. In nicotiana and petunia gametophytic self incompatibility system is present, whereas in gerbera sporophytic system of self-incompatibility is present that can be utilized in practical application of guaranteed cross pollination under open field conditions. Nicotiana and petunia produces enough seeds per pollination, which can adequately compensate the high cost of F_1 hybrid seeds. e.g. petunia, lilium.
4. **Male sterility:** Absence or atropy or mis or malformed of male sex organ (functional pollen) in normal bisexual flower. Male sterility is of three types: genetic male sterility, cytoplasmic male sterility and cytoplasmic- genetic male sterility. In Tagetes and Zinnia, genetic male-sterility facilitates efficient and economic hybrid seed production. Whenever pollen sterility is governed by a single recessive gene maintenance of the genetic stock is difficult as there will be continuous segregation of the fertile and sterile individuals in 1:1 proportion. This phenomenon is present in marigold, zinnia, delphinium, antirrhinum, calceolaria, salvia and impatiens. In ageratum, however both the male sterility and self incompatibility system are prevalent and a choice can be made on the basis of economy in the seed production. In petunia CMS has been observed but the use of this type of male sterility is not so practically common because of breakdown of male sterility in the maternal parent or malformation of flowers in F_1 plants. Male sterility in Sunflower has been used to produce ornamental varieties like Sunrich Orange (Japan) and Orit (Israel) which have no pollen grains and allergic effects like other male fertile varieties grown for their seed.
5. **Somatic embryogenesis:** Somatic embryogenesis using liquid bioreactors may be an economically viable option for hybrid seed production in flowers. The use of this technique is of preference over others, requiring less space and labour, least cost involved and faster propagation.
6. **Double flower condition:** Double flower character should be inherited as a dominant character to be manifested in F_1 hybrids. Many flower species in the

family Asteraceae assume a double form because of numerous petals but a central disc may be seen when petals are pulled aside and make a composite head such as in ageratum, aster, chrysanthemum, cornflower, dahlia, daisy, gaillardia, marigold, pyrethrum, rudbeckia and sunflower. Sunflower illustrates most clearly the satisfactory example of pollination where each floret bears only a single ovary. In most cases, double flower results from a transformation of the anthers into petals. Therefore, the double flower character can be regarded as a form of male sterility. However, in the case of gerbera, where the pistils are hidden by petals, manual hand pollination is tedious for maintenance of the female parent. Reimann-Philip (1969) developed a breeding scheme using the double flower character in garden carnation (*Dianthus caryophyllus*) as a form of male sterility. Some floral abnormalities like the Cinderella character in begonia and Femina in *Tagetes erecta* and *Zinnia elegans* results in male sterility.

7. **Triploidy:** Among the flower species in *Tagetes*, commercial triploid hybrid (2n=36) results from a cross between diploid female *Tagetes erecta* (2n = 24) and tetraploid male *Tagetes patula* (Also known as mule marigolds, triploid hybrids are a cross between African marigolds and French marigolds). These marigolds cannot produce seeds with which to germinate additional plants. Instead, each new plant must be created by crossing a French marigold plant and an African marigold plant. The triploid hybrid can rebloom two or three times during its season.Triploid cultivars are breeding crosses between *Tagetes erecta* and *Tagetes patula.* One advantage of these cultivars is that they provide the longest overall display of colour in the landscape, often lasting through the hot summer weather into August and September. However, seed germination during production may be considerably slower than for the African and French types. These plants are sterile, so they will not produce viable seed.

Some flowering plants have sterility and methods used for production of hybrid seed are given in Table 1.

Table 1: Flower crops have sterility and methods used for production of hybrid seed.

S. No.	Crop	Method being used
1.	Antirrhinum	Genetic male sterility, hand pollination.
2.	Marigold	Apetalous and double flowered sterility, hand pollination.
3.	Petunia	Self incompatibility, genetic male sterility, cytoplasmic male sterility, hand pollination.
4.	Zinnia	Apetalous male sterility, hand pollination.

BASIC PROCEDURES FOR HYBRID SEED PRODUCTION

Production of inbred lines

Inbred lines are the critical part of F_1 hybrid production accounting for much of the cost and complexity of producing F_1 hybrid seed. Inbreeding is allowing closely related plants to set seed. This, after several generations, lead to a population of very similar plants. Inbred lines lack vigour and perform poorly and are difficult and expensive to maintain. But when two inbred lines are crossed, the first generation (F_1) is uniform and vigorous due to heterosis. Breeders of F_1 hybrids aim to use parent lines whose progeny show particularly strong hybrid vigour.

Some plants are intolerant of inbreeding and self pollination. They suffer from 'inbreeding depression', causing a serious loss of vigour, fertility and stature. It is not possible to raise F_1 hybrids for species that are not readily inbred. Wild plants tend to be intolerant of inbreeding, but the process of domestication has led to populations that can tolerate inbreeding and therefore, hybrids are feasible for many crops. Wild species and standard open pollinated varieties are the source of inbred lines. In case of self pollinated crops homozygous varieties are used, whereas for cross pollinated inbred are developed.

1. Testing of combining ability of inbred lines

The value of an inbred is assessed from its performance in hybrid combination with other inbreds. The inbred lines are evaluated on the basis of their general combining ability (gca) and specific combining ability (sca). There are following methods:

i. Top cross method

This method refers to a cross between an inbred line and an open pollinated variety. This method is suitable to test the general combining ability of inbreds and in identification of promising one. Here the combining ability refers to yield performance.

A × Open pollinated variety

ii. Single cross method

This method is used to measure the specific combining ability (sca) of those inbreds which are selected on the basis of top cross performance. In this method, single crosses are made and the resulting F_1 hybrids are evaluated.

A × B

iii. Double cross method

The hybrid progeny from a cross between two single crosses is known as double cross hybrid. It involves four different inbred lines. Seeds obtained by this method are more uniform in shape and size.

(A × B) × (C × D)

iv. Three way cross hybrid

The hybrid progeny between a single cross and an inbred lines is referred to as three way cross hybrid. Such hybrids produces seeds of normal shape and size. The single cross is used as female and inbred as male.

(A × B) × C

v. Multiple cross

A cross involving more than four inbred line is referred to as multiple crosses. Multiple crosses generally have wider adaptation.

vi. Convergent method

This method can be used in partially cross pollinated species where the resulting population will be highly heterozygous. It involves combining of the desirable character from a number of inbreds.

vii. Pair cross method

It allows the use of every plant in pair crosses with every other plant in selected groups (Griffing, 1956).

Crossing between parental lines and production of F_1 in bulk

Plant breeders seek to make better plants. By crossing selected plants with different but desirable features, they produce a plant that has the best features of both parents. This is then crossed further to produce a stable plant whose seeds produce plants that are true to type. There are several techniques used to cross plants, but producing F_1 hybrids is one of the commonest and most effective methods. F_1 hybrids, which are largely annuals are produced by crossing two stable seed lines (called inbred lines) that give rise to especially uniform progeny that possess good vigour, yield and other properties. Only few crossing mechanisms have been adopted for commercial hybrid seed production. They are:

1. Hand emasculation and pollination e.g. ageratum, petunia, bellis, wall flower.
2. Hand emasculation and natural pollination e.g. gladiolus.
3. Hand elimination of male plants.
4. Self incompatibility e.g. petunia, pansy, stocks and kale.
5. Dicliny: monoecious and dioecious.

6. Male sterility: GMS e.g. marigold, zinnia, ageratum and calceolaria.

   CMS e.g. petunia.
7. Chemical emasculation- selective elimination of pollen production that is use of gameticides.
8. Use of marker genes to identify the selfs so that they can be eliminated as seedlings.

SOME ALTERNATIVES TO F_1 HYBRIDS

F_2 hybrids

These offspring of F_1 hybrids are variable but for some crops this is desirable. F_2 hybrids are relatively inexpensive. Pansy 'Joker Series' AGM are a widely offered F_2 hybrid where the variation compared to F_1 hybrids is not considered to be a drawback and whose seed is significantly less expensive than F_1 hybrid seed.

Open pollinated seeds

When a group of selected plants are grown together and allowed to freely pollinate each other, the seed is said to be open pollinated. Before the advent of hybrids, all seeds were open pollinated. Open pollinated seed is available for all crops offered as hybrids, but is usually significantly less uniform, vigorous and productive. Open pollinated seed is relatively inexpensive and can often be readily collected by home gardeners. However, seeds harvested from garden plants will not always come true to their parent, particularly if there is a related plant nearby with which it could have hybridized.

HYBRID SEED PRODUCTION IN SOME IMPORTANT ORNAMENTAL CROPS

Arctotis: 'Archnah' and 'Archley' are cultivars of *Arctotis fastuosa* derived through controlled pollination between the seed parent 'Flame' and the pollen parent 'Silver Carpet' and is submitted for the registration of plant variety rights in Australia (Anon., 2002a,b).

Ageratum: Hand emasculation and pollination is not possible due to complex nature of flowers and it is difficult to produce large quantity of seeds. Male sterility has been reported in this crop and it is used as seed parent. The qualities of F_1 hybrid seeds are superior to open pollinated varieties.

Antirrhinum: In antirrhinum cross pollination can be carried out without emasculation. The attractive flowers are pollinated repeatedly by bees in open field conditions. Therefore, hand emasculation should be done by using buds which are expected to open after two days. On the basis of plant height, flower compactness, number of branches and flowers, colour, size and duration of flowers, inbred lines are selected. F_1 hybrids are uniform and semi-dwarf and with a large number of effective tillers. A mutant of antirrhinum, open throated, penstemon types is available in which emasculation and pollination is easier at young open flower. Tetra Giant antirrhinums are very popular now-a-days because of their large deep coloured and long lasting flowers. Tetraploid snapdragons are raised from some promising F_1 hybrids.

Alyssum: Self sterility has been found in alyssum, therefore sometimes emasculation may not be needed and crossing can be done on a small cluster of flowers by pollen dusting from the male parent. It bears small flowers in compact heads. The hybrids can be quickly identified if the male parent carries a marker gene of flower or foliage. Otherwise crossed population is grown adjacent to the source of female lines and hybrids can be identified at or before maturity.

Amaryllis: By crossing between diploids and tetraploids, triploid amaryllis hybrids are produced and the resultant plants bear taller, ruffled, bright coloured and long last flowering.

Balsam: It is a self pollinated ornamental crop. Manual emasculation is carried out when flowers have just started to open. One or two days after emasculation pollination is done with anther tube.

Begonia: Begonia is a monoecious bulbous ornamental crop facilitating cross fertilization. If the female flowers are protected from the male flowers of same plant emasculation is not needed. The staminate buds are removed from the seed parent inbred before pollen dehiscence and the companion inbred supplies pollen for F_1 hybrid seeds. Reciprocal pollination increases the seed quality.

Bougainvillea: Hybrids evolution in bougainvilleas is a big problem because of the huge number of pollen and seed sterility or both. Induction of colchicines restored the fertility of selected parents. Popular varieties are available to produce colourful floriferous and improved hybrid bougainvilleas.

Carnation: In carnation stigma are larger than the anthers, therefore there is no need of emasculation. Cross pollination is beneficial when practised at the mature flower stage if the female flowers are protected from insect pollination.

China aster: Basically China aster is a self pollinated crop. The flower heads are composed of perfect disc florets and pistillate ray florets. In some flowers stamens and pistils do not mature at a time. Emasculation is done by the removal of disc florets and trimming of ray florets up to the level of stigma. Pollination is practised two to three days after pollination. Seed maturity takes place after one month of crossing.

Dianthus: Hand emasculation and hand pollination are easy and are practiced for hybrid seed production in Dianthus.

Evening Primrose: Emasculation and pollination are easy in evening primrose as both flowers and buds are large in size. Emasculation should be done in the late bud stage and pollination should be carried out when the style has split at its apex. Within 20-25 days of pollination seed will be matured.

Gazania: A new cultivar Sugaja of gazania hybrid derived from crossing seed parent Sunset Hannah with pollen parent breeding line A960022, is selected for its double flower colour and plant compactness. Cv. Sugaja is submitted for the registration of plant variety rights in Australia (Anon., 2000b).

Gypsophila: *Gypsophila paniculata* cvs. Dangysha, Dangypmini, Dangypflash and Dansferoy were derived from the hybridization of various *G. paniculata* genotypes and is submitted for the registration of plant variety rights in Australia (Anon., 2000a,c,d,e). Cv. Dangysha was selected for its erect, upright growth, narrow conical inflorescence shape and improved harvestability.

Geranium: Now a days F_1 hybrids are being produced to obtain uniform colour, plant height and simultaneous flowering annually. Prior to cross pollination self-compatible and hand emasculation is practised in the bud stage.

Hollyhock: There is a huge number of anthers in the flowers and are located surrounding the styles. Styles form a tuft in the centre of the flower but separated from one another. It is suggested not to carry out emasculation and pollination in the bud stage at least two days before the flower opening. Two important parent materials of hollyhock available, namely, 'Hungarian Rosea Annua White' from USA used as female parent and 'Supreme Mixed' from UK used as male parent for F_1 hybrid seed production.

Marigold: Marigold consists of two types of flowers. Disc flowers are having both anthers and pistils, whereas ray flowers are pistillate. Flower heads are emasculated manually by excising disc flowers before anthesis leaving only pistillate ray flowers which serve as seed parent and planted closely with pollen producing inbreds. A fully double inbred having no disc flowers can be used as seed parent but segregation may be found. There is a male sterile mutant known as 'apetalous' which segregates at a 1:1 ratio which can be used for hybrid seed production. Now a days, dwarf, prolific bearer and free flowering heavy golden yellow F_1 marigolds are produced through the use of male sterile African diploid marigold (*Tagetes erecta*) and fertile tetraploid French marigold (*Tagetes patula*).

Myosotis: Emasculation is done in the bud stage by the removal of anthers and pollination is done immediately after emasculation.

Petunia: F_1 hybrids of petunia are commercially produced by hand pollination technique because of production and huge number of seeds by single pollination. In petunia homozygous double flowers are female sterile but they have ability to produce viable pollen. Since double character is dominant, Double Grandiflora petunia hybrids are produced from the crosses of single × double. Gametophytic system of incompatibility is available in petunia and it can be overcome by bud self pollination. Inbred lines are crossed reciprocally without emasculation for the production of large quantities of seed, whereas in another method, eight or ten plants are selected and mass pollinated and the process is repeated with small families three to four generations to get uniform lines. By crossing two such lines after emasculation of the female plants in bud stage results in F_1 hybrid seeds are production.

Phlox: Two cycles of artificial selection were performed to increase autogamous fruiting in two wild populations of the self incompatible *Phlox drummondii*, to decrease autogamous fruiting in two wild populations of the self-compatible *P. cuspidata*, and to both increase and decrease autogamous fruiting in a cultivar Salmon Beauty of *P. drummondii* which is pseudo self compatible. The breeding systems were determined to be genetically quite flexible, independent of inbreeding depression and other genetic phenomena which could hinder a breeding system shift. This is especially true for increasing autogamy. Self pollen pistil compatibility seems to be the single character affected by selection. Based on the continuous variation in both autogamy and self compatibility, it is suggested that the change has been due to genes which modify the self-incompatibility reaction rather than to the simple segregation of alleles at the *S* locus (Bixby and Levin, 1996).

Primula: Heterostyly favour cross pollination in primula. Flowers are of two types 'pin' and 'thrum'. 'Pin' flowers are normally fertilized by pollen of 'thrum' anthers and short styled thrum flowers are fertilised by the pollen of 'pin' anthers and produces F_1 hybrids. Homostyly found in *Primula vulgaris* are self fertile and heterostyly found in *Primula obconica* forces cross pollination. Therefore, F_1 hybrid seeds are produced by hand pollination without emasculation.

Sweet pea: Sweet Pea flower consists of five petals, a 'standard', two 'keel petals' and one pair of 'wing petals'. Before the opening of the flower, anthers shed their pollen. Self pollination occur unless emasculation is carried out in the bud stage when the buds reached in length similar or longer than the calyx. Pollen covered brush is used for pollination. The seed matures after 25-30 days of pollination.

Verbena: Flowers of verbena are small, coloured, tubular and of head type. Hybrids are produced by crossing perennial small flowering species *Verbena tenuisecta* and large flowered annual species *Verbena hybrida* and the resultant hybrids are drought resistant, attractive, free flowering, floriferous and are easily maintained by vegetative propagation.

Viola: Open pollinated varieties of viola are uniform in height, flower size and colour and they easily set seeds. Commercially F_1 hybrid seeds can be produced by emasculation and pollination with male inbred. F_2 hybrid seeds may also be produced for obtaining mixed colours.

Wall flower: Self incompatibilty system is present in wall flower. Flower buds are emasculated at two days before opening of flowers when the sepals of each bud shift from just above the base to the tip. Pollination is done immediately after emasculation. Since inflorescence is racemose, it is possible to make many pollinations on a single branch. Male sterility has been reported in dwarf violet and creamy types of wall flowers. Therefore by using male sterility through hand pollination techniques hybrid seed can be produced.

Zinnia: Composite types of flowers are found in zinnia. Hand emasculation and pollination technique is used for the production of hybrid seed. An apetalous mutant has been reported in zinnia and is considered as the seed parent. Tetraploid hybrids are also available in zinnia. Tetraploid zinnia is developed by the colchicine treatment of diploid and they are vigorous in growth and flowers are bigger in size. Hybrid varieties of some ornamental crops are given in Table 2.

Table 2: Ornamental crops and their F_1 hybrids

S. No.	Flower	F_1 Hybrids
1.	*Acroclinum*	Blue Hawaii, White Hawaii and Royal Hawaii.
2.	*Tagetes*	Pusa Sankar1, Inca Series, Solar Red, Discovery Yellow, Durango Bolero, Durango Red, Gold Coin Series, Climax Series, Nugget, Red & Gold, Space Edge, Show Boat, Red Heaven, Red and Gold Nugget, Red Seven Star, Galore Series.
3.	*Alcea*	Pusa Pink Beauty, Pusa Yellow Beauty, Pusa Pastel Pink Supreme, Pusa Apricot Supreme.
4.	*Begonia*	Primma Donna, Gin Cocktail, Lotto Mixed, Illumination Orange, Illumination Apricot, Illumination Rose, Non-stop Mixed, Non-stop Ornament Mixed, Pin Up Flame.
5.	*Portulaca*	Sundial Peach, CalypsoMix, Tutti Frutti Mix, Margaritta.
6.	*Cineraria*	Vanezia, Jester Mix Blue, Jester Crimson, Jester Pink, Jester Pink Bicolour.
7.	*Delphinium*	Guardian, Steinchen Strain, Magic Fountains, Blue Cloud, Giant Imperial, QIS.
8.	*Impatiens*	Super Elfing, Tutu Red Bicolour, Belizzy Colour, Balls Double Mixed, Envoy Peach Butterfly, Extreme Mixed, Java Series, Borneo, Tango, Show Topper, Accent, Blitz, Minette, Carousal, Swirl.
9.	*Ageratum houstonianum*	Blue Blazer, Blue Horizon, Blue Hawaii, Blue Pearl, Blue Puffs, North Sea, Summer Snow, Blue Heaven, Summer Sky.
10.	*Petunia*	Flambe Salmon, Cascade Pink, Comanche, Red joy Improved, Orchid Mist, Double Cascade Blue, Humming Rose, Humming Scarlet, Humming Violet, Humming White, Glitters, Coral Satin, Satellite, Silver Medal, Polaris, Delight Mixed, Apple Tart, Cherry Tart, Bonanza Mixed, Colour Parade Mixed, Daddy Mixed, Super Cascade Improved Mixed, Balleriana, Pink Morn, Silver Medal, Orange Bells, Coral Bells, Summer Sun, Polo, Apricot Tart, Blue Danube, Caprice, Duet, Apple Blossom, Blue Lase, Cherry Blossom, Calypso, Chiffon Cascade, E1 Toro, Maytime, Super Cascade Red, Super Cascade White, Super Magic Pink, Fantasy Blue, Fantasy Red

(Contd.)

(*Contd. Table 2*)

S. No.	Flower	F_1 Hybrids
		And White, Fantasy Rose, Fantasy Salmon, Fantasy Scarlet, Fantasy Violet Blue, Fantasy White, Super Magic Red, Yellow Magic, Blue Flash, Pink Flash, Red Flash, Velvet Flash, White Flair, Blue Frost, Glacier, Snow Cloud, Presto, Rhapsody, Sonata, Valentine, Purple Pirouette, Bridal Bouquet, Circus, Noctume Improved, Salmon Bouquet, White Joy, Pink Joy, Purple Plum, Sugar Plum, Star Joy, Humming Blue, Allegro, Lollipop, Merlin, Polo.
11.	*Verbena*	Adonis Mango, Blue Spires, Pink Spires.
12.	*China aster*	Serene, Colour Carpet.
13.	*Celosia*	Century, Sparkler, Toreador.
14.	*Iris*	Blue diamond, Ideal, Telstar, White Wedge Wood.
15.	Larkspur	Blue Cloud, Giant Imperial, QIS.
16.	*Lavendula*	Silver Streak.
17.	*Phlox*	Patriot, Chanal.
18.	*Eschscholzia*	Mission, Bells.
19.	*Antirrhinum*	Rocket, Sonnet Mix, Butterfly Series, Kim mixed, Magic Carpet, Carioca, Coronette, Floral Carpet.
20.	*Scabiosa*	Blue Moon.
21.	*Limonium*	Misty, Excellent, Fortress Mix, Pastel Shades, Arrow, New Era, New Wonder.
22.	*Zinnia*	Fairyland Gold, Fairyland Scarlet, Sunrise, Gold Sun, Red Sun, Peter Pan, Golden Dawn, Border Beauty Yellow, Border Beauty Rose, Dasher Orange, Dasher Scarlet, Exquisite, Bonanza, Dreamland Coral, Dreamland Ivory, Dreamland Scarlet, Dreamland Yellow, Sunrise Red, Sunrise Yellow, Summer Dress Scarlet, Firecracker, Rosy Future, Torch, Yellow Zenith, Burpee's Bouquet Hybrid, Cherry Sun, Silver Sun.
23.	*Viola*	Flamenco, Black Moon, Cats White.
24.	*Helianthus*	Double Shine, Sunrich Orange, Jerusalem Gold, Little Dorrit, Harlequin, Elite Sun, Solar Eclipse, Jerusalem Seri.
25.	*Pelargonium*	Ringo 2000, Black Velvet Series, Maverick Coral, Pinto, Printer Series, Show Girl, Jackpot.
26.	*Amaranthus*	Amar Shola, Sparkler Series.
27.	*Viola*	Valentine, Pandoras Box, Sorbet, Sweeties, Penny Sun Rise, Ultima Morpho.
28.	*Primula*	Peso Series, Lira Series, Corona Series, Twilight, Sterling Series, Earnst, Benary, Joker, Finesse, Lucento, Finale, Twilly Series, Juno.
29.	*Calendula*	Coronets, Gypsy Festival Mix, Pacific Beauty Mix, Mandarin.
30.	*Dahlia variabilis*	Waverly Series.
31.	*Plumbago auriculata*	Plumbago, Escapade.
32.	*Papaver nudicaule*	Champagne, Bubbles, Wonderland Mix.

REFERENCES

Anonymous 2000a. 'Dangysha' syn Yukinko. Danziger 'Dan' Flower Farm, Moshav Mishmar Hashivar. *Plant Varieties Journal*, **13**(4): 29-30.

Anonymous 2002b. 'Sugaja'. Macquarie Fields, NSW. *Australia Plant Varieties Journal*, **15**(1): 51-52.

Anonymous 2000c. 'Dangypmini'. Danziger 'Dan' Flower Farm, Moshav Mishmar Hashivar. Israel. *Plant Varieties Journal*, **13**(4): 29.

Anonymous 2000d. 'Dangypflash'. Danziger 'Dan' Flower Farm, Moshav Mishmar Hashivar. Israel. *Plant Varieties Journal*, **13**(4): 28-29.

Anonymous 2000e. 'Danfesroy'. Danziger 'Dan' Flower Farm, Moshav Mishmar Hashivar. Israel. *Plant Varieties Journal*, **13**(4): 27-28.

Anonymous 2002a. 'Archnah'. NuFlora International Pvt. Ltd., Macquarie Fields, NSW. Australia. *Plant Varieties Journal*, **15**(4): 16-17.

Anonymous 2002b. 'Archley'. NuFlora International Pvt. Ltd., Macquarie Fields, NSW. Australia. *Plant Varieties Journal*, **15**(4): 15-16.

Bixby, P.J. and Levin, D.A. 1996. Response to selection on autogamy in phlox evolution. *Ecology,* **50**(2): 892-99.

Griffing, B. 1956. Concept of general specific combining ability in relation to diallel crossing systems, *Aust. J. Biol. Sci,* **9:** 463–93.

Chapter – 6

Intellectual Property Rights Issues

Intellectual Property Rights (IPR) are legal rights, which result from intellectual activity in industrial, scientific, literary and artistic fields. These rights safeguard creators and other producers of intellectual goods and services by granting them certain time-limited rights to control their use. Protected IP rights like other property can be a matter of trade, which can be owned, sold or bought. These are intangible and non exhausted consumption. The term "intellectual property" denotes the specific legal rights which authors, inventors and other IP holders may hold and exercise and not the intellectual work itself (Chadha, 2013). Intellectual property rights are essentially territorial in nature so, where registration is necessary, for patents, registered designs and trademarks, these will not exist in other countries unless registration has been sought and obtained there.

IP protection is crucial for a sustainable contribution of plant breeding and seed supply. Breeding new varieties of plants, which are developed after contributing number of years to the selective inheritance of traits which provide improved yields, higher quality and better resistance to such plant varieties requires a substantial investment in terms of skill, labour, material and economic resources. However, a new variety once released could in many cases be readily reproduced by others so as to deprive its breeder of the opportunity to benefit adequately from the investment made. Thus, an effective system of plant variety protection (PVP) is a key enabler for investment in breeding and the development of new varieties of plants or improving existing plant varieties, encourage importation of foreign varieties, promote exportation of plant varieties provide access to information of the created products and the methodology of creation for the enhancement of social welfare and generally benefit the market place. An international system of IPR protection for plant varieties expands these benefits by facilitating access to new varieties created in other states (Gautam *et al.*, 2012).

Intellectual property is usually divided into two branches: (i) industrial property and (ii) copyright and the rights which neighbour upon copyright. The principal categories of industrial property are: patents, trademarks, geographical indications, industrial designs and trade secrets. Industrial property according to the 1883 Paris Convention for the Protection of Industrial Property is to be understood in the broadest sense and to apply not only to industry and commerce proper but also to agricultural and extractive industries and to all manufactured or natural products including wines, grain, tobacco, leaf, fruit, cattle, minerals, mineral waters, beer, flowers and flour.

TYPES/TOOLS OF IPRs

A. Patents.

B. Trademarks.

C. Copyrights and related rights.

D. Geographical Indications.

E. Industrial Designs.

F. Trade Secrets.

G. Layout Design for Integrated Circuits.

H. Protection of New Plant Variety.

A. Patents

A patent is an exclusive right granted for an invention, which is a product or a process that provides a new way of doing something or offers a new technical solution to a problem. It provides protection for the invention to the owner of the patent. The protection is granted for a limited period, i.e. 20 years. Patent protection means that the invention cannot be commercially made, used, distributed or sold without the patent owner's consent. A patent owner has the right to decide who may or may not use the patented invention for the period in which the invention is protected. The patent owner may give permission to, or license, other parties to use the invention on mutually agreed terms. The owner may also sell the right to the invention to someone else who will then become the new owner of the patent. Once a patent expires, the protection ends and an invention enters the public domain that is, the owner no longer holds exclusive rights to the invention which becomes available to commercial exploitation by others. All patent owners are obliged in return for patent protection to publicly disclose information on their invention in order to enrich the total body of technical knowledge in the world. Such an ever-increasing body of public knowledge promotes further creativity and innovation in others. In this way, patents provide not only protection for the owner but valuable information and inspiration for future generations of researchers and inventors.

B. Trademarks

A trademark is a distinctive sign that identifies certain goods or services as those produced or provided by a specific person or enterprise. It may be one or a combination of words, letters and numerals. They may consist of drawings, symbols, three- dimensional signs such as the shape and packaging of goods, audible signs such as music or vocal sounds, fragrances or colours used as distinguishing features. It provides protection to the owner of the mark by ensuring the exclusive right to use it to identify goods or services, or to authorize another to use it in return for payment. It helps consumers identify and purchase a product or service because its nature and quality indicated by its unique trademark, meets their needs. Registration of trademark is prima facie proof of its ownership giving statutory right to the proprietor. Trademark rights may be held in perpetuity.

The Trade Marks Act, 1999 and the Trade Marks Rules, 2002 came into force with effect from 15th September 2003. The Act enlarges the definition of "Mark" and "Trademark". It also provides for Registration of "Collective Marks" and "Service Marks" in addition to goods. Article 15 to 18 of TRIPS185 provides that term of registration should not be for less than 7 years. In the Indian Act of 1999 the term of Registration is for 10 years and for renewal it is for a term of 10 years.

C. Copyrights and Related Rights

Copyright is a legal term describing rights given to creators for their literary and artistic works. The kinds of works covered by copyright include: literary works such as novels, poems, plays, reference works, newspapers and computer programs; databases; films, musical compositions and choreography; artistic works such as paintings, drawings, photographs and sculpture; architecture and advertisements, maps and technical drawings. Copyright subsists in a work by virtue of creation; hence, it's not mandatory to register. However, registering a copyright provides evidence that copyright subsists in the work and creator is the owner of the work. Creators often sell the rights to their works to individuals or companies best able to market the works in return for payment. These payments are often made dependent on the actual use of the work and are then referred to as royalties. These economic rights have a time limit, (other than photographs) is for life of author plus sixty years after creator's death. India's Copyright Act protects computer programmes, among other "works of authorship".

India has one of the most modern copyright laws in any country. In keeping with international copyright principles, the Indian Copyright Act provides that the copyright in an original book, including any "derivative work", will be owned by the individual who created it. This individual will enjoy the "economic rights" in the work, in addition to "moral rights", such as the right to attribution. The Copyright Act recognizes the concept of "work for hire", accordingly when an employee creates a work in the scope of his employment, the employer will normally be deemed to own the "economic rights" in the work. However the employee will generally retain the "moral rights". Section 18 of the Copyright Act of India provides that the owner of the copyright in an existing work or the prospective owner of the copyright in the future work may assign the copyright either wholly or partially and either generally or subject to limitations and either for the whole term of the copyright or any part thereof.

D. Geographical Indications (GI)

GI are signs used on goods that have a specific geographical origin and possess qualities or a reputation that are due to that place of origin. Agricultural products typically have qualities that derive from their place of production and are influenced by specific local factors, such as climate and soil. They may also highlight specific qualities of a product, which are due to human factors that can be found in the place of origin of the products, such as specific manufacturing skills and traditions. A geographical indication points to a specific place or region of production that determines the characteristic qualities of the product that originates therein. It is important that the product derives its qualities and reputation from that place. Place of origin may be a village or town, a region or a country. It is an exclusive right given to a particular community hence the benefits of its registration are shared by the all members of the community. Recently the GIs of goods like Chanderi Sarees, Kullu Shawls, Wet Grinders, etc. have been registered. Keeping in view the large diversity of traditional products spread all over the country. The registration under GI will be very important in future growth of the tribes/communities/skilled artisans associated in developing such products.

E. Industrial Designs

Industrial designs refer to creative activity, which result in the ornamental or formal appearance of a product and design right refers to a novel or original design that is accorded to the proprietor of a validly registered design. Industrial designs are an element of intellectual property. Under the TRIPS Agreement, minimum standards of protection of industrial designs have been provided for. As a developing country, India has already amended its national legislation to provide for these minimal standards. The essential purpose of design law it to promote and protect the design element of industrial production. It is also intended to promote innovative activity in the field of industries. The existing legislation on industrial designs in India is contained in the New Designs Act, 2000 and this Act will serve its purpose well in the rapid changes in technology and international developments. India has also achieved a mature status in the field of industrial designs and in view of globalization of the economy, the present legislation is aligned with the changed technical and commercial scenario and made to conform to international trends in design administration. This replacement Act is also aimed to enact a more detailed classification of design to conform to the international system and to take care of the proliferation of design related activities in various fields.

F. Trade Secrets

It may be confidential business information that provides an enterprise a competitive edge may be considered a trade secret. Usually these are manufacturing or industrial secrets and commercial secrets. These include sales methods, distribution methods, consumer profiles, advertising strategies, lists of suppliers and clients, and manufacturing processes. Contrary to patents, trade secrets are protected without registration. A trade secret can be protected for an unlimited period of time but a substantial element of secrecy must exist, so that, except by the use of improper means, there would be difficulty in acquiring the information. Considering the vast availability of traditional knowledge in the country the protection under this will be very crucial in reaping benefits from such type of knowledge. The Trades secret, traditional knowledge are also interlinked/associated with the geographical indications.

G. Layout Design for Integrated Circuits

Semiconductor Integrated Circuit means a product having transistors and other circuitry elements, which are inseparably formed on a semiconductor material or an insulating material or inside the semiconductor material and designed to perform an electronic circuitry function. The aim of the Semiconductor Integrated Circuits Layout-Design Act 2000 is to provide protection of Intellectual Property Right (IPR) in the area of Semiconductor Integrated Circuit Layout Designs and for matters connected therewith or incidental thereto. The main focus of SICLD Act is to provide for routes and mechanism for protection of IPR in Chip Layout Designs created and matters related to it. The SICLD Act empowers the registered proprietor of the layout-design an inherent right to use the layout-design, commercially exploit it and obtain relief in respect of any infringement. The initial term of registration is for 10 years; thereafter it may be renewed from time to time. Department of Information Technology Ministry of Communications and Information Technology is the administrative ministry looking after its registration and other matters.

H. Protection of New Plant Variety

The objective of this act is to recognize the role of farmers as cultivators and conservers and the contribution of traditional, rural and tribal communities to the country's agro biodiversity by rewarding them for their contribution and to stimulate investment for R & D for the development of new plant varieties to facilitate the growth of the seed industry. The Plant Variety Protection and Farmers Rights Act, 2001 was enacted in India to protect the New Plant Variety; the act has come into force on 30th Oct., 2005 through Authority. Initially 12 crop species have been identified for registration. i.e. Rice, Wheat, Maize, Sorghum, Pearl millet, Chickpea, Green gram, Black gram, Lentil, Kidney bean etc. India has opted for sui-generic system instead of patents for protecting new plant variety. Department Agriculture and Cooperation is the administrative ministry looking after its registration and other matters.

IPR Standards of the TRIPS Agreement

The reasons offered by economists as to why a state might confer IPR monopolies include the encouragement of invention and innovation and of technology transfer and investment. For most countries, however, IPR legislation has been introduced in the first instance for the purpose of implementing international treaty obligations or obligations under inter-governmental agreements. Until 1994, the international IPR regime was largely administered by the World Intellectual Property Organization (WIPO). The perceived failure of WIPO to deal with the growth of counterfeiting and piracy and the difficulties which were being experienced in negotiating improvements to the WIPO-administered IPR regime caused a number of countries to persuade the parties to the General Agreement on Tariffs and Trade (GATT) to assume responsibility for IPRs. This resulted in the promulgation in 1994 of the Agreement on Trade Related Aspects of Intellectual Property Rights (TRIPS) as one of the results of the Uruguay Round of the GATT (Blakeney, 2011).

Article 1.1 of the TRIPS Agreement requires WTO Members to "give effect to the provisions of this Agreement" but it points out that "Members may, but shall not be obliged to implement in their law more extensive protection than is required by this Agreement, provided that such protection does not contravene the provisions of this Agreement." In other words the norms which are prescribed by TRIPS are taken to define a legislative minimum. This provision concludes by pointing out that "Members shall be free to determine the appropriate method of implementing the provisions of this Agreement within their own legal system and practice."

In recent years bilateral Free Trade Agreements (FTAs) contain IPR chapters which either raise the minimum requirements prescribed by TRIPS, or they prescribe the form of compliance with TRIPS. For example, the USAs FTAs prescribe UPOV 1991 as the form of compliance with the obligation in Article 27.3(b) of TRIPS for Members to introduce plant variety rights protection. The "Objectives" of the TRIPS Agreement are set out in Article 7. It states that the protection and enforcement of intellectual property rights should contribute to the promotion of technological innovation and to the transfer and dissemination of technology, to the mutual advantage of producers and users of technological knowledge and in a manner conducive to social and economic welfare and to a balance of rights and obligations.

PROTECTING FLOWER CROPS VARIETIES THROUGH PROTECTION OF PLANT VARIETIES AND FARMERS RIGHTS ACT, 2001

In India, plant variety protection bill titled "Protection of Plant Varieties and Farmers Rights Bill" was passed by the Lok Sabha on 9 August, 2001, is now a PVP Act under Department of Agriculture and Co-operation of Ministry of Agriculture, GOI. This act "Protection of Plant Variety and Farmers Rights Act", 2001 was adopted to meet the obligation under Article 27(3) (b) of TRIPS Agreement to protect rights of farmers and plant breeders and to encourage the development of new plant varieties.

The Act is based on a *Sui generis* system and is unique in sense that it concurrently recognizes the rights of breeders, farmers, communities and researchers. It confers an exclusive right upon the breeder or his successor, his agent or licensee, to produce, sell, market, distribute, import or export of the registered variety. Further, the Act provides for compulsory license of a registered variety, if the seeds/propagating material is not available to the public at a reasonable price or quantity. Any person or group of persons or any organization can also claim for benefit sharing if the plant genetic material belonging to them is shared in the development of a registered variety. The researchers are conferred the right to use any registered variety for conducting experiment or research and the use of a variety by any person as an initial source of variety for the purpose of creating the other varieties. The strong protection granted to a plant breeder over his/her variety is present in the section dealing with the infringement of breeders' right where punishment in the form of substantial fines and jail term has been prescribed for those infringe the rights of the registered breeder. New plant varieties are afforded legal protection under this approach to encourage commercial plant breeders to invest the resources, labour and time needed to improve existing plant varieties by ensuring that breeders receive adequate remuneration when they market the propagating material of those improved varieties (Kannaiyan *et al.,* 2008).

The main objectives of the legislation are to (a) stimulate investments for research and development both in the public and the private sectors for the development of new plant varieties by ensuring appropriate returns on such investments and (b) facilitate the growth of seed industry in the country through domestic and foreign investment which will ensure the availability of high quality seeds and planting material to Indian farmers. Following were some of the declared advantages for developing a *Sui generis* system for plant variety protection in India (Dewan, 2011)

i. It allowed for superior planting material, leading to increased agricultural production.

ii. It facilitated investment by private sector enterprises in the development of superior plant varieties and also gave an impetus to the building up of infrastructure for the seed industry.

iii. It encouraged the competition between the private and public sector in the field of plant breeding effectively and efficiently in the larger national interest.

iv. It enabled India to meet its national obligations under the international agreement on tariffs and trade (GATT).

Protection of plant varieties in India was started by the Protection of Plant Varieties & Farmers Right Authority with 12 crop species in 2007 and presently, 57 crop species have been covered under PPV & FRA (Srivastava *et al.*, 2013). Out of these, 25 horticultural crops have been covered under PPV & FRA (Table 1).

Table 1: Horticultural crops covered under the PPVFRA.

Flowers	5	Rose (*Rosa* spp. other than *R. damascena*), Chrysanthemum, Bamboo Leaf Orchid or Boat Orchid, Spray Orchid, Vanda or Blue Orchid
Medicinal and Aromatics	5	Isabgol, Menthol Mint, Damask Rose, Periwinkle, Brahmi
Vegetables	9	Brinjal, Cabbage, Cauliflower, Garlic, Okra, Onion, Potato, Tomato, Field pea
Fruit crop	2	Mango, Coconut
Spices	4	Black pepper, Small cardamom, Turmeric, Ginger

REGISTRATION OF PLANT VARIETY

A. The Various Categories of Varieties eligible for Registration

1. **New plant variety:** A variety which conforms to the criteria of novelty i.e not been sold or otherwise disposed off in India earlier than 1 year and outside India (in case of trees and vines earlier than six years or in any other case, earlier than four years).

2. **Essentially derived variety (EDV):** A variety which has been predominantly derived from an initial variety or from a variety that itself is predominantly derived from such initial variety and while retaining the expression of the essential characteristics of the initial variety is clearly distinguishable from the initial variety.

3. **Extant notified variety:** A variety which has been notified under the Seeds Act Section 5 of the Seeds Act, 1966 (54 of 1966).

4. **Extant variety about which there is common knowledge (VCK):** An extant variety about which there is common knowledge is a variety which has not been released and notified under the Seeds Act, 1966 have been sold or otherwise disposed of in India for more than a year from the date of filing. It should be under cultivation in a state/region/country, even as "truthfully labeled" variety, finds a entry in official list/register of varieties in any country granting PVP, including filing an application for PBR, including an accession in a National/International Gene Bank and adequate description of the variety in a publication that may be considered a part of the public technical knowledge. These varieties are protected for the period of 15 year from date of notification in Gazette.

5. **Farmer's variety:** A variety which has been traditionally cultivated and evolved by the farmers in their fields or is a wild relative or land race of a variety about which the farmers possess the common knowledge.

Example variety: Illustrates the DUS characteristic and/or provides the basis for ascribing the appropriate state of expression to each variety. All DUS characteristics and their states of expression included in the DUS test guidelines of a crop species are covered by the minimum total number of example varieties. The set of example varieties for a given characteristic provides information on the range of expression of the DUS characteristic.

Reference variety: A variety has similar grouping DUS characteristics as compared to that of candidate variety for comparision with the candidate variety during the conduct of DUS trial. The reference variety may be either a notified variety or a variety registered under the Act and the number of reference variety included in the DUS trial may be one or more than one.

B. Criteria for Registration

A variety is eligible for registration under the Act if it essentially fulfills the 3 criteria

Distinctness: The variety is deemed to be distinct if it is clearly distinguishable from any other variety of which the existence at the time of submission of the application is a matter of common knowledge.

Uniformity: The variety is deemed to be homogeneous if it is having regard to the variation that may be expected from the particular features of its own reproduction, sufficiently homogeneous as regards its relevant characteristic.

Stability: The variety is deemed to be stable if in its essential characteristics remain true to its description after repeated reproduction or propagation.

These criteria are also called 'DUS Character' and the candidate variety must be distinguished by atleast one essential characteristic from a variety, which is a matter of common knowledge in any country at the time of filing application, sufficiently uniform in expression of its essential characteristics which should remain unchanged even after repeated propagation. The variety should also have a single and distinct denomination also.

C. Exceptions

Extant notified varieties: Exempted from DUS testing. Seven Extant Variety Recommendation committee (EVRC) members examine the suitability for registration.

Extant Variety about which there is common knowledge (VCK): DUS testing is done for one growing season at two DUS testing locations with half the quantity of seeds as specified for new variety.

Special tests: Special tests are to be conducted only when DUS test fails to establish the requirement of distinctiveness. The DUS testing shall be field and laboratory based, multilocational for atleast two crop seasons. The authority shall charge separate fees for conducting special test. Broadly, these tests can be classified into five main groups: physical tests, biochemical tests, molecular tests, response tests and organo-leptic tests.

D. Fee Structure

Application for registration of plant varieties should be accompanied with the fee of registration notified by the Authority (new and EDV Individual Rs. 5000/-, Institutional – Rs. 7000/-, Commercial Rs. 10000/-); Extant variety notified under the Seeds Act, 1966–1000/- and variety about which there is a common knowledge (Individual Rs. 2000/-, Institutional Rs. 3000/-, Commercial Rs. 5000/-) (The Gazette of India, 2008 and 2009e). No fee is to be paid by the farmer for registration of farmers variety. While the annual fee shall be determined on the basis of declaration given by the registered breeder or agent or licensee regarding the sales value of the seeds of the variety registered under the Act during the previous year and royalty, if any, received during the previous year from the sale proceed of seeds of the registered variety and verified by the authority. The applicant has to submit the DUS test fee (as notified for different crops) for conducting the DUS test of the candidate variety (Table 2).

Table 2: DUS test fee structure of crop species.

S. No.	Crop species	DUS test fees (Rs.)
1.	Rose, Chrysanthemum, Black pepper, Small cardamom, Ginger, Turmeric	45,000
2.	Field pea	20,000
3.	Tomato, Brinjal, Okra, Cabbage, Cauliflower, Onion, Garlic	40,000
4.	Potato	48,000
5.	Mango	30,000

After getting all the information about the variety and receiving DUS test fee and seed, the variety is send to DUS testing centres located at various locations of India for testing DUS characteristics. After DUS testing at one or two locations depending upon the category and type of crop results are submitted by the respective DUS centre to the authority which are being analyzed for the claimed characters by the applicant and subjected to issue of certificate only when atleast one of the essential character is present in the claimed variety as per DUS guidelines.

ISSUANCE OF REGISTRATION CERTIFICATE

The PPV & FR Authority has issued a total of 600 registration certificates of different crop species till now (Srivastava *et al.*, 2013). The PPV & FR Authority has also opened a National Register of "Registered Plant Varieties" having all details of the registered plant varieties and is the authentication of the plant varieties registered. As a requirement under the act, for the purpose of benefit sharing, the authority also send a copy of the certificate of registration to the National Biodiversity Authority and Indian Council of Agricultural Research (The Gazette of India, 2011).

MAINTENANCE OF REGISTERED VARIETIES

As per the Act, it is mandatory to maintain the seed samples/propagating material of the registered plant varieties up to a period of protection provided to the candidate variety and also address the issues for intellectual property of plant breeder's rights, compulsory license etc. Authority has established the National Gene Bank at National Bureau of Plant Genetic Resources (NBPGR), New Delhi for medium term storage of true samples of orthodox seeds of all registered varieties for their entire period of protection. After the expiry of protection period, the seed may be submitted to NBPGR/any public repository (Brahmi and Choudhary, 2011). For recalcitrant seeds 3 gene banks have been developed at Dapoli, Ranchi and Mashobra.

Farmers Rights

The PPV & FR Act provides exhaustive and wide ranging rights to farmers in accordance with the FAO International Undertaking on Farmers' Rights and relevant CBD Articles on conservation and sharing biodiversity and benefit sharing (Gautam *et al.*, 2011). It is one of the most important characteristics which distinguishes it from the UPOV which treats farmer's rights as a privilege and not a right in itself as compared to the rights granted to the breeders. The Act contains farmers' rights as positive rights, arising from the past, present and future contribution of farmers in conserving, improving and making available PGR, particularly those in centres of origin/diversity (Agarwal, 2011). The Act treats the farmer as plant breeder so far as the farmers' variety is concerned and they can register them under the Act without paying any fee. It protects farmers' interest by ensuring access to benefit sharing if their material is used for development of new varieties (Ravi, 2004).

REFERENCES

Agarwal, R. 2011. http://jurisonline.in/2011/11/ protection-of-farmers-and-breeders-rights.

Blackeney, M.L. 2011. Intellectual Property (IP) Issues and International Treaties Related to Genetic Resources in Agrobiodiversity in Malaysia II: Conservation and Sustainable Utilization Malaysian Agricultural Research and Development Institute (MARDI) Kuala Lumpur.

Brahmi, P and Chaudhary, V. 2011. Protection of plant varieties: systems across countries. *Plant Genetic Resources*, **9**(3): 384-91.

Chadha, R.K. 2013. Intellectual Property Rights. IPR issues in digital environment: meeting the challenges, 1-55.

Dewan, M. 2011. IPR protection in agriculture: an overview. *J. IPR*, **16**: 131-8.

Gautam, P.L., Singh, A.K., Srivastava, M. and Singh, P.K. 2012. Protection of Plant Varieties and Farmers' Rights: A Review. *Indian J. Plant Genet. Resour.*, **25**(1): 9–30.

Gautam, P.L., Srivastava, D.S., Mishra, Rajganesh, D.S. and Singh, A.K. 2011. Farmers rights under the Protection of Plant Varieties and Farmers' Rights Act of India. *Think India Journal* (submitted).

Kannaiyan, S., Parthasarathy, V.A. and Prasath, D. 2008. Intellectual Property Rights. In: Recent Initiatives in Horticulture (Eds.) K.L. Chadha, A.K. Singh and V.B. Patel. The Horticultural Society of India, New Delhi, 123-39.

Ravi, B.S. 2004. Manual on Farmers' Rights. MS Swaminathan Research Foundation. Chennai, 80p.

Srivastava, M., Bhardwaj, M., Agarwal, R.C. and Singh, B. 2013. Protecting vegetable crops varieties through Protection of Plant Varieties and Farmers Rights Act, 2001. *Proce.*: National Symp. on Abiotic and Biotic Stress Management in Vegetable Crops, 111-7.

The Gazette of India (2011) Notification of bodies to which a copy of certificate of registration issued under the PPV & FR Act, 2001 shall be sent. SO 1912 (E) August 18, 2011.

TRIPS Agreement, Article 27.3(b), http://www.wto.org/english/docs e/legal e/27-TRIPS.pdf www.cbd.int

Section II

Cut Flower Crops

Chapter – 7

Alstroemeria

Alstroemeria belongs to the family Alstromeriaceae. Alstroemeria flowers third in terms of popularity and foreign exchange earnings in Kenya after roses and statice. Alstroemeria is named after a Swedish botanist "Baron Klas von Alstroemer". The popularity of this flower is still growing in the global perspective. It is one of the quickest growing flowers with varying colours like purple, lavender, salmon, red, pink, white, orange, yellow and bicolour. It is grown for cut flowers but is also popular as a potted plant. It is grown in the higher ranges of the Western Ghats, Himachal Pradesh, Jammu & Kashmir and Karnataka in India.

ORIGIN AND DISTRIBUTION

The alstroemeria, sometimes called the Peruvian Lily, is native to Tropical America particularly Chile. There are some 30 species of alstroemeria that occurs in grasslands and Pampas of South America, Argentina, Bolivia, Brazil, Chile, Paraguay, Peru and Venezuela. In America, it flourishes in different climates from high in the Andes mountains to the deserts of Pacific coastline. Even deep into the tropics of Brazil, the plants flourish abundantly. It was introduced to Europe in 1753 when seeds were sent to Europe by Claus Alstroemeria of Sweden during his travels in South America. It was widely cultivated in England during the early 1800's and it was John Hedgewood, the founder of the Horticultural Society who first introduced the white variety in 1832. Since then, the Alstroemeria has been hybridized extensively. It is named after a Swedish botanist, Baron Klas von Alstroemer, who brought the first seeds of *A. oelegrina* from South America to Europe. Columbia is the number one in the world as far as growing area of alstroemeria is concerned. The second and third positions are taken by the

Netherlands and Japan respectively. Other producing countries are United Kingdom and Italy. Alstroemeria is a new introduction into the world floriculture scene.

Most commercial cultivars of alstroemeria are patented and protected through plant breeder's rights. Because of the fact that quite a large number of new varieties and hybrids of alstroemeria are planted in various countries in each year, royalty is paid based upon their square footage under production or number of plants leased. The new Inca Collection bred by Konst Alstroemeria B.V. is a breakthrough in the presently dwarf alstroemeria assortment. The new extended range of alstroemeria cultivar offers the opportunity to grow these top quality perennials in the garden, pots and for cut flower. These dwarf type plants reaching a height from 20-50 cm, combine with the quality and colours of the long lasting flowers. New varieties have gained popularity in California (USA), New South Wales (Australia) and Spanish south east coastline, Sicily (Italy). However, for a very long time already alstroemeria has been cherished worldwide for its exotic flowers.

In India cultivation of alstroemeria has been started around Bengaluru, Pune and Hyderabad. It has been introduced in Palampur, Solan, Srinagar and hills of Uttrakhand and Ooty, to popularize this crop among growers for domestic and export markets (Singh, 2006).

SPECIES

At present there is no consensus about the number of species that exist within the genus *Alstroemeria,* but it probably exceeds 60 (Uphof, 1952). The species *A. aurantiaca, A. ligtu* and *A. pelegrina* are probably progenitors of the modern hybrids, which produce larger flowers and are extremely lasting as a cut flower. There are about 50 known species of *Alstroemeria* which can be classified into two main groups: Orchid type and Butterfly type. *Alstroemeria psittacina*, native from SE Brazil, NE of Paraguay, Uruguay and NE of Argentina, *A. caryophyllaea*, from Brazil, *A. aurea*, from Argentina and Chile, *A. ligtu*, *A. versicolour*, *A. revoluta*, *A. pelegrina*, *A. violacea*, *A. hookeri*, all from Chile, have been used, among other, in crossings from which several commercial cultivars have been obtained.

Alstroemeria species are briefly summarized, the ones that have more potential use in breeding programme (Sanso *et al*., 2005).

***A. isabellana*:** It is found in Argentina, Uruguay, Paraguay and Brazil. This plant can be up to about 1.5 m tall, has rigid leaves and pendulous, tubular flowers, colour pink, orange red or yellow and tipped in green. Inner tepals are stripped and spotted.

***A. longistaminea*:** Occurs at semi-shaded sites in forests of Pernambuco, Sergipe and Bahia states of Brazil. This species of tender foliage can be up to 1.5 m tall. It has pendulous flowers with the outer tepals broader than inner and colour varying from yellow greenish to pink greenish.

***A. sellowiana*:** It is an odoriferous species, just as *A. caryophyllaea,* is shorter than the other Brazilian species, can be up to 0.5 m tall; the leaves are rigid, narrow and the flowers are yellow to orange with dark spots on inner tepals.

***A. foliosa*:** It is one of the showiest species from Brazil. It is about 0.4-0.6 m tall, the leaves are chartaceus and the colour varies from yellow to deep orange or red.

***A. speciosa*:** It is another showy species that grows on the shadow at forest of Serra do Mar in Sao Paulo states. It can be up to 1.5 m tall; the leaves are large and tender and the flowers are also large, 4-5 cm long, red or yellow coloured with the inner tepals dark spotted.

***A. variegata*:** It has crowded leaves and flowers ranging from red to yellowish on outside, and yellow and dark striated on inside.

***A. plantaginea*:** This species can be up to 1.2 m tall. The flowers are about 3.5 cm long, with the outer tepals colour ranging from red to orange and the inner ones with a yellow background and dark spots.

***A. gardneri*:** Plant upto 1.5 m tall, with rigid leaves. The flowers size is about 3.5-4.5 cm long and the colour varies from deep red to yellow red on the outside, yellow and spotted internally.

***A. malmeana*:** This species can be up to 0.9 m tall, the leaves are rigid, amplexicaulos and the flowers, red, pendulous with the stamens and the yellowish stigma exposed.

***A. stenophylla*:** Its height runs about 0.2-0.7 m tall. The leaves are narrow linear, rigid or chartaceus. The flowers are very similar to *A. gardneri*, reddish or orange, 2-3.8 cm long and the inner tepals are dark spotted.

***A. aurantiaca*:** It has 60-120 cm long flowering stem. About 10-30 flowers found in compound umbel, bright yellow, outer segments tipped green and inner ones spotted brown.

***A. pelegrina*:** It has 30 cm long flowering stem or less, 5 cm long flower, lilac, inner segments spotted red-purple.

***A. hookeri*:** It has 45-60 cm long flowering stem, flowers in loose heads, orange, suffused pink, upper petals spotted and blotched red and yellow.

Other species are *A. brasiliensis, A. campaniflora, A. haemantha, A. pulchella, A. violacea, A. pygmaea, A. patagonica A. andina* subsp. *venustula, A. pseudospathulata, A. spathulata, A. garaventae, A. umbellate A. graminea, A. werdermannii* and *A. polyphylla.*

FLOWER MORPHOLOGY AND BIOLOGY

Plants are herbaceous perennnials. The leaves are frequently resupinate, the inflorescence is terminal and umbeliform. The flowers are epigynous and are nearly always zygomorphic due to the position of the inner tepals. The zygomorphy is reinforced by the position of the filaments, which are located in the lower part of the flower during anthesis. In *A. pygmaea* and *A. patagonica* the flowers are actinomorphic but with some ones showing some degree of zygomorphy. Most of the floral characters have taxonomic importance. A large variation regarding flower colours and patterns is represented. Some species have tiny flowers with a diameter of about 1cm, e.g. *A. revoluta* while other such as *A. magnifica* and *A. pelegrina*, the type species of the genus, have large showy flowers with a diameter of up to 7 cm. Some Brazilian species, such as *A. isabellana* have more tubular flowers with narrower perianth segments. The pollen of all species investigated is ellipsoid, heteropolar and monosulcate. Symmetry is bilateral, six stamens (3+3), with narrow filaments and elongate pseudo-basifixed anthers, longitudinal and introrse dehiscence; trilocular or unilocular pistil with a single style, with three apical stigmatic branches, that produce an exudate on its surface, a trilocular or unilocular ovary, with distinct decurrent borders of the outer tepals and a circular scar at the position of attachment of the perigone; it contains numerous anatropous ovules on axile or parietal placentae; the fruit is a capsule which is generally loculicidal, occasionally it is indehiscent or opens explosively, flowers are often protandrous, *i.e.* the anthers shed their pollen before the stigma is receptive. Anthers dehisce in a temporal sequence. Flowers are temporally unisexual. This mechanism limits the possibilities of self-pollination within and among flowers in the same umbel (Traub, 1943; Aizen and Basilio, 1995; Snow and Grove, 1995), flowers are most probably pollinated by hummingbirds and bees (Stinson, 1942), bumble bees (Aker and Healy, 1990), and butterflies (Bayer, 1988).

GENETICS

Alstroemeria has special interest because of its large chromosomes and its asymmetric karyotypes. It may be considered chromosomally stable with a constancy of the chromosome number, 2n = 16, a unique level of ploidy, 2*x*, and an asymmetrical complement (Sanso and Hunziker, 1998 and Sanso, 2002). Evaluation of the genetic diversity of wild species of Alstroemeriaceae and alstroemeria cultivars has been attempted using chromosomal research and molecular markers. Karyotype analysis has been carried out and applied to taxonomic research. The chromosome numbers of wild *Alstroemeria* species have been determined (Strasburger 1882, Taylor 1926, Hang and Tsuchiya 1988 and Stephens *et al*., 1993). According to these studies, the somatic chromosome number is 2n = 2*x* = 16. Sanso and Hunziker (1998) revealed that the basic chromosome number of alstroemeria (n=8). Although chromosome sizes vary from species to species, they are karyotypically very similar in terms of relative size relationships between the chromosomes. Between species, there is a considerable variation in nuclear DNA content, the amount of C-banded heterochromatin (Buitendijk and Ramanna, 1996) and the presence or lack of satellites on several pairs of chromosomes. Two species differed by about a factor of two in the total chromosome length (*A. magnifica* = 116 µm; *A. pygmaea* = 224 µm). However, in alstroemeria, bimodal karyotypes occur despite these differences in the total chromosome length. It seems that in this group DNA has been added to the complements without altering the karyotype morphology too much (Sanso, 2002).

Heritability and genetic advances

The variation among and within species opens up the possibility for production of a wide range of new cultivars suited for various purposes, such as cut flowers, pot plants and garden and landscape uses. It should be possible to increase the variation and enhance the quality of the commercial cultivars by including a larger part of the genetic variation found among wild species. Many other wild taxa, that are scarcely known so far, are also of potential use in breeding programmes. Random amplified polymorphic DNA (RAPD) markers are used to study genetic variation in alstroemeria.

BREEDING OBJECTIVES

Breeders continued to broaden the assortment, with respect to the flower colour, the size of the flower, the size of the stem, and the length of the flowering period. Winter flowering or year-round flowering has remained a very important selection criterion. Additional important criteria for alstroemeria cut flowers are: the flower colour and its ornamental value, the production of high quality stems, the size and colour of the flower buds and the quality of the leaves. For pot plant cultivars breeders concentrate on plant characters such as short stems, simultaneous and uniform flowering, good leaf quality and performance of the whole plant. Important breeding objectives are as given below.

i. Novelty of colour

ii. To develop winter hardy cultivars

iii. More number of spikes per plant

iv. Dwarfness of plant for pots

v. Longer duration of flowering in field

vi. Attractive scent

vii. Longer vase life

viii. Resistance to viruses and diseases

ix. Resistance to insects and pests

BREEDING METHODS

There is always a demand for new cultivars of ornamental flower species, depending on the preferences of people worldwide. The flowers of alstroemeria exhibit particular spots on the tepal, this characteristic however does not find favour in the Japanese market. Therefore, in order to satisfy the Japanese market, Miyake (1989) introduced a breed of spotless *Alstroemeria* species without these spots on the tepal. Moreover, various cultivars have been produced by hybridization and irradiation treatments (Broertjes and Verboom, 1974). Various breeding methods described below.

Introduction

Alstroemeria varieties recommended by IHBT Palampur: Pluto, Capri, Tiara and Aladdin. Germplasm collected by Y.S. Parmar University of Horticulture and Forestry, Solan: Amor, Cinderella, Rosita, Serena, No. 14, Tiara.

Recently developed varieties are:

i. Himalaya: Snow-white flowers, each bloom opens majestically and last in many days.

ii. Nadya: Deep red colour flowers.

iii. N-Joy: Bi-colour white and pink flower.

iv. Shakira: Flower Yellow with subtle shades of green.

Selection

There are several cultivars of alstroemeria available all over the world. The development of new cultivars is very rapid and mostly done by various commercial flower companies. Less number of varieties evolved through selection (Anon., 2005u, 2012b, 2012d, 2013c). Some of them are described in Table 1.

Table 1: Varieties developed through selection.

S.No.	Variety	Breeder's name & release year	Chief characteristics
1.	Victoria	Gebr. van Zanten (1993)	Medium sized flowers, outer petals orange in colour, inner petals light yellow in colour marked end to end with dark brown longitudinal pencil stripes, plant vigorous upright growth habit with long stems, continuous production of flowers during each of its recurrent blooming seasons.
2.	Flamengo	Lezan V.O.F. (1988)	Year-around and profuse production, large flowers of Amaranth rose and Spinel red on white colouration of the uppermost petals, flowers being borne singly on strong, short peduncles comprising an umbel formation carried on strong, tall canes, flowers long lasting quality both on the plant and as cut flowers.
3.	Nevada	Koninklijke Van Zanten (1994)	Medium-size flowers having generally an ivory-white flower colour, leaves and internodes being shorter than common varieties, pedicels having a horizontal setting on the main stem and high productivity.
4.	Koncayuko	Konst Breeding B.V. (2010)	Compact and mounding plant habit; sturdy and strong plants; vigorous growth habit; red purple-coloured flowers; and good garden performance.

Hybridization

Alstroemeria has gained distinction due to the breeding work of Dr. Mark Bridgen at the University of Connecticut, U.S.A. He started the work in 1987 with his first exploration to Brazil with the late Fred Meyer. He made a collection of only the fragrant species of *Alstroemeria* and a few others. Hybridization has been proved a powerful tool of development of outstanding varieties of alstroemeria. In development of varieties remarkable work have been done by several plant breeding companies/organizations in which some companies of the Netherlands developed several varieties. These are Van Zanten Plants B.V., Aalsmeer, The Netherlands (Anon., 2005g, i, k, l, n, p, t, 2006a, b, c, w, x, y, 2007e, 2008a, b, c, d, e, 2013a, b, c, d, e, g), Wulfinghoff Alstroemeria B.V., Rijswijik, The Netherlands (Anon., 2009a, b, c, d, e, 2011a, b), Van Staaveren B.V. (Anon., 2005a, c, d, e, f, h, j, m, o, q, 2011c, 2012a,c) and Konst Breeding B.V., Nieuwveens, The Netherlands (Anon., 2005v, u, 2007a, b, c, d, 2008a, 2010a, b, c, d, 2012e, h, g, 2013h, i). Details of varieties developed by these companies become popular worldwide and are depicted in Table 2, 3, 4 and 5.

Table 2: Alstroemeria varieties developed by Van Zanten Plants B.V., Aalsmeer, The Netherlands.

S.No.	Variety	Parents	Release year	Chief characteristics
1.	Stamond	87r1345-1 × 87G1069-2	1995	Flowers of good quality having inner petals with greyed purple spots against a primarily white background with a central section with yellow intonation.
2.	Zanvedere	50665-4 × 50438-2	2003	Erect flowering stems; purple, yellow green and white-coloured flowers with dark purple spots and stripes; and excellent postproduction longevity.
3.	Staqueen	60T0040-01 × 87G1069-2	2003	Erect flowering stems; intense light and dark orange bi-coloured flowers with dark purple spots and stripes; excellent postproduction longevity.
4.	Staprioxa	92D53-1 × 86F679-1	2003	Compact and uniform plant growth habit; freely branching habit; freely flowering habit; red purple and yellow-coloured flowers with dark purple-coloured spots and stripes.
5.	Staprirange	88D1240-2 × 86F679-1	2003	Compact and uniform plant growth habit; freely branching habit; freely flowering habit; light red and yellow-coloured flowers with dark purple-coloured spots and stripes.
6.	Staprisara	93D834-16 × 93G112-2	2003	Compact and uniform plant growth habit; freely branching habit; freely flowering habit; and yellow and red purple bi-coloured flowers with dark purple-coloured spots and stripes.
7.	Staprivane	88D1240-2 × 86F679-1	2003	Compact and uniform plant growth habit; freely branching habit; freely flowering habit; purple-coloured flowers with dark purple-coloured spots and stripes.
8.	Zalsamay	95115-3PN × 93077-3PN	2004	Erect and strong flowering stems; vigorous growth habit; intense white and purple bi-coloured flowers; good postproduction longevity.
9.	Zaprijul	94186-002PN × 96123-001PN	2005	Compact, upright and outwardly spreading plant habit; purple and violet bi-coloured flowers with dark purple-coloured spots and stripes; and excellent postproduction longevity.
10.	Zalsamot	96791-001 × 96677-003	2005	Flower colour is red purple –purple.
11.	Zalsanem	96Y0255-6 × 87G1069-2	2005	Flower colour is purple.
12.	Zaprinous	97120-003PN × 95197-002D	2005	Compact and mounded plant habit; vigorous growth habit; purple-coloured flowers and good postproduction longevity.
13.	Zalsarest	95299-4 × 86021-7B	2005	Erect and strong flowering stems; vigorous growth habit; white-coloured flowers; good postproduction longevity.

(*Contd.*)

(*Contd. Table 2*)

S.No.	Variety	Parents	Release year	Chief characteristics
14.	Zalsanyx	97844-7PN × 98657-6	2008	Erect and strong flowering stems; vigorous growth habit; intense dark purple-coloured flowers; and excellent postproduction longevity.
15.	Zapriari	00-0023-01 × 87-1069-02	2009	Upright, somewhat outwardly spreading and mounded plant habit; vigorous growth habit; yellow orange-coloured flowers; and good container and garden performance.
16.	Zaprilou	88-1240-02 × 87-1069-02	2009	Upright, somewhat outwardly spreading and mounded plant habit; vigorous growth habit; red purple-coloured flowers; and good garden performance.
17.	Zalsachic	98426-2 × 98557-3	2010	Erect and strong flowering stems; vigorous growth habit; intense red-coloured flowers; excellent postproduction longevity.
18.	Zalsadon	98415-2 × 871069-2	2010	Erect and strong flowering stems; vigorous growth habit; white and yellow-coloured flowers; and excellent postproduction longevity.
19.	Zalsalan	00469-4 × 871069-2	2010	Erect and strong flowering stems; vigorous growth habit; white and yellow-coloured flowers with red purple-coloured stripes; excellent postproduction longevity.
20.	Zalsamon	9818-1 × 0023-1	2010	Erect and strong flowering stems; vigorous growth habit; bright yellow-coloured flowers; excellent postproduction longevity.
21.	Zalsaden	98441-5 × 97167-3	2010	Erect and strong flowering stems; vigorous growth habit; red purple-coloured flowers; excellent postproduction longevity.
22.	Zalsaney	1907-11 × 871069-2	2011	Flower colour is white.
23.	Zalsatal	537-2 × 20419-10	2010	Erect and strong flowering stems; vigorous growth habit; red-coloured flowers; excellent postproduction longevity.
24.	Zaprielia	04-1060-001 × 05-6213-001	2011	Compact, upright, somewhat outwardly spreading and mounded plant habit; vigorous growth habit; dark pink-coloured flowers; good for container and garden.
25.	Zapriamin	4935-1 × 56213-11	2012	Compact, upright, somewhat outwardly spreading and mounded plant habit; vigorous growth habit; bright orange-coloured flowers; good for container and garden.

Table 3: Alstroemeria varieties developed by Wulfinghoff Alstroemeria B.V., Rijswijik, The Netherlands.

S. No.	Variety	Parents	Release year	Chief characteristics
1.	Arabella	T10 × 179-6	2009	Flower colour is red.
2.	Natalie	T 10 × 465-1	2009	Flower colour is white.
3.	Tara	T19 × 231-8	2009	Suitable for pots larger than 7 cm; bushy-dwarf growth habit; red flowers with a yellow base, black stripes on all petals; and 3 weeks inflorescence longevity on the plant; 2 weeks as a cut flower.
4.	Christina	T10 × 1205-35	2009	Flower colour is yellow.
5.	Davina	T10 × 390/6	2009	Flower colour is medium pink.
6.	Gina	T39 × 1205/35	2011	Compact and mounding plant habit; sturdy and strong plants; vigorous growth habit; large red purple and yellow-coloured flowers; and good garden performance.
7.	Lucy	T38 × 1205/35	2011	Compact and mounding plant habit; sturdy and strong plants; vigorous growth habit; large pink and yellow-coloured flowers; and good garden performance.

Table 4: Alstroemeria varieties developed by Van Staaveren B.V.

S. No.	Variety	Parents	Release year	Chief characteristics
1.	Stapripur	9318-1 × 4469-1	1994	Continuous and profuse production of medium sized flowers, purple colour. Inner petals are lilac purple base portion. Stem length 30 cm. Suited for pot and garden.
2.	Stalove	84 G 447-2 × 85 G 325-1	1994	Profuse and recurrent production of very large flowers, lemon yellow colour. Crimson spots on two inner lateral petals. Erect stems branching from a tall main stem forming a strong upright bush.
3.	Stabec	R1684-1 × 87G1069-21	1995	Flowers have outer petals with light red to deep pink colouration on a white background and inner petals with similar colour but additionally including greyed red spots on a yellow background.
4.	Statiren	86R1684-1 × 87G1069-21	1995	Flowers of good quality and attractive colouration.
5.	Stasach	85T305-1 × 86F679-1	1996	Flowers of good quality of unusual red colouration.
6.	Stalona	86T1119-01 × 87G1069-02	1997	Flowers of good quality and attractive colouration.
7.	Staprizsa	91D196-1 × 7G1069-02	1998	Flowers of good quality and attractive colouration, suitable for growing in pots.

(Contd.)

(*Contd. Table 4*)

S. No.	Variety	Parents	Release year	Chief characteristics
8.	Stabelin	87R1280-1 × 87G1069-02	1998	Flowers of good quality and attractive colouration.
9.	Staprisis	91D186-3 × 86G713-1	1998	Flowers of good quality and attractive colouration, suitable for growing in pots.
10.	Staprimon	89D1577-1 × 89G1041-01	1998	Flowers of good quality and attractive colouration, suitable for growing in pots.
11.	Stapripal	91D174-1 × 86G713-1	1999	Flowers of good quality and attractive colouration.
12.	Stalauli	82R0473-6 × 86F1115-3	1999	Flowers of good quality and attractive colouration.
13.	Stalog	89W875-1 × 87G1069-2	2002	White flowers of good quality and attractive colouration.

Table 5: Alstroemeria varieties developed by Konst Breeding B.V., Nieuwveens, The Netherlands.

S. No.	Variety	Parents	Release year	Chief characteristics
1.	Konovatio	5230-2 × 5332-7	2004	Flower colour is red purple.
2.	Kogoa	6454-6 × 90-3-8	2005	Flower colour is yellow. Stem length medium to long. Inner median tepal stripes present.
3.	Koncalga	5326-1 × 5346-9	2006	Flower colour is white and yellow.
4.	Konimpa	8656-1 × 8124-7	2006	Flower colour is orange red.
5.	Konsacram	5230-2 × 6454-6	2006	Flower colour is light pink.
6.	Konzifer	5803-21 × 4495-5	2006	Flower colour is orange.
7.	Konsirak	7103-2 × 6842-18	2006	Flower colour is pink.
8.	Konamul	10056-5 × 97-0-4	2008	Flower colour is orange.
9.	Konratus	5782-14 × 7068-2	2008	Flower colour is purple pink.
10.	Konevotio	88-19-5 × 97-0-4	2008	Flower colour is medium pink.
11.	Konpulse	3150-30 × 5261-4	2008	Flower colour is red.
12.	Koncajoli	5261-4 × 8144-1	2010	Compact and mounding plant habit; sturdy and strong plants; vigorous growth habit; light red-coloured flowers with distinct red purple-coloured stripes; and good garden performance.
13.	Koncavanti	5261-4 ×7567-3	2010	Compact and mounding plant habit; sturdy and strong plants; vigorous growth habit; pink-coloured flowers with distinct dark purple-coloured stripes; and good garden performance.
14.	Konshakira	13839-2 × 9748-2	2011	Flower colour is medium yellow.
15.	Konglacier	6454-6 × 9458-2	2011	Flower colour is light yellow.
16.	Konpepper	21100-1 × 17931-1	2012	Flower colour is red.

Interspecific hybrids

Alstroemeria breeding through interspecific hybridization has been successful. Shinoda and Murata (2003) conducted detailed studies on interspecific hybridization *via* ovule culture and reported the possibilities and limitations of producing hybrids in 15 *Alstroemeria* species. Intergeneric hybridization is required to produce further variations in *Alstroemeria* species and thus produce new cultivars. Spontaneous interspecific hybridization probably occurs in nature, no specific examples have been reported so far. Several attempts to synthetically hybridize species were successful (Foster, 1948 and Duncan, 1977 and 1982). Some hybrids, such as the *A. ligtu* hybrids (Comber, 1946) and the 'Van Houtte's hybrids (Sahin, 1973) were fertile, but most other hybrids were sterile. The hybridization barriers restricted the progress in breeding, but nevertheless during the 1960's the hybrid 'Walter Fleming' was raised, and it proved to be a commercial success (Goemans, 1962). Although this diploid (2n=2x=16) cultivar was highly sterile, it was used as a parent, and the 'Parigo hybrids' were developed from it. A characteristic feature of these early hybrids is, that they possess large open flowers that resemble orchids, and for this reason they became generally known as 'Orchid type' cultivars. Most of these hybrids were highly sterile. A remarkable fact is that these early hybrids consisted of diploids, triploids (2n=3x=24) and occasionally tetraploids (2n=4x=32), and thus probably originated through the functioning of unreduced gametes (Ramanna, 1991, 1992).

Diploid interspecific hybrids between *Alstroemeria* species that have been used for the development of cultivars, has indicated the occurrence of unreduced gametes (Ramanna and Buitendijk, unpublished). It was found that the interspecific hybrids between Chilean species produced hardly any unreduced gametes, whereas those between Chilean and Brazilian species produced high frequencies of both unreduced pollen and unreduced eggs. From the latter interspecific hybrids, large numbers of sexual polyploids have been produced experimentally. Another significant development was that, besides inter Chilean species hybrids, also Chilean-Brazilian species hybrids were produced. Probably because the flowers of such hybrids resembled butterflies, they were referred to as 'Butterfly type' cultivars. Also in these hybrids spontaneous polyploidization led to triploid and tetraploid cultivars. Some important interspecific hybrids are given in Table 6.

Table 6: Interspecific hybrids of alstroemeria.

S.No.	Variety	Species involved in breeding
1.	Elisa	(*A. caryophyllae* × *A. psittacina*) × *A. pelegrina*
2.	Freedom	*A. caryophyllae* × Chilean hybrid
3.	Inca Koya	*A. caryophyllae* × Chilean hybrid
4.	Meyer's Dwarf	*A. pelegrina* × *A. psittacina*
5.	Stinson's Hybrid	*A. caryophyllae* × *A. psittacina*
6.	Sweet Laura	*A. aurea* × *A. caryophyllae*

Some other interspecific hybrids: *A. magnifica* × *A. inodora*, *A. inodora* × *A. magnifica*, *A. pelegrina* × *A. inodora*, *A. pelegrina* × *A. magnifica*, *A. aurea* × *A. ligtu*, *A. aurea* × *A. inodora*, *A. pelegrina* × *A. aurea* and *A. magnifica* × *A. Aurea*.

Mutation

The application of mutation breeding offered breeders the possibility of inducing favourable mutations in already existing cultivars (Broertjes and Verboom, 1974). After irradiation of young actively growing rhizomes with X-rays, usually a range of new forms was created, of which some proved to be improvements. Hence, the irradiation of cultivars with X-rays became a standard procedure in breeding programmes. In early studies, X-rays were used and now-a-days gamma (γ) radiation has become an efficient mutagenic agent in ornamental crops such as alstroemeria. Gamma irradiation has been widely used as a breeding technique to obtain new cultivars, where several cultivars have been obtained through rhizome radiation. Even though the biochemistry base of mutation is still not completely clear, it is thought that mutagenic treatments can generate chromosome rearrangements or produce changes in some genes to other allelic forms, explaining the phenotypic variation. *In vitro* culture techniques can be considered as an appropriate tool for inducing mutations because it can supply a large amount of homogeneous and virus free individuals to be irradiated. In addition, *in vitro* culture also provides an efficient propagation system for irradiated material allowing assessment of mutant stability and easy later selection. At the moment Alstroemeria is one of the major products in the cut flower industry and the breeding of new cultivars of this species have been based on the induction of mutagenesis through irradiation of rhizomes with x-rays and gamma rays. Alstroemeria mutagenesis has been focused in changing few traits on outstanding cultivars.

Goemans (1962) carried out work and found that mutation cannot be applied in the case of Alstroemeria and the only available method seemed therefore to be the irradiation of actively growing rhizomes of young plants. These steadily growing rhizomes carry tiny buds, in various stages of development, ultimately growing out into flowering shoots. Many of the recent hybrids are developed by him. By crossing a sport of an unpublished species with a second species and subsequently with *A. aurantiuca,* he obtained the so-called 'Parigo hybrids', such as the pink-flowered 'Ballerina', 'Pink Attraction' and 'Pink Perfection', the red-flowered 'Carmen', 'Marina' and 'Pimpernel', the yellow-flowered. 'Sussex Gold' and the pinky-yellow flowered cultivar Parigo's Charm, to mention only a few.

Since a mutated cell has its largest chance when induced in the youngest possible developmental stage of a meristem or apex, the best material to be irradiated seemed to be the actively growing rhizomes of young plants. The best time for irradiation was during March-April. First selections could often be made in May or June and promising mutants were isolated by cutting them off the rhizome. A second selection was made during the next flowering period. If the irradiation was too heavy, the first selection could be made only in spring of the next year. Plants irradiated during reduced vegetative growth generally die. Since 1970, many more rhizomes of various cultivars have been irradiated. They have produced a number of mutants, which are now being propagated and selected. Among the cultivars irradiated are also several mutants which, in turn, have produced mutants after irradiation. From cv. Canaria, for instance, a mutant has been obtained with more pronounced striping.

Apart from the fact that mutation breeding is the only way of inducing variation in sterile cultivars of vegetatively propagated crops, as in many of the best alstroemeria hybrids, it has been demonstrated that commercial mutants can be produced fairly easily and rapidly by this method. The heterozygosity, which was not expected to be large since the modern assortment derives from only a few ancestral species, turned out to be large enough to induce and obtain variability for a number of important characteristics. Some important mutant varieties are given in Table 7. There are also several mutant varieties (Anon., 2005b, 2007g, 2007f, 2010e, 2013f) developed by breeders/companies (Table 8) become popular.

Table 7: Mutant varieties of alstroemeria.

S.No.	Variety	Parent	Type of mutagen	Remarks
1.	Canaria	Orchid flower	X ray	Yellow flower colour
2.	Capitol	Carmen	X ray	Salmon-pink flower colour
3.	Fanfare	Carmen	X ray	Red flower colour
4.	Harlequin	Paringo's Charm	X ray	Orange-yellow flower colour
5.	Harmony Stabrons	Regina	X ray	Bronze flower colour
6.	Result	Carmen	X ray	Bright red flower colour
7.	Rosali Staliro	Starosa	X ray	Pink flower colour
8.	Rosita Stareza	Regina	X ray	Pink flower colour
9.	Trident	Carmen	X ray	Pink flower colour
10.	Valiant	Carmen	X ray	Light red flower colour
11.	White Wings	Orchid flower	X ray	Plant with stolons white except yellow ears
12.	Yellow Tiger	Orchid flower	X ray	Yellow flower colour, striking black stripes
13.	Zebra Stazeb	Orchid flower	X ray	Heavily stripped flower
14.	Zenith	Carmen	X ray	Orange-red flower colour

Table 8: Mutant varieties alstroemeria evolved by Breeders/Companies.

S. No.	Variety	Parents	Breeders & release year	Chief Characteristics
1.	Stabecor	Stabec	C.A. Hoogendoorn (2002)	Flowers of good quality and attractive colouration.
2.	Zapriteres	Staprivane	Van Zanten Plants B.V. (2007)	Compact and mounded plant habit; vigorous growth habit; light and dark pink-coloured flowers; and good postproduction longevity.
3.	Zaprifabi	Stapridani	Van Zanten Plants B.V. (2007)	Compact and mounded plant habit; vigorous growth habit; variegated dark green-coloured foliage with white-coloured margins; pale yellow-coloured flowers; and good postproduction longevity.
4.	Konanel	Konsacram	Konst Breeding B.V. (2009)	Flower colour is red purple.
5.	Zaprilet	Staprioxa	Van Zanten Plants B.V. (2009)	It is characterized by its upright and somewhat outwardly spreading plant habit; vigorous growth habit; red-coloured flowers; and good garden performance.

Polyploidy

Lu and Bridgen (1996) provided detailed data on the effect of genotype, developmental stage of an embryo and the culture medium on interspecific hybrids of alstroemeria In order to restore fertility, amphidiploid plants were produced by colchicine treatment (Ishikawa *et al*., 1999) from interspecific crosses of *A. ligtu* hybrid and *A. pelegrina* var. *rosea*. the amphidiploid plants that were crossed with *A. ligtu* hybrid, *A. pelegrina* var. *rosea*, *A. aurea*, *A. paupercula* and *A. psittacina* produced triploid progeny plants. Shinoda and Murata (2003) carried out great number of cross hybridizations through ovule cultures using *A. angustifolia* and *A. aurea*.

BIOTECHNOLOGY

Recently, biotechnological approaches are being applied in order to improve alstroemeria strains. Particle bombardment and *Agrobacterium*-mediated procedures were applied for genetic transformation and some transformed plants with marker genes were produced.

Tissue culture

First publications on *in vitro* cultures of alstroemeria date back to 1973 and 1974, when attempts were made to initiate rhizome cultures from greenhouse grown plants. In later years, aseptic cultures were used to eliminate viruses, to micropropagate rhizomes, to produce hybrid plants and to regenerate plants through a phase of callus or somatic embryos for genetic transformation purposes. The development of *in vitro* methods to overcome hybridization barriers has recently opened the possibility to produce hybrids on a relatively large scale. Due to these techniques many new interspecific combinations are expected to be produced. Interspecific hybrid plants have been produced by ovule cultures. By improving certain culture techniques, sexual incompatibility was overcome in some cross combinations using ovule cultures. Plant regeneration systems that involved the use of explants, immature ovules, leaves, etc., through callus cultures have been reported.

In vitro fertilization to overcome post-fertilization barriers

In the case of alstroemeria, isolation of egg cells and zygotes from ovules has been attempted in order to develop an *in vitro* fertilization technique. Generally, interspecific hybridizations fail due to post-fertilization barriers or inhibition of pollen tube elongation. By using fluorescence microscopy, de Jeu and Jacobsen (1995) observed the entry of pollen tubes into ovules and the subsequent fertilization and embryo development in interspecific diallel crossing among *A. aurea*, *A. hookeri*, *A. pulchra*, *A. ligtu*, *A. pelegrina*, *A inodora* and *A. brasiliensis*. They described the post-fertilization barrier-induced incompatibility in many interspecific hybridizations of alstroemeria. This result indicated that embryo rescue before abortion is effective in obtaining successful interspecific hybridizations. The development of culture system from leaf and stem in alstroemeria is valuable, although successful organogenesis from leaf and stem is strongly genotype-dependent at the present. This could expand the possibilities of use of culture systems for alstroemeria. By utilizing the embryogenic callus culture in alstroemeria, isolation of protoplasts, subsequent protoplast culture and regeneration of plants from protoplasts were achieved (Kim *et al.*, 2005).

Genomic *in situ* hybridization (GISH) enabled the identification of parental genomes in the following interspecific hybrids: *A. magnifica* × *A. inodora, A. inodora* × *A. magnifica, A. pelegrina* × *A. inodora, A. pelegrina* × *A. magnifica, A. aurea* × *A. ligtu, A. aurea* × *A. inodora, A. pelegrina* × *A. aurea* and *A. magnifica* × *A. aurea* (Kuipers *et al.* 1997). Kuipers *et al.* (2002) also reported the molecular cloning of highly repetitive DNA sequences from the Brazilian species *A. psittacina* and *A. inodora*. *In situ* hybridization using the repetitive DNA sequences revealed specific localization patterns in the chromosomes of *A. psittacina* and *A. inodora*.

Molecular breeding

Molecular markers have also been developed for amplified fragment length polymorphism (AFLP) by Han *et al.* (1999). By using AFLP markers, their group constructed a linkage map for *A. aurea* (Han *et al.*, 2002) and demonstrated the differences between the Chiliean and Brazilian *Alstroemeria* species. Based on the results, the *Alstroemeria* species were roughly divided into 3 subclades according to their geographical area of occurrence, *i.e.* northern Chile, central Chile and Brazil. As mentioned above, taxonomic and phylogentic analyses using molecular markers and karyotype analysis have yielded accurate data regarding the genetic relationships and diversity in Alstroemeriaceae. In addition, the nuclear content in the *Alstroemeria* species and some of their hybrids was estimated by flow cytometry using PI and DAPI (Buitendijk *et al.*, 1997). In alstroemeria, *in vitro* fertilization using isolated egg cells and zygotes from ovules has been attempted (Hoshino *et al.*, 2006).

Random amplified polymorphic DNA (RAPD) analysis was utilized to analyse the genetic relationships among alstroemeria cultivars (Dubouzet *et al.*, 1997) and between wild *Alstroemeria* species and cultivars. Aros *et al.* (2006) used RAPD markers in combination with morphological descriptors and demonstrated the genetic diversity between commercial cultivars and wild *Alstroemeria* species.

Genetic Engineering

Particle bombardment and *Agrobacterium*-mediated procedures were applied for genetic transformation, resulting in the generation of transformed plants containing the marker genes. The first attempt at genetic transformation in alstroemeria was reported by Van Schaik *et al.* (2000). They confirmed transient gene expression in embryogenic callus after particle bombardment. Foreign gene expression examined with a β-glucuronidase (GUS) assay indicated the reducible transformation conditions in which the *GUS* gene was driven by the maize ubiquitin promoter; suitable parameters for the assay equipment have been reported. Although stable transformed calluses were obtained after visual selection based on luciferase activity, no plant regeneration was observed during the 2 years of maintenance. Subsequently, Lin *et al.* (2000) succeeded in producing transgenic plants by particle bombardment. They used a *luciferase* gene from a firefly as the reporter gene under the control of the maize ubiquitin promoter (*Ubi1*) and attempted transferring this gene into friable embryogenic calluses and proembryos. Following visual selection, transgenic plantlets were regenerated through embryogenesis. Moreover, another selection by phosphinothricin (PPT) helped in obtaining *bar* and *GUS* gene-incorporated plants after particle bombardment. The successful transformation obtained by Lin *et al.* (2000) appears to be due to the selection of elite embryogenic callus lines with a high regeneration potential and maintaining these cells till abiolistic procedure was performed.

Hoshino *et al.* (2000) reported on the genetic transformation of alstroemeria *via* co-cultivation of embryogenic suspension cells with *Agrobacterium tumefaciens*. Embryogenic cell suspension cultures were established *via* ovule culture of an interspecific cross between *A. pelegrina* and *A. magenta*. Cell suspension cultures could be initiated from the callus using liquid MS media containing picloram. An *Agrobacterium* mediated transformation system was developed by using the embryogenic suspension cells. Further research is needed before useful genes can be successfully incorporated into alstroemeria and the safety of genetically modified plants can be evaluated. Subsequently, transgenic techniques will contribute in improving the plant species.

Buitendijk and Ramanna (1996) showed that Giemsa C-banding and Feulgen staining could be used to identify *A. angustifolia*, *A. aurea*, *A. ligtu*, *A. magnifica*, *A. pelegrina*, *A. philippii* and *A. psittacina*. Analysis of *A. aurea*-specific repeats using fluorescence *in situ* hybridization (FISH) revealed distinct localization patterns in the heterochromatic regions of all *A. aurea* chromosomes. FISH was also applied to the chromosomes of *Alstroemeria* species by Kamstra *et al.* (1997). Species specific repetitive DNA and highly conserved rDNA sequences were localized on metaphase chromosomes of the Chilean species *A. aurea* and the Brazilian species *A. inodora*. The FISH patterns showed obvious differences between these 2 species.

SUMMARY

In breeding programmes for alstroemeria, it is expected that the hybrid plants will possess traits such as virus and disease resistance, attractive flower colours, novel flower shapes, flower longevity, dwarfness and attractive scent. Furthermore, flowering control is important for commercial production. Further studies combining recent biotechnological techniques are necessary to produce valuable cultivars of alstroemeria. The hybrid plants are expected to have an increased ornamental value. Studies should focus on the production of hybrids by cross combinations between distantly related genera such as *Alstroemeria* and *Bomarea*. Several barriers need to be overcome before hybrid varieties of *Alstroemeria* species can be obtained. In order to overcome post-fertilization barriers, further improvements in hybridization techniques are necessary.

REFERENCES

Aizen, M.A. and Basilio, A. 1995. Within and among flower sex-phase distribution in *Alstroemeria aurea* (Alstroemeriaceae). *Canadian J. Botany*, **73**: 1986-94.

Aker, S. and Healy, W. 1990. The phytogeography of the genus *Alstroemeria. Herbertia*, **46**: 76-87.

Anonymous. 2005a. *Alstroemeria* 'Stabec'. Official Journal of Plant Breeder's Rights Office, IPAustralia. *Plant Varieties J.*, **18**(2): 306.

Anonymous. 2005b. *Alstroemeria* 'Stabecor'. Official Journal of Plant Breeder's Rights Office, IPAustralia. *Plant Varieties J.*, **18**(2): 306.

Anonymous. 2005c. *Alstroemeria* 'Stabelin'. Official Journal of Plant Breeder's Rights Office, IPAustralia. *Plant Varieties J.*, **18**(2): 306.

Anonymous. 2005d. *Alstroemeria* 'Stalauli'. Official Journal of Plant Breeder's Rights Office, IPAustralia. *Plant Varieties J.*, **18**(2): 306.

Anonymous. 2005e. *Alstroemeria* 'Stalog'. Official Journal of Plant Breeder's Rights Office, IPAustralia. *Plant Varieties J.*, **18**(2): 307.

Anonymous. 2005f. *Alstroemeria* 'Stalona'. Official Journal of Plant Breeder's Rights Office, IPAustralia. *Plant Varieties J.*, **18**(2): 307.

Anonymous. 2005g. *Alstroemeria* 'Stamond'. Official Journal of Plant Breeder's Rights Office, IPAustralia. *Plant Varieties J.*, **18**(2): 307.

Anonymous. 2005h. *Alstroemeria* 'Staprimon'. Official Journal of Plant Breeder's Rights Office, IPAustralia. *Plant Varieties J.*, **18**(2): 307.

Anonymous. 2005i. *Alstroemeria* 'Staprioxa'. Official Journal of Plant Breeder's Rights Office, IPAustralia. *Plant Varieties J.*, **18**(2): 307.

Anonymous. 2005j. *Alstroemeria* 'Stapripal'. Official Journal of Plant Breeder's Rights Office, IPAustralia. *Plant Varieties J.*, **18**(2): 307.

Anonymous. 2005k. *Alstroemeria* 'Staprirange'. Official Journal of Plant Breeder's Rights Office, IPAustralia. *Plant Varieties J.*, **18**(1): 454.

Anonymous. 2005l. *Alstroemeria* 'Staprisara'. Official Journal of Plant Breeder's Rights Office, IPAustralia. *Plant Varieties J.*, **18**(1): 456.

Anonymous. 2005m. *Alstroemeria* 'Staprisis'. Official Journal of Plant Breeder's Rights Office, IPAustralia. *Plant Varieties J.*, **18**(2): 307.

Anonymous. 2005n. *Alstroemeria* 'Staprivane'. Official Journal of Plant Breeder's Rights Office, IPAustralia. *Plant Varieties J.*, **18**(2): 307.

Anonymous. 2005o. *Alstroemeria* 'Staprizsa'. Official Journal of Plant Breeder's Rights Office, IPAustralia. *Plant Varieties J.*, **18**(2): 307.

Anonymous. 2005p. *Alstroemeria* 'Staqueen'. Official Journal of Plant Breeder's Rights Office, IPAustralia. *Plant Varieties J.*, **18**(1): 453.

Anonymous. 2005q. *Alstroemeria* 'Stasach'. Official Journal of Plant Breeder's Rights Office, IPAustralia. *Plant Varieties J.*, **18**(2): 307.

Anonymous. 2005r. *Alstroemeria* 'Victoria'. Official Journal of Plant Breeder's Rights Office, IPAustralia. *Plant Varieties J.*, **18**(2): 308.

Anonymous. 2005s. *Alstroemeria* 'Virginia'. Official Journal of Plant Breeder's Rights Office, IPAustralia. *Plant Varieties J.*, **18**(2): 308.

Anonymous. 2005t. *Alstroemeria* 'Zanvedere'. Official Journal of Plant Breeder's Rights Office, IPAustralia. *Plant Varieties J.*, **18**(1): 457.

Anonymous. 2005u. *Alstroemeria* 'Kogoa'. Official Journal of Plant Breeder's Rights Office, IPAustralia. *Plant Varieties J.*, **18**(4): 169-172.

Anonymous. 2005v. *Alstroemeria* 'Konovatio'. Official Journal of Plant Breeder's Rights Office, IPAustralia. *Plant Varieties J.*, **18**(4): 165-168.

Anonymous. 2005w. *Alstroemeria* 'Zaprinous'. Official Journal of Plant Breeder's Rights Office, IPAustralia. *Plant Varieties J.*, **18**(4): 251-254.

Anonymous. 2005x. *Alstroemeria* 'Zaprijul'. Official Journal of Plant Breeder's Rights Office, IPAustralia. *Plant Varieties J.*, **18**(4): 255-259.

Anonymous. 2005y. *Alstroemeria* 'Zalsarest'. Official Journal of Plant Breeder's Rights Office, IPAustralia. *Plant Varieties J.*, **18**(4): 260-264.

Anonymous. 2006a. *Alstroemeria* 'Zalsamay'. Official Journal of Plant Breeder's Rights Office, IPAustralia. *Plant Varieties J.*, **19**(1): 364.

Anonymous. 2006b. *Alstroemeria* 'Zalsamot'. Official Journal of Plant Breeder's Rights Office, IPAustralia. *Plant Varieties J.*, **19**(3): 326-330.

Anonymous. 2006c. *Alstroemeria* 'Zalsanem'. Official Journal of Plant Breeder's Rights Office, IPAustralia. *Plant Varieties J.*, **19**(3): 321-325.

Anonymous. 2007a. *Alstroemeria* 'Koncalga'. Official Journal of Plant Breeder's Rights Office, IPAustralia. *Plant Varieties J.*, **20**(1): 294-296.

Anonymous. 2007b. *Alstroemeria* 'Konsacram'. Official Journal of Plant Breeder's Rights Office, IPAustralia. *Plant Varieties J.*, **20**(1): 300-302.

Anonymous. 2007c. *Alstroemeria* 'Konsirak'. Official Journal of Plant Breeder's Rights Office, IPAustralia. *Plant Varieties J.*, **20**(1): 288-290.

Anonymous. 2007d. *Alstroemeria* 'Konzifer'. Official Journal of Plant Breeder's Rights Office, IPAustralia. *Plant Varieties J.*, **20**(1): 297-299.

Anonymous. 2007e. *Alstroemeria* 'Zalsanyx'. Official Journal of Plant Breeder's Rights Office, IPAustralia. *Plant Varieties J.*, **20**(1): 283-285.

Anonymous. 2007f. *Alstroemeria* 'Zaprifabi'. Official Journal of Plant Breeder's Rights Office, IPAustralia. *Plant Varieties J.*, **20**(1): 291-293.

Anonymous. 2007g. *Alstroemeria* 'Zapriteres'. Official Journal of Plant Breeder's Rights Office, IPAustralia. *Plant Varieties J.*, **20**(1): 286-287.

Anonymous. 2007h. *Alstroemeria* 'Konimpa'. Official Journal of Plant Breeder's Rights Office, IPAustralia. *Plant Varieties J.*, **20**(4): 80-82.

Anonymous. 2008a. *Alstroemeria* 'Zalsachic'. Official Journal of Plant Breeder's Rights Office, IPAustralia. *Plant Varieties J.*, **21**(1): 178-180.

Anonymous. 2008b. *Alstroemeria* 'Zalsaden'. Official Journal of Plant Breeder's Rights Office, IPAustralia. *Plant Varieties J.*, **21**(1): 169-171.

Anonymous. 2008c. *Alstroemeria* 'Zalsadon'. Official Journal of Plant Breeder's Rights Office, IPAustralia. *Plant Varieties J.*, **21**(1): 175-177.

Anonymous. 2008d. *Alstroemeria* 'Zalsalan'. Official Journal of Plant Breeder's Rights Office, IPAustralia. *Plant Varieties J.*, **21**(1): 172-174.

Anonymous. 2008e. *Alstroemeria* 'Zalsamon'. Official Journal of Plant Breeder's Rights Office, IPAustralia. *Plant Varieties J.*, **21**(1): 181-183.

Anonymous. 2009a. *Alstroemeria* 'Arabella'. Official Journal of Plant Breeder's Rights Office, IPAustralia. *Plant Varieties J.*, **22**(4): 256-258.

Anonymous. 2009b. *Alstroemeria* 'Tara'. Official Journal of Plant Breeder's Rights Office, IPAustralia. *Plant Varieties J.*, **22**(4): 259-261.

Anonymous. 2009c. *Alstroemeria*'Natalie' . Official Journal of Plant Breeder's Rights Office, IPAustralia. *Plant Varieties J.*, **22**(4): 262-264.

Anonymous. 2009d. *Alstroemeria* 'Christina'. Official Journal of Plant Breeder's Rights Office, IPAustralia. *Plant Varieties J.*, **22**(4): 265-267.

Anonymous. 2009e. *Alstroemeria* 'Davina'. Official Journal of Plant Breeder's Rights Office, IPAustralia. *Plant Varieties J.*, **22**(4): 268-270.

Anonymous. 2010a. *Alstroemeria* 'Konamul'. Official Journal of Plant Breeder's Rights Office, IPAustralia. *Plant Varieties J.*, **23**(2): 215-217.

Anonymous. 2010b. *Alstroemeria* 'Konevotio'. Official Journal of Plant Breeder's Rights Office, IPAustralia. *Plant Varieties J.*, **23**(2): 218-220.

Anonymous. 2010c. *Alstroemeria* 'Konratus'. Official Journal of Plant Breeder's Rights Office, IPAustralia. *Plant Varieties J.*, **23**(2): 221-223.

Anonymous. 2010d. *Alstroemeria* 'Konpulse'. Official Journal of Plant Breeder's Rights Office, IPAustralia. *Plant Varieties J.*, **23**(2): 224-226.

Anonymous. 2010e. *Alstroemeria* 'Konanel'. Official Journal of Plant Breeder's Rights Office, IPAustralia. *Plant Varieties J.*, **23**(2): 227-229.

Anonymous. 2011a. *Alstroemeria* 'Gina'. Official Journal of Plant Breeder's Rights Office, IPAustralia. *Plant Varieties J.*, **24**(1): 21.

Anonymous. 2011b. *Alstroemeria* 'Lucy'. Official Journal of Plant Breeder's Rights Office, IPAustralia. *Plant Varieties J.*, **24**(1): 21.

Anonymous. 2011c. *Alstroemeria* 'Statiren'. Official Journal of Plant Breeder's Rights Office, IPAustralia. *Plant Varieties J.*, **24**(2): 288.

Anonymous. 2012a..*Alstroemeria* 'Stapripur'. Official Journal of Plant Breeder's Rights Office, IPAustralia. *Plant Varieties J.*, **25**(2): 230.

Anonymous. 2012b..*Alstroemeria* 'Flamengo'. Official Journal of Plant Breeder's Rights Office, IPAustralia. *Plant Varieties J.*, **25**(2): 229.

Anonymous. 2012c..*Alstroemeria* 'Stalove'. Official Journal of Plant Breeder's Rights Office, IPAustralia. *Plant Varieties J.*, **25**(2): 230.

Anonymous. 2012d. *Alstroemeria* 'Nevada'. Official Journal of Plant Breeder's Rights Office, IPAustralia. *Plant Varieties J.*, **25**(2): 230.

Anonymous. 2012e. *Alstroemeria* 'Koncajoli'. Official Journal of Plant Breeder's Rights Office, IPAustralia. *Plant Varieties J.*, **25**(4): 359-361.

Anonymous. 2012f. *Alstroemeria* 'Koncayuko'. Official Journal of Plant Breeder's Rights Office, IPAustralia. *Plant Varieties J.*, **25**(4): 362-364.

Anonymous. 2012g. *Alstroemeria* 'Konshakira'. Official Journal of Plant Breeder's Rights Office, IPAustralia. *Plant Varieties J.*, **25**(4): 365-367.

Anonymous. 2012h. *Alstroemeria* 'Koncavanti'. Official Journal of Plant Breeder's Rights Office, IPAustralia. *Plant Varieties J.*, **25**(4): 368-370.

Anonymous. 2013a. *Alstroemeria* 'Zalsaney'. Official Journal of Plant Breeder's Rights Office, IPAustralia. *Plant Varieties J.*, **26**(1): 198-200.

Anonymous. 2013b. *Alstroemeria* 'Zapriamin'. Official Journal of Plant Breeder's Rights Office, IPAustralia. *Plant Varieties J.*, **26**(1): 201-203.

Anonymous. 2013c. *Alstroemeria* 'Zalsatal'. Official Journal of Plant Breeder's Rights Office, IPAustralia. *Plant Varieties J.*, **26**(1): 204-206.

Anonymous. 2013d. *Alstroemeria* 'Zapriari'. Official Journal of Plant Breeder's Rights Office, IPAustralia. *Plant Varieties J.*, **26**(1): 207-209.

Anonymous. 2013e. *Alstroemeria* 'Zaprilou'. Official Journal of Plant Breeder's Rights Office, IPAustralia. *Plant Varieties J.*, **26**(1): 210-212.

Anonymous. 2013f. *Alstroemeria* 'Zaprilet'. Official Journal of Plant Breeder's Rights Office, IPAustralia. *Plant Varieties J.*, **26**(1): 213-216.

Anonymous. 2013g. *Alstroemeria* 'Zaprielia'. Official Journal of Plant Breeder's Rights Office, IPAustralia. *Plant Varieties J.*, **26**(1): 217-219.

Anonymous. 2013h. *Alstroemeria* 'Konglacier'. Official Journal of Plant Breeder's Rights Office, IPAustralia. *Plant Varieties J.*, **26**(3): 216-218.

Anonymous. 2013i. *Alstroemeria* 'Konpepper'. Official Journal of Plant Breeder's Rights Office, IPAustralia. *Plant Varieties J.*, **26**(3): 213-215.

Aros, D., Meneses, C. and Infante, R. 2006. Genetic diversity of wild species and cultivated varieties of alstroemeria estimated through morphological descriptors and RAPD markers. *Scientia Horticulturae*, **108**: 86-90.

Bayer, E. 1988. Beitrag zur Cytologie der Alstroemeriaceae. *Mitteilungen der Botanischen Staatssammlung München*, **27**: 1-6.

Broertjes, C. and Verboom, H. 1974. Mutation breeding of alstroemeria. *Euphytica*, **23**: 39-44.

Buitendijk, J.H mei 1998. A cytological characterization of genomes of *Alstroemeria,* the production of interspecific hybrids, and their performance during micro propagation. Thesis, Department of Plant Breeding of Wageningen Agricultural University. The Netherlands.

Buitendijk, J.H. and Ramanna, M.S. 1996. Giemsa C-banded karyotypes of eight species of *Alstroemeria* L. and some of their hybrids. *Annals Bot.*, **78**: 449-57.

Buitendijk, J.H., Boon, E.J. and Ramanna, M.S. 1997. Nuclear DNA content in twelve species of *Alstroemeria* and some of their hybrids. *Annals Bot.*, **79**: 343-53.

Comber, H.F. 1946. *Alstroemeria* 'Ligtu hybrids'. *J. Royal Horticultural Society* LXXI-**9**: 274-77.

de Jeu, M.J. and Jacobsen, E. 1995. Early post-fertilization ovule culture in *Alstroemeria* L. and barriers to interspecific hybridization. *Euphytica*, **86**: 15-23.

Dubouzet, J.G., Murata, N. and Shinoda, K. 1997. RAPD analysis of genetic relationships among *Alstroemeria L.* cultivars. *Scientia Horticulturae*, **68**: 181-89.

Duncan, D.D. 1977. *Alstroemeria × Davisiae. Plant Life*, **33**: 71-72.

Duncan, D.D. 1982. *Alstroemeria* x *Zebrina* Duncan. *Plant Life*, **38**: 56.

Foster, M.B. 1948. Fragrant alstroemerias in Florida. *Herbertia*, **15**: 78-79.

Goemans, J.A.M. 1962. Breeding of alstroemeria. *J. R. Hort. Sot.*, **87**(6): 282-84.

Han, T.H., Van Eck, H.J., De Jeu, M.J. and Jacobsen, E. 2000. Genetic diversity of Chilean and Brazilian *Alstroemeria* species assessed by AFLP analysis. *Heredity*, **84**: 564-69.

Han, T.H., Van Eck, H.J., De Jeu, M.J. and Jacobsen, E. 1999. Optimization of AFLP fingerprinting of organisms with large-sized genome: a study on *Alstroemeria* spp. *Theoretical and Applied Genetics*, **98**: 465-71.

Han, T.H., Van Eck, H.J., De Jeu, M.J. and Jacobsen, E. 2002. The construction of a linkage map of *Alstroemeria aurea* by AFLP markers. *Euphytica*, **128**: 153-64.

Hang, A. and Tsuchiya, T. 1988. Chromosome studies in the genus *Alstroemeria*. II. Chromosome constitutions of eleven additional cultivars. *Plant Breeding*, **100**: 273-279.

Hoshino, Y., Murata, N. and Shinoda, K. 2006. Isolation of individual egg cells and zygotes in *Alstroemeria* followed by manual selection with a microcapillary-connected micropump. *Annals Bot.*, **97**: 1139-144.

Hoshino, Y., Sato, H., Murata, N. and Shinoda, K. 2000. Genetic transformation of the interspecific hybrid of alstroemeria *via* co-cultivation of embryogenic suspension cells with *Agrobacterium tumefaciens*. *Asia Pacific Conf. Plant Tissue Cult. Agribiotech.*, pp 93.

Ishikawa, T., Takayama, T. and Ishizaka, H. 1999. Amphidiploids between *Alstroemeria ligtu* L. hybrid and *A. pelegrina* L. var. *rosea* induced through colchicine treatment and their reproductive characteristics. *Scientia Horticulturae*, **80**: 235-46.

Kamstra, S.A., Kuipers, A.G.J., De Jeu, M.J., Ramanna, M.S. and Jacobsen, E. 1997. Physical localization of repetitive DNA sequences in alstroemeria: kayotyping of two species with species-specific and ribosomal DNA. *Genome*, **40**: 652-58.

Kim, J.B., Bergervoet, J.E.M., Raemakers, C.J.J.M., Jacobsen, E. and Visser, R.G.F. 2005. Isolation of protoplasts, and culture and regeneration into plants in alstroemeria. *In Vitro Cellular Developmental Biology Plant*, **41**: 505-10.

Kuipers, A.G.J., Kamstra, S.A., De Jeu, M.J. and Visser, R.G.F. 2002. Molecular characterization and physical localization of highly repetitive DNA sequences from Brazilian *Alstroemeria* species. *Chromosome Research*, **10**: 389-98.

Kuipers, A.G.J., Van, Os D.P.M., De Jong, J.H. and Ramanna M.S. 1997. Molecular cytogenetics of *Alstroemeria:* identification of parental genomes in interspecific hybrids and characterization of repetitive DNA families in constitutive heterochromatin. *Chromosome Research,* **5**: 31-39.

Lin, H.S., Van der Toorn, C., Raemakers, K.J.J.M., Visser, R.G.F., De Jeu, M.J. and Jacobsen, E. 2000. Genetic transformation of alstroemeria using particle bombardment. *Molecular Breeding*, **6**: 369-77.

Lu, C. and Bridgen, M.P. 1996. Effects of genotype, culture medium and embryo developmental stage on the *in vitro* responses from ovule cultures of interspecific hybrids of *Alstroemeria. Plant Sci.*, **116**: 205-12.

Miyake, I. 1989. Breeding spotless alstroemeria in Japan. *Herbertia*, **45**: 40-44.

Ramanna, M.S. 1991. Somatic versus meiotic doubling of chromosomes in plants: implications for polyploidy in breeding. *In:* Kimber, G. (ed.). *Proce.*: 2nd International Symposium on Chromosome Engineering in plants, pp. 118-27.

Ramanna, M.S. 1992a. The role of sexual polyploidization in the origins of horticultural crops: *Alstroemeria* as an example. *In*: Mariani A, Tavoletti S, (eds.): *Proce.* of the workshop: Gametes with somatic chromosome number in the evolution and breeding of polyploid polysomic species: Achievements and perspectives, Perugia, Italy, pp. 83-89.

Sahin, Z.K.T. 1973. Breeding and growing alstroemerias in Europe. *Plant Life*, **1973**: 74-80.

Sanso, A.M. 2002. Chromosome studies in Andean taxa of *Alstroemeria* (Alstroemeriaceae). *Bot. J. Linn. Soc.*, **138**: 451- 59.

Sanso, A.M. and Hunziker, J.H. 1998. Karyological studies in *Alstroemeria* and *Bomarea* (Alstroemeriaceae). *Hereditas*, **129**: 67-74.

Shinoda, K. and Murata, N. 2003. Cross-compatibility in interspecific hybridization of fifteen *Alstroemeria* species. *J. Japanese Soc. Horti. Sci.*, **72**: 557-561.

Singh, A.K. 2006. Flower Crops: Cultivation and Management. Pub. New India Publishing Agency, New Delhi. pp. 9-10.

Snow, A.A. and Grove, K.F. 1995. Protandry, a neuter phase, and unisexual umbels in a hermaphroditic, neotropical vine (*Bomarea acutifolia,* Alstroemeriaceae). *American J. Bot.*, **82**: 741-44.

Stephens, J. L., Tsuchiya, T. and Hughes, H. 1993. Chromosome studies in *Alstroemeria pelegrina* L. *International J. Plant Sci.*, **154**: 565-71.

Stinson, H. L. 1942. Alstroemeria breeding possibilities. *Herbertia,* **9**: 124-25.

Strasburger, E. 1882. Über den Teilunsvorgang der Zellkerne und das Verhältnis der Kernteilung zur Zeleteilung. *Archiv für Mikroskopische Anatomie*, **21**: 476-590.

Taylor, W.R. 1926. Chromosome morphology in *Fritillaria, Alstroemeria, Silphium,* and other genera. *American J. Bot.*, **13**: 179-93.

Traub, H.P. 1943. Dichogamy and interspecific sterility. *Herbertia,* **10**: 13-14.

Uphof, J.C.T. 1952. A review of the genus *Alstroemeria. Plant Life*, **8**: 36-53.

Van Schaik, C.E., Van der Toorn, C., De Jeu, M.J., Raemakers, C.J.J.M. and Visser, R.G.F. 2000. Towards genetic transformation in the monocot *Alstroemeria* L. *Euphytica*, **115**: 17-26.

Chapter – 8

Anthurium

Anthurium (*Anthurium andraeanum* Linden) belongs to the complex family Araceae, is a slow growing perennial that requires shady, humid conditions as found in tropical forests. The name anthurium is derived from Greek word "*anthos*" meaning flower and "*oura*" tail referring to the spadix. Thus, anthurium is also known as "tail flower" (Tajuddin and Prakash 1996). Anthuriums are gaining popularity due to higher returns per unit area and their beautiful and attractive long lasting flowers. Anthurium ranks ninth in the global flower trade and commands a respectable price both for its cut flower and whole plant. Commercial production has focused on two major species viz. *A. andraeanum* and *A. scherzerianum. A. andraeanum* is grown mostly for cut flower production, the main production areas being Hawaii, The Netherlands, and some other tropical and subtropical countries. *A. scherzerianum* is sold as a flowering potted plant, with main production areas located in Europe. In India, the anthurium cut flower industry is still in its infancy. At present anthuriums are mostly grown in some small gardens and nurseries. The important states cultivating anthuriums are Assam, Kerala, Tamilnadu (Salem), Karnataka (Coorg) and Mizoram where the favourable climate exists (Agasimani *et al.*, 2011).

ORIGIN AND DISTRIBUTION

Anthurium andreanum Linden Ex Andre is named after a French Horticulturist Edouard Francois Andre (1840-1911) who discovered it in 1876 and described it for the first time. The genus includes around 1000 species (Croat, 1980). Most of them are tropical herbs originating from warm areas of Central and South America (Gantait *et al.*, 2008). The distribution of this genus extends from northern Mexico and the Greater

Antilles to southern Brazil, northern Argentina, and Paraguay (Croat and Sheffer, 1983). Most anthurium species are native of tropical rain forests and are primarily epiphytic in nature. Thus, in their natural habitat, they receive ample, frequent water with good drainage. In Brazil, it is noted the occurrence of around 130 native species (Gonçalves, 2003).

In Malaysia, anthurium is mostly grown in Pahang (Cameron highlands), Selangor, Perak and Johor mainly for cut flower production. The Netherlands is the chief supplier of the anthurium in the European market. Hawaii covers 60% export share of Japan and the rest of the west coast of the USA. The Caribbean produces the anthurium for the east coast of United States and Canada. Mauritius is one of the important producer of anthurium and it exports to Japan and Italy. While some eastern countries like Taiwan, Philippines, China and Thailand mainly produce for their local market. In India, anthurium is commercially grown in Kerala, Karnataka, Tamil Nadu, West Bengal, Maharashtra and north east India. Kerala is identified as the most desirable place for growing anthurium (Singh, 2006).

SPECIES

Less than ten species of Anthurium are cultivated, although numerous hybrids and varieties exist due to the ease of interspecific hybridization. The important species of Anthurium are *A. acutangulum*, *A. acutifolium*, *A. aemulum*, *A. amnicola*, *A. andicola*, *A. andreanum*, *A. androphyoides*, *A. bakeri*, *A. bogotense*, *A. brownie*, *A. clarinervium*, *A. comtum*, *A. corrugatum*, *A. crystallinum*, *A. digitatum*, *A. ferrierense*, *A. grande*, *A. grandifolium*, *A. hacumense*, *A. harrisii*, *A. hoffmannii*, *A. hookeri*, *A. hygrophyilum*, *A. kyburzii*, *A.* × *mortefontenense* (*A. veitchii* × *A. andreanum*), *A. pendulifolium*, *A. pentaphyllum*, *A. pittieri*, *A. pulchrum*, *A. radicans*, *A. regale*, *A. regnellianum*, *A. scandens*, *A. scherzerianum* a tropical American origin is an epiphyte (Ignasse, 1984), *A. spathiphyllum*, *A. spectabile*, *A. subsignatum*, *A. splendidum*, *A. undatum*, *A. vitchii*, *A. warocqueanum* were confined to north-western South America (Boyce, 1995), *A. watemaliense*, *A. obpyriforme* from Loja province in Southern Ecuador. The species has been confused with *A. oxybelium* Schott but shows considerable morphological and ecological differences (Leimbeck and Croat, 2002). *A. leonii* E. G. Goncalves from Minas Gerais State in South eastern Brazil (Goncalves, 2005). Seven new species from central and south America with trilobed leaves includes: *A. ancuashii*, *A. chimborazense*, *A. ternifolium*, *A. pinkleyi*, *A. constrictum*, *A. moronense* and *A. rotundatum* (Carlsen and Croat, 2004). *A. viridispathum* E. G. Gonc. Sp. Nova from Espirito Santo state of Vanuatu (Goncalves, 2005a).

FLOWER MORPHOLOGY AND BIOLOGY

Flower parts

Anthuriums are cultivated for its attractive long lasting flower which is not really a flower but an inflorescence rising from the base of a bract. Anthurium flowers are small (about 3 mm) and develop crowded in a spike on a fleshy axis, called a spadix, a characteristic of the Araceae. A common feature of the Araceae is the typical, cup-shaped inflorescence. It consists of numerous flowers closely arranged in a spadix together with an outer colourful heart shaped sheath called spathe. Flowers are sessile and embedded in fleshy peduncle. They are protogynous, hermoproditic with 4 petals, 4 stamens and a bicarpellary ovary with one or two ovules.

Anthesis

It takes about 20 to 25 days from the appearance of the bud to unfurling of the spathe. The female phase starts from the second day and extends up to 20th day from the complete opening of the spathe. The sepals and petals are rudimentary; stigma appears as a rounded protuberance on the spadix when it is mature. Since, anthuriums are bisexual and protogynous, with the spadix first producing a female phase followed by, after about a month, a male phase. This reduces self-pollination, but increases the frequency of cross-pollination in plants. Pollen matures week to ten days after stigma becomes receptive to prevent self pollination. The large variations in inflorescences are due to the variations in shape, colour and the size of the spathe and spadix. The longevity of the spathe on the plant varies from 60 to 90 days, depending upon the variety. In a study on cultivars Honeymoon Red, White, Lady Jane Pink, Chilli Red and Pink, observations revealed that anthesis and anther dehiscence occurred between 8-10 am, as they are favoured by moderately low temperatre and high relative humidity (Bindu and Mercy, 1996). Bai *et al.* (2005) observed that the pollen grains were two celled at the time of anther dehiscence in cv. Arizona. They also noted that during early morning, when the inflorescences stay at their fifth development stage, is the optimum opportunity for pistil to get pollen grains, with the suggestion to collect pollen at the end of the seventh stage.

Mode of pollination

The flower structure suggest that anthurium is out breeder, cross pollination often takes place by insects such as bees, beetles and flies. The coloured spathe attract birds thereby spreading the seeds. Croat (1980) reported that in a study of about 30 species there is difference between species in the rate of flower development. Of the species with floral scent, those smelling yeasty are probably fly pollinated, and those with sweet aromas probably bee pollinated, large numbers of bees are sometimes seen on the flowers of some species. The best time for pollination is when mature stigmas on the spadix produced a sticky material.

Cyclanthura flower weevils (Coleoptera: Curculionidae: Derelomini) are identified for the first time as pollinators of multiple species of anthurium (Araceae) in Costa Rica (Franz 2007). The weevils are present on the inflorescences in small numbers during the pistillate and staminate phase of anthesis and consume plant tissues and pollen. The individuals of one species of *Cyclanthura* can visit several *Anthurium* species within the same locality. Eeuglossines bees are the common visitors of members of this aroid clade (Williams and Dressler, 1976). Other floral visitors included chrysomelids, nymphs of hemipterans, small parasitoid wasps, and thrips (but not euglossines). None of these are sufficiently behaviorally synchronized with the anthesis of anthurium to act as reliable pollinators. In contrast to this, many species of derelomine flower weevils are pollinators of such morphologically complex taxa as Arecaceae and Cyclanthaceae (Henderson, 1986; Eriksson, 1994 and Franz 2004).

Pollen viability

Pollen matures week to ten days after stigma becomes receptive to prevent self pollination. The large variations in inflorescences are due to the variations in shape, colour and the size of the spathe and spadix. The longevity of the spathe on the plant varies from 60 to 90 days, depending upon the variety.

Stigma receptivity

The stigma becomes receptive soon after unfurling of the spathe. The pollen is shed after 12-15 days after opening of the flower and production of powder like pollen can be observed on the spadix.

Seed set

After pollination, coloured berries develop on the spadix bottom upwards. Each berry contains one or two seeds that can be planted to generate heterozygous and heterogeneous seedlings.

GENETICS

The basic chromosome number of anthurium is n= 15, 16, 22 with *A. andreanum* as 2n= 30 (Sharma and Bhattacharya, 1966). However in a study for determining chromosome numbers it was found to be ranges from 2n=24 to 66, with 30 being the most common. Paleoneuploidy, polyploidy and B- chromosomes are basic features of the genus (Sheffer and Croat, 1983, Sengupta *et al.*, 1990 and Bindu and Mercy, 1998) reported that somatic chromosome number for *Anthurium andraeanum* was 2n= 30 + 2B. In *A. andraeanum*, a somatic chromosome number of 30, with one or more B chromosomes have been reported. Triploid cultivars were also reported by Ali (1979) in *A. crystallinum*.

Genetics of the major spathe colours

There are two major genes, M and O that are responsible for five major colours of *A. andraeanum*: red, orange, pink, coral and white. Gene M controls the production of cyaniding 3- rutinoside and O controls that of pelargonidin 3- rutinoside. Red and pink result when both M and O are present and orange and coral result when only O is present. While double recessive mmoo results in white colour of spathes (Kamemoto *et al*., 1988) and the spathe is purple when the genotype is M-O-pp (Wannakrairoj *et al*., 1990). The inheritance pattern of different colours is as follows.

a. Red × Red: The progenies are red or segregate into red and orange with red dominant over orange.

b. Red × Pink: The progenies are all red or segregate into red, orange and pink.

c. Red × Orange: The progenies are all red or segregate into 1:1 ratio of red and orange.

d. Pink × Pink: The progenies produce 3 type segregations, red, orange and white (di-hybrid ratio) indicating that multiple allele system does not work here.

e. Pink × Orange: The progenies show 1:1 ratio of red and orange groups, the red groups includes red and pink and orange includes orange and coral.

Flavonoid and senescence associated gene expression

Collette *et al*. (2004) observed the temporal and spatial expression of anthocyanin biosynthesis during flower development in anthurium, where a cDNA library was constructed from mRNA from the spathe and full length cDNA clones identified for the flavonoid biosynthetic enzymes. They concluded that control of anthocyanin biosynthesis in anthurium spathe differs from that described for flowers of other species. Hayden *et al*. (2004) studied the senescence regulation signaling pathways and genes for future efforts to modify leaf senescence *via* biotechnology. Results indicates that a senescence activated pathway correctly regulated the Arabidopsis sag12 cysteine protease gene and suggested the presence of sag12 protease in anthurium.

Heritability and genetic advance

In the study six F_1 hybrids were produced from three inter-specific crosses involving *Anthurium andraeanum* Gloria Angara, *A. andraeanum* Pink Exotica, *A. crystallinum* and *A. hookeri* were characterized morphologically and cytologically, and compared to each other and to their parents. It was found that suckering ability in F_1-4 and F_1-5 was inherited from the male parent, *A. andraeanum* Gloria Angara. Spathe and spadix colours varied and distinctively different from those of the parental species. F_1-1, F_1-2 and F_1-3 have black to plum purple spadices not found either in *A. hookeri* and *A. crystallium*. These interspecific hybrids do not produce pollen but F_1-1 and F_1-4 were observed to produce fruits by open pollination (Aurigue, 1993).

Histological analysis of somatic embryos derived from *in vitro* cultured laminas of *A. andraeanum* showed bipolarity with the presence of shoot and root poles connected by procambium. Storage of proteins, starch and raphides as well as suspensor like structure and an epidermis were observed in the somatic embryos (Matsumoto *et al.*, 1996).

BREEDING OBJECTIVES

i. Improved flower longevity

ii. New flower colour especially of dark colour with puckered surface

iii. Varieties with short internodes

iv. More vigorous growth

v. Varieties with more number of flowers

vi. Resistant from anthracnose disease

vii. Resistant from blight

viii. Resistant from insects pests

ix. Shortened juvenile period

x. Fast multiplication rate

BREEDING METHODS

Breeding programme is essential for any flower crop, essential to develop facilities and technologies and protocols for *in vitro* propagation of selected hybrids and varieties. A large number of species closely related to *A. andreanum* which are cross compatible are available to the anthurium breeders. Such large gene pool offers vast potential to develop new varieties.

Selection

A variety Gemini was selected from cv. White Aristocrat (Hartman and Lamb, 1996). It has glossy dark pink spathes with a contrasting white and yellow spadix, the spathes being carried above the foliage; the growth habit is moderately branched and somewhat spreading.

Another var. Northstar was also selected from cv. White Aristocrat and released. This variety has heart shaped light pink spathes which fade to white, and a pink spadix, with the spathes being held well above the foliage in the centre of the plant, inflorescence has a light sweet, mint fragrance, plant is moderately branched (Hartman and Lamb, 1998). Several varieties have been developed through selection from University of the Philippines, Los Banos and IAC, Brazil with desirable features. Some of them are listed in Table 1 and 2.

Table 1: Varieties evolved through selection from University of Philippines, Los Banos.

S.l. No.	Variety	Selected from	Characteristics
1.	UPLB White	Germplasm collection	Spathe white with occasional pink blushes at maturity, slightly ridged with medium texture, large to extra large in size, broad heart in shape, open lobes, spadix reddish pink and reclining.
2.	Eternity	Open pollinated of *A.* 'Blushing Innocence'	Spathe light pink, slightly ridged with medium texture, medium to large in size, broad heart in shape with closed lobes. Spadix reddish brown.

Table 2: Varieties evolved through selection from IAC, Brazil.

S.l. No.	Variety	Characteristics
1.	IAC Astral (IAC 154)	Spathe of strong and luminous coral colour, spadix white/yellow, plant productive and tolerant to bacteriosis. Good for pot culture, cut flower, vase life over 20 days
2.	IAC Cananeia (IAC 16772)	Large spathe, white colour (greenish on the borders of very vigorous plants), spadix rose coloured, plant vigorous with rapid growth, productive cut flower, more vase life.
3.	IAC Eidibel (IAC O-11)	Plant medium size, very productive; medium-sized, heart-shaped spathe, strong red colour, thick texture, spadix white, slightly scented; cut flower with long post-harvest durability.
4.	IAC Ômega (IAC 14021)	Large spathe of coral colour with spadix white/yellow, cut flower with long post-harvest durability
5.	Iguape (IAC 17236)	Spathe medium size, good texture, brilliant dark red colour having spadix cream colour with rose orangish shades. Cut flower, vase life 20 days
6.	Isla (IAC 14018)	Spathe roundish, white colour with greenish borders and spadix white/yellow a tall plant.
7.	Jupiter (IAC 17237)	A tall plant having spathe white with spadix pinkish.
8.	Juquia (IAC 17260)	Spathe coral with spadix white/yellow, small plant with strong sprouting and good as a pot plant.
9.	Jureia (IAC The-5)	Spathe brilliant coral colour, white/yellow spadix, plant medium vigour, cut flower, vase life 30 days.
10.	Luau (IAC N-15)	Spathe medium size, good texture, white colour; spadix almost totally white. Good as cut flower.
11.	Netuno (IAC 16770)	Plant medium vigour, medium size, dark brown, heart-shaped spathe white spadix with yellow-greenish tip.
12.	Rubi (IAC 14019)	Large spathe red colour and spadix white/yellow, tall plant, cut flower with long post-harvest durability.

Hybridization

Breeding technique

Artificial pollination: Anthurium produces bisexual flowers. Its inflorescences consist of a spadix enclosed by a spathe. The spadix is a fleshy spike covered with many small, complete flowers. The spathe unfurls and exposes the spadix several days before flowers are receptive for breeding. The unfurling of the spathe reveals numerous, uniformly sized bisexual flowers located along the entire spadix. New flowers become receptive each day, beginning at the spadix base and advancing gradually towards the tip over a two-week period. Pistillate (female) receptivity is indicated by a glistening shine of stigmatic surfaces and stickiness to the touch. Small drops of exudate are sometimes present. Flowers may remain receptive for more than one day.

The stigmatic surfaces become dry and brown before pollen is dehisence, so floral emasculation is not required. *Anthurium* pollen begins to appear along the spadix, usually at the bottom first and then proceeds towards the tip. Pollen is produced for up to two weeks because of the differences in the maturation of individual pistillate flowers. Pollen is not dispersed by wind. To transfer the pollen to the receptive female flowers, it should be collected on fingertips then gently rubbed it onto receptive stigmas.

Storage of pollen: It is always best to use fresh pollen. However, if pollen is in short supply, it can be stored in a container in a high-humidity environment in a refrigerator. Collect the pollen in a container, such as a petri dish. Place a lid over the petri dish, and set it on top of a wet paper towel or other moistened material. Enclose the wrapped petri dish within a sealed storage container. At no time should the pollen be directly in contact with the wet paper towel. Avoid splashing water droplets onto the pollen.

Seed set and production: To achieve maximum seed production, spadix needs to be pollinated more than once. To ensure seed set, no special environmental manipulation such as controlling relative humidity is needed after pollen transfer. Pollinated flowers develop mature fruits within four-to-six months. Fruit will require up to five-to-eight months to mature. As seeds mature, spadices begin to turn to a yellowish colour and soften.

Hybridization among different coloured spathe is being taken by hobbyist and growers. Improved spathe colours in shades of red, orange, white are receiving more attention. A large number of growers are introducing Hawaiian, Dutch and Mauritius hybrid for their breeding programmes. Varieties like Marian See Furth, Nitta, Lady Jane, Southern Blush etc. are freely used in breeding programme. Varieties developed by different organizations and breeders are given in Table 3, 4, 5, 6 and 7.

Table 3: Hybrid varieties of anthurium with characteristics.

S. No.	Variety	Parent	Chief characteristics
1.	Kalapana	Paradise Pink × Diamond Jubilee	Plants medium and bloom red-green, 17 cm flowers. Vigorous growth habit, produces 6 flowers/year. The spathe grows on long stems. Resistant to anthracnose and bacterial blight.
2.	Tropic Mist	Sea Pearl × Manoa Mist	A cream coloured, round spathes which grown at 30 to 45 degree angle from the stems.
3.	Tropic Ice	Anuenue × UH515	A compact variety with small foliage. Spathe white-green. About 9 flowers/year. Resistance to bacterial blight and anthracnose.

Table 4: Hybrids developed from IIHR, Bengaluru, India.

S. No.	Variety	Year	Characteristics
1.	IIHR 26	1990	Plant is erect with short stem and internodes. Very attractive with orange coloured spathe and reclining spadix. Yield 8.3 flowers/year.
2.	IIHR 139	1991	A floriferous hybrid, spath is white heart shaped with reclining yellow orange spadix. Yield 7.2 flowers/year.
3.	IIHR 243	1992	Flowers are bright red with puckered, heart shaped spathe and slightly overlapping basal lobes. The spadix is yellow reclining and very ideal for packing.
4.	IIHR 51	1992	White spathe and pink spadix. The flowers are highly fragrant spread a pleasant chocolate fragrance. Yield 11-12 flowers/year.

Table 5: Varieties of anthurium developed from University of Hawaii.

S. No.	Variety	Year	Chief characteristics
1.	Arcs	1988	A tulip type, purplish spathe and spadix. The spadix is darker than the spathe. Cultivar is a complex hybrid of *A. andreanum* cv. Uniwai, *A. kamemotoanum*, *A. formosum*, and *A. amnicola*. Resistant to anthracnose. 7 flowers/year.
2.	Ellison Onizuka	1986	The spathe length 20 cm, width 15 cm. Resistant to anthracnose.
3.	Madam Butterfly		Miniature, butterfly, red obake. Spathe about 6 cm wide and 7.5 cm long. Cultivar is also known as 'Mickey Mouse'.
4.	Marian Seefurth	1963	High yielding cultivar, 6.9 inflorescences/year (Kamenoto, 1981). Highly resistant to anthracnose, susceptible to bacterial blight.
5.	Mauna Kea	1977	Green obake, but the green is not very prominent in young plants.
6.	Midori		Glossy green spathe which has overlapping lobes, spadix green. Excellent vase life.

Table 6: Hybrid varieties of anthurium developed by some breeders.

S. No.	Variety	Breeder & release year	Parent	Chief characteristics
1.	Atwelve	Marian W. Osiecki (1999)	A2- No. 10,210 × Seedling selection No. 91-94-2	Dwarf growth habit, early and year-round flowering, very good longevity, red-coloured spathes and contrasting lighter coloured spadices.
2.	Aeighteen	Marian W. Osiecki (1999)	A2-No. 10,210 × Seedling selection No. 91-94-2	Dwarf growth habit, early flowering, purple-coloured spathes and fairly light-coloured spadices, very good longevity maintaining lavender colour for about five weeks.
3.	Red Love	Leonardus W.B.M. van Rijn (1998)	No- 924 × No- 9319	Freely branching plant habit, numerous inflorescences that are positioned high above the foliage on long, strong and erect scapes, broad glossy red spathes with auriculate base, and good post-production longevity.
4.	Orange Love	Magdalena J.M. Van Rijn (1998)	No- 924 × No- 9322	Upright, outward arching and freely branching plant habit, broad orange spathes with auriculate base, and good post-production longevity.
5.	Sugar Love	Magdalena J.M. Van Rijn (1998)	No- 923 × No- 9322.	Compact, upright, outwardly arching and freely branching plant habit, broad salmon-pink spathes with truncate base and good post-production longevity.
6.	Lady Love	Leonardus van Rijn (1999)	No- 92-4 × No- 94-13	Upright plant habit, freely clumping growth habit, glossy pink spathes with green lobes; year-round continuous flowering and good post-production longevity.
7.	Atwenty	Marian W. Osiecki (2000)	A2- No. 10,210 × Seedling selection No- 91-94-2	Dwarf growth habit, early and freely flowering habit, Large spathes in proportion to the overall plant size, salmon pink-coloured spathes with various amounts of green colouration in the lobes, bright yellow-coloured spadices and long-lasting spathes.
8.	Fresh Love	Leonardus W.B.M. van Rijn (2001)	No- 97-015 × No- 97-010	Spreading plant habit, freely clumping growth habit, white and green-coloured spathes with light orange-coloured spadices, freely flowering habit and good inflorescence longevity.
9.	Tender Love	Leonardus W.B.M. van Rijn (2003)	No- 9608 × No- 9715.	Upright and outwardly spreading plant habit, freely clumping growth habit, dark pink and green bi-coloured spathes with light orange-coloured spadices, freely flowering habit, and good inflorescence longevity.
10.	Changing Love	Leonardus W.B.M. van Rijn(2003)	No- 9715 × No- 9713	Upright and outwardly spreading plant habit, Freely clumping growth habit, greenish white, Green and light red tri-coloured spathes with grayed purple veins and grayed purple-coloured spadices, freely flowering habit and good inflorescence longevity.

Table 7: Interspecific hybrids (Source: University of Florida, IFAS Extension, 2005).

S. No.	Hybrid	Institute	Name	Chief characteristics
1.	*A.* × Bubble Gum	-	Bubble Gum	Pale pink spathes with pink spadicies. Muted leaves.
2.	*A.* × Cotton Candy	-	Cotton Candy	Light or salmon pink spade-shaped flowers with yellow spadicies, foliage green.
3.	*A.* × Favorita	-	Favorita	Solid white spathes and spadicies with deep green foliage.
4.	*A.* × Gemini	-	Gemini	Foliage dark green, flower brick-red to pink red, green or yellowish spadicies.
5.	*A.* × Kingston	-	Kingston	Paler red spathes, pale pink spadicies.
6.	*A.* × Krypton	-	Krypton	Deep red spathes and spadicies.
7.	*A.* × Lady Jane	-	Lady Jane	Pink spathes and spadicies with dark green foliage.
8.	*A.* × Lady Ruth	-	Lady Ruth	Rose coloured spathes with whitish spadicies.
9.	*A.* × Nicolien	-	Nicolien	Mint pink spathes with similar spadicies. Dark green foliage.
10.	*A.* × Nicoya	Oglesby	Nicoya	Deep pink spathes, white spadicies. Round shaped flower, crimson red with a contrasting pink to white spadix.
11.	*A.* × Pacora	Olgesby	Pacora	Red spathes, with white spadicies, pot type anthurium.
12.	*A.* × Picante	-	Picante	Dark fuchsia spathes, purple spadicies, dark foliage.
13.	*A.* × Pink Frost	Twyford International	Pink Frost	Bright pink spathes, white spadicies.
14.	*A.* × Purple Plum	-	Purple Plum	Purple spathes, spadicies, bright green foliage.
15.	*A.* × Red Hot	University of Florida	Red Hot	Excellent flower count, compact, tough foliage with a symmetrical growth habit; brilliant red tulip shape flower.
16.	*A.* × Salsa	-	Salsa	Very deep red spathe and swirled shape, yellowish flower stalks, good flower count, it branches freely, good disease resistance.
17.	*A.* × Small Talk (Pink)	-	Small Talk (Pink)	Prolific and very early pink blooming, heart shaped flowers. Good for pot culture.
18.	*A.* × Sundial	-	Sundial	Bright red spathes, yellowish spadicies. Dark foliage.
19.	*A.* × Orange Hot	-	Orange Hot	Sort from 'Red Hot', good flower count, very symmetrical foliage canopy.
20	*A.* × Tropic Fire	-	Tropic Fire	Brilliant red spathe with white spadicies.
21.	*A.* × Sarah	Florida Nursery Growers Landscapers Association	Sarah	Extra large pink obaki (bicoloured) blooms, large leaves. Grows very big.

Development of disease resistant varieties: Anthurium is affected by bacterial blight (*Xanthomonas campestris* Pv. *Dieffenbachiae*) and anthracnose is also known as black nose or spadix rot, caused by *Colletotrichum gleosporioides*. Breeding work carried out by various organization to develop disease resistant cultivars. Varieties evolved resistant to anthracnose and bacterial blight are Kalapana, Tropic Ice, whereas var. IAC Astral is blight resistant and cvs. Arcs and Tropic Ice are resistant to anthracnose.

Mutation

Since mutation breeding has been more successful in ornamental plants because changes in phenotypic characteristics like colour, shape or size of flower and chlorophyll variation in leaves can be easily detected. Mutation breeding is solely based on artificial mutations. Mutagenic agents have been used to induce useful phenotypic variations in plants for more than 70 years (Foster and Twell, 1996). *In vitro* plantlet regeneration in a number of anthurium (*Anthurium andreanum*) varieties was achieved from callus cultures derived from young, tender leaf explants on Nitsch's (1969) medium. When explants (leaves, seeds, *in vitro* plantlets) were irradiated, best response was observed with the 5 Grays (Gy) treatment in terms of callus formation and regeneration while the 15 Gy dose was lethal to the anthurium tissues. The phenotypic results indicated a boosting effect of the 5 Gy dose on the leaf tissues (Puchooa, 2005). Effect of gamma rays on mutation of anthurium Double Spathe cultures at the doses of 0, 1, 2, 3, 4 and 5 gray were studied under aseptic condition also the nodes and callus of anthurium were studied on effect of colchicines. Results shown that the dose of 5 gray caused the highest morphological abnormalities and the survival rate and growth were reduced when concentration of colchicines were increased, it also tends to cause chimera on leaves (Wichuta, 1994). Some varieties also evolved through mutation (Table 8).

Table 8: Mutant varieties of anthurium.

S. No.	Variety	Breeder &reference	Parent	Chief characteristics
1.	Lucky Leny	Leonardus W.B.M. van Rijn(2001)	Leny	Upright and somewhat outwardly spreading plant habit, dark red-coloured spathes with cream to light tan-coloured spadices that are positioned upright and beyond the foliage on strong and erect scapes and good inflorescence longevity.
2.	Exciting Love	Leonardus W. B. M. van Rijn(2003)	Red Love	Upright and outwardly spreading plant habit, Freely clumping growth habit, Grayed purple-coloured spathes with green-coloured margins and pale yellow-coloured spadices, Freely flowering habit, Good inflorescence longevity.

BIOTECHNOLOGY

Tissue Culture

Propagation by tissue culture techniques appears as an alternative to increase the production. The success of tissue culture is related to the correct choice of explants material (George *et al*., 2008). Micropropagation of anthurium is commercially used (Hamidah *et al*., 1997 and Martin *et al*., 2003) also an alternative method of vegetative propagation which is well suited for the multiplication of elite clones (Silva *et al*., 2005). Micropropagation of anthurium has been achieved with various tissues including leaf, petiole, spadix, spathe, seed, lateral bud and shoot tips. Plant regeneration of *A. andraeanum* has been achieved through adventitious shoots formation from callus and direct shoot regeneration from lamina explants. The tissue culture of anthurium was first reported by Pierik *et al*. (1974) and they used liquid culture to proliferate callus (Martin *et al*., 2003). In anthurium, the liquid or raft culture instead of solid medium was used on the regeneration of adventitious shoots from leaf explants (Teng, 1997). The method for *in vitro* production of plantlets of *Anthurium andreanum* was first developed by Pierik *et al*. (1974). The production of *in vitro* plants directly from proliferating axillary buds (Kunisaki, 1980), adventitious buds (Cen *et al*., 1993), leaf or petiole organogenic callus culture (Pierik *et al*., 1974; Pierik, 1975, 1976; Finnie and Van Staden, 1986 and Kuehnle and Sugii, 1991) and from somatic embryos derived from *in vitro* grown leaf blade explants (Kuehnle *et al*., 1992) has been reported.

Vargas *et al*. (2004) established an alternative method for regeneration of anthurium plants. They obtained *in vitro* plants from germinated seed and plantlets obtained from micro-cuttings culture. These plantlets showed callus at the stem base. Micro propagation by callus induction from leaf culture was difficult step for *in vitro* propagation. Various physical and biological factors including media play role during anthurium *in vitro* propagation (Silva *et al*., 2005). One of the important factors was genotype for the *in vitro* multiplication. Nhut *et al*. (2006) showed that genotypes of 10 anthurium cultivars had different response to callus induction and shoot regeneration. Bejoy *et al*. (2008) used 1-1.5 cm^2 lamine segments of *A. andraeanum* cv. Agnihothri on half-strength MS basal medium. They compared the regeneration efficiencies of different concentrations of plant growth regulators. Martin *et al*. (2003) cultured brown and green lamina explants of *A. andraeanum* cultivars Tinora Red and Senator and found that the explants from brown young lamina of Senator and Tinora Red showed 60% and 70% callus formation and green lamina explants showed 40% and 45% callus formation respectively. Bejoy *et al*. (2008) reported that pale green leaf explants were better and showed good callus development in the initial medium.

Nhut *et al*. (2006) studied the effects of 10 different anthurium genotypes on callus induction derived from leaf explants. 'Arizona' callus induction rate was 9.8% after 60 days and 10.7% after 100 days. They reported 65.1% explants of Pistache induced calli after 100 days of culture. Also, they realized that this process was best for Pistache, not for the other anthurium cultivars. Callus induction from leaf explants depends strongly on genotype.

Te-chato *et al.* (2006) used leaves, node and internodes of three genotypes of anthurium (Plew Thien Phuket, Sonat, Valantino). They cultured the explants Murashige and Skoog medium (MS) (Murashige and Skoog 1962), modified MS (MMS), woody plant medium (WPM) and Nitsch & Nitsch medium (NN) and observed highest callus formation of 86.6% from leaf explants in MS and MMS medium. Callus formation of cv. Valantino was higher according to the others.

Martin *et al.* (2003) cultured their regenerated *A. andraeanum* plantlets in half strength MS medium supplemented with 0.54 μM NAA and 0.93 μM Kinetin for *in vitro* rooting. *In vitro* rooting percentage of Tinora Red and Senator varieties were 100% and 99%. Feng *et al.* (2006) reported that, among the regenerated anthurium shoots, approximately 50% had roots in raft culture. Bejoy *et al.* (2008) achieved about 98% root from the newly formed shoots in 0.5 mg/l NAA in half-strength MS medium in six weeks. Vargas *et al.* (2004) achieved regeneration of *A. andraeanum* from callus tissue. Subculture of calli generates a higher number of shoots per callus. Four weeks later they observed an average of 3.6 shoots per explants produced and 6 weeks later, the calli increased in its size and 43.8 plantlets per callus were obtained. Bejoy *et al.* (2008) reported that multiplication was enhanced in the subculture of *Anthurium andraeanum* Hort. cv. Agnihotthri. They succeeded to increase the rate of shoot production in the second multiplication stage with the average of 9.7 shoots per explant in 60 days. Spadix explants were found to had a better capacity for regeneration than leaf segments on the modified Nitsch media and the plantlets derived from spadix segments were less variable (Singh *et al.*, 1991).

Seeds sterilized in 3% NaOCl for 30 min produced the highest percentage of germination while sterilizing for 15 min resulted in the highest seedling height and leaf number in a study of micropropagation of *A. andraeanum*. It was also found that multiple shoots were reduced from shoot tips and nodes on MS medium supplemented with 1.0 mg BA/litre (Somaya *et al.*, 1998). Montes *et al.* (1999) developed a methodology for micropropagation of *A. cubense* in which all explants were found to be healthy and seed germination was 100% when the seeds from physiologically ripe inflorescences were disinfected by dipping in 70% alcohol for 1 s, dipping in sodium hypochlorite for 5 minutes, then rinsing 3 times with distilled water, again dipping in sodium hypochlorite for 8 minutes and rinsing lastly for 3 times. More than 15 adventitious sprouts/explant were produced within 60 days on Murashige and Skoog medium. Transient expression of uidA reporter gene was found in *Anthurium andraeanum* variety Avo Nette when callus was induced from shoot bases and leaf blades of on modified MS media containing 0.3 mg 2, 4- D/litre and 0.5 mg benzyladenine /litre. Plant regenerated from this was obtained at a high frequency in all callus types, irrespective of their origin, in modified MS media containing 0.5 mg benzyladnine/litre (Yakandawala *et al.*, 2000).

Prabhakara *et al.* (2001) standardized media for *A. andraeanum* seed germination where subsequent growth was carried out by using 5 different media: full strength MS, half strength MS, Morel, Knudson 'C' and Nitsch. The quickest germination (10.45 days) was obtained on half strength MS medium whereas delayed germination (14.53 days) was recorded in Morel's medium. However, the earliest 3rd and 4th leaf emergence, highest shoot, longest root and maximum fresh and dry weight were recorded from Nitsch medium. Paswan and Mahanta (2001) studied the effect of the growth regulators, NAA and BAP at different concentration on germination, growth and development of *A. andraeanum* cv. Agnihotri and found that 90% germination of seeds was observed within 8-10 days in modified MS medium supplemented with BAP, which developed into 3rd leaf stage within 61 days. Yu *et al.* (1995) found callus culture to be more promising than shoot culture for micropropagation in cultivars Belinda for pot plant production and Hazrija and Ingrid for cut flower production. *In vitro* plant generation of *A. andraeanum* cultivars Pumasillo and Corallis have been achieved through adventitious shoots formation *via* callus culture. Both the varieties were responded differently to callus induction, callus induction, callus proliferation and shoot initiation treatments (Kumari *et al.*, 2011). Somatic embryogenesis was induced in *A. andraeanum* Hort cv. Pumasillo, in which young brown leaf lamina were used as explants. The germination of somatic embryos was obtained in culture media supplemented with 3 mg/l BAP and 5mg/l adenine (Kumari *et al.*, 2011).

Somatic embryogenesis

Several types of explants, growth regulators, sugars and gelling agents were tested to induce somatic embryogenesis in *Anthurium scherzerianum* and somatic embryos were converted to entire plants on a medium with 0.1 mg/litre kinetin (Hamidah *et al.*, 1995). In a study of histological analysis of somatic embryos derived from *in vitro* grown lamina of *A. andraeanum* showed bipolarity with the presence of shoot and root poles connected by procambium. Both modes of somatic embryogenesis arose from the mesophyll (Matsumoto *et al.*, 1996). In inducing cyclic somatic embryogenesis of *Anthurium scherzerianum* Schott it was found that cyclic embryogenesis occurred independently of the orientation of the explants and all embryos that reached maturation produced shoots which were phenotypically normal. The embryogenic and regeneration capacity during the 5 successive cycles, which took approximately 2 years, was unreduced (Hamidah *et al.*, 1997). Somatic embryos regenerated entire plants on a medium with 0.46 5M kinetin when leaf pieces were micro propagated in 18 5M 2, 4-D and a high sucrose concentration (6%), solidified with Gelrite (Hamidah *et al.*, 1997). Embryogenesis was induced with cvs. UH965, UH1060 and UH780 using a modified half strength MS medium supplemented with 1-4 mg 2, 4- D and 0.33- 1 mg kinetin/litre when their leaf blade explants were grown *in vitro*. Whole leaf blades on medium solidified with 0.18% Gelrite produced more somatic embryos than leaves on medium with 0.7% Bacto- agar (Kuehnle *et al.*, 1992).

Several concentrations and different types of growth regulators were tested to induce indirect somatic embryogenesis in *Anthurium andraeanum* Lind. where the differentiation of somatic embryos were achieved in the culture medium MS modified containing 4, 44 micro M of 6 benzylaminopurine under conditions of darkness (Bautista *et al.*, 2008). Raad *et al.* (2012) studied the effect of cultivar, type and age of explants, light conditions and plant growth regulators (NAA and BA) on callus formation of *A. andreanum* Linden cultivars Casino and Antadra results showed that the explants source and age, growth conditions, type of cultivar and different concentrations of growth regulators significantly influenced on callus production. Segments of lamina and petiole (micro-cuttings or explants) were cultured in MS basal medium with different concentrations of NAA (0.01, 0.1, 0.5, 1 and 2 mg/l) and BA (0.5, 1, 2 and 3 mg/l) to produce callus. Younger explants grown in dark conditions exhibited better responses to callusing than that of older explants and grown in light conditions. Callus production was better in Antadra than that of Casino. For explants types, permanent callus was produced on the lamina explants, not on the petiole explants. Plant regeneration of *A. andraeanum* cv. Rubrun was studied where, seeds from plant spadixes were germinated on a medium supplemented with 2.2 µm B.A. and after 2 weeks, 74% of the seeds germinated and four weeks later, micro-cuttings from these plantlets were subcultured on 4.4 µm B.A. and 0.5 µm NAA. Anatomical studies showed the organogenic nature of these calli (Vargas *et al.*, 2004).

Molecular breeding

Randomly amplified polymorphic DNA (RAPD) markers were utilized to determine the genetic relationships of nine morphologically similar pot plant cultivars of *Anthurium* sp. by developing DNA fingerprints (DFP). Of 25 arbitrary primers screened, nine generated DFPs that were used in computing the genetic distance (d) and similarity coefficient (C) values. All cultivars tested exhibited a high degree of genetic similarity. 'Lady Ann' and 'Lady Beth' possessed the closest relationship with d and C values of 0.06 and 0.98, respectively. The next closest genetic relationship was between 'Red Hot' and 'Southern Blush' (d = 0.33, C = 0.89). These two cultivars exhibited a more distant relationship to the other seven cultivars as indicated by higher 'd' values. However, this study showed that the nine *Anthurium* cultivars examined were genetically closely related. These cultivars share specific DNA bands with three possible parental species (*A. andraeanum* Linden ex Andre, *A. antioquens* L., and *A. amnicola* Dressler) included in this study, which may indicate similarities in their pedigree. This study shows that RAPDs can be a useful tool to distinguish *Anthurium* pot plant cultivars as well as identify their genetic relationships (Ranamukhaarachchi *et al.*, 2001).

Randomly amplified polymorphic DNA (RAPD) markers fingerprinting were used to assess the level of genetic variations among 24 cut-flower *Anthurium andraeanum* Hort. cultivars. Eight decamer primers produced a total of 98 reproducible PCR bands that were used to calculate the Nei and Li's genetic distance (GDNL) coefficients amongst the cultivars. GDNL values ranged from 0.018 to 0.163 with an average of 0.09 (representing an average genetic similarity of 91.34%). This significantly low average genetic distance among the various cultivars indicated that genetic variation among the cultivars was low. A dendrogram, produced using unweighted pair group method using arithmetic averages (UPGMA), grouped the cultivars into four main clusters. Cultivar 'Antartica' was genetically distinct from all the others. 'Midori' and 'Bourgogne' together formed a cluster whereas the remaining 21 cultivars grouped into two clusters and were closely related to each other. Clusters did not relate to cultivar provenance or origin and were independent of floral colour and spathe category. Finding correlations between these morphological traits to RAPD markers would necessitate extensive primer screening. Nevertheless, RAPD markers fingerprinting allowed a rapid assessment of the level of genetic variation that would otherwise be difficult to evaluate using the limited number of morphological markers present among these closely related anthurium cultivars (Nowbuth *et al*., 2005).

Castro *et al*. (2004) discussed the applications of RFLP (restriction fragment length polymorphism), RAPD (random amplified polymorphic DNA), micro satellite or SSR (simple sequence repeats), AFLP (amplified fragment length polymorphism), SCAR (sequence characterised amplified regions) and STS (sequence tagged site) techniques for studies of genetic polymorphism of *Anthurium* genotypes.

Ge *et al*. (2012) comparatively investigated genetic variations among a collection of anthuriums using morphological traits and amplified fragment length polymorphism (AFLP) markers. Both morphological and AFLP-based data clustered the accessions from *Anthurium andraeanum*, *Anthurium scherzerianum* and *Anthurium clarinervium* into three separate groups, with a significant though low correlation (r=0.3) observed between the morphological and AFLP-based similarity matrices. The AFLP-based principal coordinate analysis (PCoA) divided the entire accessions into three parts, reinforcing the AFLP-based clustering. Moreover, the Bayesian-based structuring assigned the anthurium accessions to three subgroups but with 27 accessions retained in the admixed groups, probably implying the highly heterozygous genome of the investigated anthuriums.

Genetic engineering

Griesbach (1994) developed a system for transfering plants through electrophoresis by introducing foreign DNA into protocorms of Calanthe sedenii using plasmid DNA (pBI-121) containing the gusA gene under control of CaMV 35S promoter. Approximately 45% of the protocorms were alive one month after treatment. Transformation was confirmed through polymerase chain reaction analysis.

Variety Paradise Pink has a pink heart-shaped spathe and Tropic Flame has a red tulip-type (cupped) spathe (Kamemoto and Kuehnle, 1996). These were engineered to produce the cecropin-like Shiva-1 lytic peptide in leaves, through *Agrobacterium*-

mediated transformation with the plasmid pBPRS1 containing the PR1 secretion signal for Shiva-1, driven by the double CaMV 35S promoter.

Development of transgenic

The first reports on the production of transgenic anthurium plant carried out by Chen and Kuehnle (1996). They reported the production of transgenic anthurium through *A. tumefaceins*- mediated transformation. Etiolated internode explants were co-cultivated with *A. tumefaceins* strain LBA404 carrying the binary vector pCa2Att. This vector contains the antibacterial gene attacin, the kanamycin resistance gene (*npt*II) for selection of transgenic events and the GUS reporter gene. Both Rudolph and UH1060 anthurium cultivars were used and produced trangenic plants. Chen *et al.* (1997) also generated transgenic plantsvia the transformation of root cuttings using the strain/vector LBA4404/ pCa2Att. Five cultivars UH1003, UH1060, Anuenue, Rudolph and Mauna Kea were tested from which only Anuenue producing transgenic shoots at an efficiency of 1.3% (Table 9).

Table 9: *Anthurium* species from which transgenic plants have been generated.

Plant species	Transformation method	References
Anthurium andraeanum	*Agrobacterium tumefaciens*	Chen and Kuehnle (1996) Chen *et al.* (1997)

Breeding for blight resistance

In vitro culture promotes accelerated multiplication of superior clones and is a pre-requisite for the improvement of plants via genetic engineering technique. Kuehnle *et al*. (1990) studied the genetic engineering of blight resistance in anthurium in which two gene transfer systems tested and found both promising for introducing for resistance to blight (*Xanthomonas campestris* pv. dieffenbachiae) into anthurium.

Kuehnle *et al*. (2001) worked out to evolve disease resistant variety in anthurium. Genome breeding using *Anthurium antioquiense* has produced blight tolerant cut flowers. However, these lack the standard heart shaped spathe due to the species influence.

REFERENCES

Agasimani, A.D., Harish, D.K., Imamsaheb, S.J. and Patil, V.S. 2011. Anthurium varieties performance in rainy and winter seasons under greenhouse. *Research J. Agril. Sci.,* **2**(2): 337-339.

Ali, R. 1979. Cytology of a natural triploid *Anthurium crystallinum* Lind & Andre. *New Botanist*, **6**: 161-166.

Aurigue, F.B. 1993. Morphology and cytology of *A. andreanum* Linden, *A. crystallinum* Linden and andre, *A. hookeri* Kunth and their hybrids. College, Laguna (Phillippines).

Bai, Jingshu., Zhu, Xiaoqing., Guo, Huihong. and Li, Zhidan. 2005. Biological features of flowering and development of male gametophyte in *Anthurium andreanum. Forestry Studies in China*, **7**(1): 20-24.

Bejoy, M., Sumitha, V.R. and Anish, N. P. 2008. Foliar regeneration in *Anthurium andraeanum* Hort. cv. Agnihotthri. *Biotechnology*, **7**(1):134-138.

Bindu, M. R. and Mercy, S. T. 1996. Pollen studies in *Anthurium andreanum* Lin. *J. Tropical Agriculture*, **34**(2): 96-98.

Bindu, M.R. and Mercy, S.T. 1998. Chromosome behaviour and karyotype analysis in *Anthurium andreanum* Lind. *South Indian Horti.*, **45**(3-4): 134-138.

Boyce, P. 1995. *Anthurium warocqueanum. Curtis's Botanial Magazine*, **12**(3): 127-130.

Carlsen, M. and Croat, T.B. 2004. New species of *Anthurium* Sect. Semaeophyllium (Araceae) from central and south America. *Novon*, **14**(4): 401- 412.

Castro, A. C. R., de Resende, L. V., Guimaraes, W. N. R. and Loges, V. 2004. Use of molecular techniques in studies of genetic diversity of *Anthurium. Revista Brasileira de Horticultura Ornamental*, **10**(1/2):6-9.

Cen, Y.Q., Jiang, R.M., Deng, Z.L. and Ni, D.X. 1993. *In vitro* propagation of *Anthurium andreanum.* Morphogehesis and effects of physical and chemical factors. *Acta Hort. Sinica*, (**2**): 187-192.

Chen, F.C. and Kuehnle, A.R. 1996. Obtaining transgenic Anthurium through *Agrobacterium* mediated transformation of etiolated internodes. *J. American Soc. Hortl. Sci.*, **121**(1): 47-51.

Chen, F.C., Kuehnle, A.R., and Sugi, N. 1997. *Anthurium* roots for micropropagation and *Agrobacterium* mediated gene transfer, *Pl. Cell Tiss. Org. Cult.*, **49**: 71-74.

Collette, V.E., Jameson, P.E., Schwinn, K.E., Umaharan, P. and Davies, K.M. 2004. Temporal and spatial expression of flavonoid biosynthesis genes in flowers of *Anthurium andraeanum. Physiologia Plantarum*, **122**(3): 297-304.

Croat, T. B. and Sheffer, R. D. 1983. The sectional groupings of *Anthurium* (Araceae). *Aroideana,* **6**: 85-123.

Croat, T.B. 1980. Flowering behaviour of the neotropical genus *Anthurium* (Araceae). *American J. Bot.*, **67**(6): 888-904.

Eriksson, R. 1994. The remarkable weevil pollination of the Neotropical Carludovicoideae (Cyclanthaceae). *Plant Syst. Evol.*, **189**: 75-81.

Finnie, J.F. and Van S.J. 1986. *In vitro* culture of *Anthurium andreanum. South African J.Bot.*, **52**(4): 343-346.

Foster, G.D. and Twell, D. 1996. Plant Gene Isolation: Principles and Practice. John Wiley and Sons Ltd. p. 215-245.

Franz, N.M. 2004. Analyzing the history of the derelomine flower weevil-*Carludovica* association (Coleoptera: Curculionidae; Cyclanthaceae). *Biol. J. Linn. Soc.*, **81**: 483-

Franz, N.M. 2007. Pollination of anthurium (Araceae) by derelomine flower weevils (Coleoptera: Curculionidae). *Rev. Biol. Trop.*, **55**:1.

Gantait, S., Mandal, N., Bhattacharya, S., Das, P.K. 2008. *In vitro* mass multiplication withpure genetic identity in *Anthurium andreanum* Lind. *Plant Tissue Cult. Biotech.*, **18**:113-122.

Ge,Y., Zhang, F., Shen, X., Yu, Y., Pan, X., Liu, X., Liu, J., Pan, G. and Tian, D. 2012. Genetic variations within a collection of anthuriums unraveled by morphological traits and AFLP markers. *Biochemical Systematics and Ecology*, **45**: 34-40.

George, E.F., Hall, M.A. and Klerk, J.D. 2008. Plant propagation by tissue culture, *The Background, Springer*, **1**: 65-75.

Goncalves, E.G. 2005. A new anthurium sect. Pachyneurium (Araceae) from Minas Gerai State, Brazil. *Novon*, **15**(1): 133-135.

Goncalves, E.G. 2005a. A new species of Anthurium (Araceae) from Espirito Santo State,eastern Brazil. *Feddes Repertorium*, **116**(1-2): 92-95.

Gonçalves, E.G. 2003. Brazilian Aroids – Diversity and potential ornamental uses. *Abst*: International Symposium on New Floricultural Crops, 5. Campinas, SP, Instituto Agronomico.

Griesbach, R. J. 1994. An improved method for transforming plants through electrophoresis. *Plant Science Limerick*, **102**(1): 81-89.

Hamidah, M., Debergh, P. and Karim, A.G.A. 1997. Cyclic somatic embryogenesis of *Anthurium scherzerianum* Schott. *Acta Hort.*, **447**: 123-124.

Hamidah, M., Debergh, P.C. and Karim, A.G.A.1995. Somatic embryogenesis of *Anthurium scherzerianum* Schott. *Mededelingen- Faculteit – Landbouwkundige-en-Toegepaste- Biologische – Wetenschappen – Universiteit- Gent* (Belgium), **60**(4): 1671-1673.

Hamidah, M., Karim, A.G.A. and Debergh, P. 1997. Somatic embryogenesis and plant regeneration in *Anthurium scherzerianum*. *Plant Cell, Tissue, Organ Cult.,* **48**: 189- 193.

Hartman, Robert D. and Lamb, Ann E. 1998. Anthurium var. 'Northstar'. Official Journal of Plant Breeder's Rights Office, IPAustralia, *Plant Varieties J.*, **18**(2): 308.

Hartman, Robert D. and Lamb, Ann E. 1996. Anthurium var. 'Gemini'. Official Journal of Plant Breeder's Rights Office, IPAustralia, *Plant Varieties J.*, **18**(2): 308.

Hayden, D.M. and Christopher, D.A. 2004. Characterization of senescence associated gene expression and senescence dependent and independent cysteine proteases differing in microsomal processing in Anthurium. *Plant Science*, **166**(3): 779-790.

Henderson, A. 1986. A review of pollination studies in the Palmae. *Bot. Rev.*, **52**: 221-259.

Ignasse, J.F. 1984. Technical data on *Anthurium scherzerianum*. *Horticulture Francaise*, **160**:10.

Kamemoto, H. and Kuehnle, A.R. 1996. Breeding anthuriums in Hawaii. University of Hawaii Press, Honolulu, p.168.

Kamemoto, H., Iwata, Ry. and Marutani, M. 1988. Genetics of the major spathe colours in anthurium. Research series, College of Tropical Agriculture and Human resources, University of Hawaii. **56**: 11.

Kuehnle A.R., Chen, F.C. and Sugii, N. 2001. Transgenic Anthurium. *In*: Bajaj, Y.P.S. (ed.), Biotechnology in Agriculture and Forestry Vol.48, Transgenic Crops III. Springer, Verlag Berlin Heidelberg, pp. 3-15.

Kuehnle, A.R. and Sugii, N. 1991. Callus induction and plantlet regeneration in tissue cultures of Hawaiian anthuriums. *Hort. Science*, **26**(7): 919-921.

Kuehnle, A.R., Chen, F.C. and Sugii, N. 1992. Somatic embryogenesis and plant regeneration in *Anthurium andreanum* hybrids. *Plant Cell Reports*, **11**(9): 438-442.

Kuehnle, A.R., Chen, F.C., Sugii, N., Jaynes, J.M., Sanford, J.C. and Alvarez, A.M. 1990. Genetic engineering of blight resistance in Anthurium. *Proc*. Third Anthurium Bliht Conference, May 23-24, at the University of Hawaii at Hilo. p. 52-54.

Kumari, S., Desai, J.R. and Shah, R.R. 2011. *In vitro* propagation of two commercial cut flower *Anthurium andraeanum* Hort. cvs. Pumasillo and Corallis. National Conference on Recent Trends and Future Prospects in Floriculture. 5- 8 March 2011. SUPUA&T, Meerut. p. 124.

Kumari, S., Desai, J.R. and Shah, R.R. 2011. RAPD analysis of morphologically distinct callus induced from leaf explants of *Anthurium andraeanum* Hort. National Conference on Recent Trends and Future Prospects in Floriculture. 5- 8 March 2011. SVPUA&T, Meerut. pp. 139-140.

Kumari, S., Desai, J.R. and Shah, R.R. 2011. Somatic embryogenesis in *Anthurium andraeanum* Hort. cv. Pumasillo. National Conference on Recent Trends and Future Prospects in Floriculture. 5- 8 March 2011. SVPUA&T, Meerut. p. 138.

Kumari, S., Desai, J.R. and Shah, R.R. 2011.Optimization of DNA and PCR protocol for RAPD analysis of *A. andraeanum* Hort using callus. National Conference on Recent Trends and Future Prospects in Floriculture. 5- 8 March 2011. SVPUA&T, Meerut. pp.139.

Kunisaki, J.T. 1980. *In vitro* propagation of *Anthurium andreanum* Lind. *Hort Science*, **15**: 508-509.

Leimbeck, R. F. and Croat, T. B. 2002. A new species of *Anthurium* (Araceae) from SouthernEcuador and a revision of the *Anthurium oxybelium* Schott complex. *Novon*, **12**(4): 474-480.

Martin, K.P., Joseph, D., Madassery J. and Philip, V.J. 2003. Direct shoot regeneration from lamina explants of two commercial cut flower cltivars of *Anthurium andraeanum* Hort. *In Vitro Cell. Dev. Biol-Plant,* **39**: 500-504.

Matsumoto, T.K., Webb, D.T. and Kuehnle, A.R.1996. Histology and origin of somatic embryos derived from *Anthurium andraeanum* Linden ex Andre lamina. *J. American Soc. Horti. Sci.*, **121**(3): 404-407.

Montes, S., Hernandez, M.M. and Varela, M. 1999. Organogenesis in *Anthurium cubense*. *Cultivos Tropicales*, **20**(1): 51-54.

Murashige, T. and Skoog, F. 1962. A revised medium for rapid growth and bio-assays with tobacco tissue cultures. *Physiol. Plant*, **15**: 473-497.

Nhut, D.T., Duy, N., Vy, N.N.H., Khue, C.D., Khiem, D.V. and Vinh, D.N. 2006. Impact of *Anthurium* spp. genotype on callus induction derived from leaf explants, shot and root regeneration capacity from callus. *J. Applied Horti.*, **8**(2): 135-137.

Nitsch, J.P., 1969. Experimental androgenesis in Nicotiana. *Phytomorphology*, **19**: 389–404.

Nowbuth, P., Khittoo, G ., Bahorun, T. and Venkatasamy, S. 2005. Assessing genetic diversity of some *Anthurium andraeanum* Hort. cut-flower cultivars using RAPD Markers. *African J. Biotech.*, **4**(10): 1189-194.

Osiecki, Marian W. 1999. Anthurium var. 'Atwelve'. Official Journal of Plant Breeder's Rights Office, IPAustralia, *Plant Varieties J.*, **18**(1): 170-74.

Osiecki, Marian W. 1999. Anthurium var. 'Aeighteen'. Official Journal of Plant Breeder's Rights Office, IPAustralia, **18**(1): 175-77.

Osiecki, Marian W. 2000. Anthurium var. 'Atwenty'. Official Journal of Plant Breeder's Rights Office, IPAustralia, *Plant Varieties J.*, **18**(1): 166-69.

Paswan, L. and Mahanta, S. 2001. Effect of growth regulators on germination of anthurium (*Anthurium andraeanum* Lind.) seeds under *in vitro* conditions. *J. Agril. Sci. Soc.North East India*, **14**(2): 274-277.

Pierik, R.L.M. 1975. Callus multiplication of *Anthurium andreanum* Lind in liquid media. *Netherlands J. Agri Sci* ., **23**: 299-302.

Pierik, R.L.M. 1976. *Anthurium andreanum* plantlets produced from callus tissues cultivated *in vitro*. *Physiol. Plant* **37**: 80-82.

Pierik, R.L.M., Steegmans, H.H.M. and Van der meys, J.A.J. 1974. Plantlet formation in callus tissues of *Anthurium andreanum* Lind. *Sci. Hort.*, **2**: 193-198.

Prabhakara, H.L., Reddy, B.S., Hegde, R.V. and Patil , B.C. 2001. Effect of different media on *in vitro* seed germination and subsequent growth of *Anthurium andraeanum* Lind. *Karnataka J.Agril. Sci.*, **14**(3): 712-715.

Puchooa, D. 2005. *In vitro* mutation breeding of anthurium by gamma radiation. *Int. J. Agric. Biol.*, **7**: 11-20.

Raad, M.K., Zanjani, S.B., Sayyad, A.R., Maghsudi, M. and Kaviani, B. 2012. Effect of cultivar, type and age of explants, light conditions and plant growth regulators on callus formation of anthurium. *American-Eurasian J. Agric. & Environ. Sci.*, **12**(6): 706-712.

Ranamukhaarachchi, D.G., Henny, R.J., Guy, C.L. and Li, Q. B. 2001. DNA Fingerprinting to identify nine *Anthurium* pot plant cultivars and examine their genetic relationship. *HortScience*, **36**(4): 758-60.

Rijn, Leonardus van. 1999. Anthurium var. 'Lady Love'. Official Journal of Plant Breeder's Rights Office, IPAustralia, *Plant Varieties J.*, **18**(1): 196-201.

Rijn, Leonardus W.B.M. van. 2001. Anthurium var. 'Fresh Love'. Official Journal of Plant Breeder's Rights Office, IPAustralia, *Plant Varieties J.*, **18**(1): 190-95.

Rijn, Leonardus W.B.M. van. 2001. Anthurium var. 'Lucky Leny'. Official Journal of Plant Breeder's Rights Office, IPAustralia, *Plant Varieties J.*, **18**(1): 213-18.

Rijn, Leonardus W.B.M. van. 2003. Anthurium var. 'Changing Love'. Official Journal of Plant Breeder's Rights Office, IPAustralia, *Plant Varieties J.*, **18**(1): 219-24.

Rijn, Leonardus W.B.M. van. 2003. Anthurium var. 'Tender Love'. Official Journal of Plant Breeder's Rights Office, IPAustralia, *Plant Varieties J.*, **18**(1): 237-42.

Rijn, Leonardus W.B.M. van.1998. Anthurium var. 'Red Love'. Official Journal of Plant Breeder's Rights Office, IPAustralia, **18**(1): 207-12.

Rijn, Leonardus W.B.M. van.2003. Anthurium var. 'Exciting Love'. Official Journal of Plant Breeder's Rights Office, IPAustralia, *Plant Varieties J.*, **18**(1): 231-36.

Rijn, Magdalena J.M. Van. 1998. Anthurium var. 'Orange Love'. Official Journal of Plant Breeder's Rights Office, IPAustralia, *Plant Varieties J.*, **18**(1): 184-89.

Rijn, Magdalena J.M. Van. 1998. Anthurium var. 'Sugar Love'. Official Journal of Plant Breeder's Rights Office, IPAustralia, *Plant Varieties J.*, **18**(1): 225-30.

Sengupta, D.K., Chettri, Rabindra and Chettri, R. 1990. Cytological studies on two ornamental aroids growing in Eastern Himalaya. *Environment & Ecology*, **7**(4): 1025-1028.

Sharma, A.K. and Bhattacharya, G.N. 1966. A cytotaxonomic study on some taxa of Araceae. *Genét. Ibér.*, **18**: 1-26.

Sheffer, R.D. and Croat, T.B. 1983. Chromosome numbers in the genus *Anthurium* (Araceae) II. *American J.Bot.*, **70**(6): 858-871.

Silva, J.A.T., Nagae S. and Tanaka, M. 2005. Effect of physical factors on micropropagationof *Anthurium andraeanum*. *Plant Tissue Culture,* **15**(1): 1-6.

Singh, A.K. 2006. Flower Crops: Cultivation and Management. Pub. New India Publishing Agency, New Delhi. 24p.

Singh, F., Sangama P.J. and Pierik, R.L.M. 1991. Micropropagation and plant conformity in *Anthurium andraeanum*. *Current Plant Science and Biotechnology in Agriculture*, **12:** 201-204.

Somaya, K.U., Narayanswamy, P. and Jayaprasad, K.V. 1998. Micropropagation studies in *Anthurium andraeanum* Lind. *Karnataka J. Agril. Sci.*, **11**(2): 466-470.

Tajuddin, E. and Prakash, R. 1996. Anthurium. The Directorate of Extension, Kerala Agricultural University, pp 5-6.

Te-chato, S., Susanon, T. and Sontikun, Y. 2006. Medium influencing embryogenesis andorganogenesis in *Anthurium* spp. 06. Cultivar, explant type and culture. *Songklanakarin J.Sci. Technol.*, **28**(4): 717-722.

Teng, W.L. 1997. Regeneration of anthurium adventitious shoots using liquid or raft culture. *Plant Cell Tissue Org. Cult.*, **49**: 153- 156.

Vargas, T.E., Mejias, A., Oropeza, M. and De, G.E. 2004. Plant regeneration of *Anthurium andraeanum* cv. Rubrun. *Electronic J. Biotech.*, **7**(3): 282-286.

Wannakrairoj, S. and Kamemoto, H. 1990. Inheritance of purple spathe in anthurium. *J. Am. Soc. Hortic. Sci.*, **115**(1): 169-171.

Wichuta, R. 1994. Effects of colchicines and gamma radiation on mutation of anthurium double spathe *in vitro*. *AGRIS*. Bangkok (Thailand).

Williams, N.H. and Dressler, R. L. 1976. Euglossine pollination of *Spathiphyllum* (Araceae). *Selbyana*, **1**: 349-356.

Yakandawala, G., Peiris, S.E. and Yakandawala, Y.L.N. 2000. Transient expression of uidA reporter gene in regenerable callus tissues of *Anthurium andraeanum* Lind. by *Agrobacterium* mediated transformation. *Tropical Agril. Res.*, **12**: 50-55.

Yu, K., Paek, K., Yu, K.J. and Paek, K.Y. 1995. Micropropagation of *Anthurium* spp. through shoot tip and callus culture. *J. Korean Soc. Hortil. Sci.*, **36**(5): 684-694.

Chapter – 9

Antirrhinum

Antirrhinum is a genus commonly known as snapdragon belongs to family Scrophulariaceae. Snapdragon is one of the flowering plant present with vast genetic diversity in India. The term Antirrhinum is derived from the Greek word "*anti*" means like and "*rhin*" means a nose, referring to the snout like shape of the flower. The cultivated *Antirrhinum majus* is grown in tropical, subtropical and temperate countries. The common name snapdragon, originates from the flowers reaction to having their throats squeezed, which causes the mouth of the flower to snap open like a dragon's mouth. Flowers are borne on terminal long spikes of many colours except blue and with numerous shades. It is one of the excellent cut flowers which have long lasting qualities. These are also used as garden plants, bedding plants, in rockeries or herbaceous borders and as potted plants. Plants can be grown for cut flowers in an open field or under protection depending on the climate. The leaves and flowers are antiphlogistic, bitter, resolvent and stimulant (Chiej, 1984).

ORIGIN AND DISTRIBUTION

The garden snapdragon (*Antirrhinum majus*) has emerged as a model organism during early studies of inheritance and mutation (Darwin, 1868) because of its diploid inheritance, ease of cultivation and variation in morphology and flower colour. The genus *Antirrhinum* includes 15 New World species in the section *Saerorhinum* and 21 Old World species in the sections *Antirrhinum* and *Orontium*. *A. majus* (Snapdragon) is a species of *Antirrhinum* native to the Mediterranean region, from Morocco and Portugal north to southern France and East to Turkey and Syria. Most species are found around the Mediterranean Sea and in North America (Stubbe, 1966). The progenitor of the modern snapdragon is from the Mediterranean region and more specifically, it is native to southern France. It was believed that the snapdragon reached Britain with the Romans at an early date, where it has been naturalized on mountainous regions. Thereafter, it spread to different parts of the world from Britain.

The scientific name of the genus *Antirrhinum* was first defined by Carl von Linne in the year 1753. In the year 1914 Professor Erwin Baur the founder of the Max-Palnck-Institut für Züchtungsforschung in Cologne published a volume called 'Einführung in die experimentelle Vererbungslehre' (Introduction to Experimental Genetics) 2nd edition. This book contained several plates with reproductions of water colour paintings of snapdragons as examples. He crossed a white line with a red line resulting in a pink flower. This pink flower then was selfed and the offspring showed the typical distribution of 1 white, 2 pink and 1 red as a demonstration of Mendelian inheritance. *Antirrhinum glandulous* is native of California. In United States of America, winter flowering cultivars were introduced in 1926 by Frank Volz of Cheviot, Ohio. In 1959, snapdragon were included in the Agricultural Censuses of Horticulturally Special Crops and ranked seventh as a cut flower crop.

The genus *Antirrhinum* offers an inexhaustible resource for genetics-based studies in developmental biology, biochemistry and evolution (Stubbe, 1966). Wide diversity of *Antirrhinum* species, combined with classical and molecular genetics, the two traditional strengths of *Antirrhinum* provides an opportunity for developmental, evolutionary and ecological approaches (Schwarz-Sommer *et al.*, 2003). Transposons responsible for flower variegation were identified in the 1980s. The Max-Planck-Institute in Cologne, Germany, allowing genes involved in flower and leaf development and in pigmentation to be isolated by transposon tagging. Population genetic studies that have been applied to *Antirrhinum* species show different population sizes, geographic distributions and breeding systems, from self-fertility to obligate out-crossing (Jimenez *et al.*, 2002 and Mateu-Andres and de Paco, 2006). It was introduced to India with the advent of British. Among them, only *A. majus* has been domesticated as an ornamental.

SPECIES

Initially it was reported that *Antirrhinum* comprises about 32 species (Chittenden and Synge, 1956). Forty to forty-two species have been reported under the genus *Antirrhinum*. Other important genera in this family are *Calaceolaria*, *Hebe*, *Penstemon*, *Verbascum*, *Nemesia*, *Veronica*, *Castilleja*, *Pedicularis*, *Digitalis* and *Minulus* (Kumar and Bhattacharjee, 2006). Description of some of the important species of *Antirrhinum* are as follows:

Antirrhinum majus: *Antirrhinum majus* is cultivated in Turkey, Spain, France, India, England and several other countries as commercial flower. It is often treated as an annual and harvested for its flowers. Lavender, pink, red and orange are commonly found for bloom colour. There are five subspecies of *Antirrhinum majus*

i. *Antirrhinum majus* subsp. *majus*

ii. *Antirrhinum majus* subsp. *cirrhigerum*

iii. *Antirrhinum majus* subsp. *linkianum*

iv. *Antirrhinum majus* subsp. *litigiosum*

v. *Antirrhinum majus* subsp. *tortuosum*

***Antirrhinum multiflorum*: ***Antirrhinum multiflorum* is also well known as Sierra snapdragons. They are native to California and a section of the Sierra Nevada. It grows on slopes and appears in burned or damaged areas frequently.

***Antirrhinum nuttallianum*:** This species is frequently found in Baja and southern California. It is often found along the near coastline and some coastal ranges. A vine is used to climb or creep on objects while it grows wild.

***Antirrhinum virga*: ***Antirrhinum virga* is California native in the north coast. It is most commonly referred to as the tall snapdragon for its height. The stems are non-climbing and it grows in serpentine soil conditions.

The species vary considerably in morphology, size and habitat.

FLOWER MORPHOLOGY AND BIOLOGY

Antirrhinums are facultative long day plant. The inflorescence is a terminal raceme. In antirrhinum inflorescences that are normally indeterminate are converted to a determinate architecture (Bradley *et al*., 1997). The flower makes up the reproductive part of a plant and is found at the end of a stalk. The base of the flower, the receptacle, has small green leaves known as sepals that provide support for the blossom and protection for the bud. The showiest parts of a flower are the petals, which are brightly coloured to attract insects to the inner reproductive parts. A flower can be male, containing only stamens; female, containing only a pistil; or have both reproductive structures, which is known as a perfect flower. Stamens are four and not winged. Many flowers are symmetrical, but the special appearance of snapdragon blossoms originates in the fact that they are bilaterally symmetrical. Flowers are of two types- Buuy shaped (5 small sepals, base of corolla is fused into tube like structure which divides into two lips, lower lip having projecting palate, that helps in closing the mouth of corolla tube) and Penstemon shaped (flower has 5 small sepals, base of corolla is fused into tube like structure with open mouth). Androecium 4 with long filament, dosified, didynamous, gynoecium is in single stigma with superior and bicarpellary ovary. The flowers are produced on a tall spike each flower is 3.5-4.5 cm long, zygomorphic. Snapdragon flowers bud all along the length of the spike. The buds near the top of the spike bloom last, with the buds located nearest to the bottom of the spike blooming first. *A. majus* use homologous mechanism in floral development (Coen and Meyerow, 1991).

The height of the plant can vary with dwarf varieties having snapdragon blossoms that are more thickly clustered. In addition, though commonly red or white, the snapdragon blossom can also be pink, lavender, yellow and orange. Viable uncontaminated pollen was obtained only from snapdragon flowers which were left after being dipped in a sterilant solution, on the stem until anther dehiscence, pollination *in vitro* failed to produce seeds (Balatkova and Tupy, 1973). Wild plants have red flowers and are homozygous and self fertile. The leaves are spirally arranged are 1-7 cm long and 2-2.5 cm broad. Leaves are lanceolate, entire dark green to purplish and mainly opposite (Bailey and Bailey, 1976). Size of a leaf increases with leaf position (Harte, 1983).

Hand pollination is required for seed setting. Although, snapdragon is often cross-pollinated in nature but selfing is a problem and therefore, the plants are pollinated by bumblebees and the flowers close over the insects when they enter and deposit pollen on their bodies. Cultivated varieties are of other colours, notably yellow, heterozygous and cross-pollinated, yellow and yellowish colours being preferred by bees which are bumble bees, which are heavy enough to open them. Small flowered cultivars may also be pollinated by other species. Flower anthesis takes place early in the morning (6-8 am), 5 days after bud appearance. Pollens are viable 5-6 days after bud appearance and maximum at about to burst stage of anthers. Stigma becomes receptive 3-6 days after bud appearance. The best time for pollination is early morning. The fruit is an ovoid capsule 10-14 mm diameter, containing numerous small seeds. In general, maximum seed setting is observed when flowers are pollinated at the time of flower opening (10-15 days after bud appearance).

GENETICS

Antirrhinum majus is amenable to classical genetics. It has a relatively short duration crop *i.e.* 4 months, is diploid with chromosome number 2n = 16 and is easily self and cross-pollinated. Tetraploid forms are also available having bigger sized flowers.

Bacterial artificial chromosome (BAC) and TAC libraries representing the antirrhinum genome are available now. A range of self-compatible and transposable element mutants of *A. majus*, were generated in which the different domains of the *S*-locus are affected in isolation. Seven related species of snapdragon *A. siculum*, *A. majus*, *A. latifolium*, *A. linkianum*, *A. litigiosum*, *A. cirrhigherum* and *A. tortuosum* were studied in order to compare levels of genetic variation and its partitioning in them and to check relationships between genetic patterns and the reproductive system. Genetic variability in *A. siculum* was found to be the lowest known in the genus. Mean values of FIT and FIS were mostly positive and not significantly different from zero. Population genetic studies that have been applied to *Antirrhinum* species show different population sizes, geographic distributions and breeding systems, from self-fertility to obligate outcrossing (Jiménez *et al.*, 2002 and De Paco and Mateu-Aandres, 2006). *Antirrhinum* species have been the subject of population genetic studies, including those aimed at assessing genetic diversity in order to inform conservation strategies for rare endemic species (Mateu-Andres, 2004). The present variability among the different germplasm of snapdragon has been studied by various workers (Singh and Arora, 1980, Sanderson and Martin, 1984 and Von-Auw, 1990 and Singh, 2005). The heritability of various characters had been studied (Singh and Arora, 1980).

Previous work suggested that the haplotype structure of the two genes is probably maintained by recombination suppression at the *S*-locus. To examine features associated with this suppression, first mapped the *S*-locus of *A. hispanicum* to a highly heterochromatic region close to the distal end of the short arm of chromosome 8. Both leptotene chromosome and DNA fiber FISH analyses showed an obvious haplotype specificity of the antirrhinum *S*-locus that is consistent with its haplotype structure. A chromosome inversion was also detected around this region between *A. majus* and *A. hispanicum*. These results revealed that DNA sequence polymorphism and a heterochromatic location are associated with the *S*-locus. Possible roles of these features in maintenance of the haplotype specificity involved in both self and non self recognition Yang *et al*. (2007). Identification of genetically superior plants is an important prerequisite for successful hybridization programme. The attraction of combining ability analysis (Sprague and Tatum, 1942) is that these provide an empirical summary of complex observation and a reasonable basis for forecasting. The performance of yet untested crossed but yet makes no genetical assumption (Simmonds, 1979). The genetic system controlling flower colour in *Antirrhinum majus* was identified by diallel analysis (Jauhari, 2006).

Male sterility and self incompatibility

Male sterility and self incompatibility are also reported in snapdragon (Sykorova, 1975 and McCubbin *et al*., 1992). Interspecific self incompatible hybrids have been successfully propagated through micropropagation from hypocotyls explants (Wang *et al*., 2004). *A. majus* and some wild species *A. valentinum*, *A. subbaeticum* and *A. siculum* are self-fertile. The majority of *Antirrhinum* species show gametophytic self-incompatibility that is determined by a single, complex *S*-locus. Individuals of self-incompatible species reject pollen carrying an *S*-allele that corresponds to one of their own alleles and are therefore obligate outbreeders. Mapping and sequence analysis of the active *S*-locus of *A. hispanicum*, followed by expression and functional studies, identified the likely pollen component to be an F-box protein involved in targeting the RNase for degradation (Lai *et al*., 2002). The inheritance of self-incompatibility was studied, putative *S*-allele products in the pistil were identified and lines were generated, which were suitable for transposon tagging (McCubbin *et al*., 1992). Cultivated varieties and wild progenitors of *A. majus* and *A. siculum* are able to self-pollinate whereas all other members of the genus show gametophytic self-incompatibility and therefore are largely or entirely out breeding.

Heritability and genetic advance

In India, enough variability can be seen in snapdragon, harnessing of its genetic potential is still to be dealt with. Heritability, genetic variability and expected genetic advance of snapdragon cultivars were assessed. Significant varietal differences were observed for all characters studied (Singh and Arora, 1980). The significant advances in the genetic and molecular study of *A. majus*, very limited work has been done on the karyotype and cytogenetic structure of the *A. majus* genome. The genetic linkage groups (LGs) have not been integrated into individual chromosomes. In recent years, extensive molecular and genomic resources have been established in *A. majus* (Schwarz-Sommer *et al.*, 2003). A standard karyotype has been established by anchoring centromeric repeats on meiotic pachytene chromosome using fluorescent *in situ* hybridization (FISH). An ideogram based on the DAPI-staining pattern of pachytene chromosomes was developed to depict the distribution of heterochromatin in the *A. majus* genome. To integrate the genetic and chromosomal maps one or two molecular markers from each LG to screen an antirrhinum transformation-competent artificial chromosome (TAC) library are selected. These genetically anchored TAC clones were labeled as FISH probes to hybridize to pachytene chromosomes of *A. majus*. As a result, the relationship between chromosomes and the LGs in antirrhinum has been established (Zhang *et al.*, 2005).

BREEDING OBJECTIVES

Breeding objectives of snapdragon are based on production problems and consumer preference.

i. Novelty of colour
ii. More number of florets per spike
iii. Compactness of spike
iv. More number of spikes per plant
v. Dwarfness of plant for bedding
vi. Longer duration of flowering in field
vii. More number of open florets at a time
viii. Longer vase life
ix. Resistance to rust and other biotic and abiotic factors

BREEDING METHODS

Classical breeding together with the most advanced techniques in molecular biology have been applied to a wide variety of species, contributing to the elucidation of molecular mechanisms in plant development. Various methods of breeding are as follows.

Introduction

Introductions of snapdragon from Australia, England and USA have been done. A few cultivars/hybrids have been introduced by NBPGR are given in Table 1.

Table 1: Introductions of snapdragon from Australia, England and USA.

S. No.	Country	Varieties/hybrids
1.	USA	Crimson Giant, Rose Bal and Highnon Sharp Yellow.
2.	Australia	F_1 hybrid Medan, Butterfly, F_1 hybrid Bright Butterfly, Stock Trisomic Mixed and Giant Column Formula Mixture.
3.	England	Black Price, Braun Blanquetii, Frosted Sunset, La Bella Red & White, Peaches & Cream, Kim Orange, Royal Bride.

Hybridization

Hybridization in *Antirrhinum majus* was first started with an introduction of F_1 hybrid Christmas Cheer by Fred and Helen Windmiller in Columbus, Ohio in 1938 (Ball, 1957). Several germplasm of snapdragon (*Antirrhinum majus*) were collected and evaluated for breeding purpose (Singh, 2005 and Jauhari and Singh, 2005), later on inbred and F_{1S} of snapdragon (*Antirrhinum majus*) were developed and evaluated for flowering and post-harvest attributes at Pantnagar (Jauhari and Singh, 2006). Its eight species, *A. siculum, A. linkianum, A. tortuosum, A. cirrhigerum, A. glultinosum, A. majus* and *A. orontium* were intercrossed in nearly all directions (Mahal, 1972). The crosses, *A. majus* × *A. cirrhigerum, A. majus* × *A. tortuosum, A. majus* × *A. glultinosum* and *A. cirrhigerum* × *A. linkianum* were successful in both the directions, whereas crosses, *A. majus* × *A. linkianum, A. majus* × *A. siculum, A. molle* × *A. tortuosum* and *A. tortuosum* × *A. linkianum* were successful only in one direction. Crosses involving *A. orontium* and *A. siculum* with other species (except *A. siculum* with *A. majus*) failed altogether. The hybrids of the successful crosses were all fertile and from their normal meiosis behaviour, it was inferred that the species concerned are very closely related. Morphologically, the hybrids produced were mostly intermediate, but dominance of some characters of one of the parent was also observed.

At meiosis, 16 chromosomes were seen to form bivalents which indicated genetic homology of the species involved. The F_2 plants were mostly balanced types, but some disharmonious gene combinations were manifested by abnormal growth and flower forms that are radically different from either of the parents which suggested evolutionary potentialities through transgressive gene action. About 18 wild *Antirrhinum* species are found in Southwest Europe and northern Africa. These display considerable variation in traits, including flower colour patterns, flower and leaf shape and size and plant growth habit. Usefully, most of the species are interfertile and so amenable to genetic analysis. This makes it feasible to identify and isolate genes that underlie species differences and to test the adaptive importance of traits. Some important varieties and hybrids of snapdragon grown for various purposes are given in Table 2.

Table 2: Some important varieties and hybrids of antirrhinum.

S. No.	Type	Varieties/hybrids
1.	F_1 Hybrid	Bizarre, Bright Butterflies, Crimson Topper, Floral Cluster, Forcing Strain, Frontier Strain, Jackpot, White Topper, Yellow Topper, Lavender Topper, Little Darling, Madam Butterfly, Rox Topper, Snow Man, Tall Rocket, Weeding Bells.
2.	F_1 Hybrid Double	Crimson Supreme, New Glamour Shades, Scarlet Supreme, Super Jet, Vanguard, White Supreme.
3.	Tall cultivars	*A. majus* Maximum, Tip Top, Rocket Hybrids, F_1 Butterfly Type.
4.	Semi-Tall cultivar	*A. majus* nanum, Tetra Ruffled, F_1 Carioca, F_1 Coronette.
5.	Dwarf cultivar	F_1 Floral Carpet, Little Darling, F_1 Kolibre, Tom Thumb, Magic Carpet.

Interspecific hybrids: The interspecific hybrids between *A. majus* and *A. molle* have been reported. There was intergrading variation in characters such as plant height, habit of branching and corolla form. From the progeny of this specific hybrid, new class of *Antirrhinum* has been developed. These hybrids have creeping habit and are highly suitable for covering the grounds as carpet and hence, the group is called as miniature or magic carpet. These hybrids are also suited to rock gardens. The fertile hybrids between *A. majus* × *A. linkianum* have also been produced. Mahal (1972) intercrossed eight species, namely, *A. siculum*, *A. linkianum*, *A. tortuosum*, *A. cirrhigerum*, *A. molle*, *A. glutinosum*, *A. majus* and *A. orontium* in nearly all directions. The crosses *A majus* × *A cirrhigerum*, *A majus* × *A tortuosum*, *A majus* × *A glutinosum* and *A. cirrhigerum* × *A. linkianum* were successful in both the directions, whereas, crosses, *A. majus* × *A. linkianum*, *A. majus* × *A. siculu*, *A. molle* × *A. tortuosum* and *A. tortuosum* × *A. linkianum* were successfully only in one direction. Crosses involving *A. orontium* and *A. siculum* with other species except *A. siculum* with *A. majus* failed altogether. Hybrids of successful crosses were all fertile.

Heterosis breeding: Hybrids were uniform with semi-dwarf habit and large number of tillers with prominently held inflorescences bearing closely set reasonably large flowers with perfect synchronous flowering. Crossing inbred lines of *A. majus* with a dwarf species *A. glutinosum* produced F_1 dwarf hybrid (Mahal, 1972). Genetic studies on diploid and tetraploid progenies in *Antirrhinum* revealed appreciable heterosis in F_1 hybrids for all the characters studied (Malik, 1979). Afterward no significant work on genetic improvement of snapdragon, so far has been done in India. The effect of light, radiation, soil temperature, air temperature and saturation deficit of the air on heterosis was studied in snapdragon (Grau, 1972, Claus *et al.*, 1974 and Edwards and Goldenberg,

1976). The estimates of heterobeltiosis and relative heterosis of the 45 crosses for 20 floricultural traits were studied in antirrhinum. Most of the crosses manifested significant heterosis over better, mid and standard parents (Jauhari, 2006).

A study was conducted with 10 inbreds and 45 crosses of snapdragon to find out heterosis (Jauhari and Singh, 2009). Heterosis for plant height, number of leaves/plant, number of primary branches/plant, leaf area index, diameter of spike, length of spike, number of seeds/pod and weight of seeds/pod were observed in positive direction, whereas days taken to flowering and days taken to seed ripening were calculated in negative direction. Standard heterosis was observed maximum (523.42%) for number of seeds/pod followed by weight of seeds/pod (284.35%) and length of spike (179.95%). Mean heterobeltiosis for plant height during both years varied from -35.95% (AG-2 × SA-1) to 33.75% (Sant-11 × AG-5). Heterobeltiosis for number of primary branches/ plant was found maximum (33.19%) in AG-1 × AG-4. Out of 45 crosses only 8 crosses showed positive heterosis over better parent leaf area index. During both years, earliest flowering occurred in AG-2 × AG-5(24.46%) over its better parents. Mean heterobeltiosis for number of leaves/plant ranged from -74.81% (SA-1 × San-11) to 23.72% (AG-1 × Sant-22).

Resistance breeding: Emsweller and Jones (1934) reported the development of cultivars resistant to rust *Puccinia antirrhini*. The backcross method was used to combine the resistance with the flower colour and plant habit of standard commercial cultivar. Therefore, resistant parent lines were used widely for F_1 hybrid production with large number of other susceptible parents. Varieties evolved resistant to rust are given in Table 3.

Table 3: Rust resistance varieties of antirrhinum.

Resistance	Varieties
Rust (*Puccinia antirrhini*)	Moon Glow, Rainbow Mixed, Rose Light, Carmine Monarch, Scarlet Monarch, Victory, White Freedom, Wisely Cheerful, Yellow Freedom.

Polyploidy

Nebel and Ruttle (1938) reported the value of tetraploid antirrhinums which were developed by the use of colchicines (0.2%). This brought revolution due to much increase in flower size. Tetraploid cultivars are widely grown and known for large flower. Triploids have been produced by crossing tetraploid and diploid. Normally, the triploids should have been tidier in appearance because of the absence of seed pods and seeds but all the triploids produce normal number of capsules. This is probably due to the fact that ovules and pollen grains are produced in such abundance, failures in fertilization passed unnoticed. Antirrhinum is well known for its lack of natural polyploidy. The genus has been divided into two sections *Antirrhinum* and *Sarrrhinum*, which are distributed in the Mediterranean region of Europe and south-western parts of North America, respectively. So far from the European element of the genus 16 species have been worked out cytogenetically. All these are diploid (n = 8). Furthermore, there is predominance of self-incompatible and cross-fertilized species. However, gene flow between species is restricted by bee specification in relation to colour, size and structure of flower and ecological isolation. As a result there is a mixed population of different species like *A. majus* and *A. glutinosum*. There is ordinarily little interspecific hybridization although the species cross readily when pollinated by hand. Fertile and segregating progenies are known to result from hybrids involving *A. majus* and *A. glutinosum.*

European species of *Antirrhinum* have an ecospecific pattern in which species differentiation is maintained at diploid level by ethological and ecogeographic isolation. Accordingly any polyploids would have auto or segmental alloploid characters. Such polyploids, unaided by other supportive mechanism, have inherent limitations and are unable to establish in nature, although autoploids of the ornamental *A. majus* are a tremendous success under garden conditions. Tetraploid snapdragon lines were also developed, which are of high performance. The autoploids are hardier, more vigorous, stouter and possess larger flowers than corresponding diploids. Induced colchiploids by treating the seeds of the variety Katrain Local with 0.4% colchicine were reported (Jain *et al.*, 1962). The most characteristic features of the induced tetraploids, which distinguished them from the diploids, were the increased size of individual flowers, which resulted in a more compact form of inflorescence and the increased length of the main floral branch. Flowers in the tetraploids had a more ruffled corolla marked by the presence of small papillae-like structure not found in the diploids. Colchiploids of *A. majus*, *A. siculum*, *A. linkianum* and two varieties of *A. orontium* were raised (Mahal *et al.*, 1968 and Mahal, 1972). These autotetraploids were hardier, more stout, slower in growth and shorter with larger flowers, pollen grains, fruits and seeds than their diploid forms. Meiotic behaviour was found to be typical autotetraploid with as many as maximum possible number of quadrivalents in a large number of cases. In all the autotetraploids, pollen and seed fertility was quite high. A plant in autotetraploid population showed cytomixis. All pollen cells in this plant had a varying number of chromosomes from the enucleate condition to a very high number (2n = 72).

The plant was totally pollen and seed fertile. Crosses between diploids and tetraploids were made in both the directions. It was found that this cross succeeds to a slight extent only when a tetraploid is used as seed parent. The triploids produced were better morphologically than both the parents, being faster in growth and longer in flowering duration. At meiosis, the three sets of chromosomes were found to form as many as eight trivalents in a number of cells. Anaphase segregation was highly irregular, but pollen and seed fertility was appreciably high. As a result of these studies, the National

Botanical Research Institute, Lucknow, India, released one variety named Tetra Giant. The resultant tetraploid is hardier, sturdier, stockier and shorter than the corresponding diploids. It has a higher number of flowering stems, with conspicuously larger, deeper coloured flowers, which are long lasting than their diploid counterparts. Polyploidy was induced in *A. majus* using an aqueous solution of colchicines, the tetraploid plant has thicker but deformed cotyledons, larger stomata and larger thicker petals, the flowering period was 1-3 weeks later than in diploid plants and on self-pollination the pollen grains were sterile due to meiotic abnormalities (Zhang and Yue, 1990). Varieties developed through polyploidy breeding are given in Table 4.

Table 4: Antirrhinum varieties developed through polyploidy breeding.

S.No.	Ploidy level	Varieties
1.	Tetraploid	Bright Scarlet, Glacier, High Noon, Rosabel, Velvet Beauty, Red Shades
2.	Triploid	Velvet Beauty

MUTATION

Baur (1924) described small mutation in snapdragon. The total mutation rate for all genes in the complement will naturally be much higher. Large differences between strains in spontaneous mutation rates in *A. majus* have also been found (Baur, 1924). A number of macro mutation have been isolated in *A. majus* (Stubbe, 1966). These mutation are: Transcendens number of stamens reduced from normal four to two; Hemiradialis - five stamens approach to radial symmetry of corolla; Neohemiradialis- six anthers; Radialis - radially symmetrical flowers; Hirzina-floral spurs, approaching *Linaria*; Fistulata-reduction of corolla limb and tubular corolla; Subconnata –one cotyledon. The effects of gamma rays on callus initiation and growth, root and embryo differentiation showed marked differences in radio-sensitivity, higher rates were inhibitory (Rao *et al*., 1976).

Tetra giant snapdragons were raised at NBRI, Lucknow from some F_1 hybrids. The resultant tetraploids are hardier, sturdier, stockier and shorter than corresponding diploids. They have generally more flowering stems with conspicuously larger, deeper-coloured flowers that are longer than their diploid counterparts (Khoshoo, 1990).

BIOTECHNOLOGY

Tissue culture

Poirer *et al.* (1974) isolated protoplast enzymatically from the leaves of Kimosy Blanc antirrhinum were cultured on a synthetic medium and their growth and development were followed *in vitro* and found that cultured protoplasts increased in size, regenerated cell walls, divided and formed small colonies and emryoids. Cell wall formation and cell division required both auxin and cytokinin in the medium. Generation and cytological observation of adventitious buds from hypocotyls explants of *Antirrhinum majus* showed that plant regeneration originated from epidermal cells of explants (Yu *et al.*, 1996). A broadly applicable method for the successful induction of root systems in a number of cultivars of *A. majus* has been determined (Atkinson *et al.*, 1991). A rapid and efficient regeneration system has been developed for interspecific self-incompatible hybrids between *A. majus* × *A. hispanicum* multiple shoots were induced from hypocotyls explants on Murashige and Skoog medium (Wang *et al.*, 2004). *In vitro* morphogenetic response and distribution of endigenous plant harmones in hypocotyls segments of *A. majus* confirmed the general principle of *in vitro* culture that a high concentration of cytokinin and a low concentration of auxin promote the induction of shoot morphogenesis (Okubo *et al.*, 1991).

Molecular breeding

The snapdragon provided the first molecular access to those genes controlling flowering meristem and individual flower organ development – the stamen, pistil, pollen and stigma – as well as controlling leaf development, organelle growth and floral asymmetry. DEFICIENS genes and AGAMOUS genes were isolated from *A. majus* and *A. thaliana*, respectively have created interest in the commercial value of cut flower with novel flower shapes through molecular manipulation of floral HOMEOTIC genes (Bowmen *et al.*, 1989). Jimenez (2002) conducted the RAPD study to evaluate the level of genetic variation present in *Antirrhinum subbaeticum* to improve its conservation. Baur (2003) recognized the potential evolutionary genetics and used snapdragons to identify and map the genes responsible for evolutionary differences in flower morphology and colour. Bey *et al.* (2004) studied the characterization of antirrhinum petal development and identification of target genes of the class B MADS Box gene DEFICIENS (DEF). The B MADS box transcription factor DEF and GLOBOSA (GLO) together control the organogenesis of petals and stamens. The first classical genetic map of *A. majus* covered a genetic distance of ~420 cM, including 57 morphological markers scattered on eight linkage groups (Stubbe, 1966). A molecular marker-based genetic linkage map has been constructed recently on the basis of an F_2 population of 92 individuals derived from an interspecific hybrid *A. majus* × *A. molle* (Schwarz-Sommer *et al.*, 2003b). These maps, together with its relatively small genome size, make positional cloning feasible in *Antirrhinum* (Lai *et al.*, 2002).

Schwarz- Sommer *et al.* (2010) created a molecular linkage map of *A. majus* based on segregation of markers in the F_2 population of two inbred lab strains of *A. majus*. The resulting map consisted of over 300 markers in eight linkage groups, which could be aligned with a classical recombination map and the *A. majus* karyotype. The distribution of recombination frequencies and distorted transmission of parental alleles differed from those of a previous inter-species hybrid. The differences varied in magnitude and direction between chromosomes, suggesting that they had multiple causes. The map, which covered an estimated of 95% of the genome with an average interval of 2 cM, was used to analyze the distribution of a newly discovered family of MITE transposons and tested for its utility in positioning seven mutations that affect aspects of plant size. They found that the current map has an estimated interval of 1.28 Mb between markers. It shows a lower level of transmission ratio distortion and a longer length than the previous inter-species map, making it potentially more useful. The molecular recombination map further indicates that the IDLE MITE transposons are distributed throughout the genome and are relatively stable. The map proved effective in mapping classical morphological mutations of *A. majus*.

Genetic engineering

Early reports on transformation of antirrhinum describe the production of transgenic tissue and plants using *A. tumefaciens* and *A. rhizogenes* (Tanaka *et al.,* 1998). However, none of these techniques were efficient enough to be used for the routine generation of transgenic plants. More recent reports have described improvements on these techniques. Two groups recent publications on generation of transgenic antirrhinum using *A. rhizogenes*. GUS-positive transgenic plants were produced from the variety Golden Monarch after transformation of hypocotyls with the *A. rhizogenes* strain LBA9402 carrying the binary vector pBI121 (Senior *et al.*, 1995). More recently techniques have been published on the use of *A. tumefaciens* for the generation of transgenic antirrhinum. These techniques allow the production of morphologically normal plants without the interference from rol genes. Heidmann *et al.* (1998) transformed highly regenerative hypocotyls with *A. tumefaciens* strain GV3101 carrying the pMP90RK helper plasmid. A recent development in vector technology called the multi-autobtransformation (MAT) vector system has also been used to transform antirrhinum (Cui *et al.,* 2000, 2001).This system uses the morphological changes caused by the oncogenes of *Agrobacterium* (ipt or rol genes) as the selectable marker and this system may also do away with the need for additional selective markers, such as antibiotic or herbicide resistance (Sugita *et al.*, 1999, 2000 and a).

Plants of *A. majus* carrying the semi-dominant Macho alleles of the plena gene display carpelloid sepals and staminoid petals, but the two inner flower whorls of stamens and carpels are normal and produce fertile gametes. In the recessive plena mutant in contrast the two outer whorls are normal whereas the stamens are largely or entirely petaloid and the carpels sepaloid, thus producing weakly male fertile or fully sterile lines (Loennig and Saedler, 1994). A new family of DNA binding proteins includes putative transcriptional regulators of the *A. majus* floral meristem identity gene SQUAMOSA, the newly identified sbp gene involved in the control of early flower development (Klein *et al.*, 1995). Apparent redundancy in myb gene function provides gearing for the control of flavonoid biosynthesis in antirrhinum flowers (Moyano *et al.*, 1996).

CYCLOIDEA (CYC) and DICHOTOMA (DICH) are paralogous genes that determine adaxial flower identity in bilaterally symmetric flowers of *A. majus*. In *A. majus*, pal-rec an allele of the pallid series is considered to be involved in anthocyanin production (Aslam, 1993). The flower meristem identity genes floricaula (flo) and squamosa (squa) promote a change in phyllotaxy from spiral to whorled in antirrhinum. The benzenoid ester, methylbenzoate is one of the most abundant scent compounds detected in the majority of snapdragon varieties. It is produced in upper and lower lobes of petals by enzymatic methylation of benzoic acid in the reaction catalysed by S-adenosyl-L-methionine: benzoic acid carboxyl methytransferase (BAMT). Cells of the petal epidermis of antirrhinum, like those of many flowering plants, have a conical shape that is dependent on activity of the MYB transcription factor encoded by *MIXTA*. These conical cells intensify flower colour by reducing reflection, as revealed by *MIXTA* mutants with flat epidermal cells and are important in attracting pollinating bees (Noda *et al.*, 1994). The delayed terminal flower phenotype is caused by a conditional mutation in the CENTRORADIALS gene of snapdragon (Cremer *et al.*, 2001).

An efficient *Agrobacterium*-mediated transformation of *Antirrhinum majus* was achieved *via* indirect shoot organogenesis from hypocotyl explants of seedlings. Stable transformants were obtained by inoculating explants with *A. tumefaciens* strain GV2260 harbouring the binary vector pBIGFP121, which contains the neomycin phosphotransferase gene (NPT II) as a selectable marker and the gene for the Green Fluorescent Protein (GFP) as a visual marker. Putative transformants were identified by selection for kanamycin resistance and by examining the shoots using fluorescence microscopy. PCR and Southern analyses confirmed integration of the GFP gene into the genomes of the transformants. The transformants had a morphologically normal phenotype. The transgene was shown to be inherited in a Mendelian manner. This improved method requires only a small number of seeds for explant preparation and three changes of medium; the overall transformation efficiency achieved, based on the recovery of transformed plants after 4–5 months of culture, reached 8–9%. This success rate makes the protocol very useful for producing transgenic *A. majus* plants (Cui *et al.*, 2003).

The recent discoveries of the mechanisms of flowering in the model plants *A. majus, A. thaliana* and *Petunia hybrida* allowed new sets of genes to be available. These include genes, which affect flowering time and flower architecture, commonly named: flowering-time genes, meristem-identity genes and organ-identity genes. The possibility of using molecular techniques to transfer genes in major ornamentals has greatly increased the resources available to plant breeders. Flower induction genes could affect flowering in species unrelated to the plant from which they were isolated. Moreover, native genes can be over expressed or suppressed. In the near future plant biotechnologists and plant breeders will continue to work alongside to improve floricultural plants (Giovannini, 2007). Study on randomly amplified polymorphic genome DNA of diploid and tetraploid *A. majus* showed that the DNA fingerprint of diploid *A. majus* and that of its autotetraploid were different with cultivars and primers (Wu *et al.*, 2005). DNA fingerprinting can offer molecular evidence for identifying new cultivars of *Antirrhinum majus* and establish the basis for protecting new cultivars. Some workers produced transgenic (Table 5).

Table 5: *Antirrhinum* species from which transgenic plants have been generated.

Plant species	Transformation method	References
Antirrhinum majus	*Agrobacterium rhizogenes.*	Senior *et al.* (1995), Hoshino and Mii (1998), Hoshino *et al.* (1998).
	Agrobacterium tumefaciens	Heidmann *et al.* (1998), Cui *et al.* (2000, 2001 and 2003).

SUMMARY

In addition to antirrhinum continuing its role as a genetic model for plant biology and development, several new experimental avenues are to be explored with antirrhinum and related species. Furthermore, given the tremendous morphological and ecological diversity in this crop, a phylogeny within the genus *Antirrhinum* with an emphasis on the New World section *Saerorhinum* may help to pinpoint interesting patterns that warrant further study. To establish a system that promotes industry oriented collaborative research to enhance germplasm utilization and conservation. The wild relatives of crops continue to be an important reservoir of genes for potential use in agriculture. Sometimes, the genes furnished have had a dramatic effect on improvement and contributed in evolutionary studies. Therefore, there is a continued urgency to conserve these wild genetic resources appropriately, both *in situ* and *ex situ* and to characterize them for future crop improvement.

REFERENCES

Aslam, K.M. 1993. The Sfm system of gene control in *Antirrhinum majus*. *Theor. Applied Gene.*, **86**(4): 401-05.

Atkinson, N.J., Newbury, H.J. and Ford-Lloyd, B.V. 1991. *In vitro* adventitious root induction in *Antirrhinum majus*. *Plant Cell, Tissue Organ Cult.*, **27**: 77-79.

Bailey, L.H. and Bailey, E.Z. 1976. Hortus Third i–xiv, MacMillan, New York, pp.1–1290.

Balatkova, V. and Tupy, J. 1973. The significance of the methods of stigmatal and placental pollination *in vitro* in *Antirrhinum majus* seed and callus formation on placentae. *Biologia Plantarum* (Prague), **15**(2): 102-06.

Ball, V. 1957. The Ball Red Book, 9th ed., Geo. J. Ball. Inc., West Chicago, Illinois.

Baur, E. 1924. Untersuchungen über das Wesen, die Entstehung und die Vererbung von Rassen-unterschieden bei *Antirrhinum majus*. *Biblio. Genet.*, **4**: 1-70.

Bey, M., Stuber, K., Fellenberg, K., Schwarz-Sommer, Z., Sommer, H., Seadler, H. And Zachgo, S. 2004. Charcterization of *Antirrhinum* petal development and identification of target genes of the class B MADS box gene DEFICIENS. *The Plant Cell*, **16**: 3197-215.

Bowman, J.L., Smyth, D.R. and Meyrowitz, H. 1989. Genes directing flower development in Arabidopsis. *Plant Cell.*, **3**: 749-58.

Bradley, D., Ratcliffe, O., Vincent, C., Carpenter, R. and Coen, E. 1997. Inflorescence commitment and architecture in Arabidopsis. *Science*, **275**(5296): 80-83.

Chiej R. 1984. Encyclopaedia of Medicinal Plants. MacDonald, London, UK, pp. 448.

Chittenden, F.J. and Synge, P.M. 1956. Dictionary of Gardening. The Clarendon Press, Oxford.

Claus, S., Grau, I. and Unger, K. 1974. Studies on the reaction norm of ideotypes demonstrated by the growth in height of *Antirrhinum majus*. *Archiv-fur-zuchtungsforschung*, **4**(1): 5-17.

Coen, E.S. and Meyerow, E.M. 1991. The war of the whorls genetic interactions controlling flower development. *Nature*, **353**(6339): 57-59.

Cremer, F., Lonnig, W.E., Saedler, H. and Huijser, P. 2001. The delayed terminal flower phenotype is caused by a conditional mutation in the centroradialis gene of snapdragon. *Plant Physio.*, **126**(3): 1031-41.

Cui, M.L., Handa, E.T. and Ezura, E.H. 2003. An improved protocol for Agrobacterium-mediated transformation of *Antirrhinum majus* L. *Mol. Gen. Genomics*, **270**: 296–302.

Cui, M.L., Takayanagi, K., Kamada, H., Nishimura, S. and Handa, T. 2000. Transformation of *Antirrhinum majus* L. by a rol-type multi-auto-transformation (MAT) vector system, *Plant Sci*. **159**: 273-80.

Cui, M.L., Takayanagi, K., Kamada, H., Nishimura, S. and Handa, T. 2001. Efficient shoot regeneration from hairy roots of *Antirrhinum majus* L transformed by rol type MAT vector system, *Plant Cell Rep.*, **20**: 55-59.

Darwin, C.R. 1868. Variation of Animals and Plants under Domestication. John Murray, London.

De Paco, I. and Mateu-Aandres, L. 2006. Genetic diversity and the reproductive system in related species of *Antirrhinum. Ann. Bot.,* **98**: 1053–60.

Edwards, K.J.R. and Goldenberg, J.B. 1976. A temperature effect on the expression of genotypic differences in flowering induction in *Antirrhinum majus. Ann. Bot.,* **40**: 1277–83.

Emsweller, S.L. and Jones, H.A. 1934. The inheritance of resistance to rust in snapdragon. *Hilgardia*, **8**: 179-10.

Giovannini, A. 2007. The genes for flowering and their involvement in ornamental applied research. *Ital. Horti.*, **14**(1): 50-58.

Grau, I. 1972. A comparison of the phenotype of two Antirrhinum breeding strains and their F_1 progeny, based on the number of flowers. *Archiv-fur-zuchtungsforschung*, **2**(3): 271-74.

Harte, C. 1983. Pheno genetics of the form of leaves in *Antirrhinum majus* 6. the final size of leaves of *Antirrhinum majus* sippe 50 and several mutants in relation to the induction of flowering. *Biologisches Zentralblatt*, **102**(3): 299-21.

Heidmann, I., Efremova, N., Saedler, H. and Schwarzsommer, Z. 1998. A protocol for transformation and regeneration of *Antirrhinum majus, Plant J.* **13**: 723-28.

Hoshino, Y. and Mii, M. 1998. Bialaphos stimulates shoot regeneration from hairy roots of snapdragon (*Antirrhinum majus* L.) transformed by *Agrobacterium rhizogenes*. *Plant Cell Rep.,* **17:** 256-61.

Hoshino, Y., Turkan, I., and Mii, M. 1998. Transgenic bialaphos-resistant snapdragon (*Antirrhinum majus* L.) produced by *Agrobacterium rhizogenes* transformation. *Scientia Hort.*, **76:** 37-57.

Jain, H.K., Rana, R.S., Alexander, M.P. and Sharma, J.N. 1962. A study of fertility and phenotype in induced tetraploids of *Antirrhinum. Indian J. Genet.*, **22**:154-59.

Jauhari, S. 2006. Hybrid breeding in snapdragon (*Antirrhinum majus* L.). Ph.D. Thesis, GBPUA & T, Pantnagar, Uttarakhand, India.

Jauhari, S. and Singh, A.K. 2005. Comparative performance of different germplasm of snapdragon. *J. Ornamental Hort.,* **8**(3): 216-218.

Jauhari, S. and Singh, A.K. 2006. Evaluation of inbred and F_{1s} for flowering and postharvest attributes in snapdragon (*Antirrhinum majus*). *J. Ornamental Hort.*, **9**(2): 91-96.

Jauhari, S. and Singh, A.K. 2009. Heterosis breeding in snapdragon. *Abst.* National Conference on Floriculture for Livelihood and Profitability, IARI, New Delhi. p. 9.

Jimenez, J.F., Sanchez, G.P., Guemes, J., Warner, O. and Rossello, J.A. 2002. Genetic variability in a narrow endemic snapdragon (*Antirrhinum subbaeticum,* Scrophulariaceae) using RAPD markers. *Heredity*, **89**(5): 387-93.

Khoshoo, T.N. 1990. Creativity in ornamentals. *Indian Horticulture*, **34**(4): 25-30.

Klein, J., Saedler, H. and Huijser, P. 1995. A new family of DNA binding proteins includes putative transcriptional regulators of the *Antirrhinum majus* floral meristem identity gene SQUAMOSA. *Molecular and General Genetics*, **250**(1): 7-16.

Kumar, V. and Bhattacharjee, S.K. 2006. Antirrhinum. *In*: S.K. Bhattacharjee (ed.) Advances in Ornamental Horticulture, Vol I. Pointer, Jaipur, Rajasthan, India, pp. 234-46.

Lai, Z., Ma. W., Han, B., Liang, L. and Zhang, Y. 2002. An F-box gene linked to the self incompatibility (*S*) locus of *Antirrhinum* is expressed specifically in pollen and tapetum. *Plant Mol. Biol*., **50:** 29-42.

Loennig, W.E. and Saedler, H. 1994. The homeotic Macho mutant of *Antirrhinum majus* reverts to wild type or mutates to the homeotic plena phenotype. *Molecular & General Genetics*, **245**(5): 636-43.

Mahal, C. 1972. Cytogenetics of *Antirrhinum* and *Lantana.* Ph.D. Thesis, Kanpur University, Kanpur, India.

Mahal, C., Pal, M. and Khoshoo, T.N. 1968. Artificial and natural polyploids in *Antirrhinum. Curr. Sci.*, **37**: 5-7.

Malik, R.S. 1979. Study on heterosis in antirrhinum (*Antirrhinum majus* L). Ph.D. Thesis, Agra University, Agra, India.

Mateu-Andres, I. 2004. Low levels of allozyme variability in the threatened species *Antirrhinum subbaeticum* and *A. pertegasii* (Scrophulariaceae): Implications for conservation of the species, genetic diversity and the reproductive system in related species of *Antirrhinum. Ann. Bot.,* **94**: 797–04.

Mateu-Andres, I. and de Paco, L. 2006. Genetic diversity and the reproductive system in related species of *Antirrhinum. Ann. Bot.*, **98**: 1053-60.

McCubbin, A., Carpenter, R., Coen, E. and Dickinson, H. 1992. Self-incompatibility in Antirrhinum. *In*: Ottaviano, E., Mulcahy, D.L., Sari, G.M. and Mulcahy, G.B. (eds.) Angiosperm Pollen and Ovules. Springer, New York, pp. 104-09.

Moyano, E., Martinez, J.F. and Martin, C. 1996. Apparent redundancy in myb gene function provides gearing for the control of flavonoid biosynthesis in antirrhinum flowers. *The Plant Cell*, **8**(9):1519-1532.

Nebel, B.R. and Ruttle, M.L. 1938. The cytological and genetical significance of colchicine. *J. Hered*., **29**: 3-9.

Noda, K., Glover, B.J., Linstead, P. and Martin, C. 1994. Flower colour intensity depends on specialized cell shape controlled by a Myb-related transcription factor. *Nature*, **369**: 661-64.

Okubo, H., Wada, K. and Uemotos, S. 1991. *In vitro* morphogenetic response and distribution of endogenous plant hormones in hypocotyls segments of snapdragon (*Antirrhinum majus*). *Plant Cell Reports*, **10**(10): 501-04.

Poirer, H.S., Rao, P.S. and Harada, H. 1974. Culture of mesophyll protoplasts and stem segments of *Antirrhinum majus* (snapdragon) growth and organization. *J. Exp. Bot*., **25**(87): 752-60.

Rao, P.S., Bapat, V.A. and Harada, H. 1976. Gamma radiation and hormonal factors controlling morphogenesis in organ cultures of *Antirrhinum majus* L. cv. Red Majestic Chief. *Z. Pflanzenphysiol.*, **80**: 144-52.

Sanderson, K.C. and Martin, W.C. 1984. Evaluation and scheduling of snapdragon cultivars. Bulletin Agricultural Experiment Station, Auburn University, No. **468**: 270.

Schwarz-Sommer, Z., Davies, B. and Hudson, A. 2003. An everlasting pioneer: the story of Antirrhinum research. *Nat. Rev. Genet.*, **4**(8): 657-66.

Senior, I., Holford, P., Cooley, R.N. and Newbury, H.J. 1995. Transformation of *Antirrhinum majus* using *Agrobacterium rhizogenes*, *J. Exp. Bot*., **46:** 1233-39.

Simmonds, N.W. 1979. Principles of Crop Improvement, Longman, London. 408 p.

Singh, A.K. 2005. Evaluation of snapdragon (*Antirrhinum majus* L.) germplasm. *Progressive Hort.*, **37**(1): 96-99.

Singh, J. and Arora, J.S. 1980. Variability studies on some quantitative characters in snapdragon (*Antirrhinum majus*). *J. Res.*, **17**(3): 258-61.

Sprague, G.F. and Tatum, L.A. 1942. General versus specific combining ability in single cross of corn. *J. Amer. Soc. Agron.*, **34**: 923-32.

Stubbe, H. 1966. *Genetik und Zytologie von Antirrhinum L. sect. Antirrhinum.* VEB Gustav Fischer, Jena, Germany.

Sugita, K., Matsunaga, E. and Ebinuma, H. 1999. Effective selection system for generating marker-free transgenic plants independent of sexual crossing, *Plant Cell Rep.*, **18**: 941- 47.

Sugita, K., Matsunaga, E. and Ebinuma, H. 2000. A transformation vector for the production of marker-free transgenic plant containing a single copy transgene at high frequency, *Plant J.* **22**: 461-69.

Sugita, K., Matsunaga, E. and Ebinuma, H. 2000a. Transgene stacking in plants in the absence of sexual crossing, *Mol. Breed.*, **6**: 239-46.

Sykorova, O. 1975. Male sterility in snapdragon (*Antirrhinum majus* L.) and its practical utilization. Genet. Slechteni., **11**(3): 189-98.

Tanaka, Y., Tsuda, S. and Kusumi, T. 1998. Physiol Metabolic engineering to modify flower colour *Plant Cell Phys.*, **39**: 1119-26.

Von-Auw, P. 1990. Contribution to the comparative study of 40 Antirrhinum bedding cultivars at the Botanic School and Garden of the city of Geneva. *Revue Horticole Suisse*, **63**(4): 89-93.

Wang, H., Ouyang, J. and Xue, Y. 2004. A facile procedure for efficient plantlets regeneration from self-incompatible hybrids in *Antirrhinum*. 2004. *Plant Cell Tissue Organ Culture*, **76**(1): 61-65.

Wu, FuChun., Hu, Xiu. and Zheng, SiXiang. 2005. Study on randomly amplified polymorphic genome DNA of diploid and tetraploid *Antirrhinum majus*. *J. Yunnan Agril. Uni.*, **20**(4): 482-85.

Yang, Q., Zhang, D., Li, Q., Cheng, Z. and Xue, Y. 2007. Heterochromatic and genetic features are consistent with recombination suppression of the self-incompatibility locus in *Antirrhinum*. *Plant J.*, **51**: 141-51.

Yu, D., Deng, G., Sheng, Y., Li, D., Yu, D.Q., Deng, Q.L., Sheng, Y.N. and Li, B.J. 1996. Generation and cytological observation of adventitious buds from hypocotyls explants of *Antrrhinum majus*. *Acta Scientiarum Naturalium Universitatis Sunyatseni*, **35**(4): 75-79.

Zhang, Dongfen., Yang, Qiuying., Bao, Weidong., Zhang, Yu., Han, Bin., Xue, Yongbiao. and Cheng, Zhukuan. 2005. Molecular cytogenetic characterization of the *Antirrhinum majus* genome. *Genetics*, **169**(1): 325-335.

Zhang, X.F. and Yue, H. 1990. The preliminary study on polyploidy of *Antirrhinum majus*. *Acta Horticulturae Sinica.*, **17**(1): 76-80.

Chapter – 10

Carnation

Carnation (*Dianthus caryophyllus* L.) is a member of the family Caryophyllaceae. *Dianthus* word was first coined by Theophrastus from Greek word "*dios*" meaning divine and "*anthos*" meaning flower. It is a half hardy, perennial plant, propagated commercially by terminal cuttings. It is being produced as a cut flower in the world. Carnation cut flower trade contributed about 15% of the world flower market. Till now, new carnation varieties have been produced mainly through conventional breeding method. The breeding procedure typically consists of hybridization, self-pollination and selection (Holly and Baker, 1992). The breeding programme did not get enough momentum due to heterozygosity, limited gene pool and inadequate knowledge of genetic make up. However genetic engineering and molecular breeding have opened the way to achieve novel colour and good vase life of flowers.

ORIGIN AND DISTRIBUTION

It is native of Mediterranean region and its distribution is restricted to Greece, Italy, Sicily and Sardinia (Tutin *et al.*, 1993). According to Pliny, the centre of origin of carnation is Spain. The generic name *Dianthus* comes from writing of the Theophrastus who lived about 300 BC; *dios,* divine and *anthos,* flower; the flower of the gods. The species name *caryophyllus* is coined from Greek word *Caryon*, nut and *phyllon*, leaf. Linnaeus chose the species name, *caryophyllus,* after the genus of clove, as the fragrance from carnation is reminiscent of clove. The common word carnation probably comes from a Greek word coronation because of its use as decorating the crowns of Greek athletes during ancient times. Carnation has been cultivated for over 2000 years and today commercial cultivation is the result of 200 years of improvement and breeding.

During sixteenth century its improvement work was started in various countries. The gardens of Italy, France, Germany, Holland and England have contributed much to the development of cultivars. The natural climates for carnation occur near 30° N or S latitudes of equator and on the western edges of the continents.

Carnation growing countries are Spain, Kenya, Columbia, Israel, Ceylon, Canary Islands, Australia, Valparaiso, Chile, USA and South Africa. Other suitable carnation growing areas modified by altitude are Bogota, Columbia, mountain area of Mexico and Central America and some parts of Kenya in Africa. An altitude of 2800 m near equator is near ideal as any place in the world. In Europe, carnations are being produced in open fields in lesser cost.

In India carnation is grown intensively in Nasik, Pune, Srinagar, Bengaluru, Coimbatore, Delhi and Kolkata. Places having congenial climate like Kalimpong, Shimla, Solan, Kullu, Ooty, Nainital, Chaubattia, Yercaud, Kodaikanal etc. are most suitable areas for the production of cut flowers. From Himachal Pradesh and Ooty, rooted cuttings are supplied by the growers to the different parts of the country (Singh, 2006).

SPECIES

Bailey (1969) reported 300 wild species of carnation. However, Galbally and Galbally (1997) and Jurgens *et al.* (2003) reported more than 300 identified species. The commercially important species and their origin have been mentioned in Table 1.

Table 1: Commercially popular members of *Dianthus* species

S. No.	Botanical name	Common name	Origin
1.	*D. alpinus* L.[a]		Austrian Alps.
2.	*D. arenarius* L.[ab]		Northern and eastern mountains of Europe.
3.	*D. armeria* L.[b]	Deptford pink	
4.	*D. arvernensis*[a]	Finnish Pink	Auvrgne mountains of France.
5.	*D. barbatus L.*[ab]	Sweet William	Grown in Britain as early as 1573.
6.	*D. carthusianorum L.*[b]	Cluster-head Pink	
7.	*D. caryophyllus* L.[ab]	Carnation	Mediterranean.
8.	*D. chinensis* L.[ab]	Indian Pink or Rainbow Pink	Hills of eastern Asia.
9.	*D. deltoides* L.[ab]	Maiden Pink	Europe and Asia and was reported in Britain in 1581.
10.	*D. erinaceus* Boiss.[a]		Dwarf, alpine form, mountains of the Middle East.
11.	*D. fragrans* M.F. Adams[b]	Fragrant Pink	
12.	*D. freynii* Vandas[a]		Native of Hungary and Bosnia
13.	*D. gratianopolitanus* Vill.[ab]		Southwestern France, introduced to

(*Contd.*)

S. No.	Botanical name	Common name	Origin
			Britain in 1792.
14.	*D. haematocalyx* Boiss. & Heldr.[a]		Greece.
15.	*D. knappii* Asch. & Kanitz[a]		Hungary.
16.	*D. microlepis* Bioss.[a]		Mountains of Bulgaria
17.	*D. myrtinervius* Grisch[a]		Alpine meadows of Macedonia
18.	*D. neglectus* Loisel.[abc]		Swiss and Italian Alps.
19.	*D. nitidus* Waldst & Kit.[ab]		Mountains of Macedonia.
20.	*D. plumarius* L.[ab]	Feathered Pink	Southern Russia.
21.	*D. repens* Willd.[b]		Eastern European grasslands.[d]
22.	*D. seguieri* Vill.[b]		Temperate regions of the eastern Mediterranean.[d]
23.	*D. squarrosus* M. Bieb[a]		Southern Russia
24.	*D. superbus* L.[ab]		Central Europe
25.	*D. sylvestris* Wulfen[ab]	Woodland Pink	Alpine part of southern Europe.

a. From Galbally and Galbally (1997).
b. Listed in Global biodiversity information data portal (GBIF): www.gbif.net. The relevant authority is identified as Catalogue of life partnerships: Integrated Taxonomic Information System.
c. *D. neglectus* was identified by Galbally and Galbally (1997) and its alternative name *D. pavonius* is also mentioned. The GBIF portal lists it as *D. pavonius*.
d. *Source*: Rock garden plant database: (www.kadel.cz 2004).

Some other important species in the improvement of carnations are *D. anatolicus*, *D. caesius*, *D. callizonus*, *D. capitatus*, *D. cinnabarinus*, *D. cruentus*, *D. diadematus*, *D. fimbriatus*, *D. giganteus*, *D. glacialis*, *D. grandiflorus*, *D. heddewigii*, *D. hybridus*, *D. lalifolius*, *D. monspessulanus*, *D. nobilis*, *D. pancicii*, *D. petraeus*, *D. versicolour*, *D. viscidus*, *D. winteri*, *D. neglactus* var. *roysii*, *D. allwoodii*, *D. deffoides* and *D. pontederae*. In India about 12 species of *Dianthus* genus are found. Out of these 9 species are wild (Anon., 1952). Development of modern carnation includes *D. caryophyllus*, *D. chinensis*, *D. barbatbus*, *D. knappi*, *D. plumarius*, *D. winteri*, *D. heddewigii*, *D. laciniatus*, *D. nobilis*, *D. superbus*, *D. arenarius*, *D. gratianopolitanus and D. arboreus*. *D. barbatus* and *D. chinensis* are winter annual species propagated by seeds. The yellow colour in carnation is believed to have been contributed by *D. knappi*.

FLOWER MORPHOLOGY AND BIOLOGY

Flower morphology and biology were studied by various workers (Bhat *et al.*, 1993 and Cuellar *et al.*, 1999). According to the study of Bhat *et al.* (1993) carnation flowers are pedicellate, complete, hermaphrodite, actinomorphic, pentamerous, hypogynous and borne in terminal or solitary regions of the growing stem. Flowers are protandrous type. Petals and stamens are variable in number (5-50 and 10-30, respectively).

Classification of carnation according to floral morphology

On the basis of floral morphology, there are different kinds of carnations which are described below.

Chaubad or Marguerette: This is the result of crossing *D. chinensis* × *D. caryophyllus*. These annual flowers may be single or double with fringed petals and shorter vase life.

Border and Picotee: In border carnation, single stem has been produced from the base which becomes bushy in the subsequent years. The flowers are smooth edged, frilled with open centre. Border flowers are further classified on the basis of colour.

Bizarres- Flowers have clear ground, marked and flaked with 2 or 3 colours.

Flakes- Flowers are with clear ground, flaked with one colour.

Selfs- There is only single clear colour of any shade.

Fancies- Flowers have a yellow or white background or mottled or flaked and spotted with various colours.

Picotees- Flowers are always either white or yellow and have a narrow band near the margin of each petal.

Malmaison: Plants are bushy with broader leaves, the buds are rounded and the calyx is shorter and apt to burst. Flowers are fragrant.

Perpetual flowering: Plant flowers all the year round. The flowers are of self colour with longer vase life.

The florists carnations are grouped into two classes.

1. **Standard:** They produce large blooms on longer flower stalks. The plants perform better under cooler climate.

2. **Sprays:** Spray carnation produce many flowers of small size with weaker stems. Sprays are suitable for warm climate.

Anthesis

The stigma tends to be higher than the anthers within the pinks and carnation. The pistil consisted of 2-5 carpels. The gamosepalous calyx forms a tube like structure enclosing the corolla and essential organs before anthesis.

Anthesis starts early in the morning at 6.00 am. Maximum numbers of flowers open between 9.00 and 10.00 am. Carnation is highly cross-pollinated crop (Watts, 1980). Anther dehiscence commences immediately after anthesis, which is maximum between 9.00–10.00 am.

Pollen viability

Pollen viability increases gradually until just open flower stage, time at which anthesis occurs and then it decreases in fully open flower stage. Pollen production is also affected by temperature and temperature of 23°C proved to be optimum (Kho and Baer, 1973). In case of var. Nora Barlo pollens were viable only in just open flower and fully open flower stage. In Kaly variety pollens do not remain viable at any developmental stage. The highest percentage of pollen germination has been observed in cv. Casterlaro with 73.33% (Cuellar *et al.*, 1999). Pollens can be stored at 20-25°C for 10 days with 31.94% viability, while at 4°C temperature for 60 days with 54.70% viability.

Stigma receptivity

Stigma becomes receptive after 48 hours of anthesis and remains viable upto 9 days with a peak at 72 hrs after anthesis (Bhat *et al.*, 1993). However, optimal moment of receptivity was determined at the fully open flower stage (Cuellar *et al.*, 1999). Cross pollination in carnation improves seed set resulting in increase of seven folds in seed production (Sparnaaij and Beeger, 1973). The highest number of seeds/capsule has been obtained with hand-cross pollination compared with hand self-pollination. An average of 39 days is required for capsule maturity (Bhat *et al.*, 1993).

In the context of *Dianthus barbatus*, the inflorescence initiation takes approximately at 160 days, whereas the flower bud takes an average of 16 days for full development. The maximum flower anthesis occurs between 10-12 am. Similar to flower anthesis, maximum anther dehiscence has been recorded between 10-12 am. Anther dehiscence is faster on sunny days than on rainy days and high humidity. The maximum pollen viability (96%) has been reported in freshly dehisced pollen grains. Maximum germination of pollens and pollen tube growth has been measured in 6% sucrose media. The stigma becomes receptive after 48 hrs of anthesis as pollination after 48 hrs of anthesis gives maximum seed set (Gupta *et al.*, 2004).

GENETICS

Dianthus caryophyllus is a diploid (2n = 30) with basic chromosome number x = 15. *Dianthus knappi* is also a diploid species. Most of the species of *Dianthus* genus are diploid but tetraploid and hexaploid have also been reported. Tetraploids have larger flowers and more rigid stems than diploid but they are less productive and therefore are seldom grown. *D. chinensis* is tetraploid (2n = 60). *D. gratianopolitanus* is another species which is tetraploid (Howard, 1968 and 1971) but also exists in hexaploid form. *D. allowoodii* develoed by Mr. Allwood in England is a hexaploid (2n = 60). Triploid carnations were produced for commercial purpose, but the resulting plants were mostly aneuploid.

The doubleness in flowers is due to the conversion of stamens into petals. Thus most of the varieties are male sterile.

Inheritance of flower type and doubleness

Various characteristics of carnation are governed by qualitative genes; however some are governed by quantitative genes. Doubleness and singleness is a complex of qualitative genes. Commercial carnations grown by florists is an unified hybrid obtained from a cross between single and super double and the bullhead type is incompletely dominant to the single and the hybrid between two is intermediate, which is known as commercial double. Doubleness in carnation is a monogenic character governed by dominant gene.

The doubleness has been further investigated by Dumlao (1969). Carnation is grouped into three groups *viz.*, single (4-6 petals), semi-double (30-60 petals) and double (100-350 petals). Single flowers are of two types, the basic single (dd) and variable single (Ee). Basic single gives single flowers, whereas variable single type segregates after selfing into 3 singles (EE or Ee) to 1 semi-double (ee). Doubleness is governed by single recessive gene (e). Semi-double is also having two types of inheritance, one is of lower petals and another one is of higher petals. Lower petaled inheritance is further divided into two types, the recessive genotype ee produces 8-15 petals and FF and Ff genotypes produce 16-45 petals. The F gene is incompletely dominant to f. While doubleness in higher petaled is governed by D gene. After selfing Dd genotype, gives semi-double (DD or Dd) and single in 3:1 proportion. Thus, true doubles can be obtained by selection after continuous selfing. Unlinger and Lindgren (1984) released cv. N 74133 a germplasm and the progenies had resulted in large double flower with lavender and pink colours, those were dominant.

Inheritance of flower colour

Mehlquist (1940) studied the inheritance of colour and found that 6 factors are responsible for expression of self colour. The 'Y' factor controls yellow colour. 'I' controls production of ivory anthoxanthin. 'I' is epistatic to 'Y'. 'A' controls the anthocyanin production. But 'A' will express only in the presence of 'Y' and 'J'. 'S' controls the production of scarlet-red anthocyanin, its recessive allele 'r' produces faded red. 'R' is dominant and produces carmine red, whereas its recessive permits the production of scarlet red anthocyanin only. 'M' controls the colours of red series. The colour inheritance in F_1 and F_2 progenies are given in Table 2 as described by Mehlquist (1940).

Table 2: Inheritance of colour in the carnation.

S. No.	Parents	F_1	F_2
1.	Salmon × red	Red	3 red : 1 salmon
2.	Salmon × crimson	Crimson	9 crimson : 3 red: 3 lavender pink : 1 salmon
3.	Salmon × light pink	Light pink	3 light pink : 1 salmon
4.	Salmon × deep pink	Dark pink	9 deep pink : 3 light pink : 3 red : 1 salmon
5.	Red × crimson	Crimson	3 crimson : 1 red
6.	Red × deep pink	Deep pink	3 deep pink : 1 red
7.	Crimson × deep pink	Purple	9 purple : 3 deep pink: 3 crimson : 1 red
8.	Light pink × deep pink	Deep pink	3 deep pink : 1 light pink
9.	Light pink × red	Deep pink	9 deep pink : 3 light pink : 3 red : 1 salmon
10.	Deep pink × purple	Purple	3 purple : 1 deep pink

Carnation flowers also exhibit incomplete dominance in the expression of flower colour when homozygous red flowers are crossed with homozygous white flowers, the offspring, all heterozygous have pink colour. Earlier workers revealed the biochemistry of flower colour that chemical effect, accompanied alterations in the genetic constitutions with respect to factors 'S', 'R' and 'M'. It was found that the degree of nuclear hydroxylation of the anthocyanin pigments of the flower was affected by factor 'R'. Cyanic forms that contained 'R' in homozygous or heterozygous condition contained cyanidin glycosides. The homozygous recessive (rr) contained pelargonidin derivatives and in the white and yellow only rr specimens were found (which had no anthocynins) but flavanol, kaempferol, the hydroxylation pattern of which was same as that of pelargonidin. Itoh *et al*. (2002) studied the excision of transposable elements, responsible for change in colour. In the Rhapsody cultivar of the carnation, which bears white flowers variegated with red flecks and sectors, a transposable element, dTdic 1, belonging to the *Ac/Ds* superfamily, was found within the dihydroflavonol 4-reductase (*DFR*) gene. The red flecks and sectors of cv. Rhapsody may be attributable to a reversion to *DFR* activity after the excision of *dTdic 1*. The yellow colour of the carnation petals is attributed to the synthesis and accumulation of chalcone 2'-glucoside. In several carnation cultivars that bear yellow flowers variegated with white flecks and sectors, both the chalcone isomerase (*CHI)* and *DFR* genes are disrupted by *dTdic 1*. The inheritance of various characters in carnation is given in Table 3.

Table 3: The inheritance of various characters in carnation.

S.No.	Character	Gene	Ratio
1.	Lutescent seedling vs normal	*lut* recessive gene, monogenic showing greenish yellow colour	3 normal: 1 lutescent
2.	Club-neck vs normal	*cb* recessive gene, monogenic	3 normal: 1 club neck
3.	Virescent vs normal	*vir* recessive gene, monogenic	3 normal: 1 virescent
4.	Lazy virescent vs normal	*laz* recessive gene, monogenic	3 normal: 1 lazy virescent
5.	Dwarf vs normal	*dw* recessive gene, monogenic	3 normal: 1 dwarf
6.	Thin vs normal	*single* recessive gene	3 normal: 1 thin

Heritability and genetic advance

Heritability of various characters has been studied by several workers. Diameter of flowering stem and petal number showed high heritable values and higher genetic advance in case of plant height and petal number (Anon., 1986). Whereas, Kaicker (1988) reported that medium and low heritability was found for number of petals, duration of flowering and cut flower life, which can be utilized for exploitation of heterosis. In a half diallel method, Sparnaaij and Demmink (1987) found significant additive variance for resistance against fusarium wilt. Further trial confirmed the predominance of additive inheritance (Boayen *et al.*, 1991). High heritability with high genetic advance was observed for flowering duration followed by stem strength and vase life. Identification of these traits can be useful in breeding programme through selection. These characters are governed by additive gene action (Misra and Gupta, 2003).

BREEDING OBJECTIVES

i. More number of branches/plant
ii. Compact plant growth
iii. Winter hardiness
iv. Strong stem
v. Fragrance
vi. Large flower size
vii. Attractive and novel colour
viii. Longer vase life
ix. Long duration of flowering
x. Early flowering
xi. High yield and productivity of flowers
xii. Calyx splitting resistance
xiii. Resistance to diseases particularly *Fusarium*, *Phialophora*, etc.

BREEDING METHODS

Introduction

Due to lack of gene pool, breeding of carnation can be restricted. NBPGR has introduced 77 varieties as well as species from Australia, Cyprus, France, Germany, Japan, UK and USA. Some important introductions from different countries to India are given in (Table 4).

Table 4: Varieties and species of carnation introduced from different countries.

S. No.	Country	Varieties and species
1.	Australia	Chaubad Giant Mixed, F_1 Hybrid Dwarf Mixed Colonis, Genes of France, *Dianthus plumarius*, *D. patraeus*, Perpetual Giant Mixed.
2.	Cyprus	Yellow Dusty, William Sim, Lena, White Sim and Arther Sim.
3.	France	Picotee Mixed, Grenddin, King of Blacks, Sunflod, Triompha Vauk, Grenadin Roi Des Roses, Vauie Selection Clause and Grenddin Range.
4.	Germany	Laddie Sim, Pink Manic, Red Sim, Arthur Sim, Dusty Sim, Scania.
5.	Japan	*D. chinensis* var. *laciniatus* Regel and *D. chinensis* var. *heddewigii* Regal.
6.	UK	Chaubad Allwoods, *D. spiculifolius*, *D. armeria*, *D. balbissii*, *D. carthusianorum*, *D. deltoids*, *D. pinifolius* var. *lilacinus*, *D. pulmarius*, *D. turkestianicus*, *D. superbus*, *D. praecox*, *D. sternbergii* and *D. fragrans*.
7.	USA	Red Wonder, Scarlet Wonder, Pixie Delight and Scarlet Luminette.

The standard varieties like White Sim, Dusty Sim, Laddie Sim and Arthur Sim and spray varieties *viz*., White Elegance, Pride, Scarlet Elegance and Exquisite have performed well in Ludhiana while variety Arthur Sim performed well in Delhi, Pune and Solan. The performance of cv. Scania was found to be the best under Solan and Pune conditions (Gill, 1990).

Selection

Method of improvement through seedling selection was mostly utilized in carnation. One seedling selection of perpetual carnation had larger and more beautiful flowers in France during 1857 and was called as Malmaison Giant. The plants are much stiffer type with broader leaves and massive appearance and have very large double flowers which are well filled with perfume.

Two varieties of carnation *viz.* Linlong 2 and Linlong 3 were selected from hybrid population of *Dianthus caryophyllus*. These cultivars are characterized by large and fragrant flowers, long flowering season, strong adaptability to desert and easy propagation. These are good ornamental plants for cut flower (Sun and Lin, 2007). A variety named 'Arka Flame' was developed by IIHR, Bengaluru and released in 2008.

Hybridization

Emasculation and pollination: The technique of crossing two varieties together is easy. The flowers must be emasculated at the bud stage as anther dehiscence starts after flower opening. Bagging is done after emasculating buds. The ripened stigma shows numerous minute hairs with a glistening drop of fluid at the tip, after 3-4 days of anthesis. The fresh pollen, collected from newly opened flowers, is transferred on the ripe stigma. The stigma should remain dry and protected from moisture for at least six hours. If the pollen has taken, the petals collapse in a typical way 12 to 48 hrs after pollination. As soon as this happens, the calyx should be torn and open on the underside to prevent moisture which might rot the seed pod. The petals should be pulled out as soon as they have shrivelled enough to pull out easily and the calyx should be removed entirely. When the browning extends about a third of the way down the pod, the pod should be gathered, placed in a paper bag and kept for a week in a dry place. After extracting seeds from dried pods, seeds must be stored in packet until they are sown.

Heterosis: It was found that mean number of laterals per plant, flower size and increased number of petal of F_1 hybrids exceeded the parental means. Appreciable amount of heterosis per flower size (54.5%) has been observed in some combinations, though earliness to flowering was highly variable (Kaicker, 1988).

The earliest improvement work through hybridization in carnation began in the 16th century. In 1717, Thomas Fairchild developed Mule Pink by crossing *D. barbatus* × *D. caryophyllus*. The hybrid was double flowered and sterile (Crane and Lawrence, 1952). A French gardener, M. Dalmais obtained a constant blooming carnation Atim by crossing Eillet de Mahon and Eillet Biohon. The perpetual-flowering race of carnation, which has been developed by crossing of *D. caryophyllus* and *D. chinensis* is originated in France. The cross of *D. chinensis* and *D. caryophyllus* has given rise to Marguerite carnations, which have broader and greener leaves than *D. caryophyllus* (Gorer, 1970). Another cross of Mrs. T.W. Lawson with progeny of Pink Perpetual and Border carnation, Britannia was obtained with early flowering of scarlet colour that does not loose its colour during winter months. Mr. Martin Smith developed a cross between Malmaison and Border carnation, named as Mrs. Martin Smith, which grew 150 cm high and had flowers of 20 cm diameter. In 1951 Mr. Allwood prepared a race called Royal carnation by combining Malmaison with Perpetual carnations. Allowood made around 300 crosses and only found 32 fertile seeds. They were further crossed with Picotee and the resultant hybrids were of perpetual flowering Picotees. These Picotees are characterized by different coloured petal margin. *Dianthus allwoodii* is a hybrid between *D. plumerius* and perpetual flowering carnation. They are having perpetual flowering with very hardy plants, scented, large double and semi-double flowers. In a further improvement programme Sweet Wivelsfield was developed by crossing of *D. allwoodii* and *D. barbatus*, whereas a new race was evolved by hybridizing Sweet Wivelsfield with deeply fringed *D. superbus* cultivar. Another named series Delight was resulted from crossing Sweet Wivelsfield and *D. neglectus* var. *roysii* (Gorer, 1970).

Some series like Diantinis, developed by Dr. Nobbio, Italy shows attractive colour pattern; micro-carnations, bred by Mr. Brea. Italy is a very compact type with button like flowers produced in large quantities; multiflora carnations developed by Dutch firm of Van Staaveren, Holland having smaller flowers than spray, larger than micros, cross of Dianntini × border carnation, known as Pico which provides flowers of wider colour range and colour pattern with dark centres; F_1 Knight series, developed by Goldsmith, California, produces 15-20 flowers/stem and due to strong stems needing no support under Delhi condition; F_1 Pixie Delight and Scarlet Luminette by Burpee Seed Co. are also one of the best achievements of classical hybridization (Sparnaaij and Demmink, 1983 and Kaicker, 1988).

Some important hybrids are Allwood's Yellow, Allwood's Cream, Allwood's Pink, Emily Allwood (pearl white) and George Allwood (white fringed and scented). In Chaubad Giants Series, Marie Chaubad (sulphur yellow), Legiond Honour (blood red), Enticelant (red), Nero (purple) and Giroflee (violet pink) are the important hybrids which all are strongly scented and very beautiful. Other F_1 hybrids Pixie Delight, Scarlet Luminette and Scarlet Knight are also very popular for cut flowers.

Information regarding development of modern races through hybridization in carnation is very less due to secrecy in trade. There are only 50 cultivars of standard and spray types which are being maintained by AICFIP at various centres of India.

Varieties Statas (Van Staaveren, 1997) and Statropur (Van Staaveren, 1998) have been developed by hybridization. Cv. Statas is the cut flower type producing medium to long stems and large double and pink flower. This is the hybrid of Barlow × Klekopi. Stratropur is the variety of miniature spray type with pink purple coloured flowers. It is the cross of 83000-1J × 84200-2J. Onozaki *et al.* (2001) adopted repeated crossing of Pallas, Sand Rosa, Candy, Tanga, White Sim and Scania and selection for 3 generations to improve the vase life of carnation flowers. All 39 selected lines showed significantly longer vase life (11.1 to 17.5 days) than White Sim (5.4 days). A new cultivar Yundieyi of *Dianthus caryophyllus* was developed through hybridization from parentage Rendiz-Vors and Odino (Gui *et al.,* 2011).

Details of varieties of carnation developed through hybridization are given in Table 5.

Table 5: Varieties of carnation developed through hybridization.

S. No.	Variety	Parent	Year	Characteristics
1.	Statas	Barlow × Klekopi	1997	Flower colour pink. Flowers are large and double. Stem medium to large. Good for cut flower.
2.	Stratropur	83000-1J × 84200-2J	1998	Spray type. Flowers pink purple coloured. It is a miniature type variety.
3.	Yundieyi	Rendiz-Vors × Odino	2011	Flowers are big, good flower shape and fragrance. Long branches stand erect. Propagation coefficient is high. Longer vase-life. Resistance to transport.

Breeding of Spray type carnations started in Bulgaria during1986. After import of 14 spray carnation varieties from the company Van Staveren, i.e. Red Baron, Red Ministar, Danilo, Silvery Pink, Salmony, Sam's Pride, Mirna, Jolivette, Tony, Albivette, White Lilli Ann, Exquisitte, Scarlet Elegance and Casino breeding work was started. At a later stage, 10 more varieties grown in Bulgaria, were included: Bianca, Evelin,

Festival, Gold Star, Hellisal, Pink Hellisal, Red Barbara, Royalette, Starlight and Super Gold. After a comprehensive breeding 10 varieties of carnation were released (Table 6). The blooms of the varieties Naslada, Elmaz, Feja, Ira and Biljana are strongly fragrant, those of Regina - slightly fragrant and Janita, Rossitsa, Krassina, Niki and Russalka - non-fragrant. Varieties Naslada, Elmaz, Janita, Regina, Rossitsa and Ira have the highest yield and best quality of the cut flower. Varieties Elmaz, Ira, Regina and Niki showed the highest resistance to *Fusarium* wilt.

Table 6: Spray type carnation through hybridization in Bulgaria

S.No.	Variety	Parentage	Characteristics
1.	Krassina	Casino × Exquisite	Double, bicolour flowers. Petals red at base with a lighter stripe towards edge. Petals are slightly serrated, almost even-edged, with strongly pleated leaf blade that accounts for the originality of the flower.
2.	Biljana	Danilo × Albivette	White double flowers with radial red hatches, strongly fragrant. The petals are slightly serrated.
3.	Regina	Silvery Pink × Casino	Dark purple, double and slightly fragrant flowers with slightly serrated petals. Resistant to *Fusarium* wilt.
4.	Janita	Mirna × Salmony	Double, deep pink, non-fragrant flowers and deeply serrated petals.
5.	Niki	Red Ministar × Sam's Pride	Bright red, double, non-fragrant flowers and slightly serrated petals. Resistant to *Fusarium* wilt.
6.	Feja	Salmony × Casino	Double, light pink, strongly fragrant flowers with serrated petals.
7.	Naslada	Casino × Salmony	Large, double, dark red, strongly fragrant flowers and deeply serrated petals.
8.	Russalka	Red Ministar × White Lilli Ann	Very large, double, non-fragrant flowers. Pale yellow colour wit radial red hatches and deeply serrated petals.
9.	Elmaz	Barbara × *D. gratianopolitanus*	Large, double, light pink, strongly fragrant flowers and deeply serrated petals. Resistant to *Fusarium* wilt.
10.	Ira	Biljana × Russalka	Double non-fragrant flowers, white with thin and thick red hatches and strongly serrated petals. Resistant to *Fusarium* wilt.

Breeding for resistance to *Fusarium* and bacterial wilt

Breeding for resistance to diseases curtails the dependency of pesticides. The first work in this direction was reported by Guba (1951). Various varieties like Waltham Pink, Regal Pink, Spicy Rose and Mrs. E.F. Guba had been developed for multiplying disease resistance (wilt, rust and blight) through hybridization. The history of extensive breeding for resistance to vascular wilts in carnation started in 1960's at the IVI to develop resistance to *Phialophora cinerescens*. This fungus also causes a vascular wilt similar to that of *Fusarium oxysporum* but slower to develop. After extensive screening and about six generations of back crossing with commercial varieties, a high level of resistance had been achieved inbreeding population (Sparnaaij and Demmink, 1976, 1977). In Sim group var. Rose and Arthur Sim showed high level of resistance against rust (Semina and Shestachenko, 1981). cv. Raveda (Red) and Novada (Pink) are also resistant to *Fusarium* wilt (Demmink *et al*., 1982).

A variety Duca was reported to be resistant against all the pathotypes of *F. oxysponem* f. sp. *dianthi* (Garibaldi, 1983). Frank (1985) stated that the best quality yellow flower cv. Angela had the greatest resistance to *Fusarium oxysporum* whereas Salome and Vanya were disease tolerant. Onozaki *et al*. (1998) reported that the resistance of *Dianthus* wild species to bacterial disease (*Pseudomonas caryophylli*) was inherited by interspecific hybridization. Homozygous dihaploid lines (DHLS) from standard carnation clones, resistant to *F. oxysporum* f sp. *dianthi* have been developed. These lines can be further utilized to accelerate breeding for new resistant varieties (Dolcet-Sanjuan *et al*., 2001).

Fusarium wilt resistant cv. Jasujung was developed from the cross of red standard cultivar 'Francesco' and purple standard cultivar 'Saturn' and recommended to grow over 8°C at night and under 25°C at days (Joung *et al*., 2009). Similarly cv. Lyra was made by crossing between spray cultivar Ballantyne and red spray cultivar Chateau, which is early in flowering, multi branching, fragrance and long flower stalk, with medium resistance to *Fusarium* wilt (Kwon *et al*., 2010). A new spray carnation of early flowering, having white colour with red peak edge 'Lepus' developed a cross of 'Diamond' and '99257-7 Line'. It has medium fragrance, many flowers and attractive flower arrangement. In particular, flowering was 10 days early (Kwon *et al*., 2012).

Bacterial wilt caused by *Burkholderia caryophylli* is one of the most damaging diseases of carnations in Japan. Yagi *et al*. (2012) performed repeated crossing and selection for resistance to produce a new cultivar having resistance derived from *Dianthus capitatus* ssp. *andrzejowskianus*, a wild species highly resistant to bacterial wilt. They succeeded in developing the world's first wilt-resistant carnation cultivar namely Karen Rouge. It is a deep yellowish-red standard-type cultivar selected from a cross between carnation cultivar Miracle Rouge and resistant line 4AZ31-5. Cv. Karen Rouge has a high degree of resistance to bacterial wilt, and its productivity and other traits desirable for cut flowers are similar to those of modern commercial cultivars. In 2004, they initiated marker-assisted selection with the STS-WG44 marker, which is linked to a QTL with a large effect for bacterial wilt resistance derived from *D. capitatus* ssp. *andrzejowskianus*. Details of varieties of carnation resistant to *Fusarium* and bacterial wilt are given in Table 7.

Table 7: Varieties of carnation resistant to *Fusarium* and bacterial wilt.

S. No.	Variety	Parent	Year	Characteristics
1.	Jasujung	Francesco × Saturn	2000	Standard type. Flower purple colour. Recommended to grow over 8°C at night and under 25°C at days. Good for cut flower, has good fragrance and long stem. Strong resistances to *Fusarium* wilt.
2.	Lyra	Ballantyne × Chateau	2003	Spray carnation. Flower bright pink colour. Recommended to grow over 8°C at night and under 25°C at days. Early flowering, multi branching, fragrance and long flower stalk. Good for cut flower. Medium resistances to *Fusarium* wilt.
3.	Lepus	Diamond × 99257-7 Line	2004	Spray type. Flower white colour with red peak edge. Strong flower stalks. Produce many flowers and good for flower arrangement. Medium resistances to *Fusarium* wilt.
4.	Karen Rouge	Miracle Rouge × 4AZ31-5	2012	Standard type. Flower colour deep yellowish red. Good for cut flowers. High degree of resistance to bacterial wilt (*Burkholderia caryophylli*).

Interspecific hybridization

An investigation has been done to find out the possibility of crossing compatibility of *D. caryophyllus* with some wild species (Atanasova, 1998). It was observed that crosses between *D. silvestris*, *D. knappi*, *D. sequierii*, *D. carthusianorum* and *D. caryophyllus* were successful and able to produce seeds and hybrid plants crosses involving *D. chinensis*, *D. plumarius*, and *D. gratianopolitanus* were successful when the spray carnation cultivars were used as the maternal plants. Crosses between cultivated carnation and *D. gianteus*. *D. barbatus* and *D. deltoids* did not produce seeds.

Dhananjaya and Tejaswini (2004) have developed interspecific hybrid CS-1 by crossing *D. caryophyllus* and *D. chinensis*, which was found to have commercial potential to be grown as a miniature spray carnation in open field. Guobao *et al.* (1995) have obtained 30 new varieties by interspecific hybridization of *D. caryophyllus* × *D. chinensis* and *D. caryophyllus* × *D. barbatus*. They were being successful in deriving most of the characters of *D. chinensis* and *D. barbatus* through 3-4 generations of back crossing and developed 130 promising hybrid forms which may be used as parent material. Some important species of Dianthus have strong characteristics against various biotic and abiotic stresses as well as plant architecture may be used for interspecific hybridization are described in Table 8.

Table 8: Important species used in interspecific hybridization as a source of different Characteristics.

S. No.	Species	Characteristics	Reference
1.	*D. giganteus,D. cruentus* and *D. carthusiana*	Strong stem and compact vegetative growth	Sparnaaij and Demmink (1983)
2.	*D. chinensis*	Higher productivity and yield distribution	Sparnaaij and Demmink (1983), Sparnaaij and Koehorst-Van Putten (1990)
3.	*D. chinensis*, *D. pancicii* and *D. pontederae*	Compact growth and small flowers	Demmink (1982)
4.	*D. knappii*	More number of branches	Sparnaaij and Demmink (1983)
5.	*D. knappii* and *D. cruentus*	Resistance against	*Fusarium* and *Phialophora* Sparnaaij and Demmink (1983)
6.	*D. chinensis*	Early flowering	Sparnaaij and Koehorst (1990)

Mutation

About 25% of the contribution of varieties in The Netherlands was estimated through bud sports (Wasscher, 1956). An impressive number of bud sports of various flower colour and high quality strains were obtained from William Sim through selection (Holly and Baker, 1963). These mutations occurred spontaneously.

Induced mutations also have paramount role in breeding as a diagnostic tool in carnation. For the very first time, induced floral colour and single flowered mutants were obtained by Richter and Singleton (1955). Cv. White Sim is a very much quoted example of induced mutation, which resulted from rooted cuttings, treated with ã rays and was introduced in 1962 in USA by Mehlquist. Pink Sim and White Sim are the mutants of cv. William Sim. Rooted cuttings of William Sim, White Sim and Pink Sim were irradiated with X-rays of 2500 to 5000 R and a change in colour from pink to red and white to red was observed (47-91%), whereas the reverse change occurred with very less frequency (0-4%). This change is due to alteration of the chimeric nature of Pink Sim and White Sim as a result of a radiation induced morphological alteration of shoot apex (Sagawa and Mehlquist, 1957).

Generally the Sim carnations are sterile, but irradiation has also been found effective in restoring the male fertility in single flowered mutants derived from William Sim (Mehlquist and Sagowa, 1962). In contrast to normal pink coloured flowers, rose pink and variegated pink mutants have been isolated. The most important work for commercial mutation breeding, however, was carried out by Sparnaay and co-workers (Sparnaay, 1974, Sparnaay and Demmink, 1971 and Sparnaay *et al.*, 1974). They observed higher rooting in X-ray mutated unrooted cuttings with 2.5 kr to 5 kr and they also found the higher flower production, time of flowering and yield from the induced pre-selected plants on the basis of clonal yield data. Chimeral variations by using gamma rays and from the variations homogenous plants *in vivo* were obtained (Dommergues and Gillot, 1973).

However, carnation has the ability to maintain stable chimeras. Pereau-Leroy (1974 and 1975) has shown that orange flowering cv. Jacqueline which was genetically yellow L1 and genetically red L11 and L111 also produced identical cyto-chimeras in L11 and L111 after colchicine treatment. The double and semi-double forms of flowers have been shown to be caused by a dominant mutation at a single locus (Scovel *et al*., 1998) that shows meristic effects and homeotic conversion of stamens to petals.

Samata *et al*. (1979) have also obtained colour variation in flower after application of γ rays. During *in vitro* mutagenesis, 5 k rad dose of gamma-rays was found to be lethal, whereas irradiated plants of 1 krad and 3 krad showed pink streak on white petals of the flower as compared to white coloured petals of White Sim (Oradee and Chaichoompon, 1985). More than 35 new varieties have been developed by petal colour mutation from the French cv. Londorgo (Ferrero *et al*., 1987). Chemical mutations have also been reported in carnation by various workers. Vosamae and Kask (1972) treated seeds of cv. Chabaud with nitrosoethyl urea at 2.5 mM and found changes in flower colour in second and third generations. The treatment of EMS at 1.0 or 2.5% also provided several changes in colour and Enzett Barther. Fruiting was selected from a shoot of cv. Arthur Sim and cv. Enzell Folklore and Dione from cv. William Sim were isolated (Hentrich and Glawe, 1982).

Induced change in flower colour was observed through *in vitro* application of EMS to axillary bud explants of cv. Espana (Singh *et al*., 2000). Some another study *in vitro* mutants have been evolved by applying physical (γ rays) and chemical (nitroso-methyl urea and ethyl methane sulphonate) mutagens. Electrophoretic studies of these mutants reflected the variability at genetic level (Singh *et al*., 2002). Higher yield of flowers can also be obtained by mutations. Seed treatment with 0.0125% EMS for 3 hrs produced highest number of flower buds per plant and number of stigma (Bhattacharya, 2003). Okamura (2006) investigated the efficiency of ion-beam irradiation in two different cultivars Vital and Star which were irradiated with carbon ions and cultured until shoot regeneration occurred. A wider mutation spectrum was obtained in plants irradiated with carbon ions than with gamma rays. In cultivar Star, new yellow mutants were obtained through both ion-beam irradiation and chronic gamma irradiation. Okamura *et al*. (2012) developed a systematic and directed method to create novel glittering mutants in carnation (*Dianthus caryophyllus*) by combining the advantages of ion-beam breeding and genomic information and created two highly novel carnations with the most glittering colour. Induced mutant varieties are given in Table 9.

Table 9: Induced mutant varieties of carnation.

S. No.	Mutant	Parent	Year
1.	Pink Sim	William Sim	1962
2.	White Sim	William Sim	1962

BIOTECHNOLOGY

Tissue culture

Successful *in vitro* propagation of Gaudina was achieved by inoculating axillary bud explants in Gamborg basal medium + 1.0 mg benzyladenine/litre + 1.0 mg NAA/ litre for shoot induction and proliferation followed by transferring the shoots to half-strength MS medium + 0.80 mg IBA/litre for rooting (Wankhede *et al.*, 2007). Pepo and Toth (2007) reported that the capability of different cultivars for *in vitro* multiplication varied significantly from each other. Carnation cv. Lipstick was micropropagated using shoot tips of greenhouse-grown plants and nodal segments of *in vitro*-grown shoots cultured in MS medium supplemented with various growth regulators. High rates of multiplication of shoot tips were recorded in the medium containing basal MS salts supplemented with 0.5 mg NAA, 1.5 mg kinetin and 0.5 mg GA_3 litre^{-2} (Bora *et al.*, 2007). Ali *et al.* (2008) found that best rooting response was obtained on MS medium containing 1.0 mg/l NAA when apical and nodal meristems were used. A protocol has been established for callus multiplication and plant regeneration in carnation. Maximum shoots were obtained on the MS medium in combinations of 0.5 mg/l NAA and IAA each (Sharma *et al.*, 2010).

Somaclonal variation

Due to presence of sterility in most of the carnation varieties, biotechnology is an attractive approach for creating novel characters in the crop. Somaclonal variants have been isolated from cv. Cerise Royal as a result of disruption of the chimeral structure of the meristem (Leshem, 1985, 1986). Adventitious shoot regeneration from carnation petals was performed to assess its potential for the production of somaclonal variants (Vainstein *et al.*, 1992). Two out of 100 regenerants significantly differed from White Sim. One somaclonal variants had a large number of red stripes while the other one showed round edged petals. Some salt tolerance variants have been isolated by Haouala *et al.* (2003). A dwarf somaclonal variant from cv. Espana and many floral variations were obtained from cv. Candy (Kumar *et al.*, 2004). Embryo culture has been successfully employed to overcome the incompatibility of species hybridization and obtained interspecific hybrids of the combinations of *D. caryophyllus* × *D. chinensis* and *D. caryophyllus* × *D. barbatus* (Guobao *et al.*, 1995).

Diploid plants were regenerated from unpollinated ovules of a tetraploid F_1 plants through gynogenesis (Demmink *et al.*, 1987). These diploids from F_1 hybrid of *D. trifasciculatus* (tetraploid) and Diantini types (tetraploid interspecific hybrid) did not produce any pollen and were used as male sterile. Sato *et al.* (2000) developed double haploids by pseudo fertilized ovule culture. The emasculated flower buds of carnation were pollinated with pollen inactivated by X-ray irradiation and after 2-3 weeks the ovaries were explanted.

Molecular breeding

Various molecular markers have been extensively used in carnation for identification and characterization at gene or genes. DNA finger printing using mini and micro satellites has used to identify genotypes and establish genetic distance. RAPD was also used to identify a marker linked to bull head flowers in carnation (Vainstein *et al.*, 1995). RAPD and RFLP markers were found to be tightly linked to the locus controlling carnation flower type and RFLP marker discriminated with 100% accuracy between the semi-double and double flower phenotype in carnations of both Mediterranean and American groups (Scovel *et al.*, 1998). Thirteen SSR (simple sequence repeat) markers were successfully developed in carnation using an enriched genomic library method (Kimura *et al.*, 2009).

Genetic engineering

Development through genetic modification has been most widely occurred in carnation. Several workers have developed transformation systems for carnation by using *A. tumefaciens* and microprojectile bombardment. Zuker *et al.* (1995) produced transgenics by micro projectile bombardment of highly regenerative stem segments. Bombardment of cv. White Sim stem segments was performed with a plasmoid containing UidA and bar genes and around 3% of the plantlets exhibited strong and stable expression of genes. The suitable conditions of particle bombardment for the introduction of foreign genes to carnation shoot apices were studied (Hasegawa, 2003). The optimum pressure and amount of gold particles per bombardment must be 1350 psi and 0.5 mg respectively. The optimum concentration of geneticin for transgenic shoot selection was found to be 10 mg/litre.

Vainstein, *et al.* (1993) generated transgenic carnations via *A. tumefaciens* mediated transformation with bacterial GUS and CAT reporter genes using the binary vectors and findings revealed that the use of CAT rather than GUS reporter genes may prove advantageous in establishing an efficient carnation transformation system. Transgenics developed through *Agrobacterium*-mediated transformation of leaf (Van Altvorst *et al.*, 1995 and Lin *et al.*, 2003), petal (Van Altvorst *et al.*, 1996 and Nakano *et al.*, 1999) and stem (Zuker *et al.*, 1997). The efficiency of gene transfer depends on the carnation genotype, explant age and co cultivation time. A higher transformation frequency can be observed with the binary vector pCGN 7001 than with the p35S GUSint vector. Leaf explant, whilst giving reduced transformation efficiency, are more suitable for the regeneration of transgenic plants, whereas from stem explant regenerated plants showed 30% stable GUS expression throughout the plant. Ovadis *et al*. (1999) used a combination of biolistic and *Agrobacterium* mediated transformation method.

Shoot apices of carnation (*Dianthus caryophyllus*) cultivars were wounded using a Biolistics PDS 1000/He system and subsequently infected with *Agrobacterium tumefaciens* strain AGLO carrying uidA and pAMrolC. Combination of the two techniques led to reproductively high gene transfer frequencies as evidenced by the transient GUS expression assay in stem explants. Transgenic plants expressing rolC exhibited increased axillary bud breakage and development and also better rooting ability.

Zuker *et al*. (2001) established a unique and efficient transformation with main features of high efficiency (approximately 2 transgenes per 10 explants) and suitability to numerous cultivars. The established transformation procedure was used to generate carnation with novel agronomic and ornamental traits.

Among various factors in different explants, co cultivation medium is one of the most important factors for the successful genetic transformation of carnation node explants, which has been studied by Nontaswatsri *et al*. (2004).

Breeding for development of novel colour

For developing blue colour in carnation, the missing F32 52 h in carnation was introduced into this species lead to the production of carnation flowers with bluish hues (Tanaka *et al*., 1998). By introducing the DFR and F32 52 H genes derived from petunia into the carnation in white colour in which anthocyanidin synthesis never takes place due to the lack of dihydroflavonol 4-reductase (DFR) activity, delphinidin is produced in the petals and the flower colour changes to purple-violet. In addition, the produced delphinidin is transformed to delphinidin 3, 5-(malyl) diglucoside by the flavonoid 3- glucosyl-transferase (3GT) contained in the delphinidin produced. In some carnations that contain methyltransferase (MT), petunidin is produced.

Anthocyanins are the major pigments contributing to carnation flower colouration. Most carnation varieties are sterile and hence molecular breeding is an attractive approach to creating novel colours in this commercially importantcrop. Characterization of anthocyanins in the flowers of the modern carnation cv. Eilat revealed that only the orange pelargonidin accumulates, due to a lackof both flavonoid 32, 52 -hydroxylase and flavonoid 32 -hydroxylase activities. To modify flower colour in cv. Eilat, antisense suppression was used (Zuker *et al*., 2002) to block the expression of a gene encoding flavanone3-hydroxylase, a key step in the anthocyanin pathway. The transgenic plants exhibited flower colour modifications ranging from attenuation to complete loss of their original orange/reddish colour. In the latter, only traces of pelargonidin were detected. Dramatic suppression of flavanone 3-hydroxylase level/activity in these transgenes was confirmed by northern blot, RT-PCR and enzymatic assays. The new phenotype has been stable for over 4 years of vegetative propagation. Moreover, transgenic plants with severe colour modification were more fragrant than control plants. GC-MS headspace analyses revealed that transgenic anti-*f3h* flowers emit higher levels of methyl benzoate. The possible interrelation between pathways leading to anthocyanin and fragrance production is discussed.

Breeding for fragrance

The most modern cultivars lack distinct fragrance. Although a considerable research has been conducted under conventional breeding; molecular breeding also holds promise for improving this character. It is very interesting that in carnation blocked F3h expression led to the production of flowers with not only novel colours but also strongly increased fragrance. Analysis of fragrance compounds in the transgenic plants revealed increased levels of benzoic acid derivatives (Zuker *et al.*, 2001). Transgenic plants were generated by expressing the linalool synthase gene which catalyses the production of monoterpenes from *Clarkia breweri* (Lavy *et al.*, 2002).

Breeding for post harvest life

For improving vase life cv. Yellow Star was transformed with antisense genomic DNA of ACC oxidase (1-amino cyclopropane -1 carboxylase oxido reductase) enzyme. The vase life of transgenic plants was 4 days longer than that of control plants (Yu and Bao, 2004).

Breeding for disease resistance

To obtain fungal resistance, transgenic carnation with osmotin, PR-1 and/or chitinase genes were generated. A high level of resistance in these transgenes to a major carnation pathogen (*Fusarium oxysporum* f. sp. *dianthi*, race 2) was obtained. The rolC gene from *Agrobacteruium rhizogenes*, driven by a CaMV 358 promoter was harnessed to generate carnation plants with improved performance. Transgenic lines exhibited dramatically improved rooting ability and production yield (in terms of both number of stem cuttings and number of flowering stalks per mother plant). Moreover, these traits were stable following 2 years of greenhouse testing. A successful transformation procedure using *Agrobacterium* mediated gene transfer was established for the bacterial wilt resistant *Dianthus* species, *D. acicularis* (Tomonori and Masaliro, 2005).

Development of transgenic

Two primary objectives of this developmental work are the extension of vase life through the modification of ethylene biosynthesis and perception and the generation of new colours. Two transgenic varieties of mauve colour cv. Moondust and cv. Moonshadow have been developed. Both are now in commercial production and being sold in Australia, Japan and USA and it is the first ornamental crop to have GM cultivars in commercial production (Table 10).

Table 10: Transgenic varieties of carnation.

S. No.	Mutant	Flower colour	Description
1.	Moondust	Mauve	These are first transgenic varieties in ornamental crop to have GM cultivars in commercial production. Developed through the modification of ethylene biosynthesis to enhance vase life. Varieties are now in commercial production and being sold in Australia, Japan and USA.
2.	Moonshadow	Mauve	

REFERENCES

Ali, A., Afrasiab, H., Naz, S., Rauf, M. and Iqbal, J. 2008. An efficient protocol for *in vitro* propagation of carnation (*Dianthus caryophyllus*). *Pakistan J. Bot.*, **40**(1): 111-121.

Annoymous. 1986. Annual Report, Division of Floriculture and Landscaping, IARI, New Delhi.

Anonymous, 1952. The Wealth of India, III: D-E, CSIR, New Delhi, pp. 52-53.

Atanasova, B. 1998. Investigation of the possibility of crossing carnation (*D. caryophyllus* f. spray) with some wild species of the genus *Dianthus*. *Rosteniev dni-Nauki*, **35**(3): 200-205.

Bailey, L.H. 1969. Manual of Cultivated Plants. MacMillan Co, N.Y. USA.

Bhat, M.S., John, A.Q. and Abdullah, M. 1993. Floral biology and seed setting studies in carnation (*Dianthus caryophyllus* L.). *Progessive Horti.*, **23**(1-4): 60-67.

Bhattacharya, C. 2003. Effect of ethyl methane sulphonate on carnation (*Dianthus caryophyllus* L.). *Environment & Ecology*, **21**(2): 301-305.

Boayen, R.P., Sparnaaij, L.D., Jansen, J. and Niemann, G.J. 1991. Inheritance of resistance components in carnation against Fusarium wilt. *Netherlands Erlands J. Pl. Path.*, **97**: 73-86.

Bora, A., Singh, S., Talukdar, M.C. and Hazarika, B.N. 2007. An efficient method for *in vitro* plant regeneration in carnation. *Indian J. Horti.*, **64**(4): 439-443.

Crane, M.B. and Lawrence, W.J.C. 1952. Genetics of Garden Plants., MacMillan, London.

Cuellar, S., Bilboo, O., Castro, C., Nustez. C., Rodriguez, L.E. and Mosquera, T. 1999. Evaluation of the phenotypic characteristics of F_1 hybrids from crosses involving 10 carnation varieties (*Dianthus caryophyllus* L.). *Acta Horticulrurae*, **482**: 291-295.

Demmink, J.F. 1982. Dianthini, a new group of spray carnations. *Vakblad voor de Bloemisterij*, **37**(4): 28-29.

Demmink, J.F., Custers, J.B.M. and Berger Voet, J.H.W. 1987. Gynogenesis do bypass crossing barriers between diploid *Dianthus* species. *Acta Hort.*, **216**: 343-344.

Demmink, J.F., Karsten, M. Giessen, A.C. Van-der. 1982. First wilt resistant cluster carnations coming along (cultivars, Fusarium, molds, Phialophora, Netherlands). *Vakble-Bloemisterij*, **37**(25): 24-27.

Dhananjaya, M. and Tejaswini. 2004. Interspecific hybrid: An attempt to combine advantage of carnation and pinks. *J. Orna. Hort.*, **7**(3-4): 1-6.

Dolcet-Sanjuan, R., Claveria, E., Leaurado, M., Ortigosa, A. and Arus, P. 2001. Carnation (*D. caryophyllus* L.) dihaploid lines resistant to *Fusarium oxysporum* f sp. *dianthi*, *Acta Hort.*, **560**: 141-144.

Dommergues, P. and Gillot, J. 1973. Genetically homogenous clones in all their ontogenetic layers obtained from an American pink chimera (*Dianthus caryophyllus*). *Ann. Amelior. Plant*, **23**(2): 83-93.

Dumlao, R.C. 1970. Inheritance of doubleness in carnation (*Dianthus caryophyllus* L.) Ann Arbor: University Microfilms.

Ferrero, F., Silvy, A., Jay, M. and Ledems, P. 1987. Experiment evidence proving the mutational origin of carnation cultivars obtained from the 'londorgo' genotype. *Acta Hort.*, **216**: 205-214.

Frank, K. 1985. Fusarium resistant carnation cultivars. *Kerteszete-es- Szoleszet*, **34**:4-7.

Galbally, J. and Galbally, E. 1997. Carnations and pinks for garden and greenhouse. Timber Press, Portland, Oregon, USA. 310 p.

Garibaldi, A., 1983, Resistenza di cultivar di garofano nei confronti di 8 patatipo di *Fusarium oxysporum* f. sp. *dianthi*. *Revista della Orthoflorafrutticoltura Italiana*, **67**: 201-270.

Gill, A.P.S. 1990. Commercial growing of Carnations under north Indian conditions. *In*: X Workshop All Indian Coordinated Floriculture Improvement Project held at KAU, Vellanikkara, Trichur.

Gorer, R. 1970. The development of garden flowers. Eyre and Spottiswoode Pub. Ltd., U.K.

Guba, E.F. 1951. Breeding for disease resistant in carnation. *Phytopathology*, **42**: 112-113.

Gui, M., Chen, M., Lu, Z.H., Zhou, X.H., Long, J., Song, J. and Mo, X.J. 2011. A new *Dianthus caryophyllus* cultivar Yundieyi. *Acta Horticulturae* Sinica, **38**(8): 1625-1626.

Guobao, W., Jingping, Y., Xiaomei, Z., Gong Cheng, F., Dawei, L. and Zhen, Y. 1995. Studies on carnation (*Dianthus caryophyllus*) species hybridization. *Acta Hort.*, **404**: 82-90.

Gupta, Y.C., Rana, N., Dhiman, S.R. and Jain, R. 2004. Floral biology and seed setting studies in *Dianthus barbatus* L. *J. Orna. Hort.*, **7**(3-4): 11-17.

Haouala, F., Hannachi, C. and Zid, E. 2003. Exploitation of somaclonal variability for research of carnation (*Dianthus caryophyllus* L.) tolerant to salinity. *Tropicultura*, **21**(1): 16-21.

Hasegawa, T. Manago, N. and Kato, Y. 2003. Production of transgenic carnation introduced sarcotoxin gene through particle bombardment. *Research Bulletin of the Aichi Ke Agricultural Research Center*, **35**: 143-147.

Hentrich, W and Glawe, M. 1982. Breeding carnations using EMS application to the leaf axils of young plants. *Archiv fur Zuchtungsforschung*, **12**(3): 197-207.

Holly, W.D. and Baker, R. 1963. Carnation production including the history, breeding, culture and marketing of carnations. William C. Brown, Dubugene, Jowa, pp. 142.

Holly, W.D. and Baker, R. 1992. Breeding for better varieties. *In:* W.D. Holly and R. Baker (eds.). Carnation Production II. Colourado State University, pp. 21-30.

Howard G.S. 1968. Polyploid carnation: Their identification and breeding behavior. *J. Hered.*, **59**: 77-80.

Howard, G.S. 1971. Polyploid carnations: five quantitative character estimates for new hybrid populations. *J. Amer. Soc. Hort. Sci.*, **96**(5): 673-674.

Itoh, Y., Higeta, D., Suzuki, A., Yoshida, H. and Ozeki, Y. 2002. Excision of transposable elements from the chalcone isomerase and dihydroflavons 4- reductase genes may contribute to the variegation of the yellow flowered carnation (*Dianthus caryophyllus). Plant Cell Physio.*, **43**(5): 578-585.

Joung, HyangYoung, Lee, YoungRan, Choi, SeongYoul, Kim, HyoungDeug and Kim, TaeIl. 2009. Breeding of purple standard carnation Jasujung with fragrance for cut flower. *Korean J. Breed. Sci.*, **41**(4): 507-509.

Jurgens, A., Witt, T. and Gottsberger, G. 2003. Flower scent composition in *Dianthus* and *Saponaria* spp. *Biochemical Systematics and Ecology*, **31**: 345-357.

Kaicker, U.S. 1988. Breeding of carnation (*Dianthus caryophyllus* L.). A review. *Haryana J. Hort. Sci.*, **17**(3-4): 166-176.

Kho, Y.O. and Baer, J. 1973. The effect of temperature on pollen production in carnations. *Euphytica*, **22**: 467-470.

Kimura, T., Yagi, M., Nishitani, C., Onozaki, T., Ban, Y. and Yamamoto, T. 2009. Development of SSR markers in carnation (*Dianthus caryophyllus*). *J. Japanese Soc. Hortil. Sci.*, **78**(1): 115-123.

Kumar, A., Singh, S.K., Sharma S.K. and Kumar, S. 2004. Regeneration of somaclonal variants in carnation raised through indirect organogenesis. *Orissa J. Hort.*, **32**(2): 1-6.

Kwon, OhKeun., Kim, KwangJin., Joung, HyangYoung., Choi, SeongRoul. and Lee, JungSoo. 2010. Breeding of pink colour spray carnation 'Lyra' for cut flower. *Korean J. Breed. Sci.*, **42**(6): 645-648.

Kwon, OhKeun., Shin, HakKi., Choi, SeongRoul., Joung, HyangYoung. and Lee, JungSoo. 2012. Breeding for spray carnation 'Lepus' of white colour with red edge. *Korean J. Hortil. Sci. Tech.*, **30**(1): 90-92.

Lavy, M., Zuker, A., Lewinsohn, E., Larkov, O., Ravid, V., Vainstein, A. and Weiss, D. 2002. Linalool and Linalool oxide production in transgenic carnation flower expressing the *Clarkia breweri* linalool synthase gene. *Molecular Breeding*, **9**: 103-111.

Lin, R.C., Chen, L.Q., Bao, M.Z. and Sun, Z.Y. 2003. Establishment of efficient *Agrobacterium tumefaciens* mediated geneticf transformation system of carnation (*Dianthus caryophyllus* L.). *Forest Research*, **16**(2): 123-128.

Mehlquist, G.A.L. 1940. Inheritance in the carnation (*Dianthus caryophyllus*) I. Inheritance of flower colour. *Amer Soc.Hort. Sci.*, **37**: 1019-1021.

Mehlquist, G.A.L. and Sagowa, Y. 1962. The effect of gamma radiation on carnations. *Proc*: 16th Int. Hort. Cong. Brussels, **1**: 293.

Misra, Sanyat and Gupta, Y.C. 2003. Genetic variability in carnation. *J. Orna. Hort.*, **6**(1): 20-23.

Nakano, M., Koike, Y., Watanabe, Y. and Suzuki, S. 1999. Production of transgenic plantlets of carnation via *Agrobacterium* mediated transformation of petal explants. *Bull. Faculty Agri., Niigata University*, **51**(2): 105-114.

Nontaswatsri, C., Fukai, S. and Goi, M. 2004. Revised cocultivation conditions produce effective *Agrobacterium* mediated genetic transformation of carnation (*D. caryophyllus* L.). *Plant Science*, **166**(1): 59-68.

Okamura, M. 2006. Flower breeding in quantum beam technology and its commercialization. *Gamma Field Symposium*, **45**: 77-89.

Okamura, M., Umemoto, N. and Onishi, N. 2012. Breeding glittering carnations by an efficient mutagenesis system. *Plant Biotech.*, **29**(3): 209-214.

Onozaki, T., Ikeda, H. and Yamaguchi, T. 2001. Genetic improvement of vase life of carnation flowers by crossing and selection. *Scientia Horticulturae*, **87**: 107-120.

Onozaki, T., Yamaguchi, I.T. and Kage, T. 1998. Introduction of bacterial wilt (*Pseudomonas caryophylli*) resistance in *Dianthus* wild species to carnation. *Acta Horti.*, **454**: 127-132.

Oradee, S. and Chaichoompon, S. 1985. Effects of gamma irradiation on 'White Sim' carnation (*Dianthus caryophyllus* L.) Shoot tip culture. *Proc*: 23rd National Conference, Volume 1. Bangkok. pp. 205-214.

Ovadis, M., Zuker, A., Tzfira, T., Abroni, A., Shklarman, E., Scovel, G. Itzhaki, H., Ben-Meir, H. vainstein, A., Altmanm, A., Ziv, M, and Izhar, S. 1999. Generation of transgenic carnation plants with novel characteristics by combining microprojectile bombardment with *Agrobacterium tumefaciens* transformation. *Curr. Plant Sci. Biotech. Agri.*, **36**:10.

Pepo, P. and Toth, S. 2007. *In vitro* cloning of carnation (*Dianthus caryophyllus* L.) genotypes. *Kertgazdasag – Horticulture*, **39**(1): 47-52.

Pereau-Leroy, P. 1974. Comparison of the physiological and genetic effects of neutron and gamma radiation on carnation (*Dianthus caryophyllus* L.) pipings. IABA. Viennao, Munich- Neuterberg. pp. 441-449.

Pereau-Leroy, P. 1975. Genetic interaction between the tissues of carnation petals as periclinal chimeras (dianthus). Thesis Universite Clermont-Ferr and Ser. E. No. 199. p. 169.

Richter, A. and Singleton, W.R. 1955. The effect of chronic gamma radiation on the production of somatic mutation in carnations. *Proc*. Natl. Acad. Sci., USA, **41**(5): 295-300.

Sagawa, Y. and Mehlquist, G.A.L. 1957. The mechanism responsible for some X-ray induced changes in flower colour of the carnation (*Dianthus caryophyllus*). *American J. Bot.*, **44**(5): 397-403.

Samata, Y., Isuyuki, Y., Inazu, K. and Iizuka K. 1979. Induction of flower colour changes in carnation (*Dianthus caryophyllus*) by gamma ray irradiation. Bull. Fac. Agric. Tamagawa Univ., Japan, No. 19, pp. 29-41.

Sato, S. Katoh, N., Yoshida, H., Iwai, S. and Hagimori, M. 2000. Production of doubled haploid plants of carnation (*Dianthus caryophyllus*) by pseudofertilized ovule culture. *Scientia Horticulturae*, **83**(3-4): 301-310.

Scovel, G., Ben-Meir, H., Ovadis, M., Itzhali, H. and Vainstein, A. 1998. RAPD and RFLP markers tightly linked to the locus controlling carnation (*Dianthus caryophyllus*) flower type. *Theor. Appli. Genet.*, **96**: 117-122.

Semina, S.N. and Shestachenko, G.N. 1981. Rust resistance of carnation *Dianthus caryophyllus* infected with *Uromyces caryophyllus*. *Mikologiya I Fitopatologia*, **15**: 238-240.

Sharma, C., Kaur, R., Chandel, S. and Aggarwal, G. 2010. *In vitro* callus multiplication and shoot regeneration in carnation cv. Raggio de Sole. *J. App. Biosci.*, **36**(1): 55-57.

Singh, A.K. 2006. Flower Crops: Cultivation and Management. Pub. New India Publishing Agency, New Delhi. pp. 47-59.

Singh, K.P., Choudhary, M.L., Kumar, P.A. and Suchitra. 2002. Characterization of *in vitro* induced mutants of carnation by means of electrophoretic protein analysis. *Indian J. Hort.*, **59**(4): 427-430.

Singh, K.P., Singh B., Raghava, S.P.S. and Kalia, C.S. 2000. Induced flower colour mutations in carnation through *in vitro* application of chemical mutagen, *Indian J. Gen. Pl. Br.*, **60**: 535-539.

Sparnaaij, L.D. and Beeger, G.W. 1973. The improvement of seed production for breeding purposes in the greenhouse carnation *Dianthus caryophyllus* L. *Euphytica*, **22**: 274-278.

Sparnaaij, L.D. and Demmink, J.F. 1976. Breeding for resistance to *Phialophora cinerescens* in glass house carnations (*Dianthus caryophyllus*). *Euphytica*, **25**: 329-338.

Sparnaaij, L.D. and Demmink, J.F. 1977. Progress towards *Fusarium* resistance in carnations. *Acta Hort.*, **71**: 107-113.

Sparnaaij, L.D. and Demmink, J.F. 1983. Carnations of the future. *Acta Hort.*, **141**: 17-23.

Sparnaaij, L.D. and Demmink, J.F. 1987. Current research in the Netherlands on improvement of breeding and selection methods for resistance of carnations to *Fusarium oxysporum* f. sp *dianthi*. *Acta Hort.*, **216**: 111-124.

Sparnaaij, L.D. and Koehorst–van Putten, H.J.J. 1990. Selection for early flowering in progenies of interspecific crosses of ten species in the genus *Dianthus*. *Euphytica*, **50**: 211-220.

Sparnaay, L.D. 1974. Progress report on clonal selection in carnation after irradiation. *In*: Meeting of the Mutation Breeding contact Group. C. Broetjes (ed.). Wageningen, pp. 28-31.

Sparnaay, L.D. and Demmink, J.F. 1971. Mutation breeding as a means of improving productivity in commercial carnation cultivars. *In*: Eucarpia Meeting on Ornamental, Wageningen, pp. 38-44.

Sparnaay, L.D., Demmink, J.F. and Gaeretsen, F. 1974. Clonal selection in carnation. *In*: Eucarpia Meeting on Ornamentals. *Frejus Inst. Hortic. Plant Breed.*, Wageningen, pp. 39-50.

Sun, Q. and Lin, D.W. 2007. Two new cultivars Linlong2 and Linlong3 of Dianthus. 2007. *Acta Horticulturae* Sinica, **34**(5): 1339- 1340.

Tanaka Y. Tsuda, S. and Kasumi, T. 1998. Metabolic engineering to modify flower colour. *Plant Cell Physiol.*, **39**: 1119-1126.

Tomonori, S. and Masahiro, M. 2005. *Agrobacterium tumefaciens* mediated transformation of highly regenerable (cell suspension cultures in *Dianthus acicularis*. *J. Hort. Sci. Biotech.*, **80**(4): 393-398.

Tutin, T.G., Bergs, N.A., Chater , A.O., Edmondson, J.R., Heywood, V.H., Moore, D.M., Valentine, D.H., Walters, S.M. and Webb, D.A. 1993. Flora Europia, Cambridge University Press, Cambridge. pp. 227-246.

Unlinger, R.D. and Lindgren, D.T. 1984. N. 74133 *Dianthus* germplasm, *HortScience*, **19**(6): 904.

Vainstein, A, Fisher, M. and Ziv, M. 1993. Applicability of reporter genes to carnation transformation. *Hort. Sci.*, **28**(11): 1122-1124.

Vainstein, A., Ben-Meir, H., Zuker, A., Watad, A.A., Scovel, G., Ahroni, A., Ovadis, M. and Weiss, D. 1995. Molecular markers and genetic transformation in the breeding of ornamentals. *Acta Hort.*, **420**: 65-67.

Vainstein, A., Fisher, M. and Ziv, M. 1992. Shoot regeneration from petals as a basis for genetic variation and transformation. *Acta Hort.*, **314**: 39-43.

Van Altvorst, A.C., Koehorst, H., Dejong, J. and Dons, H.J.M. 1996. Transgenic carnation plants obtained by *Agrobacterium tumefaciens* mediated transformation of petal explants. *Plant Cell Tiss. Org. Cult.*, **45**: 169-173.

Van Altvorst, A.C., Riksen, T., Koehorst, H. and Dons, H.J.M. 1995. Transgenic carnations obtained by *Agrobacterium tumefaciens* mediated transformation of leaf explants. *Transgenic Res.*, **4**(2): 105-113.

Van Staaveren, 1998. Variety Statropur Syn Gypsy. *Plant Varieties Journal*, **11**(3): 17.

Van Staaveren, BV. 1997. Variety statas syn Tasman. *Plant Varieties Journal*, **10**(4): 23.

Vosamae, H. and Kask, K. 1972. Experiments on the use of N-nitroso-N-alkylureas in the breeding of Chabaud carnations M2-M3. *Biologia*, **21**(2): 178-180.

Wankhede, M., Patil, S. and Kalamkar, V. 2007. *In vitro* propagation of carnation cv. Gaudina. *J. Soils Crops*, **17**(1): 46-51.

Wasscher, J. 1956. The importance of sports in some florist's flowers. *Euphytica*, **5**: 163-170.

Watts, L. 1980. Flower and Vegetable Plant Breeding. Grower Book Pub, London. p.184.

Yagi, M., Tanikawa, N., Shibata, M. and Onozaki, T. 2012. Breeding of carnations (*Dianthus caryophyllus* L.) for resistance to bacterial wilt (*Burkholderia caryophylli*). *Acta Hort.*, **937**: 427-431.

Yu, Y.X. and Bao, M.Z. 2004. Carnation flower vase life prolonged by transformation with antisgense ACC Oxidase Gene. *Acta Hort.*, **31**(5): 633-636.

Zuker, A., Ahroni, A., Vainstein, A., Allman, A. and Ziv, M. 1997. A highly efficient method for carnation transformation. *Acta Hort.*, **447**: 373-375.

Zuker, A., Chang P.L., Ahroni, A., Cheah K., Woodson, K., Woodson, W.R., Bressan, R.A., Wated, A.A., Hasegawa, P.M. and Vainstein A. 1995. Transformation of carnation through microprojectile bombardment. *Scientia Horticulturae*, **64**(3): 177-185.

Zuker, A., Shklaraman, E., Scovel, G., Ben, M.H., Ovadis, M., Neta, Shari, I., Ben, Y.Y., Weiss, D., Watad, A., and Vainstein, A. 2001. Genetic Engineering of agronomic and ornamental traits in carnation. *Acta Hort.*, **560**: 91-94.

Zuker, A., Tzfira, T., Ben-Meir, H., Ovadis, M., Shklarman, E., Itzhaki, H., Forkamann, G., Martens, S., Netashanir, I., Weiss, D. and Vainstein, A. 2002. Modification of flower colour and fragrance by antisense suppression of the flavanone 3-hydroxylase gene. *Mol. Breed.*, **9**: 33-41.

Chapter – 11
Gerbera

Gerbera (*Gerbera jamesonii*), a tender, stemless, perennial flowering plant, belongs to family Asteraceae. Gerbera was named in honour of the German botanist and naturalist "Traugott Gerber" who was a friend of Carolus Linnaeus. Gerbera is very popular and widely used as cut flower or as a decorative garden plant. It was discovered in 1878 near Barberton, eastern Transvaal (South Africa), therefore, referred as the Barberton daisy or Transvaal daisy. It is ideal for beds, borders and pots and is the fifth most important cut flower in the world.

ORIGIN AND DISTRIBUTION

Gerbera is originated from South Africa. The native distribution of genus *Gerbera* comprising 30 species, extends to Africa, Madagascar, tropical Asia and South America (Hansen, 1985 and Bremer, 1994). Whereas, it was also reported that gerbera is found in the temperate and tropical parts of Asia and Africa and about 7 species are recorded in India. This plant is distributed in the temperate Himalayas from Kashmir to Nepal. Gerbera is commonly seen on the dry slopes near Shimla and on hillsides above Gulmarg in Kashmir (Anon., 1956). The *G. jamesonii* is found naturally in grassland in sandy soils in Mpumalanga and Barberton district. The genus *Gerbera* has approximately 14 southern African species, some of which are found only in Western Cape. *G. aurantica* is endemic to the mist belt grassland region of KwaZulu-Natal and is known from approximately 9 scattered populations from Helahela in the south to Babanango in the north. This species is under threat of extinction due to habitat fragmentation and degradation. In India they are distributed in temperate Himalayas at altitudes of 1300-3200 metres and observed from Kashmir to Nepal.

The genus *Gerbera* is named in honour of German naturalist Traugott Gerber who travelled in Russia in 1743 and started first botanical garden in Moscow. However, for species i.e. *jamesonii*, it is a matter of different opinion. The *jamesonii* was named for Dr. L. S. Jameson, a British colonial statesman in South Africa who was one of the discoverers of this species, whereas in some literature, the species was named after Robert Jameson who collected live specimens while on a prospecting expedition to the Barberton district in 1884, even though the species collected on three earlier occasions by other people. In 1888, Medley Wood, the curator of the Durban Botanical Garden sent plants to Kew, which subsequently flowered. This is native to South Africa, Madagascar, Asia, South America and Indonesia.

The first official description of South African species *G. jamesonii* was made by J.D. Hooker in Curtis Botanical magazine with coloured illustration in 1889. An active society on this subject is to be found in Australia under the name W.A. Gerbera Society. The society has used a classification system of gerbera. The breeding of gerbera started at the end of the nineteenth century in Cambridge, England, when two South African sp. were crossed. *G. jamesonii* native to Natal and Transvaal and *G. viridifolia* from western Cape were crossed together by R. Irwin Lynch in 1887, after both these species were introduced to Cambridge Botanic Gardens by Jameson and the hybrid was named as G. × *cantabrigensis.*

The majority of the present commercially cultivated varieties originated from the crossing of progenies of two species i.e. *G. jamesonii* and *G. viridifolia*. Natural hybrids of the two species have not been found. It is possible that also other wild gerbera species have been used in breeding, but for that hardly any information exists. Large scale commercial breeding of gerberas took place mostly in the Netherlands. At the turn of the century, gerberas were cultivated in England, Belgium, USA, Germany, Australia, Italy and India. Today, gerbera is known as an important article of trade and it belongs to the important cut flower in the world.

SPECIES

The genus *Gerbera* has approximately 40 species, was earlier reported by Bailey (1963). However, presently it has 84 species i.e. *G. aberdarica*, *G. abyssinica*, *G. ambigua*, *G. anandria* : Ghostly Daisy, *G. anandria* var. *anandria*, *G. anandria* var. *densiloba*, *G. anandria* var. *integripetala*, *G. anandria* var. *bonatiana*, *G. aspleniifolia*, *G. aurantiaca*, *G. bojeri*, *G. bonatiana*, *G. bracteata*, *G. brevipes*, *G. burchellii*, *G. burmanni*, *G. candollie*, *G. canaleriei*, *G. chilensis*, *G. cineraria*, *G. connata*, *G. conrathii*, *G. cordata*, *G. coronopifolia*, *G. curvisquama*, *G. delavayi*, *G. discolour*, *G. diversifolia*, *G. elegans*, *G. elliptica*, *G. emirnensis*, *G. ferruginea*, *G. flava*, *G. galpinii*, *G. glandulosa*, *G. henryi*, *G. hieracioides*, *G. hirsuta*, *G. hypochaeridoides*, *G. integripetala*, *G. jamesonii*, *G. knorringiana*, *G. kokanica*, *G. kraussii*, *G. kunzeana*, *G. lacei*, *G. lagascae*, *G. lanuginosa*, *G. lasiopus*, *G. latiligulata*, *G. leandrii*, *G. leiocarpa*, *G. leucothrix*, *G.*

lijiangensis, G. lynchii, G. macrocephala, G. nepalensis, G. nervosa, G. nivea, G. parva, G. peregrina, G. perrieri, G. petasitifolia, G. piloselloides, G. plantaginea, G. plicata, G. podophylla, G. pterodonta, G. pulvinata, G. pumila, G. randii, G. raphanifolia, G. ruficoma, G. saxatilis, G. semifloscularis, G. serotina, G. speciosa, G. tanantii, G. tomentosa, G. tuberosa, G. uncinata, G. viridifolia, G. welwitschii, G. wrightii.

The habitat of the horticultural significant species *Gerbera jamesonii* H. Bolus ex Hook f. is restricted to the eastern part of Mpumalanga and the southern part of the Limpopo Province in South Africa is widely used in improvement programme. Important species of Gerbera are described below.

***G. jamesonii*:** Most important species. Plant is hairy throughout. Leaves are wooly on their undersides and radiate out from a central growing point in rosette fashion. Many solitary orange scarlet heads. The plants produce flower stalks upto 40-45 cm tall, topped with single orange flower heads. Individual blossoms are durable and may last up to 4 weeks on the plant. Single or double flowered cultivars and hybrids are available.

***G. aurantiaca*:** The species is a low growing herb with thick, thong like roots and rosettes of elliptical, dark green, lanceolate to oblong, acute and 12-15 cm length. Flower heads are pink or orange and anthers yellow. Flowering takes place in spring from September to November.

***G. ambigua*:** It is as called Pink and White gerbera and is a strikingly beautiful grassland and garden plant. The flowers are not single flowers but a head made up of many tiny florets massed together. The petals (ray florets) are usually pink, orange or white.

***G. piloselloides*:** Leaves are ovate. Flower colours are pink, white, and yellow. It is grown at 500-1900m above msl. Bloom in spring and summer. It prefers cold, semi-shady site with a humus rich, moist soil.

***G. viridifolia*:** Leaves elliptical or oblong, obtuse, green on both sides and smooth or nearly so. Flower stalk short. Flower heads dirty white and small.

Other important species are *G. kraussii*, *G. kunzeana* and *G. tomentosa*.

FLOWER MORPHOLOGY AND BIOLOGY

The gerbera is characterized by head-like inflorescence (capitula). More than one morphological flower type can exist in a single inflorescence. In gerbera, three flower types are distinguished: the marginal ray florets, the central disc florets and the intermediate trans florets. Each ray, disc and trans florets of gerbera usually have 5 petals, 5 stamens and 1 pistil made of 2 joined carpels to begin with. Mature ray and trans, however, have only female sex organs. That is because anthers arrest at intermediate stages of development resulting in the formation of non functional staminoides in ray and trans florets. Disc florets are bisexual and contain fertile stamens with inferior ovary. The florets are actinomorphic. Ray florets produce infertile stamens and have zygomorphic corollas.

The gerbera petals can be divided in 3 different flower types: dorsal, lateral and ventral petals. It is not so very usual to have gerbera florets-especially disc florets with fused corolla, either. Some fused flowers may have even 6-12 petals. The base of corolla is surrounded by pappus bristles, specialized organs that are involved in seed dispersal.

Mode of pollination

Since the flowers are protogynous, self-pollination is almost impossible. Insects of all types can also work to pollinate the plants. Butterflies and moths are native pollinators that frequently pollinate the gerbera flower. The purpose of the brightly coloured single petals is to attract pollinating bees and other insects. The single-petal flowers are infertile and the center flowers are the fertile blossoms. The bees who wander freely over the flower, tracking pollen as they collect nectar. Hummingbirds, another pollinator drawn by vivid colour, may brush against the tops of the flowers, spreading the pollen from the swollen stamens to the sticky stigma with the beat of a wing.

Pollen viability

Study carried out to estimate the pollen viability and the possibility of crossings among different accessions. Root tip and pollen collected from the cultivars and non commercial accessions were stained with acito-carmine at 45%. Pollen viability estimated by counting the number of viable and non viable pollen grains. The accessions displayed high pollen viability, varying from 87.67-99.27%. The seed formation was 4.46% in the crossings among cultivars, and 50% among cultivar and non commercial accessions. The genomic compatibility among the accessions, the high pollen viability and the success in the seed formation among commercial and non-commercial accessions reveal the possibility of hybrid production with new allelic combination and transference of desirable traits from the non-commercial to the commercial accessions (Cardoso *et al.*, 2009).

Pollination and seed set

After pollination, when stigma has been fertilized, it is no longer receptive to other pollen. If some self-pollination occurs, it would be resulted in the production of fewer viable seeds and of a high proportion of plants with poor characteristics (Schiva, 1974). Seed setting could take place both autogamously and heterogamously (Luckana and Somjit, 1984). Hand pollination resulted in higher seeds/inflorescence in first year and seeds/plant in the second year (Franceschetti, 1975). Among the various methods of pollination, natural and artificial pollination, both showed around equal seed setting i.e. 63.31 and 61.44 seeds/flower, respectively (Luckana and Somjit, 1984). Cross pollination could give more achenes than self-pollination (Luckana and Somjit, 1985). Time is also very crucial for seed setting. After pollination in March or in spring season, seed production has been found higher in summer (Franceschetti, 1975 and Luckana and Somjit, 1985).

GENETICS

The basic chromosome number of *Gerbera jamesonii* is n = 25 and any change in number alters the phenotype. Most of cultivars are diploid (2n = 50) hybrids of *G. jamesonii*. Polyploidy have been reported to show desirable characters. Pryor (1972) reported that the size of plant as well as flower is more in tetraploids than diploid.

A study was conducted to confirm the chromosome number of 6 *Gerbera hybrida* cultivars (Cariba, Amazone, Pink Elegance, Terra Fame, Azteca and Tennessee), to determine the chromosome number in 7 non commercial accessions (A7, A8, A9, A10, A11, A12 and A13) of *Gerbera* sp. The chromosome counting was performed in intact metaphase cells. All the accessions presented 50 chromosomes, indicating that the morphological variation in the capitulum (simple, semidouble and double) is not due to the chromosome number mutation or polyploidy (Cardoso *et al.,* 2009).

Inheritance of flower characters

Productivity of gerbera is determined by additive effect of genes (Mutseniets *et al.*, 1978). The inheritance of flower characters is also governed by the genes with additive effect. Earlier, Maurer and Horn (1967) reported that flower colour is governed by four factors: A - causing intense red; a - causing dull red; H - causing A and a to be lighter colour and B which became evident in the absence of A, causing ivory or yellow colour.

In gerbera, anthocyanin pigmentation is controlled according to all important morphological levels of the inflorescence (flower types, flower organs, within organs). Chalcone synthase (CHS) is a key enzyme in flavonoid biosynthesis, whereas dihydroflavonol-4-reductase (DFR) catalyzes the reduction of dihydroflavonols to luccoanthocyanidins. In order to identify genes that the plant can regulate according to these anatomical patterns, Helariutta *et al*. (1995) analysed gene expression affecting two enzymatic steps, CHS and DFR of the anthocyanin pathway in 5 gerbera varieties with spatially restricted anthocyanin pigmentation patterns. The *dfr* expression profiles vary at the levels of floral organ, flower type and region within corolla during inflorescence development according to the anthocyanin pigmentation of the cultivars. In contrast, *chs* expression, although regulated in a tissue specific manner during inflorescence development, varies only occasionally.

Helariutta *et al*. (1995a) studied the diversity of *chs*-like genes by analyzing the structure, expression pattern and catalytical properties of the corresponding enzymes of 3 genes that are active during corolla development of *Gerbera hybrida*. The expression patterns demonstrate that *chs*-like genes are representatives of 3 distinct genetic programmes that are active during organ differentiation in gerbera. The *gchs 1* and *gchs 3* code for typical CHS enzymes and their gene expression pattern temporally correlates with flavonol (*gchs 1*, *gchs 3*) and anthocyanin (*gchs1*) synthesis during corolla development. Whereas, *gchs 2* is evidently not involved in anthocyanin synthesis and encodes an enzyme with novel catalytic properties (Elomaa *et al*., 1995).

Heritability for flower characters

Heritability studies on different characteristics were investigated by various workers. Heritability was high for inflorescence diameter and number of ray florets and low for disc diameter and stem length (Schiva, 1972). Heritability estimation results suggest that most genetic variance for flower yield is additive (Harding *et al.*, 1981). Harding *et al.* (1991) estimated additive, dominance, maternal and paternal components of variance for yield, scapes, flowers, disc, ray and trans florets, leaves and branching of *Gerbera hybrida*. They observed that reciprocal differences do not seriously affect estimates of additive variance or heritability. Estimation of heritability for important flowering traits has been done by various workers (Drennan *et al*., 1986, Yu *et al*., 1993 and Huang *et al*., 1995). High heritability was observed for 15 important characters except flower number/plant by Aswath *et al.* (1998). Similarly, Mahanta *et al.* (1998) observed broad sense heritability for 11 characters except size of flower and vase life. In another study, high heritability combined with high genetic advance was observed for weight of ray florets, disc diameter, fresh weight of stalk and inflorescence and stalk length (Anuradha and Gowda, 1999) and heritability estimates and genetic advance were highest for cut flower yield and leaf area, which indicate additive gene action (Nair and Shiva, 2003).

Heritability for post-harvest longevity

Development of cultivars with genetically superior post-harvest longevity is an important aspect. Estimates for heritability of vase life have been reported. Tesi (1978), using clones from a population of *Gerbera jamesonii* with semi-double flower, concluded that vase life is strongly influenced by environmental factors. Harding *et al.* (1981a) concluded that since narrow sense heritability ($h^2 = 0.24$ and 0.38) and broad sense heritability ($H^2 = 0.36$ and 0.46) were moderately low for two successive generations, either intense selection or selection over a large number of generations would be required to increase mean vase life. Wernett *et al.* (1996a) reported that genetic variation may be largely controlled by additive gene action since broad sense and narrow sense heritability were nearly equal. Therefore, in a fixed model experiment, progeny means obtained from a top-cross mating would effectively determine parents with good combining ability for increased vase life. In a study of genetic investigations in dry vase life of gerbera, Aswath (2004) observed that high degree of environmental influence was found to influence most of the characters. Dominant and epistatic gene action was found on characters like per cent inner conduit.

BREEDING OBJECTIVES

i. Uniform and improved plant growth
ii. More number of flowers per plant
iii. Novel form
iv. Fancy colour
v. Structural stem strength
vi. No stem bending and remain upright in vase
vii. More water uptake in vase
viii. Longer keeping quality
ix. Compactness and uniformity of growth, uniformity and strength of flower stem for pot culture
x. Resistance against disease especially *Phytophthora*
xi. Resistance against insect especially white flies

BREEDING METHODS

Introduction

Gerbera varieties introduced from various countries by different nurseries, private companies and institutes for commercial production and improvement are Alegro, Alesmara, Aruba, Black Heart, Blue Bay, Brasel, Camilla, Cango, Caressa, Carrona, Casva, Citella, Crossid, Darolnella, Delama, Devas Memory, Diablow, Dingo, Dokevita, Evening Bell, Funda, Fravoury Rahman, Gateway, General Kaisar, Glaria, Glory, Goddongak, Golden Gate, Gold Spot, Greenish Yellow, Lilla Billa, Mella, Mistique, Orange Glame, Ornella, Optima, Pink Elegance, Pink Star, Popular, Racko, Regilla, Reward, Rosabella, Rosna, Rubian, Ruby Red, Saffron, Salman Spray, Sangria, Savannah, Sunanda, Sunway, Thalassa, Tiramigu, Tiramigu, Tonneke, Twiggy, V. Record, Xamena, etc.

Selection

In India, very little work has been done in gerbera improvement. Selection has been done from varieties collected from other countries and germplasm existing in India. Varieties available at various centres are Calcutta Pink, Calcutta Red, Calcutta White, Calcutta Yellow, G.S. Lal, Kalimpong Pink, Kalimpong Red, Kalimpong Yellow, Pride of Sikkim, Priyadarshini (Anon., 2002).

Hybridization

Hybridization has been successfully utilized for the development of improved cultivars of gerbera. The breeding of gerbera started at the end of 19th century in Cambridge, England when 2 South African species *G. jamesonii* and *G. viridifolia* were crossed by R.I. Lynch. He named hybrid as *G. cantabrigiensis*, known today as *G. hybrida*. The majority of the present commercially cultivated varieties originated from the crossing progenies of these two species. Natural hybrids of these two species have not been found (Hansen, 1985). R.I. Lynch also created the variety of gerbera 'Brilliant' as a cross between a variety from Natal known as 'Sir Michael Forstor' and *G. jamesonii*. Presently about thousand cultivars of gerbera exist. They vary greatly in shape and size, colours include white, yellow, orange, red and pink.

Combining ability: De Jong and Garretsen (1985) analysed combining ability for post-harvest longevity in gerberas using a diallel mating scheme with 12 parents. In this study 3 characteristics of vase life i.e. days to witting or folding; percent folding and stem curvature were examined. General combining ability was significant for each characteristic. They made several conclusions *viz.*, folding results in shorter vase life than wilting; higher curvature may increase the incidence of folding, thereby resulting in shorter vase life; no relationship exists between curvature and percent late folding and late folding is difficult to select against early folding. Large intraplant variation for both days to wilt and per cent folding was concluded to be major obstacle in breeding to improve post-harvest longevity.

Improving post-harvest longevity in gerbera by breeding was considered possible for many years. Smith and Nelson (1967) noted differences among gerbera cultivars and they suggested selection and breeding could minimize this variation. Kohl (1968) proposed selecting cultivars with an increased ability for water uptake and cultivars having structural stem strength are requisite for maintaining the popularity of gerbera. Von Wilberg (1973) recorded differences in the frequency of stem bending cultivars and noted the need to breed stems that remain upright. Van Meeteren (1978) observed that increased ion leakage in ligule of gerbera cut flower depended on the cultivar. Later van Meeteren (1981) suggested pressure potential of ligule from recent harvested flowers might be a good indicator for vase life and a possible selection criterion in breeding programme to improve the post-harvest longevity.

De Jong (1986) tested 12 parents and 80 progenies for days to bending or wilting of petals of gerbera. Large difference in days to wilt was noted. It is suggested that parent choice is the major factor in vase life.

Heterosis: Various important traits were considered in breeding programme by several workers. In crossing experiment, Schiva (1974) observed that the heritability of ligule length and stalk length showed a high level heterosis. Ligule length was found to be little affected by environmental conditions and to be well correlated with certain other flower characters; it could therefore be used in breeding programmes for the rapid early choice of plants for commercial use. Statistics were also developed to measure the relative importance of genotype × environmental (GE) interaction. Estimates indicated that as much as half of the genetic variance for cut-flower yield in the Davis Population of gerbera may be attributed to GE interactions. It is suggested that response equations that do not take interaction bias into account will overestimate selection potential (Byrne *et al.*, 1987).

Breeding for flower quality and post-harvest life: A total of 69 crosses were made of same colour, in order to obtain hybrids with colour uniformity. One hybrid (K1/1 × Brigitte) was selected in which all inflorescences were yellow (Oriecki, 1991). Parent forms of gerbera were chosen on the basis of combining ability for cut-flower quality and yield. Experiments were also carried out with tetraploid forms. Cvs. Jautriite, Ruusinjs, Mikus, Raunis, Cilda and Toms produced high quality cut-flowers. Good ornamental value was characteristic of cvs. Zeltene, Aime, Aira and Liina. Inbreeding depression was marked in cvs. Aime, Bille, Lelde and Teiksma. No segregation for flower colour was seen in cvs. Ede (dark red) and all white cultivars Delphi, Symphonie and clones 35-5, 55-3 and 83-6 (Buivids, 1992).

A top-cross mating scheme was used as a screening method to determine longest vase life among 31 parents. Cv. Apple Blossom was used as the male donor because of its excellent vase life rating and low intraplant variation. A 5 × 5 diallel cross was performed. Progeny population resulted an increase in vase life at 3.4 days (Wernett *et al.*, 1996).

A cultivar Runa was developed from a cross between cvs. Ansofie and Sunset. It is yellow, large flower and double type cultivar. Average yield of this cultivar is 49 flowers/ plant in greenhouse (Chung, 2004). A new pink gerbera (*Gerbera hybrida* Hort.) cultivar Party Time was developed from a cross between 'Hope' and 'Sunnyeo', followed by seedling and line selections at the Flower Research Institute of Gyeongnam Agricultural Research and Extension Services (GARES), South Korea in 2007. Cultivar Party Time (Grant No. 2691) was registered to the Korea Seed and Variety Service (KSVS) for commercialization in 2009 (Chung *et al.*, 2013). Another new gerbera cultivar 'Purple Queen' of cross between 'Selma' with long peduncle height and 'Marathon' with pinkish purple and semi-double flower in 2007. This variety was released by the National Institute of Horticultural & Herbal Science (NIHHS), South Korea in 2011(Park *et al.*, 2012). Details of these varieties are given in Table 1.

Table 1: Gerbera varieties developed through hybridization.

S. No.	Hybrids	Parentages	Year	Chief characteristics
1.	Runa	Ansofie × Sunset	2004	Yellow colour, double type and large flower. It has good harmony in ray floret colour and flower center colour with stable flower shape. The vase life is 12.2 days.
2.	Party Time	Hope × Sunnyeo	2009	Semi-double type large flowers in a good harmony with pink ray floret and brown center. It has a good, stable flower shape and strong peduncles. Vase life 12.7 days. Average flower yield is 47.3 stems/plant/year.
3.	Purple Queen	Selma × Marathon	2011	Purple flower colour dark brown center, flower semi-double, 11.4 cm in diameter. Inner ray floret 6.1 cm, disc diameter 2.3 cm. It has long peduncle (50.3 cm), and width 5.0 mm. Vase life 12.2 days, good for cut flower. Average flower yield is 51.4 stems/plant/year.

Development of *Phytophthora* resistant cultivar: Resistance to soil borne wilting disease caused by *Phytophthora cryptogea*, was the subject of one of the earliest resistant breeding programme. Hybridization followed by selection has resulted in development of 12 new varieties, among which Aktur, Migar and Atair showed resistance to *Phytophthora* (Danhardt, 1966). Sparnaaij *et al.* (1975) selected surviving plants from a seriously affected commercial greenhouse from *Phytophthora*. These plants were cloned and compared in a screening trial. Out of these 6 showed widely different resistance levels, were inter-crossed for a diallel analysis. The result indicated mainly additive resistance levels. Percentage of healthy plants varied from 5 to 84%.

Through hybridization a gerbera cultivar Liangfen, highly resistant to *Phytophthora cryptogea* was developed (Li *et al.*, 2008). It is also suitable for protected cultivation. Another *Phytophthora* resistant cultivar Autumn Day was also evolved by Li *et al.* (2011). It has good ability for storage and transportation. Details description of these varieties are given in Table 2.

Table 2: Gerbera varieties resistant to *Phytophthora cryptogea* disease.

S. No.	Hybrids	Parentages	Year	Chief characteristics
1.	Liangfen	Aruba True × Love	2008	It belongs to wide-petal standard cut flower cultivar with pink petal, blank centre, semi-double. Diameter of flower head 10-12 cm. Peduncle length 50-55 cm. Vase life 12-15 days. Flower yield 30-35 stems/plant/year.
2.	Autumn Day	Sombero × Dongri	2011	It belongs to a standard cut flower cultivar with orange-red petal black centre and semi-double type. D10-12 cm. Peduncle length is 50–55 cm, diameter of peduncle 0.55-0.65 cm. Suitable for protected cultivation. Vase life 12-15 days. Flower yield 23-25 stems/plant/year.

Polyploidy

A tetraploid of *Gerbera jamesonii*, obtained from colchicine seed treatment is characterized by increased plant and flower size. It was fertile and produced tetraploid seedlings following self-pollination (Pryor, 1972). Tetraploids were created by colchicine treatment to reduce quantitative partial resistance against *Phytophthora* in gerbera (Sparnaaij, 1979). In this case resistance to *Phytophthora* was not investigated because of rapid change in clonal propagation but it was observed that uniformity within progeny was remarkably high for all external characters.

Tosca *et al.* (1995) treated haploid gerbera plantlets with colchicine and oryzalin for 48 hours and founded that at comparable doses; oryzalin is as efficient as colchicine for chromosome doubling of gerbera, but less phytotoxic. They concluded, oryzalin may be considered superior to colchicine because of its lower phytotoxicity and absence of long term effects. They suggested that longer oryzalin treatments could probably induce the diploidization of a larger number of cells and reduce the problem of chimeric plants.

Mutation

Li *et al.* (2009) successfully induced tetraploids from *in vitro* plantlets of diploid *Gerbera jamesonii* Bolus 'Cabana' and 'Dalma' (2n=2x=50) by dipping shoot tips with colchicines at different concentrations. The most efficient condition for inducing tetraploids was treated with 0.10% colchicine for 48 h. About 200 and 300 stable tetraploids of 'Cabana' and 'Dalma' were respectively obtained.

BIOTECHNOLOGY

Tissue culture

Rapid *in vitro* shoot multiplication on MS basal medium supplemented with 0.7 mg l^{-1} thidiazuron was induced in *Gerbera jamesonii* var. Pink Star. Healthy shoots were developed in large number bearing deep green leaves. Maximum shoot elongation was discernible on the same medium in comparison to other treatments. Cent percent rooting of the shootlets was observed on MS supplemented with 0.5 mg l^{-1} NAA. Addition of 10 micro M $AgNO_3$ in combination with 0.5 mg l^{-1} NAA and IBA, respectively, showed development of fewer roots but displayed excellent plant health (Mandal *et al.,* 2010).

In vitro propagation in three varieties of gerbera from flower bud and flower stalk explants was carried out (Akter *et al.*, 2012). It was observed that flower bud and flower stalk were suitable and superior for callus induction and subsequent regeneration of *in vitro* shoots when cultured on MS supplemented with 5.0 mg/l BAP and 1.0 mg/ l NAA. Variety with red petal showed the best response in producing multiple shoots among the three varieties.

Somaclonal variation

Somaclonal variation may arise due to genetic or epigenetic changes (Jain *et al.*, 1998). Seeds of European strain gerbera were germinated *in vitro*. The shoots were then treated with some mutagens such as gamma rays and colchicine. At 1- 4 krad doses, plant survived but growth was decreased as compared to control. At dose 1 krad the survival of flowers was 100% and treatments of doses at 2, 3, 4 and 5-10 krad caused 25, 46, 50 and 100% death, respectively. Irradiation with 1 krad of gamma rays for 1-3 times resulted in some abnormal characters like stunt shoots, divided leaves at apex, rough or thick, narrow and thin leaves. Some florets had red strips on the yellow ray florets and some flowers were larger than control but not different in chromosome number. Shoots treated with colchicines at 0.025 and 0.05% solution for 24 hours. Some plants survived at 0.025% treatment. It was found that one plant was tetraploid (4x = 100), 2 plants were mixoploid and rest were diploid (2x = 50) from cytological studies (Rongrong, 1985).

Somatic mutagenesis

In somatic mutagenesis study, shoots of *in vitro* raised plantlets of cv. Rebecca were irradiated with a single dose of 20 Gy gamma rays and then micropropagated for 2 cycles. Irradiation treatment reduced a 25% reduction in propagation rate in each cycle. Of the initial shoots, 52% produced a clonal progeny showing flower variations and among the MV_2 plants, 15% were variants. Variation was observed on flower number, length and width of ligules and colour. In MV_2, only 3% plants showed stable mutation and remaining 97% showed chimerism at various levels. In this study 13 morphological and 6 colour type variations were observed (Laneri *et al.*, 1990). They further concluded that more than 2 micropropagation cycles are necessary to increase the percentage of stable mutants.

Jerzy *et al.* (1994) irradiated *in vitro* regenerated plants with gamma rays (5-25 Gy). They observed that chimeras from leaf explants forming adventitious shoots significantly increased the percentage of stable mutants with uniform inflorescence colour change in the VM_2. Hasbullah *et al.* (2012) observed gradual decline in the number of shoots regenerated from irradiated petiole explants when explants were exposed to 20 Gray of irradiation dose. Gradual decline was observed based on plant height as the dose of gamma irradiation increased.

Molecular breeding

RAPD analysis involving decamer primers confirm true to-type-nature of the micropropagated plantlets at genomic level. Profuse microtillering and subsequent whole plant development as gleaned from this study offers ample scope to develop large number of clones from a stock of choice for *in vitro* conservation, characterization and screening for diverse characters (Mandal *et al.*, 2010).

The genetic fidelity of *in vitro* raised gerbera clones was assessed by using random amplified polymorphic DNA (RAPD) and inter-simple sequence repeat (ISSR) markers. Out of 35 RAPD and 32 ISSR primers screened, only 12 RAPD and 10 ISSR primers produced clear, reproducible and scorable bands. The 12 RAPD primers produced 54 distinct and scorable bands, with an average of 4.5 bands per primer. The number of scorable bands for ISSR primers varied from 3 (ISSR-14) to 9 (ISSR-07), with an average of 5.5 bands per primer. The number of bands generated per primer was greater in ISSR than RAPD. All banding profiles from micropropagated plants were monomorphic and similar to those of the mother plant. A similarity matrix based on Jaccard's coefficient revealed that the pair-wise value between the mother and the in vitro-raised plantlets was 1, indicating 100% similarity. This confirmed the true-to-type nature of the in vitro-raised clones (Bhatia *et al.*, 2012).

Genetic engineering

In India, Nagaraju *et al.* (1998) used *A. tumefaciens* strain LBA4404 carrying the binary vector pBI 121 (GUS and *nptII*) to generate transgenetic plants from the cultivar RCGH-164. They tested petiole, leaf and shoot tip explants harvested from *in vitro* grown shoots and obtained transformation efficiencies of 3%, 0.5% and 17%, respectively. They reported that all the transgenics were normal. Sugita *et al.* (2000) reported that newly developed MAT vector system using the shooty gene from *Agrobacterium* for enhanced regeneration may be very useful in gerbera transformation. Korbin *et al.* (2002) transformed gerbera plants with tomato spotted wilt virus (TSWV) nucleoprotein gene. After 10 months, they selected 74 (5.4%) transgenic gerbera plants. The level of nucleoprotein estimated by ELISA was low in transformed tissues, but mechanically inoculated, transgenic plants demonstrated resistance to TSWV.

Elomaa *et al.* (1998) isolated a promoter for one of the *dfr* genes, *Pgdfr 2*. Gerbera plants transgenic for a *Pgdfr 2-uidA* construct reveal that the activity of the *Pgdfr 2* promoter from one variety follows the pigmentation in other varieties which have different colour patterns. They have also isolated a cDNA (*gmyc 1*) homologous to the characterized genes encoding bHLH-type regulators of the anthocyanin pathway. The major role of *gmyc 1* is in regulalting *dfr* activity in corolla and carpel, but not pappus and stamen. FNS catayzes the conversion of flavanones to flavones via the introduction of a double bond between C-2 and C-3 of the flavanones (Heller and Forkmann, 1993). FNS II has been characterized in *Gerbera hybrida* (Martern and Forkmann, 1998).

Development of transgenic

First transgenic plant in gerbera was produced from the cultivar Terra Regina carrying the antisense chalcone synthase gene (*gchs*) for the alteration of flower colour. Pre-cultured petioles were transformed using *Agrobacterium tumefaciens* strain C 58 carrying the disarmed Ti plamid pGV2260. Two binary vector were used, pHTT370 and pHTT377 carrying antisense *gchs 1* and *gchs 2*, respectively (Elomaa *et al.*, 1993 and 1996).

Four transgenics were generated in the first transformation investigation and 17 in second experiment. A biotechnology company Florigene transformed gerbera by using *A. tumefaciens* strain LBA4404 (Robinson and Firoozabady, 1993). They noticed that transformation efficiency is dependent on cultivar and explant.

In an another study various *A. tumefaciens* strains (LBA4404, EHA101 and EHA105) were used with binary vector containing GUS and either the *nptII* or *bar* selective marker genes for the transformation of the gerbera cv. Mariola (Orlikowska *et al.*, 1997). They observed that very young petioles to be the effective explant. The highest transformation efficiencies were achieved using EHA101 and the *bar* gene for selection of transformants. About 10 stable transgenics were regenerated from 200 explants.

REFERENCES

Akter, N., Hoque, M.I. and Sarker, R.H. 2012. *In vitro* propagation in three varieties of gerbera (*Gerbera jamesonii* Bolus.) from flower bud and flower stalk explants. *J. Plant Tissue Culture Biotech.*, **22**(2): 143-152.

Anonymous, 2002. Passport Data of Flower Crop Germplasm. AICRP on Floriculture, ICAR, New Delhi.

Anonymous. 1956. The Wealth of India. CSIR, New Delhi, **4**: 287.

Anuradha, S. and Gowda, J.V.N. 1999. Quantitative genetic studies in gerbera. *Crop Res. Hisar*, **18**: 60-63.

Aswath, C. 2004. Genetic investigations in dry vase life of gerbera. *Abst*: National Symposium on Recent Trends and Future Strategies in Ornamental Horticulture, UAS, Dharwad, Karnataka, pp. 29.

Aswath, C., Parthasarthy, V.A. and Bhowmik, G. 1998. Genetic studies in economic characters of gerbera. *J. Maharastra Agril. Uni.*, **23** (2): 140-142.

Bailey, L.H. 1963. The Standard Cyclopedia of Horticulture, MacMillan Co., New York, Vol. I, A-E, 1200 p.

Bhatia, R., Singh, K.P., Sharma, T.R. and Jhang T. 2012. Evaluation of the genetic fidelity of in vitro-propagated gerbera (Gerbera jamesonii Bolus) using DNA-based markers. *Plant Cell Tissue Organ Cult.,* **104**(1): 131-135.

Bremer, K. 1994. Asteraceae: Cladistics and Classification. Timber Press, Portland, Oregon.

Buivids, K., Ozolinya, L., Muceniece, G. and Marga, G. 1992. Main trends in breeding Chrysanthemum and gerbera for cut flowers in Latvia. *Eksperimentine Biologija*, **3 & 4**: 61-62.

Byrne, T., Drennan, D. and Harding, J. 1987. Genotype × environment interaction bias in heritability estimates for cut-flower yield of greenhouse gerberas. *J. American Soc. Hortil. Sci.*, **112**(4): 724-727.

Chung YongMo, Hwang JuChean, Chin YoungDon, Lee JungSoo, Kim SuKyeong, Ro ChiWoong, Yi YoungByung and Kwon OhChang. 2013. Breeding of a new pink *Gerbera hybrida* 'Party Time' with strong peduncle for cut flower. *Korean J. Hortil. Sci. Tech.,* **31**(1): 128-132.

Chung, Y.M., Hwang, J.C., Yi, Y.B., Kin, J. and Kwon, O.C. 2004. A new high yielding yellow gerbera cut flower cultivar with strong peduncle, Runa. *Korean J. Breeding*, **36**(1): 67-68.

Danhardt, W. 1966. Hybridization in *Gerbera jamesonii*. *Arch. Gartenb*., **14**: 17-24.

De Jong, J. 1986. Breeding for keeping quality in gerbera. *Acta Horticulturae*, **181**: 353-357.

De Jong, J. and Garretsen, F. 1985. Genetic analysis of cut flower longevity in gerbera. *Euphytica*, **34**: 779-784.

Drennan, D., Harding, J. and Byrne, T.G. 1986. Heritability of inflorescence and floret traits in gerbera. *Euphytica*, **35** (1): 319-330.

Elomaa, P., Helaniutta, Y., Kotilainen, M. and Teeri, T.H. 1995. Transformation of antisense cDNAs of the chalcone synthase super-genefamily into *Gerbera hybrida:* Differential effect on the expression of family members. *Mol. Breed.*, **2**: 41-50.

Elomaa, P., Honkanen, J., Puske, R., Sepponen, P., Helariutta, Y., Mento, M., Kotilainen, M., Nevalainen, L. and Terri, T.H. 1993. *Agrobacterium*-mediated transfer of antisense chalcone synthase cDNA of *Gerbera hybrida* inhibits flower pigmentation. *Biotechnology*, **11**: 508-511.

Elomaa, P., Mehto, M., Kotilainen, M., Helariutta, Y., Neralainen, L. and Teeri, T.H. 1998. A bHLH transcription factor mediates organ, region and flower type specific signals on dihydroflavonol-4-reductase (*dfr*) gene expression in the inflorescence of *Gerbera hybrida* (Asteraceae). *Plant J.*, **16**: 93-99.

Franceschetti, U. 1975. Pollination and seed production in gerbera. *Colture Protette*, **4** (1): 19-23.

Hansen, H.V. 1985. A taxonomic revision of the genus *Gerbera* (Compositae, Mutisieae) sections Gerbera, Parva, Piloselloides (in Africa) and Lasiopus. *Opera Bot.*, **78**: 1-36.

Harding, J., Byrne, T.G. and Nelson, R.L. 1981. Estimation of heritability and response to selection for cut flower yield in gerbera. *Euphytica*, **30** (2): 313-322.

Harding, J., Byrne, T.G. and Nelson, R.L. 1981a. Heritability of cut flower vase longevity in gerbera. *Euphytica*, **30**: 653-657.

Harding, J., Huang H. and Byrne, T.G. 1991. Maternal, paternal, additive and dominance components of variance in gerbera. *Theo. Appl. Genetics*, **82** (6): 756-760.

Hasbullah, N.A., Taha, R.M. and Azani, S.N.M. 2012. Irradiation effect on *in vitro* organogenesis, callus growth and plantlet development of *Gerbera jamesonii*. *Horticultura Brasileira*, **30**(2): 252-257.

Helariutta, Y., Kotilainen, M., Elomaa, P. and Terri, T.H. 1995. *Gerbera hybrida* (Asteraceae) imposes regulation of several anatomical levels during inflorescence development on the gene for dihydroflavonol-4-reductase. *Plant Mol. Biol.*, **28**: 935-941.

Helariutta, Y.K., Elomaa, P., Kotilainen, M., Griesbach, R.J., Schroder, J. and Teeri, T.H., 1995a. Chalcone synthase-like genes active during corolla development are differentially expressed and encode enzymes with different catalytic properties in *Gerbera hybrida* (Asteraceae). *Plant Mol. Biol.*, **28**: 47-60.

Heller, W. and Forkmann, G. 1993. Biosynthesis of flavonoids. *In:* J.B. Harborne (ed.). The Flavonoids: Advances in Research since 1986. Chapman and Hall Ltd., London, pp. 399-425.

Huang, H., Harding, J., Byrne, T. and Famula, T. 1995. Estimation of long term genetic improvement for gerbera using the best linear unbiased predication (BLUP) procedure. *Theo. Appl. Genetics*, **91**(5): 790-794.

Jain, S.M., Vitti, D., Tucci, M., Grassotti, A., Rugini, E., Saccardo, F. 1998. Biotechnology and mutagenesis in gerbera improvement. *Advances Hortil. Sci.*, **12** (1): 47-53.

Jerzy, M., Zalewska, M. and Garczewska, A. 1994. Induction of mutations and separation of chimeras *in vitro* in *Gerbera jamesonii. Zeszyty Problemowe Postepow Narek Rolniczych*, **414**: 343-352.

Kohl, H.C. 1968. Gerberas: Their culture and commercial possibilities. *S. Flor. Nursery*, **28**: 24-26.

Korbin, M., Powyszynska, M., Komorowsha, B. and Wawrzynczak, D. 2002. Transformation of gerbera plants with tomato spotted wilt virus (TSWV) nucleoprotein gene. *Acta Horticulturae*, **572**: 149-157.

Laneri, U., Franconi, R. and Altavista, P. 1990. Somatic mutagenesis of *Gerbera jamesonii* hybr.: irradiation and *in vitro* culture. *Acta Horticulturae*, **280**: 395-402.

Li, Han, Yan, Bo, Zhang, Ting, Jiang, YaLian, Zhang, Hao, Yu, LiXia and Li, ShenChong. 2009. Preliminary studies on polyploidy mutation of cut flower *Gerbera jamesonii* Bolus. *Acta Horticulturae Sinica*, **36**(4): 605-610.

Li, Shen-chong, Cao, Hua, Li, Shu-bin, Jiang, Ya-lian, Yang, Chun-mei, Shan, Qin-li and Wang, Ji-hua. 2011. A new cut flower gerbera cultivar Autumn Day. *Acta Horticulturae Sinica*, **38**(11): 2235-2236.

Li, Shen-chong, Li, Jin-ze , Wu, Li-fang , Jiang, Ya-lian , Cao, Hua, and Li, Han. 2008. A new cut flower gerbera cultivar Liangfen with resistance to *Phytophthora cryptogea*. *Acta Horticulturae Sinica*, **35**(3): 466.

Luckana, P. and Somjit, B. 1984. Effect of natural and artificial pollination on gerbera seed production. Agricultural Research and Extension Research Report (Thailand), pp 97-106.

Luckana, P. and Somjit, B. 1985. Study for suitable time of gerbera seed production. Agricultural Research and Extension Research Report (Thailand), pp. 1-11.

Mahanta, P., Choudhury, S., Paswan, L. and Talukdar, M.C. 1998. Studies on variability and heritability of some quantitative characters in gerbera (*Gerbera jamesonii*). *South Indian Hort*., **46**: 43-46.

Mandal, A.B., Das, Sonali and Biswas, Anusrita. 2010. *In vitro* propagation of gerbera and assessment of clonal fidelity of the mericlones. *Indian J. Hort*., **67**(3): 357-361.

Martern, S. and Forkmann, G. 1998. Genetic control of flavone synthase II activity in flowers gerbera hybrids. *Phytochemistry*, **49**: 1953-1958.

Maurer, J. and Horn, W. 1967. Colour inheritance studies in gerbera. *Gertenwelt*, **67**: 63-64.

Mutseniets, G.Y., Rashal, I.D. and Dishler, V.Y. 1978. Inheritance of quantitative characters in *Gerbera jamesonii. Genetica*, **14**: 238-241.

Nagaraju, V., Srinivas, G.S.L. and Sita, G.L. 1998. *Agrobacterium*-mediated genetic transformation in *Gerbera hybrida*. *Current Sci*., **74**(7): 630-634.

Nair, S.A. and Shiva, K.N. 2003. Genetic variability and path coefficient analysis in gerbera. *J. Orna. Hort*., **6**(3): 180-187.

Oriecki, M. 1991. Study of a F_1 hybrid of gerbera (*Gerbera jamesonii)* with yellow inflorescences and its greenhouse performance. *Prace Instytuta Sadownictwa I kwiaciarstwa w Skierniewicach Seria B Rosliny Ozdobne*, **16**: 23-30.

Orlikowska, T., Nowak, E., Altman, A. and Ziv, M. 1997. Factors affecting transformation of gerbera. *Acta Horticulturae*, **447**: 619-621.

Park, SangKun, Shin, HakKi Kwon, YoungSoon, Jung, JeaA, Choi and SeongYoul. 2012. A new purple and semi-double gerbera cultivar Purple Queen with long and thick peduncle for cut-flower. *Korean J. Breed. Sci*., **44**(4): 617-620.

Pryor, R.L. 1972. Cytogenetical investigation in gerbera. *Hort. Science*, **7**: 197-198.

Robinson, K.E.P. and Firoozabady, E. 1993. Transformation of floriculture crops. *Sci. Hort*., **55**: 83-99.

Rongrong, V. 1985. *In vitro* mutation breeding of gerbera (*Gerbera jamesonii* Hort.). Bangkok, Thailand, Leaves, 81.

Schiva, T. 1972. Analyses of biometric characteristics of commercial importance in *Gerbera jamesonii* and the consequences of inbreeding. *Annali della Facolta di Scienze Agrarie della Universita degli Studi di Torino*, **8**: 355-366.

Schiva, T. 1974. Breeding gerberas. Genetic analysis of quantitative characters of commercial importance and the results of inbreeding. *Annali dell' Istituto Sperimentale per la Floricoltura*, **5** (2): 27-36.

Smith, D.E. and Nelson, R.L. 1967. Gerbera propagation. *Calif. Agri*., **21** (12):7.

Sparnaaij, L.D. 1979. Polyploidy in flower breeding. *Hort. Science*, **14**: 496-499.

Sparnaaij, L.D., Garretsen, F. and Bekker, W. 1975. Additive inheritance of resistance to *Phytophthora cryptogea in Gerbera jamesonii. Euphytica*, **24**: 551-556.

Sujita, K., Kasahara, T., Matsunaga, E. and Ebinuma, H. 2000. A transformation vector for the production of marker-free transgenic plants containing a single copy transgene at high frequency. *Plant J*., **22**: 461-469.

Tesi, R. 1978. Variation of some characters of flowers in clones of *Gerbera jamesonii hybrida. In:* L. Quagliotti and A. Baldi (eds.). Genetics and breeding of carnation and gerbera. Proc. Eucarpia Mtg., Alassio, Italy. pp. 227-232.

Tosca, A., Pandolfi, R., Citterio, S., Fasoli, A. and Sgorbati, S. 1995. Determination by flow-cytometry of the chromosome doubling capacity of colchicines and oryzalin in gynogenetic haploids of *gerbera*. *Plant Cell Reports*, **14**: 455-458.

Van Meeteren, U. 1978. Water relations and keeping quality of cut gerbera flowers. II. Water balance of ageing flowers. *Scientia Hort*., **9**: 189-197.

Van Meeteren, U. 1981. Role of pressure potential in keeping quality of cut gerbera inflorescences. *Acta Horticulturae*, **113**: 143-150.

Von Wilberg, B. 1973. Physiological investigations on the problem of freedom from stem break as prerequisite in breeding gerbera forms which remain upright when used as cut flower. *Z. pflanzenzuchtung*, **69**: 107-114.

Wernett, H.C., Sheehan, T.J., Wilfret, G.J., Marousky, F.J., Lyrene, P.M. and Knauft, D.A. 1996. Post-harvest longevity of cut-flower gerbera. I. Response to selection for vase life components. *J. Amer. Soc. Hort. Sci.*, **121**(2): 216-221.

Wernett, H.C., Wilfret, G.J., Sheehan, T.J., Lyrene, P.M., Martin, F.G., Inhite, T.L., Powell, G.L. and Wilcox, C.J. 1996a. Post-harvest longevity of cut-flower gerbera. II. Hertability of vase life. *J. Amer. Hort. Sci.*, **121**(2): 222-224.

Yu, Y., Harding, J., Byrne, T. and Famula, T. 1993. Estimation of components of genetic variance and heritability for flowering time and yield in gerbera using Derivative-Free Restricted Maximum Likelihood (DFRML). *Theo. Appl. Genetics*, **86**: 234-236.

Chapter – 12

Gladiolus

Gladiolus (*Gladiolus* spp.) belongs to family Iridaceae and subfamily Ixioideae. Gladiolus is a Latin word, the diminutive of "*gladius*" meaning sword, sometimes called as sword lily. The plant is herbaceous perennial and semi-hardy. It is commonly grown from rounded, symmetrical corm, which is enveloped in several layers of brownish, fibrous tunics. The cultivated gladiolus is a complex assemblage of cultivars originating from complex ancestry of natural selection or hybridization involving several species. Modern garden gladiolus is a complex of at least twelve species (Raghava, 2000). It is an important commercial cut flower in both domestic and international markets.

ORIGIN AND DISTRIBUTION

There are two main centres of distribution of *Gladiolus* species, one in Mediterranean basin and the other in southern and central Africa (Lewis *et al.*, 1972 and Delpierre and Du Plessis 1974). Most of the species are native to Mediterranean region and tropical parts of the South Africa, particularly the region of the Cape of Good Hope (Raghava, 2000).

The name gladiolus was coined by Pliny the Elder (A.D. 23-79) to describe the shape of the leaf which resembles that of sword (Latin word gladius meaning sword). It is native to South Africa. Most of the species scattered throughout South Africa, Zambesi River, Rhodesia, Mt. Kilimangero, Tanganyika, Natal and Cape of Good Hope, Republic of South Africa, Madagascar and Abyssinia (Ethiopia). Gladioli are said to be cultivated since the days of ancient Greece. As evidenced by a record in Lyte's Nieeve Herball, it is known since 1578, first introduced in France and soon after spread to England, Germany, Holland and North America. No evidence is available that any species of gladiolus of South African origin was ever grown in Europe prior to 1700. Cornutus described *Gladiolus aethiopion* bearing light red-scarlet flowers and is from Cape region. In the middle of eighteenth century the South African species were introduced into Europe. *Gladiolus tristis* closely related to *G. grandis* was first introduced in Chelsea Physii Garden, UK in 1745 from Cape region and Walat. *G. illyricus* is native to Mediterranean region. In 1749, *Gladiolus grandis* was naturalized in Spain from South Africa.

By 1816, France, Holland and Belgium started raising hybrids, the first being Ghant Gladiolus, *Gladiolus gandavensis*, in 1841, by crossing *G. cardinalis* and *G. psittacinus* and vice-versa. *Gladiolus primulinus* having central petals with hooded characteristics was collected from its native place on the banks of Zambesi River near Victoria Falls in 1902. Victor Lemoine of France used *G. primulinus* for producing 'Primulinus Hybrids' during 1906-1908 while A.E. Kunderd of Indiana, U.S.A. in 1915 produced a similar race 'Primulinus Grandiflora Hybrids' by using same parentage as Lemoine. In 1907, first ruffled gladiolus with fimbriated and frilled petals was produced by the efforts of Kunderd.

Gladiolus is a potential flower crop for floriculture industry in India. Gladiolus is commercially grown in West Bengal, Maharashtra, Uttar Pradesh, Uttarakhand, Punjab, Haryana, Sikkim, Jammu and Kashmir, Karnataka, Gujarat, Himachal Pradesh, Tamil Nadu, Madhya Pradesh, Delhi and Rajasthan. The area of the states of U.P., Uttarakhand, Haryana, Punjab, Rajasthan and H.P. and surrounding Delhi grow gladiolus mostly for Delhi flower market (Singh, 2006).

SPECIES

The genus *Gladiolus* contains about 260 species i.e. *G. abbreviatus* Andrews, *G. acuminatus* F. Bol., *G. aequinoctialis* Herb., *G. alatus* L. (sect. Hebea), *G. aleppicus* Boiss. var. *aleppicus, G. anatolicus, G. andringitrae* Goldblatt, *G. angustus* L. (sect. Blandus) : Long-tubed Painted Lady, *G. antandroyi* Goldblatt, *G. antholyzoides* Baker, *G. appendiculatus* G. Lewis, *G. aquamontanus* Goldblatt & Vlok, *G. arcuatus* Klatt, *G. atropurpureus* Baker, *G. atroviolaceus* Boiss., *G. aurantiacus* Klatt, *G. aureus* Baker : Golden Gladiolus, *G. bellus* C.H. Wright, *G. benguellensis* Baker (sect. Ophiolyza), *G. bilineatus* G.J. Lewis, *G. boehmii* Vaupel (1882), *G. bojeri* (Baker) Goldblatt, *G. brachylimbus* Baker (1893), *G. brachyphyllus* F. Bolus or Bolus f., *G. brevifolius* Jacq. (sect. Linearifolius), *G. brevitubus* G. Lewis, *G. buckerveldii* (L. Bolus) Goldblatt, *G. bullatus* Thunb. ex G. Lewis : Caledon Bluebell, *G. caeruleus* Goldblatt and J.C. Manning, *G. calcaratus* G. Lewis, *G. calcicola* Goldblatt, *G. callianthus* Mosais : Abyssinian Gladiolus, *G. canaliculatus* Goldblatt, *G. candidus* (Rendle) Goldblatt, *G. cardinalis* Curtis (sect. Blandus), *G. carinatus* Aiton, *G. carmineus* C.H. Wright (sect. Blandus) : Cliff Lily, *G. carneus* (sect. Blandus) : Large Painted Lady, *G. caryophyllaceus*

(Burm. f.) Poiret, *G. cataractarum* Oberm., *G. caucasicus* Herb., *G. ceresianus* L. Bolus, *G. citrinus* Klatt, *G.* × *colvillei* : Colville's Gladiolus, *G. communis* L. (sect. Gladiolus) : Cornflag (type species), *G. communis* subsp. *byzantinus* (sect. Gladiolus), *G. communis* subsp. *communis* (sect. Gladiolus), *G. conrathii* Baker (1898), *G. crassifolius* Baker, *G. crispulatus* L. Bolus, *G. cruentus* T. Moore (sect. Ophiolyza), *G. curtifolius* Marais, *G. cuspidatus, G. cylindraceus* G. Lewis, *G. cymbarius* Baker, *G. dalenii* Van Geel (sect. Ophiolyza), *G. debilis* Ker Gawler (sect. Homoglossum) : Small Painted Lady, *G. decaryi* Goldblatt, *G. decipiens* Vaupel, *G. decoratus* Baker, *G. densiflorus* Baker, *G. deserticolus* Goldblatt, *G. dolomiticus* Oberm., *G. dracocephalus* Hook.f., *G. dregei* Klatt, *G. dubius, G. dzavakheticus, G. ecklonii* Lehm., *G. edulis* Burchell ex Ker Gawler, *G. elliotii* Baker (sect. Ophiolyza), *G. emiliae* L. Bolus, *G. engysiphon* G. Lewis, *G. equitans* Thunb. (sect. Hebea), *G. erectiflorus* Baker, *G. exiguus* G. Lewis, *G. flanaganii* Baker : Suicide Gladiolus, *G. floribundus* Jacq., *G. fourcadei* (L. Bolus) Goldblatt and De Vos, *G.* × *gandavensis* (sect. Ophiolyza) [= *G. dalenii* × *G. oppositiflorus*], *G. garnierii* Klatt, *G. geardii* L. Bolus, *G. goetzii* Harms, *G. gracilis* Jacq. (sect. Homoglossum) : Reed Bells, *G. gracillimus* Baker, *G. grandiflorus* (sect. Blandus), *G. gregarius* Welw. ex Baker (sect. Densiflorus), *G. griseus* Goldblatt and J.C. Manning, *G. gueinzii* Kunze, *G. guthriei* F. Bol. (sect. Linearifolius), *G. halophilus, G. harmsianus* Vaupel, *G. heterolobus* Vaupel, *G. hirsutus* Jacq. (sect. Linearifolius) : Small Pink Afrikaner, *G. hollandii* L. Bolus, *G. horombensis* Goldblatt, *G. huillensis* (Welw. ex Baker) Goldblatt, *G. hyalinus* Jacq., *G. illyricus* W.D.J. Koch, *G. imbricatus*, *G. inandensis* Baker, *G. incospicus* Baker, *G. inflatus* Thunb., *G. inflexus* Goldblatt and J.C. Manning, *G. insolens* Goldblatt and J.C. Manning, *G. intonsus* Goldblatt, *G. invenustus* G. J. Lewis, *G. involutus* (sect. Hebea), *G. iroensis* (A. Chev.) Marais, *G. italicus* P. Mill. (sect. Gladiolus) : Italian Gladiolus, *G. johnstoni* Baker (s. d.), *G. jonquilliodorus* Ecklon ex G. Lewis, *G. junodi* Baker, *G. kamiesbergensis* G. Lewis, *G. karendensis* Baker, *G. katubensis* De Wild., *G. klattianus* Hutch., *G. kotschyanus* Boiss, *G. kubangensis* Harms, *G. lapeirousioides* Goldblatt, *G. laxiflorus* Baker, *G. lemoinei, G. leptosiphon* Bolus f., *G. liliaceus* Houtt. (sect. Homoglossum), *G. linearifolius* Vaupel, *G. linearis* N.E.Br., *G. longanus* Harms, *G. longicollis* Baker (sect. Homoglossum), *G. longicollis* subsp. *longicollis* (sect. Homoglossum), *G. longicollis* subsp. *platypetalus* (sect. Homoglossum), *G. loteniensis* Hilliard & Burtt, *G. louiseae* L. Bolus, *G. lundaensis* Goldblatt, *G. luteus* Lam., *G. lyalinus, G. macneilii* Oberm., *G. macowani* Baker, *G. macowanii* Baker (s. d.), *G. macrospathus* Goldblatt, *G. maculatus* Sweet, *G. magnificus* (Harms) Goldblatt, *G. malangensis* Baker (1879), *G. malvinus* Goldblatt and J.C. Manning, *G. marlothii* G. Lewis, *G. martleyi* L. Bolus (sect. Homoglossum), *G. meliusculus* (G. Lewis) Goldblatt & J.C. Manning, *G. melleri* Baker (sect. Ophiolyza), *G. micranthus* Baker (1901), *G. microcarpus* G. Lewis, *G. microsiphon* Baker, *G. milleri* Ker Gawler, *G. mirus* Vaupel, *G. monticola* G. Lewis ex Goldblatt and J.C. Manning, *G. mortonius* (sect. Densiflorus), *G. mostertiae* L. Bolus, *G. muenzneri* F. Vaup, *G. murielae* (sect. Acidanthera), *G. natalensis* (Eckl.) Hook., *G. nerineoides* G. Lewis, *G. newii* Baker, *G. nigromontanus*

Goldblatt, *G. niveus* Goldblatt and J.C. Manning, *G. nyasicus* Goldblatt, *G. oatesii* Rolfe, *G. ochroleucus* Baker (sect. Densiflorus), *G. odoratus* L. Bolus, *G. oliganthus* Baker, *G. oppositiflorus* Herbert (sect. Ophiolyza), *G. orchidiflorus* Andrews (sect. Hebea), *G. oreocharis* Schltr., *G. pallidus* Baker, *G. paludosus* Baker, *G. palustris* : Marsh Gladiolus, *G. papilio* Hook. f. (sect. Densiflorus): Goldblotch Gladiolus, *G. pappei* Baker (sect. Blandus), *G. pardalinus* Goldblatt and J.C. Manning, *G. parvulus* Schltr., *G. patersoniae* F. Bolus or Bolus f., *G. pavonia* Goldblatt and J.C. Manning, *G. permeabilis* Delaroche (sect. Hebea), *G. perrieri* Goldblatt, *G. persicus* Boiss., *G. pillansii* G. Lewis, *G. pole-evansii* Verd., *G. praecostatus, G. praelongitubus* G.J. Lewis, *G. pretoriensis* Kuntze, *G. priorii* (N. E. Br.) Goldblatt and De Vos, *G. prismatosiphon* Schltr., *G. pritzelii* Diels, *G. psittacinus, G. psittacinus* Hook., *G. puberulus* Vaupel, *G. pubigerus* G. Lewis, *G. pulchellus* Klatt, *G. pulcherrimus* (G. Lewis) Goldblatt and J.C. Manning, *G. punctulatus* Schrank, *G. pusillus, G. quadrangularis* (Burm. f.) Ker Gawler, *G. quadrangulus* (Delaroche) Barnard, *G. ramosus, G. recurvus* (sect. Homoglossum), *G. rehmannii* Baker, *G. remotifolius* Baker, *G. rigidifolius* Baker, *G. robertsoniae* F. Bolus or Bolus f., *G. rogersii* Baker, *G. roseovenosus* Goldblatt and J.C. Manning, *G. rubellus* Goldblatt, *G. rudis* Lichtst. ex Roem. & Schult., *G. rupicola* F. Vaupel, *G. saccatus* (Klatt) Goldblatt and M.P. de Vos, *G. salteri* G. Lewis, *G. saundersii* Hook. f. : Saunders' Gladiolus, Lesotho Lily, *G. schlechteri* Baker, *G. schweinfurthii* Baker, *G. scullyi* Baker, *G. serapiiflorus* Goldblatt, *G. serenjensis* Goldblatt, *G. sericeovillosus* Hook. f., *G. serpenticola* Goldblatt & J.C. Manning, *G. somalensis, G. speciosus* Thunb., *G. spectabilis* Baker (s. d.), *G. splendens* (Sweet) Herbert, *G. stefaniae* Oberm., *G. stellatus* G. Lewis, *G. subcaeruleus* G. Lewis, *G. sufflavus* (G. Lewis) Goldblatt and J.C. Manning, *G. sulcatus* Goldblatt, *G. taubertianus* Schltr., *G. tenellus* Ecklon, *G. tenuis, G. teretifolius* Goldblatt & De Vos, *G. trichonemifolius* (sect. Homoglossum), *G. tristis* (sect. Homoglossum), *G. tristis* var. *concolor* (Salisb.) Baker: Ever-flowering Gladiolus, *G. tritoniaeformis, G. tritoniiformis* Kuntze, *G. uitenhagensis* Goldblatt and Vlok, *G. undulatus* L. (sect. Blandus): Waved-flowered Gladiolus, *G. unguiculatus* Baker, *G. usambarensis* Marais ex Goldblatt, *G. uysiae* L. Bolus ex G. Lewis, *G. vaginatus* F. Bolus or Bolus f. (sect. Homoglossum), *G. validissimus* Vaupel, *G. vandermerwei* (L. Bolus) Goldblatt and De Vos, *G. varius* F. Bolus or Bolus f., *G. velutinus* De Wild., *G. venustus* G. Lewis (sect. Hebea), *G. vernus* Oberm., *G. vinoso-maculatus* Kies, *G. violaceo-lineatus* G. Lewis, *G. virescens* Thunb. (sect. Hebea), *G. viridiflorus* G. Lewis, *G. viridis* Aiton, *G. watermeyeri* (sect. Hebea), *G. watsonioides* Baker, *G. watsonius* Thunb. (sect. Homoglossum), *G. woodii* Baker, *G. zimbabweensis* Goldblatt. Out of these, 250 species are growing in sub-Saharan Africa, mostly South Africa (Goldblatt *et al.*, 2001). About 10 species are growing in Eurasia. There are 160 species of gladiolus endemic in southern Africa and 76 in tropical Africa.

FLOWER MORPHOLOGY AND BIOLOGY

The individual flower is broad and funnel shaped and called floret. The floret opens in an acropetal succession i.e. from base to the top. Each floret is enclosed by 2 green bracts and has 3 main parts i.e. tepals, stamens and pistil. Sepals and petals are collectively called tepals. A group of tepals is known as perianth. The perianth segments are variously coloured, 6 in number and are arranged in 2 whorls. The upper 3 are large and the lower 3 are small and often have blotches or markings on the throat. Stamens are 3, each stamen has a long filament inserted on the tube. Pistil consist of 3 parts i.e. stigma, style and ovary. The ovary is present at the base of the floret. Style is single and supports 3-lobed stigma above the anthers. Flowers are hermaphrodite and often protandrous.

When the plant reaches 6-7 leaf stage, the flowering spike starts emerging through the leaves (Anon., 2002). Flowers start opening after the sunset and open fully in the morning, except few exceptions for example cultivar Setting Sun opens its flowers by 11.00 am in cloudy weather, whereas as certain species as *Gladiolus floribundus* closes its segments at night while *G. tristis* open fully in the evening (Lewis *et al.*, 1972).

Mode of pollination

The flower structure and positional barrier between anthers and stigma suggest that gladiolus is essentially an outbreeder (Ohri and Khoshoo, 1981). Studies on mode of pollination by Goldblatt *et al.* (2001) indicated that the primary or sole pollinators include long-tongued aphid and anthophorine bees (Apidae), short-tongued halietid and andrenid bees (Halictidae, Andrenidae), Sometimes in combination with hopliine beetles (Scarabaeidae), long-proboscid flies (Nemestrinidae, Tabanidae), large butterflies (Satyridae), moths (mostly Noctuidae and Sphingidae) and subirds (Passerinae). In gladiolus moth-pollinated flowers are usually large, long-tubed and pale-coloured or mottled dull purple to brown, are usually richly scented, often open fully only at night and produce relatively concentrated nectar that is sucrose rich. Butterfly pollinated flowers, in contrast, are fully open during day, close partially or completely at night, are often bright crimson to scarlet, usually with prominent white splashes on the lower tepals, but are also large, have a long tube and produce quantities of relatively dilute nectar, either sucrose rich or hexose rich (Goldblatt and Manning, 2002).

Pollen viability

Asynchronous flowering of different cultivars is a big hurdle to the breeders. Pollen storage is one of the methods to overcome this problem. Cvs. Darshan, Dhiraj, Sagar, Sapna and Shobha and *G. callianthus* were stored at -65º C and -196º C (in liquid nitrogen) for 3 days and 9 months.

The dehiscence started at 8.00 am in several cultivars and at 8.30 am in some cultivars. The peak period of dehiscence was observed between 8.00 to 9.30 am (Roychoudhary, 1980 and Shah *et al.*, 1988). However, Singh and Singh (1985) observed dehiscence between 7 to 10 am and 8 to 11 am in cvs. George Mazure and Ratna Butterfly, respectively. It suggests that dehiscence of a variety or species should be observed before breeding programme.

Anthesis and stigma receptivity

Shah *et al.* (1988) studied floral biology of gladiolus in ten cultivars and observed that the anthesis was completed on same day. It started between 4:00 to 5.00 am and continued by 11.30 am in seven cultivars, whereas in cvs. Spite, Carmine and Gospel Sog anthesis started earlier (3.00 to 4.00 am and up to 10.30 am), however the peak period was observed between 8 to 10 am in all the cultivars. Similar observations have also been made by Roychoudhary (1980) and Singh and Singh (1985). However, Choudhary (1991) recorded anthesis in gladiolus cv. Melody between 5.00 and 6.35 hrs.

Normally after 1-2 days of opening of florets stigmas become thick and hairy and their surface becomes sticky. The hairs of the stigma help to catch the pollens that fall on it. This is the stage when stigma becomes receptive (Anon., 2002). The stigma became receptive one day before anthesis and remains receptive up to 2 days after anthesis, however, the maximum stigma receptivity was observed on the day of anthesis (Grabowska, 1969, Roychoudhary, 1980 and Shah *et al.*, 1988), whereas in contrast Singh and Singh (1985) observed stigma receptivity for 72 hours after anthesis. Studies were made on the floral biology of gladiolus on cvs. American Beauty, Green Willow, Oscar and White Oak and *G. callianthus* var. Murielae by Mahawer and Misra (1993a) and no significant differences were observed on anthesis, anther dehiscence and stigma receptively between the cultivars.

GENETICS

The modern garden and cultivated varieties of gladiolus originated from complex of at least a dozen species (Buch, 1972 and Everett, 1981). Repeated cycles of hybridization among elemental species (both diploid and polyploid) lead to recombination between multiple sets of genomes and polysomic inheritance (Ohri and Khoshoo, 1983). Original chromosome counts confirm that x=15 is probably the ancestral basic number for the genus *Gladiolus* but *G. atropurpureus* has x=12, *G. serapiflorus*, *G. gregarious* and *G. pseudospicatus* have x=11 (2n=22); *G. unguiculatus* has x=13 and *G. actinomorphanthus* has 2n=28 (Goldblatt *et al.*, 1993). The chromosomes of gladiolus ranges from 2n=30 to 120 (diploid, triploid, pentaploid, hexaploid, octaploid and hyperaneauploid (Misra and Singh, 1989). Chromosome numbers according to Ohri (2000) varied between 2n=30 and 180. The Mediterranean species (*G. communis*, *G. segetum*, *G. blandus*) are polyploids (2n=60 to 130) have not participated in producing the modern gladiolus cultivars. South African species which are primarily diploids (2n=30) have been hybridized to produce outstanding hybrids (Halevy, 2000). After crossing desirable strains obtained in F_1 generation, were vegetatively propagated without being segregated in the following generation, hence cultivars of today may be F_2, F_3, F_4 or F_7 or so of a particular was further blended with some extra parents at nearly every generation and they are not allowed to segregate freely in further generation due to asexual propagation. Therefore, now the modern cultivars have become more complex that the offspring obtained by crossing them, even two seedlings, do not appear similar (Misra, 1975).

Variability, heritability and genetic advance

In the study of inheritance of some qualitative characters, Grabowska (1983) observed that in the F_1 hybrids coloured florets dominated over white ones. Also the parents with creamy white and salmon florets occurred in F_1 hybrids of yellow cultivars. Plain edges of floret petals dominated over wavy edges. An alternating floret arrangement predominated in the inflorescences of F_1 hybrids and it was independent from that of the parents. Observations on flowering date, plant height, spike length, floret number, ornamental value of spike and overall value of 6 clones and 15 intercrosses were done. A large amount of variation among crosses was due to general combining ability. Heritability, estimated by mid-parent-offspring regression, was high in the case of first 4 characters and very low for the last 2 characters was observed (Cohat, 1988).

Genetic variability in cultivars showed high value of heritability. Number of florets/spike, spike length and spike weight showed high heritability along with high genetic advance for additive gene effect. Plant height, number of leaves and days taken to flowering showed lower genetic advance with non additive gene effect (Gowda, 1989), whereas in another study Anuradha and Gowda (1990) observed that number of cormels and cormel weight showed high heritability along with high genetic advance showing additive gene effects.

High heritability for weight of corms observed by Deepti (2000). Variability studies of 35 (11 Indian and 24 exotic) genotypes for 27 characters showed high heritability and genetic advance on number of shoots/spikes, daughter corms and cormels/plant and weight of cormels (Balaram *et al*., 2000a). Nazir *et al*. (2003) observed that rachis length, plant height and durability of spike showed significant positive correlation with number of florets/spike and suggested that these characters are of primary importance for improvement of gladiolus. In a variability and heritability study Basavaraja *et al*. (2003) stated that selection of individual plant would be done based on days to first flower and plant height which showed moderate to high genotypic coefficient of variation, heritability and genetic advance over mean.

A study was under taken involving 30 diverse genotypes of gladiolus collected from different parts of India to determine genetic variability, heritability and genetic advance. Ten most important growth and flowering related traits were analyzed for this purpose. Significant differences among genotypes were observed for all the characters under study. The cultivars Sunny Boy, Rose Supreme, Rose Spire, White Prosperity, Shobha, Suchitra, Hunting Song, Praha and Bis Bis were found outstanding for vegetative and flowering characters. Most of the characters under study exhibited moderate to low variability. Phenotypic co-efficient of variation was higher than genotypic co-efficient of variation for most of the characters under study. The highest phenotypic and genotypic co-efficient of variation was recorded for number of cormels per plant. This trait also exhibited high heritability along with high genetic advance (Bhatia and Grewal, 2009).

In a study significant variations were recorded for the various characters. Phenotypic and genotypic coefficients of variation were highest for number of cormels and weight of cormels produced per plant, indicating presence of sufficient genetic variability for selection in these traits. High heritability and high genetic advance for number of cormels per plant, weight of cormels per plant, leaf area, number of spikes per plot, number of corms per plot, number of florets remaining open at a time, number of spikes per plant, leaf width, spike diameter, weight of corm, rachis length, vase life of spike, number of corms per plant and plant height indicated the presence of additive gene effects in these traits and their amicability for direct selection. The non additive gene effects were evident in spike length, size index of corms, number of florets per spike, number of days to slipping thus warranting use of heterosis breeding for these characters. The selection on the basis of cormels per plant will be more effective for further breeding programme (Chaudhary *et al.*, 2012).

Inheritance of flower colour

There are different patterns of colouring in gladiolus. Six anthocyanidins (pelargonidin, cyanidin, peonidin, delphinidin, petunidin and malvidin) were identified by Cohen *et al.* (1986) and they observed that pigments of each group may appear individually or in conjunction with pigments from either or both groups, all pigments contributing to the visible colour in the petals. Specific parental clones containing a single pigment produced offspring lines having no pigmental variation and little or no difference in the visible colour. This opens the possibility for F_1 hybrid production.

BREEDING OBJECTIVES

i. Improved plant growth

ii. Stem type such as slenderness and flexibility to high wind

iii. Better spike quality specially length and strength, symmetrical arrangement and facing of the florets in one direction, no side spikes

iv. New flower colour

v. Desired form (i.e. compactness, texture and frilling) and size of floret

vi. More number of florets/spike

vii. More number of florets opened at a time

viii. Fragrant variety

ix. Good substance of petals (i.e. it can withstand the rigours of handling and transport)

x. More vase life

xi. High rate of corm and cormel multiplication

xii. Resistance to diseases (i.e. *Fusarium* wilt and *Botrytis* blight)

ICAR identified various Coordinated Centres for breeding and germplasm collection of gladiolus as given below.

For collection, maintenance and assessment of germplasm

- Acharya N.G. Ranga Agricultural University, Hyderabad
- Birsa Agricultural University, Kanke, Ranchi
- Dr. Y.S. Parmar University of Horticulture and Forestry, Solan
- Maharana Pratap University of Agriculture and Technology, Udaipur
- Punjab Agricultural University, Ludhiana
- Mahatma Phule Krishi Vidyapeeth, Pune
- G.B. Pant University of Agriculture and Technology, Pantnagar

For breeding and testing new cultivars

- Indian Institute of Horticultural Research, Bangaluru
- Indian Agricultural Research Institute, New Delhi
- National Botanical Research Institute, Lucknow
- Dr. Y.S. Parmar University of Horticulture and Forestry, Solan
- Punjab Agricultural University, Ludhiana
- Mahatma Phule Krishi Vidyapeeth, Pune
- G.B. Pant University of Agriculture and Technology, Pantnagar
- Sher-e-Kashmir University of Agricultural Sciences and Technology, Srinagar

BREEDING METHODS

Introduction

Hundreds of varieties of gladiolus have been introduced to India evolved or originated from USA, Russia, Canada, The Netherlands, Egypt, England, Australia and Israel. All these introduced varieties were evaluated in different agroclimatic zones for commercial production. List of important exotic varieties is given in Table 1.

Table 1: Gladiolus varieties introduced from different countries.

S. No.	Country	Variety introduced
1.	Russia	Apache, Autumn Charm, Autumn Rose, Blue Spie, Dawn Watch, Day Dream, Inspiration WIR 821, Inspiration WIR 842, Melody, Banana Split, Oscar.
2.	USA	American Beauty, Angel Fire, Annabel and Arrows, Any Bellsele, Astro, Atom Dragon, Ave, Chorus Girl, Christian Jane, Creamy Green, Departing Sun, Everglow, Twenty Four Karat, Friendship, Gambier Pearl, Green Lilac, Green Willow, Happy Birthday, Heady Wine,
3.	Israel	Nova Lux, Rose Prince, Trader Horn, White Projiate.
4.	The Netherlands	Australian Fir, Bis-Bis, Donald Duck, George Mazure, Melody, Oscar, Peter Pears, Pfitzers Sensation, Scarlet Queen, Snow Princes, Spring Song, Sylvia.
5.	England	Anglia, Cliffs of Dover.
6.	Australia	The Barton.
7.	Canada	Green Bay, Regency

Selection

As gladiolus is a cross pollinated crop, many insects and birds visit the flowers (Goldblatt *et al.*, 2001 and Goldblatt and Manning, 2002). The resulted seed setting provides options for evolution of new combinations. After evaluation of seedling raised from open pollinated seeds outstanding one can be selected. Description of some Indian varieties evolved from seedlings are mentioned in Table 2.

Table 2: Important varieties of gladiolus evolved through selection.

S. No.	Variety	Parentage	Chief characteristics
Varieties evolved at IARI, New Delhi			
1.	Agnirekha (1980)	Sylvia	Floret scarlet and fire red with saffron yellow blotch, spikes 85 cm, diameter 9.3 cm, with 18 florets, mid-season.
2.	Mayur (1980)	Sylvia	Florets lilac-purple with dark purple throat, spikes 75 cm long, florets 16 and 12.0 cm across, mid-season.
3.	Pusa Suhagin (1987)	Sylvia	Florets ruby-red with barium yellow streaks on the lower tepal, spikes 91 cm long, florets 16 and 8.5 cm across, late season.
4.	Bindiya (1997)	Ratna Butterfly	Florets yellowish cream with fan-shaped red coloration on 2 side falls, spikes 94 cm long, with 18 florets per spike, floret size 11.0 cm, mid-season.
5.	Sarang (1997)	White Oak	Florets purple-red, spikes 114 cm long with 21 florets of 13.0 cm diameter, mid-season.
6.	Sukanya (2000)	Salmon Queen	Floret colour white with scarlet ring in the lip, spikes 98 cm long with 18 florets with diameter 9.7 cm, early mid-season.
7.	Swapnil (2000)	Viola	Floret colour red with white lip, spikes 109 cm long with 17 florets of 11.4 cm diameter, early blooming.
8.	Urvashi (2002)	Red Soft Glow	Florets soft red purple with deep purple, spike 80 cm long with 17 florets of 11 cm diameter, mid season bloomer.
9.	Pusa Red Valentine	Regency	Florets brick/blood red, spike long, mid flowering.
10.	Pusa Subham (2009)	-	Florets cream or light yellow with red-purple marking on yellow base of the throat. Spikes 85 cm, early variety, good multiplier.
Varieties evolved at NBRI, Lucknow			
1.	Basant Bahar	Unias Challenge	Florets yellow with magenta specks in throat, spikes 50 cm long with 14 florets per spike.
2.	Gazel	White Friendship	Florets pink with darker tips and linear shading, having yellow throat, spikes 35 cm with 14 florets per spike.
3.	Jwala	Psittacinus hybrid	Florets vermillion with blotched yellow throat, spikes 65 cm long with 14 florets per spike .
4.	Kalima	Sylvia	Florets red with 2 side falls blotched yellow, spike 80 cm long with 15 florets per spike.
5.	Priyadarshini	Lavanesque	Florets mauve and throat white, spikes 45 cm long with 12 florets per spike.
6.	Sada Bahar	Sylvia	Florets speckled purple with sulphur yellow petal mid-ribs, spikes 50 cm long with 14 florets per spike.
7.	Smita	Lavanesque	Florets rose with darker margins, spikes 45 cm long with 14 florets per spike.

Hybridization

Most of the modern cultivars of gladiolus have been developed through hybridization. To avoid foreign pollen, emasculation is done preferably in the afternoon.

It can be done in morning before 7.30 am. Pollination is more successful in the morning. Pollen is dusted on the sticky stigmatic surface. The process is done on the following day of emasculation to ensure pollination. About 4-5 florets should be used for hybridization and remaining portion of spike should be removed. Female floret should be covered before emasculation and after pollination. Style will begin to wilt within 48 hours. The pods generally mature after 4-5 weeks of pollination. Pod should be picked when become brown and dried. Approximately 90-100 seeds are expected from a pod (Randhawa and Mukhopadhyay, 2001). Hybridization work was started in 1806 by William Herbert (England) while in India hybridization was started in 1956 by Bajrang Bahadur Singh Bhadri, the then Lieutenant Governor of Himachal Pradesh (Negi and Raghava, 1995) and developed some 150 varieties suitable for cultivation in hills of Himachal Pradesh and other places (Bhadri, 1963). Gladiolus varieties developed through hybridization by different organizations are given in Table 3.

Table 3: Gladiolus varieties developed through hybridization.

S. No.	Variety	Parent	Chief characteristics
Varieties evolved at IARI, New Delhi			
1.	Suchitra	Sylvia × Jo Wagenaar	Florets camellia rose with vermillion and purple blotch, spikes 80-85 cm long, florets 16-18 and 9-10 cm across, mid season.
2.	Neelam (1995)	Sylvia × Patricia	Mid season variety with deep mauve florets, spikes 65 cm long with 13 florets of 9.0 cm diameter.
3.	Anjali (1997)	Sancerre × Rose Spire	Florets pinkish white, spikes 80 cm long, florets 17 and 8.5 cm across, early mid-season.
4.	Pusa Archana (1997)	Creamy Green × American Beauty	Florets scarlet pink with yellow dusting on falls, spike 67 cm long, florets 16 and 11.5 cm across, mid-season.
5.	Chirag (1997)	Cygnet × Little Fawn	Florets orange in colour with deeper throat, spikes 99 cm long with 18 florets, floret size 11.0 cm mid-season.
6.	Shweta (1997)	Wind Song × Pink Frost	Florets frilled white with green-yellow throat, spikes 90 cm long with 17 florets of 10.0 cm diameter, mid-season.
7.	Sunayna (1997)	George Mazure × Melody	Florets pink with red throat, spikes 85 cm long with 17 florets, floret diameter 12.5 cm, early season.
8.	Vandana (1997)	George Mazure × Eurovision	Orange coloured, 95 cm long spikes, 16 florets of 13.0 cm diameter, early mid-season.
9.	Chandni (1997)	Green Woodpecker × White Butterfly	Florets greenish white, spike 80 cm long with 14 florets, floret size 10.0 cm, early season.
10.	Dhanvantri (1995)	Jr. Prom × Lucky Star	Florets light yellow, spikes 100 cm long with 16 florets, floret diameter 11.3 cm, mid-season.
11.	Gunjan (2000)	White Oak × Oscar	Floret soft orange with deeper splashing on 2 side falls, spikes 87 cm long, 16 florets/spike, floret diameter 11.0 cm, early mid-season.

(*Contd.*)

(*Contd. Table 3*)

S. No.	Variety	Parent	Chief characteristics
12.	Kamini (2000)	Ave × Christian Jane	Floret orange-red, purple red lip, light yellow base on 2 side falls, spikes 88 cm long, 14 florets/spike, floret diameter 11.0 cm, early mid-season.
13.	Lohit (2000)	Creamy Green × American Beauty	Floret colour red with white mid-ribs on 2 side falls, spikes 125 cm long with 16 florets per spike, floret diameter 11.8 cm, early mid-season
14.	Mohini (2000)	Ave × Christian Jane	Floret colour red-purple, spike length 92 cm with 15 florets/spike, floret diameter 9.5 cm, early mid-season.
15.	Rangmahal (2000)	Red Bantam × Flaura Belli	Florets ruffled, red-purple and compactly arranged, spikes 84 cm long with 16 florets per spike, floret diameter 10.5 cm, mid-season.
16.	Urmil (2000)	Tinker Belle × Break O'Dawn	Florets violet with creamy throat, spikes 91 cm long with 13 florets per spike, floret diameter 11.5 cm, early blooming.
17.	Urmi (2002)	Berlew × Heady Wine	Florets red purple, pale dusted throat, spikes 88 cm long, 16 florets per spike, floret diameter 10 cm, good multiplier.
18.	Pusa Kiran /Ave Open (2009)	-	Florets white with red lines on the lower tepals, long spikes with 16 florets per spike, early flowering.
19.	Pusa Vidushi (2011)	Melody × Berlew	Florets purplish white with greyed purple spots on base of the throat, spikes straight, 15-16 florets, early and mid maturing.
20.	Pusa Manmohak (2011)	Mayur × Hunting Song	Florets saffron red with thin whitish stripes on throat, compact arrangement of florets, mid maturing.
21.	Pusa Unnati (2013)	Berlew × Heady Wine (Tall)	Height 158.66 cm. Spike length 141.00 cm.] Rachis length 71.00 cm.
22.	Pusa Srijana (2013)	Berlew × Heady Wine	Height 103.22 cm. Spike length 85.25 cm. Rachis length 49.55 cm. No. of florets 16.66 per spike.
Varieties evolved at IIHR, Hessaraghatta, Bengaluru			
1.	Aarti (1980)	Shirley × Melody	Florets poppy-red with purple-red and canary-yellow blotch, spikes 63.6 cm long, 11 florets, mid-season.
2.	Apsara (1980)	Black Jack × Friendship	Florets ruby-red with barium yellow flecks in throat, spikes 97.6 cm long, 18 florets, floret diameter 11 cm.
3.	Darshan	Watermelon Pink × Shirley	Floret colour ruby-purple with red-purple margins having white blotch.
4.	Kum Kum (1993)	Watermelon Pink × Lady John	Florets red with yellow blotch.
5.	Meera (1979)	G.P. 1 × Friendship	Florets white, spikes 90 cm long with 18 florets.
6.	Nazrana (1979)	Black Jack × Friendship	Florets cardinal-red with barium yellow flash in throat, spikes 104 cm long with 18 florets.
7.	Poonam (1979)	Geliber Herald × R.N. 121	Florets yellow, spikes 98 cm long with 17 florets.

(*Contd.*)

(*Contd. Table 3*)

S. No.	Variety	Parent	Chief characteristics
8.	Sagar (1994)	Melody × Wild Rose	Florets red with yellow blotch, spikes 119 cm long with 20 florets per spike.
9.	Sapna (1979)	Green Woodpecker × Friendship	Florets greenish yellow, spikes 83 cm long with 17 florets per spike.
10.	Arka Amar (2010)	Watermelon Pink × Aarti	Florets red with red margin and yellow blotch. Florets wavy, arranged in double rows, 1-2 spikes/plant.
11.	Arka Naveen (2010)	74-39-1 × Tropic Seas	Florets purple violet, having red purple margin with yellow green blotch, 1-2 spikes/plant.
12.	Arka Gold (2010)	Green Bay × Gold Medal 412	Florets yellow with red blotch. Florets slightly ruffled and wavy, 1-1.5 spikes/plant.
Varieties evolved at NBRI, Lucknow			
1.	Archana (1984)	Sylvia × Friendship	Floret purple with yellow blotch and white mid-rib, spikes 70 cm long 18 florets/spike, mid-season.
2.	Arun (1984)	Sylvia × Fancy	Florets vermillion, spikes 70 cm long with 13 florets per spike, late bloomer.
3.	Dhiraj (1993)	Beauty Spot × Psittacinus hybrid	Florets purple with deeper and yellow blotch.
4.	Hans (1985)	Friendship × *G. tristis*	Florets white with falls having green mid-rib, spike length 70 cm with 16 florets per spike.
5.	Indrani (1985)	Friendship × *G. tristis*	Florets crimson with white mid-rib, spikes 80 cm long with 16 florets per spike.
6.	Manhar (1983)	Friendship × *G. tristis*	Florets yellow with rosy tips and yellow throat, spikes 60 cm long with 14 florets per spike.
7.	Manisha (1983)	Friendship × *G. tristis*	Florets yellow with purple splashes at tips, spikes 80 cm long with 14 florets per spike, late bloomer.
8.	Manmohan (1982)	Friendship × *G. tristis*	Florets yellow with purple splashes at tips, Spikes of 80 cm length with 14 florets, Late bloomer.
9.	Manohar (1982)	Friendship × *G. tristis*	Florets purple and throat yellow, spikes 80 cm long with 14 florets per spike, late bloomer.
10.	Mohini (1982)	Friendship × *G. tristis*	Florets white splashed with rose and throat yellow, spikes 60 cm long with 12 florets per spike, late bloomer.
11.	Mridula (1985)	Friendship × *G. tristis*	Florets purple specked at edges with yellow throat and white mid-ribs, spikes 70 cm long with 14 florets per spike, mid-season bloomer.
12.	Mukta (1981)	Friendship × *G. tristis*	Florets sulphur-yellow splashed with purple, spike 64 cm long, 15 florets/spike, late bloomer.
13.	Pitamber (1985)	Friendship × *G. tristis*	Florets light-green, purple streak in throat, spikes 64 cm long, 15 florets, mid-season bloomer.
14.	Sanyukta (1984)	Friendship × *G. tristis*	Florets rose with primrose-yellow throat, spikes 80 cm long, 18 florets per spike, mid-season bloomer.
15.	Triloki (1984)	Friendship × *G. tristis*	Florets rose with yellow throat, spikes 75 cm long, 14 florets, mid-season bloomer.
16.	Kohran	Sylvia × King Lear	Florets azalea pink with shading on both sides with pansy violet, spikes 35 cm long, 14-18 florets/spike.

(*Contd.*)

(*Contd. Table 3*)

S. No.	Variety	Parent	Chief characteristics
17.	Neelima (2006)	Tropic Sea × Snow Princess	Florets bluish purple, spike 90 cm long, 15 florets, used as cut flower and decoration.
18.	Urvashi (2006)	Eurovision × Snow Princess	Florets light creamy orange, spike 95 cm long, 16 florets per spike, used as cut flowers.
19.	Suverna (2008)	Fedelio × Hall Mark	Florets in combination of three colours (orange red, red, and creamy yellow).
20.	Roshni (2010)	Pink Friendship × Red Beauty	Florets pinkish red with orange red throat, spike 75 cm with 12-15 florets/spike.
21.	Jamuni	Lavender Puff × Tropic Sea	Florets bicolour of purple and yellow white, spike 95 cm, 16-18 florets, compact type.
Varieties evolved at HETC, Chaubattia, Ranikhet, Almora			
1.	Chaubattia Arunima	Oscar × Motherfisher	Mandarin red flowers with white strips, spike length 89.2 cm.
2.	Chaubattia Ankur	Oscar × Friendship	Flower colour Dutch vermillion, spike length 85.8 cm.
3.	Chaubattia Tripti	Sunny Boy × Oscar	Flower chartreuse green with azalea pink strips, spike length 88.9 cm.
4.	Chaubattia Shobhit	Meria × Tropic Sea	Early season hybrid with rose purple flowers and an average spike length 86.2 cm.
Varieties evolved at PAU, Ludhiana			
1.	Punjab Dawn	Suchitra × Melody	Spikes 60-70 cm long, flowering in 90-100 days, 12-15 florets/spike, floret size 8.5-9.5 cm of light pink colour
2.	Punjab Morning	Sancerre × White Prosperity	Spikes 60-70 cm long, 12-15 florets per spike, takes 90-100 days to flower. Floret diameter 8.5-9.0 cm of white colour with creamish throat.
3.	Shan-e-Punjab	Suchitra × Melody	Spikes 70-80 cm long with 12-15 florets, takes 90 days to flower, floret size 9-9.5 cm with pinkish orange colour with dark pink throat.

Breeding for disease resistance (*Fusarium, Botrytis, Stromatinia*)

Most of the commercial varieties of gladiolus are susceptible to diseases caused by *Fusarium oxysporum* and *Botrytis gladiolorum*. Therefore, breeding for resistance to these dreaded diseases is of great importance. Chandra *et al*. (1985) tested 41 cultivars and 14 hybrids and screened Australian Fair and Mansoer varieties tolerant to *Fusarium oxysporum* and suggested that these varieties could be used in breeding programme to develop wilt tolerant cultivars. Screening of 50 cultivars against *Fusarium* wilt was done under field condition by Kaur *et al*. (1989) and Beauty Crest, Camellia, *G. callianthus*, George Mazure, Mayur, Melody, R.N. 121, Snow Princess and Sylvia were rated as resistant.

Van Eijk *et al.* (1990) tested several cultivars and species to find out resistance to dry rot disease (*Stromatinia gladioli*) and found corms, cormels and seeds of *G. italicus* show a high degree of resistance when planted or sown in the soil infested with *Stromatinia gladioli.* Hence, this species is recommended in breeding programme as parent.

Negi *et al.* (1991) made several possible crosses from 5 susceptible with 4 tolerant varieties to *Fusarium oxysporum*. They found a resistant hybrid (Beauty Spot × Psittacinus Hybrid) and 2 tolerant (Watermelon Pink × Lady John and Watermelon Pink × Mansoer) hybrids. Variety Morning Mist originated from a cross between GC 354 (Blue Baron × Baverly Ann) and GC 143 (China Blue × Baverly Ann) was claimed as tolerant to *Fusarium oxysporum* by Wilfret (1993). Guo-Tai Jun *et al.* (2001) also developed Chenghongjiao variety resistant to diseases. Rao *et al.* (2003) bred Indian and exotic varieties and IIHR, Hybrid No. 86-17-6 was found very promising. The corms of this hybrid were inoculated artificially with *Fusarium oxysporum* f. sp. *gladioli* race 1. The Hybrid No. 86-17-6 was found to be moderately resistant to *Fusarium* wilt disease. This variety was named as 'Arka Kesar'. A hybrid selection of the cross of cvs. Watermelon Pink and Aarti was made and a variety Arka Amar was developed. This variety was also found resistant to *Fusarium* wilt (Anon., 2013). Description of these varieties are given in Table 4.

Table 4: Disease resistant varieties of gladiolus.

S. No.	Variety	Parent	Chief characteristics
1.	Arka Kesar	Vink's Glory × Sagar	Floret colour yellow orange, having orange margin. Orange-red blotch with yellow-green line on tepals. Moderate resistant to *Fusarium* wilt.
2.	Arka Amar	Watermelon Pink × Aarti	Floret colour red, having red margin and white line on tepals with yellow blotch. Resistant to *Fusarium* wilt.

Breeding for fragrance: Some South African *Gladiolus* spp. have fragrance which varies from Apple Blossom type to rose and dirty. First fragrant variety Lucky Star was developed by crossing cv. Filigree × *Acidanthera murielae*. Other scented cultivars are Pink Fragrance, Spice, Yellow Rose (Mukhopadhyay, 2002). Wild species of gladiolus can be utilized in breeding programme to develop fragrant varieties has been advocated by Rao and Janakiram (1992) and they produced interspecific hybrids between gladiolus cultivars and *G. callianthus* with mild fragrance.

Mutation

Gladiolus mutant are generally lighter in colour than the original parent. Cv. White Friendship is a very popular mutant obtained from its known parent Friendship. IIHR, IARI, NBRI and other organization did some significant contribution in this line. Various workers have been adopted different doses of gamma irradiation for improvement of gladiolus. In general higher doses were found to be lethal and doses from 0.5 to 5.0 krad are advisable for the gamma irradiation in gladiolus. Misra and Bajpai (1983) studies the effects of gamma irradiation. Gamma rays at 3 krad level proved very favourable for production of more shoots/plant. When corms of 3 varieties were subjected to doses of 1-15 krad, the LD_{50} proved to be between 10 and 15 krad, doses 10 krad adversely affected vegetative and floral traits. Sensitivity to radiation varied between varieties and between traits. A stable mutant with shell pink floret was isolated form var. Wild Rose in the 1 krad treatment and released as a variety under the name Shobha (Raghava *et al.*, 1988).

Sedel-Nikova (1988) irradiated corms of cvs. Blue Isle, Dixieland, Frisky and White Frosting and found out some interesting mutation. The new forms showed crimson and light red flower colouration. The colour changes were evidently associated with chromosome reorganization in the apical L2 layer of the main shoot of the corm, resulted quantitative and qualitative changes in anthocyanin synthesis. Whereas Rao *et al.* (1990) studied the effect of gamma irradiation on anther calli. Young undifferentiated green calli from anther of gladiolus cv. Picardy was irradiated with 0.5, 1.0, 5.0 and 10.0 krad of gamma rays. LD_{50} was found between 1.0 and 5.0 krad. In 5.0 and 10.0 krad treatment calli turned brown. Shoot buds and shoots were obtained in 0.5 and 1.0 krad. Banerji *et al.* (1994) irradiated dormant corms of cv. White Friendship with 250, 500, 750, 1000 and 1250 Gy gamma rays. Morphological abnormalities in foliage and florets and chromosomal aberration during root tip mitosis increased with increase in dose. Flowering ceased after treatment with 1250 Gy. MV_2 and MV_3 generations also followed the abnormalities. Pink flower colour was detected in a few plants as a sectorial chimeric form in MV_2 population. In the MV_3 from the 750 Gy treatment one plant produced a spike with light pink florets. This mutant has been isolated in a pure form. Cvs. Fidelio, Rose Supreme and Spic & Span were irradiated at 0, 500, 1000 and 1500 rad of gamma rays. Sensitivity to radiation was observed only in cv. Fidelio. In vM_1 and vM_2 generations the mutant exhibited white coloured tepals with slight ruffling at lowest dose (500 rad). This mutant produced desirable qualitative and quantitative traits (Bhuj and Singh, 2002) and later released as variety Shubhangini.

Many workers have been used chemical mutagens for further improvement. Banerji and Gupta (1982) treated *G. psittacinus* cv. Hookeri cultivar orange with 0, 0.05, 0.1 and 0.2% aqueous solution of methyl methanesulphonate for 8 hrs. Reduction in important growth and flower traits was observed in treated corms. Floral abnormalities increased with increase in concentration of mutagen. LD_{50} on survival basis lies at 0.2% concentration. Banerji and Datta (1986) treated corms of *G. psiltacinus* var. Hookeri cultivar red and orange with 1, 2 and 4% formalin for 6, 12 and 24 hours. LD_{50} was found to be above 4% in formalin. In vM_2 somatic mutation in flower shape was detected and was isolated in pure form 4% formalin for 12 hours duration in the cultivar Red. Corms of var. Sylvia were treated with Ethyl methanesulphonate (0.2-1.0%) and ethylamine (0.1%), nitrosomethyl urea (0.2-0.4%) in combination with dimethyl sulphonate. Treatments alone were not found effective, however, 0.2% NMU in combination with 2 krad gamma radiation gave highly attractive colour mutant without any change in size, number and form of florets (Kaicker and Singh, 1983). Descriptions of some varieties developed through mutation breeding are given in Table 5.

Table 5: Description of mutant varieties of gladiolus.

S. No.	Mutant	Parent	Chief characteristics
1.	Shakti (1980)	Wild Rose	Florets pink with yellow throat, spikes 97 cm long with 18 florets/spike.
2.	Shobha (1981)	Wild Rose induced mutant	It requires 50 days to flowering. Spike 97 cm and rachis 62 cm long, florets shell pink with empire yellow throat and 6 florets open at a time.
3.	Swarnima (2000)	Dhanvantari spontaneous mutant	Florets coppery yellow, spikes 102 cm long with 16 florets/spike, floret diameter 11.3 cm, mid-season.
4.	Subhangini (2002)	Fidelio induced mutant	Spikes are 95-100 cm long with 15-18 florets, florets are white, slightly ruffled and 12 cm across, very good multiplier.

BIOTECHNOLOGY

Tissue culture

Callus formation and organogenesis were observed in root sections of gladiolus cormels cultured in MS medium supplied with NAA with or without benzyladenine (BA). Regeneration rate was highest in MS medium supplemented with 5 mg NAA and BA/litre. Root and cormel development were observed on regenerated shoots transferred in MS medium. The dormancy of the corms were broken by low temperature. The regenerated plants exhibited the same characteristic flower shape and colour as the mother plant. Varietal differences were observed in terms of adventitious bud formation (Kasumi *et al.*, 2005).

Gladiolus imbricatus, a rare species spread in the hill region of Europe, is resistant to abiotic and biotic stress being one of the most cold-tolerant in the genus. The aim of such study is to develop *in vitro* technologies for micropropagation, multiplication, corm development, somatic embryogenesis and medium-term storage of this endangered species, Red listed in South-Eastern and Central European countries. Initiation of *in vitro* cultures was started from wounded seeds or excised mature embryos. Micropropagation and multiplication was achieved on MS solid medium with 30 g L^{-1} sucrose and 1 mg L^{-1} N-6-benzyl adenine (BA). The callus developed from wounded seeds on a richer MS medium proved to be embryogenic. Increased concentrations of sucrose promote corm development as also shown for other genotypes of cultivated gladioli. Further on a two-step culture protocol, on solid and liquid shaken MS based media, each for 6 months was done to evaluate the effects of acetic acid and ancymidol on corm development. Acetic acid + sucrose 9%, alone or in combination with ancymidol stimulated corm formation and yield. For medium-term conservation maintenance of the *in vitro* cultures in low temperature in the dark proved to be the best in reducing the growth rate of the shoots after 3, 6 and 12 months. The recovery of plant growth was evaluated after 12 months, by the transfer to normal growth conditions. 25% of the plants were recovered after one year storage (Rakosy-Tican, 2012).

Somatic embryogenesis

In a somatic embryogenesis study Tomotsune *et al.* (1994) observed highest rate of somatic embryogenesis. 224.1 somatic embryos/g was obtained when callus was induced in medium containing 5 mg NAA/l was placed on growth regulator free medium. Stefaniak (1994) placed corm slices, young leaf bases and plantlets without root of 4 gladiolus cultivars on MS medium supplemented with 2, 6 or 10 mg 2, 4-D or NAA/l. Friable embryogenic callus and somatic embryos were obtained from cvs. Blue Isle, Jenny Lee and Rose Supreme with 2 mg 2,4-D and from cv. Peter Pears with 10 mg NAA/l. Somatic embryos transferred on MS hormone free medium regenerated into plantlets. For all cultivars, a higher percentage of corm slices produced embryogenic callus that regenerated plantlets. All plantlets obtained through embryogenesis did not differ phenotypically from the parental clones, whereas, primary somatic embryo development was observed on MS medium supplemented with zeatin or 0.25 5M BA in gladiolus cv. Peter Pears. Combinations of zeatin and BA or 0.5 5M BA in the regeneration medium induced secondary somatic embryos on previously formed immature primary somatic embryos. Histological studies showed that secondary somatic embryos had a well defined suspensor and small coleoptiles. Among the regenerated plantlets some albino shoots were recovered, indicating that *in vitro* culture may induce a certain degree of somaclonal variation. Regenerated plantlets were rooted on hormone free MS medium and successfully transferred to soil (Remotti, 1995).

Somatic embryogenesis from the stem tips of cormels resulted in flower colour variations among regenerated plants from 3 gladiolus cultivars. The highest frequency (51.1%) of somatic embryos was induced in MS medium supplemented with 5 mg NAA/l. Cvs. Topaz and Traveler formed more somatic embryos than cv. Hector. Flower colour variants originated among regenerated plants grown in the field. No chimeric change occurred in the flower i.e. all florets of a given plant displayed one colour. The flower colour of variants changed into pale to deep colour from that of original cultivar. These findings reveal that some variations in flower colour induced through somatic embryogenesis are useful in developing new gladiolus cultivars (Kasumi *et al.*, 1999).

Kumar *et al.* (2002) observed that heat shock (50º C for 1h) induced somatic embryogenesis or shoot bud differentiation in 1 year old cytokinin dependent and auxin dependent callus culture of gladiolus cvs. Her Majesty and Aldebaran grown on high (> 174 mM) or low (< 116 mM) sucrose medium, respectively.

Mutagenesis

Gamma irradiation to cormel explants on callus formation, somatic embryogenesis and flower colour was investigated by Kasumi *et al.* (2001) in gladiolus cvs. Traveler and Topaz. The total doses of 100-200 Gy at 10 Gy/hour caused a 50% reduction in the rate of callus formation and somatic embryogenesis (LD_{50}). Flower colour variants occurred in cv. Traveler but not in cv. Topaz. The frequency of colour variants increased, resulting in the appearance of extremely deep colours and morphological variants when tissue culture was used in combination with gamma irradiation. Most flower colour variants were solid (i.e. nonsectorial chimeric) variants. In a study Zang and Wang (2008) revealed that LD_{50} of gladiolus radiation by electron beam varied greatly between the two cultivars i.e. Beauty Queen possessed higher resistance to radiation as compared to Rose Supreme. LD_{50} of the corm of gladiolus Rose Supreme was found to be around 240 Gy, and it was more than 240 Gy for Beauty Queen.

Concentrations of colchicine on gladiolus culture were used at 0, 50, 100 and 200 ppm on the duration of 6, 12 24, and 36 hours soaking. Increasing concentrations of colchicine decreased weight, size of callus and height stem but increased thickness of leaf. Duration of soaking had effect on weight of callus but not on size of callus, stem height, leaf thickness, size of stomata and weight and size of cormel. Chromosome number remained unchanged (Tiwa Raknim, 1990).

Molecular breeding

Takatsu *et al.* (2001) obtained information on the genetic relationship among wild *Gladiolus* species through randomly amplified polymorphic DNA (RAPD) analysis. Out of the 140 tested primers, 32 amplified a total of 133 RAPD bans in 33 *Gladiolus* species. The genetic distance was calculated from the data of these RAPDs, and a dendrogram was generated. Interspecific crosses were carried out in seven combinations within or between clusters, and F_1 seedlings were obtained from most combinations. The RAPD analysis showed that these F_1 seedlings were real hybrids. Results suggest that RAPD markers are useful for detecting genetic relationships in *Gladiolus* species and for interspecific crosses in breeding programmes.

Pathania and Misra (2001) used random amplified polymorphic DNA markers to confirm the mutability of cv. Eurovision mutants. Out of 7 random primers used for PCR amplification, one primer OP × 02 (52 -TTCCGCCACC-32) produced polymorphic banding pattern. A monomorphic band of 1 kb was present, both in parent cultivar and its mutants. However a single band (0.8 kb) difference between 5 mutants, ERM-3, ERM-4, ERM-5, ERM-6 and ERM-7 was obtained. This distinguishes the mutants from cultivar Eurovision at molecular levels. The mutants with red-and-white colour developed from the gladiolus cv. Spic and Span which was induced by 75 GY were compared with the control by AFLP molecular marker technology. Results indicated that the differences between the dual colour flower mutant and the control were related to the changes of genetic material, some of them may be related to the genes involved in flower colour formation (Liu *et al.*, 2009). When the treated corm bud of *G. hybridus* Hort with EMS of different consistency was analyzed by RAPD (Wang *et al.* 2009) and ISSR (Gong *et al.*, 2010) results showed that EMS was a very effective mutagenic agent, with the increase of consistency the mutagenic range increased first then decreased, among which 0.6% EMS treatment had the biggest influence.

Genetic engineering

Cormel slices of gladiolus cvs. Jenny Lee, Peter Pears and Florida Flame were co-bombarded with M10 tungsten particles coated with plasmid p355 Ac, containing the phosphinothricin acetyltransferase (PAT) gene under the control of the CaMV 355 promoter and pAct 1D, containing the GUS (vid A) gene under the control of rice actin promoter. Plantlets were regenerated from 14% of bombarded cormel slices selected on 8 mg phosphinothricin (glufosinatel/litre). The integration of the PAT and GUS genes was confirmed by PCR amplification. Following selection of transgenic plantlets on 4 mg phosphinothricin/litre, stable transformation was confirmed by PCR amplification, Southern hybridization and histochemical staining. GUS expression was observed primarily in callus cells and root meristems (Kamo *et al.*, 1995).

More than 100 transgenic gladiolus plants were recovered after particle bombardment of regenerable suspension cells and callus. For transformation, gladiolus callus and suspension cells were co-bombarded with phosphinothricin acetyltransferase (PAT) and GUS expressing plasmids. Stably transformed calluses were selected on medium containing either phosphinothricin (glufosinate) (PPT) or bialaphos (bilanafos) followed by transfer to a regeneration medium to recover transgenic plants. Stable transformation was confirmed by detection of the PAT gene by DNA gel blot analysis and by enzymatic assays to measure GUS activity. In general, particle bombardment of regenerable suspension cells rather than callus resulted in the largest number of transformants. The rate of co-expression for GUS in PPT resistant plants was high (70%). Promoters that are typically more efficient in dicotyledonous plants were very active in gladiolus, a monocotyledonous bulb plant (Kamo *et al.*, 1995a). Babu and Chawla (2000) observed transient expression of GUS after *Agrobacterium* mediated transformation of gladiolus tissue. They used the *Agrobacterium* strain LBA 4404 carrying either the binary vector pBI141 (*act*-GUS) or pTOK233 (355-GUS). Transgenic gladiolus transformed with the bean yellow mosaic virus (BYMV) coat portion (CCP) gene in either sense or antisense (AS) orientation were developed using biolistics. This is the first report of developing a floral bulb crop with antiviral genes to BYMV (Kamo *et al.*, 1997 and Kamo *et al.*, 2005).

REFERENCES

Altaee, A.H.Y. and Bashi, B.Z.K. 2011. Response two *Gladiolus* cv. to propagation *in vitro*. *Diyala Agril. Sci. J.*, **3**(1): 134-155.

Anonymous. 2002. Gladiolus. AICRP on Floriculture, Tech. Bulletin no.14. Pub. Project Coordinator, ICAR, IARI, New Delhi. pp. 100.

Anonymous. 2010. IIHR News, Indian Institute of Horticultural Research, Bengaluru, **31**(1): 4.

Anonymous. 2013. IIHR News, Indian Institute of Horticultural Research, Bengaluru, **34**(1): 5-6.

Anuradha, S. and Narayana Gowda, J.V. 1990. Genetic variability in gladiolus. *Prog. Hort.*,**22**(1-4):55-59.

Babu, P. and Chawla, H.S. 2000. *In vitro* regeneration and *Agrobacterium* mediated transformation in gladiolus. *J. Hort. Sci. Biotech.*, **75**: 400-404.

Balaram, M.V., Janakiram T., Vasantha Kumar, E., Choudhary, M.L., Ramachandran, N. and Ganeshan, S. 2000a. Genetic variability among gladiolus genotypes. *Proc*: National Conference on Gladiolus, NBRI, Lucknow, p. 17.

Balaram, M.V., Janakiram, T., Ganeshan, S., Choudhary, M.L. and Vasantha Kumar, E. 2000. Pollen storage feasibility studies in *Gladiolus* spp. *Proc*: National Conference on Gladiolus, NBRI, Lucknow, p. 58.

Banerji, B.K. and Datta, S.K. 1986. Improvement of ornamental plants by induced mutation: Gladiolus. *Floriculture*, **6**: 3-5.

Banerji, B.K. and Gupta, M.N. 1982. Effects of methyl methansulphonate on *Gladiolus psittacinus* var Hookeri cultivar Orange. North American Gladiolus Council Bulletin, **150**: 64-66.

Banerji, B.K., Datta, S.K. and Sharma, S.C. 1994. Gamma irradiation studies on gladiolus cv. White Friendship. *J. Nuclear Agri. Bio.*, **23**(3): 127-133.

Basavaraja, N., Hemla Naik, B. and Mohamad Ali, A. 2003. Genetic variabiligy studies in gladiolus (*Gladiolus hybridus* L.). *Abst*: National Symposium on Recent Advances in Indian Floriculture. ISOH, Kerala. p.73.

Bhadri, B.B.S. 1963. Possibilities gladiolus culture in Himachal and other hills. Raj Niwas, Simla.

Bhatia, R. and Grewal, H.P.S. 2009. Genetic variability and heritability studies in gladiolus. *Indian Agriculturist*, **53**(1/2): 1-6.

Bhuj, B.D. and Singh, R. 2002. Induced mutation in gladiolus. *Abst*. National Symposium on Indian Floriculture in the New Millennium. ISOH, Bangalore. p. 58.

Buch, P.O. 1972. The Species in the World of the Gladiolus Council, Edgerton Press. Md. pp. 2-7.

Chandra, K.J., Negi, S.S., Raghava, S.P.S. and Sharma, T.V.R.S. 1985. Evaluation of gladiolus cultivars and hybrids for resistance to *Fusarium oxysporum* f. sp. *gladioli*. *Indian J. Hort.*, **42**(3-4): 304-305.

Choudhary, M., Moond, S.K., Anop Kumari, Beniwal, B S. 2012. Genetic variability in quantitative characters of gladiolus (*Gladiolus × hybridus* Hort.). *International J. Agril. Sci.*, 8(1): 138-141.

Choudhary, M.L. 1990. Study on pollen viability of gladiolus. *Adv. Pl. Sci.*, **3**(2): 291- 295.

Choudhary, M.L. 1991. Response of sucrose and boric acid on pollen viability of *Gladiolus* spp. *Haryana J. Hort. Sci.,* **20**(1-2): 73-76.

Clarke, A.E., Considine, J.A., Ward, R. and Knox, R.B. 1977. Mechanism of pollination in Gladiolus: roles of the stigma and pollen-tube guide. *Ann. Bot.,* **41**(171): 15-20.

Cohat, J. 1988. Estimates of heritability of some characters in gladiolus (*Gladiolus grandiflorus* Hort.). *Agronomie*, **8**(3): 179-185.

Cohen, A., Akavia, N. and Umiel, N. 1986. The identification of anthocyanin pigments in the petals as an aid to the breeding of gladiolus. *Acta Horticulturae*, **177**(2): 375-383.

Deepti. 2000. Coefficient of variation, heritability and expected genetic advance in gladiolus. *Proc*: National Conference on Gladiolus, NBRI, Lucknow, p. 70.

Delpierre, G.R. and Du Plessis, N.M. 1974. The Winter-Growing Gladioli of South Africa. Tafeberg, Cape Town.

Dhaduk, B.K., Singh, B. and Dadlani, N.K. 1987. Effect of different methods of pollination on seed set in gladiolus. *South Indian Hort.*, **35**(3): 260-265.

Emek, Y. and Erdag, B. 2007. Somatic embryogenesis from leaf explants of *Gladiolus anatolicus* (Boiss.) Stapf. *Pakistan J. Bio. Sci.*, **10**(8): 1190-1194.

Erdag, B.B., Emek, Y.C. and Aktas, L.Y. 2009. *In vitro* somatic embryogenesis from cormel-derived callus cultures of Gladiolus anatolicus (Boiss.) Stapf. *Propagation Ornamental Plants*, **9**(4): 176-180.

Everett, T.H. 1981. New York Botanical Garden Illustrated Encyclopedia of Horticulture, Vol. 5 Garland, N.Y., pp. 1485-1490.

Goldblatt, P. and Manning, J.C. 2002. Evidence for moth and butterfly pollination in gladiolus (Iridaceae Crocoideae). *Ann. Missouri Bot. Garden*, **89**(1): 110-124.

Goldblatt, P., Manning, J.C. and Bernhardt, P. 2001. Radiation of pollination systems in gladiolus (Iridaceae: Crocoideae) in Southern Africa. *Ann. Missouri Bot. Garden*, **88**(4): 713-34.

Goldblatt, P., Takei, M. and Razzaq, Z.A. 1993. Chromosome cytology in tropical African gladiolus (Iridaceae). *Ann.Missouri Bot. Garden*, **80**(2): 461-470.

Gong, ShuFang., Fu, HuiJan. and Wang, JinGang. 2010. ISSR analysis of M_1 generation of *Gladiolus hybridus* Hort treated by EMS. *J. Northeast Agril. Uni.*, **17**(2): 22-26.

Gowda, J.V.N. 1989. Genetic and phenotypic variability and correlation in quantitative and qualitative characters in gladiolus. *Vortraege-fuer-Pflanzenzuechtung,* **15**(18): 4.

Grabowska, B. 1969. Investigations on the biology of flowering in garden gladioli, *Gladiolus hybridus* Hort. and its intervarietal hybrids. Part I, Studies on the biology of flowering. *Hodowla Roslin, aklimatyzacia I Nasiennictwo*, **13**: 483-407.

Grabowska, B. 1983. Studies on breeding of gladioli (*Gladiolus hybr.* Hort.). Part II. Inheritance of some qualitative characters. *Prace-Instytutu-Sadownictwa-i-Kwiaciarstwa-w-Skierniewicach,-B-Rosliny-Ozdobne*, **8**: 35-46.

Guo, Taijun., Jiao, Peijuan., Zhang, Yafeng., Li, JiHai. and Lu, Wenpeng. 2001. Chenghongjiao - a new variety of gladiolus. *Acta Horticulturae Sinica*, **28**(2): 184.

Halevy, A.H. 2000. Gladiolus. *In*: A.H. Halevy. (ed.) CRC Handbook of Flowering, Vol III. CRC Press, Florida, US, pp. 63-70.

Kaicker, U.S. and Singh, H.P. 1983. Somatic mutations in gladiolus. *Abst:* XV International Congress of Genetics, New Delhi. p. 471.

Kamo, K., Blowers, A., Smith, F. and Van Eck, J. 1995. Stable transformation of gladiolus by particle gun bombardment of cormels. *Plant Science Limerick,* **110**(1): 105-111.

Kamo, K., Blowers, A., Smith, F. and Van Eck, J. 1995a. Stable transformation of gladiolus using suspension cells and callus. *J. American Soc. Hort. Sci.*, **120**(2): 347-352.

Kamo, K., Gera, A., Cohen, J., Hammond, J., Blowers, A., Smith, F. and Van Eck, J. 2005. Transgenic gladiolus plants transformed with the bean yellow mosaic virus coat-protein gene in either sense or antisense orientation. *Plant Cell Reports*, **23**(9): 654-663.

Kamo, K., Hammond, J. and Roh, M. 1997. Transformation of gladiolus for disease resistance. *J. Korean Soc. Hort. Sci.*, **38**(2): 188-193.

Kasumi, M. Takatsu, Y. Suzuki, K. Gonai, T. Nogi, M. Yamada, T. Manabe, T. 2005. Callus formation and plant regeneration from root explant of gladiolus (*Gladiolus* x *grandiflora* hort.). *Horticultural Research* (Japan), **4**(1): 7-10.

Kasumi, M., Takatsu, Y., Manabe, T. and Hayashi, M. 2001. The effects of irradiating gladiolus (*Gladiolus* × *grandiflora* Hort.) cormels with gamma rays on callus formation, somatic embryogenesis and flower color variations in the regenerated plants. *J. Japanese Soc. Hort. Sci.*, **70**(1): 126-128.

Kasumi, M., Takatsu, Y., Tomotsune, H. and Sakuma, F. 1999. Somatic embryogenesis from cultured cormel stem tips and flower color variation among regenerated plants of gladiolus. *J. Japanese Soc. Hort. Sci.*, **68**(1): 168-175.

Kaur, S., Arora, J.S. and Khanna, K. 1989. *Fusarium* wilt is a limiting factor in commercial cultivation of gladiolus. *Indian Horticulture*, **34**: 21-22.

Kumar, A., Palni, L.M.S., Sood, A., Sharma, M., Palni, U.T. and Gupta, A.K. 2002. Heat-shock induced somatic embryogenesis in callus cultures of gladiolus in the presence of high sucrose. *J. Hort. Sci. Biotech.*, **77**(1): 73-78.

Lewis, G.J., Obermeyer, A.A. and Barnard, T.T. 1972. Gladiolus- a revision of South African species. *J. South. Afr. Bot.,* **10**: 316.

Liu, ChangMing., Zhang, Xian Xu. and Qing, Xian Feng. 2009. AFLP analysis on dual colour flower gladiolus mutants induced by radiation. *Acta Botanica Boreali-Occidentalia Sinica*, **29**(1): 69-74.

Mahawer, L.N. and Misra, R.L. 1993. Studies on blossom biology of gladiolus under Delhi condition. *J. Orna. Hort.*, **1**(2): 16-20.

Mahawer, L.N. and Misra, R.L. 1993a. Studies on gladiolus pollination with regard to stages of anther and stigma. *J. Orna. Hort.*, **1**(2): 51-54.

Misra, R. L. 1975. Gladiolus Breeding, Association Newsletter, **12**:2-5.

Misra, R.L. and Bajpai, P.N. 1983. Effects of mutagens on shooting, leaf number, heading, plant height and spike length in gladioli. *Indian J. Hort.*, **40**: 107-111.

Misra, R.L. and Singh, B. 1989. Gladiolus. *In*: T.K. Bose and L.D. Yadav (eds.) Commercial Flowers. Calcutta, Naya Prokash. pp. 267-253.

Misra, R.L., De, L.C. and Saini, H.C. 2001. Manifestation of heterosis in exotic × indigenous crosses of gladiolus. *J. Orna. Hort.*, **4**(1): 1-6.

Mohamed Yasseen, Y. 2000. *In vitro* somatic embryogenesis and plant regeneration from gladiolus root explants. *Ann. Agril. Sci.*, **45**(2): 647-657.

Mukhopadhyay, A. 2002. Gladiolus. Publications and Information Division, Indian Council of Agricultural Research. New Delhi, p. 83.

Nazir, M., Dwivedi, V.K. and Bhat, K.L. 2003. Correlation and path coefficient analysis in gladiolus over different environments. *Abst:* National Symposium on Recent Advances in Indian Floriculture. ISOH, Kerala. p. 81.

Negi, S.S. and Raghava, S.P.S. 1995. Gladiolus Breeding. *In:* K.L. Chadha and S.K. Bhattacharjee (eds.). Advances in Horticulture: Ornamental Plants, Vol. 12-II. Malhotra Publishing House, New Delhi. pp. 441-460.

Negi, S.S., Raghava, S.P.S., Chacko, C.I., Rao, T.M., Ramachandran, N., Prakash, J. and Pierik, R.L.M. 1991. Breeding for quality and resistance to *Fusarium* wilt in gladiolus. *Proce*: International Seminar on New Frontiers in Horticulture, Bangalore, Indo-American Hybrid Seeds. p. 25.

Ohri, D. 1982. Self-incompatibility in gladiolus. *Incompatibility Newsletter.* **14**: 111-115.

Ohri, D. 2000. Cytogenetical evolution of garden gladiolus. *Proce*: National Conference on Gladiolus, Lucknow, NBRI. p. 11.

Ohri, D. and Khoshoo, T.N. 1981. Cytogenetics of garden gladiolus I. Pollination mechanism and breeding system. *Proc. Indian Natn. Sci. Acad.* **47**: 510-515.

Ohri, D. and Khoshoo, T.N. 1983. Cytogenetics of garden gladiolus. *Zeitschrift-fur-Pflanzenzuchtung*, **91**(1): 46-60.

Pant, C.C. and Lal, S.D. 1993. Heterosis in gladiolus-II (*Adiolus primulinus* Baker.). *Prog. Hort.*, **25**(1-2): 1-4.

Pant, C.C., Lal, S.D. and Shah, D. 2000. Combining ability studies in gladiolus-II. *J. Orna. Hort.*, **3**(1): 43-47.

Pathania, N.S. and Misra, R.L. 2001. Characterization of gladiolus mutants using RAPD markers. *J. Orna. Hort.*, **4**(2): 65-68.

Raghava, S.P.S. 2000. Genetic improvement of gladiolus in India. *Proc*: National Conference on Gladiolus, Lucknow, NBRI. p. 1.

Raghava, S.P.S., Negi, S.S., Sharma, T.V.R.S. and Balakrishnan, K.A. 1988. Gamma ray induced mutants in gladiolus. *J. Nuclear Agri. Bio.*, **17**(1): 5-10.

Rakosy-Tican, E. Bors, and B. Szatmari, A. M. 2012. *In vitro* culture and medium-term conservation of the rare wild species *Gladiolus imbricatus. African J. Biotech.*, **11**: 14703-14712.

Randhawa, G.S. and Mukhopadhyay, A. 2001. Floriculture in India. New Delhi, Allied Publishers Ltd. pp. 376-83.

Rao, T.M. and Janakiram, T. 1992. Breeding for fragrance in gladiolus utilizing wild species. *Abst:* National Seminar on Commercial Floriculture in India: Present & Potential, Bangalore. p. 36.

Rao, T.M., Janakiram, T., Negi, S.S. and Ramachandran, N. 2003. Development of gladiolus cultivars moderately resistant to *Fusarium* wilt through hybridization. *Abst*: National Symposium on Recent Advances in Indian Floriculture. ISOH, Kerala. p. 81.

Rao, T.M., Negi, S.S. and Dore Swamy, R. 1990. Effect of gamma irradiation on anthers calli of gladiolus. *Abst:* VII. Int. Congr. Plant Tissue Cell culture, Amesterdam, pp. 4-82.

Remotti, P.C. 1995. Primary and secondary embryogenesis from cell suspension cultures of gladiolus. *Plant Science Limerick*, **107**(2): 205-214.

Roychoudhury, N. 1980. Flower biology of gladiolus. *Mysore Hort. Sci.*, **20**(4): 36-38. Sedel-Nikova, L.L. 1988. Mutations in flower colour of gladioli from gamma irradiated corms. *Byulleten Glavnogo Botanicheskogo Sada*, **148**: 55-60.

Sedel-Nikova, L.L. 1989. Polystigmaty in gladiolus. *Byulleten2 Glavnogo Botanicheskogo Sada*, **154**: 80-83.

Shah, A., Pant, C.C. and Lal, S.D. 1988. Floral biology of some gladiolus cultivars under temperate conditions. *Prog. Hort.*, **20**(1-2): 74-76.

Singh, A.K. 2006. Flower Crops: Cultivation and Management. Pub. New India Publishing Agency, New Delhi. pp. 147-166.

Singh, B. and Singh, M. 1985. Studies of floral biology in gladiolus. *Prog. Hort.*, **17**(2): 134-35.

Stefaniak, B. 1994. Somatic embryogenesis and plant regeneration of gladiolus (*Gladiolus* Hort.). *Plant Cell Reports*, **13**(7): 386-389.

Takatsu, Y., Kasumi, M., Manabe, T., Hayashi, M., Inoue, E. and Marubashi, W. 2001. Temperature effects on interspecific hybridization between *Gladiolus grandiflora* and *G. tristis. Hort Science*, **36**(2): 341-43.

Takatsu, Y., Miyamoto, M., Inoue, E., Yamada, T., Manabe, T. and Kasumi, M. 2001a. Interspecific hybridization among wild Gladiolus species of southern Africa base don randomly amplified polymorphic DNA markers. *Scientia Horticulturae*, **91**(3-4): 339-348.

Timchenko, O.D. 1990. Study of varieties of Gladiolus × hybridus with a view to using them in breeding work. *Ukrains2 kii Botanichnii Zhurnal*, **47**(4): 96-98.

Tiwa Raknim. 1990. Effect of colchicine on mutation of gladiolus *in vitro*. Bangkok. 54p.

Tomotsune, H., Kasumi, M. and Takatsu, Y. 1994. Propagation of gladiolus by somatic embryogenesis. IPPS Japan potential region. *Proce*: International Plant Propagators Society Combined Proceedings, **44**: 239-244.

Van Eijk, J.P., Van Zaayen, A. and Eikelboom, W. 1990. Developing a test method for selection on resistance to dry-rot disease (*Stromatinia gladioli*) in gladiolus. *Acta Horticulturae*, **266**: 365-371.

Wang, JinGang, Guo, Ying, Che, DaiDi, Liu, ShenKui and Yang, ChuanPin. 2009. RAPD analysis of M_1 generation of *Gladiolus hybridus* Hort. treated by EMS. *J. Northeast Agril. Univ.*, **16**(1): 5-9.

Wilfret, G.J. 1993. Morning Mist gladiolus. *Hort Sci.*, **28**(7): 752-753.

Zhang, ZhiWei. and Wang, Dan. 2008. Effect of electron beam irradiation on growth and development of induced gladiolus. *J. Northeast Forestry Uni.*, **36**(1): 26-27.

Chapter – 13
Heliconia

Heliconia belongs to family Heliconiaceae is popular as ornamental plants and cut flowers because of their brilliant colours and exotic appearance. The genus name "*Heliconia*" is derived from "*Helicon*", a mountain in southern Greece regarded by the ancient Greeks as the home of the Muses, thus suggesting the relationship between these plants and the bananas, genus *Musa* (Berry and Kress, 1991). Unique features of the Heliconiaceae are (1) medium to large erect herbs rising from underground rhizomes; (2) each erect shoot is composed of a stem and leaves, whereas the stem is made up of an axis covered by overlapping sheathing leaf bases, technically called a pseudostem; (3) inverted flowers and (4) the presence of a single staminode. Heliconia has long been popular horticulturally because of their showy inflorescences. It was so attractive that early explorers of the tropics returned to Europe with several species that became prized green house specimens. Heliconia, grown in nearly all of the tropical regions of the world, including Africa and Asia. In India, heliconia is grown in North-East region and Kerala.

ORIGIN AND DISTRIBUTION

Heliconias are native only to Central and South America and some of the islands of the South Pacific, for horticultural and commercial popularity. Berry and Kress (1991) stated that they are native from Tropical America and found at different altitudes, from sea level up to 2,000 meters in Central and South America, and up to 500 meters in the South Pacific Islands (Criley and Broschat, 1992).

Most of the *Heliconia* species are distributed primarily in Neotropical areas from the North of Mexico to the South of Brazil (Santos, 1978, Dahlgren *et al.*, 1985 and

Kress, 1990). A small Paleotropical group, about eight species, occurs in islands of the South Pacific (Tomlinson, 1969 and Kress, 1985). There are approximately 40 Brazilian species with two primary areas of distribution: the Amazon basin and the Atlantic coastal forest (Kress, 1990). About 100 or more species, native mainly to tropical and subtropical South and Central America (Cronquist, 1981).

Several attempts have been made over the years to sort out the taxonomy of the Old World *Heliconia*. Baker (1893), followed by Schumann (1900) and Winkler (1930), dealt with the problem of identity and origin of these taxa by assigning them to the neotropical species *Heliconia bihai* L. Others such as Ridley (1908a, b), Backer (1920) and Green (1969) recognized the distinctive and endemic features of the paleotropical taxa that separated them from the neotropical species. This ornamental rhizomatous herbaceous plant from the *Heliconia* genus, belongs to the Musaceae family, now constitutes the Heliconiaceae family in the Zingiberales order. The various species of *Heliconia* are subdivided into five subgenera: *Heliconia, Taeniostrobus, Stenochlamys, Heliconiopsis* and *Griggisia*; and 28 sections (Kress *et al.*, 1993).

SPECIES

In *Heliconia* genus, the number of species ranges from 250 to 300 and there are also a great number of cultivars and 23 natural hybrids (Santos, 1978, Dahlgren *et al.*, 1985, Kress, 1990, Berry and Kress, 1991 and Castro *et al.*, 2007), these plants can be found either in shaded places, such as forests or at full Sun areas, such as forest edges and roadsides (Castro and Graziano, 1997).

Species and cultivar identification is primarily based on morphological differences and colouration of the flowers and inflorescences. Significant variations exist within and between species of heliconia. However, natural variation among individuals and populations of heliconias has caused confusion (Berry and Kress, 1991). Variation may be due to geographical isolation, environmental influence, particularly light and nutrients. Clones of *H. psittacorum* growing close together tend to vary in vegetative and floral characteristics, e.g., size and colour of bracts. In addition, postharvest characteristics can vary from clone to clone (Donselman and Broschat, 1986). Following are some common species of *Heliconia*.

H. humilis: Native to Trinidad and Brazil, having shiny green leaves and erect flower heads with salmon red boat shaped bracts changing to green towards tip and ridges and greenish yellow flowers.

H. latispatha: Native of Central and South America. Showy plants with green broad oblong leaves and erect inflorescence with well separated boat shaped bracts orange yellow at the base and red towards tip.

H. bihal: A large perennial herb oblong smooth textured pointed green leaves, greenish yellow flowers clustered in the axils of large stiff boat shaped crimson red flattened bracts with pointed tip and arranged in two ranks on erect inflorescence.

***H. distans*:** Native of Tropical America having slender bracts, lemon yellow at base and crimson on top.

***H. collinsiana* (Pendula):** Robust tropical perennial herb with erect inflorescence having boat shaped orange red bracts becoming darker as they mature and creamy white flowers. The seeds are first orange, later turning blue.

FLOWER MORPHOLOGY AND BIOLOGY

Heliconia plants are erect, from 0.5 m to 10.0 m high with various sized leaves. The stem is formed by overlapping of the blades, of the leaves. Each leaf is composed of two halves separated by a tapering midrib. Leaves in different positions on the plant have a different absorption potential of sunlight for photosynthesis when exposed to different degrees of sunlight (He *et al.*, 2006). Flowers produce ample nectar that attracts pollinators, most prevalent of which are hummingbirds (Bruna *et al.*, 2004). The actual flower is the tiny little that emerges from the large showy bracts. The flower group (called an inflorescence) is almost always at the end of long, leafy shoots. The stem is either erected or drooping, with respect to the leafy shoot from which it emerges. The inflorescence is made up of the peduncle and modified leaf-like structure known as bracts. The rachis connects adjacent bracts. Within each bract there is a coil of inconspicuous flowers.

In a study performed by Eunoh (1992) the minimum day length required for flower bud initiation in *H. angusta* was 13 hr for 7 weeks. Anatomical analysis of shoot apex also confirmed transformation of vegetative apex to reproductive apex under day length 13 hr or longer. In day length longer than 13 hr, the differences in day length had no significant effect on flowering time which was 16 weeks after long day treatment. Day lengths of 12, 14, 16 and 18 hr had number effect on time to flower after initiation or number of flowers, but plant grown in 18 hr produced longer flower stalks. Night temperature had a significant effect on the number of flowers or flower quality. Harvesting of flowers very early in the day at low temperature condition was found to enhance germinability (Salih 1994). Kirchoff (1992) states that the flowers of the Heliconiaceae are bisexual and zygomorphic, each subtended by an opaque or membranous, variously coloured floral bract, persistent or decomposing after anthesis. The two trimerous perianth whorls are united at the base into short floral tubes. At the top of this tube the median sepal becomes free from the other perianth members, while the remaining members are united into a lip through the adnation of the sepals to the petals. The androecium consists of six androecial members arranged in two whorls. There are five pollen-bearing stamens and a single staminode, inserted opposite the free sepal. The ovary is inferior and trilocular with one basally inserted anatropous ovule per locule. Gynopleural nectaries are present beginning below and extending through the locular region of the ovary. Phenetic and cladistic analyses using 231 polymorphic RAPD markers demonstrated that the genus *Heliconia* is monophyletic

(Marouelli *et al*., 2010). In Heliconia, thread like structure connecting the pollen grains are described by Rose and Barthlott (1995). These threads decay products of walls separating the pollen chambers and products of rupture of the mature anthers in the stomium region.

Heliconia acuminata is an understory herb native to the nonflooded forests of Amazonia and the Guyanas. It is a self-incompatible hermaphroditic perennial with limited vegetative reproduction (Bruna and Kress, 2002.). Flowering begins in late January and continues through April; during this time it is the most abundant flowering plant in the understory. Most reproductive plants have one inflorescence with a total of 20-25 flowers. Each flower is open for only one day, after which the perianth and the style abscise.

Sanjeev *et al.* (2010) carried out a study on heliconia to study the floral biology of 15 heliconia species and varieties for sixteen floral characters. The study revealed that under South Kerala condition. There was no uniformity in flowering behaviour of different varieties. Considering the flowering pattern, some of the varieties showed continuous flowering throughout the year, whereas some others were significantly seasonal.

The fruits also develop within the bracts. The bracts are often filled with water and house a distinctive aquatic micro- ecosystem. Fruits are blue-purple when ripe and primarily bird dispersed (Uriarte *et al*., 2011). Post dispersal seed survival was studied and had many results (Hoii and Lulow, 2006). The study showed that seed size was not a determinant of post dispersal seed survival. The highest amount of seed predation came from mammals.

Pollination

The Hispaniolan mango, *Anthracothorax dominicus*, is the principal pollinator of both *H. bihai* and *H. caribaea* on Hispaniola (Rodriguez *et al*., 2011). Many insects groups (Arthropoda, Insecta) are associated to *Heliconia*, such as beetles (Coleoptera), ants (Hymenoptera, Formicidae), aphids (Hemiptera, Aphidoidea) and scale insects (Hemiptera, Coccoidea). Some *Heliconia* species accumulate exudates, water and floral parts in these bracts that form a micro habitat called phytotelmata which favours the occurrence of these insects (Oliveira *et al.,* 2010). Stein and Hensen (2011) founds that a stingless bee species (*Trigona* sp.) appeared to be the most abundant visitor to the ornithophilous flowers of *H. angusta,* along with four different species of hummingbirds and two species of butterflies. Most flower visitors were recorded between 9.00 am and 1.00 pm with a higher number visiting under semi-shaded conditions than in full shade. Hummingbird numbers increased with flower abundance while the other visitor group numbers were not affected. Although heliconia are almost exclusively pollinated by hummingbirds, some bat pollination has been found to occur. *H. solomonensis* is pollinated by the macroglosine bat (*Melonycteris woodfordi*) in the Solomon Islands. *Heliconia solomonensis* has green inflorescences and flowers that open at night, which is typical of bat pollinated plants. The macroglosine bat is the only known nocturnal pollinator of *H. solomonensis* (Kress, 1985). Hummingbirds are the main pollinators of heliconia in many locations. At La Selva research station in Costa Rica it was found that specific species of Heliconia have specific hummingbird pollinators (Stiles, 1975). It was found that many species of *Heliconia*, even the newly colonized species, all had many pollinators visit (Feinsinger, 1978).

Pollinators

Heliconia is pollinated by various types of agents. Comparison of pollinator related characters in bat, honeyeater and humming bird pollinated heliconia is given in Table 1.

Table 1: Comparison of pollinator related characters in heliconia.

S. No.	Character	Bat pollinated taxa	Honeyeater pollinated taxa	Hummingbird pollinated taxa
1.	Inflorescence orientation	Pendent and erect	Erect	Pendent and erect
2.	Flower colour	Green	Green or yellow	Yellow
3.	Perianth length	50-68 mm	45-75 mm	28-65 mm
4.	Flower longevity	< 24 hr	< 24 hr	< 24 hr
5.	Anthesis	Evening	Morning	Morning
6.	Breeding concentration	Self compatible and self incompatible	Self compatible	Self compatible
7.	Species diversity	4	2	225

GENETICS

The chromosome number 2n = 24 was determined in 31 taxa of *Heliconia*, representing nearly the entire morphological spectrum of the genus. The karyotype is a graded series of rather small chromosomes ranging between 0.7 and 1.7 urn. One pair has satellites. It is concluded that morphological chromosome differentiation played no great part in the diversification of the genus.

Six cultivars of *Heliconia psittacorum* were selected by Lee *et al.* (1994) for studies on their natural fruit bearing ability, pollen formation and pollination under tropical climatic conditions at Singapore. Three of them namely, Tay, Andromeda and Lady Di were partially fertile with a very low rate of fruit set, ranging from 2.8 to 4.7 %. They were found to be diploid with 2n = 24 chromosomes. Whereas, the other three namely, Petra, Sassy and Iris were completely sterile. Their pollen grains were of variable sizes and appear to be abnormally fragmented. All three were confirmed to be triploid (2n = 3x =36).

There are taxonomic confusions and uncertainties about the number of species and the relationships among them. Molecular studies are therefore necessary for better understanding of the species boundaries of these plants. Genetic variability and the phylogenetic relationships of 124 accessions of the genus Heliconia based on RAPD markers was examined. Phenetic and cladistic analyses, using 231 polymorphic RAPD markers, demonstrated that the genus Heliconia is monophyletic. Groupings corresponding to currently recognized species and some subgenera were found, and cultivars and hybrids were found to cluster with their parents. RAPD analysis generally agreed with morphological species classification, except for the position of the subgenus Stenochlamys, which was found to be polyphyletic (Marouelli, 2010).

There are some cold hardy species as given below may play important role in improvement of heliconia.

Cold hardy species

***H. angusta*:** This hardy species from southern Brazil comes in 3 colours, yellow, red and orange. It is cold hardy compact upright and wind resistant. Common name Holiday or Christmas heliconia.

***H. spissa*:** There are 2 forms. The Mexico Red is the most common. It has upright blooms, pinkish red. Similar to *H. schiedeana,* but fuller and showier. The plant grows 5-6 feet and has naturally lacerated weeping leaves that give it a split leaf philodendron look which is very attractive.

***H. latispatha*:** There are many forms of *H. latispatha* and they vary in size, hardiness, and colour. They all have a deep keel to the bracts which have a long elegant taper.

***H. bourgeana*:** This plant from Mexico is amazingly hardy and wind resistant for having such huge leaves.

***H. subulata*:** This may be the hardiest.

***H. champneiana* 'Maya Blood':** Although this species from Mexico. It is a hardy species.

Genetic variability and heritability

Twelve heliconia varieties were evaluated to ascertain genetic parameters of variability. Phenotypic coefficients of variation (PCV) were higher than genotypic coefficients of variation (GCV) for all the characters studied. However, higher PCV and GCV estimates were found for flower yield, bract size and leaf area. High heritability with high genetic advance was observed for flower yield; leaf area and bract size (Sheela *et al.*, 2005). Isaza *et al.* (2012) characterized the genetic variability of 67 genotypes of cultivated heliconias belonging to *Heliconia caribaea*, *H. bihai, H. orthotricha*, *H. stricta*, *H. wagneriana* and *H. psittacorum* as well as that of several interspecific hybrids such as *H. psittacorum* × *H. spathocircinata* and *H. caribaea* × *H. bihai*. They further concluded that *heliconia* genus are highly variable and diverse.

A genetic study of seven cultivars of *H. psittacorum* and *Heliconia interspecific* hybrids was carried out. The heritability estimate and genetic variation coefficient were highest for stem diameter (99.32% and 56.90%, respectively) and for CVg/CVe (1.85), indicating a favourable situation for selection. The genetic correlations of stem diameter with days to inflorescence emergence (0.64), period from shoot emergence to stem cut (0.63) and stem weight (0.96) showed that the time from inflorescence emergence to cut is longer and the stem weight is greater for genotypes with larger stem diameter. Inflorescence length, stem diameter and days to inflorescence emergence were the most important traits, accounting for 99.55% of the total variation. For stem diameter and inflorescence length, the repeatability values exceeded 0.60 and for stem diameter, stem weight, days to inflorescence emergence and inflorescence length the coefficients of determination exceeded 93% (Rocha *et al.*, 2010).

BREEDING OBJECTIVES

i. Higher flower yield
ii. Winter blooming proclivity
iii. Longer vase life
iv. New and different types of flowers
v. New colour variety
vi. Resistance to diseases like bacterial wilt and fungal root rot

BREEDING METHODS

Introduction

Some promising varieties were introduced by College of Tropical Agriculture and Human Resource (CTAHR) of University of Hawaii (1999) as described in Table 2.

Table 2: Heliconia varieties introduced in Hawaii.

S. No.	Variety	Introduced from	Characteristics
1.	Kauai Morning Sun	CTAHR	Erect, compact, light-bodied inflorescence with light rosy-red, relatively non pubescent bracts. Upper margins and tips of bracts are green, accented by a narrow strip of black. Pale yellow rachis is medium in length. Plant habit is musoid, compact, clumping with green, glabrous leaves on slightly pubescent petioles. Flowering heaviest from October to February. Vase life 26 days.
2.	Kauai Christmas	CTAHR	Erect, medium-bodied inflorescences, bright red, pubescent bracts with green upper margins and tips accented below by a narrow, black border. Rachis is long and white. Plant habit musoid, clumping, medium in height. Petioles slightly pubescent, glabrous leaves. Flowering from December to February. Vase life 19 days.
3.	Kauai Sunset	CTAHR	Looks like deepening sunset. Erect, stout-bodied inflorescences. Bracts deep red and pubescent, with green upper margins and tips accented below by a flat-black border. Rachis light yellow and compact. Plant habit musoid, clumping, medium-short in height. Petioles pubescent, with glabrous leaves. Flowering from January to March. Vase life 19 days.

Selection

A study was carried out to see the potential of *Heliconia* brazilian species as cut flower by Merida *et al.* (2013). Study consists of 28 species and varieties, 25 of which are naturally occurring in Brazil: *H. episcopalis* Vell, *H. hirsute* L., *H. bihai* (L.) L. 'Nappi Yellow', *H. stricta* Huber, *H. spathocircinata* Aristeg, *H. lourteigiae* Emygdio & Santos, *H. farinosa* Raddi, *H. rivularis* Emygdio & Santos, *H. pseudoaemygdiana* Emygdio & Santos, *H. acuminate* C. Rich, *H. angusta* Vell. 'Aurorea', *H. angusta* Vell. 'Flava', *H. angusta* Vell. 'Holiday', *H. psittacorum* L., *H. metallica* Planch. & Linden, *H. subulata* R & P., *H. chartaceae* Lane ex Barreiros 'Sexy Pink', *H. chartaceae* Lane ex Barreiros 'Sexy Scarlet', *H. pendula* Wawra, *H. marginata* (Griggs) Pittier, *H. rauliniana* Barreiros, *H. rostrata* R. & P., *H. densiflora* Verlot, *H. aemegydiana* Burle Marx, *H. tenebrosa* MacBride, and *H. richardiana* Miq. Based in the morphological and agronomic characterization data and the post-harvest evaluation, it was found that *H. bihai* 'Nappi Yellow', *H. stricta, H. lourteigiae, H. farinosa*, and *H. angusta* 'Aurora' are very suitable for use as cut flower.

Heliconia is a cross pollinated crop and attract various insects, birds and bats, therefore possibility of some outstanding selections from open pollinated. Following are some varieties that were selected and developed through a partnership with DPIFM

Crops, Forestry and Horticulture Division and the NTHA cut-flower growers group. Some seedling selection of heliconia is given in Table 3.

Table 3: Varieties evolved through selection.

S. No.	Variety	Parent	Characteristics
1.	Pink Blush	*H. psittacorum* Parakeet	Unique large inflorescence form with thick long stem, excellent cut flower. Vase life 22 days.
2.	Flamenco	*H. psittacorum* Parakeet	Beautiful coloured large pink inflorescence form, vase life 25 days.
3.	Pearl	*H. psittacorum* Parakeet	It is a rare pale lemon to cream coloured. *Psittacorum* inflorescence ideal for floral wedding arrangements. Prolific producer of flowers even during the dry season. Vase life 21 days.
4.	Fantasia	*H. psittacorum* Parakeet	Attractive small inflorescence, red-orange bracts, highlighted by a bright yellow band at the base. Vase life 21 days.
5.	Rosie Q	*H. psittacorum* Lady Di	Pink, apricot coloured. Vase life 18 days.
6.	Red Devil	*H. psittacorum* Lady Di	Reddest of all the nine new *Psittacorums*, Good producer of flowers. Vase life 18 days.
7.	Rose Red	*H. psittacorum* Lady Di	Rosy-red coloured inflorescence. Vase life 14 days.
8.	Exotic	*H. psittacorum* Suzi	Variegated leaves. Height 1 to 1.5 m, early red flower. Good production.
9.	Cracker Jack	*H. psittacorum* Lizette	Firey-red to orange coloured inflorescence, small, good as filler in arrangements. Vase life 24 days.

Hybridization (Interspecific)

Hybridization is the most common method of plant improvement but in heliconia artificial hybridization is unsuccessful due to pre-fertilization barriers. Even though there are some natural hybrids, artificial hybrids are not reported in heliconia. Most species of heliconia are selfcompatible but cross fertilization is unsuccessful (Berry and Kress, 1991, Kress, 1983 and Lee *et al.*, 1994). Existence of pre-fertilization crossability barriers (stigmatic, stylar or ovarian) act as the ultimate mechanism for preventing hybridization in heliconia (Kress, 1983).

Artificial hybridization among 14 species of neotropical Heliconia was studied by Kress (1983) at two sites in Costa Rica. Observation of pollen tube growth by means of fluorescence microscopy and seed set were used to determine the extent of crossability. Crossability barriers between the majorities of species were strong and foreign pollen tubes were inhibited at the stigmatic surface within the stylar tissue or within the ovary. Some interspecific hybrids reported by Berry and Kress (1991). Hybrid varieties of heliconia are given in Table 4.

Table 4: Hybrid varieties of heliconia.

S. No.	Variety	Interspecific cross	Chief characteristics	Inventor (year)
1.	Cinnamon Twist	*H. bihai* × *H. spathocircinata*	Erect inflorescence, bracts red with yellow, rachis scarlet, sepals greenish yellow, height 1.5 to 3.4 m.	D. Carli (1988)
2.	Alan Carle	*H. psittacorum* × *H. spathocircinata*	Erect inflorescence, bracts reddish, rachis greenish to yellow, sepals yellowish height 0.9 to 1.5 m.	F. Berry (1989)
3.	Bushmaster	*H. mariae* × *H. pogonantha holerythra*	Pendent inflorescence, bracts red, rachis red, sepals faint yellow, height 4.6 to 6 m.	F. Berry and J. Kress (1991)
4.	Carib Flame	*H. bihai* × *H. caribaea*	Erect inflorescence, bracts red with orange infusion, rachis green, height 2.4 to 4.6 m.	J. Criswick
5.	Criswick	*H. bihai* × *H. caribaea*	Erect inflorescence, bracts red with greenish tip, rachis red, sepals green distally, height 2.4 to 5.8 m.	F. Berry (1989)
6.	Crocodile	*H. curtispatha* × *H. mariae*	Pendent inflorescence, bracts red with black lip, rachis red, sepals yellow, height 6.5 m.	C. Black & A. Rodríguez (2004)
7.	Grand Etang	*H. bihai* × *H. caribaea*	Erect inflorescence, bracts dark red to maroon, bract mostly green, Rachis red, sepals light green, height 2.7 to 4.6 m.	F. Berry and A. Carle (1989)
8.	Golden Torch	*H. psittacorum* × *H. spathocircinata*	Erect inflorescence, bracts, golden or yellow, rachis golden, sepals golden with faint green tip, height 0.8 to 2.7 m.	Will and H. Donselman (1977)
9.	Golden Torch Adrian	*H. psittacorum* × *H. spathocircinata*	Erect inflorescence, spiral bracts, maroon to dark red at base, rachis maroon to dark red, sepals golden with faint green distally, height 1.1 to 1.5 m.	D. Carli (1989)
10.	Grenadier	*H. bihai* × *H. caribaea*	Erect inflorescence, bracts red with thin yellowish lip, bracts green, rachis red, sepals green, height 2.4 to 4.2 m.	J. Criswick

(*Contd.*)

(*Contd. Table 4*)

S.No.	Variety	Interspecific cross	Chief characteristics	Inventor (year)
11.	Guyana	*H. psittacorum* × *H. spathocircinata*	Erect inflorescence, bracts, red-orange, rachis red-orange, sepals pale orange, height 1.5 to 1.8m.	J. Nelson (1989)
12.	Harvey Bullis	*H. imbricata* × *H. sarapiquensis*	Erect inflorescence, bracts red with orange flush, bract with green keel, rachis yellow, sepals yellow to green, height 2.1 to 3.6 m.	F. Berry (1989)
13.	Johnson Beharry V.C.	*H. psittacorum* × *H. spathocircinata*	Erect inflorescence, bracts uniform deep crimson, rachis deep crimson, sepals yellow, height 1.3 to 1.9 m.	J. Criswick (2005)
14.	José Abalo	*H. imbricata* × *H. latispatha*	Erect inflorescence, spiral bracts, gold with orange tint, rachis yellow with green tint, sepals pale olive green, height 1.8 to 5.5 m.	F. Berry (1989)
15.	Pedro Ortiz	*H. bourgeana* × *H. collinsiana*	Distichous bracts, bright solid red, rachis red, sepals golden yellow, height 1.8 to 2.4 m.	F. Berry and B. Ballantyne (1991)
16.	Richmond Red	*H. bihai* × *H. caribaea*	Erect inflorescence, bracts red with maroon, rachis dark red, sepals green height 2.7 to 4.6 m.	I. Bannochie
17.	Toucan	*H. clinophila* × *H. secunda*	Pendent inflorescence, spiral bracts, red coloured, basal bract with green cheek, rachis light yellow or red, sepals light yellow, height 1.2 to 1.8 m.	F. Berry (1989)
18.	Vermillion Lake	*H. bihai* × *H. caribaea*	Erect inflorescence, bracts red, bract orange, rachis red above and yellowish below, height 2.7 to 4.9 m.	J. Criswick (1987)
19.	Yellow Dolly	*H. bihai* × *H. caribaea*	Erect inflorescence, bracts red, rachis red above and yellow below, height 2.7 to 4.2 m.	J. Criswick (1989)
20.	Yellow Parrot	*H. psittacorum* × *H. spathocircinata*	Erect inflorescence, bracts solid yellow, height 0.9 to 1.5 m.	K. Vinzant (2002)

Mutation

The high economic importance of ornamental plants represents a stimulus for breeders in the research for new cultivars. Among the several employed techniques, gamma radiation induced mutation can bring promising results in different *Heliconia* species. Rodrigues *et al.* (2005) in a study found that the LD_{50} for the *in vitro* culture was ranged between 30 and 40 Gy when different doses i.e. 0, 10, 20, 30, 40 and 50 Gy gamma rays were used in 25 shoot apices. Sangwanangkul *et al.* (2007) found that Heliconia 'Keanae' showed high sensitivity to 250 Gy in a study to find the sensitivity of cut flowers and foliage towards electron beam irradiation. In a study it was found that *Heliconia psittacorum* and *Heliconia rostrata* were not tolerant to gamma and e-beam radiation of 300 Gy. The most evident damage caused by irradiation was the browning, followed by drying up (Kikuchi, 2003).

BIOTECHNOLOGY

Biotechnology has been considered helpful to solve some of the problems related to the domestication of heliconias. Multiplication of heliconias was the first problem because of the slow germination of its seeds (about three months to three years). Multiplication by rhizome (the underground stem of a heliconia) is also time consuming and therefore hardly an alternative. Tissue culture is a good tool of multiplication of heliconia.

Tissue culture

Talukdar *et al.* (2002) conduct an *in vitro* propagation of shoot explants from *H. psittacorum* in modified MS medium comprising of different treatments *viz*. 0.25 mg IAA/ litre and benzyladenine at either 4, 6, 8 or 10 mg/litre. Zeng *et al.* (2004) obtained new buds by dark culture from the rhizomes of ten species or cultivars and hybrids of heliconia which were used as explants. The best results for rooting were observed on MS medium supplemented with 0.5 mg/ NAA plus 0.5% active carbon. The plantlets showed a survival rate of more than 90%. Bora and Paswan (2003) founds a successful establishment of threefold increase in length and more than double in breadth of *H. psittacorum* axillary bud in MS medium when supplemented by 2.5 mg benzyladenine, 0.2 mg IAA, and 2.5 mg GA_3/litre which was ideal for initiation of multiple shoots and their multiplication rates and MS medium containing 0.75 mg IAA and 0.05 mg NAA/ litre gave maximum rooting response and earliest root initiation.

Since, disinfection of explants is a major problem with micropropagating species where underground rhizomes are the only source of axillary buds. Pedersen and Brandt (1992) develop a procedure for the disinfection of rhizome tips, based on trimming of scale leaves three times altering with short (1- 10 minutes) immersions in disinfectants was used with *Heliconia psittacorum* cv. Choconiana Dwarf. Results shows that rhizome tips shorter than 4 cm, from fast growing stock plants, resulted in the lowest contamination and the percentage of growing uninfected explants obtained after three weeks of culture on MS medium supplemented with growth regulators was 60. *In vitro* propagation of *Heliconia psittacorum* cv. Rhizomatosa was done by Shiau Yihjuh *et al.* (1999) using terminal shoot tip explants and found that well developed shoots were successfully obtained after 45 days of culture of shoot tip explants on MS medium supplemented with 4- 8 mg/litre BA and 0.5 mg/litre NAA.

Pollination barrier and *in vitro* pollination and fertilization

Most species of heliconia are selfcompatible but cross fertilization is unsuccessful (Berry and Kress, 1991, Kress, 1983 and Lee *et al*., 1994). Existence of pre-fertilization crossability barriers (stigmatic, stylar or ovarian) act as the ultimate mechanism for preventing hybridization in heliconia (Kress, 1983). The technique of *in vitro* pollination and fertilization is very promising for overcoming pre-fertilization barriers due to incompatibility and for raising new genotypes through seed propagation. The purpose of *in vitro* pollination and fertilization is to bring pollination and embryogenesis under artificial control so as to produce hybrid embryos among plants that cannot be crossed by conventional methods of plant breeding. The specific objective of this study was to evolve a protocol for the *in vitro* pollination and fertilization technique for heliconia.

To overcome from this barrier an experiment was conducted to standardize an appropriate *in vitro* pollination and fertilization technique for heliconia and thereby evolve a suitable protocol for a hybridization programme. Three cultivars of *Heliconia psittacorum, viz*. 'Lady Di', 'Andromeda' and 'Parakeet' were selected for the study. ME3 medium with 12% PEG was identified as the suitable medium for the *in vitro* pollen germination of heliconia. Among the different *in vitro* pollination techniques, modified placental pollination was found ideal for heliconia cultivars. The pollinated ovules, cultured in the medium of MS + NAA 2 mg l^{-1} + BA 2.5 mg l^{-1} + Yeast extract 250 mg l^{-1} showed best results. The *in vitro* seeds were smaller than *in vivo* seeds. In the *in vitro* seeds, the embryo was at the chalazal end and three progressive stages of embryos were identified, viz. globular, reniform and elongated. The embryos from *in vitro* formed seeds dissected 40 days after pollination showed positive response on embryo culture in MS medium with 3% sucrose (Dhanya and Sheela, 2008).

Somaclonal variation

Rodrigues (2008) in a study, investigated the occurance of somaclonal variation in micropropagated changes of *Heliconia bihai* cv. Lobster Claw I in which the occurrence of three types of somaclonal variants was observed. Variation of the chlorophyll in the leaf, low stature variant and pseudostem and petiole colour variant, the latter with ornamental potential. The somaclonal variation rate for *Heliconia bihai* cv. Lobster Claw I, under the proposed conditions, was 61.40%. A protocol was developed for the *in vitro* induction of embryogenic callus in *Heliconia chartacea* cultivars Sexy Pink and Sexy Scarlet from ovary sections in which Culture media were supplemented with IAA (0, 1 and 2 mg/litre) and 2, 4-D (0, 5, 10, 15 and 20 mg/litre). Results indicate that Sexy Pink presented a higher frequency of callus development compared to Sexy Scarlet (Ulisses *et al.*, 2007).

Molecular breeding

A protocol was developed to extract DNA from heliconia leaves and to analyse genetic variation using RAPD. The percentage genetic similarities among *Heliconia* species, cultivars and hybrids were determined which indicates that the RAPD technique is useful in distinguishing species and cultivars of heliconia. It was also found that 2 triploid cultivars of *H. psittacorum*- Iris and Petra showed identical RAPD profile which 10 different primers in agarose and polyacrylamide gels, suggesting that they are of the same genotype.

There are taxonomic confusions and uncertainties about the number of species and the relationships among them. Molecular studies are therefore necessary for better understanding of the species boundaries of these plants. Genetic variability and the phylogenetic relationships of 124 accessions of the genus Heliconia based on RAPD markers was examined. Phenetic and cladistic analyses, using 231 polymorphic RAPD markers, demonstrated that the genus Heliconia is monophyletic. Groupings corresponding to currently recognized species and some subgenera were found, and cultivars and hybrids were found to cluster with their parents. RAPD analysis generally agreed with morphological species classification, except for the position of the subgenus Stenochlamys, which was found to be polyphyletic (Marouelli, 2010).

Recent developments in molecular biology have made DNA fingerprinting of plants feasible and an array of techniques are available today to characterize DNA sequence/ polymorphism. Several of these have been used in phylogenic studies, genetic diversity analysis, forensic science, paternity determination, plant varietal characterization, and the like (Williams *et al.*, 2004). In heliconia, a protocol to extract DNA from leaves and analyze the genetic variations using RAPD has been developed (Kumar *et al.*, 1998).

Molecular characterization of heliconia through RAPD assay in seventeen species and varieties was performed by Sheela *et al.* (2006), in which eight primers that produced the highest number of bands were used for DNA amplification. The genetic similarity matrix showed the highest value between Petra Orange and Parakeet, while the lowest was between Golden Torch and *H. humilis*. Thus implying a strong parallelism between genetic and morphologic/taxonomic variability of *Heliconia* genotypes. The present RAPD analysis is also useful in detecting genetic variations and similarities among the heliconia genotypes. Similar work was reported earlier by Goh *et al.* (1995) in heliconia.

DNA markers, which allow the access to variability at DNA level, emerge as an efficient alternative for plant species characterization by quantifying diversity and determining its genetic structure (Bruns *et al.*, 1991). Since, molecular marker is important from the genetic variability assessment point of view, because the rDNA mutltigenic family once subjected to a rapid evolution in concert event, allows greater precision in the reconstruction process of the relationship between species based on sequencing, since this phenomenon increases the intragenomic uniformity (Baldwin *et al.*, 1995). Guimaraes *et al.* (2012) in a study, found that there was no repetition of genetic material among the cultivars and interspecific hybrids of *H. psittacorum* evaluated, indicating the necessity to use other regions that could provide potentially informative characters. Molecular analysis using amplified fragment length polymorphism (AFLP) markers revealed a total of 170 bands in 67 genotypes of cultivated heliconias of *Heliconia caribaea, H. bihai, H. orthotricha, H. stricta, H. wagneriana* and *H. psittacorum* as well as that of several interspecific hybrids such as *H. psittacorum* × *H. spathocircinata* and *H. caribaea* × *H. bihai*. The lowest percentage of polymorphism was found in *H. psittacorum* (17.65%) and the highest was in *H. stricta* (55.88%). Thus, study revealed the monophyletic origin of the heliconiaceae family, and identified the *Heliconia* subgenus as monophyletic (Isaza *et al.*, 2012).

REFERENCES

Anderson, L. 1984. The chromosome number of *Heliconia* (Musaceae). *Nordic J. Bot.*, **4**(2): 191-193.

Backer, C.A. 1920. *Heliconia indica* Lamarck, an insufficiently known species of the East-Indian Archipelago. *Bull. Jard. Bot. Buitenzorg*, **3**(2): 315-319.

Baker, J.G. 1893. A synopsis of the genera and species of Musaceae. *Ann. Bot.*, **7**: 189-222.

Baldwin, B.G., Sanderson, M.J., Porter, J.M., Wojcichowiski, M.F. Campbell, C.S. and Donoghue, M.J. 1995. The ITS Region of Nuclear Ribosomal DNA: A Valuable Source of Evidence on Angiosperm Phylogeny. *Ann. Missouri Bot. Gard,.* **82**: 247-277.

Berry, F. and Kress, W.J. 1991. Heliconia: An Identification Guide. Smithsonian Institution Press, Washington and London, pp. 1-334.

Bora, S. and Paswan, L. 2003. *In vitro* propagation of *Heliconia psittacorum* through axillary bud. *J. Orna. Hort.*, **6**(1): 11-15.

Bruna, E.M. 2002. Effects of forest fragmentation on *Heliconia acuminata* seedling recruitment in central Amazonia. *Oecologia,* **132**: 235-243.

Bruna, E.M. and Kress, W.J. 2002. Habitat fragmentation and the demographic structure of an Amazonian understory herb (*Heliconia acuminata*). *Conservation Biology,* **16**: 1256-1266.

Bruna, E.M., Kress, W.J. Marques, F. and da Silva, O.F. 2004. *Heliconia acuminata* reproductive success is independent of local floral density. *Acta Amazonica,* **34**(3): 467-471.

Bruns, T.D., White, T.J. and Taylor, J.W. 1991. Fungal molecular systematics. *Ann. Rev. Ecol. Sys.*, **22**: 525-564.

Castro, C.E.F. and Graziano, T.T. 1997. Especies do Genero *Heliconia* (Heliconiaceae). R. Bras. Hortic. Ornam. Campinas, Brazil, **3**: 15–28.

Castro, C.E.F., Goncalves, C. and May, A. 2007. Atualizacao da nomenclatura de especies do genero *Heliconia* (Heliconiaceae). *R. Bras. Hortic. Ornam. Campinas*, Brazil, **13**: 38–62.

Criley, R.A. and Broschat, T.K. 1992. Heliconia: Botany and Horticulturae of new floral crop. *Hortic. Review*, **14**: 1-55.

Cronquist, A. 1981. An Integrated System of Classification of Flowering Plants. Columbia University Press, New York, pp. 1-1262.

Dahlgren, R.M.T., Clifford, H.T. and Yeo, P.F. (1985), The families of the monocotyledons. *Springer-Verlag*, Berlin, pp. 350-358.

Dhanya, A. and Sheela, V.L., 2008. Standardization of *in vitro* pollination and fertilization technique for *Heliconia. Heliconia Soci. International Bulletin,* **14**(2): 1-11.

Donald P., Watson R. and Smith, R. 1974. Ornamental Heliconias. Co-operative Extension Service. University of Hawaii. Circular, 482.

Donselman, H. and Broschat, T.K. 1986. Production of *Heliconia psittacorum* for cut flowers in South Florida. *Bul. Heliconia Soc. Intl.*, **1**: 4-6.

Eunoh, K.1992. Flowering in *Heliconia angusta*. University of Hawaii.

Feinsinger, P. 1978. Ecological interactions between plants and hummingbirds in a successional tropical community. *Ecological Monographs,* **48**: 269-287.

Goh, C.J., Kumar, P.P. and Yau, C.K. 1995. Genetic variations detected with RAPD markers in *Heliconia. Acta Hort.*, **420**: 72-74.

Green, P.S. 1969. Notes on Melanesian plants: II. Old World *Heliconia* (Musaceae). Kew Bull., **23**: 471-478.

Guimaraes, W.N.R. Ferraz, G.M.G., Martins, L.S.S., Resende, L.V., Burity, H.A. and Loges, V. 2012. Genetic Diversity Analysis of *Heliconia psittacorum* Cultivars and Interspecific Hybrids Using Nuclear and Chloroplast DNA Regions. Agricultural Science, Dr. Godwin Aflakpui (Ed.), In Tech. pp. 11-22.

He, J., Chee, C. and Goh, C. 2006. Photoinhibition of Heliconia under natural tropical conditions: The importance of leaf orientation for light interception and leaf temperature. *Plant Cell and Environment,* **19**: 1238-1248.

Hoii, Karen and Megan Lulow. 2006. Effects of species, habitat, and distance from edge on post-dispersal seed predation in a Tropical Rainforest. *Biotropica,* **29**: 459-468.

Isaza, L., Marulanda, M.L. and Lopez, A.M. 2012. Genetic diversity and molecular characterization of several *heliconia* species in Colombia. *Genet. Mol. Res.*, **11**(4): 4552-4563.

Kikuchi, O.K. 2003. Gamma and electron-beam irradiation of cut flowers. *Radiation Physics Chemistry,* **66**: 77-79.

Kirchoff, B.K. 1992. Ovary structure and anatomy in the Heliconiaceae and Musaceae (Zingiberales). *Canadian J. Bot.*, **70**: 2490-2508.

Kress, W.J. 1983. Crossability barrier in Neotropical Heliconia. *Ann. Bot.,* **52**: 131-141.

Kress, W.J. 1985. Bat Pollination of an Old World Heliconia. *Biotropica*, **17**(4): 302-308.

Kress, W.J. 1990. The diversity and distribution of *Heliconia* (Heliconiaceae) in Brazil. *Acta Bot. Bras.*, **4**: 159-167.

Kress, W.J., Betancur, J. Roesel, C.S. and Echeverry, B.E. 1993. Listapreliminar de las Heliconias de Colombia y cinco espécies nuevas. *Caldasia*, **17**:183-197.

Kumar, P.P., Nathan, M.J. and Goh, C.J. 1996. Invlovement of ethylene on growth and plant regeneration in callus cultures of *Heliconia psittacorum* L. *Plant Growth Regulation*, **19**(2): 145-156.

Kumar, P.P., Yau, J.C.K. and Goh, C.J. 1998. Genetic analysis of *Heliconia* species and cultivars with randomly amplified polymorphic DNA (RAPD) markers. *J. Am. Soc. Hort. Sci.*, **123**: 91-97.

Lee, Y.H., Ng, N.Y. and Goh, C.J. 1994. Pollen formation and fruit set in some cultivars of *Heliconia psittacorum. Scient. Hort.,* **60**: 167-172.

Marouelli, L.P., Inglis, P.W., Ferreira, M.A. and Buso, G.S.C. 2010. Genetic relationships among heliconia (Heliconiaceae) species based on RAPD markers. *Genetic Molecular Research*, **9**(3): 1377-1387.

Merida, D., Ferreira de Castro, C.E., Gonçalves, C., Soares Pimenta, R., Sant'Ana Tucci, M.L. and Rocha Moreira, S. 2013. Potential use of *Heliconia* brazilian species as cut flower. *Acta Hort.*, **1000**: 139-145.

Nathan, M.J., Goh, C.J. and Kumar, P.P. 1992. *In vitro* propagation of Heliconia psittacorum by bud culture. *Hort. Sci.,* **27**(5): 450-452.

Nathan, M.J., Kumar, P.P. and Goh, C.J. 1993. High frequency plant regeneration in *Heliconia psittacorum* L. *Plant Science*, **90**(1): 63-71.

Oliveira, T.R.S. de., Sena, D.C. de A., Loges, V., Dacamara, C.A.G. and Reis, A.C. 2010. *Revista Brasileira de Horticultura Ornamental*, **16**(2): 174-178.

Pedersen, C. and Brandt, K. 1992. A method for disinfection of underground rhizome tips of Alstroemeria and Heliconia. *Acta Horticulturae*, **325**: 499-504.

Ridley, N. H. 1908b. Heliconias. *Gard. Chron. Ser.*, **7**: 13.

Ridley, N.H. 1908a. Heliconias. *Agric. Bull.*, **6**: 129-132.

Rocha, F.H.A., Loges, V., da Costa, A.S., de Aragao, F.A.S. and Santos, V.F. 2010. Genetic study with *Heliconia psittacorum* and interspecific hybrids. *Crop Breeding Applied Biotech.,* **10**: 282-288.

Rodrigues, P.H.V. 2008. Somaclonal variation in micropropagated *Heliconia Bihai* cv. Lobster Claw I plantlets (Heliconiaceae). *Sci. Agric.*, **65**(6): 681-684.

Rodrigues, P.H.V., Tulmann N.A. and Dutra, M. de F.B. 2005. Gamma radiation LD 50 determination in *Heliconia bihai* (Heliconiaceae) explants. *Revista Brasileira de Horticultura Ornamental*, **11**(1): 79-81.

Rodriguez, M.S., Kress, W.J., Temeles, E.J. and Melendez-Ackerman, E. 2011. Plant pollinator interactions and floral convergence in two species of *Heliconia* from the Caribbean Islands. *Oecologia*, **167**(4): 1075-1083.

Rose, M.J. and Barthlott, W. 1995. Pollen connecting threads in Heliconia (Heliconiaceae). *Plant Systematics and Evolution*, **195**(1-2): 61-65.

Salih, J. 1994. Pollen Studies on Heliconia. B.Sc. Desertification. National Univ. of Singapore, Singapore.

Sangwanangkul, P., Saradhuldhat, P. and Paull, R.E. 2007. Survey of tropical cut flower and foliage responses to irradiation. *Post harvest Bio. Tech.*, **48**: 264-271.

Sanjeev, S.J., Sheela, V.L. and Lekshmi, P.R.G. 2010. Floral biology and correlation studies in *Heliconia. J. Orna. Hort.*, **13**(1): 8-19.

Santos, E. 1978. Revisão das espécies do gênero *Heliconia* L. (Musaceae *s.l.*) espontâneas na região fluminense. *Rodriguésia*, **30**: 99-201.

Schumann, K. 1900. Musaceac. *In*: A. Engler, (ed.). Pflanzenr. IV. **45**: 1-45.

Sheela, V.L., Lekshmi, P.R.G., Nair, C.S.J. and Rajmohan, K. 2006. Molecular characterization of *Heliconia* by RAPD assay. *J. Tropi. Agri.,* **44**(1-2): 37-41.

Sheela, V.L., Rakhi, R., Nair, C.S., Jayachandran, Sabina, G.T. 2005. Genetic variability in heliconia. *J. Ornamental Hort.*, **8**(4): 284-286.

Shiau, Yihjuh., Wang, Tsai Yih. and Tsay, HsinSheng. 1999. Studies on tissue culture of *Heliconia psittacorum* cv. Rhizomatosa. III. Effect of cefotaxime and plant growth regulators on growth of shoot tip explants *in vitro*. *J. Agril. Res. China,* **48**(1): 42-51.

Stein, K. and Hensen, I. 2011. Potential pollinators and robbers: A study of the floral visitors of *Heliconia angusta* (Heliconiaceae) and their behaviour. *J. Pollination Ecology*, **4**: 39-47.

Stiles, G. 1975. Ecology, flowering phenology and hummingbird pollination of some Costa Rican *Heliconia* species. *Ecology*, **56**: 285-301.

Talukdar, M.C. 2002. *In vitro* propagation of heliconia. *J. Orna. Hort.*, **5**(1): 45-46.

Tomlinson, P.B. 1959. An anatomical approach to the classification of the Musaceae. *J. Linn. Soc. (Bot.)*, **55**: 779-809.

Ulisses, C., Camara, T., Willadino, L. and Albuquerque, C. 2007. Induction of somatic embryogenesis in *Heliconia chartacea* Lane ex Barreiros cv. Sexy pink and cv. Sexy Scarlet from sections of ovaries. *Revista Brasileira de Ciencias Agrarias*, **2**(4): 281-284.

Uriarte, M.A., Da Silva, M.T.B., Rubim, P., Johnson, E. and Bruna, E.M. 2011. Disentangling the drivers of reduced long-distance seed dispersal by birds in an experimentally fragmented landscape. *Ecology,* **92**(4): 924-937.

Williams, K.J., Kress, W.J. and Prince, L.M. 2004. A new classification of the gingers (Zingiberaceae) based on DNA evidence. Bull. Heliconia Soc. International, Miami, Florida, **11**: 3-9.

Winkler, H. 1930. Musaceae. *In*: A. Engler., and K. Prantl (eds). Die Natürlichen Pflanzenfamilien, ed. 2, Bd. **15**: 505-541.

Zeng, SongJun., Guo, Shao Cong., Wu, Kun Lin., Zhang, YiQi., Chen, GuoHua and Duan Jun. 2004. *In vitro* propagation of heliconia plants. *J. Tropical and Subtropical Bot*., **12**(2): 153-158.

Chapter – 14

Kangaroo Paws

Kangaroo Paws (*Anigozanthos* spp.) belongs to family Haemodoraceae. It is commonly known as Sword-like or Iris-like. The word Anigozanthos was coined by French botanist Jacques de La Billardiere. This is derived from Greek words "*anoigo*" meaning to "expand or unequal" and "*anthos*" meaning "flower or irregular corolla". It can be used as cut flower, pot plant, landscape plants or dry flower (Gui *et al*., 2005). The kangaroo paws is gaining popularity among gardeners, horticulturists, nursery man also in world market due to its beautiful, unique, colourful pattern of flowers and also being perennial in nature.

ORIGIN AND DISTRIBUTION

It is originated in Australia. Louden's Hortus Britanicus, A Catalog of a Plants Endemic, Cultivated or Introduced to Britain, written in 1830, states that in 1808 *Anigozanthos manglesii* was introduced by Robert Mangles who had possibly received seed from his brothers in New Holland. *Anigozanthos rufus* was introduced into cultivation in 1824. The Royal Horticultural Society Dictionary of Gardening dates of the Botanical Magazine Plates are listed as *A. flavida* 1808 (BM 1151), *A. manglesii* 1833 BR, *A. pulcherrimus* 1844 (BM 4507). The first report of introduction of 8-10 species of *Anigozanthos* under green house from Australia in USA was reported in 1902 edition of Bailey's Cyclopedia of American Horticulture. The first hybrid Bush Gems was introduced from Australia by Mr. Meyer, he had also made good selection of *A. rufus and A. pulcherrimus* these were propagated by Oglesby tissue culture lab. Many of them were released for sale at the Huntington Gardens annual plant sale in 1985.

All species of Kangaroo paw are restricted in distribution to the southern areas of Western Australia. *Anigozanthos flavidus* occurs in the extreme south west of Western Australia growing from Augusta east to Two Peoples Bay and north to Waroona. It can be found on roadsides, river banks, swamps, shallow water and in eucalypt forests, tolerating light shade. The species *A. manglesii* is widely distributed throughout the Southwest Australian bio geographic regions, preferring white, yellow or grey sand or sandy loam. The occurrence is confirmed in the northernmost part of its range in the Geraldton Sandplains and the Swan Coastal Plain near Perth. Extensive records also exist of occurrence in Jarrah Forest and Warren region to the south but not reaching the southern coasts. It extends inland to the Avon Wheatbelt.

SPECIES

The taxonomy of Kangaroo Paws is problematic and has been the subject of numerous studies (Hopper and Campbell, 1977, Hopper, 1980, Anderberg and Eldenas, 1991, Simpson, 1990 and Hopper *et al.*, 1999, 2009). In particular, the validity of Macropidia as a monotypic genus has been discussed in detail with some authors supporting its monotypic status (Hopper and Campbell 1977, Hopper 1980 and Hopper *et al.*, 1999, 2009) while others have argued in favour of it being lumped within *Anigozanthos* (Simpson, 1990 and Anderberg and Eldenas, 1991). Relationships between closely related taxa within *Anigozanthos* are also unclear (Hopper, 1980). Contributing to this taxonomic obscurity is the ease with which some species within *Anigozanthos hybridise* (Hopper 1977, 1977a, 1978, 1980, Hopper and Burbidge 1978 and Shchori *et al.*, 1995).

Botanically the Haemodoraceae is divided into two subfamilies, the Haemodoroideae and the Conostylidoideae, both of which contain Australian species. The Conostylidoideae are best known horticulturally as this subfamily includes the famed kangaroo paws of the genus *Anigozanthos*. The genus comprises of 11 species restricted to southwest corner of Western Australia. Geerinck (1968) and Robertson (1976) noted the Haemodoraceae as uncertain, different combination of tribes and genera first time Robert Brown (1810) recognized the Haemodoraceae as a formal taxonomic unit composed of three south African genera, *Dilatris*, *Lanaria* and *Wachendorfia* and four Australian genera *Anigozanthos, Conostylis, Haemodorum* and *Phlebocarya* (Simpson, 1990). The Table 1 indicates the Australian members of the family and it comprises of 13 genera (Griesbach, 1990).

Table 1: Genus, number of species and its distribution.

Genus	No. of Species	Common Name	Distribution
Subfamily Haemodoroideae			
Haemodorum	20	Bloodroots	South-western, northern and eastern Australia, including Tasmania and New Guinea
Subfamily Conostylidoideae			
Anigozanthos	11	Kangaroo paws/catspaws	Southwest Western Australia
Blancoa	1	Red bugle	Southwest Western Australia
Conostylis	46	Cone flowers	Southwest Western Australia
Macropidia	1	Black kangaroo paw	Southwest Western Australia
Phlebocarya	3	None	Southwest Western Australia
Tribonanthes	5	None	Southwest Western Australia

Many of the most successful cultivars grown at the gardens are hybrids which have *Anigozanthos flavidus* as a parent. The following kangaroo paws species are grown in most temperate parts of Australia and most have been successfully cultivated at the gardens (Table 2).

Table 2: Important species of *Anigozanthos* with characteristics.

S. No.	Species	Characteristics
Branched Inflorescence		
1.	*A. flavidus* - Tall Kangaroo Paw	Most common of all of the Kangaroo Paws in cultivation. It is also the toughest, able to take wet conditions, heavier soils, full sun to light shade and is resistant to disease. It is 183 cm tall and flowering in a range of colours from yellow, pink to red.
2.	*A. pulcherrimus* Golden Kangaroo Paws	Attractive species with brilliant golden yellow flowers on reddish flower stems. Requires very good drainage and full sun. This species is very susceptible to ink spot but is a good co-parent to *A. flavidus* in producing showy hardy yellow strains.
3.	*A. onycis* Branched Cat's Paw	A low growing kangaroo paws that has only recently been described and is virtually unknown in cultivation. The specific name onycis is from the Greek word onyx which means claw and refers to the claw-like appearance of the open flowers. The flowers can be red or yellow suffused with red and differs from *A. humilis* by having a branched inflorescence. It is noted to be one of the most floriferous species in the wild.
4.	*A. rufus* Red Kangaroo Paw	Showy species with deep red flowers. It can tolerates moderate frosts and a wider range of soils than other kangaroo paws. Its susceptibility to Ink disease and its attractiveness to snails has been noted in Australia.

(*Contd.*)

(*Contd. Table* 2)

S. No.	Species	Characteristics
Unbranched Inflorescence		
5.	*A. bicolor* Little Kangaroo Paw	Similar to *A. manglesii* but with smaller stature and flowers. Tolerates winter wet moisture if drainage is adequate. This species can be severely damaged by frost. It has been used to create Emerald Gem, Bush Opal and Bush Nugget.
6.	*A. grabrielae*	A species not in cultivation in the U.S. and was considered up until 20 years ago a subspecies of *bicolor* (*A. bicolor* var. *minor*). As this species is even smaller than the little kangaroo paw it is mostly suited to the rock garden. It reportedly has a propensity to flower and can be a very attractive sight.
7.	*A. humilis* Cat's Paw	An attractive, small kangaroo paw, flowers are yellow to orange. Although it has proven to be short lived in cultivation, *A. humilis* has been an essential ingredient in the Bush Gem plants, Bush Nugget, Bush Gold, Bush Glow and Bush Flame.
8.	*A. kalbarriensis*	Very attractive small red flowering plant which has close ties to *A. humilis*. It has proven difficult to propagate and grow, which has led to its absence in cultivation.
9.	*A. manglesii* Red and Green Kangaroo Paw	It is the floral emblem of Western Australia. The combination of red hairs that clothe the stems and flower bases with the green perianth and reflexed petals makes for a slightly unreal effect. Although susceptible to the dreaded Ink disease, this plant's life span can be increased by good cultural practices such as good drainage, good air circulation, drip irrigation, snail and slug control, and clump division every 2-3 years. *A. manglesii* is a parent to Bush Emerald and Emerald Gem.
10.	*A. preissii* Albany's Cat's Paw	This Kangaroo Paw has the largest individual flowers of the genus, with the orange to red flowers held above the foliage. Although more tolerant of wetter soils than most other kangaroo paws and listed as long lived in the wild, it is listed as difficult to maintain in cultivation. Its slender upright flower stems, with the flowers clustered toward the top has made it a good parent for the cut flower variety Regal Claw.
11.	*A. viridis* Green Kangaroo Paw	Small, emerald-green flowering kangaroo paw with fine deep green leaves. It has been used in hybridizing work to create Bush Magic and Bush Emerald.

FLOWER MORPHOLOGY AND BIOLOGY

It is a rhizomatous herbaceaous perennial with flat strap-shaped leaves that folded at the midrib. The rhizome makes plant fire and drought resistant as it is able to re-sprout once conditions improve. It is clumping plants occasionally reaching 1 metre wide and high, but usually much smaller. The leaves arise from an underground stem, known as a rhizome and flowers occur in terminal clusters on stems arising from the base of the leaves. These flowering stems may be shorter or longer than the leaves, depending on the genus. Flowers have little or no fragrance. The electron microscopy study of the flower morphology and initiation of flowering of two species *A. manglesii* and *A. flavidus* (Motum and Goodwin, 1987) and it was found that in both the species the first signs of transition to flowering are swelling of the apical dome, a change in leaf phyllotaxy from alternate to radial and commencement of rachis elongation. In branched inflorescences, pedicels are initiated spirally and acropetally. The spike of kangaroo paws is composed of flowers in a sequence of diminishing size formed because one flower primordium continues to enlarge and differentiate organs whilst the other divides again to form the next flower until the specific number have been formed. Floral organs are initiated in alternate fashion in whorls on a radially symmetric disc from the outer edge towards the centre. Elongation of the floral tube and zygomorphy take place late in the floral process.

Gathe and Watson (2008) described the morphology and biology of kangaroo paw's flowers. Flowers aggregated in inflorescences in racemes or in spikes. The terminal inflorescence unit to cymose or racemose. Inflorescences scapiflorous, terminal, scape simple or forked or paniculate, inflorescence pubescent, few or many flowered. Flowers are subsessile, very irregular, zygomorphic, 3 merous, cyclic, tricyclic or tetracyclic. Perigone tube present. Perianth of tepals are 6 in 1 whorl, 1 -whorled are isomerous, petaloid, similar in the two whorls, green, or yellow, or orange, or red, or black. Androecial members adnate; all equal; free of one another; 1-3 -whorled. Androecium exclusively of fertile stamens. Stamens 6 in 1-3 series all more or less similar in shape; isomerous with the perianth, or diplostemonous; near the base of the perianth lobes; alterniperianthial. Filaments appendiculate, or not appendiculate. Anthers basifixed; versatile, or non-versatile; dehiscing via longitudinal slits; introrse; appendaged or unappendaged. Gynoecium 3 carpelled. The pistil 3 celled. Gynoecium syncarpous; eu-syncarpous; partly inferior, or inferior. Ovary plurilocular; 3 locular. Gynoecium stylate. Styles 1, when 3 they are partially joined; attenuate from the ovary, or from a depression at the top of the ovary; apical; much longer than the ovary at anthesis. Stigmas 1; capitate. Placentation axile. Ovules 2-100 per locule, non-arillate, orthotropous to hemianatropous. Fruits are non-fleshy, dehiscent and capsule. Capsules are loculicidal and fruits are 3 celled. Seeds copiously endospermic. Endosperm oily. Seeds are winged, or wingless. Number of cotyledons are l and testa without phytomelan.

Rowley and Rowley (1996) studied the pollen wall structure in *Anigozanthos viridis*. The exine of *Anigozanthos viridis* pollen is divided into two zones by a plane which has been referred to as a "commissural line" or a junction plane. Following early stages there is a pronounced reduction in thickness of the inner exine, probably due to stretching of the exine without the addition of new material as development progresses. The fibrillar zone under the exine in microspores is described as oncus-like. Since this onciform zone precedes microspore mitosis and intine formation, the fibrillar zone is divided into two layers in mature and is not a two-layered intine.

Mode of pollination

The flower is pollinated by birds but there are no bird pollinated plants in Europe and the hummingbird is the only pollinating bird in the US. But in Australia, about 17% of the plants are pollinated by birds ranging from the tiny honey eaters to the fairly large wattle bird. The bird lands on the kangaroo paw stem and seeks nectar inside the tubular flower. When it does the five stamens that make up the "paw" of the flower smack the bird on top of the head leaving pollen behind. Interestingly, each species of Kangaroo Paw hits the bird in a slightly different place. This does not prevent cross pollination as there are other characteristics that do that but it does help conserve pollen by not wasting it where it cannot be utilized.

The main pollination vectors used by Australian members of the Haemodoraceae are birds, small marsupials and insects. Often the flower structure of particular species gives a guide to the particular pollinators for that species. For example, a kangaroo paw flower is long, narrow and tubular-shaped with protruding, pollen-bearing structures and a protruding pollen-accepting structure. These seem ideally designed for low-beaked birds which seek out the nectar deep from within the flower and at the same time are either dabbed with pollen from the anthers or deposit pollen from another flower on the stigma. Kangaroo paws are also often pollinated by small nectar-feeding marsupials such as the small honey possum. On the other hand, the open flowers of *Conostylis* species seem less likely to require such specialised pollination. Observations have confirmed that these are generally insect pollinated. The plant is bird pollinated as flower are tubular and split in 6 lobes lengthwise which resembles claw shape, the flowers are brilliantly coloured and in few species hair like structure is present on flower may be of same colour or contrasting, hair may cause irritation in skin and eye when come in contact.

Anthesis and pollen viability

Flower inducation requires short days therefore flower scapes start appearing in late autumn. Low temperatures also influence induction and flower formation. Flowers start to open from mid winter, a flower branch can be picked with closed or open flowers.

The viability of *A. manglesii* pollen was estimated (Sukhvibul and Considine, 1993) by germination *in vitro* and by fluorescence and the more reliable method was then used to assess the effect of storage conditions on pollen longevity. Pollen germination was a more accurate test than staining for pollen of low and intermediate viability. Effective storage could be achieved in a number of ways depending on the duration required. Dehydrated pollen retained viability longer in storage than did hydrated pollen. Pollen from intact anthers could be stored for up to 5 weeks at room temperature with 50% or better germination provided it was stored on silica gel. Pollen isolated from anthers and stored at 2° C had reduced viability in comparison with that stored at room temperature or at -80º C. Pollen stored at -196º C lost only 5% viability after 20 weeks and it was concluded that this would be the most suitable method for long term storage for breeding or conservation purposes.

GENETICS

The chromosome number of kangaroo paw is 2n=12. The haploid number is n= 6 of *A. flavidus* (Stenar, 1927, Green, 1961 and Simpson, 1990), *A. humilis*, *A. manglesii*, *A. viridis* and *A. bicolour* (Green, 1961 and Simpson, 1990). Cytology n = 4-8, 11, 14, 16, 21, 28 (*Conostylidoideae*) and n = 12, 15, 19-21, 24 (*Haemodoroideae*). DNA is 5 bp insertion absent from the plastid gene *mat*K.

Inheritance of flower colour

The variance in colour tunes and the colour combinations of ovary and tube gives the breeder a large range of combinations. The use of plants of the same colour and different origins might show a segregation in offspring colour. This might be the result of different complementary genes that are responsible for different pathway deficiencies in synthesis of anthocyanin (Shchori and Kochba, 1998).

Self incompatibility

A. manglesii is an outcross and self incompatible plant (Hopper 1987). Obtaining homozygous purelines by selfing is impossible due to self incompatibility. Homozygosity can be achieved by sib crosses only. Sib cross fertility might be limited due to their high self incompatibility and inbreeding depression. For the hand crosses efficiency self incompatibility is helpful since no emasculation is needed. Seed yield of interspecific crosses depend on their genetic distance (Shchori *et al*., 1995). Within the *A. manglesii* species fertility might be the result of self incompatible gene combination.

BREEDING OBJECTIVES

i. Improved pot and garden performance.

ii. Seed propagated hybrid varieties.

iii. High flower quality.

iv. Increased disease resistance particularly to ink disease.

v. Better natural branching.

vi. Repeat flowering nature.

vii. Neater upright foliage.

viii. Attractive flower colours.

ix. Earliness in flowering then repeat flower.
x. Selection of tougher plants that require little maintenance and less water.

BREEDING METHODS

Selection

In breeding programme of *Anigozanthos* first step is selection from wild population. Though selection parents were choosen for production of hybrids of *A. manglesii* Fira Vol., Ohir Vol., Shani Vol., and Chen Vol (Shchori *et al.*, 1997). Selective Breeding has produced some exceptional Kangaroo Paws that thrive in different climatic conditions like the 'Bush' series. 'Bush Eclipse' kangaroo paws produced by tissue culture, 'Bush Eclipse' is a selected form of *Macropidia fuliginosa* and is from the Bush Gem series. The evergreen species *Anigozanthos flavidus* is the longest lived, most vigorous and most disease resistant of all other *Anigozanthos*. Plant breeders refined this promising species by collecting its seed from the wild, growing it, then selecting and cross pollinating the best of those seedlings. By repeating this process again and again, they eventually created several very magnificent cultivars of *A. flavidus*. *Anigozanthos flavidus* also became the main species used for hybridizing with other more difficult to grow species. This created stronger growing kangaroo paws hybrids with a wider range of colours and sizes. Some varieties develop through selection (Table 3) such as *A. flavidus* 'Red' and 'Pink' ('Pink Joey' in the trade) are available and the species has been used to create several hybrids.

Table 3: Varieties developed through selection.

S. No.	Varieties	Selection from	Characteristics
1.	*A. flavidus* 'Red'	*A. flavidus*	Plants tall 1.8 m. Flower red, resistant to diseases.
2.	*A. flavidus* 'Pink' ('Pink Joey' trade name	*A. flavidus*	Plants tall, grown in full sun to light shade and is resistant to disease. Flower pink.
3.	Lilac Queen (Anon., 2009a)	Open pollination *A. flavidus*	Plant height is tall to very tall, ramification present, leaf attitude spreading, Perianth tube colour is green, Perianth lobes are reflexing and weak and time of beginning of flowering is medium.

Hybridization

Kangaroo paws are relatively easy to hybridize and depending on the parent species can set more than 40 seeds per flower head. The pollen is easily accessible and presented at the end of the flower on the anthers close to the stigma. In some species such as *A. gabrielae* care must be taken when pollinating that self pollen is not transferred to the same flower. This risk can be overcome by removing the anthers of a flower prior to its pollen being released. Primary hybrids between all the *Anigozanthos* species have been produced as well as a few secondary and tertiary hybrid (Griesbach, 1990). Natural hybridization occurs in the *Anigozanthos* genus between most of the species in different rates of fertility. Fertility in different interspecific and intraspecific crosses of *A. manglesii*, produced by hand pollination was measured (Shchori *et al.*, 1995) by the rate of seed setting of the crosses and by seed yield/cross. Seed yield ranged from 0 to 280 seeds/cross, with the highest yielding combination being *A. viridis* × *A. manglesii.* In sub genus Haplanthesis most of the interspecific crosses are fertile and easily obtained. Some other interspecific crosses are obtained with more difficulty but still enable gene exchange (Hopper, 1987). The gene exchange can be obtained in some of the species by a direct cross or by an indirect cross using a more easily crossed species as an intermediate (Shchori *et al.*, 1995). The use of embryo rescue technique might be a helpful tool in hardly achieved combinations. There are several varieties developed through hybridization (Table 4).

Table 4: Varieties developed through hybridization in kangaroo paw.

S. No.	Varieties	Parentage	Characteristics
1.	Big Red	[*A. Manglesii* × *A. flavidus* (4n)] × [*A. rufus* × *A. flavidus*]	Bright red, evergreen perennial, 122-183 cm in height, good disease resistance. Developed by Keith Oliver at Lakkari native plant nursery.
2.	Bush Dawn	*A. flavidus* × *A. pulcherrimus*	Yellow flower, evergreen perennial, 122-183cm in height. Hybridized by late Merv Turner.
3.	Bush Devil	Bush Ranger 4n hybrid	Red flower, evergreen perennial, 91-122 cm height.
4.	Bush Games	*A. manglesii* × *A. Flavidus* × *A. Viridis* × *A. humilis*	Red green flower, evergreen perennial, 122-152 cm in height.
5.	Bush Garnet	*A. perissii* × *A.* 'Bush Twilight'	Bear burgundy red inflorescence, plant is about 30-91 cm in height.
6.	Bush Gold	*A. flavidus* × *A. humilis*	Golden yellow flower, height 61-91 cm. Developed by Angus Stewart.
7.	Bush Lantern	*A. humilis* × *A. bicolour* × *A. flavidus*	Yellow fuzzy, dwarf flowers, narrow linear light green foliage.
8.	Bush Pearl	*A. humilis* × *A. Bicolour* × *A. flavidus*	Pink fuzzy, dwarf flower, perennial, about 30-60 cm height, narrow green linear leaves.
9.	Bush Ranger	*A. humilis* × *A. flavidus*	Red flower, dwarf. Evergreen perennial, 30-60 cm height.
10.	Bush Sunset	*A. flavidus* × *A. rufus*	Dark red flower, evergreen perennial, 122-152 cm height.
11.	Coral Pink	*A. flavidus* × *A. humilis*	Coral pink flower. Evergreen perennial, 30-91 cm height.
12.	Gold Velvet	(*A. humilis* × *A. flavidus*) × (*A. Pulcherrimus* × *A. flavidus*)	Golden large flower, evergreen perennial, 122-183 cm height. Developed by Keith Oliver and first released by Lullfitz Nursery.

(*Contd.*)

(*Contd. Table 4*)

S. No.	Varieties	Parentage	Characteristics
13.	Harmony	*A. flavidus* × *A. pulcherrimus*	Red-yellow flower, evergreen perennial, 122-183 cm height.
14.	Kanga Burgundy (Rambueleg)	*A. flavidus* 'H80' × *A. flavidus*	Red orange flower, evergreen perennial.
15.	Orange Cross	*A. preissi* × *A. flavidus* hybrid	Orange flower, evergreen perennial.
16.	Pink Joey	*A. flavidus* hybrid	Dwarf pink flower, evergreen perennial, 61-91cm height.
17.	Red Cross	*A. rufus* × *A. flavidus*	Dark burgundy flower, evergreen perennial, 122-183 cm height.
18.	Regal Claw Hopper hybrid	*A. flavidus* × *A. preisii*	Orange red large flower, evergreen perennial, 122-152 cm height.
19.	Dwarf Delight Hopper hybrid	*A. onycis* × *A. flavidus*	Red flower buds which turn orange and apricot at flower opening. Multiple branching inflorescence. 61-91 cm height. Pollens are not completely sterile therefore has potential in future breeding.
20.	Red Cross Hopper hybrid	*A. rufus* × *A. flavidus*	Multi-branched stems with rich burgundy colouration on both flower and stem.
21.	Bush Baby	*A. humilis* × *A. flavidus*	Pale orange flower that are tinged burgundy at the tips. Branched flower stems, plant height 46 cm suitable for rock garden.
22.	Bush Emerald	*A. manglesii* × *A. viridis*	Emerald green flower, unbranched red stems, foliage grey-green colour.
23.	Bush Dawn	*A. pulcherrimus* × *A. flavidus*	Bright yellow flower tinted green, good cut flower.
24.	Bush Glow	*A. humilis* × *A. flavidus*	Double ranked golden flower with red bases along a bronzy-red stem.
25.	Bush Nugget	*A. humilis* × *A. Flavidus* × *A. humilis* × *A. bicolor*	Red stem with orange flowers.
26.	Bush Ranger	*A. humilis* × *A. flavidus*	Bright red flower on branched, plant height 46 cm.
27.	Autumn Mystery	*A. manglesii* × *A. flavidus*	Flowers and stems usually hairy - clumping plant. Plant is 1 m tall, Flowers having red tip yellow bases.
28.	Autumn Sunrise	*A. pulcherrimus* × *A. flavidus*	Flowers red yellow and green Flowers and stems usually hairy - clumping plant. Plant is 1.75 m tall.
29.	KLEAC11213 (Anon., 2012c)	Rambubona × Bush Emerald	Plant height is medium, medium to strong leaf glaucosity, yellow colour perianth tube and very late flowering.
30.	KLEAC11212 (Anon., 2012d)	Bush Aura × Bush Emerald	Plant height medium, very few numbers of inflorescence, very weak leaf glaucosity and green colour perianth tube.
31.	KLEAC11211 (Anon., 2012e)	Rambubona × Bush Reverly	Short to medium plant height, very weak leaf glaucosity and yellow colour perianth tube.

(*Contd.*)

(*Contd. Table 4*)

S. No.	Varieties	Parentage	Characteristics
32.	Amber Velvet (Anon., 2011a)	(*A. humilis* × *A. flavidus*) × (*A. pulcherrimus* × *A. flavidus*)	Plant height medium, six numbers of anthers at top of perianth, early flowering.
33.	Gold Velvet (Anon.,2011)	(*A. humilis* × *A.flavidus*) × (*A. pulcherrimus* × *A. flavidus*)	Plant height is medium, leaf length medium to long and early flowering.
34.	Rambudan (Anon., 2009b)	H0061 × Emerald Gem	Plant height short to medium, semi-erect leaf, ramification absent and Perianth lobes reflexing very strong.
35.	Rambubona (Anon., 2009)	190/1×150/1-3	Plant height short, leaf semi-erect, ramification present, Perianth tube predominant colour yellow, Perianth lobes reflexing medium and mid season flowering.
36.	Regal Velvet (Anon., 2008)	*A. manglesii* × *A. flavidus*	Plant height medium, semi-erect leaf, inflorescence ramification present. Perianth tube colour green, and perianth lobes reflexing strong.
37.	Rambueleg (Anon., 2012f)	H80 × Flashpoint	Plant height short, ramification present, perianth tube colour red, perianth lobes reflexing weak and very early flowering.
38.	Ramboramp	believed to be *A. humilis* × *A. flavidus* hybrid	Plant height medium, semi-erect leaf, ramification present, perianth tube colour orange-red, perianth lobes reflexing medium and very early flowering.
39.	Rambozazz (Anon., 2012g)	Joey Lipstick × A02-1744	Plant height short, leaf attitude semi-erect, ramification present, perianth lobes reflexing weak and very early flowering.
40.	Ramboblitz (Anon., 2012h)	A02-1697(Serenity) × A02-1686	Plant height short to medium, leaf erect, ramification present. Perianth tube orange-red, Perianth lobes reflexing very weak to weak and early flowering.
41.	Ramboball (Anon., 2012)	Unknown	Plant height short, semi-erect leaf, ramification present, perianth tube colour red, lobes reflexing weak to medium. Early flowering.
42.	Bush Spark (Anon., 2012a)	H35 × *A. flavidus*	Plant height medium short to medium, semi-erect leaf. Perianth tube colour red, lobes medium weak. Early flowering.
43.	Bush Inferno (Anon., 2012b)	H35 × *A. flavidus*	Plant height short to medium, leaf semi-erect, ramification present, perianth tube colour red, lobes reflexing weak to medium. Early flowering.

Breeding for hardiness

Haemodoraceae have the ability to hybridize readily. Numerous hybrids of the genus *Anigozanthos* have become commercially available in the last 15 or so years. The main aim of hybridization is to combine the hardiness of *A. flavidus* with the showiness of some of the other species, including the famed *A. manglesii* in the form of the hardier Bush Emerald. When selecting kangaroo paw hybrids the best indicator of hardiness lies in the leaf type. Those hybrids with a leathery glossy, mid-green leaf generally have *A. flavidus* as one parent which generally makes them a much hardier garden plant. Hybrids with a grey green leaf are usually derived from less hardy species and generally will only thrive for one or two seasons in the garden.

Table 5: Hardy *Anigozanthos* hybrids and cultivars based on *A. flavidus*.

S.No.	Cultivar or Hybrid	Parentage	Characteristics
1.	*A. flavidus* forms	*A. flavidus*	Greenish-yellow flowers, tall stem, hardy. A number of more attractive colour forms are available (red, orange, pink).
2.	Big Red	(*A. manglesii* × *A. flavidus*) × (*A. rufus* × *A. humilis*)	Deep red flower, moderately vigorous plant about 1 - 1.5 m.
3.	Bush Haze	*A. flavidus* × *A. pulcherrimus*	Yellow flowers suffused with red on stems to about 1 - 1.5 m. Moderately vigorous.
4.	Bush Ranger	*A. flavidus* × *A. humilis*	Bright red flowers on stems to about 0.5 m. One of the hardiest of small-growing cultivars.
5.	Gold Fever	*A. flavidus* × *A. pulcherrimus*	Moderately vigorous plant with yellow/orange flowers on stems to about 1 - 1.5 metres.
6.	Pink Joey	Probably a form of *A. flavidus*	Bright pink flowers on stems to about 0.5 metres.
7.	Red Cross	*A. flavidus* × *A. rufus*	Deep red flower, moderately vigorous plant about 1 metre.
8.	Yellow Gem	*A. flavidus* × *A. pulcherrimus*	Bright yellow flower, moderately vigorous plant about 1 - 1.5 metres.

Development of hybrid kangaroo paws in Australia

Hopper hybrids

In the mid 1970's, Dr. Stephen Hopper began crossing different species of *Anigozanthos* at the Western Australian Wildlife Research Centre in Wanneroo, WA. Primary interest was to study the phylogenetic relationships between the different species. Some of these crosses turned out to have horticultural value and were tissue cultured by Dr. Jenny McComb, of Murdoch University. They came into cultivation in the US in the early 1980's because of their value in the cut flower industry. Although some of these plants trickled into the nursery trade. With the ability to mass produce *Anigozanthos* through tissue culture, Hoppers three hybrids, *A.* 'Red Cross', *A.* 'Dwarf Delight' and *A.* 'Regal Claw' were soon spread around to many nurseries in southern California.

Oliver hybrids

In the mid 1980's, work was being done on the hybridizing of *Anigozanthos* for the horticultural industry. By this time there were quite a few Kangaroo Paws on the market

Keith Oliver, at Lakkari Native Plant Nursery was working on a breeding programme to produce good quality ornamental plants. Several hybrids were produced in the Australian plant market including *A.* 'Early Spring', *A.* 'Big Red', *A.* 'Emerald Glow' and *A.* 'Harmony'. All of these plants showed hybrid vigour but unfortunately were still susceptible to Ink Disease.

Resistance breeding

Professor Merv Turner of Monbulk, Victoria embarked on a programme to cross different species and then test their susceptibility to Ink disease and develop a hybrid after 10 years first by careful selection and breeding and then by exposing young plants to diseased stock. In March 1986, they were the first hybrids to be bred for disease resistance and were called the Bush Gems. These hybrids are recognizable for the name "Bush" or "Gem" in their cultivar name.

Selection for disease tolerance and compact forms is likely to require several generations to achieve more desirable outcomes as the species such as *A. humillis* or *A. gabrielae* that have the best compact forms are also the most susceptible to diseases such as ink spot and rust. Crossing for disease resistance is shown by hybridising a selection of *A. manelesii* with ink spot tolerance with a selection of rust tolerant *A. viridis* 'Green Dragon'. This cross has resulted in a range of hybrids with varying tolerance for each disease. Those hybrids showing the greatest disease tolerance are then selected for further breeding with the end result being a hybrid or hybrids that have multi-gene expression of the characteristic being sought.

Mutation

Ornamental plants appear to be ideal systems for mutation breeding as many characters of economic interest, i.e. flower traits or the growth habit are easily monitored after mutagenic treatment. Furthermore, many ornamental species are heterozygous and often propagated vegetatively, thus allowing the detection, selection and conservation of mutants within the M_1-generation. Appropriate strategies in mutation induction as application of fractionated doses and recurrent irradiation combined with *in vitro* culture techniques may lead to rapid success also in homozygous, polyploid species. Spontaneous mutations and unusual genetic combinations in the production of ornamental *Anigozanthos* are described. The unusual hybrids include *A. rufus* × *A. pulcherrimus* and *A. pulcherrimus* × *A. rufus*. A green and yellow sport of the usually red and yellow *A. manglesii* and its possible multiplication using seed and *in vitro* culture are described (Beng, 1998).

Some mutant varieties either through sport or spontaneous mutation has been evolved (Table 6).

Table 6: Varieties developed through mutation.

S. No.	Varieties	Type of mutation	Characteristics
1.	Bush Diamond	Bush Pearl sport	Red Green flower. Evergreen perennial, 60-91 cm height.
2.	Bush Ember (Anon., 1995a)	Spontaneous mutation in Bush Ranger	Plant height averages 266.6 mm, produces 30-31inflorescences/plant, 12 flowers/in florescence, free flowering and compact growth habit.
3.	Bush Splendour (Anon., 1995)	Spontaneous mutation in Bush Ranger	Plant height 279 mm, produce upto 20 inflorescences/plant and 11-12 flowers/ inflorescence, free flowering and compact growth habit.
4.	Sunglow (Anon., 1996)	Spontaneous mutation in the orange-flowered cross *A. pulcherrimus* × *A. rufus*,	Yellow flower, tall growing, upright, perennial, 8-9 flowers/inflorescence.
5.	Rambodiam (Anon., 2012i)	Spontaneous mutation in A02-1769	Plant height short, ramification present, perianth tube colour greenish with reflexing medium. Early flowering.

Polyploidy

There are several interspecific hybrids with improved flower colours, heat tolerance and growth habit. These hybrids are sterile due to divergent evolution of the parent species. Colchicine was used to double the chromosome number of an important sterile hybrid from *A. humilis* × *A. flavidus*. This hybrid is ever blooming, dwarf, and heat tolerant. The resulting allotetraploid is fertile and progeny are now being evaluated (Griesbach, 1990).

BIOTECHNOLOGY

Tissue culture

It has become one of the most promising methods of propagation for many plants of Anigozanthos. Many of the new hybrids becoming available have been propagated by tissue culture both in Australia and by companies in the United States. The key technologies of rapidly propagating Anigozanthos such as the induction of explants and screening of enrichment media as well as rooting, acclimatizing and transplanting were studied and a system of rapidly propagating Anigozanthos through tissue culture was set up, which made it possible extensively growing Anigozanthos seedlings through tissue culture and rapid propagation (Qian *et al.*, 2005).

Johnson, (1997) excised leaf segments from plants of *Anigozanthos* sp. *Blandfordia grandiflora* and *Thysanotus juncifolius* were cultured on Murashige and Skoog medium supplemented with picloram or 2,4-D at 5, 10 or 20 micro M and 3% sucrose. Initiated, nodule-like callus was transferred to medium containing picloram and BAP [benzyladenine], both at 8 micro M and incubated in the dark for 35 days. Embryogenic callus was induced on all tested media for the 3 genera. Callus initiation took 2 months for *Anigozanthos* and 4-6 months for the other 2 genera. Germination of somatic embryos on MS medium with BAP and/or 2iP was very slow and the rate of germination was very low. Cryostorage of shoot apices of the perennial monocotyledonous species *Anigozanthos viridis* subsp. *viridis* was investigated (Turner *et al.*, 2000) by using a modified vitrification protocol. The highest post-thaw survival of shoot apices (41.4%) involved preculturing shoot apices on 0.4 M sorbitol for 48 h followed by incubation in a vitrification cryoprotective solution (PVS2) for 25 min at 0°C. The level and type of cytokinin used in the culture stage was also found to influence cryostorage success with post-thaw survival decreasing from 41% with zero to low levels of cytokinins to below 5% for cytokinin levels that are typical of plant tissue applications (2.5 micro M) for Australian plant species. Five Haemodoraceae taxa (*Anigozanthos* spp. and *Conostylis* spp.) were successfully cryopreserved with this modified protocol; however, a sixth taxon, *Macropidia fuliginosa* [*A. fuliginosus*], remained unresponsive to this vitrification technique.

The effect of plant growth regulators on somatic embryogenesis was examined by Johnson (2000) in 3 Australian monocotyledonous species used as cut flowers or pot plants: *Anigozanthos* cv. Harmony, *Blandfordia grandiflora* cv. Queen Anne and *Thysanotus juncifolius*. Embryogenic callus was established. Cultures were maintained at 24 + or -2°C, 16 h light or in the dark, on Murashige and Skoog supplemented with 6-benzylaminopurine (BAP) and NAA. Nodular callus of *B. grandiflora* and *T. juncifolius* formed somatic embryos on MS supplemented with BAP at 0.5 micro M, and with BAP and NAA at 1 micro M. *Anigozanthos* embryogenic callus only formed roots only under conditions similar to those applied to the other 2 species.

The tissue culture of *Anigozanthos* sp. showed that the optimum sterilization of explants was obtained (Liu and Ding, 2003) in the treatment with 0.1% HgCl2 for 7-8 minutes. The concentrations of NAA and 6-BA [benzyladenine] affected the callus formation, shoot induction and rooting of plantlets. Increasing 6-BA concentration resulted in increased callus formation. The BA/NAA ratio determined the shoot induction. Half-strength MS medium was the best for rooting, while 6-BA treatment inhibited rooting. The best media for the different culture stages were: MS + 6-BA at 1 mg/litre + NAA at 0.1 mg/litre + charcoal at 3% for the callus stage; MS + 6-BA at 0.5 mg/litre + NAA at 0.5 mg/litre for the shoot induction stage; and half-strength MS + IBA at 0.2 mg/litre for the rooting stage.

Somatic mutagenesis

Trimble and Membrey (1991) developed the dwarf compact variety with medium green and upright leaves. It has a light green perianth tube and light yellow ovary, both covered in light yellow hairs. It arose as a sport in a population of tissue cultured bicentennial [(*A. humilis* × *A. bicolor*) × *A. flavidus*], which is taller, more spreading and has dark red hairs on its floral organs.

Molecular breeding

In an effort to resolve relations within the family Haemodoraceae, nucleotide sequence data for the *trn*L-F region of plastid DNA were analysed by Hopper *et al.* (1999). Results supported that Haemodoraceae are monophyletic and that there are two subfamilies Conostylidoideae and Haemodoroideae. It is unlikely that Haemodoraceae are of Gondwanan origin and the phylogenetic pattern indicates a largely relictual distribution with a recent radiation in Western Australia. In Conostylidoideae, *Tribonanthes* is supported as a monogeneric tribe but tribal status is not supported for *Phlebocarya. Macropidia* is sister to *Anigozanthos. Blancoa* and *Macropidia* should be recognized as distinct genera. The two clades in Haemodoroideae are *Dilatris* (*Lachnanthes* + *Haemodorum*) and Xiphidium (Schiekia + Wachendorfia).

Roux *et al.* (2010) constructed the first molecular phylogeny for all species of the Kangaroo Paw group. As previously determined by taxonomists working on herbarium specimens, naturalized populations were identified as *A. flavidus*. In addition, they also identified a second species, *A. rufus*. Relative genome size estimates for *Anigozanthos* species indicated that small inter-specific differences in genome sizes are positively correlated to hybrid fitness. *Anigozanthos flavidus* and *A. rufus* have relatively compatible genomes and may produce fertile hybrids under field conditions. However, for species whose genome size differ more than 30%, there is little inter-specific compatibility and consequently a very low risk of producing fertile hybrids.

REFERENCES

Anderberg, A.A. and Eldenas, P. 1991. A cladistic study of Anigozanthos and Macropidia. *Australian Systematic Bot.*, **4**: 655-64.

Anonymous, 1995a. Variety: 'Bush Ember'. Application no: 94/065. *Plant Varieties J.*, **8**(3): 15-16.

Anonymous, 1995. Variety: 'Bush Splendour'. Application no: 94/061. *Plant Varieties J.*, **8**(3): 18-19.

Anonymous, 1996. Variety: 'Sunglow'. Application no: 93/227. *Plant Varieties J.*, **9**(2): 26-27.

Anonymous, 2008. Kangaroo paw (*Anigozanthos* hybrid). Variety: 'Regal Velvet'. Application Number 2006/012. *Plant Varieties J.*, **20**(4): 195-96.

Anonymous, 2009a. Kangaroo paw (*Anigozanthos* hybrid). Variety: 'Lilac Queen'. Application Number 2004/262. *Plant Varieties J.*, **22**(2): 201-02.

Anonymous, 2009b. Kangaroo paw (*Anigozanthos* hybrid). Variety: 'Rambudan'. Application Number 2007/293. *Plant Varieties J.*, **23**(4): 139-41.

Anonymous, 2009. Kangaroo paw (*Anigozanthos* hybrid). Variety: 'Rambubona'. Application Number 2007/295. *Plant Varieties J.*, **23**(4): 142-43.

Anonymous, 2011a. Kangaroo paw (*Anigozanthos* hybrid). Variety: 'Amber Velvet'. Application Number 2005/047. *Plant Varieties J.*, **24**(1): 189-90.

Anonymous, 2011. Kangaroo paw (*Anigozanthos* hybrid). Variety: 'Gold Velvet'. Application Number 2005/048. *Plant Varieties J.*, **24**(1): 191-93.

Anonymous, 2012a. Kangaroo paw (*Anigozanthos* hybrid). Variety: 'Bush Spark'. Application Number 2004/139. *Plant Varieties J.*, **25**(2): 160-64.

Anonymous, 2012b. Kangaroo paw (*Anigozanthos* hybrid). Variety: 'Bush Inferno'. Application Number 2004/076. *Plant Varieties J.*, **25**(2): 165-68.

Anonymous, 2012c. Kangaroo paw (*Anigozanthos* hybrid). Variety: 'KLEAC11213'. Application Number 2011/269. *Plant Varieties J.*, **25**(4): 316-19.

Anonymous, 2012d. Kangaroo paw (*Anigozanthos* hybrid). Variety: 'KLEAC11212'. Application Number 2011/268. *Plant Varieties J.*, **25**(4): 320-22.

Anonymous, 2012e. Kangaroo paw (*Anigozanthos* hybrid). Variety: 'KLEAC11211'. Application Number 2011/267. *Plant Varieties J.*, **25**(4): 323-25.

Anonymous, 2012f. Kangaroo paw (*Anigozanthos* hybrid). Variety: 'Rambueleg'. Application Number 2007/294. *Plant Varieties J.*, **25**(2): 119-21.

Anonymous, 2012g. Kangaroo paw (*Anigozanthos* hybrid). Variety: 'Rambozazz'. Application Number 2010/040. *Plant Varieties J.*, **25**(2): 125-27.

Anonymous, 2012h. Kangaroo paw (*Anigozanthos* hybrid). Variety: 'Ramboblitz'. Application Number 2008/119. *Plant Varieties J.*, **25**(2): 128-30.

Anonymous, 2012i. Kangaroo paw (*Anigozanthos* hybrid). Variety: 'Rambodiam'. Application Number 2008/118. *Plant Varieties J.*, **25**(2): 131-33.

Anonymous, 2012. Kangaroo paw (*Anigozanthos* hybrid). Variety: 'Ramboball'. Application Number 2008/120. *Plant Varieties J.*, **25**(2): 134-36.

Beng, T. 1998. Kangaroo paws: not your average kangaroo paw. *Australian Plants*, **19**(154): 267-70.

Brown, R. 1810. Prodromus Florae Novae Hollandiae et Insulae Van-Diemen exhibens characteres plantarum quas annis. Taylor: London. pp. 1802-05.

Gathe, J. and Watson, L. 2008. *Anigozanthos* Labill. *Relation du Voyage a la Recherche de la Perouse*, **1**: 409-10.

Geerinck, D. 1968. Considerations taxonomiques au sujet des Haemodoraceae et de Hypoxidaceae (Monocotyledones). *Bull Soc. Roy. Bot. Belgique*, **101**: 265- 78.

Green, J.W. 1961. The genus Constylis. II. Taxonomy. *Proc.*: Linn. Soc. New South Wales **85**: 334-373.

Griesbach, R.J. 1990. A fertile tetraploid *Anigozanthos* hybrid produced by *in vitro* colchicine treatment. *HortScience*, **25**(7): 802-03.

Gui, M., Xiong, L., Mo, X.J., Nong, J. and Jian, H.Y. 2005. Introduction experiment of a new flower Kangaroo paws (Anigozanthos). *Southwest China J. Agril. Sci.*, **18**(3): 321-24.

Hopper, S.D .1977a. The reproductive capacity of *Anigozanthos manglesii* D. Don, *A. humilis* Lindl. and their hybrids in a wild population. *Aust J. Bot.*, **25**: 423-28.

Hopper, S.D. 1977. The structure and dynamics of a hybrid population of *Anigozanthos manglesii* D.Don and *A. humilis* Lindl. (Haemodoraceae). *Australian J. Bot.*, **25**: 413-22.

Hopper, S.D. 1978. Nomenclatural notes and new taxa in the *Conostylis aculeata* group (Haemodoraceae). *Nuytsia*, **2**: 254-64.

Hopper, S.D. 1980. A biosystematic study of the Kangaroo Paws, Anigozanthos and Macropidia (Haemodoraceae). *Australian J. Bot.*, **28**(6): 659-80.

Hopper, S.D. 1987. Blancoa, Anigozanthos, Macropidia. *Flora of Australia*, **45**: 110-28.

Hopper, S.D. and Burbidge, A.H. 1978. Assortative pollination by red wattle birds in a hybrid population of *Anigozanthos* Labill. (Haemodoraceae). – *Aust. J. Bot.*, **26**: 335-50.

Hopper, S.D. and Campbell, N.A. 1977. A multivariate morphometric study of species relationships in kangaroo paws (Anigozanthos Labill. And Macropidia Drumm. ex Harv. Haemodoraceae). *Australian J. Bot.*, **25**: 523-44.

Hopper, S.D., Fay, M.F., Rossetto, M. and Chase, M.W. 1999. A molecular phylogenetic analysis of the bloodroot and kangaroo paw family, Haemodoraceae: taxonomic, biogeographic and conservation implications. *Botanical J. Linnean Society*, **131**(3): 285-99.

Hopper, S.D., Smioth, R.J., Fay, M.F., Manning, J.C. and Chase, M.W. 2009. Molecular phylogenetics of Haemodoraceae in the Greater Cape and Southwest Australian floristic regions. *Mol. Phylogen. Evol.*, **51**: 19-30.

Johnson, K.A. 1997. Development of somatic embryogenesis for three Australian monocotyledonous plant genera. *Acta Horticulturae,* **447**: 531-33.

Johnson, K.A. 2000. Induction of somatic embryogenesis in Australian monocots. *Acta Horticulturae*, **520**: 67-74.

Liu, C. and Ding, M. 2003. The studies of the tissue culture of *Anigozanthos* (Kangaroo paws). *Acta Horticulturae Sinica*, **30**(1): 113-14.

Motum, G.J. and Goodwin, P.B. 1987. Floral initiation in Kangaroo paw (*Anigozanthus* spp.): A scanning electron microscope study. *Scientia Horticulturae*, **32**(1-2): 115-22.

Qian, X.W., Zhu, Q. and Li, B.Y. 2005. Rapid propagation of A*nigozanthos* seedling by tissue culture. *Acta Agriculturae Shanghai*, **21**(1): 1-3.

Robertson, K.R. 1976. The genera of Haemodoraceae in the Southeastern United States. *J. Arnold Arbor*, **57**: 205-16.

Roux, J.J., le Geerts, S., Ivey, P., Krauss, S., Richardson, D.M., Suda, J. and Wilson, J.R.U. 2010. Molecular systematics and ecology of invasive Kangaroo Paws in South Africa: management implications for a horticulturally important genus. *Biological Invasions*, **12**(12): 3989-4002.

Rowley, J.R. and Rowley, J.S. 1996. Pollen wall structure in *Anigozanthos viridis* (Haemodoraceae). *Acta Societatis Botanicorum Poloniae*, **65**(1/2): 83-90.

Shchori, Y. and Kochba, M. 1998. Breeding of red and green kangaroo paw (*Anigozanthos manglesii)*. *Acta Horticulturae*, **454**: 111-14.

Shchori, Y., Kochba, M. and Afgin, L. 1995. Hybrid fertility in different crosses of *Anigozanthos manglesii*. *Acta Horticulturae*, **420**: 147-49.

Shchori, Y., Kochba, M., Barzilay, A. and Afgin, L. 1997. Breeding of *Anigozanthos manglesii*. *Acta Horticulturae*, **453**: 99-100.

Simpson, M.G. 1990. Phylogeny and Classification of the Haemodoraceae. *Ann. Missouri Botanical Garden*, **77**(4): 722-84.

Stenar, H. 1927. Zur Entwicklungsgeschichte der Gat-tung Anigozanthos Labill. *Bot. Not.*, 1927: 104-14.

Sukhvibul, N. and Considine, J.A. 1993. Medium and long term storage of *Anigozanthos manglesii* (D. Don) pollen. *New Zealand J. Crop Hort. Sci.*, **21**(4): 343-47.

Trimble, R. and Membrey, S. 1991. Kangaroo paw (*Anigozanthos* hybrid). Variety: 'Lemon Whizz'. Application no. 90/099. *Plant Varieties J.*, **4**(3): 18.

Turner, S.R., Touchell, D.H., Dixon, K.W. and Tan, B. 2000. Cryopreservation of *Anigozanthos viridis* ssp. *viridis* and related taxa from the south-west of Western Australia. *Australian J. Bot.*, **48**(6): 739-44.

Chapter – 15

Lilium

Lilium is a popular genus of ornamental bulbous plants, belonging to family Liliaceae. Lilium is originated from a Greek word "*li*" meaning whiteness. It is an herbaceous perennial plant having scaly bulb. Despite the range of forms and their popularity as garden plants, there are two main groups of commercial importance as cut flowers i.e. Asiatic and Oriental Hybrids. They are the result of interspecific hybridization. However, intraspecific and interspecific hybridization face the hurdles like interspecific incompatibility, embryo abortion, degeneration of ovule, poor seed set and pollen sterility. There are several breeding methods and techniques which have been used to overcome these hurdles.

ORIGIN AND DISTRIBUTION

All species of *Lilium* are native to northern hemisphere. The northern limit extends up to Canada and Siberia, whereas in south, this extends up to Florida and the Nilgiri mountains of India. Majority of lilies known to cultivation are native to the northern hemisphere, chiefly South Canada, Siberia, Florida and India. It is distributed to east coast of Asia, the west coast of North America and the Mediterranean region. Several species i.e. *Lilium nepalensis, L. wallichianum* and *L. polyphyllum* are native of Himalayan region and *L. niligiriensis* have originated from Nilgiri hills. Few species have also been found in north east India. The Far East is the home of almost half of the world lilies and most garden lilies grown in the world have originated there.

The Assyrian monuments dating back to 1000 BC with sculptured lily forms is the first authentic record. The Palestinian coin makers were stamping lily on shekels before 143 BC. *Lilium longiflorum* popularly known as Easter lily became a florist flower which existed in the Japanese gardens before trade opened the orient to the west. After that it was brought to the England by the Royal Horticultural Society in 1819. Lilies reached to Bermuda from England where it flourished well. Lily depicted an ancient Greek pottery, featured in literature and religions as diverse as pagan myths. Today lilies are enjoying a renaissance of recognition. In Italy and France they are among the most fashionable of flowers. In the USA their popularity has increased tremendously. A number of lily bulbs were imported by the US from the Netherlands for both professional forcing and garden usage. In India, lilies are mostly grown in the hills but *Lilium longiflorum* do well in the plains, especially under subtropical regions. There are about 7000 registered varieties of lily. The International Lily Register is maintained, under international agreement by the Royal Horticultural Society in London.

Improvement and breeding of *Lilium* started about 200 years back (Shimizu, 1987). Although significant contribution in this field is about 50 years old when breeding of Asiatic hybrids stared and some improved varieties evolved (McRae, 1998). Since 1970 lilium has become one of the most important bulbous flower for its cut flower production. Presently companies involve in the breeding of lilium belonging to the Netherlands are dominating worldwide. Since last two decades more than 100 new cultivars released every year for its marketing.

Presently, due to outstanding flower shape and fragrance, Oriental hybrids have become the most important group. During last 10 years, several Oriental hybrid varieties have been released with early flowering, novel flower colours and large flower size. However, Asiatic hybrids were the leading group until 1980. The Netherlands is leading in bulb production followed by Japan, USA, Australia, Chile and South Africa.

SPECIES

There are 400 species in genus *Gerbera* (Bailey, 1963). The genus is divided into 7 subgenera and these are:

i. Eulirion (true lilies, trumpet or funnel shaped lilies)

ii. Archelirion (open flowered lilies)

iii. Martagon (turk's cap or turban shaped lilies)

iv. Pseudo-Martagon (bell flowered lilies)

v. Isolirion (erect or upright flowered lilies)

vi. Cardiocrinum (heart shaped leaved lilies)

vii. Notholirion (fritillaria like lilies)

The most important species for lily improvement are *L. tigrinum*, *L. bulbiferum*, *L. maculatum*, *L. dauricum*, *L. davidii*, *L. lancifolium*, *L. concolor*, *L. duchartrei*, *L. auratum*, *L. speciosum*, *L. rubellum*, *L. longiflorum*, *L. japonicum and L. nobilissimum.*

There are several other species which are involved in breeding of lily.

In recent years, the much improved forms and colours in varieties have been developed from highly concentrated and specialized breeding programme in many countries. The several hundreds of crosses between species were made and hybrids were evolved. These hybrid groups are described below along with wild species.

Asiatic hybrids: The Asiatic hybrids were derived from a species that originated in Asia. They bloom in early summer, usually flowering for over a month. Their 10 to 15 cm blossoms, which may face up, out or down, come in shades of red, orange, yellow, pink, lavender and white. The plants produce compact growth and range from 30 to 150 cm in height. They are great for growing in containers and for cutting and are quite disease resistant.

Oriental hybrids: They were mostly derived from *L. auratum* and *L. speciosum.* These beautiful plants range in height from 60 to 240 cm. They produce huge flowers up to 30 cm across. The pleasantly scented blossoms are borne in late summer. They are bowl shaped with recurving petals. They come in white, pinks, deep reds and bicolours. These are great for growing in containers, but may need staking because of their size.

Martagon hybrids: These hybrids are derived from *L. martagon* (Martagon lily and Turk's cap lily). They produce their 5 to 10 cm, nodding flowers in early summer. The flowers come in white, yellow, orange, tangerine, mahogany, brown, lavender or lilac colours. Martagon hybrids can reach heights ranging from 90 to 180 cm.

Candidum hybrids: Candidum hybrids were derived from *L. candidum* (Madonna lily or White lily), produce their fragrant, 10-12 cm flowers in early summer. These plants grow from 90 to 120 cm high.

American hybrids: These hybrids are derived from native North American lilies. They grow from 120 to 240 cm in height and produce their 10-15 cm flowers in late spring and early summer.

Longiflorum hybrids: These are derived from the popular *L. longiflorum* (Easter lily). These hybrids produce aromatic, white, trumpet-shaped flowers. They grow to about 90 cm high and bloom in mid summer, although they can be forced to bloom indoors for Easter.

Trumpet hybrids: The Trupmpet hybrids were derived from Asiatic species. These lilies grow from 120 to 180 cm high and produce large, 15-25 cm flowers in the summer.

Wild species: These are native to North America, Europe and Asia. Wild species are the kind that still occur growing in the wild and which have not yet been affected by hybridization. *L. auratum* (Gold banded lily) is one of the most beautiful of the wild species. It bears up to thirty five, star shaped, highly aromatic flowers per stem, usually in August and sometimes in September. The large flowers can reach a diameter of 25-30 cm. They are white with a golden yellow stripe down the centre and numerous tiny red speckles. *L. lancifolium* (Tiger lily) is a well-known native of Japan and China. Its flowers are borne in late summer. They consist of reflexed, bright red or orange-red petals covered with little speckles. An interesting characteristic of this flower is that bulbils appear in the leaf axils. This lily has played an important part in the development of the Asiatic hybrids and other lilies.

FLOWER MORPHOLOGY AND BIOLOGY

The lily flower has a long perianth tube which comprises of two fused whorls and 6 isomorphic perianth segments. Flower of lily is stalked and is subtended by bract. There are prominent 6 anthers, dehiscing introrsely and produce orange coloured sticky pollen. The anthers are borne on long filaments (5 cm) attached near the base of the perianth tube. The ovary is superior and trilocular. It bears a style and 3 lobed stigma. Fruit is an oblong capsule borne above the base of the perianth segments. It is 6 ribbed which divide into 3 cells. Each cell closely packed with flattened, brown and soft coated seed.

Anthesis and stigma receptivity

Stigma remains receptive for more than 1 week. For pollen germination, anther dehiscence and pollen drying are essential. The anthers were collected 1 to 2 days before anthesis and dried before being incubated with culture medium showed increase in germination (55%) than that of without drying (0 to 15%). Pollens remain viable for several days after anthesis. The best germination was exhibited by pollen whose moisture was 15% or less and was harvested 1 day or longer after anthesis (Lin and Dickinson, 1984).

Mode of pollination

Lily is a cross pollinated crop. Wild species of lilies have large trumpet-shaped corollas of various appealing colours, such as white, red and yellow, and long stamens and pistils. These floral traits appear to be adaptive to pollination by animals. Since Brantjes and Bos (1980) report that both diurnal and nocturnal hawkmoths visit *Lilium martogon,* many other efforts have been made to describe pollinators of wild lilies in various contexts. The most intensive study was conducted by Skinner (1988), who observed pollinators of 13 west American species of *Lilium* and recognized five types of pollination: butterfly pollination (*L. humbolodtii*, *L. kelleyanum* and *L. kelloggii*), butterfly and hummingbird pollination (*L. wigginsii*, *L. pardalinum* and *L. pardalinum* ssp. *vollmeri*), hummingbird pollination (*L. columbianum*, *L. occidentale* and *L. bolanderi*), hummingbird and bumblebee pollination (*L. parvum* and *L. maritimum*) and hawkmoth pollination (*L. washingtonianum* and *L. parryi*). Additional studies reported various flower visitors, including fritillaries to *L. concolor* var. *pulchellum* (Hayashi and Kawano, 2007), nocturnal hawkmoths to *L. formosanum* (Rodger *et al.*, 2010) and *L. japonicum* (Chiba and Shimizu, 2004), and hawkmoths and swallowtails to *L. auratum* (Hayashi and Kawano, 2007 and Morinaga *et al.*, 2009). These studies have shown that flowers of *Lilium* are mostly pollinated by diurnal and nocturnal Lepidoptera, and less frequently by hummingbirds and bumblebees.

GENETICS

Most of the *Lilium* spp. are diploid (2n = 24). The vegetative cells are diploid, the germ cells are haploid and the endosperm in *Lilium* is pentaploid. All the diploid cultivars are cross-fertile and the seedlings obtained by hybridization are highly variable, that shows the immense heterozygosity of the parents. *L. japonicum* and *L. lancifolium* are normally triploid and do not set seeds because of the irregular meiosis (Noda, 1978). In diploid species, bulbils are formed before flowering. Morphological differences in chromosomes were recorded among the species (Sen, 1978). On the basis of karyotype analysis, Noda (1987) classified the Japanese originated *Lilium* species into 5 groups and are given in Table 1.

Inheritance of flowering

Flowering time in lilies is genetically controlled (Freeman, 1980). The interspecific hybridization of early flowering species of Asiatic lilies like *L. dauricum*, *L. bulbiferum* and *L. concolor* gave dominantly early flowering plants. Similarly late flowering species i.e. *L. lichtlinii* var. *maximowiczii*, *L. tigrinum* var. *diploid* and *L. wilsonii* produced in the high majority of late flowering plants. The genetic base for the flowering time in Asiatic lilies, in the first generation (F_1) is inherited, conditioned and segregated as a complete dominance effect of the parent plants (Mynett, 1996).

Genotypic variation has been studied for post harvest flower longevity of 63 Asiatic lily hybrids and considerable genotypic variation with high degree of genotypic determination was observed (Vander Meulen-Muisers *et al.*, 1998).

Diallel crosses were carried out among 10 Asiatic lily cultivars varying in flower longevity (Lim *et al.*, 2002). The number of flowers, water uptake and earliness were moderately correlated to the individual flower longevity. Water uptake of stems at post harvest showed negative correlation with individual flower longevity.

Inheritance of flower colour

In order to produce a desirable colour in the next generations, understanding of the inheritance of flower colour is needed. The genetics of flower colour has been determined differently in the hybrid groups. In the Asiatic hybrids, orange flowers are the dominant wild type, while yellow is determined by a single-recessive gene. Another gene determines the absence of carotenoid flower colours resulting in white or pink

Table 1: Origination of *Lilium* species.

S. No.	Species	Chromosome set (n=12)												Group
		a	b	c	d	e	f	g	h	i	j	k	l	
1.	*L. medeoloides*			*	*		*				*			A
2.	*L. maculatum*	*	*				*	*						
3.	*L. dauricum*	*	*				*	*						
4.	*L. lancifolium*	*	*					*				*		B
5.	*L. leichtlinii*		*				*	*				*		
6.	*L. collosum*	*	*				*	*				*		
7.	*L. concolor*	*	*				*	*				*		
8.	*L. longiflorum*			*	*			*						C
9.	*L. speciosum*	*			*							*		D
10.	*L. alexomdrae*	*			*							*		E
11.	*L. nobilissimum*	*												
12.	*L. auratum*	*			*							*		
13.	*L. japonicum*		*		*		*			*				
14.	*L. rubellum*	*	*		*							*		

* The specific place of nucleolar constriction.

flowers. In Orientals, dark pink is dominant over white. In this group, the carotenoid colours have a minor role (Banba, 1975).

Abe *et al.* (2002) investigated the segregation of the two traits among 96 F_1 plants derived from a cross between commercial cultivars Montreux and Connecticut King in order to understand the genetic background of two floral anthocyanin pigmentation traits, anthocyanin pigmentation in the flower tepals and spot formation, in the Asiatic hybrid lily (2n = 24). Cv. Montreux has anthocyanin pigmentation in the tepals with many spots, and cv. Connecticut King has flowers with carotenoid pigmentation without spots. The F_1 plants with or without anthocyanin pigment in the tepals segregated with a 1:1 segregation ratio, indicating that a single gene controls anthocyanin pigmentation in the tepals. The number of spots per square centimeter of all tepals showed continuous distribution in the F_1 plants. To map the loci for the two anthocyanin pigmentation traits, molecular linkage maps in the Asiatic hybrid lily were constructed using a double pseudo-testcross strategy, with the same F_1 plants used for phenotypic evaluation, and 212 PCR-based DNA markers. The trait for anthocyanin pigmentation in tepals was used as a trait marker. The map of cv. Montreux comprised 95 markers in 26 linkage groups, and the map of cv. Connecticut King used 119 markers in 24 linkage groups. The total map lengths were 867.5 and 1,114.8 cM, respectively. The trait locus for anthocyanin pigmentation in the tepals was between markers ASR35-180 and P506-40 in linkage group 1 of the cv. Montreux map with a map distance of 1.2 cM and 2.6 cM, respectively. A single-point analysis of quantitative trait loci (QTLs) for tepal spot number identified two putative QTLs in linkage groups 1 and 19 of the 'Connecticut King' map. One putative QTL in linkage group 19 explained 64% of the total phenotypic variation. Because both putative QTLs were mapped on the linkage map of 'Connecticut King' that has no spots, dominant alleles of them might suppress spot formation.

To clarify the genetic basis for carotenoid pigmentation in flower tepals of Asiatic hybrid lily (*Lilium* sp.), Nakano *et al.* (2005) evaluated the segregation of a tepal-carotenoid content among F_1 plants derived from a cross between cv. Montreux (having a small amount of carotenoids) and cv. Connecticut King (having a large amount of carotenoids), and mapped genetic loci for the carotenoid pigmentation onto the molecular linkage maps of cv. Montreux and cv. Connecticut King that they constructed previously. The tepal-carotenoid content among the F_1 plants showed continuous segregation, indicating that several genes are associated with this trait. Quantitative trait loci (QTL) analysis identified one QTL, qCARmon6, on the sixth linkage group of the cv. Montreux map. qCARmon6 explained 58.2% of the total phenotypic variation, that is, this locus had a large effect on the carotenoid accumulation. The result that qCARmon6 was mapped on the linkage group of cv. Montreux which has a small amount of carotenoid pigments in tepals indicates that this locus has a dominant negative effect on carotenoid pigmentation.

A breeding programme was begun by Balode (2007) to obtain new varieties with a wide range of colours. For each plant, starting with the anthesis of the first flower, the colouring was analyzed separately. The 'Royal Horticultural Society Colour Chart' was used for evaluating the colour of the different cultivars. Significant differences between and within combinations were detected. Colour analysis allows conclusion on heredity of progeny traits. It was found that cultivars in orange-red produced hybrids in orange-red as the dominant colour over white, yellow and violet. In cross combination, 'Olga' (apricot-white with brown spots) × 'Arabeska' (salmon-orange), high variability was observed. As inherited from the female parent and considered the dominant traits, intermediates in shades of rosy-pink, apricot, salmon-orange with brown spots in the centres were observed. Hybrids similar to parents by colour were produced by the Trumpet hybrid lilies in cross combination of 'Zemgale' (creamy white, unspotted) × 'White Henryi' (white with yellow-orange centres). Altogether 353 plants were estimated in 14 cross combinations. Of those cases, 41% showed co-domination over the corresponding traits of a cultivar and in colour 36% were female and 23% male parents.

Ploszaj-Witkowska and Falkiewicz (2012) studied the seedlings obtained from the hybridization of cultivars and clones of lilies which belong to the Asiatic hybrids, group such as Abiza' × Connecticut King (ACK 49 indiv.) and Altair × Miss Alice (AMA 26 indiv.), and verified, them according to the COBORU methodology, with respect to their tolerance to the climatic conditions prevailing in the region. During the vegetative season, measurements were taken of the following morphological traits (as recommended by the COBORU): beginning of the vegetative growth, beginning of flowering, end of flowering, height and length of the inflorescence, assessment of the shoots and foliage, measurements of single flowers, determination of the colour of flowers according to the colour chart RHSCC. Most of the ACK hybrids obtained in this experiment inherited the colour of the perigon and spots from the clone 'Abizag'. In turn, 4 colours: yellow, orange, bright- and dark-red, could be distinguished among the AMA hybrids.

Inheritance of pollenlessness

Pollenless hybrid was obtained from a cross of cvs. Montreux and Geneva and backcrossing of this F_1 hybrid with either of the parents resulted pollenless phenotype and it is concluded that cytoplasmic gene is related to the pollenless trait. F_1 plants showing pollenless phenotype had shorter tepals than F_1 plants with pollen grains indicating that pollenless genes reduce flower size directly or indirectly (Yamagishi, 2003).

Self incompatibility

Self incompatibility has been reported in *Lilium* spp. (Horn, 2002). A heteromorphic self incompatibility system is inherited and controlled by a single locus with two alleles designated S/s. The interspecific cross incompatibility has been observed in *Lilium*. Specific proteins and lipids of the exudates from the stigma inhibit or accelerate the pollen germination. *L. longiflorum* pistils secrete large amount of exudates. The stylar canal is also covered with the same exudates. This is composed of a gel like solution containing protein, polysaccharide, phenolic compounds and lipids (Labarca *et al*., 1970).

Self compatibility was studied in hybrid *L. formolongi* (*L. formosanum* × *L. longiflorum*) and concluded that a single dominant gene may control self compatibility (Anderson and Dunn, 2003).

Heritability and genetic advance

The correlation and combining abilities of growth and flowering characteristics in the F_1 hybrids of *L. formolongi* and *L. formosanum* diallel crosses were studied (Song *et al*., 2004). Plant height was positively correlated with stem diameter or internode length and negatively correlated with flower height. The flower diameter was highly positively correlated with the number of leaves and negatively correlated with internode length or flower height. The mean squares of general (GCA) and specific combining ability (SCA) were highly significant for all the growth and flowering characteristics, except internode length. The mean square values of GCA were higher than those of SCA for plant height, leaf length, leaf width, internode length, days to flowering and flower height indicating additive gene actions. The lines B and C for plant height and D for flower diameter showed relatively high GCA effects. The A × D and B × D crosses showed high SCA effects on increasing flower size and decreasing plant height and days to flowering. The A × B, B × C and C × D crosses showed high SCA effects on increasing plant height, stem diameter, petal length, and flower height. The broad sense heritability was generally higher than the narrow sense heritability. Both plant height and leaf length showed the highest heritability in the broad and narrow sense.

BREEDING OBJECTIVES

i. Large size of flower

ii. Strong stem

iii. Higher flower longevity

iv. Good forcing ability

v. Good bulb growth of *L. longiflorum*

vi. Long duration of flowering

vii. Early flowering

viii. Less pollen variety

ix. Resistance against *Fusarium oxysporum*, *Botrytis elliptica* etc. diseases.

x. Resistance against Cucumber mosaic virus (CMV), Lily symptomless virus (LSV) and Lily mottle virus (LMoV), Lily mottle virus (LMoV) diseases.

BREEDING METHODS

Selection

Due to highly cross pollinated crop, selection from segregants is more common in lily. However, some varieties have been selected during breeding programme for different traits. In one of the improvement programme, a close relationship has been found between the low incidence of flower bud withering and male sterility and a yellow flowered cv. Yellito was selected, which produces little or no pollen and showed promising trait for growing in poor light conditions (Van Tuyl, 1985). *Fusarium oxysporum* is the most devastating fungi of lily. Resistance against this fungus has been found both in cultivated varieties and in wild species of lily (Straathof *et al.*, 1993 and Straathof and Van Tuyl, 1993). Besides, a research work has also been done in lily to develop seedling tests for optimizing selection for resistance against *Fusarium* (Straathof and Leffeer, 1994).

Hybridization

Several hundreds of lily cultivars are the result of concerted efforts of hybridization. The crosses between species were made and a number of hybrid groups have been evolved. These groups are Asiatic hybrids, Martagon hybrids, Candidum hybrids, American hybrids, Longiflorum hybrids, Trumpet hybrids and Oriental hybrids. Out of these groups Asiatic and Oriental hybrids are commercially important groups. Asiatic hybrids derived from 8 species of *L. tigrinum*, *L. bulbiferum*, *L. maculatum*, *L. dauricum*, *L. davidii*, *L. lancifolium*, *L. concolor*, and *L. duchartrei*. The Oriental hybrids originated from cross of *L. auratum*, *L. speciosum* and *L. rubellum*. Japan has a long history of lily breeding. It is evident that the development of *L.* × *elegans* Thumb is relatively old because the strains had been mentioned in a book 300 years ago (Shimizu, 1960). Mr. Nishimura, in Nagano made cross between *L. formosanum* with *L. longiflorum* around 1928 and developed *L. formolongi*. It combines the characters of flowering within a year. There is no problem with the infection of virus. Hybrids obtained by the cross of *L. longiflorum* display vigorous growth, have large foliage like *L. longiflorum* and have the ability to produce good bulb. Lariptol, one of LA hybrids has big pink flowers, 16-17 cm in length, on the stem (130-140 cm) with dark green leaves. Abe and Kawada (1966 and 1970) obtained hybrids by crossing *L. speciosum* with *L. auratum*, *L. rubellum*, *L. japonicum* and *L. mobilissinum* and developed 'Pacific Hybrids'. *Lilium* spp. has fragrance or resistant to various biotic and abiotic stresses may utilize for breeding a new variety are given in Table 2.

Table 2: Species having fragrance and resistant to various biotic and abiotic stresses.

S. No.	Species	Characteristics
1.	*L.* × *fialkovaja*	Fragrance
2.	*L. dauricum*	Earliness and *Fusarium* resistance
3.	*L. longiflorum*	Fragrance
4.	*L. henryi*	Virus and *Fusarium* resistance
5.	*L. candidum*	Low temperature and low light tolerance

Cultivar Mona Lisa was developed from a hybridization made by the inventor Floris Vletter in a controlled breeding programme in Rijnsburg, The Netherlands in 1979. The parents were Stargazer × F 4431-77. Hybrid was Oriental type, pink and white coloured flowers. Pink at the centre and white at the tips with red markings on the inner side (Anon., 1990).

Cheng and Mattson (1972) found self incompatibility with complete 4 × 4 diallel crosses of 4 Mid-Century hybrid lilies. For all crosses normal stigmatic pollination produced seed while for selfing intra stylar pollination methods resulted in a slightly greater seed set. The external feature of lily hybrids derived from 15 crosses between different *Lilium* spp. was exhibited. The crosses between *L. speciosum* × *L. henryi* and *L. shikayoma* × *L. henryi* showed deeper colour than their parents. Severe inbreeding depression has been reported by selfing of *L. formolongi* hybrid. This included malformed flowers with petals/tepals becoming leaves (Anderson and Dunn, 2003).

A cultivar Yun Xia is *Lilium* Oriental hybrid derived from Oriental × Oriental. It is characterized by large, upright and star flowers. The colour is pink, the spot on the petal is very few. It has green-white style, orange-brown pollen, and strong plants (Wu *et al.*, 2008). Rhee *et al.* (2009) developed a new cultivar Heugjinju by crossing between *Lilium* Asiatic hybrid Nerone, black red coloured, and *L.* Asiatic hybrid 'La Toya', purple coloured. Several varieties of lilium developed through hybridization by breeders. In general parentage are not disclosed, since varieties are under breeders right's act (Anon., 2005a, b, c, d, 2006a, b, c, e, 2007a, c, d, e, f, g, 2008a, b, 2009, 2010a, b, c, d). These varieties are given in Table 3.

Table 3: Lilium varieties evolved through hybridization by several breeders.

S. No.	Variety	Parents	Breeder's & release year	Chief Characteristics
1.	Veronese	PH 96-59 × RH 96-4	Vletter & Den Haan Beheer B.V. (2004)	Flower colour is light yellow.
2.	Halifax	92-80 × RM-96-2	Vletter & Den Haan Beheer B.V. (2004)	Flower colour is red purple.
3.	Valparaiso	PG 95-048 × RH 95-017	Vletter & Den Haan Beheer B.V. (2004)	Flower colour is light yellow.
4.	Vina Del Mar	PG 95-048 × Aubade	Vletter & Den Haan Beheer B.V. (2004)	Flower colour is light yellow.
5.	Brisbane	Acapulco × Unnamed seedling	Sande B.V.(2002)	Flower main colour of inner side of inner tepal is red-purple.
6.	Montezuma	Unknown	Vletter & Den Haan Beheer B.V. (2005)	Flower colour is dark purple-red.
7.	Zanlorvenna	Unknown	Van Zanten Flower bulbs B.V. (2005)	Flower colour is red-purple.
8.	Zanlortrofeo	Unknown	Van Zanten Flower bulbs B.V. (2005)	Flower colour is red-purple.
9.	Corso	Unknown	Voort Cees A Van Der (2002)	Large erected flowers with long petals and curly edges, soft red-purple colour with contrasting yellow/ green mid-vein, vigorous growth and long lasting quality.
10.	Genova	Unknown	Van Der Voort Cees A. (2002)	Erected, light-pink flowers, contrasting orange pollen and markings on the inner side of the throat, vigorous growth and long storage capabilities.
11.	Fenice	Unknown	Vletter & Den Haan Beheer B.V. (2007)	Flower colour is light to medium pink.
12.	Argentina	Unknown	Vletter & Den Haan Beheer B.V. (2007)	Flower colour is medium red purple.
13.	Belladonna	PG 96-034 × Devotion	Vletter & Den Haan Beheer B.V. (2007)	Flower colour is yellow.
14.	Lido	93-040 × RW97-001	Vletter & Den Haan Beheer B.V. (2007)	Flower colour red purple.
15.	Giacondo	Gentile × Benevento	Vletter & Den Haan Beheer B.V. (2007)	Flower colour is whitish pink.
16.	Mothers Choice	Siberia × OR-90	Mak 't Z and B.V. (2005)	Flower colour is group white.

(Contd.)

(*Contd. Table 3*)

S. No.	Variety	Parents	Breeder's & release year	Chief Characteristics
17.	Zanlorsanna	Unknown	Van Zanten Flower bulbs B.V. (2004)	Flower colour is light blue-pink.
18.	Catalonie	Simplon × RW 94-020	Vletter & Den Haan Beheer B.V. (2007)	Flower colour is white. Plant height is medium to tall. Inflorescence compactness compact.
19.	Dordogne	Unknown	Vletter & Den Haan Beheer B.V. (2006)	Large, upright, hot pink flowers with yellow and white center colours, and vigorous growth.
20.	Tarragona	Unknown	Vletter & Den Haan Beheer B.V. (2005)	Creamy white flowers with yellow center colour and contrasting orange pollen, and vigorous growth.
21.	Lombardia	Unknown	Vletter & Den Haan Beheer B.V. (2002)	Light pink flowers, contrasting yellow mid-vein, and contrasting orange pollen.
22.	Bernini	Unknown	Vletter & Den Haan Beheer B.V. (2006)	Large, upright, pink flowers, curled axis of the outer tepals, dark-pink spots on the tepals, and vigorous growth habit.

At IARI Regional Station, Katrain (H.P.) inter-varietal lilium hybrids were developed first time in India and evaluated. Hybrids PKLH 1, PKLH 3 and PKLH 8 were found promising in respect of vegetative and flowering characters (Anon., 2012). Details of these varieties are described in Table 4.

Table 4: Lilium hybrids developed at IARI Regional Station, Katrain (H.P.).

S. No.	Hybrids	Parentages	Chief characteristics
1.	PKLH 1	Pollyanna × Prato	Flower orange, compact, size about 15.5 cm, long and sturdy spikes, suitable for cut flower production.
2.	PKLH 3	Shiraj × Sumplon	Flower pink with yellow tinted center, size about 17.6 cm, compact, short and sturdy stem, good flower longevity, suitable for pot culture.
3.	PKLH 8	Navona × Brunello	Flower orange red, suitable for cut flower production.

Interspecific hybridization

Successful crosses were obtained between *L. pumilum* and Asiatic hybrids and between *L. henryi* and Oriental hybrids. Eleven interspecific hybrids were developed by the intrastylar pollination techniques and their fertility was investigated (Asano, 1980) in backcrosses, the hybrids as female parents were used. More successful results were obtained where the species originally used as pollen parents were again used as pollen parents. For breeding of low light requirement cultivars, Van Tuyll and Van Groenestijn (1982) selected cultivars Connecticut King, Enchantment and Nutmegger as the most light sensitive cultivars (<10% flowering with 6000 lux or natural light) and cultivars Pirate and Uncle Sam the least sensitive (>92% flowering with 6000 lux). In a diallel cross of these cultivars, Tobasco and Pirate indicated a short forcing period for their progeny and cultivars Uncle Sam, Nutmegger and Connecticut King a long forcing period.

Interspecific hybridization between *L. primulinum* var. *burmaniacum* and *L. longiflorum*, *L. elegans* or *L. speciosum* was successful (Kriangkrai and Oradee, 1991). In the above crosses *L. longiflorum* cv. Teppoyuri, *L. elegans* cvs. Apeldoorn, Bellona, Connecticut King, Fastival, Mont Blanc, Sterling Star and *L. speciosum* cvs. Cassablanca and Star Gazer were used.

To breed for late flowering in Asiatic lilies, the late flowering species *L. leichtlinii* var. *maximowiczii*, *L. tigrinum* var. *diploid* and *L. wilsonii* were crossed reciprocally and with the very late cv. Yellow Blaze. Three of the best late flowering hybrids have been selected (Mynett *et al.*, 1997). Hybrids between *L. collosum* and *L. regale* showed 26% fruit set and hybrid seedlings had strong and vigorous roots (Yang *et al.*, 1998). Some new LA hybrids were obtained from the crosses *L. longiflorum* × Asiatic hybrids and *L. formolongi* × *L. rubellum* (or *L. elegans*). LA hybrids have nicely shaped pastel coloured flowers and faintly scented (Fiedler, 1999).

Resistance to *Fusarium oxysporum* and *Botrytis elliptica*

Fusarium oxysporum var *lilii*, *Botrytis* and viruses are the most severe pathogens for lilies. Research has been performed at Plant Research International (The Netherlands) for the evaluation of disease resistance for *Fusarium* and *Botrytis* and useful screening techniques has been developed (Straathof and Van Tuyl, 1994). A high degree of *Fusarium* resistance was found in some Asiatic cultivars and in some species e.g. *L. dauricum.* In contrast, all Oriental hybrids were relatively susceptible. On the other, hand Oriental hybrids were highly resistant to *Botrytis*, while many Asiatic hybrids and some species e.g., *L. pumilum*, *L cernuum* were very susceptible.

Straathof and Van Tuyl (1994) studied the genetic variation to *Fusarium* resistance in 82 *Lilium* genotypes. A high level of resistance was found in the Asiatic hybrids (section *Sinomartagon*) and to a lesser extent in cultivars of *L. longiflorum* (section

Leucolirion). The Oriental hybrids (section *Archelirion*) showed only moderate resistance. In the sections *Sinomartagon, Leucolirion* and *Martagon* species were detected with high levels of resistance. An accession of *L. dauricum* (section *Sinomartagon*) was the most resistant source. Possibilities for introgression of *Fusarium* resistance into cultivars by interspecific hybridization, using the resistance found in *Lilium* genotypes are discussed.

Research on lilium has been emphasized on the importance of breeding for resistance to the fungus *Botrytis elliptica* by early screening of Asiatic *Lilium* that showed wide variability to infection. *Fusarium oxysporum* alone or in combination with other fungi causes basal rot, the most serious disease of Asiatic *Lilium*. A method for assessing resistance was developed and is being used in a breeding programme. Detection of viruses remains an important method of control in the absence of genetically engineered resistance and identified virus resistance in breeding programmes. Tissue culture combined with chemical treatment was successfully used to produce virus-free *Lilium* (Lawson and Hsu, 1996).

Crosses between cultivars with different levels of resistance to *Botrytis elliptica* have been obtained. Estimation of parent plants and the produced hybrids have been carried out by two years of testing on plants grown in natural environment. The level of *Botrytis* resistance was rated visually from 0-4 grades (0=healthy - 4=very susceptible). When *Botrytis*-resistant parents were crossed, the hybrids were characterized by degree of resistance to the disease; when *Botrytis*-susceptible cultivars were crossed with cultivars resistant to the disease, the hybrids are intermediate or susceptible. Under field conditions, 10 perspective hybrids have been tested for resistance to the grey mould. After two years of testing, no significant differences in disease level have been found (Balode, 2009).

Some *Fusarium* resistant varieties evolved after rigorous breeding. A strong resistant to *Fusarium* wilt *Lilium* Oriental cv. Guiyang Red was evolved (Zheng *et al*., 2010). A new Oriental lily cultivar Xiyangyang was selected from Cai74 × OT-62. It is strongly resistant to *Fusarium oxyspoyum* f. sp. *lilii*. and has strong adaptability to cold (Wu *et al*., 2013). Another resistant to wilt cultivar Luona was evolved (Luo *et al*., 2013). Chief characteristics of these varieties are given in Table 5.

Table 5: Lily varieties resistant to *Fusarium oxyspoyum* disease.

S. No.	Variety	Parentage	Year	Characteristic
1.	Guiyang Red	Caiyun 3 × Ruyi	2010	Flower colour bright red with white edge. Stigma colour is milk white. The stem colour is green with purple spots. It has strong adaptability and easy propagation. Excellent cut flower.
2.	Xiyangyang	Cai74 × OT-62	2013	Flower major colour is white, second colour is red which is distributed on midribs. Many spots on tepals. Stigma colour slightly green, stem colour is green with purple spots. It has strong adaptability to cold and easy propagation
3.	Luona	Cai74 × Sorbone	2013	Flower major colour is white, second colour is red which is distributed on midribs. Stigma colour is slightly greenÿ the stem colour is green with purple spots. It has strong adaptability to cold and easy propagation.

Breeding for *Phytophthora* blight resistance

Yang *et al*. (2011) first reported the *Phytophthora* blight of lily in Yunnan (China). They isolated the pathogen causing lily *Phytophthora* blight from diseased plants. The fungi was identified as *Phytophthora nicotianae* according to morphological characteristics, pathogenicity and the sequence of ribosomal DNA-ITS. The sequence of ribosomal DNA ITS of isolate P1 was showed a sequence identity of 98%-99% with *Phytophthora nicotianae*. PCR detection using *P. nicotianae* - specific primers indicated that the pathogen was *P. nicotianae*.

Resistance to Cucumber mosaic virus, Lily symptomless virus and Lily mottle virus

One of the most serious threats in the lily cultivation is basal rot caused by the soil-borne pathogen *Fusarium oxysporum* f.sp. *lilii.* Breeding for resistance of lily is limited by the relative long juvenile phase, the relative slow propagation of clones and the lack of reliable screening assays during this phase. The Asiatic hybrid lily (*Lilium* L.) 'Connecticut King' harbours several beneficial traits such as partial resistance to *Fusarium* and complete resistance to tulip breaking virus (TBV). The variation in resistance to *Fusarium* was determined in four different greenhouse tests in four different years on the same 100 descendants of a backcross population. There was considerable variation in *Fusarium* resistance levels over the years.TBV resistance was determined and segregated, as expected, like a monogenic trait (1:1). In order to identify quantitative trait loci (QTLs) for *Fusarium* resistance and also to identify linked markers to the

TBV resistance a QTL mapping approach was carried out in this backcross population. Segregation of 399 polymorphic AFLP fragments was determined. Two different enzyme combinations were used, *Pst*I/*Mse*I and *Eco*RI/*Mse*I, in the latter one extra selective nucleotide was added to the *Mse*I primer. A paternal ('Connecticut King') linkage map was calculated. This map consisted of 24 linkage groups (*Lilium* 2n=2x=24) with 251 markers covering 1367 cM. About 25% of the markers remained unlinked. Four QTLs for *Fusarium* resistance mapped to linkage groups 1, 5, 13 and 16 respectively. The resistance gene for TBV was placed on linkage group 9 (Van Heusden *et al.*, 2002).

Li *et al.* (2013) carried out the study to establish effective ways to detect and eliminate three major viruses in lilies namely, *Cucumber mosaic virus* (CMV), *Lily symptomless virus* (LSV) and *Lily mottle virus* (LMoV). *In vitro* plantlets of *Lilium longiflorum* cv. Raizan 1 were used as subject material. Specific primer pairs were designed according to the gene sequence of the three viruses and 18S rRNA, and used for multiplex RT-PCR. Several reaction components (dNTPs, Mg^{2+} and *Taq* DNA polymerase), annealing temperature and cycle numbers were optimized to conduct multiplex RTPCR effectively. The results illustrated that culturing shoot tips of <=1 mm size under light condition was the most effective way to eliminate all the three viral strains. High virus-free rate of 93.3% was observed in this method. Additionally, short culture cycles, easy handling and enhanced virus-elimination rate makes it an ideal method for producing virus-free lilies.

Virus free planting material through the culture of *Lilium* × *Enchantment* was produced and around 41% of success has been achieved (Nimi *et al.*, 2001). Incompatible as well as intergeneric pollinations for *Gasteria verrucosa* and *L. longiflorum* were applied and resulted in micropylar penetration (Willemse *et al.*, 1995).

Breeding for abiotic stress

Very limited work on abiotic stress has been done. Many attractive colour varieties are not able to tolerate cold and also poor multiplier. A *Lilium* Oriental cultivar East Red was developed through hybridization. It has strong adaptability to cold (Zheng *et al.*, 2009). A pot Oriental lily cultivar Joyous Event was evolved (Cheng and Qui, 2011). It has strong adaptability to cold and resistance of leaves burning. Xi *et al.* (2012a, b) developed two Asiatic varieties Water Lily Dew and Mid-May. These varieties have strong growing vigour and high complex resistance to main diseases. Chief characteristics of varieties are given in Table 6.

Table 6: Lily varieties have strong adaptability to cold.

S. No.	Variety	Parentage	Year	Characteristic
1	East Red	Co Amore × Acapulco	2009	Flower colour pink which is distinguished with the parents. It has excellent characteristics such as strong adaptability to cold and easy propagation. Good for cut flower.
2	Joyous Event	Acpulco × Miami	2011	Flower is pink, buds are large, flowering period is long. Stem is thick and rigid. Plant is small, height 52 cm. It has strong adaptability and resistance of leaves burning. Bulbs breeding cycle is short. Good for pot culture.
3	Water Lily Dew	Golden Horn × Brunello	2012	Flower colour bright orange-red flowers with thick petals and some purple-brown speckles. Filaments and styles are orange. Height 75-80 cm, flower diameter 14-15 cm. Strong growing vigour and high complex resistance to main disease.
4	Mid-May	Golden Horn × Brunello	2012	Flower colour bright orange, no speckle. Blossoms early and has high flower yield. Filaments and styles are orange. Height 65-70 cm, flower diameter 16-17 cm. Strong growing vigour and high complex resistance to main disease.

Breeding for pollenless variety

Lily cultivars that do not release pollen grains have been in demand since pollen grain stains are not easily removed from clothes. Yamagishi (2003) obtained one F_1 plant showing a pollenless phenotype (designated as a PL clone) from the cross between two Asiatic hybrid lily cultivars Montreux and Geneva. The PL clone showed short filaments with immature anthers and did not release pollen grains. The PL clone was crossed with four Asiatic hybrid lily cultivars, and segregation of plants with the pollenless phenotype among F_1 plants was investigated. All F_1 plants showed the pollenless phenotype when the PL clone was backcrossed with cv. Geneva and Montreux, suggesting cytoplasmic gene(s) is related to the pollenless trait. However, F_1 plants with pollen grains appeared from the crosses of the PL clone _ 'Connecticut King' and the PL clone _ 'Sancerre'. From these results, a genetic model for the pollenless trait is proposed. F_1 plants showing the pollenless phenotype had shorter tepals than F_1 plants with pollen grains, indicating that pollenless genes reduce flower size directly or indirectly.

Mutation

In *Lilium*, solid mutants can be developed by irradiating bulb scales immediately after scaling the bulbs or even better by scaling after irradiation of the bulbs.

Heterozygosity and easy vegetative propagation are two important positive circumstances for successful mutation breeding (Broertjes, 1969 and Broertjes and Alkema, 1970). In late sixties, bulb scales of cvs. Enchantment and Connecticut King were irradiated with 2.5 Gy X rays, which resulted in a change in flower colour and also produced low light tolerant mutants (Van Groenestijn and Van Tuyl, 1983). The LD_{50} of gamma rays for the lily plantlets was 500-600 R (Wang *et al*., 1989). They also reported that the combined inhibiting effect of colchicine and gamma rays on the proliferation of plantlets was greater than their separate impact. Variation in leaf thickness, leaf colour, bulb size, etc. occurred in plantlets treated with colchicine alone or in combination with irradiation. The optimal treatment for variations was 200-400 R gamma rays and 2.0 mg/l colchicine. A mutant variety Zanlotriumph from parent Zanlophator and release in 2005 (Anon., 2006d). Two mutants varieties i.e. Mies Bouwman and TX 68-1 from the Netherlands have been commercialized (Table 7).

Table 7: Lilium varieties developed through induced mutation.

S. No.	Variety	Treatment	Characteristics
1.	Mies Bouwman	Bulb scales, 2.5 Gy X rays	Orange flower colour, excellent forcing quality.
2.	TX 68-1	Bulb scales, 2.5 Gy X rays	Orange flower colour, good forcing quality.
3.	Zanlotriumph	-	Flower colour is white. Evolved by Van Zanten Flower bulbs B.V.

Polyploidization

The reasons for using polyploidy in lily breeding are the larger flowers and the stronger stems of polyploid plants. Polyploidization in lilium is very helpful in overcoming the sterility owing to lack of chromosome pairing during meiosis in the interspecific F_1 hybrids. This sterility hampers further breeding. Somatic chromosome doubling may induce homologous pairing of chromosomes and restores fertility. Colchicine was used successfully to produce fertile allotetraploids in lilium (Asano, 1982, Van Tuyl *et al*., 1990 and Van Tuyl, 1993). Restoration of fertility in the tetraploid forms was demonstrated in the hybrids *L. longiflorum* × Mont Blanc and *L. longiflorum* × *L. candidum* by treating the bulb scales with colchicine. Tetraploid fertile plants produced by treating the scales of interspecific hybrid of cross *L. auratum* cv. Shikayama × *L. henryi* for 2-8 hrs with 0.1–0.2% colchicines (Van Tuyl and Kwakkenbos, 1992). To avoid the toxic effect of colchicine, oryzalin (0.01% and 0.005%) a herbicide with anti-mitotic activity was also used successfully in doubling the chromosome number (Van Tuyl *et al*., 1992b).

Thus, several tetraploid hybrids i.e. *L. henryi* × *L. candidlum* and of *L. longiflorum* × *L. henryi* have come into flower and showed a fully restored fertility. Triploid progenies were obtained by using interspecific hybrids Shikayama × *L. henryi* and *L. auratum* ×

L. henryi as pollen parents. Back crossing these triploids with *L. auratum* × *L. henryi* gave a number of aneuploids with a chromosome number between 36 and 48 (Van Tuyl and Van Holsteijn, 1996). Arzate Fernandez *et al.* (1998b) produced triploid by successful hybridization of *L. concolor* and *L. longiflorum*. The studies of allopolyploids revealed that in lily 2n gametes can be most useful for the introgression of alien genes and chromosomes into cultivars and the introgression can be achieved through recombination due to genetic crossing over between alien chromosome as well as addition of alien chromosome, which is extremely difficult or impossible to achieve in the case of colchicine induced allopolyploids (Ramanna and Jacobsen, 2003).

The mechanism of viable pollen production in allopolyploids developed through mitotic and meiotic polyploidization was reviewed by Van Tuyl and Lim (2003). Mitotic polyploidization possesses one homologous chromosome set. They undergo normal meiotic division like diploid cells. However, meiotic polyploidization often shows irregular chromosome division resulting in two of unreduced chromosomes number instead of reduced chromosome number in tetrads. There are two significant mechanisms, FDR and SDR, for the formation of 2n gametes. The FDR gametes increase heterozygosity while SDR gamete increases homozygosity. Application of meiotic polyploidization in interspecific breeding programme may be of great importance for the introgression of characters from diploids to tetraploids. Induction and selection of tetraploids were also performed to create stock material for breeding of tetraploid lilies. Colchicine pretreatment and *in vitro* culture of bulb scales yielded polyploids of *L. monadelphum*, *L. candidum* and *L. pseudotigrinum* f. *unicolor*, cultivars of Oriental hybrids and Asiatic hybrids. The polyploids are having broad fleshy bulb scales (Prosceviscius *et al.*, 2003). A triploid of *Lilium* Oriental hybrid 'Yahe' is obtained by a method of triploid production. The average flower diameter is 18.69 cm. The colour is pink, the spot on the petal is small and very few. It has a strong adaptability to cold, and easy propagation, the new cultivar is good for cut flower (Zheng *et al.*, 2007). Several varieties developed through polypolidization are given in Table 8.

Table 8: Lilium varieties developed through polypolidization.

S. No.	Variety	Ploidy level	Chief Characteristics
1.	Avignon	Tetraploid	Red orange coloured flower, up-facing
2.	Bold Knight	Tetraploid	Red coloured flower, out-facing
3.	Apricot Supreme	Tetraploid	Apricot coloured flower, down-facing
4.	Pari Sienne	Triploid	Pink coloured flower, up-facing
5.	Tetra Connecticut King	Triploid	Yellow coloured flower, up-facing
6.	Hornback's Gold	Triploid	Yellow coloured flower, down-facing
7.	Eros	Triploid	Pink coloured flower, down-facing
8.	Rosepoint Lace	Triploid	Flowers white, pink margins down-facing
9.	Red Velvet	Triploid	Deep red coloured flower, out-facing
10.	Jasper	Triploid	Pink coloured flower, down-facing
11.	Embarrassment	Triploid	Pink coloured flower, down-facing
12.	Pink Giant	Triploid	Light pink coloured flower, down-facing
13.	Tiger Babies	Triploid	Buff to pink coloured flower, down-facing

BIOTECHNOLOGY

Tissue culture

Tissue cultures of *Lilium auratum* and *L. speciosum*, derived from bulbscales, appeared to differentiate organs (Takayama, 1979). The best material for bulblet formation was bulb scales of intact or *in vitro* produced bulblets. The optimum temperature was 20°C and optimum pH was 6. Effect of irradiance on organ formation was not obvious but leaf emergence was stimulated. Higher kinetin concentrations stimulate the formation of bulbscalcs. High NAA concentrations induce roots. On the other hand kinetin inhibits the NAA effect on root formation. A high sucrose concentration stimulated organ formation. The formation of bulblets and their growth were stimulated at increasing strength of MS medium, but the length of roots was inhibited. Interaction of strength of MS medium and sucrose concentration was examined. High concentration of both components stimulated bulblet growth, but the second strength of MS medium containing 90 or 120 g/l sucrose stimulated callus induction and inhibited the growth of bulblets. Maximum growth took 100 days for bulblets and about 50 days for roots.

Novak and Petru (1981) studied the effect of growth regulators NAA and BAP on bulblet regeneration from bulb scale explant of Lilium oriental hybrid Crimson Beauty. The combination of 5 M BAP and 1 M NAA induced maximum bulblet/plant and on the same medium bud differentiation on the ovary and leaf derived explants were successfully induced.

Potential of lily virus elimination by tissue culture *in vitro* was assessed. The medium used for explants was based on MS medium. Plantlets regenerated from explants were examined by electron microscopy. Plantlets from basal parts of the petals and of basal pieces of leaves used as explants resulted in the best virus elimination. In conclusion, tissue culture is applicable for virus elimination in lily cultivars and hybrids but the success depends on the type of explants (Dapkuniene *et al.*, 2004).

For *in vitro* propagation of *Lilium longiflorum* (meristem and twin scales) showed different responses toward different type of media. The best shoot formation response was obtained on MS medium supplemented with 2.0 mg/l of BAP. At this concentration 84% shoot formation response was obtained after 12.2 days of inoculation. When MS medium was supplemented with different concentrations of auxin and cytokinins the best shoot initiation response was gained at 2.5 mg/l of BAP with 0.5 mg/l of NAA i.e. 88% shoot initiation was obtained after 21 days of explants inoculation. Maximum shoot multiplication response was obtained at 2.5 mg/l of BAP i.e. 4.6 shoots per culture were observed after 15 days of inoculation. MS medium supplemented with 1.5 mg/l of NAA was proved to be optimum for rooting i.e. an average of 5.2 roots were formed after 12.2 days of inoculation. Temperature of 26° C and photoperiod 16 hours & 8 hours dark in 24 hour cycle induced the best response for *in vitro* shoot formation. Well developed *in vitro* plants were shifted to pots containing different media including soil, peat, soil + sand and soil + sand + peat. 100% hardening response was obtained in peat (Ali and Yasmin, 2012).

Strategies to overcome pre and post fertilization barriers

Lot of work has been done in lilium improvement through various techniques to overcome pre and post-fertilization barriers in interspecific crosses as well as to incorporate quality traits for commercial production of flower.

Self incompatibility is a noticeable pre-fertilization barrier. A positive effect of high temperature in overcoming self incompatibility and incongruity has been detected and applied in breeding programmes of lily by heating the style (Ascher and Peloquin, 1968) or by pollinating at high temperatures (Van Tuyl *et al.*, 1982). The use of mentor pollen (compatible pollen genetically inactivated by irradiation but still capable of germination, is reported to overcome inhibition in the style) has been found effective in overcoming self incompatibility but not in interspecific crosses (Van Tuyl *et al.*, 1982).

In a comparison of several pollination methods, it was shown that pre-fertilization barriers in lily can be circumvented by using cut style technique (Van Tuyl *et al.*, 1991 and Janson *et al.*, 1993). By this cut style technique, many pollen tubes grew normally into the ovary. Thus, pollen circumvents stylar and stigmatal barriers which can inhibit pollen tube growth. However, low seed set is found as a complexity with this method. A majority of the pollen tubes either grow past the inner integument or did grow along but not into the micropyle after cut style pollination. Despite the low seed set, a large number of unique interspecific lily hybrids were obtained by retaining longer part of style on the ovary (Asano and Myodo, 1977, Asano 1980, Okazaki *et al.*, 1992 and Van Creij *et al.*, 1993). The stylar graft technique was applied successfully in order to improve the cut style technique (Van Tuyl *et al.*, 1991). In this method, pollen grains are deposited on a compatible stigma. After one day, the style of the pollen donor is cut 1-2 mm above the ovary and grafted on the ovary of another plant. Style and stigma are joined *in vitro* using a piece of straw filled with *L. longiflorum* stigmatic exudates or are stuck together with only the exudates. This technique gives a better seed set as compared to cut style pollination. Various kinds of *in vitro* techniques have been developed to overcome post-fertilization barriers in lily. When abortion occurs in a very young stage and maternal tissue has no negative influence on the development of seeds, ovary culture can be applied. Van Tuyl *et al.* (1992 and 1993) developed new hybrids between *L. longiflorum*, *L. henryi*, *L. concolor*, *L. dauricum* and *L. candidum* through ovary culture. In case of large ovaries, ovary slice culture is a better option for rescuing the young seedlings *in vitro*. Following cut style pollination, ovary slice culture was successfully applied in lilium (Straathof *et al.*, 1987, Kanoh *et al.*, 1988 and Van Tuyl *et al.*, 1991). Arzate-Fernandez *et al.* (1998) adopted this technique to produce interspecific hybrids between *Lilium* × *elegance* and *L. longiflorum*. In this method ovaries were harvested 7-40 days after intrastylar pollination and after surface sterilizing, sliced into 2 mm thick disks. Seed germination occurred after 30 days of planting. This method is also reported in obtaining diploid and triploid interspecific hybrids by crossing of *L. concolor* and *L. longiflorum*. This cross took 20 days for embryo germination, while its reciprocal *L. longiflorum* × *L. concolor* germinated 25 days after planting (Arzate-Fernandez *et al.*, 1998). To overcome fertilization barrier in the cross of Asiatic lilies and *L. pumilum*, cut style pollination was found to be successful (Proscevicius, 2003).

Cultivar Apricot Marriage derived from crossing of *L.* × *formolongl* with *L. elegance* via ovule culture showed vigorous growth characteristics and light yellowish orange petal colour (Mastsumoto *et al.*, 1998).

Embryo culture can be applied successfully in hybrids in which pollinated flowers can stay on the plant for a notable time, before natural abscission occurs. In lily this method has been widely adopted for successful interspecific hybridization (Asano and Myodo, 1977, Asano, 1980, Van Tuyl *et al.*, 1991, Kriangkrai and Oradee, 1991, Kriangkrai, 1993, Okazaki *et al.*, 1994 and 1995, Kim, 1998 and Proscevicius, 2003). The report on embryo culture was first published by Nakajima (1940) who developed the hybrids between *L. speciosum* and *L. auratum* and also between *L. speciosum* and *L. japonicum*. Kriangkrai and Oradee (1991) obtained 6 out of 4196 hybrid seeds from embryo culture of the crosses between *L. longiflorum*, *L. elegans* or *L. speciosum* and *L. primulinum* var. *burmaniacum*. In nature, the crosses between these species were unsuccessful. Kriangkrai (1993) made a cross between Lily Doi (*L. primulinum*) used as female and Longiflorum hybrids, Asiatic hybrids and Oriental hybrids as male plant. To increase the variation in flower size of *L.* × Asiatic hybrids and *L. dauricum* were reciprocally crossed with *L. concolor* and the embryo rescued hybrids were evaluated for cut flower production and promising progenies were selected to serve as mother plants for further breeding (Okazaki *et al.*, 1995). Kim (1998) successfully applied embryo rescue technique to obtain a hybrid Nellie White of *L. longiflorum* and *L. callosum*.

Another type of post-fertilization barrier can be found after raising the hybrids that show F_1 sterility using mitotic (chromosome doubling) and meiotic polyploidization techniques. Amphidiploids with restored fertility can be produced from these diploid hybrids (Van Tuyl *et al.*, 1992b and Van Tuyl, 1993).

Another specific post-fertilization barrier is lack of introgression, while using amphidiploid interspecific hybrid. Genomic *in situ* hybridization techniques provide an insight into the parental genome composition and homologous recombination break points (Karlov *et al.*, 1999 and Lim *et al.*, 2000).

Introgression of alien chromosome segments from donor species into recipient cultivar through backcrossing of F_1 hybrid was studied using *in situ* hybridization technique and the use of 2n gametes found to be the most promising way for the introgression of desirable characters (Van Tuyl *et al.*, 2002).

Using electric pulses, a high frequency of cell fusion was induced between pollen protoplasts and generative cell (Ueda *et al.*, 1990). Somatic hybridization *via* electro-fusion was performed in combination of Oriental hybrid lilies cvs. Acapulco and Shirotae and *Lilium* × *Formolongi* cv. Hakucho. This is the first report to describe the successful realization of lilium somatic hybridization *via* protoplast fusion (Horita *et al.*, 2003).

Molecular breeding

The use of molecular markers in lily has opened new ways to confirm the hybridal status and to obtain the information about the parental genotypes. This confirmation in *Lilium* was performed for earlier isozyme polymorphism (Fernandez *et al.*, 1996), via the morphology of leaves and chromosomes (Okazaki *et al.*, 1994), carbohydrate and flavonol content (Roh *et al.*, 1996) and by the RAPD markers (Obata *et al.*, 2000). Wiejacha *et al.* (2001) developed RAPD fingerprints of the genotypes of lilium that can be used in breeding programme in order to nominate band markers useable for the selection of hybrids at the *in vitro* stage. Using the molecular marker techniques *Fusarium* resistance of the Asiatic cultivar Connecticut King is being marked at the lily genome but there must be efficient screening method at seedling level (Van Tuyl and Holsteijn, 1996).

Yamagishi *et al.* (2002) examined the PCR-based DNA markers in Asiatic hybrid lily. Random primers with various lengths (10-, 12-, 15- and 20-base) were tested in randomly amplified polymorphic DNA (RAPD) reactions with total DNA from two cultivars Montreux and Connecticut King. Thirty-three out of 61 (54%) 15-base primers and 14 out of 21 (67%) 20-base primers generated polymorphic fragments whereas 17 out of 145 (12%) and 4 out of 24 (17%) of the 10- and 12-base primers, respectively, amplified them. The efficiency of the RAPD reaction increased with increasing primer length and examined inter-simple sequence repeat (ISSR) reactions using 3'-anchored simple sequence repeat (ASSR) primers, and found 33 out of 63 (52%) primers that amplified polymorphic bands between the two cultivars. They repeated all RAPD and ISSR reactions at least two times and obtained the same DNA fingerprints, indicating high reproducibility obtained with these reactions. All polymorphic fragments were inherited into their F_1 hybrids, indicating that these DNA markers are genetically stable.

Genetic engineering

Lilies are monocotyledonous plants that have been generally recalcitrant to molecular genetic manipulation because of limitations that restrict utilization of transformation technologies that are routinely applied to dicotyledonous plants. Genetic transformation mediated by bombardment with microscopic metal particles carrying target gene is the preferred method in monocotyledonous plants.

Incorporation of genes via microprojectile bombardment-mediated transformation in *L. longiflorum* was studied by Watad *et al.* (1997). Transgenic lily plants were obtained *via* the above method with plasmids containing the Uid A reporter gene and phosphinothricin acetyl transferase gene that detoxifies the herbicide 'Basta' known as PAT gene. After Southern analysis of 3 plants it was indicated that all contained the PAT gene. A particle gun was also used for the introduction of the NPT 11 gene (for kanamycin resistance) into pollen of lily (*L. longiflorum*) cv. Gelria (Vander Leedge *et al.*, 1997). Lipsky *et al.* (2002) developed cucumber mosaic virus resistant lily plants by particle bombardment method. This method has been proved to be the most preferred method for gene transformation (Cohen *et al.*, 2004). However, preliminary studies have been carried out in the development of an *Agrobacterium*-mediated transformation system for lily (Langeveld *et al.*, 1995). A system for production of transgenic plants has been developed for the Oriental hybrid lily, lilium cv. Acapulco by *Agrobacterium*-mediated genetic transformation. Six hygromycin resistant culture lines were obtained by Hoshi *et al.* (2004).

Breeding for flower colour

Nakatsuka *et al.* (2003) isolated cDNA clones of three chalcone synthase (CHS) genes (LhCHSA, LhCHSB and LhCHSC) and one dihydroflavonol 4-reductase (DFR) gene (LhDFR) and analyzed the spatial and temporal expression of these genes for two Asiatic hybrid lily cultivars Montreux (pink tepals with spots) and Connecticut King (yellow tepals without spots). Pink tepals and tepal spots contain the same kind of anthocyanin, and yellow tepals have no anthocyanin. In both cultivars, the expression of all four genes was not detected in unpigmented leaves, stems and white bulbscales, but was detected in anthocyanin pigmentation-induced bulbscales, indicating that these genes are transcriptionally active in both cultivars. In flower organs, three CHS genes were expressed in anthocyanin-pigmented tepals, filaments and pistils of cv. Montreux, and in unpigmented filaments and pistils of 'Connecticut King'. Pigmented anthers accumulated only LhCHSC mRNA in both cultivars. LhDFR was abundantly expressed in pigmented tepals, anthers, filaments and pistils of cv. Montreux, and in pigmented anthers of cv. Connecticut King. These results indicate that different expression of these genes, especially of the LhDFR gene, controls the distinct pigmentation pattern in flower organs in both cultivars. In three F_1 plants with many spots on yellow or white tepals, derived from a cross between cvs. Montreux and Connecticut King, anthocyanin accumulation in tepal spots was accompanied by the LhCHSA and LhDFR expression, indicating that the expression of CHS and DFR genes is independently controlled between tepals and tepal spots.

To clarify mechanisms of regulation of anthocyanin pigmentation, Yamagishi *et al.* (2010) isolated the two full-length cDNAs of R2R3-MYB (LhMYB6 and LhMYB12) from the anthocyanin-accumulating tepals of cultivar Montreux. Analysis of the deduced amino acid sequences indicated they have homology with petunia AN2, homologous sequences of which had not been isolated in species of monocots. Yeast two-hybrid analysis showed that LhMYB6 and LhMYB12 interacted with the *Lilium* hybrid basic helix-loop-helix 2 (LhbHLH2) protein. Transient expression analysis indicated that co-expression of LhMYB6 and LhbHLH2 or LhMYB12 and LhbHLH2, introduced by a microprojectile, activated the transcription of anthocyanin biosynthesis genes in lily bulb scales. Spatial and temporal transcription of LhMYB6 and LhMYB12 was analyzed. The expression of LhMYB12 corresponded well with anthocyanin pigmentation in tepals, filaments and styles, and that of LhMYB6 correlated with anthocyanin spots in tepals and light-induced pigmentation in leaves. The results indicated that LhMYB6 and LhMYB12 positively regulate anthocyanin biosynthesis and determine organ- and tissue-specific accumulation of anthocyanin.

Yamagishi *et al.* (2010) compared the content and composition of carotenoids and transcriptional changes of carotenoid biosynthetic genes between tepals of yellow, red, pink and white flowers in order to elucidate the regulatory mechanisms of carotenoid accumulation in tepals of Asiatic hybrid lily (*Lilium* spp.) Yellow and red tepal cultivars accumulated high amounts of carotenoids, and cultivars with pink and white tepals accumulated small amounts of carotenoids. Most of the carotenoids in yellow petals were antheraxanthin, (9*Z*)-violaxanthin, *cis*-lutein and violaxanthin. Capsanthin was accumulated in red tepals. Transcription of phytoene synthase, phytoene desaturase, æ-carotene desaturase, carotenoid isomerase and β-ring hydroxylase in tepals increased as tepals grew. Among yellow and red tepal cultivars, the amounts of carotenoids were well associated with the transcription levels of biosynthetic genes, indicating that transcription levels regulate carotenoid accumulation in these cultivars. However, in pink and white tepals, the transcription levels of biosynthetic genes were similar to those in red tepals, indicating that small amounts of carotenoids are not explained by down-regulation of biosynthetic genes. Another factor should reduce carotenoid amounts in these tepals.

Lai *et al.* (2011) investigated the effects of elevated temperature on anthocyanin accumulation in the tepals of a pink-flowered cultivar Marrero and to clarify the flower stage most susceptible to elevated temperature, the flowers were divided into seven developmental stages. Potted plants with flowers at these stages were incubated at 35°C or 20°C for 2 days. The elevated temperature at 35°C caused poor colouration in the tepals at stage (St) 2 and 3. At 20°C, anthocyanin accumulation began at St 2 and anthocyanin content increased at St 3. The elevated temperature did not affect anthocyanin colouration at other stages. To know the mechanism for the poor colouration at St 2 and 3 at 35°C, the transcription of *LhMYB12* (which regulates anthocyanin biosynthesis), *chalcone synthase* (*CHS*), *flavanone 3-hydroxylase* (*F3H*), and *dihydroflavonol 4-reductase* (*DFR*) as well as endogenous sugar content was evaluated in tepals. The transcription of *LhMYB12*, *CHS*, *F3H*, and *DFR* was suppressed at St 2 and 3 under the elevated temperature. The change in sugar content in the tepals at 35° C was not correlated with the reduced anthocyanin accumulation. Thus they concluded that elevated temperatures inhibit anthocyanin biosynthesis at St 2 and 3 *via* the suppression of *LhMYB12* transcription.

Anthocyanins are the predominant visible pigments in flowers of the Oriental hybrid lily (*Lilium* spp.). To understand the regulatory mechanisms of flower colouration in this hybrid lily, full-length cDNA of the *R2R3-MYB* gene, *LhSorMYB12*, was isolated from anthocyanin-accumulating tepals of the cultivar Sorbonne. The deduced amino acid sequence of *LhSorMYB12* showed 85% (209/246) identity, including a gap of seven amino acids, with an amino acid sequence in *LhMonMYB12* that regulates anthocyanin biosynthesis in the Asiatic hybrid lily. Transcripts of *LhSorMYB12* accumulated in anthocyanin-accumulating tepals of three cultivars but not in white tepals of two other cultivars. The white-tepal cultivar Rialto showed *LhMYB12* transcription in tepals, but the Rialto sequence contained amino acid replacements in the R2 repeat, resulting in reduced transcription of anthocyanin biosynthesis genes and in no pigmentation. Spatial and temporal transcription of *LhSorMYB12* showed that transcription of *LhMYB12* corresponded well with transcription of anthocyanin biosynthesis genes in tepal regions where anthocyanins accumulated and in anthocyanin spots on tepals. These results indicate that *LhSorMYB12* definitely regulates anthocyanin biosynthesis and is responsible for the tepal colour difference between Oriental hybrid lily cultivars (Yamagishi, 2011).

In Asiatic hybrid lilies (*Lilium* spp.) a single dominant locus determines anthocyanin biosynthesis in the tepals; however the gene that determines this trait has not been previously reported. Furthermore, anthocyanin colour hue in tepals varies between cultivars, but the mechanisms underlying this variation are not known. Masumi *et al.* (2012) showed that *LhMYB12*, which is homologous to petunia *An2*, determines anthocyanin pigmentation in tepals and that the level of *LhMYB12* transcription affects the quantity of anthocyanin pigments, producing colour hue variation in tepals. The pink-tepal cultivars Montreux and Renoir were heterozygous for the *LhMYB12* gene. Their F_1 population segregated 3:1 for the presence or absence of anthocyanin in tepals, and *LhMYB12* cosegregated perfectly with the presence of anthocyanin, indicating that *LhMYB12* determines anthocyanin biosynthesis in tepals. Among seven cultivars, a single anthocyanin cyanidin 3-*O*-β-rutinoside was detected in tepals displaying light pink, pink–red, dark red, or chocolate brown hues and they suggested that colour hue variation in tepals is strongly related to anthocyanin quantities. The accumulation of *LhMYB12* transcripts in tepals varied between seven cultivars in the same order, and transcription levels were positively correlated with pigment quantities. This is due the levels of transcription of *LhMYB12* vary among the different cultivars and subsequently affect the quantities of pigment in tepals. The above results indicated that *LhMYB12* determines anthocyanin pigmentation and generates the variation in anthocyanin colour hue in lily tepals.

REFERENCES

Abe, H., Nakano, M., Nakatsuka, A., Nakayama, M., Koshioka, M. and Yamagishi, M. 2002. Genetic analysis of floral anthocyanin pigmentation traits in Asiatic hybrid lily using molecular linkage maps. *Theor. Applied Genetics,* **105**: 1175-182.

Abe, S. and Kawada, J. 1966. Interspecific hybrids in Lilium. *Sinkaki*., **51**: 21-25.

Abe, S. and Kawada, J. 1970. *Lilium* Pacific Hybrids. *Nou kou to lengei*, **7**: 11-12.

Ali, A. and Yasmin, S. 2012. Tissue culture studies for micropropagation of *Lilium*: *In vitro* propagation of *Lilium*. LAP LAMBERT Academic Publishing, pp.124.

Anderson, N.O. and Dunn, B. 2003. Inbreeding depression in seed propagated *Lilium* × Formolongi inbreds. *Acta Horticulturae*, **624**: 43-49.

Anonymous. 1990. Patent No. Plant 7379.

Anonymous. 2005a. *Lilium*'Veronese'. Official Journal of Plant Breeder's Rights Office, IPAustralia. *Plant Varieties J.*, **18**(1): 287-289.

Anonymous. 2005b. *Lilium* 'Halifax'. Official Journal of Plant Breeder's Rights Office, IPAustralia. *Plant Varieties J.*, **18**(1): 290-292.

Anonymous. 2005c. *Lilium* 'Valparaiso'. Official Journal of Plant Breeder's Rights Office, IPAustralia. *Plant Varieties J.*, **18**(1): 293-295.

Anonymous. 2005d. *Lilium* 'Vina Del Mar'. Official Journal of Plant Breeder's Rights Office, IPAustralia. *Plant Varieties J.*, **18**(1): 296-298.

Anonymous. 2006a. *Lilium* 'Brisbane'. Official Journal of Plant Breeder's Rights Office, IPAustralia. *Plant Varieties J.*, **19**(1): 323-327.

Anonymous. 2006b. *Lilium* 'Montezuma'. Official Journal of Plant Breeder's Rights Office, IPAustralia. *Plant Varieties J.*, **19**(1): 346-351.

Anonymous. 2006c. *Lilium* 'Zanlorvenna'. Official Journal of Plant Breeder's Rights Office, IPAustralia. *Plant Varieties J.*, **19**(3): 306-310.

Anonymous. 2006d. *Lilium* 'Zanlotriumph'. Plant Varieties Journal, Official Journal of Plant Breeder's Rights Office, IPAustralia. *Plant Varieties J.*, **19**(3): 311-315.

Anonymous. 2006e. *Lilium* 'Zanlortrofeo'. Official Journal of Plant Breeder's Rights Office, IPAustralia. *Plant Varieties J.*, **19**(3): 316-320.

Anonymous. 2007a. *Lilium* 'Corso'. Official Journal of Plant Breeder's Rights Office, IPAustralia. *Plant Varieties J.*, **20**(2): 344.

Anonymous. 2007c. *Lilium* 'Fenice'. Official Journal of Plant Breeder's Rights Office, IPAustralia. *Plant Varieties J.*, **20**(3): 170-172.

Anonymous. 2007d. *Lilium* 'Argentina'. Official Journal of Plant Breeder's Rights Office, IPAustralia. *Plant Varieties J.*, **20**(3): 173-176.

Anonymous. 2007e. *Lilium* 'Belladonna'. Official Journal of Plant Breeder's Rights Office, IPAustralia. *Plant Varieties J.*, **20**(3): 177-180.

Anonymous. 2007f. *Lilium* 'Lido'. Official Journal of Plant Breeder's Rights Office, IPAustralia. *Plant Varieties J.*, **20**(3): 181-184.

Anonymous. 2007g. *Lilium* 'Giancondo'. Official Journal of Plant Breeder's Rights Office, IPAustralia. *Plant Varieties J.*, **20**(3): 185-188.

Anonymous. 2008a. *Lilium* 'Mothers Choice'. Official Journal of Plant Breeder's Rights Office, IPAustralia. *Plant Varieties J.*, **21**(1): 163-165.

Anonymous. 2008b. *Lilium* 'Zanlorsanna'. Official Journal of Plant Breeder's Rights Office, IPAustralia. *Plant Varieties J.*, **21**(3): 261-264.

Anonymous. 2009. *Lilium* 'Catalonie'. Official Journal of Plant Breeder's Rights Office, IPAustralia. *Plant Varieties J.*, **22**(1): 195-197.

Anonymous. 2010a. *Lilium* 'Dordogne'. Official Journal of Plant Breeder's Rights Office, IPAustralia. *Plant Varieties J.*, **23**(1): 295.

Anonymous. 2010b. *Lilium* 'Tarragona'. Official Journal of Plant Breeder's Rights Office, IPAustralia. *Plant Varieties J.*, **23**(2): 345.

Anonymous. 2010c. *Lilium* 'Lombardia'. Official Journal of Plant Breeder's Rights Office, IPAustralia. *Plant Varieties J.*, **23**(2): 345.

Anonymous. 2010d. *Lilium* 'Benini'. Official Journal of Plant Breeder's Rights Office, IPAustralia. *Plant Varieties J.*, **23**(2): 345.

Anonymous. 2012. Annual Report 2011-12. Indian Agricultural Research Institute, New Delhi.

Anonymous.2007b. *Lilium* 'Genova'.Plant Varieties Journal, Official Journal of Plant Breeder's Rights Office, IPAustralia. *Plant Varieties J.*, **20**(2): 344.

Arzate-Fernandez, A.M., Nakazaki, T. and Tanisaka, T. 1998. Production of diploid and triploid interspecific hybrids between *Lilium concolor* and *L. longiflorum* by *in vitro* ovary slice culture. *Plant Breeding*, **117**(5): 479-484.

Arzate-Fernandez, A.M., Tanisaka, J., Nakazaki, T. and Ikehashi, H. 1998. Efficient hybridization between *Lilium* × *elegance* and *L. longiflorum* through *in vitro* ovary slice culture. *Japanese J. Breed.*, **48**(1): 71-75.

Asano, Y. 1980. Studies on crosses between distantly related species of lilies, IV. The culture of immature hybrid embryos 0.3-0.4 mm long. *J. Japan. Soc. Hort. Sci.*, **49**: 114-118.

Asano, Y. 1982. Overcoming interspecific hybrid sterility in *Lilium*. *J. Japan. Soc. Hort. Sci.*, **51**: 75-81.

Asano, Y. and Myodo, H. 1977. Studies on crosses between distantly related species of lilies. I. For the intrastylar pollination technique. *J. Japan Soc. Hort. Sci.*, **46**(1): 59-65.

Ascher, P.D. and Peloquin, S.J. 1968. Pollen tube growth and incompatibility following intra - and inter-specific pollinations in *Lilium longiflorum*. *Amer J. Bot.*, **55**: 1230-1234.

Bailey, L.H. 1963. The Standard Cyclopedia of Horticulture. The Macmillan Company, New York. pp. 1862-1878.

Balode, A. 2007. Colour analysis of flowers in *Lilium* sp. breeding. *Acta Horticulturae.*, **755**: 213-217.

Balode, A. 2009. Breeding for resistance against *Botrytis* in lily. *Acta Horticulturae.*, **836**: 143-48.

Banba, H. 1975. Pigments of lily flowers i. survey of anthocyanin. *Yearbook North America Lily Soc.*, **28**: 44-51.

Brantjes, N.B.M. and Bos, J.J. 1980. Hawkmoth behaviour and flower adaptation reducing self pollination in two Liliflorae. *New Phytolorogist,* **84***:* 139–143.

Broertjes, C. 1969. Mutation breeding in vegetatively propagated crops. *In*: G.C. Chisci and G. Houssmann (eds.), *Proc:* 5th Eucarpia Congress, Milan, 1968. *Genet. Agrar.*, **23**: 139-165.

Broertjes, C. and Alkema, H.Y. 1970. Mutation breeding in flower bulbs. *Proce*: First Int. Symp. on Flower Bulbs, Noordivijk/Lisse, vol. II. pp. 407-411.

Cheng, I.H. and Mattson, R.H. 1972. Effect of intrastylar pollination methods on seed set of *Lilium* × Mid Century' hybrid lilies. *J. American Soc. Hort. Sci.*, **97**(5): 591-592.

Cheng, Q.F. and Qui, M. 2011. A new cultivar of *Lilium* Oriental hybrid Joyous Event. *Acta Horticulturae Sinica*, **38**(12): 2421-2422

Chiba, S. *and* Shimizu, T. 2004. Life history of the near threatened *Lilium japonicum* Thunb. in Nagano and pollinators. *Bulletin Botanical Soc. Nagano,* **37***:* 1–8.

Cohen, A., Lipsky, A. Arazi, T., Lon, A., Star, R., Sandler-Ziv, D., Pintea, C., Barg, R., Salts, Y., Shabtai, S., Gaba, V. and Gera, A. 2004. Efficient genetic transformation of *Llilium longiflorum* and *Ornithogalum dubium* by particle acceleration followed by prolonged selection in liquid medium. *Acta Horticulturae*, **651**(2): 131-138.

Dapkuniene, S., Indribiunaite, G., Juodkaite, R., Navalinskiene, M. and Samuitiene, M. 2004. Tissue culture for elimination of lily viruses depending on explant type. *Acta Universities Latviensis, Biology*, **676**: 163-166.

Fernandez, A.M., Nakazaki, T. and Tanisaka, T. 1996. Development of diploid and triploid interspecific hybrids between *L. longiflorum* and *L. concolor* by ovary slice culture. *Plant Breeding*, **115**: 167-171.

Fiedler, H. 1999. Flowering ability and health were more important breeding objectives. *TASPO-Gartenbaumagazin*, **9**(10): 30-31.

Freeman, Mc. R.J. 1980. Genetics for lily breeder. The Lily Yearbook of the North American Lily Society, Inc., **33**: 98-110.

Hayashi K. *and* Kawano S. 2007. *Lilium dauricum* Ker-Gawler. *In:* Kawano S. *(ed.). Life-History Monographs of Japanese Plants III. Hokkaido University Press, Sapporo, pp.* 1–23.

Horita, M., Morohashi, H. and Komai, F. 2003. Production of fertile somatic hybrid plants between Oriental hybrid lily and *Lilium* × *fermolongi* Planta *Berlin*, **217** (4): 597-601.

Horn, W. 2002. Breeding methods and breeding research. *In*: A. Vainstein (ed.). Breeding for Ornamentals: Classical and Molecular Approaches. Kluwer Academic Publishers, The Netherlands, pp. 47-83.

Hoshi, Y. Kondo, M., Mori, S., Adachi, Y., Nakano, M. and Kobayashi, H. 2004. Production of transgenic Lily plants by *Agrobacterium* mediated transformation. *Plant Cell Reports*, **22**(6): 359-364.

Janson, J., Reinders, M.C., Van Tuyl, J.M. and Keijzer, C.J. 1993. Pollen tube growth in *Lilium longiflorum* following different pollination techniques and flower manipulations. *Acta Botanica Neerlandica*, **42**(4): 461-472.

Kanoh, K., Hayashi, M., Serizawa, Y. and Konishi, T. 1988. Production of interspecific hybrids between *Lilium longiflorum* and *L.* × *elegance* by ovary slice culture. *Japan J. Breed.*, **38**: 278-282.

Karlov, G.I., Khrustaleva, L.I., Lim, K.B. and Van Tuyl, J.M. 1999. Homoeologous recombination in 2n-genete producing interspecific hybrids of *Lilium* (Liliaceae) studied by genomic *in situ* hybridization (GISH). *Genome*, **42**(4): 681-686.

Kim, S.H. 1998. Immature embryo culture for interspecific hybridization of lilies. *J. Korean Soc. Hort. Sci.*, **39**(3): 824-826.

Kriangkrai, A. 1993. Improvement of Lily Doi (*Lilium primulinum* Balc. var. Burmaniacum Stearn) through embryo culture. Bangkok, pp. 86.

Kriangkrai, A. and Oradee, S. 1991. Interspecific hybridization of lily. *Proce*: The 29th Kasetsart University: Plant Science, Bangkok. pp. 415-422.

Labarca, C., Kroh, M. and Loewus, F. 1970. The composition of stigmatic exudates from *Lilium longiflorum*. Labelling studies with myo-inositol, D-glucose and L-proline. *Plant Physiology*, **46**: 150-156.

Lai, Y., Yamagishi, M. and Suzuki, T. 2011. Elevated temperature inhibits anthocyanin biosynthesis in the tepals of an Oriental hybrid lily via the suppression of *LhMYB12* transcription. *Scientia Horticulturae*, **132**: 59-65.

Langeveld, S.A., Gerrits, M.M., Derks, A., Boonekamp, P.M. and Bol, J.F. 1995. Transformation of lily by *Agrobacterium*. *Euphytica*, **85**: 97-100.

Lawson, R.H. and Hsu, H.T. 1996. Diseases and their control. *Acta Horticulturae*, **414**: 175-187.

Li, C.J., Yuan, X.X., Jin, X.Y., Zhao, L.X. and MingFang, Y.Y. 2013. Study on molecular detection and elimination of three lily (*Lilium longiflorum*) viruses. *J. Agril. Biotech.*, **21**(4): 489-97.

Lim, K., Vander Muisers, J.J.M., Van Tuyl, J.M. 2002. Breeding for flower longevity enhancement of Asiatic hybrids lilies. *Acta Horticulturae*. **570**: 409-413.

Lim, K.B., Chung. J.D., Van Kronenburg, B.C.E., Ramanna, M.S., De Jong, J.H. and Van Tuyl, J.M. 2000. Introgression of *Lilium rubellum* Baker chromosomes into *L. longiflorum* Thunb. a genome painting study of the F_1 hybrid, BC_1 and BC_2 progenies. *Chromosome Res.*, **8**(2): 119-125.

Lin, J.J. and Dickinson, D.B. 1984. Ability of pollen to germinate prior to anthesis and effect of dessiceation on germination. *Plant Physiol.*, **74**: 746-748.

Lipsky, A., Cohen, A., Gaba, V., Kamo, K., Gera, A. and Watad, A. 2002. Transformation of *Lilium longiflorum* plant for cucumber mosaic virus resistance by particle bombardment. *Acta Horticulturae*, **568**: 209-214.

Luo, Z.Y., Wu, L.Y., Cheng, J.Q., Mao, S.S., Huang, L.P., Fan, F.F. and Ling, W.B. 2013. A new *Lilium* Oriental cultivar Luona. *Acta Horticulturae Sinica*, **40**(1): 197-198.

Mastsumoto, O., Okafuji, Y., Mitsunaga, T. and Nakamura, T. 1998. Breeding of new lily cultivar Apricot Marriage by ovule culture. Bulletin of the Yamaguchi Agricultural Experiment Station, **49**: 1-4.

McRae, E.A. 1998. Lilies: A Guide for growers and Collectors. Timber Press, Portland, Oregon, p. 392.

Morinaga, S.I., Kumano Y., Ota A., Yamaoka R. and Sakai S. 2009. Day–night fluctuations in floral scent and their effects on reproductive success in *Lilium auratum. Population Ecology,* **51***:* 187–195.

Mynett, K. 1996. Research production and breeding of lilies in eastern European countries. *Acta Horticulturae*, **414**: 47-53.

Mynett, K., Lilien, K.H., Borochov, A. and Halevy A.H. 1997. Breeding for late flowering in Asiatic lilies. *Acta Horticulturae*, **430**: 531-537.

Nakajima, Y. 1940. On the utility of the stored *Lilium* pollen and abnormal seed in crosses of *L. speciosum* × *L. auratum*, *L. speciosum* × *L. makinoi*. *Bot. Mag. Tokyo*, **54**: 473-483.

Nakano, M., Nakatsuka, A., Nakayama, M., Koshioka, M. and Yamagishi, M. 2005. Mapping of quantitative trait loci for carotenoid pigmentation in flower tepals of Asiatic hybrid lily. *Scientia Horticulturae,* **104**: 57-64.

Nakatsuka, A., Izumi, Y. and Yamagishi, M. 2003. Spatial and temporal expression of chalcone synthase and dihydroflavonol 4-reductase genes in the Asiatic hybrid lily. *Plant Science,* **165**: 759-67.

Nimi, Y., Han, D.S. and Fujisaki, M. 2001. Production of virus free plantlets by anther culture of *Lilium* × Enchantment. *Scientia Horticulturae*, **90**(34): 325-334.

Noda, S. 1978. Chromosomes of diploid and triploid forms found in the natural populations of Tiger lily in Tsushima. *Bot. Mag. Tokyo*, **91**: 270-283.

Noda, S. 1987. Lily road toward the Japanese Islands- A cytological viewpoint *In*: M., Shimizu (ed.), The Lilies of Japan- species and Hybrids, Seibundoshinkkosha Ltd. Tokyo.

Novak, F.S. and Petru, Ev. A. 1981. Tissue culture propagation of lilium hybrids. *Scientia Horticulture* **14**(2): 191-199.

Obata, Y., Nimi, Y. Nakano, M., Okazaki, K. and Miyajima, J. 2000. Interspecific hybrids between *L. nobilissimum* and *L. regale* produced *via* ovules with placenta-tissue culture. *Scientia Hortic.*, **84**: 191-204.

Okazaki, K. Kawada, J., Kunishige, M. and Murakami, K. 1995. Introduction of the characteristics of *Lilium concolor* into *L.* × Asiatic hybrids by crossing through style cutting pollination and embryo culture. *J. Japan. Soc. Hort. Sci.*, **63**(4): 825-833.

Okazaki, K., Asano, Y. and Oosawa, K. 1994. Interspecific hybrids between Lilium Oriental Hybrid and L. Asiatic hybrid produced by embryo culture with revised media. *Breeding Sci.*, **44**: 59-64.

Okazaki, K., Umada, Y., Urashima, O., Kawada, J., Kunishige, M. and Murakami, K. 1992. Interspecific hybrids of *Lilium longiflorum* and *L* × *formolongi* with *L. rubellum* and *L. japonicum* through embryo culture. *J. Jap. Soc. Hort. Sci.*, **60**: 997-1002.

Ploszaj-Witkowska, B. and Falkiewicz, D. 2012. Morphological verification of clones of asiatic hybrid lilies. *Acta Horticulturae*, **953**: 331-36.

Proscevicius, J. 2003. Incongruous crosses in breeding of hardy and resistant lilies. Sodinin kyste ir Darzininkyste, **22**(3): 560-567.

Proscevicius, J., Sliesaravicius, Z. and Ranceliene. 2003. Selection of stock polyploid forms for breeding of lilies. *Sodinin Kiyste, ir Darzininkyste*, **22**(3): 553-559.

Ramanna, M.S. and Jacobsen, E. 2003. Relevance of sexual polyploidization for crop improvement : A review, *Euphytica*, **133**(1): 3-18.

Rhee, HyeKyung., Cho, HaeRyong., Lim, JinHee., Kim, MiSeon. and Choi, SeongYeol. 2009. Breeding of an Asiatic hybrid lily 'Heugjinju' with unspotted black red petals. *Horticulture, Environment and Biotechnology*, **50**(6): 593-595.

Rodger J.G., van Kleunen M. *and* Johnson S.D. 2010. Does specialized pollination impede plant invasions? *International J. Plant Sci.,* **171***:* 382–391.

Roh, M., Griesbach P.J. and Gross, K. 1996. Production and identification of interspecific hybrids of *Lilium longiflorum* and *L. elegans*. *Acta Horticulturae*, **414**: 93-100.

Sen, S. 1978. Intra specific differentiation in karyotype of *Lilium*. *Cytologia*, **43**: 305-315.

Shimizu, M. 1960. Sukashi-Yuri (*Lilium elegans* Thunb.) and its origin. The Lily Yearbook of the North American Lily Society., **13**: 105-120.

Shimizu, M. 1987. The Llilies of Japan: Species and Hybrids (Japanese). Seibundo Shinkosha, Tokyo, pp.148–165.

Skinner, M.W. 1988. Comparative pollination ecology and floral evolution in Pacific coast *Lilium. Ph.D. dissertation. Harvard University, Cambridge.*

Song, CheonYoung, Park, SangChul, Lee, JongSuk, Kim, YoungA, Kim, JongHwa. 2004. Correlation and combining ability of plant growth and flowering in F_1 hybrids by diallel cross in *Lilium formolongi* and *L. formosanum. Korean J. Hortil. Sci. Tech.*, **22**(1):107-13.

Straathof, T.P. and Leffeer, H.J.M. 1994. Screening resistance in seedling populations of Asiatic hybrid lily. *Euphytica*, 78: 43-51.

Straathof, T.P. and Van Tuyl, J.M. 1993. Breeding for resistance against *Fusarium* in tetraploid *Lilium*. The Lily Yearbook of the North American Lily Society, **43**: 23-27.

Straathof, T.P. and Van Tuyl, J.M. 1994. Genetic variation in resistance to *Fusarium oxysporum* f.sp. *lilii* in the genus *Lilium. Annals Applied Biology*, **125**(1): 61-72.

Straathof, T.P., Jansen. J. and Loffler, H.J.M. 1993. Determination of resistance to *Fusarium oxysporum* in *Lilium. Phytopathology*, **83**: 568-572.

Straathof, T.P., Van Tuyl, J.M., Keizer, C.J., Wilms, H.J., Kwakkenbos, A.A.M. and Van Dien, M.P. 1987. Overcoming post-fertilization barriers in Lilium by ovary and ovule culture. *Plant Cell Incompatibility Newsletter*, **19**: 69-74.

Takayama, S. 1979. Differentiation in lilium bulbscales grown *in vitro. Physiologia Plantarum*, **46**(2): 184-190.

Ueda, K., Miyamoto, Y. and Tanaka, I. 1990. Fusion studies of pollen protoplasts and generative cell protoplasts in *Lilium longiflorum. Plant Science Limerick*, **72**(2): 259-266.

Van Creij, M.G.M., Van Raamsdonk, L.W.D. and Van Tuyl, J.M. 1993. Wide interspecific hybridization of Lilium: preliminary results of the application of pollination and embryo rescue methods. The Lily Yearbook of the North American Lily Society, **43**: 28-37.

Van Groenestijn, J.E. and Van Tuyl, J. 1983. Straks green teliebelichiting meer noding. Hoopvolle resultaten van veredeling op lichtbehoe fte. *Vakblad- voor- de. Bloemisterij*, **38**(6): 30-31.

Van Heusden, A.W., Jongerius, M.C., Van Tuyl, J.M., Straathof, T.P. and Mes, J.J. 2002. Molecular assisted breeding for disease resistance in lily. *Acta Hortic*., **572**: 131-38.

Van Tuyl, J.M. 1985. Breeding work on lilies. Attention to vase life, scent and pollen. *Vakblad- voor- de. Bloemisterij*., **40**(44): 36-37.

Van Tuyl, J.M. 1993. Surgery of research on mitotic and meiotic polyploidization at CPRO-DLO. The Lily Yearbook of the North American Lily Society, **43**: 10-18.

Van Tuyl, J.M. and Kwakkenbos, T.A.M. 1992. Research on polyploidy in interspecific hybridization of lily. The Lily Yearbook of the North American Lily Society, **42**: 62-65.

Van Tuyl, J.M. and Lim, K.B. 2003. Interspecific hybridization and polyploidization as tools in ornamental plant breeding. *Acta Horticulturae*, **612**: 13-22.

Van Tuyl, J.M. and Van Groenestijn, J.E. 1982. Breeding Asiatic lilies for low light requirement. The Lily Yearbook of the North American Lily Society, **35**: 106-111.

Van Tuyl, J.M. and Van Holsteijn, H.M.C. 1996. Lily breeding research in the Netherlands. *Acta Horticulturae*, **414**: 35-45.

Van Tuyl, J.M., Marcucci, M.C. and Visser, T. 1982. Pollen and pollination experiments. VII. The effect of pollen treatment and application method on incompatibility and incongruity in *Lilium. Euphytica*, **31**: 613-619.

Van Tuyl, J.M., Mass, I.W.G.M. and Lim, K.B. 2002. Introgression in interspecific hybrids of lily. *Acta Horticultureae*, **570**: 213-218.

Van Tuyl, J.M., Meijer, H. and Van Dien, M.P. 1992b. The use of oryzalin as an alternative for colchicine in *in vitro* chromosome doubling of *Lilium* and Nerine. *Acta Horticulturae*, **325**: 625-630.

Van Tuyl, J.M., Van Dien, M.P., Van Creij, M.G.M., Van Kleinwee, T.C.M., Franken, J. and Bino, R.J. 1991. Application of *in vitro* pollination, ovary culture, ovule culture and embryo rescue for overcoming incongruity barriers in interspecific *Lilium* crosses. *Plant Sci.*, **74**: 115-126.

Van Tuyl, J.M., Van Holsteijn, H.M.C. and Kwakkenbos, A.A.M. 1990. Research on polyploidy in interspecific hybridization of lily. *Acta Horticulturae*, **266**: 323-329.

Van, Tuyl, J.M., Van Creij, M.J.M., Van Dien, M.P., Saniewski, M., Beijersbergen, J.C.M. and Bogatko, W. 1992. *In vitro* pollination and ovary culture as a breeding tool in wide hybridization of *Lilium* and Nerine. *Acta Horticulturae*, **325**: 461-466.

Vander Leedge-Plegt, L.M., Van Tuyl, J.M., Van Tunen, A.J., Lilien Kipnis, H., Borochov, A. and Halevy, A.H. 1997. Transgenic lilies via pollen-mediated transformation. *Acta Horticulturae*, **430**: 529-530.

Vander Meulen- Muissers, J.J.M., Van Oeveren, J.C. and Van Tuyl, J.M. 1998. Genotypic variation in post-harvest flower longevity of Asiatic hybrid lilies. *J. American Soc. Hort. Sci.*, **123**(2): 283-287.

Wang, Y., Xi, Y. and Wei, Z. 1989. Effects of gamma rays and colchicine on mutagenesis in somaclones of *Lilium davidii* var. Willmottiae. *Jiangsu J. Agril. Sci.*, **5**(2): 31-37.

Watad, A.A., Yun, D.J., Matsumoto, T., Niu, X., Wu, Y., Kononowicz, A.K., Bressan, R.A. and Hasegawa, P.M. 1997. Microprojectile bombardment mediated transformation of *Lilium longiflorum*. *Plant Cell Reports*, **17**(4): 262-267.

Wiejacha, K., Maraselk, A. Sabala, J. and Orlikowska, T. 2001. Molecular markers in detection of distant hybrids in *Lilium*. *Acta Horticulturae*, **546**: 281-283.

Willemse, M.T.M., Plyushch, T.A. and Reinders, M.C. 1995. *In vitro* micropylar penetration of the ovule of *Gasteria verrucosa* (Mill. H. Duval and *Lilium longiflorum* Thunb: conditions, attraction and applications. *Plant Sci.,* **108**(2): 201-208.

Wu, L.Y., Ling, W.B., Cheng, J.Q., Gong, J.H., Mao, S.S., Huang, L.P. and Fan, F.F. 2013. A new variety cultivar Xiyangyang. *Acta Horticulturae Sinica*, **40**(2): 399-400.

Wu, X.W., Zhang, Y.P., Qu, Y.H., Xiong, L., Wang, J.H., Qu, S.P., Wu, L.F., Cui, G.F., Wang, Q.G., Wang, X.N. and Zhang, P.F. 2008. A new cultivar of *Lilium* Oriental hybrid Yun Xia. *Acta Horticultrae Sinica*, **35**(8): 1245.

Xi, M.L., Wu, Z.H., Fu, W., Qi, F.J., Zhang, R.J. and Shi, J.S. 2012a. A new cultivar hybrid Water Lily Dew. *Acta Horticulturae Sinica*, **39**(4): 811-812.

Xi, M.L., Wu, Z.H., Fu, W., Qi, F.J., Zhang, R.J. and Shi, J.S. 2012b. A new *Lilium* cultivar Mid-May. *Acta Horticulturae Sinica*, **39**(5): 1013-1014.

Yamagishi M., Shimoyamada, Y., Nakatsuka, T. and Masuda, K. 2010. Two *R2R3-MYB* genes, homologues of petunia *AN2*, regulate anthocyanin biosyntheses in flower tepals, tepal spots and leaves of Asiatic hybrid lily. *Plant & Cell Physiology*, **51**: 463-74.

Yamagishi, M. 2003. A genetic model for a pollenless trait in Asiatic hybrid lily and its utilization for breeding. *Scientia Horticulturae*, **98**(3): 293-297.

Yamagishi, M. 2011. Oriental hybrid lily 'Sorbonne' homologue of LhMYB12 regulates anthocyanin biosyntheses in flower tepals and tepal spots. *Molecular Breeding,* **28**: 381–89.

Yamagishi, M., Abe, H., Nakano, M. and Nakatsuka, A. PCR-based Molecular Markers in Asiatic Hybrid Lily. 2002. *Scientia Horticulturae*, **96**: 225-34.

Yamagishi, M., Kishimoto, S. and Nakayama, M. 2010. Carotenoid composition and changes in expression of carotenoid biosynthetic genes in tepals of Asiatic hybrid lily. *Plant Breeding*, **129**: 100-07.

Yamagishi, M., Yusuke,Y. and Nakayama, M. 2012. The transcription factor LhMYB12 determines anthocyanin pigmentation in the tepals of Asiatic hybrid lilies (*Lilium* spp.) and regulates pigment quantity. *Molecular Breeding,* **30**(2): 913-25.

Yang, L.P., Zhang, X.F., Ding, B. and Zhao, L.J. 1998. Cultivation of hybrid from *Lilium callosum* × *L. regale*. *J. Northeast Forestry Univ.*, **26**(2): 37-40.

Yang, X., Qu, S., Wang, L., Liu, X., Peng, L. and Wang, J. 2011. Identification and PCR detection of *Phytophthora* blight of lily. *Acta Horticulturae Sinica*, **38**(6): 1180-184.

Zheng, S.X., Huang, J., Zhao, Yan., Guan, W.L., Wu, H.Z., Zhang, H.L., Wu, F.C. and Liu, W.L. 2007. A new cultivar of *Lilium* Oriental hybrid Yahe. *Acta Horticulturae Sinica*, **34**(5): 1338.

Zheng, S.X., Luo, F.X., Li, Y., Huang, J., Zhang, H.L., Zhao, N., Zhang, C.J. 2009. A new *Lilium* Oriental L. cultivar East Red. *Acta Horticulturae Sinica*, **36**(9): 1402.

Zheng, S.X., Zhao, Y., Liu, F.H., Wu, H.Z., Luo, F.X., Zhang, C.J. and Shi, L.J. 2010. A new *Lilium* Oriental L. cultivar Guiyang Red. *Acta Horticulturae Sinica*, **37**(12): 2043- 2044.

Chapter – 16

Lisianthus

Lisianthus is among top ten cut flower crop of the world. In short duration of 20-30 years it become top ranking flower crop of world due to its beautiful flowers which bears beauty of different top ranking old commercial flower crops with variation in flower colours, shape, size with long vase life. It looks like tulips as they began to open and as poppies when fully opened, gaps between the petals appears like daisy, ruffled petals resemble carnations and the most important one is the flower buds and opened double type lisianthus resemble rose buds and open flower respectively and there is no doubt that this is a primary reason for consumer acceptance. Double flowered lisianthus are more demanded in U.S. as cut flower about 80% of lisianthus market whereas, single flowered are more dominant in Japanese and European markets. In Japan over 129 million cut stems were sold in 2001 placing it on number one position in Europe, over 122 million cut stems were sold in 2001 and placing it in top ten ranking and in the United States over 14 million cut stems were sold in 2002, it is not only popular as a cut flower but also as a bedding plant and a pot plant in market (Grueber *et al*., 1985, Halevy and Kofranek, 1984, Roh and Lawson, 1987, Tjia and Sheehan, 1984, Tija and Sheehan 1986, Skrzypczak *et al*., 1993 and Harbaugh, 2007), a plant bears 20-30 flowers per plant (Skrzypczak *et al*., 1993).

Lisianthus (Eustoma) is herbaceous perennial in some cases annual and biennial (Farina, 1989, Halevy and Kofranek, 1984, Wilkins and Grueber, 1983 and Harbaugh, 2007) belongs to family Gentianaceae (Oren-shamir *et al*., 1999) and genus "*Lisianthus*", which is also used as the common name for *Eustoma grandiflorum*, the plant was commonly known as *Lisianthus russellianus* when first listed in seed catalogues in the early 1980's in the United State, later it was given scientific name *Eustoma grandiflorum*

(Raf.) Shinn. (synonyms *Eustoma andrewsii*; *E. russellianum*; *Lisianthus russellianus*) (Bailey and Baily, 1976, Everett, 1981 and Harbaugh, 2007). However, Lisianthus is also known as the Prairie gentian or Texas bluebell (Markham, 1996 and Harbaugh, 2007) and Prairie Rose (Harbaugh, 2007). *Eustoma* is named after the two Greek words Eu (beautiful, good, well) and stoma (mouth) (Uddin *et al.*, 2013). It is a self compatible (Harbaugh, 2007 and Markham, 1996), diploid crop (Markham, 1996). The buds are colourless is a great disadvantage as an ornamental growth. The range of flower colours available in commercial lines is limited to white and purple, shades of cream, pink, mauve and purple (Markham, 1996 and Harbaugh, 2007).

E. grandiflorum has a set of pharmacologically active compounds and therefore can be a valuable therapeutic material, or can serve as a source of pure chemical compound, apart from its strictly ornamental uses (Skrzypczak *et al.*, 1993).

ORIGIN AND DISTRIBUTION

Lisianthus is a gentian native to the central and southern United States (Bailey 1949, Shinners 1957, Deroles *et al.*, 1993 and Hecht *et al.*, 1994). Its natural habitat is the prairies of the plain states (Harbaugh, 2007). It has been observed in the northern part of Mexico, Texas, Oklahoma, Kansas, Nebraska, Colorado, Wyoming, and South Dakota (Shinner, 1957, Wood and Weaver, 1982 and Harbaugh, 2007). *E. exaltatum* is found along the sandy coastal areas of the Gulf of Mexico, such as Florida, Louisiana, and Texas, but has also been reported in the Florida Keys and islands of the West Indies (Harbaugh, 2007). Lisianthus was first commercially available in Japan in 1933 with its seed (Azrak, 1984, Ohkawa, 1994 and Harbaugh, 2007). It was first come in notice commercially worldwide when Sakata Seed Company offered lisianthus as a new cut flower in the United States in the early 1980's (Sakata, 1982 and Harbaugh, 2007), later company reintroduced new and improved F_1 hybrids in its catalogues in early 1980's (Roh and Lawson, 1984 and Harbaugh, 2007). This was the probable reason why it was considered native to Japan but later it was found in US as wild plant and it's native was changed. Dr. Ohkawa introduced seed of *E. exaltatum* in 1986 (Harbaugh, 2007). The typical wild plants have blue-purple flowers but intensive breeding in Japan over the last 30 years has produced varieties with white, pink, plum and mauve flowers (Deroles *et al.*, 1993). To enhance the popularity of lisianthus, new varieties need to developed (Deroles *et al.*, 1993). The Disney World displayed lisianthus in its gardens in 1983 and were successful in getting much attention (Parke, 1986 and Harbaugh, 2007) this proved it as landscape garden plant. Initially, most of the breeding efforts in the United States were made for pot or bedding plant (Harbaugh, 2007). Most of the selection and breeding improvements were done by Japanese breeders and in the advancements of it within 50 years of breeding before re-introducing Lisianthus (Harbaugh, 2007). The first hybrids were not uniform in flowering or vegetative characteristics and some speculated that these were not true F_1 hybrids (Roh *et al.*, 1989 and Harbaugh, 2007).

SPECIES

E. grandiflorum comprises several forms (f.) given by Skrzypczak *et al*. (1993) f. *grandiflorum*-blue-purple flowers with a dark center, f. *fisheri*-white flowers, f. *bicolour*-white flowers with purple streaks, f. *roseum*-pink flowers, f. *flaviflorum*-yellow flowers. *E. exaltatum* is commonly known as coastal dunes lisianthus, seaside Gentian and catchfly Gentian (Harbaugh, 2007). It has very small flowers with petals about 2.5 to 3.8 cm across (Harbaugh, 2007).

FLOWER MORPHOLOGY AND BIOLOGY

It is herbaceous plant with wide variation in colour, form, height, leaves etc., which make it suitable for different uses like cut stem, bedding plant, pot plant, spray type. The cut type bear single stem with few or no basal branches, its height ranges between 61-91cm depending on the season of growth and cultivar it may be as high as 121-152cm. The height of dwarf plant ranges 15-30 cm with more basal branches (Harbaugh, 2007).

Leaf characteristics vary significantly among cultivars. Leaves can be very attractive to undesirable. Leaf shapes typically are from ovate to oblong, but some selections have lanceolate (almost grassy in appearance) or oblanceolate leaves. Leaf colour ranges from gray-green (similar to carnation), light green, dark green to almost a blue-green. Leaves are opposite, sessile and glaucous, largest leaves are often 12.7cm or more in length and 7.6 cm wide, this decreases with height, upper leaves are <5 cm in length and 1.3 cm wide.

Inflorescence may appear linear on some cultivars. The first cut lisianthus cultivars was monopodial with flowers arranged in a racemose fashion. The order of blooming is acropetally resulting in open flowers displayed in layers or in a tiered fashion. The pedicel was of similar length and the internode differences were relatively large, these original cut flowers had a very open inflorescence.

The flower petals are very smooth. However, ruffled or wavy edges are also present which gives them entirely different look. In single flowers corolla consists of five petals united at the base with 5 stamens attached on the corolla throat however the first open flower on stem has 6 petals. Double lisianthus typically have two to five rows of petals. The flower colours in its natural prairie habitat range from the predominantly blue-purple hues to white, pink, and a bicolour blue or pink rim. The continuous breeding efforts developed fifteen colours (vivid blues, light blue, light to very dark purples, light to dark pink, rose, lilac, cream to pure white, yellow, wine red, peach and green). The bicolour flowers are available in white petal with a blue margin (usually termed a 'picotee' or 'rim'). The petals may have a small speck of colour at the tip, a large blotch or the petal margin completely covered with a 1mm to 5 mm (0.2 in.) wide strip of colour. Other bicolour flowers include white with just a blush of blue or pink or white with blue or pink speckles over the entire petal surface.

The eye spot is frequently a dark shade of blue/purple appearing almost black in blue flowers and burgundy to a shade of brown in pink flowers. The eye spot is usually inconspicuous and relatively small. It generally blends into the base colour of the petals. However, in white flowers the eyespot can be a pure blue, pink, or green/yellow and large that the flower appears to be bicoloured. It can be considered reverse Picotee, since the petal margins are white but the base of the petals have colour.

The most common size of flowers are about 7.6 cm across on the native lisianthus. Breeding lines from The University of Florida may bear as large as 12.7 cm for both single and double flowers, small flowers are about 5.1 cm in diameter when fully opened. Single and double flowers are classified into small (mini), normal, or large on the basis of flower size. It bears 20-40 flowers per plant which last or around 25 days (Skrzypczak *et al.*, 1993).

The ovary is 1-chambered and the stigma is two lobed positioned above the anthers. The calyx is made up of five inconspicuous green sepals that are thread like at the tip (less than 1-mm wide; 0.04-inch). In the bud stage the sepals extend beyond the petals and may contain a slight colouration of blue or pink confusing with the petal colour. While a bell shaped flower is the typical flower shape, there are many variations in flower shape. In double type the large number of petals and in some petals are numerous that they are difficult to count. They generally do not have normal stigmas and also numerous deformed stigmas and/or they are devoid of stamens.

It was found that pink, blue and white colour flowering plants can flower under eight hours photoperiod (Halevy and Kofranek, 1984) when they attain a certain age or size (Azrak, 1984) therefore, it was considered as a day-neutral plant (Halevy and Kofranek, 1984, Azrak, 1984 and Harbaugh, 2007). However, other researchers considered lisianthus a facultative long-day plant (Grueber *et al.*, 1984, Roh and Lawson, 1984, Roh *et al.*, 1989 and Harbaugh, 2007). 'Yodel White' is classified as a facultative long day plant (Harbaugh, 2007).

GENETICS

The most recent estimate of chromosome number of lisianthus is 2n = 36 (Griesbach and Bhat, 1990). Genetically defined, simply inherited markers are lacking in lisianthus. Isozyme loci have been used in ornamental species for germplasm characterization (Kobayashi *et al.*, 1987, Messeguer and Ares, 1985, Sumanasinghe, 1986 and Wendel and Parks, 1982) since they often can be evaluated in young seedlings; exhibit Mendelism inheritance, codominance and complete penetrance; and lack pleiotropic and epistatic interactions (Weeden and Wendel, 1989). Screening cultivated lisianthus germplasm has revealed variation in banding patterns for three isozyme systems: esterase (EST; E.C. 3.1.1.-), diaphorase (DIA; E.C. 1.6.99.–), and glucose-6-phosphate isomerase (GPI; E.C. 5.3. 1.9) (Rubino, 1991). Segregating lisianthus [*Eustoma grandiflorum*

(Griseb.) Shinn.] progeny were evaluated to determine the inheritance of esterase (EST), diaphorase (DIA) and glucose-6-phosphate isomerase (GPI) isozymes. Phenotypic data supported the hypotheses that EST is monomeric and controlled by one locus (Est1) with at least three alleles, DIA is tetrameric and controlled by one locus (Dia2) with at least two alleles and GPI is controlled by one locus (Gpil) with at least two alleles. These three isozymes are the first simply inherited, unlinked biochemical markers identified in lisianthus (Rubino, 1993).

Inheritance

Blue petal colour is dominant over white or pink and pink dominant over white (Blue>Pink>White). White flowers may be result of a simple recessive gene or epistasis. In the former, the white is generally a cream colour, while in the latter, a pure white results. The rim colour is very unstable and expression varies with environmental conditions. For example, flowers on the same stem that range from pure white with only a very small blue tip to completely blue. Plants exposed to cold temperatures bear blue flowers and under warmer temperatures rim flowers. To develop homozygous bicolour lines with stable colour patterns over a range of environmental conditions eight to nine generations of inbreeding are required, in comparison to five generations for simple flower colour or vegetative traits. The three major anthocyanidins, pelargonidin (Pg), cyaniding (C) and delphinidin (Dp) of Eustoma follow mendelian inheritance (Uddin *et al.*, 2002).

BREEDING OBJECTIVES

i. New colour, form, size.

ii. Throughout year flowering.

iii. Double flower.

iv. Early flowering.

v. Long vase life.

vi. Impart fragrance in flowers.

vii. Development of F_1 hybrids and interspecific hybrids which are capable in producing viable seeds.

viii. Russet tolerant variety (so can be cultivated in wide range of environment and temperature range).

ix. Resistant or tolerant varieties to biotic and abiotic strees.

x. Develop cultivar to over come herbicide susceptibility.

Use based specific breeding objectives

i. *Fusarium* sp. resistance in bedding plant variety.
ii. Synchronous and uniform flowering in pot plants.
iii. Inflorescence architecture (flat-topped for cut flowers, tiered or spray for potted/ bedding plants).
iv. Basal branching for pot and bedding plant types.
v. Dwarf cultivars for spray type and pot plants.
vi. Day neutral and heat-tolerant plants for cut flowers year round.
vii. Early colouration in cut varieties.

BREEDING METHODS

Lisianthus is an outcrossing and self-compatible species but subjected to inbreeding depression and comparatively easy to breed in comparison to other floricultural crops. Emasculation, bagging and crossing for hybrids production is comparatively easy as pollen dehiscence start after one or two days of flower opening and stigma becomes receptive after three to five days of opening thus accidental selfing is avoided. The buds are emasculated two days prior natural opening of flower. Pollination must be done when stigma becomes sticky or receptive or when pollen easily adhere to it. Pollens can be collected and stored at room temperature (22-26°C). It bears pod which contains at least 1,000 seeds or more (several thousand) in larger pods. Seeds mature in 8 to 10 weeks during the summer but may require more time in the winter. Maturity indices of seed ripening is when pods begin to turn yellow at the apex and this split longitudinally when fully ripe so must be harvested before splitting of pod to avoid seed loss especially in case of hybrid production. In addition, under humid conditions it is common to see mycelium growing inside an open seed pod, resulting in poor germination. Properly dried seed can be stored for many years at 18-20°C in envelopes in sealed plastic bags with dehiscent.

It is well known that the crop has rosen very fast in the world and today, many varieties are available for commercial purposes and for different uses. This crop is easy to breed which enable breeders to develop so many varieties in short duration of time. The first developed variety was single flowered (Harbaugh, 2007). The Pan American Seed Company has made extensive efforts in its development, Dr B.K. Harbaugh with his associates developed and improved many cultivars by selections, hybridization followed by selection and hybrid series. In early 1990's, three dwarf lisianthus cultivars were released (Rubino, 1992 and Harbaugh, 2007) these are:

- Blue Lisa from Pan American Seed Co., West Chicago, Illinois.
- Little Belle Blue from U.S. Dept. of Agr., Agr. Res. Serv., Beltsville, Maryland.
- Mermaid Blue from Sakata Seed America, Morgon Hill California.

University of Florida's Gulf Coast Research and Education Center, Bradenton, Florida started breeding program on lisianthus in 1985 (Harbaugh, 2007). They emphasised on breeding heat-tolerance and basal branching plants for pot and bedding types.

GCREC-Blue, flowered as a day-neutral plant when seedlings were grown at 28° C, but when seedling were grown at 12° C, plants flowered normally under long-days than short-days (Harbaugh, 2007). Most commercial lisianthus cultivars have a tendency to rosette when seedlings are exposed to temperatures above 25° C (Harbaugh *et al.*, 1992; Ohkawa *et al.*, 1991 and Harbaugh, 2007) the crop is sensitive to rosette formation from germination period onwards. The flower changes its colour in different environmental conditions. It is observed that there is a great variability in phenotypic expression of same genotype under different environmental conditions. In production of commercial lisianthus crop successfully not only varietals character are important but cultural practices and environmental factors are also of the same importance.

Selection

Many selection were made to breed new cultivars of lisianthus in respect to colour, form, suitable for pot, bedding and cut flower production etc. the varieties developed through selection procedure are described below in Table 1 with their parentage and characters.

Table 1: Lisianthus varieties from selection.

S. No.	Varieties	Parentage	Characteristics
1.	UF03-596	F_3 selection of UF01-94	Bicoloured flowering plant.
2.	UF03-573	F_4 selection of UF00-690 × UF0-625	Pink white bicoloured flowering plants.
3.	UF00-579	F_7 selection of UF-94-237 × UF94-393	Pink-white bicoloured flowers, basal branching and floriferousness.
4.	UF00-689	F_4 of UF96-393 × UF96-255	Pink-white bicoloured flowers.
5.	UF00-690	F_4 of UF96-393 × UF96-255.	Pink white bicoloured flower.
6.	UF00-625	F_5 of selection of UF96-419 × UF96-426	Blue-white bicoloured flowers, basal branching and floriferousness.
7.	UF7-47	S_1 of Blue Poppy	Dwarf, low branching habit, sutability to flower in summer (35°C in day).
8.	UF7-53	S_1 of Blue Poppy	Basal branching habit, summer heat tolerance.
9.	GCREC-8-21	Selection made by Selfing in Blue Poppy for three generation	Flower at high temperature (35°C day) basal branching, lower branching.
10.	GCREC-9-105	Selection made by Selfing in Blue Poppy	Flower at high temperature (35°C day) basal branching, lower branching.

(*Contd.*)

(*Contd. Table 1*)

S. No.	Varieties	Parentage	Characteristics
		for three generation	
11.	UF94-404	F_3 selection of UF92-17 × Blue Lisa	Dwarf habit, heat tolerance and white flowers with a blue eyespot.
12.	UF94-34	F_4 selection of UF8-21 × Mermaid Blue	Lower branching, heat tolerance and floriferousness.
13.	UF94-393	F_2 selection of F_1 of unknown tall parent × Marmade	Blue with Double Echo Misty Blue Dwarf plant habit and blue-white bicoloured flowers.
14.	UF96-393	F_2 selection of UF94-226 × UF94-230	Both parents are source of heat tolerant and dwarf plant habit.
15.	UF8-57	S_1 of Tosen	Short basal branching habit.
16.	UF8-21	Selected after three generations of self-pollination of Blue Poppy-1 and Mermaid Blue	Lower branching dwarf characteristics.
17.	UF99-49	F_6 selection of UF94-404 × UF94-46	Heat tolerance, compact growth habit, tulip shaded white flowers, floriferousness.
18.	UF99-16	F_6 selection of UF94-505 × UF94-34	White flowers with a vivid violet-blue center, heat tolerance, lower branching.
19.	UF9-105	Selection after three generations of self-pollination of Blue Poppy-1	Ability to flower at high temperatures.
20.	UF95-309	F_6 selection of a cross between two Yodel Pink plant	Dwarf plant, lower branching habit and vivid pink flower colour.

Hybridization

In 1977 Sakata Seed Company started breeding for development of F_1 hybrids (Harbaugh, 2007). Charles Weddle of Weddle Native Plants, Palisade, Colorado was perhaps the first breeder in the United States to formally release F_1 hybrids for cut flowers and in cooperation with Colorado State University the hybrid 'Colorado Blue Bell' was released in 1984 (Azrak, 1984 and Harbaugh, 2007).

Inbreeding depression is a serious problem in lisianthus. After a few generation of selfing much vigour is lost (Harbaugh, 2007). The use of backcrossing techniques and/or sibling crosses are necessary considerations for developing viable/strong homozygous parents to be used in making F_1 hybrids (Harbaugh, 2007).

In order to develop superior varieties of lisianthus in cut flower long cut stem large flowers, more colour forms, resistant to heat, biotic and abiotic stess and to combine characters of two different superior varieties crossing extensively practiced in development of lisianthus. Some F_1 hybrids are discussed below in Table 2.

Table 2: F_1 hybrids alongwith their parentage and characteristics.

S.No.	Hybrids	Parentage	Characteristics
1.	Twinkle series ($7\ F_1$ hybrid)	-	The series bear small flowers, spray type with many branches.
2.	TG3 (Tresoa Red)	T3 strain × G20 strain	Superior than parents, flowers are dark red colour, bicoloured, mid flowering.
3.	TG26 (Tresoa Purple)	T3 strain × G20 strain	Dark purple flowering, bicoloured, medium size flower spray type and late flowering.
4.	Florida Blue	GCREC-46× GCREC-34	First semi dwarf, tolerate 28-31°C temperature, flower petals are violet blue on the adaxial petal surface, diffusing to a variegated 1.6 to 2.0 cm white band at the base of the petals.
5.	UF01-94	UF00-579 × UF00-689	Bright pink white bicoloured flowers and basal branching.

Interspecific hybridization

Not much work has been done on interspecific hybridization. However, *E. grandiflorum* was crossed with *E. exaltatum* the first generation has intermediate sized flowers that are about half the size of *E. grandiflorum* (Harbaugh, 2007).

Rosette and heat resistant breeding

Lisianthus seedlings typically bolt after they develop three to four true leaf pairs if temperature condition is ideal (Harbaugh, 2007). However, lisianthus is sensitive to high temperatures from the time seeds absorb water during germination through the development of three to four leaf pairs (Harbaugh, 2007). Commercial production of heat tolerant and rosette free cultivars were needed in this respect many varieties were developed to overcome this constrains of production. The temperature above 25^{0}C and high temperature for 3 days during germination favours rosette formation (Okhawa *et al.*, 1991). To avoid rosette formation in lisianthus it is important to develop heat tolerant varieties. Maurine Blue and Florida Blue are the first heat tolerant cultivars of lisianthus. Seedlings can be grown at 28-31^{0}C without rosetting (Harbaugh and John, 1998). The heat resistant varieties are discussed below in Table 3.

Table 3: Heat tolerant hybrid varieties.

S. No.	Variety	Parentage	Characteristics
1.	Maurine Pink Lilac	UF94-214 × UF94-253	Dark pink adaxial petal surface with a lilac hue, diffusing to a variegated 1.8 to 2cm white band at the base of the petals, abaxial petal surface is slightly lighter shade of pink, dark red purple eyespot, heat tolerant.
2.	Maurine Pink	UF94-190 × UF94-214	Light pink petals on adaxial petals surface, abaxial petal surface is a lighter pink, dark red-purple eyespot, Heat tolerant.
3.	Maurine White	lines UF94-147 × UF94-237	White colour at the petal margin diffusing into a creamy yellow-white toward the petal's base red-purple eyespot, Heat tolerant.
4.	Maurine White on Blue	UF94-170 × UF94-163	Heat tolerant.
5.	Maurine Light Blue		UF94-163 × UF94-147 Violet-blue petals on adaxial surface, diffusing to a variegated 1.8-2cm white band at the base of petals, abaxial petal surface is lighter violet-blue, dark violet-blue eyespot. Heat tolerant.
6.	Maurine Lilac	UF94-253 × UF94-147	Pale lilac petals on the adaxial surface, diffusing to a variegated 1.8-2cm white band at the base of the petals, pale violet abaxial surface, dark red-purple eyespot and heat tolerant.
7.	Maurine Dawn	UF03-596 × UF-03-573	Heat tolerant.
8.	UF96-419	UF94-237 × UF94-393	Blue-white bicoloured flower and basal branching, heat tolerant.
9.	UF96-426	UF94-237 ×UF94-404	Bright white flowers and branching habit, heat tolerance.
10.	Florida Silver	UF99-16 × UF99-49	Flower petals white on the adaxial and abaxial petal surface. The eyespot is a violet-blue. The only dwarf or semidwarf cultivar that has a violet-blue eyespot, heat tolerant.
11.	Florida Pink	UF95-309 × UF95-321	Semi-dwarf, heat-tolerant, low rosette-forming, petals light pink. The eyespot is a dark red-purple.

Future breeding for disease and insect-pests resistance

Common disease problems in lisianthus are Gray mold (*Botrytis* blight), *Curvilaria* leaf blotch (Jones and Harbaugh, 1995) *Fusarium* crown and stem rot (Koike *et al.*, 1996, McGovern and Harbaugh, 1977 and McGovern *et al.*, 1978), *Pythium* root rot, and *Rhizoctonia* crown and stem rot. *Impatiens* necrotic spot virus or INSV (McGovern *et al.*, 1997), as well as at least six other viruses have been reported to cause infection in lisianthus. Fungus gnats (*Basidysia* spp.) is more damaging than a pest responsible for crop damage in many crops, they were found to be a major pest of lisianthus (Price *et al.*, 1998 and Harbaugh, 2007). Other major pests that need to be controlled (Ozaki, 1992) include whiteflies (*Trialeurodes vaporariorum*), aphids (*Myzus persicae* and

Aphis gossypii), beet armyworms (*Spodoptera exigua*) and thrips (several species, including *Thrips palmi.* Breeding resistant varieties for these diseases and pest can help to grow on large scale in landscaping where it is prone to *Fusarium* fungus infection.

Mutation

The dose of 100-250Gy gamma rays induced mutants were selected for heat and low rosseting mutants of *Eustom grandiflorum* which can be utilized in breeding summer varieties (Miyashita *et al.*, 2005). Gamma irradiation of seed and somaclonal variation present in *in vitro* culture are being exploited for breeding of lisianthus (Deroles *et al.*, 1993). $4He^{2+}$ion irradiation at 10Gy and H^+ ion beam at 15Gy completely inhibited shoot formation and as dose increases from 3-7Gy it decreases growth in Mickey Soft Pink but this was not observerd in White Palace (Ohki *et al.*, 2000).

Polyploidy

The tetraploid form of *Eustoma* Blue Poppy was horticulturally superior to the diploid form having thicker stems and a shorter plant that could better support the flowers (Griesbach and Bhat 1990). A tetraploid form of lisianthus was produced by using colchicine and open a scope of improvement through autoploidy. Whereas, the protoplast derived DNA cultured plants form may offer a novel method of improving lisianthus by means of producing tetraploids and aneuploids (Lindsey, *et al.*, 1994). During a study cvs. Fresh White and Hakusen developed by protoplast-derived DNA culture which showed large flower, leaves and longer guard cells. The nuclear DNA content was measured 6.33 ± 0.04pg and 3.26 ± 0.10pg DNA per DNA diploid cell suggesting aneuploids (Lindsey *et al.*, 1994).

BIOTECHNOLOGY

Most of the work has been done in the crop improvement by the private sector and therefore most of the work on improvment is unpublished or unrevealed. Recently many investigators have reported transformation of lisianthus by transgenic means (Deroles *et al.*, 1993 and Handa and Deroles, 2000). Transformation with objective to modifying flowering time, new colour and early colouration, anthocynine production gene from antirrhinum (Schwinn *et al.*, 1997 and Markham, 1996), flower and introducing fragrance (Zaccai *et al.*, 2001 and Aranovich *et al.*, 2007).

The study on pigment and co-pigment chemistry (Asen *et al.*, 1986, Markham and Ofman, 1993 and Markham, 1996) together with an extensive study of the factor influencing their expression in flower colour (Markham and Ofman, 1993 and Markham, 1996) has shown that the lisianthus pigment are different from other ornamentals with the presence of galactose residues at the 3-position of the anthocyanins and co-pigments, rather than the common glucose (Markham, 1996). For production of novel anthocyanins in lisianthus transgenic lines are made and one of them i.e. TL851 has shown anthocyanin presence. For production of transgenic lines cDNA from *Antirrhinum majus* which encodes UDP-glucose: flavonoid- 3-O-glucosyltransferase (UFGT) (Markham, 1996) was transferred in lisianthus.

The control of flowering time is essential in most of the ornamental flowers to manipulate flowering time in lisianthus. Zaccai *et al*. (2001) tested *LEAFT* (LFY) cDNA of *Arabidopsis* under the control of 35S CaMV promoter and it was found that the F_1 progenies from two different transformed plants was normally bearing flower and produced normal fertile plants but one to two weeks earlier than control, the number of flowers per inflorescence and nodal length was reduced. Study suggests that for altering flowering time genetic transformation can be used successfully.

Tissue culture

Tissue culture is important tool in obtaining new genes that could be used to further improvement of lisianthus (Anderson, 2007). Establishment of technique for the *in vitro* propagation of lisianthus now focused on the *Agrobacterium* mediated transformation to produce novel cut flower varieties (Deroles *et al*., 1993). For commercial scale multiplication selected clones can be produced by *in vitro* culture of *E. grandiflorum* (Griesbach *et al*., 1987 a and b and_Semeniuk and Griesbach, 1987). Leaf segments use in culture is convenient for micropropagation (Zenkteler and Zenkteler, 1987 and Harbaugh, 2006).

Methods for the production of plants from shoot tips and leaf and internodal stem segments via a callus phase was developed (Semeniuk and Grieshbash, 1987). This technique provided rapid method for clonal propagation production of secondary metabolites (Skrzypczak and Wesolowska, 1993 and Deroles *et al*., 1993) and a source of somaclonal variants (Griesbach and Semeniuk, 1987 and Deroles *et al*., 1993).

For rapid propagation of Polestar Yellow variety of *E. grandiflorum* by using the aseptic seedling leaves as explants. The results showed that for the induction of the adventitious buds from leaves the combination of 6-BA and IBA was better than that of 6-BA and NAA. The best bud induction medium for Polestar yellow was MS + 6-BA 0.6 mg/L + IBA 0.04 mg/L. This provided technique for large scale production of Polestar Yellow of *Eustoma grandiflorum* (Ming *et al*., 2008).

Ordogh *et al*. (2004) studied the effect of different concentration of BA and NAA on the shoot multiplication in four sports of Echo cultivar (Echo White, Echo Rose, Echo Blue, Echo Blue Picotee) in MS base media and Jambor Benczur and Marta basal medium. It was found that medium containing 0.10mg/L BA developed the highest number of shoots and the smallest leaves (in Echo White) where as absence of BA in medium gives opposite results however, in other three it developed large leaves. The higher concentration of NAA (2.0 and 3.0 mg/L) reduce number of roots. Medium containing 0.25mg/L BA gives tallest plants with longest leaves (in Echo Blue Picotee).

Somaclonal variation

Genetic variability in plants particularly somaclonal variation can be increased by *in vitro* culture (Lindsey *et al*., 1994). For genetic transformation however, an *in vitro* regeneration method is required that gives rise to shoots directly from an explant thus reducing the potential for somatic variation by removing the callus phase (Deroles *et al*., 1993). Plants regenerated from callus, frequently exhibit polyploidisation or polysomaty (Pijnacker *et al*., 1989 and Lindsey *et al*., 1994). Semeniuk and Griesbach (1987) discovered somaclonal variants in a tissue-cultured population of lisianthus. These variants are dwarf miniature and branched compared to the tall single stem parent which were not been reported in commercial seeds of cultivars (Griesbach *et al*., 1987a, b, Griesbach *et al*., 1988 and Harbaugh, 2007). The traits of the new variants were stable and progeny produced from these plants indicated genetic transmission of the traits (Harbaugh, 2007).

Molecular breeding

Molecular breeding is being used to engineer lisianthus with altered flower colours, modified flowering times and fragrance (Harbaugh, 2007). Future, classical and molecular breeding focus on disease resistance (particularly *Fusarium*), day neutrality, heat-tolerance, flowering, earliness, and post-transplant survival.

Genetic engineering

First methods for *Agrobacterium*-mediated transformation of *E. grandiflorum* (Handa, 1992) was obtained by using hairy roots obtained by *Agrobacterium rhizogenes* infection for formation of transgenic, however failed in producing fertile transgenic (Takahashi *et al*., 1998). *Agrobacterium*-mediated transformation is useful for transformation of monocot plants but that time they were not sure if it can be applied in genotypes of *E. grandiflorum* (Hiei *et al*., 1994 and Takahashi *et al*., 1998). Semeria *et al*. (1995) were successful in production of transgenic plants by bombardment-mediated transformation of suspension cultured cells of *E. grandiflorum* but were not successful to produce viable examples using leaf tissues (Takahashi *et al*., 1998). However *Agrobacterium* mediated transformation resulted in development of single whorl of mauve petals in line 1, cream –white petals in line 42, purple petals in line 54 and pink petals in line 65 from the original Lisianthus lines have now been transformed (Deroles *et al*., 1993). *Argobacterium rhizogenes* infected plants of *E. grandiflorum* show hairy roots, become dwarf, wrinkled leaves and cup shape corolla than normal and plant become less fertile (Handa *et al*., 1994). The infected plants are genetically transformed by *ROL* gene this was confirmed by Southern analysis of infected plant which showed the presence of Ri-T-DNA introduction (Handa *et al*., 1994).

Lisianthus is a beautiful and appealing flower and posses most of all the characters that a perfect flower should have except fragrance, to enhance lisianthus in this respect

(Aranovich *et al*., 2007) work done on transgenic lisianthus (*E. grandiflorum* cv. Royal Pink) using the benzyl alcohol acetyltransferase (BEAT) gene from the *Clarkia breweri* in presence of the CaMV35S promoter. It catalyses the synthesis of the volatile compound benzyl acetate also reported by Semeria *et al*. (1996) and Zaccai *et al*. (2001). For producing transgenic lisianthus plasmid was introduced in *Agrobacterium tumefaciens* strain EHA105 and used for plant transformation. The transformed plants were similar in phenotype with normal. Study found that an external supply of benzyl alcohol induce five to seven times higher production of benzyl acetate in detached flowers and leaves of transgenic lisianthus plants, compared to non-transformed plants. No benzyl acetate was detected in tissues of both control and transgenic plants fed with water. The plant feded with additional alcoholic compounds, i.e. hexanol, benzyl alcohol, isoamyl alcohol, phenethyl alcohol, and cinnamyl alcohol, supposed to be used as substrates by BEAT, transgenic *in vitro*-grown lisianthus plantlets produced significantly higher levels of acetates than control plants. Study demonstrates the possibility of producing substrate-dependent acetates in transgenic lisianthus plants which might lead to induction of new aromas (Aranovich *et al*., 2007).

REFERENCES

Aranovich, D., Lewinsohn, E. and Zaccai, M. 2007. Post-harvest enhancement of aroma in transgenic lisianthus (*Eustoma grandiflorum*) using the *Clarkia breweri* benzyl alcohol acetyl transferase (BEAT) gene. *Postharvest Bio. and Tech*., **43**(2): 255-260.

Asen, S., Griesbach, R.J., Norris, K.H. and Leonhardt, B.A. 1986. Flavonoids from *Eustoma grandiflorum* flower petals. *Phytochemistry*, **25**(11): 2509-2513.

Aylsworth, J.D. 1977. Plant breeder extraordinaire. *Greenhouse Grower*, **15**(3): 14-15.

Azrak, M.F. 1984. Cultural studies of greenhouse grown *Eustoma grandiflorum*. MS Thesis, Colorado State Univ., Fort Collins.

Bailey, L. 1949. Mannual of cultivated plants. MacMillian, New York.

Bailey, L.H. and Baily, E.Z. 1976. Hortus Third. MacMillan, New York. Corr, B. and P. Katz. (1997). A grower's guide to lisianthus production. Flora Culture International May: 16-20.

Chen, Y.T., Fang, Q.S., Chiang, C.H., Yeh, S.D., Wu, H.W. and Yu, T.A. 2010. Transgenic *Eustoma grandiflorum* expressing the *bar* gene are resistant to the herbicide Basta. *Plant Cell Tiss. Organ Cult*., **102**: 347-356.

Deroles, S.C., Ledger, S.E., Miller, R.M., Davis, K.M. and Given, N.K. 1993. Transformation in *Eustoma grandiflorum* (Lisianthus). Plang protoplast and genetic Engineering III Biotech. *Agri. and Forestry*, **22**: 202-212.

Dooner, H.K., Robbins, T.P. and Jorgensen, R.A. 1991. Genetic and developmental control of anthocyanin biosynthesis, *Ann. Rev. Genet*., **25**: 173-199.

Everett, T.H. 1981. The New York Botanical Garden Illustrated Encyclopedia of Horticulture, vol. 6. Garland Pub. Co., New York, London.

Farina, E. 1989. The cultivation of *Lisianthus* (*Lisianthus russelianus* Hook) for cut flower in annual or biennial cycle: Effect of planting date and plant age on productivity. *Acta Horticulturae*, **252**: 257-261.

Fascella, G., Agnello, S., Delmonte, F., Sciortino, B. and Giardina, G. 2009. Crop response of Lisianthus (*Eustoma grandiflorum* Shinn.) hybrids grown in soilless culture. *ActaHort*., **807**: 559-564.

Griesbach, R. J., Semeniuk, P., Roh, M. and Lawson R.H. 1987b. Genetic engineering as a tool to improve lisianthus. *Greenhouse Grower,* **5**(3): 18-20.

Griesbach, R.J. and Bhat, R.N. 1990. Colchicine-induced polyploidy in *Eustoma grandiflorum. HortScience*, **25**(10): 1284-1286.

Griesbach, R.J., Semeniuk, P., Roh, M. and Lawson, R.H. 1988. Tissue culture in the improvement of *Eustoma. HortScience,* **23**(4): 1, 791.

Griesbach, R.J., Semeniuk, P. Roh, M. and Lawson, R.H. 1987a. The use of somaclonal variation in the improvement of lisianthus. *J. Hered.*, **78**: 114-116.

Grueber, K.L., Corr, B.E. and Wilkins, H.F. 1985. Evaluation of *Eustoma grandiflorum* as a bedding plant. *Minn. State Flor. Bull.*, **34**(1): 16-18.

Halevy, A.H. and Kofranek, A.M. 1984. Evaluation of lisianthus as a new flower crop. *HortScience*, **19**: 845-847.

Handa, T. 1992. Regeneration and characterization of prairie gentian (*Eustoma grandiflorum*) plants transformed by *Agrobacterium rhizogenes. Plant Tissue Culture Letters*, **9**(1): 10-14.

Handa, T. 1996. Transformation of prairie gentian (*Eustoma grandiflorum*) with *Agrobacterium rhizogenes* harboring beta-glucuronidase (GUS) and neomycin phosphotransferase II (NPT II) genes. *J. Japanese Soc. Hortil. Sci.*, 64.

Handa, T. And S.C. Deroles. 2000. Transgenic *Eustoma grandiflorum* (Lisianthus). Biotechnology in agriculture and forestry: Transgenic Crops III. ED. Y.P.S. Ohkawa, K. (1992). The breeding and culture of *Eustoma grandiflorum.* Seibundo-Shinkosha, Japan.

Handa, T., Sugimura, T., Kato, E., Kamada, H. and Takayanagi, K. 1994. Genetic transformation of *Eustoma grandiflorum* with rol genes. *Genetic Improvement Hortil. Crops by Biotech.*, **392**: 209-218.

Harbaugh, B.K., Scott, J.W. and Rubino, D.B. 1996. Florida Blue 'Semi-Dwarf Lisianthus (*Eustoma grandiflorum* (Raf.) Shinn.). *HortScience*, **31**(6): 1057-1058.

Harbaugh, B.K. 2007. Lisianthus *Eustoma grandiflorum. In*: Neil O. Anderson's (Ed.). Flower Breeding and Genetics: Issues, Challenges and Opportunities for the 21st Century Part II Crop-specific Breeding & Genetics. pp. 645-664.

Harbaugh, B.K. and Scott, J.W. 1996. 'Maurine Blue' lisianthus (*Eustoma grandiflorum* (Raf.) Shinn). *HortScience*, **31**(6): 1055-1056.

Harbaugh, B.K. and Scott, J.W. 1998. Six heat-tolerant cultivars of lisianthus. *HortScience*, **33**: 164-165.

Harbaugh, B.K. and Scott, J.W. 1999. 'Florida Pink' and 'Florida Light Blue'- Semi-dwarf Heat-tolerant cultivars of Lisianthus. *HortScience*, **34**: 364-365.

Harbaugh, B.K. and Scott, J.W. 2001. Florida Silver'- A Semi-dwarf Heat-tolerant Lisianthus. *HortScience*, **36**(5): 988-989.

Harbaugh, B.K., Roh, M.S., Lawson, R.H. and Pemberton, B. 1992. Rosetting of lisianthus cultivars exposed to high temperatures. *HortScience,* **27**: 885-887.

Hecht, M., Ecker, R., Ran, S. and Watad, A.A. 1994. Differential expression *in vitro* of heterosis in Lisianthus (*Eustoma grandiflorum*) at various benzyladenine and Gibberellic acid levels. *In Vitro Cell Dev. Biol.*, **30**: 136-139.

Jones, J.P. and Harbaugh, B.K. 1995. *Curvularia* blotch of lisianthus. *Proc. Fla. State Hort. Soc.*, **108**: 60-62.

Kobayashi, R.S., Brewbaker, J.L. and Kamemoto, H. 1987. Identification of *Anthurium andraeanum* cultivars by gel electrophoresis. *J. Amer. Soc. Hort. Sci.*, **112**: 164-167.

Koike, S.T., Gordon, T.R. and Lindow, S.E. 1996. Crown rot of *Eustoma* caused by *Fusarium avenaceum* in California. *Plant Dis.*, **80**: 1429.

Lindsay, G.C., Hopping, M.E. and O'Brien, I.E.W. 1994. Detection of protoplast-derived DNA tetraploid Lisianthus (*Eustoma grandiflorum*) plants by leaf and flower characteristics and by flow cytometry. *Plant Cell Tissue Organ Cult.*, **38**(1): 5

Markham, K.R. 1996. Novel anthocyanins produced in petals of genetically transformed lisianthus. *Phytochemistry*, **42**(4): 1035-1038.

Mato, M., Sato, T., Himori, Y., Shibata, H. and Asari, Y. 2004. The clarification of the inheritance on flower colours and flower types for breeding of lisianthuses (*Eustoma grandiflorum*). *Bulletin Akita Agril. Experiment Station,* pp. 3-55.

McGovern, R. J. and Harbaugh, B.K. 1977. Finding fungicides for Fusarium. *Greenhouse Grower,* **15**(10): 44-45.

McGovern, R. J., Harbaugh, B.K. and Price, J. F. 1978. Potted lisianthus: Secrets of success for controlling diseases. *Greenhouse Grower,* **16**(3): 28-36.

McGovern, R.J., Polston, J.E. and Harbaugh, B.K. 1997. Detection of a severe isolate of *impatiens* necrotic spot virus infecting lisianthus in Florida. *Disease Notes*, **81**(11): 1334.

Messeguer, R. and Ares, P. 1985. Electrophoretic identification of carnation cultivars. *HortScience,* **20**: 372-373.

Ming-xia, G. 2008. Study on rapid propagation technology of *Eustoma grandiflorum in vitro* (J). *J. Anhui Agril. Sci.*, **29**: 018.

Miyashita, C., Minami, H. and Emori, K. 2005. Selection of heat-tolerant, low rosette-forming lines induced by gamma irradiation in *Eustoma grandiflorum. Bulletin Tokyo Metropolitan Agril. Experiment Station*, **33**: 53-57

Ohkawa, K. 1994. *Eustoma* (Lisianthus). *In*: K. Konishi (ed.). Horticulture in Japan, 24th International Horticultural Congress. Asakura Publishing Co., Ltd., Tokyo. pp. 159-161.

Ohkawa, K., Kano, A., Kanematsu, K. and Korenaga, M. 1991. Effects of air temperature and time on rosette formation in seedlings of *Eustoma grandiflorum* (Raf.) Shinn. *Scientia Horticulturae*, **48**(1): 171-176.

Ohki, S., Sakai, Y., Ooba, K. and Chatani, K. 2000. Use of ion beams for *in vitro mutation* breeding of ornamental plants- application in *Eustoma grandiflorum*. *Acta Hort.*, **508**: 75-82.

Ordogh, M., Jambor-Benczur, E. and Tilly-Mandy, A. 2004. Micropropagation of Echo´ cultivars of *Eustoma grandiflorum*. *In*: V International Symposium on *In Vitro* Culture and Horticultural Breeding, **725**: 457-460.

Oren-Shamir, M., Shaked-Sachray, L., Nissim-Levi, A. and Ecker, R. 1999. Anthocyanin pigmentation of lisianthus flower petals. *Plant Science*, **140**(1): 81-86.

Ozaki, K. 1992. Lisianthus: Control of Diseases and Insects, *In*: Ohkawa, K. (ed.). The breeding and culture of *Eustoma grandiflorum*. Seibundo-Shinkosha, Japan, pp.101-105.

Parke, M. 1986. New look of lisianthus. Horticulture. March: 32-34.

Pijnacker, L.P., Sree Ramulu, K., Dijkhuis, P. and Ferwerda, M.A. 1989. Flow cytometric and karyological analysis of polysomaty and polyploidisation during callus formation from leaf segments of various potato genotypes. *Theor. Appl. Genet.*, **77**: 102-110.

Price, J.F., Harbaugh, B.K. and McGovern, R.J. 1998. Potted lisianthus: Secrets of success for controlling insects. *Greenhouse Grower*, **16**(2): 22-27.

Roh, M.S. and Lawson, R.H. 1987. Research and development on new crops in the United States Department of Agriculture. *Acta Horticulturae*, **205**: 39-48.

Roh, M.S. and Lawson, R. H. 1984. The lure of lisianthus. *Greenhouse Mgr*., **2**(11): 103-104.

Roh, M.S., Halevy, A.H. and Wilkins, H.F. 1989. *Eustoma grandiflorum*. *In*: A.H. Halevy (ed.). Handbook of Flowering. Vol. VI. CRC Press, Boca Raton, Fla. Roh, pp. 322-327.

Rubino, D.B. 1991. Isozyme variation in cultivated *Exacum affine* and *Eustoma grandiflorum* germplasm. *HortScience*, **26**: 775.

Rubino, D.B. 1992. Trials rate lisianthus cultivars. *Greenhouse Manager*, **10**(1): 65-69.

Rubino, D.B. 1993. Inheritance of Esterase, Diaphorase, and Glucose-6-phosphate Isomerase in Lisianthus. *HortScience,* **28**(6): 661-663.

Sakata, T. and Co. 1982. Sakata's Reliable Seed Catalog 1982/83. Yokahama, Japan.

Schwinn, K.E., Davies, K.M., Deroles, S.C., Markham, K.R., Miller, R.M, Bradley, J.M., Manson, D.G. and Given, N.A. 1997. Expression of an *Antirrhinum majus* UDP- glucose: flavinoid-3-O-glucosyltrasnsfrease trans gene alters flavinoid glycosylstion and acylation in lisianthus (*E. grandiflorum* G Rise.). *Plant Sciences*, **125**(1): 53-61.

Semeniuk, P and Griesbach, R.J. 1987. *In vitro* propagation of prairie gentian. *Plant Cell, Tissue Org. Cult.*, **8**: 249-253.

Semeria, L., Ruffoni, B., Rabaglio, M., Genga, A., Vaira, A. and Accotto, G. 1996. Genetic transformation of *Eustoma grandiflorum* by *Agrobacterium tumefaciens*. *Plant Cell Tissue Org. Cult.*, **47**: 67–72.

Semeria, L., Vaira, A.M., Accotto, G.P. and Allavena, A. 1995. Genetic transformation of Eustoma grandiflorum Griseb by microprojectile bombardment. *Euphytica*, **85**: 125-130.

Shinners, L. H. 1957. Synopsis of the genus *Eustoma* (Gentianaceae). The Southwestern Naturalist, 38-43.

Skrzypczak, L., Wesoowska, M. and Budsianowski, J. 1993. *Eustoma grandiflorum* Shinn (Texas Bluebell): Callus culture, micropropagation and the production of gentiopicroside and other secondary metabolites. *In*: Medicinal and Aromatic Plants V. Springer Berlin Heidelberg. pp. 192-201.

Sumanasinghe, V.A. 1986. Electrophoretic, cytogenetic, crossability and morphological studies of exacum (Gentianaceae). Ph.D. Thesis. The Pennsylvania State Univ., University Park.

Takahashi, M., Nishihara, M. and Yamamura, S. 1998. Stable transformation of *Eustoma grandiflorum* by particle bombardment. *Plant Cell Reports*, **17**: 504-507.

Tjia, B. and Sheehan, T.J. 1984. Lisianthus (*Eustoma*) as a bedding plant. *Florida Ornamental Grower's Association Newsletter*, **6**(4): 1.

Tjia, B. and Sheehan, T.J. 1986. Chemical height control of *Lisianthus russellianus*. *HortScience*, **21**(1): 147-148.

Uddin, A.F.M.J., Hashimoto, F. and Sakata, Y. 2002. Mode of inheritance and allele expression of three major anthocyanadins in *Eustoma grandiflorum* cultivars. *In*: 26th International Horticultural Congress: Elegant Science in Floriculture, pp. 51-59.

Uddin, A.F.M.J., Islam, M.S., Mehraj, H., Roni, M.Z.K. and Shahrin, S. 2013. An evaluation of some Japanese Lisianthus (*Eustamo grandiflorum*) varieties grown in Bangladesh. *The Agriculturists*, **11**(1): 56-60.

Uddin, A.F.M.Jamal, Hashimoto, F., Miwa, T., Ohbo, K. and Sakata, Y. 2004. Seasonal variation in pigmentation and anthocyanidin phenetics in commercial Eustoma flowers, *Scientia Horticulturae*, **100**(1–4): 103-115.

Weeden, N.F. and Wendel, J.F. 1989. Genetics of plant isozymes, *In:* D.E. Soltis and P.S. Soltis (eds.). Isozymes in plant biology. Dioscorides Press, Portland, Ore. pp. 46-72.

Wendel, J.F. and Parks, C.R. 1982. Genetic control of isozyme variation in *Camellia japonica* L. *J. Hered.*, **73**:197-204.

Wilkins, H. F. and Grueber, K.L. 1983. *Eustoma grandiflorum*. Minnesota State Flor. Bull., **32**(6): 10-11.

Wood, D.E. and Weaver, R.E. 1982. The genera of Gentianaceae in the Southeastern United States. *J. Arnold Arbor.*, **63**: 441-487.

Zaccai, M., Lewinsohn, E. and Pichersky, E. 2001. Modifying lisianthus traits by genetic engineering. *Acta Hortic.*, **552**: 137–142.

Zaccai, M., Lewinsohn, E. and Pichersky, E. 2001. Modifying lisianthus traits by genetic engineering. *In*: 20th International Eucarpia Symposium, Section Ornamentals, Strategies for New Ornamentals-Part I 552. pp. 137-142.

Chapter – 17

Tulip

Tulip cut flower is first among the bulbous ornamentals due to its attractive colours and exquisite flowers. Tulip is a member of Liliaceae family, commonly grown from bulbs. The name tulip is derived from the Persian word "toliban" meaning "turban", which is an appropriate term to describe the flower shape of certain tulips. The mother bulb dies and is replaced by several daughter bulbs, each of which is an annual unit of growth borne in the axils of the mother bulb scale. Tulips are basically grown for three purposes i.e. bulb production, forcing as cut flower and potted plant and for landscaping (Singh, 2006). The world tulip flower industry is dominated by the Netherlands. Another important tulip producing country is France.

ORIGIN AND DISTRIBUTION

The primary gene centre of the genus *Tulipa* is located in the Pamir Alai and Tien Shan mountain ranges of central Asia. Diversification in this region, resulting in a distribution from Morocco to Western Europe and eastward to Western China. A second gene centre has been found in the Caucasus (Hoog, 1973). However, Bailey (1963) reported that the tulip is native of Siberia, Turkey, Asia Minor, China, Japan and Mediterranean countries of Europe, the species are particularly abundant in central Asia, in the Bokhara region.

Tulips were first seen in Turkey. The vast majority of species occurs in central and western Asia in the countries such as Turkey, Afghanistan, Iran, Kazakhistan, Turkmenistan, Uzbekistan and Georgia. East of this center of distribution, the range of a couple species extends into northern India and far western China but on limited basis. One species of tulip i.e. *Tulipa stellata* is found wild in the mid and high hills of

Himachal Pradesh and *T. aitchisonii* is native of Himalaya. There are around a hundred species of *Tulipa* originated from Turkey to the Pamir and Hindu Kush mountains in Asia, which have been used for breeding purposes. In fact, a small number of species can be found in the wild in Europe and those are all restricted to southern parts, primarily in the Mediterranean region as far west as Portugal. There is not a single species of tulip native to Holland, the country frequently associated with these colourful and well known flowers.

Records show that tulips were in cultivation in Turkey as early as 1000 A.D. In 1578, first tulip appeared in England. Carolus Clusius planted the first tulip in 1593 in the botanical garden in Leiden (The Netherlands). The tulips appear in France in 1598. During 1600-1650 tulip become an exclusive garden plant and it was planted in strategic places in various gardens. Commercial cultivation of tulip started during 1600 in South Haarlem. The famous first parrot tulip was described in 1630. The famous tulip firm namely Voorhelm was established in 1650 in the Haarlem. An inventory of 126 different tulips was released in 1661. The bulbs of tulip had been sold to the U.S. from Holland as early as the eighteenth century. During nineteenth century, several hybrids were developed and cultivable area expanded largely. A fragrant variety Prins van Ooslenrijk and a double early tulip Murillo were developed in 1860. The worldwide popular Darwin tulip group was introduced in 1889. The first classified list of the tulip name including cultivation practices and flower management was released during 1914-1915. The most of the modern tulip complex is based on *T. gesneriana* with some important hybrids with species tulip. There are approximately 100 of these wild or species of tulips that occur naturally in the world and several thousands of distinctly different hybrids, selection and mutants that are gracing gardens and becoming remunerative to the growers around the world. In India its cultivation is limited to amateur gardeners of hilly areas in Kashmir and Himachal Pradesh. Srinagar, Kangra, Kullu, Mandi, Solan, Shimla, etc. are most suitable areas. Tulip cultivation can be extended in new areas like hills of Uttarakhand in India.

SPECIES

The genus includes about 160 species and important ones are *Tulipa gesneriana*, *T. biflora*, *T. chrysantha*, *T. clusiana*, *T. haferi*, *T. kaufmanniana*, *T. greigii*, *T. eicheri*, *T. festeriana*, *T. montana*, *T. persica*, *T. prastans*, *T. tarda* and *T. turkestanica* (Bhalla and Sulthan, 2003). One tulip species i.e. *T. stellata* is found in wild in the mid and high hills of Himachal Pradesh. Tulips were introduced from Turkey into Europe. They flowered first time in the Netherlands in 1594. Around 1630, tulips were extremely popular. The introduced tulips have been grown and bred for a long time. This has resulted in a wide range of diversity in growth, flowering, vigour, flower colour and shape. These tulips, whose original species have not been determined, are grouped together and are called *T. gesneriana*. Earlier, Hall (1940) also reported that most of the garden tulip complex is based on *T. gesneriana* with some important hybrids and tulip species. This species is presently playing a vital role in tulip improvement programme. Henceforth, the current commercial assortment consists, mainly of cultivars from *T. gesneriana*. The second groups of cultivars, the Darwin hybrids, have been obtained from interspecific crosses between cultivars of *T. gesneriana* and genotypes of *T. fosteriana*. *T. gesneriana* and Darwin hybrids consist of more than 1100 cultivars (Van Scheepen, 1996). Other important species in commercial cultivation are *T. fosteriana*, *T. greigii* and *T. kaufmanniana*.

FLOWER MORPHOLOGY AND BIOLOGY

Flower parts

Tulip flowers are of rich and brilliant colours and ornamentally called by different names. Flowers in solid colours are 'selfs' or 'breeders'; those with mixed colours are called 'broken'; when the under-colour is white and markings lilac or purple, they are 'bybloems' and when the under colour is yellow and marking red to brown, they are known as 'bizarres' (Bailey, 1963). Tulip flowers are comprised of 2 whorls each of 3 perianth parts. Perianth deciduous, campanulate or slightly funnel shaped. Stamens are 6 in 2 whorls and hypogynous, shorter than perianth. Filament longer or shorter than anthers, alternate or filiform and dehiscing laterally. Ovary sometimes narrowed at collar, comprises of style and tripartite stigmatic surface with fringed edges. The ovary has 3 carpels, each containing 2 rows of ovules. One ovary contains between 150 ovules (*T. turkestanica*) to 210-270 ovules (most species) to 300-450 ovules (*T. gesneriana*).

Pollination

Attractive colour of flower influence pollination by attracting insects. A study in this line has been carried out by Weryszko Chmielewska *et al.* (2002). Six cultivars of *Tulipa gesneriana* namely Attila, Rijnland, Giewont, Hocus Pocus, Esperanto and Rewizor were selected to find out flower attractiveness for insect pollination. Cv. Hocus Pocus was most attractive from the pollination ecology point of view.

Pollen viability

Pollen viability greatly varied according to cultivars. Twenty one cultivars were selected for this study and sterile pollen varied from 10.5% to 98.2%, whereas number of insect pollinators was highest in cultivars with yellow perianths in spite of high level of pollen sterility (Weryszko Chemielewska *et al.*, 2003). For storage, anthers can be collected from flowers, placed in plastic containers and stored in exicator with silicagel at 5°C. In this way pollen can be stored upto 1 year without losing its germination power (Jensen, 1964).

In pollination study it is observed that a central bundle of pollen tubes develop in the ovarian cavity after compatible pollination. Pollen tubes bend sideways and grow towards and into the ovules. At 15°C, they reach the first ovules at 1-3 days after pollination. This temperature is optimal for most tulip crosses (Kho and Baer, 1971). The lowest ovules are reached about 7-11 days after pollination. The number of ovules penetrated by a pollen tube increases from 3 to 9 days after pollination. On an average 68-83% of the ovules is ultimately penetrated by a pollen tube (Van Creij *et al.*, 1997a). After pollination it takes about 12 weeks for harvesting of mature seeds (Van Creij *et al.*, 1997b).

Seed setting

Seed setting is a barrier specially in interspecific crossing in tulip. To overcome with this problem, hormone treatment of ovaries, cut-style, grafted-ovary and placental pollination were evaluated for producing interspecific crosses between *T. gesneriana* and 8 other tulip species. Seed setting was noticed with treating ovaries at 0.1% benzyladenine at 12 days after crosses between *T. gesneriana* and *T. agenensis*. Treating ovaries with 1% BAP gave a negative effect on seed set in compatible crosses. The cut-style and grafted-ovary method was not found effective in pollen tube penetration. After placental pollination, pollen tube penetration percentage was similar to stigmatic pollination; however, most of the penetrated ovules with pollen tube penetration showed embryo germination (Van Creij *et al.*, 1997).

Self incompatibility and pollen germination

Self incompatibility is reported in tulip which may be overcome by early planting or partially overcome by excision of stigma in some varieties (Keiichi and Murakami, 1992). Self incompatibility expressed by pollen tube penetration into stigma and style was studied by Moue and Urashima (1996). Pollen germination was observed in all cultivars within 4 hours to 1 day after pollination but seed setting was high to low in different cultivars. Conclusion was that degree of self-incompatibility may be related to the distance of penetration of the pollen tube into the transmitting tissue. Rate of morphological abnormalities in pollen tube was 69-100%; spiral and swollen pollen tubes was observed in the transmitting tissue from 1 to 7 days after pollination in this study.

GENETICS

The majority of the tulips (about 85%) are diploid (2n = 24), 14% are triploid (2n= 36) and 1% are tetroploids (Rees, 1985). Chromosome number of some tulip cultivars were studied by Mokhono and Yakovleva (1976) and they observed that out of 10 cultivars, one cultivar was tetraploid (2n = 48), 2 were triploid and rest 8 cultivars were diploid. They further reported that the triploids have a high ornamental value and the tetraploids and diploids were of breeding interest.

Flower colour

Flower colour is determined by the composition and proportion of different flower colour pigments. Yellow colour contains only carotenoids, while orange flowers contain at least carotenoids and cyanidin. Red flowers contain cyanidin and mostly also pelargonidin, whereas purple flower contain a combination of cyanidin and delphinidin. Pink flowers contain the largest variation in flower colour pigments but the level is mostly lower compared to the other colours. White flowers contain almost no colour pigments. In addition to single coloured tulips there are also multi-coloured tulips e.g. pink-red and purple-red. Flowers with approximately the same colour differed in the pigment composition and its vice-versa. Furthermore, in several seedlings flower pigments were found which were not present in either parents. For example, a yellow × yellow can result in some red flower plant.

Blue colour in flowers is due mainly to anthocyanins and a considerable part of blue colouration can be attributed to metal-complexed anthocyanins. Momonoi *et al.* (2009) identified a vacuolar iron transporter in *T. gesneriana* (*TgVit1*), and characterized the localization and function of this transporter protein in tulip petals. The amino acid sequence of TgVit1 is 85% similar that of the *Arabidopsis thaliana* vacuolar iron transporter AtVIT1, and also showed similarity to the AtVIT1 homolog in yeast, Ca^{2+}-sensitive cross-complementer 1 (CCC1). The gene *TgVit1* was expressed exclusively in blue-coloured epidermal cells, and protein levels increased with increasing mRNA expression and blue colouration. Transient expression experiments revealed that TgVit1 localizes to the vacuolar membrane, and is responsible for the development of the blue colour in purple cells. Expression of *TgVit1* in yeast rescued the growth defect of *ccc1* mutant cells in the presence of high concentrations of $FeSO_4$. Our results indicate that TgVit1 plays an essential role in blue colouration as a vacuolar iron transporter in tulip petals. These results suggest a new role for involvement of a vacuolar iron transporter in blue flower colour development.

Inheritance of flower colour

In order to investigate the inheritance of flower colour, Nieuwhof *et al.* (1998) crossed and selfed 7 tulip cultivars in a diallel and relative amounts of various pigments were determined in parents and F_1 hybrids. For delphinidin, quercetin and kaempferol only additive gene action was determined, whereas for carotenoids, cyanidin and pelargonidin were due to non-additive gene action and plasmic differences. They further observed that biosynthesis of delphinidin or pelargonidin was only found when cyanidin was present, while the synthesis of dephinidin and pelargonidin were antagonistic to each other.

Inheritance of flower shape

As with flower colour inheritance, prediction of the inheritance of flower shape in relation to their parents appeared to be difficult. At Plant Research International (PRI), the Netherlands, crosses made between lily-flowered cultivar, gave almost no pure lily-flowered, double flowered, edge-coloured, single-coloured and base-coloured cultivars to find out inheritance of flower shape. The lily-flowered cv. Dyanito combined with a non lily-flowered cultivar gave almost no pure lily-flowered plants. Populations with *T. accuminata* as a parent, however, gave lily-flowered plants. The double-flowered cv. Abodement gave hardly any pure double-flowered seedlings. Crosses between edge-coloured cultivars gave high percentage of edge-coloured plants and when combined together with non-edged cultivars the offspring resulted in approximately 50% of edge-coloured plants.

Inheritance of *Fusarium* resistance

The inheritance of *Fusarium* resistance was investigated and it was found that the use of one resistant parent can result in resistant descendants (Van Eijk *et al*., 1979). Van Eijk and Eikelboom (1990) have also reported *Fusarium* resistance to be largely based on additive gene action.

Genetic variability and heritability

Jhon *et al.* (2006) evaluated 40 genotypes of tulip (*Tulipa* sp.) in India under 2 different environments viz. open field and polyhouse for various traits. Increased plant height, scape length and earlier flowering was recorded under polyhouse whereas higher bulb number and bulblet weight plant was recorded in open conditions. Phenotypic coefficients of variation were higher than genotypic coefficients of variation for all characters studied under both the environments. All the characters showed high heritability, except bulb sprout and bulb number plant^{-1}. High heritability with high genetic gain was observed for plant height, scape length and bulblet weight plant^{-1} under both the environments.

BREEDING OBJECTIVES

i. Shortened juvenile period.

ii. Fast multiplication rate.

iii. More yearly increase in main bulb diameter.

iv. Variety required shorter cold period.

v. Variety required less forcing period.

vi. Improved flower longevity.

vii. New flower shape.

viii. New flower colour.

ix. Resistance to *Fusarium oxysporum* (bulb rot) and other fungi i.e. *Botrytis tulipae*, *Pythium* spp., *Rhizoctonia tuliparum/solani*.

x. Resistant to Tulip Breaking Virus (TBV), Tobacco Necrosis Virus (TNV) and Tobacco Rattle Virus (TRV).

xi. Resistance to mites.

xii. Resistance to nematodes (*Trichodoridae*, *Pratylenchus penetrans* and *Ditylenchus dipsaci*).

BREEDING METHODS

Introduction

India has different types of agro-climatic condition. Tulip required cool climate. Several improved and exotic varieties introduced from different countries. Tulip cultivars Apeldoorn, Apeldoorn's Elite, Barcelona, Blenda, Candela, Cassini, China Pink, Christmas Marvel, Don Quichotte, Golden Apeldoorn, Golden Melody, Golden Oxford, Mrs. Jhon T. Scheepers, Keen Nelis, Lucky Strike, Orange Emperor, Paul Richter, Princes Victoria, Shirly, Striped Banllon, Sweet Heart, Valentine and West Point were introduced and evaluated at SKUA&T, Srinagar, J&K, India (Jhon and Khan, 2002). Apart from these cultivars, some other var. Arabian Mystery, Burgundy Lace, Cape Cod, Charles, Copland's Purple, Double Yellow, Estella Rijinveld, Flaming Parrot, Golden Harvest, Inzell, Menton, Orange Supreme, Orange Monarch, Parade, Pink Supreme, Primvera, Princess Margaret Rose, Prominence, Red Emperor, Show Winner and Yokohama were introduced and evaluated under open and polyhouse conditions (Jhon and Khan, 2003).

Selection

Tulip has long juvenile period and slow propagation rate, hence much attention was initially paid to pre-selection at Plant Research International (PRI), the Netherlands. Pre-selection is the selection at the juvenile stage from non-flowering seedlings. The main goals of pre-selection are bulb production, forcing for cut flower production and later disease resistance. Pre-selection was based on bulb production and forcing. Best population was selected on the basis of annual increment in main bulb size as well as number of years taken from seed to flowering plant for bulb production. However, pre-selection for forcing, besides earliness, other characteristics such as stem length, firmness of leaves and stem, leaf position and number, flower size, bud presentation and the proportion between flower and leaves are important. In 1978 and 1979, plant material obtained from the pre-selection research was released by PRI. The cvs. Debutante and Silver Dollar showed the best known results and they also have long vase life.

Screening for improved post harvest life: The life of cut flower is the most important characteristic of tulip for the consumer. Van Eijk and Eikelboom (1976) and Van der Meulen *et al.* (1997) developed screening tests in which the vase life of a genotype could be estimated. A good correlation was found between longevity of flower still attached to the bulb and the vase life of the flower (Van Eijk and Eikelboom, 1976). In general variation in flower longevity among cultivars varied from 8-16 days. Studying segregation of population derived from cultivars, the flower longevity varied from 6-22 days (Van der Meulen *et al.*, 1997). This indicated that selection for longer flower longevity, based on additive effects of several genes, is a promising method.

Screening for disease resistance: Tulip is affected by several diseases. For *Fusarium* resistance, Van Eijk *et al.* (1979) produced reliable screening tests for clones as well as for juvenile seedlings at pre-selection stage. In these tests, bulbs are planted in *Fusarium* infected soil and grown under standardized conditions for entire season. After harvest, bulbs are examined for *Fusarium* infection. Almost absolute resistance was found within the *T. gesneriana* assortment. An assay for screening tulips for resistance to *Botrytis tulipae* has been developed (Straathof *et al.*, 2002). Absolute resistance was found in *T. tarda*. Research on tulip breaking virus (TBV) resulted in reliable screening tests at the seedling as well as clonal level (Romanov *et al.*, 1991, Straathof *et al.*, 1997 and Eikelboom *et al.*, 1992). Leaves of flowering plants are inoculated by viruliferous aphids. Flower breaking was observed one year after inoculation. Absolute TBV resistance was found in several *T. fosteriana* cultivars, e.g. Cantata and Princeps.

A study on assessment and screening of varieties of tulip resistance to tulip breaking virus (TBV) was carried out by Juodkaite *et al.* (2012) by applying methods of test-plants, electron microscopy, DAS-ELISA. By resistance to *Tulip breaking virus*, tulip cultivars were arranged into 4 groups: (1) resistant; (2) moderately resistant; (3) non-resistant; (4) excessively non-resistant. In the study, the collection was analyzed according to classification groups. Triumph tulips and Single late tulips dominate. Among the investigated 15 classification groups of tulips grown, Darwin hybrid tulips, Kaufmanniana, Fosteriana, Greigii, varieties and hybrids and other tulip species and their varieties and hybrids are the most resistant to Tulip breaking virus. Most cultivars of the Double early tulips and Parrot tulips were moderately resistant to Tulip breaking virus. Single early tulips, Triumph tulips, Single late tulips, Lily-flowered tulips and Fringed tulips were rather susceptible to viral diseases.

Hybridization

It is a well known fact that tulip is the most important ornamental bulb crop in the world but again it has been observed since last several decades that production of flowers and bulbs of an optimal quality can be seriously affected by many pathogens. Therefore, disease resistances are important breeding objectives now. Beside this, a shorter cold requirement and forcing period, improved flower longevity, new flower shapes and colours are important goal for tulip breeding.

T. gesneriana is the most favourite species of the breeders. Each year several thousand flowers are hand pollinated. After growing the seeds till adult plants, most breeders select around 0.1%-1.0% of the best plants (on the basis of plant habit, colour, etc.) during forcing as cut flower in the greenhouses. After vegetative propagation further selection takes place for other characters such as disease resistance, bulb production, etc. Wide genetic variation in tulip can be achieved through interspecific hybridization. Although crossing barriers have prevented the formation of hybrids in many interspecific crosses, yet *T. gesneriana* has been crossed with 12 out of approximately 55 tulip species by adopting conventional breeding methods (Van Eijk *et al*., 1991 and Van Raamsdonk *et al*., 1995). Hogenboom (1973) reported that crossing barriers impeding sexual reproduction in interspecific tulip crosses are due to incongruity between the crossed species. After seed setting and obtaining viable seed, 5-6 years are needed to obtain a flowering bulb. However, this period can be shortened from 5-6 years to 3-5 years by applying optimal field and storage conditions. Subsequently, it takes another 10-15 years to screen the tulip for desirable characters and to multiply the bulbs for commercial release. Despite many efforts to develop a rapid propagation system *in vitro* (Hulscher and Krijgsheld, 1995, Chanteloube *et al*., 1995 and Kuijpers and Langens-Gerrits, 1997), the production of tulip bulbs still occurs primarily by propagation in field.

To enrich the commercial assortment with desirable traits, many crosses between *T. gesneriana* and other tulip species have been done. However, incongruity barriers impede sexual reproduction in several interspecific crosses either in part or in whole. These barriers can prevent or diminish the formation of viable seeds at pre-fertilization, during fertilization or post-fertilization stage. A comprehensive review has been well documented by Liedl and Anderson (1993). Pre-fertilization and post-fertilization barriers have been observed in interspecific crosses with *T. gesneriana* and other 8 section of genus *Tulipa* (Van Creij *et al*., 1997a and 1997b). In hybridization, besides varieties of *T. gesneriana*, several cultivars of the Darwin hybrids (crosses between *T. gesneriana* and *T. fosteriana*) are generally triploid and use of these triploids for further breeding is not possible due to F_1 sterility. In contrary, *T. gesneriana* was found to be compatible with other species of same section of genus *Tulipa*. Studies of pollen tube in the pistil and pollen tube penetration in the ovules in reciprocal crosses between *T. gesneriana* and *T. didieri* resulted in pollen tube growth percentages comparable to intraspecific crosses of *T. gesneriana* (Van Creij *et al*., 1997a). However, hybrids have been produced in several crosses between *T. gesneriana* and representatives of the section *Eichleres*. Crosses between *T. gesneriana* (pistillate parent) and *T. kaufmanniana* and *T. fosteriana* exhibited high pollen tube penetration (Van Creij *et al*., 1997a). Small number of seeds was obtained from these crosses (Van Raamsdonk *et al*., 1995).

However, no seeds were obtained from *T. gesneriana* and *T. praestans*, despite relatively high percentage of ovules with pollen tube penetration (Van Creij *et al.*, 1997b). In this case, post-fertilization barriers hinder or prevent the formation of viable seeds.

Van Raamsdonk and De Vries (1995) reported that crosses between *T. gesneriana* and species from section *Tulipanum* produced no verified hybrids, except for *T. systola*. Relatively high percentage of pollen tube penetration were found in the cross *T. gesneriana* × *T. agenensis*. Pollen tubes penetration percentage greatly varied from 7% (several species) to 31% (*T. pulchella*) or 87% (*T. clusiana*) in reciprocal crosses (Van Creij *et al.*, 1997a).

In order to investigate the possibility of using species in bridge crosses, some tulip section can often be crossed (Van Raamsdlonk *et al.*, 1995). F_1 bulblets were produced from crosses between species of the section *Eichleres* and *Clusianae*. Nevertheless, *T. systola* might be used in bridge crosses between species of the section *Tulipanum* and *T. gesneriana*. The species of the section *Eichleres*, which can be crossed with *T. gesneriana* can only be used in bridge crosses with species of some section.

Several factors proved to influence the rate of pollen tube growth, pollen tube penetration and the number of seeds obtained in tulip crosses. Cultivar effects, effects of accessions and year effects were found to influence seed set in interspecific crosses (Van Eijk *et al.*, 1991 and Custers *et al.*, 1995).

Breeding for *Fusarium oxysporum* f. sp. *Tulipae* resistance

Most of the bulbous flowering plants are affected by *Fusarium oxysporum*. It is a very important fungal, soil born disease and fungus occurs all over the world. Van Eijk *et al.* (1978) initiated breeding work on *Fusarium oxysporum* resistance. Within *T. gesneriana* cultivars large differences in resistance were found. Subsequently these results were in juvenile and adults of the same genotype and in crossing populations to study the genetics of the resistance (Van Eijk *et al.*, 1979 and Van Eijk and Eikelboom, 1983). *Fusarium oxysporum* resistance in tulip is assessed by counting the numbers of affected plants which needs the evaluation of large numbers of plants during a whole growing season.

Much emphasis has been given on tulip to develop *Fusarium* resistant cultivars. Van Eijk and Eikelboom (1990) carried out a comprehensive study to find out resistant cultivars. To determine the effectivity of early selection, starting from one-year-old seedling bulbs in 62 progenies of the diallel, were determined and compared with adult bulbs of some progenies. The pre-selection was generally effective. In some of the least resistant progenies no effect occurred, e.g. 21 (Topscore × General Patton), 62 (Due de Berlin × General Patton), 65 (Due de Berlin × Bartigon) and 70 (Snowstar × Peerless Pink). They also determined the effect of *Fusarium* in adult phase. For that, resistant plants were selected from same progenies and propagated as clones in infested soil. High percentage of resistant clones (73% in juvenile phase and 63% in adult phase) was observed. In this study the most resistant parent cultivars showed generally high general combining ability (GCA) i.e. they transmit a high degree of resistance to their progenies. They concluded that, for breeding on resistance, at least one of the parents must have a high degree of resistance and only progenies which show a high degree of resistance during the juvenile phase may be used for further propagation. Cultivars Lucky Strike, Black Parrot and Aristocrat were found resistant to *Fusarium*, whereas cvs. Due de Berlin, Snowstar and Lustige Witwe showed low degree of resistance (Van Eijk and Eikelboom, 1990).

Tsuji *et al.* (2005) harvested 955 seedlings from a cross between cv. Don Quichotte and cv. Ben van Zanten and after 6 years evaluation developed a variety Arisa. It has adaptability of forcing and resistance against bulb rot caused by *Fusarium oxysporum* and tulip mild mottle mosaic virus. 'Arisa' has a single flower and it flowers in the middle of April in an ordinary year. The flower colour is wine red, edged purplish pink.

Breeding for *Botrytis tulipea* resistance

Second important pathogen which causes economical loss is *Botrytis tulipae*. It is a necrotropic pathogen which is able to kill its host tulip actively and is also able to live on dead tissue subsequently (Staats, 2007). As a leaf pathogen, it severely affects tulip bulb production. Straathof *et al.* (2002) reported that *T. tarda* is resistant to this pathogen, however, this species cannot be crossed with the cultivars of *T. gesneriana*. Some cultivars of *T. gesneriana* and *T. kaufmanniana* were found partial resistant to *Botrytis tulipae*. At Plant Research International (PRI) breeding for *Botrytis* resistance is in progress.

Breeding for Tulip breaking virus (TBV) resistance

Tulip breaking virus (TBV), the causal agent of flower breaking, is one of the most important pathogens. Apart from the breaking of flower colour and it reduces bulb yield (Eikelboom *et al.*, 1992). It transmitted through aphids in a non-persistent way, therefore, spread through the field in a short period of time.

A comprehensive research work on TBV screening and resistance was carried out by Straathof *et al.* (1997). An incomplete diallel was made between the susceptible *T. gesneriana* cvs. Christmas Marvel, Kees Nelis and Lustige Witwe; the partial resistant *T. fosteriana* cvs. Juvan and Madame Lefeber and the absolute resistant cvs. Cantata and Princeps. Using the percentage of healthy plants, they calculated GCA, SCA and reciprocal effects. Significant correlation in GCA values were found between 3 years. In all test, they found cvs. Cantata and Princeps had significantly better GCA value than other cultivars. They further advocated that since back crossings of the resistant hybrids are necessary to obtain suitable cultivars for cut flower production, attention must be paid to the possible F_1 sterile in the *T. gesneriana* × *T. fosteriana* (Darwin) hybrids. A cultivar Yukitsubo has evolved from cross between Albino and Fujiyama. Flower colour of this variety is yellowish white. It has single flower, dwarf stem, adaptability of forcing and found suitable for bedding and pot culture. Infection rate of TBV and *Fusarium oxysporum* is low (Urashima *et al.*, 2000).

For the future, it would be a good effort to develop resistance to the 3 main diseases of tulip in order to have multi-resistant tulips.

Mutation

Tulip being a clonally propagated crop during cultivation several natural mutants (sports) were selected (Van Scheepen, 1996). Parrot tulips are famous because of their bizarre flower shape. In the past, the shape was obtained only by spontaneous mutation of cultivars. From an inbreeding Programme at PRI, a genetic parrot cv. Amethyst was found in a selfing of cv. Cordell Hull. Mutations can exhibit differences in flower shape (parrots, fringed and double) and flower colour. Besides these, mutations also occur in colour of leaf edge, plant height and bulb production. For mutation breeding, X-rays at various doses were applied (Broertjes and Alkema, 1970 and Van Harten and Broertjes, 1989). Mutation doses varied from 350-550 rad. Generally radiation can be applied early in the planting season (August) or late (November). Many mutants are known from specific cultivars like Bartigon, William Copland, Murillo, Apeldoorn, etc. In 1975, 1977 and 1978 several radiation mutants of Preludium and Lustige Witwe were released. However, private breeders are using mutation breeding (Straathof and Eikelboom, 1997).

Polyploidy

Zeilinga and Schouten (1968) described use of laughing gas for production of tetraploids. By placing plants, which were pollinated a week previously, for a day in a cylinder with laughing gas (N_2O) and 5 to 6 atmospheric pressure. These seedlings were crossed mutually to improve the fertility of these tetraploids. Highly fertile tetraploid cultivars of *T. gesneriana* and tetraploids of *T. fosteriana* and *T. kaufmanniana* were released to Dutch tulip companies in 1989 and 1991 (Straathof and Eikelboom, 1997). Interploidy crosses between diploid (2x = 24), triploid (3x = 36) and tetraploid (2x = 48) of varieties of *T. gesneriana*, *T. fosteriana* and their hybrids. In the 2x × 3x crosses 92.6% were diploids and 7.4% were aneuploids, while in the 3x × 2x crosses 60% were diploids and 40% were aneuploids. In crosses involving Darwin hybrids (interspecific varieties between *T. geseneriana* and *T. fosteriana)*, 2 crosses among 62 cross combinations produced progenies and resulted seedlings chromosome number ranging from 24 to 36. These seedlings are useful in the introgression of genes from *T. fosteriana* into *T. gesneriana* (Okazaki *et al.*, 2000). By making crosses between tetraploids, new tetraploids can be obtained.

Several tetraploid were released by PRI. Of which 'Judith Leyster' is best known. A problem with existing tetraploids is that their progeny flowered late in the season, which made them unsuitable for early forcing. Crossing tetraploids with diplolids can result in strong growing triploids. Unfortunately, these triploids are mostly sterile and can not be used in further breeding. Tetraploid *T. gesneriana* cultivars can also be used in combination with diploid *T. fosteriana* to obtain Darwin hybrids with a better vase life and suitable for early forcing. The cv. World's Favourite released, bred from a tetraploid *T. gesneriana* seedling (Denbola × Lustige Witwe) crossed with diploid *T. fosteriana* seedling.

BIOTECHNOLOGY

Tissue culture

The F_1 seeds of interspecific hybrids of *Tulipa gesneriana*, cv. Negrita × *T. buhseana*, *T. gesneriana*, cv. Negrita × *T. tianschanica* and *T. gesneriana*, cv. Negrita × *T. iliensis* germinated after 36-67 days of culture under 4°C and formed subsequently bulblets after 3-5 months of culture under 15°C. The germination rates of F_1 seeds of Negrita × *T. buhseana*, Negrita × *T. tianschanica* and Negrita × *T. iliensis* were 60.37-63.89%, 80.28-81.25% and 3.75-4.45%, respectively. There was significant difference among them. The F_1 bulblets of Negrita × *T. buhseana* proliferated in 1/2MS medium supplemented with 2.0 mg/litre 6-benzyladenine, 1.0 mg/litre KT and 0.1 mg/litre TDZ, meanwhile, the F_1 bulblets of Negrita × *T. tianschanica* proliferated in 1/2MS medium supplemented with 5.0 mg/litre 6-benzyladenine and 0.4 mg/litre KT. There were 41.67% of germinated Negrita × *T. tianschanica* showed abnormal morphology, which could induce bulblets in 1/2MS medium supplemented with 2.0 mg/litre 6-benzyladenine and 0.4 mg/litre KT (Wang *et al.*, 2009).

Somaclonal variation

Podwyszynska *et al.* (2010) did a comprehensive investigation to find out variants through somaclonal variation and evolve a novel variety of tulip. Somaclonal variation (SV) was evaluated in tissue-culture (TC) derived plants of the tulip cultivar 'Giewont' and genotype "Bs6", the mutant selected from among the micropropagated plants of the cultivar 'Blue Parrot' in 2004. The plant material of both genotypes derived from the long-term cultures maintained *in vitro* for a period of 2, 4 or 6 years. The plants were planted outdoors in an insect-proof tunnel in 1999-2004. Phenotypic evaluation was done in 2006 and 2007. The plants of the mutant "Bs6" have parrot flowers which are 1.5-2 cm longer with stems which are 7-10 cm longer than the original 'Blue Parrot' plants. Further the flower colour changed from purple-violet to red-purple. In 2008, selected "Bs6" true-to-type plants have given start new cultivar Agalia.

Somaclonal variation frequency in 'Giewont' and "Bs6" ranged from 5.9 to 15.3%, including 1.5-6.8% SV frequency resulted from instable variants, characterized with a flower fasciation or a partly green tepal, accreted with a last leaf. Some off-type plants of genotype "Bs6" had minor changes, i.e. the flower shape was altered to non-parrot type. Variants with major changes, having highly malformed flowers, occurred regularly in both genotypes. They had always leaves with thicken, vitreous venation. Such leaf malformation was also found within juvenile plants. Similar variants were previously observed in TC-derived plants of tulip 'Prominence'. DNA analysis of 'Giewont', "Bs6" and 'Prominence' variants (both the flowering and juvenile plants) with ISSR (inter simple sequence repeats) technique confirmed that the leaf malformation resulted from genetic changes. The results indicate that the trait of leaf thicken, vitreous venation can be considered as a morphological marker for early detecting the major genetic changes within juvenile plant material (Podwyszynska *et al.*, 2010) .

Somatic embryogenesis

The effects of 3 reagents such as picloram on tulip callus induction and somatic embryogenesis were studied by using the optimal design of "311-B" regression. The results showed that the callus inductivity and the somatic embryogenesis had a very significant correlation with picloram concentration. Picloram played a decisive role in tulip callus induction, and the callus could occur through 20 d culture. The optimal medium was MS+6 mg/L picloram+0.5 mg/L BA, the callus inductivity reached 100% and each piece of the callus averagely produced 10 embryoids. Picloram and BA had a positive interaction and could be used together. But picloram had a negative interaction with hymexazol and both did not have to be used together (Zhou *et al.*, 2011).

In situ hybridization

Interspecific hybridization was attempted in wildly grown *Tulipa clusiana* f. *stellata* (Hook.) Dasgupta & Deb. and garden tulips for the development of new varieties and utilization of desirable characteristics of wild germplasm. Germination of hybrid seeds developed by this process was hampered by embryo breakdown at later stages of embryo development, resulting in failure to obtain interspecific hybrid seedlings. Embryo rescue was attempted as a means to overcome post fertilization barrier in interspecific *Tulipa clusiana* f. *stellata* × garden tulip crosses. The combination of ideal time and cultural condition for the rescue of developing embryo by culturing the fertilized ovules were studied. Optimal sprouting was observed when fertilized ovule were cultured at 30 days after pollination (DAP) in half strength MS medium with 4% sucrose, kept in dark at 4ºC initially for 8 weeks followed by 15ºC for 9 to 32 weeks. Shoot proliferation was noticed after 8 to12 weeks in MS medium with 4% sucrose, 6-benzylaminopurine (BAP) 1mg/l and Naphthalene acetic acid (NAA) 1mg/l at 2ºC. Microbulb formation occurred in 3 out of 12 cultures after 10 to 14 weeks in MS medium containing 4% sucrose, GA_3 2 mg/l and BAP 1 mg/l at 20ºC. The chromosome number of the putative hybrids was found to be 2n=36, with many chromosomal abnormalities noticed during anaphase and telophase. This technique offers possibilities for the introduction of variation in flower characters from garden tulip cultivars into *Tulipa clusiana* f. *stellata* (Karthigeyan *et al.*, 2006).

Hybridization between *Tulipa gesneriana* cultivars (G, 2n=2x=24) and Darwin hybrids (*T. gesneriana* × *T. fosteriana*; 2n=2x=24) was made by Marasek-Ciolakowska *et al.* (2011) with the aim of transferring desirable horticultural traits from *Tulipa fosteriana* (F) to *T. gesneriana* as well as its implications for breeding virus resistant tulips. The genome composition and the extent of intergenomic recombination in a number of diploid progenies (2n=2x=24) was evaluated by genomic *in situ* hybridization technique (GISH) using differently labelled total genomic DNAs of 'Princeps' (*T. fosteriana*) and 'Ile de France' (*T. gesneriana*). In all progenies analysed, it was possible to distinguish chromosomes from both parental genomes as well as the recombinant chromosomes. The number of *T. fosteriana* chromosomes as well as the number and type of recombinant chromosomes differed among progenies. The total number of intact *T. fosteriana* chromosomes in hybrids ranged from zero to three whereas the maximum number of recombinant *T. fosteriana* chromosomes was five. Among recombinant chromosomes, most contained a combination of a single *T. gesneriana* and a single *T. fosteriana* fragment. The maximum number of recombinant segments per chromosome was two and their position ranged from distal to highly interstitial. The Darwin hybrids turned out to be very useful intermediate parents for introgressing the *T. fosteriana* germplasm into the *T. gesneriana* assortment.

Assessment of intergenomic recombination through GISH analysis of F_1, BC_1 and BC_2 progenies of *Tulipa gesneriana* and *T. fosteriana* made by Marasek-Ciolakowska *et al.* (2012a) through using 23 F_1 hybrids, 14 BC_1 and 32 BC_2 progenies. The genome composition of Darwin hybrid tulips was analyzed through genomic *in situ* hybridization (GISH) of somatic chromosomes. They explained the fact that, unlike the situation in most F_1 hybrids of other plant species, certain genotypes of Darwin hybrid tulips behave like normal diploid plants producing haploid gametes and give rise to mostly diploid sporophytes.

In vitro polyploidization

Polyploids, especially triploids and tetraploids are widely used in the breeding programmes of ornamental plants, due to their desirable traits such as vigorous growth, large leaves or flowers. Tetraploids *per se* can constitute the new cultivars but, first of all, they can be utilized for further breeding. The cultivar Fringed Black and three diploid breeding clones (Pol-D 21, Pol-D 32 and Pol-D 50) of tulip were used for *in vitro* polyploidization. The initial explants (flower stem fragments), isolated from the cooled bulbs, were treated with antimitotic agents, oryzalin and amiprophos methyl, by the incubation of explants for seven or fourteen days on the regeneration medium containing above mentioned chemicals at the concentration of 5 and 10 mg L^{-1}. Subsequently explants were transferred on the same medium without antimitotic agents and subcultured three times every four, six and eight weeks, respectively. Then, bulb formation was induced by cooling of the regenerated shoots for 14 weeks at 5°C. The obtained bulbs were rooted *ex vitro*. Ploidy evaluation, using flow cytometry (performed just before cooling), revealed that tetraploids were obtained for the two genotypes, four tetrapoids (28.6%) in genotype Pol-D 21 and sixteen ones (33.3%) in genotype 'Fringed Black'. Both oryzalin and amiprophos methyl induced tetraploid production (Podwyszynska, 2012).

In a study, $2n$ gamete producing F_1 hybrids of Darwin hybrids (*T. gesneriana* × *T. fosteriana*) have been selected and used as pollen donor with the aim of producing of polyploid tulips. In total, 308 one-year-old BC_1 seedlings resulting from crosses between diploid and triploid *T. gesneriana* cultivars and diploid $2n$ gametes producers have been tested by flow cytometry analysis. The progenies from crosses at diploid level ($2x \times 2x$) were mostly diploids, whereas a few seedlings were triploids. In crosses $3x \times 2x$, 81 genotypes were tetraploids and 25 seedlings were pentaploids. Cytological analyses using GISH of the sexual polyploid progenies have shown considerable amounts of intergenomic recombination in the BC_1 generations which is desirable for introgression breeding (Marasek-Ciolakowska *et al.*, 2012b).

Strategies to overcome post fertilization barriers

For overcoming crossing barriers after interspecific hybridization, biotechnological tools have been found successful. Methods for bypassing post-fertilization barriers focus on the survival of the hybrid embryos and on restoring the fertility of F_1 hybrids. Work on this line by embryo culture, ovule culture, ovary slice culture and ovary culture have been developed to enable hybrid embryos to survive *in vitro* is well reviewed (Williams *et al.*, 1987 and Sharma *et al.*, 1996). The application of embryo-rescue techniques have been well documented by various workers (Van Tuyl *et al.*, 1990, Custers *et al.*, 1992, 1995 and Van Creij *et al.*, 1999, 2000). Fertility of tulip hybrids can be restored by chromosome doubling. In this line, tetraploid tulip cultivars have been produced after treating tulip stems from bulbs *in vitro* with oryzalin or colchicine (Eikelboom *et al.*, 2001 and Van Tuyl *et al.*, 2002). A procedure for *in vitro* pollination in tulip has been described (Van Creij, 1997) by using compatible intraspecific *T. gesneriana* crosses as model. It offers good prospects for tulip breeding.

Molecular breeding

Tulip being a long juvenile period crop is highly suitable for using molecular marker techniques and this makes pre-selection possible. Van Heusden *et al.* (2002) did work on molecular assisted breeding by using AFLP-markers in tulip. AFLP-markers were found over the entire tulip genome (12 linkage groups). Various AFLP-markers were also found to be linked with TBV-resistance. Further AFLP method was found to be successful for the fingerprinting of the tulip genome (Bondrea *et al.*, 2008). The application of PCR-derived markers for various characters will speed up the breeding process of tulips in future. Luan *et al.* (2008) carried out the RAPD analysis of tulip cultivars and its wild species.

Kiani *et al*. (2012) analyzed the genetic relationship among 39 *Tulipa* accessions using inter-simple sequence repeat (ISSR) primers and found the usefulness of DNA fingerprinting for identification of *Tulipa* accessions, and it is imperative to collect and characterize more genetic variability from the other distribution areas of this genus.

In populations of four species of tulips (*Tulipa biebersteiniana*, *T. patens*, *T. scytica* and *T. riparia*) genetic diversity was studied by means of morphological and AFLP analysis. A morphological analysis of seven quantitative and two qualitative criteria was carried out. Three selective *Eco*RI/*Mse*I primer pairs allowed one to genotype 81 individuals from 13 tulip populations with 87 loci. The low level of variability by AFLP loci were revealed in all species, including *T. biebersteiniana* (P=20.41%, UH_e=0.075), *T. patens* (26.97%, 0.082), *T. scytica* (27.53%, 0.086), and *T. riparia* (27.72%, 0.096). According to results, the variability proportion that characterizes the differences between the four *Tulip* species was lower (F_{CT}=0.235) than between populations within species (F_{ST}=0.439). *Tulipa patens* is well differentiated by means of Nei's distances, coordination, and analysis in the structure programme. An analysis in the structure revealed four genetic groups of tulips that are not completely in accordance with the analyzed species. This acknowledges the presence of complicated genetic process in the tulip population (Kutlunina *et al*., 2013).

In a study, *Tulipa fosteriana* was found to contain 45S rDNA repeat units of 9.7 and 9.5 kb, in which at least 7 types of 45S rDNAs were identified by restriction site analysis (Mizuochi *et al*., 2007). For 5S rDNA, repeat units ranging from 364 bp to 396 bp were identified. The diploid cultivars (2n=2x=24) 'Christmas Dream' and 'Queen of Night,' representing the horticultural group *T. gesneriana*, and 'Red Emperor', belonging to *T. fosteriana*, were compared cytogenetically using cloned 5S and 45S rDNAs. Fluorescence *in situ* hybridization (FISH) analysis identified many rDNA sites located on each chromosome in the diploid genomes. They identified 71 sites of 5S rDNA and 10 sites of 45S rDNA in 'Red Emperor'. Additionally, FISH analyses enabled construction of karyotypes for these cultivars. Karyotype comparison of *T. gesneriana* cultivars showed conservation of repetitive rDNA unit positioning. A clear difference in chromosome size and signal pattern was observed between *T. gesneriana* and *T. fosteriana* cultivars. They further demonstrated the unique nature of the highly repeated 5S rDNA units in these *Tulipa* species and the usefulness of FISH karyotyping with cloned 5S and 45S rDNAs to clearly distinguish between chromosomes from *T. gesneriana* and *T. fosteriana*.

The production of tulip flowers and bulbs of an optimal quality was recently affected by many pathogens. Therefore, breeding of tulip is now concentrated to develop varieties free from diseases especially bulb rot (*Fusarium oxysporum*), fire (*Botrytis tulipae*) and Tulip Breaking Virus (TBV) and second priority was given on forcing quality, flower colour and form. The inheritance of *Fusarium* resistance was investigated and found that use of one resistant parent can result in resistant descendants. An assay for screening tulips for resistance to *Botrytis tulipae* has been developed and work in this line is in progress at Plant Research International (PRI), The Netherlands.

Genetic engineering

Genetic modification of tulip has been investigated by Wilmink *et al.* (1992 and 1995). They obtained gene transfer to 7 cultivars and 7 species of *Tulipa* by particle bombardment and *Agrobacterium* transformation. Using the Particle Delivery System (Du Pont), transient expression of the GUS reporter gene for beta-glucuronidase was demonstrated and it was shown that the CAMV 35S and TR2 promoters were active in flower stem explants. Various wildtype and disarmed *Agrobacterium* strains, harbouring the 35S GUS intro gene on a binary plasmid was used for infection of flower stem explants. The GUS intro gene was expressed in 9 of the genotypes despite the fact that tulip tissue did not produce detectable amounts of virulence-inducing substances. *Agrobacterium rhizogenes* appeared to be most effective in gene transfer to tulip tissue. Seven years after selecting GUS-positive and PPT-resistant explants only one complete transgenic tulip plant flowered. It proved that stable integration was established. In a study on selecting agents and marker genes for use in genetic transformation of *T. gesneriana*, Chauvin *et al.* (1997) observed that endogenous beta-glucuronidase activity was detected in some of the tulip explants even when sampling was done *in situ*. In another study Wilmink *et al.* (1995a) compared the activity of monocot-derived promoters and 35S promoters in tulip. They observed that the tulip tissue behaved in a similar manner to *Nicotiana* tissue with all promoters tested generating a high level of expression (35S, emu, act1, emu). In addition, the inclusion of the adh1 intron with 35S promoter (35Si) significantly reduced promoter activity. Hence, it seems that tissue of tulip behaves more similarly to dicot tissue than to other monocots.

SUMMARY

Several hundred promising varieties have been developed through interspecific hybridization. *Tulipa gesneriana* is most important species used in breeding programme but pre-fertilization and post-fertilization barriers have been observed in interspecific hybridization. Now methods have been developed for overcoming these barriers. This crop has long juvenile period. It takes 20-25 years to develop new cultivar. Therefore, most of the novel varieties are under breeder's right act. Juvenile period can be reduced by pre-selection. For the future, the application of molecular assisted breeding techniques can be an important tool for reaching goal of developing novel and disease resistant assortment.

REFERENCES

Bailey, L.H. 1963. The Standard Cyclopedia of Horticulture, Vol. III: P-Z. Mac Millan Co. New York. pp. 3393-3404.

Bhalla, R. and Sulthan, B. 2003. Tulips, *In*: Bulbous Ornamentals. P.K. Rajeevan, K.P. Singh, P.K. Valsalakumari and C.K. Geetha, (eds.). ISOH, New Delhi, pp. 86-93.

Bondrea, I.O., Pamfil, D., Heusden, A.W., van Tuyl, J., van Bondrea, M., Meda, L. and Taoutaou, A. 2008. AFLP markers as a powerful tool for fingerprinting and breeding *Tulipa* genus. Lucrari Stiintifice - Universitatea de Stiinte Agronomice si Medicina Veterinara Bucuresti. Seria B. *Horticultura*, **51:** 627-629.

Broertjes, C. and Alkema, H.Y. 1970. Mutation breeding in flower bulbs. *Acta Horticulturae*, **23:** 407-412.

Chanteloube, F., Courduroux, J.C., Tort, M. and Le Nard, M. 1995. Micropropagation of *Tulipa gesneriana* L.: Regeneration of bulblets on growing floral stem segments cultured *in vitro*. *Acta Bot. Gallica*, **142:** 301-307.

Chauvin, J.E., Hamann, H., Cohat, J., Nard, M., Le Nard, M., Lilies Kipnis, H., Borochov, A. and Halevy, A.H. 1997. Selective agents and marker genes for use in genetic transformation of *Gladiolus grandiflorus* and *Tulipa gesneriana*. *Acta Horticulturae*, **430:** 291-297.

Custers, J.B.M., Eikelboom, W., Bergervoet, J.H.W. and Van Eijk, J.P. 1992. In ovulo embryo culture of tulip (*Tulipa* L.); effects of culture conditions on seedling and bulblet formation. *Scientia Hortic.*, **51:** 111-122.

Custers, J.B.M., Eikelboom, W., Bergervoet, J.H.W. and Van Eijk, J.P. 1995. Embryo rescue in the genus *Tulipa* L.; successful direct transfer of *T. kaaufmanniana* Regel germplasm into *T. gesneriana* L. *Euphytica*, **82:** 253-261.

Eikelboom, W., Straathol, Th.P., and Van Tuyl, J.M. 2001. Tetraploide "Christmas marvel" methoden om tetraploide tulpen te verkrijgen. *Bloembollencultuur*, **112**(12): 22-23.

Eikelboom, W., Van Eijk, J.P. Peters, D. and Van Tuyl, J.M. 1992. Resistance to Tulip Breaking Virus (TBV) in tulip. *Acta Horticulturae*, **325:** 631-636.

Hall, A.D. 1940. The Genus *Tulipa*. Royal Horticultural Society, London.

Hogenboom, N.G. 1973. A model for incongruity in intimate partner relationships. *Euphytica*, **22:** 219-233.

Hoog, M.H. 1973. On the Origin of *Tulipa*, Lilies and other Liliaceae. Royal Horticulture Society, London, England, pp. 47-64.

Hulscher, M. and Krijgsveld, H.T. 1995. Micropropation of tulip. *Acta Horticulturae*, **420:** 104-106.

Jensen, C.J. 1964. Pollen Storage under Vaccum. Royal Agric. University Yearbook. pp. 133-146.

Jhon, A.Q. and Khan, F.U. 2002. Evaluation of exotic tulip genotypes under Kashmir valley conditions. *J. Ornamental Hort.*, **5**(2): 27-30.

Jhon, A.Q. and Khan, F.U. 2003. Evaluation of tulip cultivars under polyhouse and open conditions. *J. Ornamental Hort.*, **6**(1): 42-45.

Jhon, A.Q., Khan, F.U. and Rather, Z.A. 2006. Genetic variability studies in tulip. *Applied Biological Research*, **8**(1/2): 37-39.

Juodkaite, R., Dapkuniene, S., Skridaila, A., Stukeniene, G., Baliuniene, A., Motiejunaite, O., Naujalis, J.R., Navalinskiene, M. Samuitiene, M. 2012. Assessment of resistance to tulip breaking virus (TBV) in the tulip collections of the Botanical Garden of Vilnius University. *Acta Horticulturae*, **937:** 977-984.

Karthigeyan, S., Dhyani, D., Raja Ram, Singh, M.K., Mukherjee,D. and Ahuja, P.S. 2006. Interspecific hybridization in himalaya wild tulip [*tulipa clusiana* f. *stellata* (hook.) Dasgupta & Deb.] x garden tulips and embryo rescue for its potential exploitation as cut flower. *Abst*. National Symposium on Ornamental Bulbous Crops, SVBPUA&T, Meerut. pp. 3-4.

Keiichi, O. and Murakami, K. 1992. Effects of flowering time (in forcing culture), stigma excision and high temperature on overcoming of self-incompatibility in *tulip*. *J. Japanese Soc. Hort. Sci.*, **61**(2): 405-411.

Kho, Y.O. and Baer, J. 1971. Incompatibility problems in species crosses of tulips. *Euphytica*, **20:** 30-35.

Kiani, M., Memariani, F. and Zarghami, H. 2012. Molecular analysis of species of *Tulipa* L. from Iran based on ISSR markers. *Plant Systematics and Evolution*, **298**(8): 1515-1522.

Kuijpers, A.M. and Langens-Gerrits, M. 1997. Propagation of tulip *in vitro*. *Acta Horticulturae*, **430:** 321-324.

Kutlunina, N.A., Polezhaeva, M.A. and Permyakova, M.V. 2013. Morphologic and AFLP analysis of relationships between tulip species *Tulipa biebersteiniana* (Liliaceae). *Russian J. Genetics*, **49**(4): 401-410.

Liedl, B.E. and Anderson, N.O. 1993. Reproductive barriers: identification, uses and circumvention. *Plant Breed Rev.*, **11**: 111-154.

Luan, QiFu, Ouyang, Tong, Jiang, YanCheng and Wang, CaiXia. 2008. Tulip RAPD analysis of cultivars and wild species in Xinjiang. *Acta Agriculturae Universitatis Jiangxiensis*, **30**(4): 656-660.

Marasek-Ciolakowska, A., He, H., Bijman, P., Ramanna, M.S., Arens, P., van Tuyl, J.M. 2012a. Assessment of intergenomic recombination through GISH analysis of F1, BC1 and BC2 progenies of *Tulipa gesneriana* and *T. fosteriana*. *Plant Systematics Evolution*, **298**(5): 887-899.

Marasek-Ciolakowska, A., Ramanna, M.S., van Tuyl, J.M. 2011. Introgression of chromosome segments of *Tulipa fosteriana* into *T. gesneriana* detected through GISH and its implications for breeding virus resistant tulips. *Acta Horticulturae*, **886:** 175-182.

Marasek-Ciolakowska, A., Xie, S., Ramanna, M.S., Arens, P. and van Tuyl, J.M. 2012b. Meiotic polyploidization in Darwin hybrid tulips. *Acta Horticulturae*, **953:** 187-192.

Mizuochi, H., Marasek, A. and Okazaki, K. 2007. Molecular cloning of *Tulipa fosteriana* rDNA and subsequent FISH analysis yields cytogenetic organization of 5S rDNA and 45S rDNA in *T. gesneriana* and *T. fosteriana*. *Euphytica*, **155**(1/2): 235-248.

Mokhono, V.S. and Yakovleva, G.V. 1976. Chromosome numbers of some tulip cultivars. *Tsitologiya -i-Genetika*, **10**(2): 123-124.

Momonoi, K., Yoshida, K., Mano, S., Takahashi, H., Nakamori, C., Shoji, K., Nitta, A. and Nishimura, M. 2009. A vacuolar iron transporter in tulip, TgVit1, is responsible for blue colouration in petal cells through iron accumulation. *Plant Journal*, **59**(3): 437-447.

Moue, T. and Urashima, O. 1996. Pollen tube growth in stigma, style and upper seedless part of valve after self-pollination in genus *Tulipa*. *Bulletin Faculty Agri. Niigate University*, **49**(1): 19-29.

Nieuwhof, M., Van Eijk, J.P., Keijzer, P., Eikelboom, W. and Van Eijk, J.P. 1998. Inheritance of flower pigments in tulip (*Tulipa* L.). *Euphytica*, **38**(1): 49-54.

Okazaki, K., Nishimura, M. and Van der Plas, L.H.W. 2000. Ploidy of progenies in crosses between diploid, triploid and tetraploid in tulip. *Acta Horticulturae*, **522**: 127-133.

Podwyszynska, M. 2012. *In vitro* tetraploid induction in tulip (*Tulipa gesneriana* L.). *Acta Horticulturae*, **961:** 391-396.

Podwyszynska, M., Kuras, A. and Korbin, M. 2010. Somaclonal variation in micropropagated tulips as a source of novel genotypes - field and molecular characteristic. *Acta Horticulturae*, **855:** 225-232.

Rees, A.R. 1985. Tulipa. *In*: CRC Handbook of Flowering. A.H. Halevy. (ed.). Vol I. CRC Press, Florida, US, pp. 272-277.

Romanov, L.R., Van Eijk, J.P., Eikelboom, W., Van Schadewijk, A.R., and Peters, D. 1991. Determining levels of resistance to tulip breaking virus (TBV) in tulip (*Tulipa* L.) cultivars. *Euphytica*, **51**(3): 273-280.

Sharma, K.D., Kaur, R. and Kumar K. 1996. Embryo rescue in plants- a review, *Euphytica*, **89:** 325-337.

Singh, A.K. 2006. Flower Crops: Cultivation and Management. Pub. New India Publishing Agency, New Delhi. pp. 371-387.

Staats, M. 2007. Botrytis species on flower bulb crops: Phylogeny, genetic variation and host specificity. Ph.D. Thesis, Wageningen University, Wageningen, The Netherlands.

Straathof, T.P., Eikelboom, W., Van Tuyl, J.M. and Peters, D. 1997. Screening for TBV resistance in seedling populations of *Tulipa* L., *Acta Horticulturae*, **432:** 391-395.

Straathof, T.P., Mes, J.J., Eikelboom, W. and Van Tuyl, J.M. 2002. A greenhouse screening assay for *Botrytis tulipae* resistance in tulips. *Acta Horticulturae*, **570:** 415-421.

Tsuji, T., Imai, T., Ikegawa, S., Kanamori, M., Umada, Y., Koizumi, M., Kizu, M., Iimura, N. and Urashima, O. 2005. The production and characteristics of the new tulip cultivar Arisa. *Acta Hort.*, **673**: 555-558.

Urashima, O, Murakami, K., Toyoda, T., Nishii, K., Tsutsui, K., Bezu, M., Hirata, Y., Kawata, J., Kunishige, M., Amano M., Kanamori, M., Imai, F., Umada, Y., Okazaki, K., Imai, T. and Tsuji, T. 2000. Breeding process and characteristics of tulip. (*Tulipa*) cultivar Yukitsubo. *Bulletin Toyama Agril. Res. Center (Japan)*, **19:** 53-64.

Van Creij, M.G.M. 1997. Interspecific hybridization in the genus *Tulipa* L., Ph.D. Thesis.

Van Creij, M.G.M., Kerckhoffs, D.M.F.J. and Van Tuyl, J.M. 1997. Application of three pollination techniques and of hormone treatments for overcoming interspecific crossing barriers in *Tulipa*. *Acta Horticulturae*, **430:** 547-557.

Van Creij, M.G.M., Kerckhoffs, D.M.F.J. and Van Tuyl, J.M. 1997a. Interspecific crosses in the genus *Tulipa* L.: Identification of pre-fertilization barriers. *Sex. Plant Reprod.*, **10:** 116-123.

Van Creij, M.G.M., Kerckhoffs, D.M.F.J. and Van Tuyl, J.M. 1999. The effect of ovule age on ovary-slice culture and ovule culture in intranspecific and interspecific crosses with *Tulipa gesneriana* L. *Euphytica*, **108:** 21-28.

Van Creij, M.G.M., Kerckhoffs, D.M.F.J., De Bruijn, S.M., Vreugdenhil, D. and Van Tuyl, J.M. 2000. The effect of medium composition on ovary-slice culture and ovule culture in intraspecific *Tulipa gesneriana* crosses. *Plant Cell Tiss. Org. Cult.*, **60:** 61-67.

Van Creij, M.G.M., Van Went, J.L. and Kerckhoffs, D.M.F.J. 1997b. The Programmeic phase, embryo and endosperm development in an intraspecific *Tulipa gesneriana* L. cross and in the incongruent interspecific cross *T. gesneriana* × *T. agenensis* DC. *Sex. Plant Reprod.*, **10:** 241-249.

Van der Meulen, J.J.M., Van Oeveren, J.C. and Van Tuyl, J.M. 1997. Breeding as a tool for improving postharvest quality characters of lily and tulip flowers. *Acta Horticulturae*, **430:** 569-575.

Van Eijk, J.P. and Eikelboom, W. 1976. Possibilities of selection for keeping quality in tulip breeding. *Euphytica*, **25:** 353-359.

Van Eijk, J.P. and Eikelboom, W. 1983. Breeding for resistance to *Fusarium oxysporum* f. sp. *tulipae* in tulip (*Tulipa* L.). 3 genotypic evaluation of cultivars and effectiveness of pre selection. *Euphytica*, **32:** 505-510.

Van Eijk, J.P. and Eikelboom, W. 1990. Evaluation of breeding research on resistance to *Fusarium oxysporum* in tulip. *Acta Horticulturae*, **266:** 357-363.

Van Eijk, J.P. Van Raamsdonk, L.W.D., Eikelboom, W. and Bino, R.J. 1991. Interspecific crosses between *Tulipa* gesneriana cultivars and wild *Tulipa* species: a survey. *Sex. Plant Reprod.*, **4:**1-5.

Van Eijk, J.P., Bergman, B.H.H. and Eikelboom, W. 1978. Breeding for resistance to *Fusurium oxysporum* f. sp. *tulipae* in tulip (*Tulipa* L.). 1. Development of a screening test for selection. *Euphytica*, **27:**255-261.

Van Eijk, J.P., Garretsen, F. and Eikelboom, W. 1979. Breeding for resistance to *Fusarium oxysporum* f. sp. *tulipae* in tulip (*Tulipa* L.). 2. Phenotypic and genotypic evaluation of cultivars. *Euphytica*, **28:** 67-71.

Van Harten, A.M. and Broertjes, C. 1989. Induced mutations in vegetatively propagated crops. *Plant Breed Rev.*, **6:** 55-91.

Van Heusden, A.W., Jongerius, M.C., Van Tuyl, J.M., Straathof, T.P. and Mes, J.J. 2002. Molecular assisted breeding for disease resistance in lily. *Acta Horticulturae*, **572:** 131-138.

Van Raamsdonk, L.W.D. and De Vries, T. 1995. Species relationships and taxonomy in *Tulipa* subgenus *Tulipa* L. *Pl. Syst. Evol.*, **195:** 13-44.

Van Raamsdonk, L.W.D., Van Eijk, J.P. and Eikelboom, W. 1995. Crossability analysis in subgenus *Tulipa* of the genus *Tulipa* L. *Bot. J. Linn. Soc.*, **117:** 147-158.

Van Scheepen, J. 1996. Classified list and International Register of tulip names. Royal General Bulb Growers Association KAVB, Hillegom, The Netherlands, 623 p.

Van Tuyl, J.M., Bino, R.J. and Custers, J.B.M. 1990. Application of *in vitro* pollination, ovary culture, ovule culture and embryo rescue in breeding of *Lilium*, *Tulipa* and *Nerine*, *In*: De Jong, J. (ed.). Integration of *in vitro* techniques in ornamental plant breeding, Proceedings of the Eucarpia symposium, section ornamentals, Wageningen, pp. 86-97.

Van Tuyl, J.M., Lim, K.B., and Ramanna, M.S. 2002. Interspecific hybridization and introgression. *In*. Vainstein, A. (ed.). Breeding for ornamentals: Classical and molecular approaches, Kluwer Academic Publishers Dordrecht/Boston/London. pp. 85-103.

Wang, CaiXia, Ouyang, Tong and Jiang, YanCheng. 2009. Aceptic germination and bulblet formation of interspecific hybrids between wild species and cultivar of tulip (*Tulipa* L.). *Plant Physiology Communications*, **45**(11): 1098-1100.

Weryszko Chemielewska, E., Hetman, J., Laskowska, H. and Michonska, M. 2002. Flower attractiveness of six tulip (*Tulipa gesneriana* L.) cultivars and some traits of flower stem structure. *Annales Universitatis Mariae Curie Skodowska Sectio EEE Horticultura*, **10:** 73-81.

Weryszko Chemielewska, E., Hetman, J., Laskowska, H., Banach, B. and Michonska, M. 2003. Pollen grains characteristics of selected tulip (*Tulipa gesneriana* L.) cultivars. *Annales Universutatis Mariae Curie Skodowska Sectio EEE. Horticultura*, **13:** 79-84.

Williams, E.G., Maheswaran, G. and Hutchinson, J.F. 1987. Embryo and ovule culture in crop improvement. *Plant Breeding Rev.*, **5:** 181-236.

Wilmink A., Van de Ven, B.C.E. and Dons, J.J.M. 1992. Expression of the GUS-Gene in the monocot tulip after introduction by particle bombardment and *Agrobacterium*. *Plant Cell Reports*, **11**(2): 76-80.

Wilmink A., Van de Ven, B.C.E., Custers, J.B.M., Van Tuyl, J.M., Eikelboom, W. and Dons, J.J.M. 1995. Genetic transformation in *Tulipa* species (tulips). *In:* Y.P.S. Bajaj (ed.). Biotechnology in Agric. and Forestry. Vol. 34, Plant Protoplasts and Genetic Engineering VI, II. **13:** 289-298.

Wilmink, A., Vandever, B.C.E. and Dons, J.J.M. 1995a. Activity of constitutive promoters in various species from Liliaceae. *Plant Mol. Biol.*, **28:** 949-955.

Zeilinga, A.E. and Schouten, H.P. 1968. Polyploidy in garden tulips II. The production of tetraploids, *Euphytica*, **17:** 303-310.

Zhou, Yin, Zhang, ZhiQi, Zhang, JianJun and Gu, JiaQing. 2011. Effects of picloram on tulip callus induction and somatic embryogenesis. *Acta Agriculturae Shanghai*, **27**(1):7-10.

Chapter – 18

Zantedeschia

Zantedeschia commonly known as Calla lily, Richardia or Pig lily belongs to family Araceae. It is the only genus in the tribe Zantedeschiae which is the sub family philodendroideae. The name Zantedeschia was given as a tribute to Italian botanist "Giovanni Zantedeschi" (1773-1846) by the German botanist Kurt Sprengel (1766-1833). It is considered by many people in the Republic of Ireland to be a symbol of Irish Republicanism. The Romans valued them so much that they often decorated the edges of the bloom with filaments of gold. These spectacular flowers are beginning to rival the rose in popularity for bridal bouquets. In India there has been initiation in the production of tubers of Calla and few tissue culture labs have started production for New Zealand and Dutch Companies. It is used as potted plants, callas are very attractive. It is used for decoration throughout the spring and early summer months focal point for a spectacular arrangement or bouquet.

ORIGIN AND DISTRIBUTION

The *Zantedeschia albomaculata* are native to areas of South Central Africa which are subjected to periodic drought (Letty, 1973). *Zantedeschia* was originally placed in the *Calla* genus by Carl Linnaeus, *Zantedeschia aethiopica* was moved in 1818 by Carl (Charles) Sigismund Kunth to the *Richardia* genus, until it was finally classified as *Zantedeschia*. Although *Zantedeschia* did not remain in the *Calla* genus, even after all this time this wonderful bulb is still referred to by its common name of calla lily.

The name "*aethiopica*" probably denotes South Africa, the country of the origin. Formerly its name was Calla or Richardia. This plant was introduced in Europe very early, apparently before Van Riebeeck had established the refreshment station at Cape.

It is also illustrated in an account of Royal Garden in Paris in 1664. It was sent as one of the interesting plant of Cape to Europe by Simon van der Stal sometime before 1697. The genus is restricted to African continent with seven recognized species *Z. aethiopica, Z. albomaculata, Z. ellottiana, Z. jucunda, Z. odoratum, Z. petlandii,* and *Z. rehmannii.*

The common calla (*Z. aethiopica*) is found from the western Cape through the eastern Cape, Kwa Zullu-Natal, Mpumalanga and in to the Northern Province. In Australia calla lily is naturalized in moist niches of the cooler parts of the New South Wales, southern and south western Victoria, moist valley in Adelaide hills and north west Tasmania. They are serious weeds of pastures and natural bush land in south west of western Australia.

The calla lily is used throughout the world as an ornamental for its flowers. It is mainly grown in temperate and sub tropical climates of the world. They also thrive well under tropical conditions. New Zealand is the topmost producer followed by The Netherland. The original summer flowering calla species were introduced in New Zealand in the early twentieth century. Finland and Scandinavia are also big producer of callas and exported to Sweden and Norway. New Zealand offers a favourable climate to grow callas outdoors and for exports it is grown under controlled conditions. In India it is a potential emerging crop and requires improvement programme for commercialization.

SPECIES

The genus *Zantedeschia* is endemic to Southern Africa and consists of two sections (Letty, 1973 and Singh *et al.*, 1996). The section *Zantedeschia* comprises of *Z. aethiopica* and *Z. odorata* while the section *Aestivae* consists of around six species: *Z. rehmannii, Z. jucunda*, *Z. elliotiana*, *Z. pentlandii*, *Z. albomaculata* and *Z. valida.* (Singh *et al.*, 1996). More than 10 species have been proposed for the genus (Traub, 1948) but most recent revision of genus (Letty, 1973) reduces the number of species to 6 *viz.*, *Zantedeschia aethiopica, Zantedeschia rehmanii, Zantedeschia jucunda, Zantedeschia elliottiana, Zantedeschia pentlandii, Zantedeschia albomaculata* which is further divided into three subspecies, *Zantedeschia albomaculata* subsp. *Albomaculata, Zantedeschia albomaculata* subsp. *macrocarpa* and *Zantedeschia albomaculata* subsp. *valida* (Letty, 1973). A seventh species *Zantedeschia odorata* has been described by Perry (1989). All species occur in the summer-rainfall areas, except for *Z. odorata* at Nieuwoudtville (winter-rainfall) and *Z. aethiopica* which is widespread and commonly found in marshy areas. Most prominent commercial *Zantedeschia* are described below.

***Zantedeschia aethiopica*:** The flower spathes of *Z. aethiopica* are white tinged with yellow with a yellow spadix and it produces glossy, arrow shaped leaves. This species grows 60-90 cm high. The leaves are fleshly stalked and arise directly from the rhizome and are succulent, broad, dark glossy in colour. The petioles are hairy, leaf lamina is sagittate, hastate or lanceolate or cordate-ovate in shape. The peduncles may be longer than leaves. The spathe is staminate above and pistillate below.

***Zantedeschia elliotiana*:** These are deciduous, medium to tall plants up to 0.6 m high, growing singly or forming small clumps. Their leaves are large, broadly ovate, deep green and speckled with white spots. The spathe is a cup-shaped funnel, bright golden-yellow and dark purple at the base within. It flowers November to January.

***Zantedeschia rehmannii*:** These are small to medium plants 0.2–0.6 m high. Their leaves are lance-shaped, dark green and very rarely speckled with white spots. The spathes are cylindrical, colour ranging from white through shades of pink to dark maroon (almost black). They flower September to February, with a peak in November to January.

***Zantedeschia albomaculata*:** These are deciduous, medium to tall plants up to 0.7 m high, growing in small clumps. They have oblong or triangular arrow-shaped leaves, often speckled with white spots. Spathes are cylindrical, white, cream or pale yellow and frequently dark purple at the base within.

***Zantedeschia albomaculata* subsp. *albomaculata*:** Leaves oblong, spathe tapering to apex, berries many, medium-sized, up to 18 mm in diameter. It flowers October to April. It is found along rocky hillsides, forest margins and stream banks.

***Zantedeschia albomaculata* subsp. *macrocarpa*:** Leaves triangular, spathe shortened at apex, berries few, large, mostly close to 20 mm in diameter. It flowers from November to April with a peak in December.

***Zantedeschia odorata*:** These are deciduous, medium to tall plants up to 0.75 m high. Their leaves are broadly ovate and not speckled. Spathes are cylindrical and white. They flower in late July and August.

***Zantedeschia jucunda*:** These are deciduous, medium to tall plants, up to 0.8 m high, growing solitary or in small clumps. Their leaves are triangular-hastate, deep green and speckled with white spots. The spathe is a cup-shaped funnel, deep or cream-yellow and dark purple at the base within. They flower from November to January.

***Zantedeschia pentlandii*:** These are deciduous, medium to tall plants, up to 0.6 m tall. Leaves broadly ovate, glaucous green and very rarely speckled with white spots. The spathe is a cup-shaped funnel, lemon-chrome yellow and dark purple at the base within. They flower from November to December.

***Zantedeschia valida*:** These are deciduous, robust plants, up to 0.75 m tall. Leaves broadly ovate, plain, not speckled or very speckled with white spots. The spathe is a cup-shaped funnel dark purple at the base within. They flower from October to March with the peak in November.

FLOWER MORPHOLOGY AND BIOLOGY

Plants of the genus *Zantedeschia* are deciduous perennials up to 1 m high and survive the dry season through rhizomes or tubers. They have many leaves that are ovate or arrow-shaped and plain green or white-speckled. Petioles are hairy or bristly below and spongy throughout. Leaf lamina is sagittate lanceolate to cordate-ovate in shape. Leaves are blotched with white semi transparent spots. The inflorescence consists of a spathe, a large modified leaf around a central axis called the spadix that carries minute flowers densely congested, female flowers at base and males above. Spathes are funnel-shaped or cylindrical, white, cream, yellow, pink or red, with or without a dark purple marking at the base inside, mostly persistent and green when fruits are ripe. It takes 10-12 weeks to come to flowering from the time of planting.

Flowers are unisexual and without a perianth. The striking arum lily flower is actually many tiny flowers arranged in a complex spiral pattern on the central column. The tiny flowers are arranged in male and female zones on the spadix. The top 7 cm are male flowers and lower 1.8 cm are female flower. The female flowers have an ovary with a short stalk above it which is the style. The spadix is surrounded by the white or coloured spathe. The flowers are faintly scented and this attracts various crawling insects and bees which are responsible for pollinating the flowers. Cross pollination occur as the anther of each flower ripen before the ovaries. The spathe changes its colour after flowering and covers the ripening berries. The succulent yellow berries attract birds which are responsible for seed dispersal. Beetles are the most likely pollinators of *Zantedeschia* flowers. Beetles belonging in the genera *Anoplachilius* and *Leucocelis* use the spathes in *Z. aethiopica* for mating. Pollen is white and released in long, fine threads. The stigma is broad and produces a sticky shiny secretion, possibly as a sugar reward to insect visitors. Fruit are berries, clustered, each with few to many seeds, green or orange when ripe, surrounded by the partial or entire spathe. Seeds with mucus, round or oval, cream or brown. It is propagated by seeds, rhizomes or tuber and micropropagation. Most cultivars are propagated vegetatively, although some are F_1 hybrid cultivars, thus propagated by seeds. A dark purple marking is constant in spathes of *Z. jucunda*, *Z. pentlandii* and *Z. valida*. It probably mimics the remains of dead flesh and aids in attracting carrion beetles that effect pollination. The purple marking is, however, absent in spathes of *Z. aethiopica* and *Z. odorata*. Scent is present in the inflorescences of these species which possibly attracts pollinators.

GENETICS

The basic chromosome number in zantedeschia is 16 and their normal diploid chromosome number is 2n=22. All *Zantedeschia* spp. are diploid with 32 chromosomes (Yao *et al.*, 1994b). Plastome-genome incompatibility (PGI) exists between species of different sections of the genus *Zantedeschia*, as well as between species within the section *Zantedeschia* (Yao *et al.*, 1994a and 1995). The plastome of *Z. odorata* is incompatible to the genome of *Z. aethiopica*, resulting in albino plants. The plastome of *Z. aethiopica* is partially incompatible to the genome of *Z. odorata*, resulting in virescence (Yao *et al.*, 1994a). Although functional interspecific hybrids can be obtained after artificial hybridization among species of the section *Aestivae* (Wolff, 1919, Shibuya, 1956 and Horn, 1962), it is possible that PGI exists. An indication for PGI was delivered by New and Paris (1967), who mentioned hybrid variegation and a non-Mendelian segregation of virescence in interspecific hybrids. If PGI exists among the section *Aestivae*, then it could have a major impact on genetic studies and breeding. Knowledge of breeding barriers and PGI within the section *Aestivae* is a prerequisite of genetic analyses and breeding. Directions of PGI between species of the section *Aestivae* must therefore be determined to identify the most promising parents and crossing directions.

Plastid inheritance

In most genera combinations of plastomes and genomes function to form normal plants, irrespective of the taxonomical distances between the plastidial genome and the nuclear genome. In some genera plastomes and genomes have co-evolved to such a degree that they can only function within specific combinations. Plastomes and genomes of five taxa (*Z. albomaculata* subsp. *albomaculata*, *Z. albomaculata* subsp. *macrocarpa*, *Z. elliotiana*, *Z. pentlandii* and *Z. rehmannii*) were combined by interspecific hybridisation. Plastomes were differentiated using plastome specific CAPS-markers. Degrees of plastome-genome incompatibility existed between the hybrid genomes of *Z. rehmannii* and *Z. albomaculata*, *Z. rehmannii* and *Z. elliotiana*, *Z. rehmannii* and *Z. pentlandii* and the plastomes of *Z. albomaculata*, *Z. elliotiana* and *Z. pentlandii*, respectively. The plastome of *Z. rehmannii* appeared compatible to all tested hybrid genomes in section *Aestivae*. It was only fully incompatible with the hybrid genome with *Z. aethiopica* (section *Zantedeschia*). The plastomes of *Z. albomaculata* and *Z.* elliotiana appeared compatible to the hybrid genome of *Z. albomaculata* and *Z. elliotiana*. Four plastomes were differentiated among the section *Aestivae*. Biparental plastid inheritance (3 to 90%) was observed among all crossing combinations, which produced more than 1600 interspecific offspring. Biparental inheritance of plastids thus appeared to be a general phenomenon among interspecific hybrids within the section *Aestivae*. A literature review shows that biparental plastid inheritance is not uncommon in interspecific hybrids. The genus *Zantedeschia* shows an unusual wide range of degrees of biparental plastid inheritance that is only recorded in genera that also show plastome-genome incompatibility (Snijder *et al.*, 2007). The mode of plastid inheritance between *Z. albomaculata* subsp. *albomaculata* and *Z. albomaculata* subsp. *macrocarpa* was also investigated by Brown *et al.* (2005). Subspecies specific CAPS markers were developed and used to demonstrate that biparental plasmid inheritance occurs between these subspecies. Thus, biparental plastid inheritance exists within *Zantedeschia* spp. and is not a result from the unusual genomic background of interspecific crosses.

BREEDING OBJECTIVES

i. New colour
ii. More number of flower
iii. Varied coloured and attractive spathe
iv. Development of genetically dwarf varieties for the pot
v. Development of cultivars with new shapes
vi. Homogenity in breeding material
vii. Resistance against insects and pests
viii. Resistance to soft rot caused by *Erwinia carotovora* subsp. *carotovora*

BREEDING METHOD

Selection

In zantedeschia several varieties were selected from *Zantedeschia aethiopica* species. In general, the selected varieties were produced by open pollination, seedling selection from improved varieties. Priority was given to evolved dwarf variety. Some important varieties are given in Table 1.

Table 1: Varieties developed through selection.

S.No.	Varieties	Parentage	Characteristics
1.	Child's Perfection	Dwarf selection of *Z. aethiopica*	It grows to about 50 cm tall in moist soil and partial shade. The arachnids on the flowers are Opiliones, commonly called harvestmen or daddy-long-legs.
2.	Crowborough	Improved selection of *Zantedeschia aethiopica*	Deciduous perennial upto 1m height, rich green, arrow-shaped leaves and funnel-shaped, white spathes to 15 cm length, prominent yellow spadix.
3.	Rod's Red	Selection of Rod Saunders	It has pure white flowers in the inside but bright cherry on the outer surface.
4.	White Giant	Selection of *Z. aethiopica*	Spotted foliage up to about 1.2 m tall and flower spikes up to 1.8 m tall
5.	Goldilocks (Anon., 2005b)	Open pollination, "dwarf" lines.	Beverley Joe was the inventor of this variety. Characterized by its high number of yellow flowers and the good proportions of the plant and flowers. It is an excellent pot-plant variety.
6.	Flamingo (Anon., 2010b)	Open pollination of *Z. aethiopica* 'Pink Mist'.	Upright plant habit; glossy medium green-coloured leaves; large pink-coloured

(*Contd.*)

(*Contd. Table 1*)

S.No.	Varieties	Parentage	Characteristics
			spathes that are positioned above and beyond the foliage on erect scapes and good inflorescence longevity. It was developed by Theo C.M. Klaver.
7.	Red Desire (Anon., 2003)	Cross pollination of a *Zantedeschia aethiopica* selection.	Upright and somewhat outwardly spreading plant habit; white-coloured spathes with dark pink-coloured throats and dark pink-coloured spadices and freely flowering habit. Petrus M. M. Hoff. Invented this variety.

Hybridization

Species of genus *Zantedeschia* fall into 2 groups: (i) Plants with foliage that does not die down in the winter in their native habitat staminoides interspersed with the female flower in the lower section of the spadix and fruits that turn orange and become mucilaginous when ripe, (ii) Plants with foliage that dies to the ground in nature during the winter no staminodes among the represented by *Zantedeschia aethiopica* with all the remaining species in the second group (Letty, 1973). Hybridization between *Zantedeschia aethiopica* and other species has not been successful while crossing among species of the second group is possible (Traub, 1948). Therefore, the commercially available zantedeschia hybrids have characteristics of second group. These species and hybrids are grown both for cut flowers and as flowering potted plants. There are several cultivars of this species. Hybrids with characters from sections are therefore very attractive for plant breeders. Hybridization between two group of *Zantedeschia* produced only albino hybrids. These hybrids survived only on artificial medium.

Table 2: Varieties developed through hybridization.

S. No.	Varieties	Parentage	Characteristics
1.	Pot Black (Anon., 2004a)	Seed parent a dark purple seedling × Chianti	Invented by Andrew G.F. Warren, Jacobus J.P. Spaans. Good as a potted plant particularly distinguished by dark purple/black spathe and lightly maculated leaves.
2.	Hot Chocolate (Anon., 2004c)	Cameo × Dark purple seedling	It bear apricot salmon colour spathe and has medium green arrow-shaped leaves while the pollen parent, the unnamed dark purple seedling has more oval shaped and darker green leaves.
3.	Pink Pot (Anon., 2004d)	Pink seedling unnamed female parent × Pink Persuasion.	Calla lily potted plant particularly distinguished by pink spathe and lightly maculated leaves.
4.	Jack of Hearts (Anon., 2005a)	Sunglow × unknown seedling with red spathes.	It bear pink red spathe and foliage with medium maculation.
5.	Captain Cupido (Anon., 2006a)	Zaailing 60 × Zaailing 05	Characterized by green leaves and yellow-green coloured flowers.
6.	Captain Reno (Anon., 2006b)	*Z. spreng* 'Zaailing 60'× *Z. spreng* 'Zaailing 35' selected by breeder in 2001 as single plant progeny.	Antonius S. J. Kapiteijn invented it. It is characterized by green leaves with light spots and purple violet coloured flowers.
7.	Purple Heart (Anon., 2007a)	Purple Seedling × Pacific Pink.	Flower colour is purple. The flower shape is fluted. The leaf shape is cordate. It is invented bySpaans, J.P. and Warren, A.G.F.
8.	Vermeer (Anon., 2007c)	Selection from cross of *Z. sprengeri* seedling 95845 × *Z. sprengeri* seedling 95846.	Round, white and purple spathe. Fast forcing time. High flower production. Invented by Cecilius Jan-Jochem Randag.
9.	Picante (Anon., 2007d)	Pot of Gold × Neroli	Invented by Spaans, J.P. and Warren, A.G.F. Calla lily plant particularly distinguished by orange flowers and sagittate leaves.
10.	Odessa (Anon., 2008)	Selection from a cross of C02029 × C02021 *Z. sprengeri*	Dark purple spathe, Fast forcing time and high flower production.
11.	Hot Cherry BLZ (Anon., 2009a)	Cross of female with pink spathe and the mae 'Majesty Red' with red/ burgundy spathe.	A purple-red spathe. Low density maculation of the leaves that is visible from both sides and produces 2 to 4 inflorescences per 3-cm tuber.
12.	Merlot BLZ (Anon., 2009b)	Majestic Red × *Zantedeschia*	Red-purple spathes, Medium-density maculation of the leaves on the upper surface

(Contd.)

(*Contd. Table 2*)

S.No.	Varieties	Parentage	Characteristics
		having pink spathes.	that is visible from the lower surface and produces 2 to 4 inflorescences per 3 cm tuber.
13.	Rose BLZ (Anon., 2009c)	Cross of the female having red-purple spathe and male having a pink spathe (both *Zantedeschia* spp.)	Distinguished by a red-purple spathe with darker red-purple veins, medium density maculation of the leaves that is visible from both sides and which produces 2 to 3 inflorescences per 3 cm tuber.
14.	Yellow Jacket BLZ (Anon., 2009d)	Sunglow × Pot of Gold	Yellow spathes. High density maculation of the leaves that is visible from both sides. Produces 2 to 4 inflorescences per 3 cm tuber.
15.	Captain Amigo (Anon., 2010a)	*Zantedeschia* sp. 45×05	Green leaves with white spots and yellow orange coloured inflorescences. Invented by Anthonius Simon-Johannes Kapiteijn.
16.	Captain Aguila (Anon., 2010c)	4131× 6484	Large yellow flowers, prolific flower production, green foliage spotted with white and a short plant height. Suitable for indoor plant production. Invented by Anthonius Simon Johannes Kapiteijn.
17.	Captain Ventura (Anon., 2010d)	44wit × 7849.	Commercially suitable for indoor plant production. Characterized by large flowers, prolific flower production, strong peduncles, and green foliage occasionally spotted with white.

Interspecific hybridization

Interspecific hybrids are readily achieved between species within section 2; however, because of post fertilization barriers, no successful crosses have been achieved between species in groups 1 and 2 (Traub, 1948). Yao *et al.* (1995) obtained fertilized embryos of these incompatible crosses by using an embryo-rescue technique, but all hybrids were albino and viable only *in vitro*. Yao *et al.* (1995) obtained albino hybrids by using only 4 cross-combinations. The hybrids, viable autotrophically, may be obtained from the crosses of more combination patterns. Incompatibility between the genome and plastome could be one of the factors to cause this dysfunction (Michishita *et al.*, 2002). Interspecific hybrids between genotypes of the *Aethiopica* and the *Aestivae* section obtained using embryo rescue techniques deliver up till now only albino plants. An *in vitro* disease test was developed for screening these heterotrophic albino hybrids. An interspecific variety evolved is given in Table 3.

Table 3: Variety developed through interspecific hybridization.

S. No.	Variety	Parentage	Characteristics
1.	Apricot Glow (Anon., 1991)	*Z. elliottiana* × *Z. rehmannii*	At the opening edges of the spathe, pink-red blush while remainder of spathe Yellow Orange Blush. Background colour is very light over blush. Spathe is trumpet shaped, not cylindrical, opening to a sharp tip, with point. In center of spathe, cylinderical in shape arising from the floral stem terminal, surrounded by the spathe.

Breeding for soft rot (*Erwinia carotovora* ssp. *carotovora*) resistance

A huge problem in growing *Zantedeschia* is the occurrence of soft rot caused by *Erwinia carotovora* ssp. *carotovora*. At Plant Research International, a research programme was started for finding and developing *Erwinia*-resistant genotypes of *Zantedeschia*. There is variation for susceptibility/resistance within section *Aestivae*. Finding the genetics behind the resistance within this section is the next step in the breeding programme. Soft rot caused by *Erwinia carotovora* subsp. *carotovora* is an important disease of *Zantedeschia* spp. and the major disease in the section *Aestivae* (Funnell, 1993, Kuehny, 2000 and Wright *et al*., 2002). There are several possibilities to breed for resistance to soft rot. The technique to introgress resistance into genotypes of the section *Aestivae* depends on the source of the resistance gene. The genetic variation of *Zantedeschia* spp. for resistance to soft rot has not been studied, but *Z. aethiopica* is presumed to be more resistant than cultivars of the section *Aestivae* (Funnell, 1993 and Yao *et al*., 1995), although this difference has not been determined scientifically. The introgression of genes from the section *Zantedeschia* into genotypes of the section *Aestivae* is not possible due to severe breeding barriers (Wolff, 1919, Traub, 1948 and Yao *et al*., 1994a). Therefore, if conventional breeding is to be successful, resistance sources must be located within the section *Aestivae*. If no resistance exists within the section *Aestivae*, it might be possible to introduce resistance genes from outside the *Zantedeschia* germplasm by genetic modification.

Conventional breeding starts with the assessment of genetic variation in the cultivated species and its crossable species. The potential of resistance breeding is unknown, because variation for resistance to soft rot has not been determined in the genus *Zantedeschia*. Moreover, selection of resistant genotypes requires sensitive disease tests to determine resistance of plants. However, these have not been described yet. Knowledge from other crops could be used as a basis for breeding *Zantedeschia* spp. for resistance to soft rot. Therefore, the perspectives to develop a resistance test for *Zantedeschia* spp. are very good. So it should be possible to breed for resistance to soft rot in *Zantedeschia* spp. as well. The occurrence of a major breeding barrier prevents the introgression of genes from the section *Zantedeschia* into cultivars of the section

Aestivae (Wolff, 1919, Traub, 1948 and Yao *et al.*, 1994a). This barrier results in albino plants that can only survive heterotrophically and is caused by reciprocal incompatibility between plastomes and genomes (Yao *et al.*, 1994a). Plastid development is probably blocked in an early stage in these albino plants (Yao and Cohen, 2000). In this line some efforts have been made by Snijder *et al.* (2004a) to identify genetic variation in *Zantedeschia* for resistance to soft rot. They developed non-destructive resistance test, and variation in aggressiveness was observed among five isolates of *Erwinia carotovora* subsp. *carotovora* without interactions between the isolates and three *Zantedeschia* accessions. Within eleven accessions of *Z. aethiopica*, variation was observed from almost complete to moderate resistance, while the *Z. odorata* accession was susceptible. All 21 *Aestivae* cultivars were susceptible. Within the *Aestivae* species, *Z. elliotiana* and *Z. pentlandii* were also susceptible, but within twelve accessions of *Z. albomaculata*, as well as in six accessions of *Z. rehmannii*, variation was found from susceptible to moderately resistant. Hence, new sources of resistance were identified that show good potentials for resistance breeding.

The pattern of heredity of resistance to *E. carotovora* (PD1784) subsp. *carotovora* in *Zantedeschia* spp. is investigated. Four species with different resistance levels (*Z. albomaculata*, *Z. elliotiana*, *Z. pentlandii* and *Z. rehmannii*) were compared to their reciprocal offspring. The occurrence of plastome-genome incompatibility (PGI) affected plant resistance in all families. Therefore, plants that suffered from PGI were omitted from genetic analyses. Resistance was quantitative and the correlation between resistance levels of parents and offspring (h^2=0.33; r^2=0.66) indicated a genetic basis of resistance. *Z. rehmannii* and *Z. albomaculata* contributed more resistance genes than *Z. elliotiana* or *Z. pentlandii*. Transgression among some of the offspring of *Z. rehmannii* and *Z. albomaculata* indicated the presence of complementary resistance genes in these two species and good potential for resistance breeding (Snijder *et al.*, 2004b).

Snijder and van Tuyl (2002) obtained plant material from commercial sources in The Netherlands to determine resistance of zantedeschia to soft rot caused by *Erwinia carotovora*. It was found that resistant control cultivar *Zantedeschia aethiopica* 'Crowborough' scored most resistant in all three tests. Within the section *Aestivae*, degrees of susceptibility were identified that were in agreement with each other and with field observations indicating reliability of two of the methods in which tubers were used. The correlation coefficient of these two tests was high. A new non-destructive test method was developed for use on seedlings which involved immersion of leaf disks in a bacterial suspension. The percentage of decayed leaf area was a measure of resistance and results were in general agreement with the other tests. These methods will be useful for breeding for soft rot resistance and performing genetic analyses.

Snijder and van Tuyl (2004) determined the potential of breeding *Zantedeschia* cultivars for resistance to soft rot caused by *Erwinia carotovora*. Hybrids of six cultivars (Black Magic, Galaxy, Pink Persuasion, Sensation, Treasure and Florex Gold) were screened using a non-destructive resistance test (leaf disc test) and scored different levels of susceptibility. Some hybrid families were less susceptible than the very susceptible reference 'Florex Gold', while all hybrid families were more susceptible than the partial resistant reference *Z. rehmannii*. The half-sib family of 'Pink Persuasion' was more resistant, but less vigorous than the half-sib family of 'Florex Gold'. Hence, it may be concluded that resistance and vigour may be negatively correlated in the deployed cultivars and that resistance should be bred from other sources than the used cultivars.

Breeding for flower colour

Zantedeschia species feature a range of cultivars with spathe (flower) colour from maroon/dark red through pink, to orange, yellow, cream and white. In the absence of information on the pigments that make up these colours in *Zantedeschia*, Lewis *et al.* (2002) analysed the flavonoid and carotenoid profiles in several commercial summer flowering cultivars that were representative of the colour range. Anthocyanins were the main pigment associated with red and pink cultivars, while carotenoids were the main pigment in the yellow cultivars. The orange cultivars were a mix of anthocyanins and carotenoids, although in different cell layers. The anthocyanin pigments were present in the epidermis and the sub-epidermis, overlaying the carotenoids. The latter were distributed more evenly throughout the sub-epidermis and mesophyll. The main pigments observed were the anthocyanin cyanidin 3-glucoside and the carotenoids lutein and b-carotene. Other flavonoids, predominantly 7-methylated luteolin *C*-glycosides, were also detected in the spathe tissue. Variation in spathe colour appears primarily due to changes in the pigment group produced and pigment concentration rather than the production of different individual pigments (Lewis *et al.*, 2002).

Mutation

Ornamental plants appear to be ideal systems for mutation breeding as many characters of economic interest, i.e. flower traits or the growth habit are easily monitored after mutagenic treatment. Limited variability in the genus makes breeding and development of new varieties challenging. New colour and forms are being developed using mutation breeding and intergeneric hybridization. Furthermore, many ornamental species are heterozygous and often propagated vegetatively thus allowing the detection, selection and conservation of mutants within the M_1-generation. Appropriate strategies in mutation induction as application of fractionated doses and recurrent irradiation combined with *in vitro* culture techniques may lead to rapid success also in homozygous, polyploid species. Inspite of its accidental character, mutation induction continues to be an attractive method for creating genetic variability in ornamentals whenever the desired characters can be expected within the genetic scope of a given species. A mutant variety Edge of Night was discovered in 2004 by Terra Nova Nurseries, Inc. from parent variety Schwarzwalder (Anon., 2006c). Some varieties developed through mutation are given in Table 4.

Table 4: Varieties developed through mutation.

S. No.	Variety (release year)	Parentage and description
1.	Edge of Night (2006)	It is evolved from parent variety Schwarzwalder. Extremely dark purple, almost black, flowers or spathe. Very dark petioles and leaf margins. Leaves spotted white and pink. Excellent vigour.
2.	Majestic Star (2007)	Naturally occuring whole plant mutation of *Zantedeschia sprengeri* var. Black Star. It was discovered by breeder in year 2003.
3.	Snow White (2009)	It was discovered in 2003 as a naturally occurring branch sport of its parent. It is dwarf but bear large white and yellow inflorescence than parent.

Polyploidy

In *Zantedeschia* all natural species and hybrids are diploid and no induced polyploidy plant has been reported. The usefulness of polyploidy in Calla lily improvement is unknown. Triploid hybrid plants were produced from crosses between diploid and tetraploid plants of the winter-dormant section in the genus *Zantedeschia.* In five different crosses, the endosperm became degenerate and embryo development was restricted to the globular stage because of post-fertilization incompatibility. Using an embryo rescue technique, hybrid plants were produced from two crosses. These plants were confirmed to be triploids by counting chromosomes in root-tip cells. The triploid plants flowered in a greenhouse and produced no pollen. As hybridizations among diploid plants in the winter dormant section are generally compatible, severe incompatibility in the diploid-tetraploid crosses may be the result of the imbalance of chromosome numbers (Yao, 1996).

All *Zantedeschia* species are diploid with 32 chromosomes (Earl, 1957 and Yao *et al.,* 1994b). Tetraploid plants produced from winter dormant *Zantedeschia* cultivars using an *in vitro* chromosome doubling method (Yao and Cohen, 1996). Change in ploidy level of one species may improve the "crossability" with an otherwise incompatible species, known as manipulation of endosperm balance number (Johnston *et al.*, 1980 and Johnston and Hanneman, 1982). Triploids have been widely used in breeding of many crops. The first step to test the breeding value of triploid *Zantedeschia* plants, crosses were made between diploid and tetraploid plants in the winter-dormant section to produce triploid plants. Although hybridizations between diploid plants in the winter dormant sections are generally compatible and produce viable hybrids (New and Paris, 1967), but there are post-fertilization incompatibility barriers in diploid-tetraploid crosses within the section and the production of triploid hybrids can only be achieved using an embryo rescue technique.

The sexual polyploidization through 2n gametes, which can result in intergenomic recombination to produce new genetic variations is paid more and more attention by plant breeders. Hence in a study, the young flowers of coloured *Zantedeschia* diploid cultivars were treated by the chemical method (colchicines) to induce 2n gamete and the cytological characteristics of 2n male gamete (2n pollen) formation were observed. The result showed that 2n pollen of coloured *Zantedeschia* was induced successfully by colchicine and 2n pollen grain was larger than that of normal 1n pollen. The best concentration of colchicines was 0.15%. There were obvious differences on cytological matters during 2n pollen formation. The abnormal phenomena like parallel spindles and fused spindles at metaphase, dyads and triads at the tetrad stage, multinuclear cell with abnormal meiosis, nuclei fusion in pollen mother cell, etc. were observed. The seeds of the treated pollen crossing with normal 1n female gametes were harvested. The chromosome identification showed that the triploid (2n=3x=48) progeny was obtained successfully. The results imply that 2n pollen induced artificially can be a powerful method to create the sexual polyploids of coloured *Zantedeschia* (Wu *et al.*, 2011).

BIOTECHNOLOGY

Tissue culture

A number of techniques are employed in tissue culture. One of these methods is to select a small section from any part of a healthy plant and place it in a vial of agar. Many sections may be taken from one plant. The agar is a gel which holds specific nutrients vital for growth. The tissue should then form a callus, which will consist of indistinguishable cells that are neither roots nor shoots. Cell development will happen very quickly due to the perfectly maintained growing conditions. Once the growth has reached a suitable size it can be removed and divided into separate sections and more separate vials. The process can be repeated again with the new sections, or they can be provided with different nutrients and hormones, which will cause the sections to develop into small plants. These new plants are grown in sterile laboratories until they are large enough to be handled. Tissue culture is a fascinating form of propagation, as large number of new plants can be grown in a short period of time. The plants grown are more vigourous and generally free from disease. By selecting good "mother" stock the young material is of a predictably high quality as each new plant is a clone of the mother bulb. This propagation method allows professional bulb growers to provide a wonderful number of healthy, strong and beautiful bulbs.

Diploid plants in all *Zantedeschia* spp. contain 32 chromosomes (Yao *et al.*, 1994a). Various modern cultivars of calla lily are developed by crosses from different species and propagated vegetatively, therefore, have a high degree of heterozygosity. The juvenile phase of calla lily is quite long, lasting for 1 to 2 years. These characterizations hamper genetic analysis of important traits and efficient breeding of the ornamental. Breeding efficiency could be improved by inbreeding programme, but it is time consuming to obtain pure lines by numerous cycles of self pollination. Compared with conventional inbreeding, haploid technology allows obtaining of homozygous doubled haploid plants (DHs) in a single generation; thus, tremendously facilitating breeding programmes (Ferrie *et al.*, 1995 and Forster and Thomas, 2005). Haploid plants are also of benefit in genetic transformation and mutation studies (Ferrie *et al.*, 2008). However, data on haploid plant production are rare in the Araceae family. Ko *et al.* (1996) earlier reported the obtaining of haploid plant using anther culture in classical white calla lily, *Zantedeschia aethiopica*. Only one or two haploid origin plants were obtained from hundreds of anthers or ovules. On the other hand, these methods also give rise of diploids that could be somatic origin. In order to confirm that they are valuable DHs, laborious genetic analyses are required. Alternatively, haploid plants can be produced through isolated microspore culture. This method offers an advantage of using unique haploid cells as source of regenerants and has potential to increase yield. Nevertheless, there are a number of factors that influence the success of microspore culture, including microspore developmental stage, stress treatment of culture, culture medium and carbohydrate source provided in medium. However, no data has been published on isolated microspore culture in the Araceae family. In this study, effect of various factors on isolated microspore culture was investigated and a successful microspore culture procedure was reported in *Z. aethiopica*. Moreover, this study revealed an essential and novel role of monosaccharides in the microspore culture of this ornamental.

Haploid plant production

Diploid plants in all *Zantedeschia* spp. contain 32 chromosomes (Yao *et al.*, 1994a). The genus consists of two sections, Zantedeschia and Aestivae. *Zantedeschia aethiopica*, the classic white calla lily, belongs to the section of Zantedeschia. This species gives white flowers, has a rhizomatous storage organ and is an evergreen. The species of the section Aestivae give coloured flowers, have a tuberous storage organ and require a period of dormancy. Various commercial cultivars have been developed from crosses of different species. Modern cultivars of calla lily are propagated vegetatively by tissue culture and generally have a high degree of heterozygosity. Calla lily is also characterized by a long juvenile phase of 1–2 years. The heterozygosity and long juvenile period hamper genetic analysis of important traits and efficient breeding of the ornamental. Haploid plant production would be of great benefit to breeding and studying genetics in calla lily. One publication of *Z. aethiopica* on haploid production by anther culture was reported (Ko *et al.*, 1996), but it is a long and quite inefficient procedure. Only one haploid plantlet was obtained from 296 cultured anthers (0.34%).

Zhang *et al.* (2011) describe a faster and more successful anther culture procedure for haploid production in *Z. aethiopica*. Their report describes advances in the anther culture of *Zantedeschia aethiopica*. Important factors for improvement as compared to the earlier procedure were: (1) using flowers from inflorescences developed at relatively low temperature during winter, (2) high temperature stress treatment at 32 °C for 2 days in the beginning of the culture, (3) use of Gamborg B5 as anther culture medium, and (4) addition of sucrose at high concentration of 8% in the culture medium. Plants were obtained via a callus phase. Frequency of anthers producing calli was around 4–5%. About 87% of the calli gave regenerants, of which 52% were haploid, 36% were diploid and the rest had other ploidy levels. In addition to chromosome counting, cytological examination of the microspore development and amplified fragment length polymorphism (AFLP) analysis of the regenerants showed that haploid as well as diploid plants originated from the microspores. Finally, 12 doubled haploid (DH) plants could be produced from each inflorescence. One quarter of the DHs equaled the original cultivar in growth vigour, while more than one third showed good fertility, indicating that inbreeding depression was not so severe in this heterozygous species. The improved protocol now enables production of sufficient number of DHs for application of haploid technology in genetic improvement and breeding of *Z. aethiopica*.

Molecular breeding

A RAPD-based identification system for calla lily cultivars (*Zantedeschia* spp.) was constructed after screening 60 arbitrary 10-mer primers. Using this method, 2 or 3 sequential PCR reactions enabled clear identification of 12 cultivars to be made within several days (Hamada and Hagimori, 1996).

A new set of microsatellite or simple sequence repeat (SSR) markers from expressed sequence tags (ESTs) was developed for arum lily (*Zantedeschia aethiopica*), which is one of the most iconic and widely recognized ornamental plants in the world. Using 2175 unigenes derived from 4283 random ESTs in arum lily, 166 primer pairs were designed and tested for amplification in 24 accessions from Asia, Europe and Africa. A total of 43 loci were polymorphic, with the number of alleles per locus ranging from 2 to 10. The observed heterozygosity, expected heterozygosity and polymorphism information content ranged from 0.2313 to 0.8480, 0.3034 to 0.8648, and 0.1015 to 0.7364, respectively. Hence, it was concluded that these novel polymorphic EST-SSR markers will facilitate future studies of genetic variation and molecular-assisted breeding systems in arum lily (Wei *et al*., 2012).

Genetic engineering

Recombinant *Zantedeschia aethiopica* agglutinin (ZAA) was expressed in *Escherichia coli* as N-terminal His-tagged fusion. After induction with isopropylthio-beta-D-galactoside (IPTG), the recombinant ZAA was purified by metal-affinity chromatography. The purified ZAA protein was applied in anti-fungal assay and the result showed that recombinant ZAA had anti-fungal activity towards leaf mold (*Fulvia fulva*), one of the most serious phytopathogenic fungi causing significant yield loss of crops. This study suggests that ZAA could be an effective candidate in genetic engineering of plants for the control of leaf mould (Lin *et al*., 2009).

An efficient protocol for the *Agrobacterium tumefaciens*-mediated transformation of calla lily (*Zantedeschia elliottiana* (W. Wats.) Engl. cultivar 'Florex Gold') is described. Shoot basal discs were co-cultivated with *A. tumefaciens* C58C1 carrying a plasmid containing neomycin phosphotransferase (nptII) and plant ferredoxin-like protein (pflp) genes. After *Agrobacterium* co-cultivation, the shoot basal discs were exposed to 100 mg l^{-1} kanamycin for selection. Twenty-eight out of 260 discs (10.8%) were found to have survived and produced shoot clusters. Twenty-six of these were confirmed to contain the pflp transgene by PCR, ending up in 10% transformation efficiency. The disease resistance investigation revealed that 18 transgenic plants exhibited resistance to soft rot disease caused by *Erwinia carotovora* subsp. *carotovora*. The presence of pflp gene was demonstrated by PCR and its accumulation and activity was confirmed by Western blot and disease resistance assay. This was the first report to show the successful transformation and resistance to a bacterial pathogen in zantedeschia. The protocol is useful for the quality improvement of calla lily through genetic transformation (Yip *et al*., 2007).

A new lectin gene was isolated by using genomic walker technology and revealed to encode a mannose-binding lectin. Analysis of a 2233 bp segment revealed a gene including a 1169 bp 5' flanking region, a 417 bp open reading frame (ORF) and a 649 bp 3' flanking region. There are two putative TATA boxes and eight possible CAAT boxes lie in the 5' flanking region. The ORF encodes a 15.1 kDa precursor, which contains a 24-amino acid signal peptide. One possible polyadenylation signal is found in the 3'-flanking region. No intron was detected within the region of genomic sequence corresponding to zaa (Zantedeschia aethiopica agglutinin) full-length cDNA, which is typical of other mannose-binding lectin gene that have been reported. The deduced amino acid sequence of the lectin gene coding region shares 49-54% homology with other known lectins. The cloning of this new lectin gene will allow to further study its structure, expression and regulation mechanisms (Chen *et al*., 2004).

Using RNA extracted from *Zantedeschia aethiopica* young leaves and primers designed according to the conservative regions of Araceae lectins, the full-length cDNA of *Z. aethiopica* agglutinin (ZAA) was cloned by rapid amplification of cDNA ends (RACE). The full-length cDNA of zaa was 871 bp and contained a 417 bp open reading frame (ORF) encoding a lectin precursor of 138 amino acids. Through comparative analysis of zaa gene and its deduced amino acid sequence with those of other Araceae species, it was found that zaa encoded a precursor lectin with signal peptide. Secondary and three-dimensional structure analyses showed that ZAA had many common characters of mannose-binding lectin superfamily and ZAA was a mannose-binding lectin with three mannose-binding sites. Southern blot analysis of the genomic DNA revealed that zaa belonged to a multi-copy gene family (Chen *et al*., 2005).

Epifluorescence microscopy of mature pollen grains of *Turnera ulmifolia* and *Zantedeschia aethiopica* stained with 4',6-diamidino-2-phenylindole demonstrated the presence of fluorescent cytoplasmic DNA aggregates in the male reproductive cells of both species. Double staining of the cells with 4',6-diamidino-2-phenylindole and 3,3'-dihexyloxacarbocyanine iodide in Technovit resin sections showed that the mitochondria of these cells did not correspond to the fluorescent cytoplasmic DNA aggregates. Electron microscopy studies revealed both plastids and mitochondria in the cells of these species. In addition, immunoelectron microscopy using an anti-DNA monoclonal antibody showed clear labeling of plastids but not mitochondria. These data provide cytological evidence for biparental plastid inheritance and maternal mitochondrial inheritance in these species (Ji *et al.*, 2004).

REFERENCES

Anonymous, 1991. Zantedeschia plant 'Apricot Glow' United States Patent. Publication Number US PP7653 P.

Anonymous, 2003. Zantedeschia plant named 'Red Desire' United States Patent. Publication Number US PP13837 P2

Anonymous, 2004a. Calla lily plant named 'Pot Black' United States Patent. Publication Number US PP14513 P2.

Anonymous, 2004b. Zantedeschia plant named 'Hot Chocolate' United States Patent. Publication Number US PP15294 P2

Anonymous, 2004c. Zantedeschia plant named 'Hot Lips' United States Patent. Publication Number US PP15260 P2

Anonymous, 2004d. Zantedeschia plant named 'Pink Pot' United States Patent. Publication Number US PP14531 P2

Anonymous, 2005a. Zantedeschia plant named 'Jack of Hearts' United States Patent. Publication Number US PP0166295 P1

Anonymous, 2005b. Zantedeschia plant named 'Goldilocks' United States Patent. Publication Number US20050204441 P1.

Anonymous, 2006a. Zantedeschia plant named 'Captain Cupido' United States Patent. Publication Number US PP17019 P2.

Anonymous, 2006b. Zantedeschia plant named 'Captain Reno' United States Patent. Publication Number US PP17052 P3.

Anonymous, 2006c. Zantedeschia 'Edge of Night'. Official Journal of Plant Breeder's Rights Office, IPAustralia. *Plant Varieties J.*, **19**(3): 359.

Anonymous, 2007a. Calla lily plant named 'Purple Heart' United States Patent. Publication Number US PP17760 P3.

Anonymous, 2007b. Variety of calla lily named 'Majestic Star' United States Patent. Publication Number US PP18110 P3.

Anonymous, 2007c. Variety of calla lily named 'Vermeer' United States Patent. Publication Number US PP18281 P3.

Anonymous, 2007d. Zantedeschia plant named 'Picante' United States Patent. Publication Number US PP17811 P3

Anonymous, 2008 Calla Lily plant named 'Odessa' United States Patent. Publication Number US PP18833 P2.

Anonymous, 2009a. Calla lily plant named 'Hot Cherry BLZ' United States Patent. Publication Number US PP20071 P3.

Anonymous, 2009b. Calla Lily Plant Named 'Merlot BLZ' United States Patent. Publication Number US PP19867 P2.

Anonymous, 2009c. Calla lily plant named 'Rosa BLZ' United States Patent. Publication Number US PP20509 P3.

Anonymous, 2009d. Calla lily plant named 'Yellow Jacket Blz' United States Patent. Publication Number US 20090205094 P1.

Anonymous, 2009e. Zantedeschia plant named 'Snow white' United States Patent. Publication Number US PP19830 P2.

Anonymous, 2010a. Zantedeschia plant named 'Captain Amigo' United States Patent. Publication Number US PP20736 P2.

Anonymous, 2010b. Zantedeschia plant named 'Flamingo' United States Patent. Publication Number US PP21134 P2

Anonymous, 2010c. Zantedeschia plant named 'Captain Aguila' United States Patent. Publication Number US 20100223704 P1.

Anonymous, 2010d. Zantedeschia plant named 'Captain Ventura' United States Patent. Publication Number US 20100223703 P1.

Brown, F.S., Snijder, R.C. and van Tuyl, J.M. 2005. Biparental plastid inheritance in *Zantedeschia albomaculata* (Araceae). *Proce.* 9th Intl. Symp. on Flower Bulbs. *Acta Hort.*, **673**: 463-468.

Chen, Z., Pang, Y., Liu, X., Wang, X., Deng, Z., Sun, X., and Tang, K. 2005. Molecular cloning and characterization of a novel mannose-binding lectin cDNA from *Zantedeschia aethiopica. Biocell.*, **29**(2): 187-93.

Chen, Z., Sun, X. and Tang, K. 2004 . Genomic cloning and characterization of a novel lectin gene from *Zantedeschia aethiopica*. *Biosci. Rep.*, **24**(3): 225-34.

Earl, G. 1957. Chromosome numbers and speciation in the genus *Zantedeschia*. *Plant Life.*, **13**: 140-46.

Ferrie, A.M.R., Palmer, C.E. and Keller, W.A. 1995. Haploid Embryogenesis. *In:* Thorpe T. A. (eds.), *In Vitro* Embryogenesis in Plants. Kluwer Academic Publishers, Dordrecht, The Netherlands, pp. 309-44.

Ferrie, A.M.R., Taylor, D.C., Mackenzie, S.L., Rakow, G., Raney, J.P. and Keller, W.A. 2008. Microspore mutagenesis of *Brassica* species for fatty acid modifications: a preliminary evaluation. *Plant Breed.*, **127:** 501-06.

Forster, B.P. and Thomas, W.T.B. 2005. Doubled haploids in genetics and plant breeding. *Plant Breed. Rev.*, **25:** 57-88.

Funnell, K.A. 1993. *Zantedeschia. In*: M. LeNard. The physiology of flower bulbs. Elsevier, Amsterdam, pp: 683-03.

Hamada, K. and Hagimori, M. 1996. RAPD-based method for cultivar-identification of calla lily (*Zantedeschia* spp.). *Scientia Horticulturae*, **65**(2/3): 215-218.

Horn, M. 1962. Breeding research on South African plants: III. Intra and interspecific compatibility I *Ixia* L., *Sparaxis* Ker, *Watsonia* Mill and *Zantedeschia* Spreng. *J. South African Bot.,* **28**(4): 269-77.

Ji, X., Zhang Q., Liu Y. and Sodmergen. 2004. Presence of plastid and absence of mitochondrial DNA in male reproductive cells as evidence for cytoplasmic inheritance in Turnera ulmifolia and *Zantedeschia aethiopica*. *Protoplasma*, **224**(3-4): 211-6.

Johnston, S.A. and Hanneman, R.E. 1982. Manipulations of endosperm balance number overcome crossing barriers between diploid *Solanum* species. *Science*, **217:** 446-48.

Johnston, S.A. and Hanneman, R.E. 1995. The genetics of triploid formation and its relationship to endosperm balance number in potato. *Genome,* **38:** 60-67.

Johnston, S.A., Den Nijis, T.P.M., Peloquin, S.J. and Hanneman, R.E. 1980. The significance of genetics balance to endosperm development in interspecific crosses. *Theor. Appl. Genet.*, **57:** 5-9.

Ko, J.A., Kim, Y.S. and Eun, J.S. 1996. Embryogenesis and plant regeneration by the anther culture of *Zantedeschia aethiopica* spp. *J. Korean Soc. Hort. Sci.*, **37:** 468-474.

Kuehny, J.S. 2000. Crop reports: Calla history and culture. *Hort. Tech.*, **10:** 267-74.

Letty, C. 1973. The genus *Zantedeschia. Bothalia,* **11**(1&2): 5-26.

Lewis, D.H., Arathoon, H.S., Swinny, E.E., Huang, S.C. and Funnell, K.A. 2003. Anthocyanin and carotenoid pigments in spathe tissue from selected *zantedeschia* hybrids. *Acta Hort.*, **624**: 147-54.

Lin, L., Liu, X., Hu, L., Zhou, Y., Sun, X. and Tang, K. 2009. Expression and purification of *Zantedeschia aethiopica* agglutinin in *Escherichia coli. Mol. Biol. Rep.,* **36**(3):437-41.

Michishita, A., Ureshino, K. and Miyajima, I. 2002. Plastomegenome incompatibility of *Rhododendron serpyllifolium* (A. Gray) Miq. to evergreen *Azalea* species belonging to series Kaempferia. *J. Japan. Soc. Hort. Sci.*, **71:** 375-81.

New, E.H. and Paris, C.D. 1967. Virescence in *Zantedeschia. Proce*: American Society for Horticultural Science, **92:** 685-95.

Perry, P. L. 1989. A new species of *Zantedeschia* (Araceae) from the western Cape. *South African J. Bot.,* **55**(4): 447-51.

Shibuya, R. 1956. Intercrossing among pink calla, white-spotted calla and yellow calla. Tokyo. pp. 97.

Singh, Y., Baijnath, H. and van Wyk, A.E. 1996. Taxonomic notes on the genus *Zantedeschia* Spreng (Araceae) in southern Africa. *South African J. Bot.,* **62**(6): 321-24.

Snijder, R.C. and van Tuyl, J.M. 2002. Evaluation of tests to determine resistance of *Zantedeschia* spp. (Araceae) to soft rot caused by *Erwinia carotovora* subsp. *Carotovora. European J. Plant Pathology,* **108:** 565-571.

Snijder, R.C. and van Tuyl, J.M. 2004. Partial resistance to *Erwinia carotovora* subsp. *carotovora* and plant vigour among F_1 hybrids of *Zantedeschia* cultivars. *Acta Horticulturae*, **651**(2): 63-67.

Snijder, R.C., Cho, H.R., Hendriks, M.M.W.B., Lindhout, P., van Tuyl, J.M. 2004a. Genetic variation in *Zantedeschia* spp. (Araceae) for resistance to soft rot caused by *Erwinia carotovora* subsp. *carotovora. Euphytica*, **135**(1): 119-128.

Snijder, R.C., Lindhout, P. and van Tuyl, J.M. 2004b. Genetic control of resistance to soft rot caused by *Erwinia carotovora* subsp. *carotovora* in *Zantedeschia* spp. (Araceae), section Aestivae. *Euphytica*, **136**(3): 319-325.

Snijder, R.C., Santiago Brown, F. and van Tuyl, J.M. 2007. The role of plastome-genome incompatibility and biparental plastid inheritance in interspecific hybridization in the genus *Zantedeschia* (Araceae). *Floriculture Ornamental Biotech.*, **2:** 150-57.

Traub, H.P. 1948. The genus *Zantedeschia. Plant Life,* **4:** 9–32.

Wei, ZunZheng, Luo, LaiBo, Zhang, HuaLi, Xiong, Min, Wang, Xian and Zhou Di. 2012. Identification and characterization of 43 novel polymorphic EST-SSR markers for arum lily, *Zantedeschia aethiopica* (Araceae). *American J. Botany*, **99**(12): 493-497.

Wolff, H. A. 1919. Vooruitgang in de Richardia's. *Weekblad voor Bloembollencultuur*, **30** (45): 196.

Wright, P.J., Burge, G.K. and Triggs, C.M. 2002. Effects of cessation of irrigation and time of lifting of tubers on bacterial soft rot of calla (*Zantedeschia* spp.) tubers. *J. Crop Hortil. Sci.,* **30:** 265-72.

Wu, HongZhi, Shi, JingFeng, Zheng, SiXiang and Zhang, JingLi Zhou Di. 2011. Induction of 2n pollen in coloured *Zantedeschia* hybrid and obtaining of its triploid. *J. Agril. Biotech.*, **19**(4): 662-668.

Yao, J. 1996. Production of triploid *Zantedeschia* hybrids using embryo rescue. *New Zealand J. Crop Hortil. Sci.,* **24:** 297-301.

Yao, J.L. and Cohen, D. 2000. Multiple gene control of plastome-genome incompatibility and plastid DNA inheritance in interspecific hybrids of *Zantedeschia. Theor. Appli. Gene.,* **101**(3): 400-06.

Yao, J.L., Cohen, D. and Rowland, R.E. 1994a. Plastid DNA inheritance and plastome-genome incompatibility in interspecific hybrids of *Zantedeschia* (Araceae). *Theor. Appli. Gene.,* **88**(2): 255-60.

Yao, J.L., Cohen, D. and Rowland, R.E. 1995. Interspecific albino and variegated hybrids in the genus *Zantedeschia. Plant Science,* **109**(2): 199-06.

Yao, J.L., Rowland, R.E. and Cohen, D. 1994. Karyotype studies in the genus *Zantedeschia* (*Araceae*). S. Afr. J. Bot., **60**: 4–7

Yip, M., Huang, H., Ger, M., Chiu, S., Tsai, Y., Lin, C. and Feng, T. 2007. Production of soft rot resistant calla lily by expressing a ferredoxin-like protein gene (pflp) in transgenic plants. *Plant Cell Rep.*, **26**(4): 449-57.

Zhang, X., Wu, Q., Li, X., Zheng, S., Wang, S., Guo, L., Zhang, L. and Custers, B.W.M. 2011. Haploid plant production in *Zantedeschia aethiopica* 'Hong Gan' using anther culture. *Scientia Horticulturae*, **129**(2): 335-42.

Section III

Cut and Loose Flower Crops

Chapter – 19

China Aster

China aster (*Callistephus chinensis* L. Nees) belongs to family Asteraceae. It is a half hardy and branching annual. Its generic name *Callistephus* is derived from two Greek words "*kalistos*" meaning beautiful and "*stephos*" meaning a crown alluding the large colourful flower heads. Plant is erect having hispid-hairy branches (Bailey, 1963). It is propagated through seed. It is cultivated as cut flower both in the field and under protected condition in temperate countries (Post, 1956 and Warren Auman, 1980). Whereas, in India it is widely grown in Karnataka, Tamil Nadu, West Bengal and Maharastra for cut as well as loose flower production (Janakiram *et al*., 2001).

ORIGIN AND DISTRIBUTION

China aster is originated from China. Several literatures also stated that China aster is native to China and Japan (Anon., 1950, Bailey, 1963 and Cockshull, 1985). The evolution of China aster has remarkable variations. The present day aster has been developed from single form of wild species, *C. chinensis*, a plant native to China. The plant was introduced in Europe in 1731 by Jesuit Missionary R.P. d Incarville.

The original plant had single flowers with two or four rows of blue, violet or white ray florets. The first change in the flower type had been the prolongation or development of central florets which lead to quilled flowers. Much original work to improve the plant was done by French. Later the British followed suit with breeding programmes. Most of the recent improvements are the work of American breeders. French florists contributed much in development of double forms. German scientists also evolved many doubled cultivars. These asters were known as German asters because the first great advances in the evolution of the plant were made in Germany and the seeds then used came largely from that country. The comet types introduced in 1886 or 1887 subsequently replaced the quilled types. With the introduction of new branching types by James Vick in 1893, the USA became the main centre of development of this plant as well as for the production of seeds. Since its introduction into Europe, it is grown in almost all the countries of the globe as a garden annual. A good amount of work on China aster on improvement and growing practices has been done in India (Singh, 2006).

SPECIES

Callistephus, a monotypic genus represented by *C. chinensis*. So far, no other species is reported. The genus *Callistemma* also erected by Cassini, is older than *Callistephus*. Cassini described the China aster as *C. hortensis*. It was first named by Linnaeus, however, as *Aster chinensis* and needs subsequently transferred this name to *Callistephus*, so that the plant now would better bear the name *Cellistephus chinensis* (Bailey, 1963). Many races of the species showing a wide range of variation in the size, shape and colour are known under cultivation.

FLOWER MORPHOLOGY AND BIOLOGY

A single inflorescence, which is commonly called head is a capitulum (Cockshull and Hughes, 1969). The original plant had single flowers with two or four rows of blue, violet or white ray florets. The first change in the flower type had been the prolongation or development of central florets which lead to quilled flowers. A China aster head often holds several hundred individual florets. The individual flowers on florets are of 2 types, female ray florets with an extended corolla forming a colourful strap-shaped ligule and a central group of small yellow hermaphrodite disc florets. The outer ray florets of the head produce no pollen. The disc florets at the centre of head are perfect and produce both, male and female organs. Stamens and pistil both do not mature simultaneously in the individual flowers. The stigma of the individual flower unfolds after the pollen is discharged from the flower therefore, it is called as geitonogamous.

From breeding point of view, it should be remembered that the florets of a head mature from the outside row of the ray flowers inward to the central disc flowers. Successive rings of florets are mature each day by which several days may elapse between the time when the first and last floret mature.

The single and semi-double varieties are predominantly cross-pollinated, whereas, the double varieties are generally self-pollinated. However, for an effective breeding programme in China aster, care and precaution to be taken against occasional cross-pollination occurring due to insects or other pollination agents. Novotna (1989) observed that tubular flowers produced more seeds than the ligulate flower in the same anthodium. This was due to poor pollination in the ligulate flowers.

GENETICS

Most of the China asters are diploid (2n=18) (Huziwara, 1954 and Cockshull, 1985). Raghava and Negi (1992) made an attempt to study the inheritance of growth habit. The genetics of drooping and erect growth habit was studied in 5 combinations, involving 6 parents. Growth habit of China aster was controlled by one major gene. Drooping was incompletely dominant over the erect habit. Gene Dp governed drooping habit, while its allele dp controlled erect habit.

Inheritance of quantitative characters

Inheritance of quantitative characters like plant height, number of main branches, laterals and flowers/plant, plant spread, days to flower, flower stalk length, flower size, flower weight, number of ray florets/flower head, duration of flowering and cut flower life in a 12×12 diallel set of crosses (excluding reciprocals). The results indicated the involvement of both additive and non-additive gene actions in the inheritance. Additive gene action was noticed in the inheritance of number of ray florets/flower head and flower size, whereas, other characters were found to be governed by non-additive gene action (Raghava, 1984).

Inheritance of flower type and doubleness

Genetic architecture in the expression of doubleness and flower types by crossing 9 parents was studied (Raghava, 1984 and Raghava and Negi, 2001). Doubleness of flower was reported to be monogenically dominant over singleness. Pompon type flower with one row of ray florets and tubular disc florets was dominate over both chrysanthemum and fluffy types. White chrysanthemum types was dominant over fluffy type. These 3 types were found to be governed by 3 independent genes Tu, D and C. Another genotype having pompon flower type with four rows of ray florets and tubular disc florets was dominant over pompon with one row of ray florets and tubular disc florets. When this pompon type was crossed with chrysanthemum and fluffy types, it was found that the tubular (showy) disc florets were incompletely dominant over normal non-showing disc florets. Thus, in this pompon type, there is a dominant allele C of the gene C, which does not allow the conversion of ray florets into tubular disc florets. Tu alone or C, or C' alone fail to produce tubular disc florets.

Inheritance of colour

As per literature available, Wit (1937) first reported that the inheritance of colour is governed by multiple alleles. Mode of inheritance of flower colour was studied by Negi and Raghava (1990). They observed that violet colour was dominant over all other colours, blue was dominant over creamy white, deep pink was dominant over pink and white, but incompletely dominant over light pink and pink was dominant over light pink and white. Four genes, designated C, B, R and P governed flower colour. It was proposed that P has a dominant allele PD (for deep pink) and recessive allele P (light pink).

Heritability and genetic advance

Heritability analysis in China aster has been done by many workers. Negi *et al.* (1983) observed that flower weight, stalk length, flower size and plant height exhibited high value of heritability in broad sense, however, number of flowers and branches/ plant and days to flower had medium heritability. In a heritability and correlation studies, Ashwath and Parthasarathy (1993) observed that plant height, stem scar diameter and number of flowers showed high frequency of genotypic correlations. The estimate of heritability and co-heritability showed that all characters were controlled by additive gene effects. In another study, heritability estimates for plant spread, leaf area, flower weight and flower yield were relatively high indicating that these characters can be selected for production of flower. High heritability with medium genetic advance was also observed for number of leaves and number of flowers/plant (Patil and Kumar, 2003).

BREEDING OBJECTIVES

i. Vigorous plant growth

ii. More number of flowers/plant for loose flower

iii. More stalk length for cut flower

iv. Improved size and shape of flower

v. Improved and new colour of flower

vi. Horizontal placement of flower

vii. Good keeping quality for cut flower varieties

viii. Resistance against China aster yellow disease

viii. Resistance against wilt disease

BREEDING METHODS

Much original work to improve the plant was done by French. Later the British followed suit with breeding programmes. Most of the recent improvements are the work of American breeders. French florists contributed much in development of double forms. German scientists also evolved many doubled cultivars. These asters were known as German asters because the first great advances in the evolution of the plant were made in Germany and the seeds that used came largely from that country. The comet types introduced in 1886 or 1887 subsequently replaced the quilled types. With the introduction of new branching types by James Vick in 1893, the USA became the main centre of development of this plant as well as for the production of seeds. Since its introduction into Europe, it is grown in almost all the countries of the globe as a garden annual. A good amount of work on improvement of China aster has been done in India and some cultivars have been evolved through selection and hybridization is now in commercial production.

Introduction

Revolutionary work on selection and hybridization has been done in other countries and several new types developed. Therefore, about 50 collections of aster have been done from Australia, Canada and USA through NBPGR, New Delhi. These collections are given in Table 1.

Table 1: Introduction of China aster varieties from Australia, Canada and USA.

S. No.	Country	Collection
1.	Australia	Giant Orego, Grande Fleur Varie, Perfection White, Perfection Rose.
2.	Canada	Dwarf Green Bright Red, Reines Marguerite Dwarf, Double Mum Flowered Blue Mumster, Reines Marguerite Dwarf Double Mum, Dwarf semi-Double Start Blend, Flowered Rose Mumster, Dwarf Double Mum Flowered Copper Mumster, Pinocchio Mixed, Creg. Finest Mixed, Dwarf Queen Yellow, Reines Marguerite Dwarf Queen White, Reines Marguerite Dark Blue, Dwarf Queen Rose Pink, Reines Marguerite Giant Princess Mixed, Reines Marguerite Ball Mixed, Ladies Formula Blend, Reines Marguerite Peach Lady, Reines Marguerite Crimson Lady, Dwarf Queen Mixed, Red Mumster, Pink Mumster, Dwaft Queen Light Blue, Rose Mini Lady, Blue Mini Lady, Mumster Formula Blend, Mixed Powerpuffs, American Branching Mixed, Pink Lady, Purple Lady, Blue Lady, Rose Lady, White Lady, Heart of France, Single California Giant Finest Mixed, Early Charm Mixed, Sky Blue Star, Dwarf Queen Blood Red.
3.	USA	Crimson Perfection, Rose, Pink, Mid Blue, Early Bird White Dark Blue, Early Bird Red, Blue Wonder, Mixed Colours, Scarlet Beauty, Blue Beauty.

Selection

Wide variation and diversity in plant growth, flower colour, form and size were observed (Anon., 1950 and Bailey, 1963). Various types of germplasm were collected at the IIHR, Bengaluru. In this collection several varieties/genotypes were found in heterozygous condition. Therefore, it was felt that single plant selection would be a best possible approach for improvement on the basis of desired characters. At this centre 50 pure lines were developed. Afterwards performance of these purelines was assessed in replicated trials for 3 consecutive years along with 3 standard.local varieties. On the basis of these testing, 2 selections, AST-1 and AST-2 performed better than several other entries. Flower colour of this selection is white and they have erect plant growth and pompon type flower head (Anon., 1977, 1981, 1982, 1983). Weigelt (1984) made selection for resistance to *Fusarium oxysporum*. Plants were grown in beds and artificially inoculated with pathogen. The survivors served as parents for next generation. This selection method resulted in highest disease resistance. A cultivar of China aster (*Callistephus chinensis*) was evolved after the selection for six generations. In 3-year field tests of contrast, plant height was 61.63% lower, while diameter of flower and number of flower were 14.91% and 26.93% higher than the control. It was released as cv. Jinongda Blue for cultivation in Jilin Province of China (Hou *et al.*, 2009).

Some improved varieties of China aster were developed from NARP Research Station of MPKV, Ganeshkhind, Pune. The details of these varieties are given below.

Phule Ganesh White: It is medium tall, erect, growing late variety. It produces snow white flower. The flowers are double chrysanthemum type with better vase life and suitable for cut flowers and garlands. Being white colour there is good demand during X-mas. The stalk length is 48.6 cm. The yield potential is 47.07 lakh flowers/ hectare.

Phule Ganesh Violet: It is early, semi erect and superior in growth and flower yield characters. It produces medium size violet flowers. The flowers are double chrysanthemum type suitable for cut flower and loose flower production. It gives 60.47 lakh flowers/hectare.

Phule Ganesh Pink: It is early, semi-erect and superior quality variety. The flowers are attractive double chrysanthemum type big size rosy pink colour and suitable for cut flower production. It produces 43.03 lakh flowers/hectare.

Phule Ganesh Purple: It is semi-erect plant with bright purple colour flowers which is preferred in the basic in the artistic floral managements and pasting work. It produces 46.82 lakh flowers/hectare.

Hybridization

Before crossing, it is very important to select the desirable male and female parents based on selective horticultural characters. First, anthers of disc florets of female parent are clipped before anthesis (dehiscence). Disc florets are bagged with butter paper bags to avoid natural pollination. Long ray florets of female plants are trimmed with scissors or sharp blade till the stigma gets exposed. Third day after emasculation, pollination is done. The flowers, which will provide the pollen are put in the sun for 1-2 hours and then beaten out on the trimmed flower. After that all crosses are properly labelled. Seeds are collected when the capitulum is properly dried. Generally seed get matured after one month of crossing. Seeds are dried and stored in dry place or cold storage.

In India, systematic hybridization work on China aster was started during early seventies at IIHR, Bengaluru followed by other institutes. Studies on combining ability has been done by several workers for selecting potential parents for hybridization as well as characterizing nature and magnitude of gene action. Raghava and Negi (1993) studied combining ability in a 12 × 12 diallel set excluding reciprocals. Plant height, main and lateral branches/plant, plant spread, days to flower, flowers/plant, stalk length, flower size and weight, ray florets/ flower head, flowering duration and cut flower life were studied. All the characters except flower size and number of ray florets/flower head were under the major influence of non-additive gene action, where heterosis breeding may be followed. Based on SCA effects, 5 specific cross combinations AST-20 × AST-16, Shell Pink × Azure Blue, AST-8 × AST-20, AST-8 × Crego Giant Pink and Giant Branching Comet or Crego Light Blue × Azure Blue were recommended for heterosis breeding. Combining ability study was also done by Patil and Rane (1994) on 12 components in 9 lines and their 72 F_1 hybrids. Significant differences due to GCA, SCA and reciprocal combining effect (RCA) for all the characters were there except SCA and RCA for vase life. Greater GCA than SCA and RCA effects indicated the role of additive gene action. Parents P5 (Aladi Selection) and P7 (Perfection Purple Selection) were good combiner for yield. In another combining ability study, high magnitude of GCA mean square was observed than the SCA mean square, which indicated predominant role of additive generation. Parents recorded significant GCA effects for 13 characters out of 15 characters. They observed that parents PG Pink and PG White were good general combiners for yield and yield attributes and cross combination PG Pink × PG Violet and PG Violet × PG White for earliness (Kumar *et al.*, 2004). Various workers studied heterobeltiosis and heterosis breeding for evolving new hybrids in China aster. In a study, 66 F_1 hybrids were evaluated to study the magnitude of heterobeltiosis. Appreciable heterobeltiosis was recorded for all the important characters except days to flowering. It indicated that the best performing hybrid was better than the best parent (Raghava *et al.*, 1988). Information on heterosis is derived from 12 yield components in 9 *Callistephus chinensis* lines and their 72 F_1 hybrids. Heterobeltiosis for flower weight was highest (99.25%) in Aladi Selection × Kashmir Local Selection and for flowers/plant (74.8%) in Rose Selection × Otur Selection (Patil and Rane, 1994a). In a diallel set involving 6 parents (excluding reciprocal), 15 hybrids were evaluated to study the magnitude of heterosis. Appreciable heterosis was observed for all the characters and the performance of F_1 hybrids was better than the top parents except the average weight of flower. Considering the economic traits like number of flowers/plant, stalk length, flower weight and earliness, the crosses PG Violet × PG White, Kamini × PG Violet, PG Pink × PG Violet combinations were found to be the best and are recommended for commercial exploitation (Kumar *et al.*, 2004a). First time in India, 4 varieties namely Kamini, Poornima, Shashank and Violet Cushion were released from IIHR, Bengaluru. The description of these varieties developed through hybridization are given in Table 2.

Table 2: China aster varieties evolved through hybridization.

S. No.	Variety	Parentage	Chief characteristics
1.	Kamini	AST-6 × AST-36	Develop through pedigree method of breeding, flower colour is deep pink, more attractive than cv. Local Pink, plant about 60 cm tall and takes about 138 days to flower, flower diameter is 6 cm and individual flower weight is 2 g, produces about 50 flowers/plant, flower stalk length is 30 cm and vase life is about 8 days.
2.	Poornima	AST-29 × AST-3	Develop through advanced pedigree selection, bearing pure white flowers of powdery puff type, takes about 105 days to flower and grows to about 50 cm, flower diameter is 5 cm, produces 25 flowers/plant and individual flower weight is 3.5 g, stalk length of 25 cm and vase life of 7 days.
3.	Shashank	Local Pink × AST-2	Flowers are powdery puff type and are creamy white in colour, bears flower 124 days after sowing on 55 cm long plants, flower size is 6 cm and weight is 2.5 g, produces 45 flowers/plant, which is more than double of Local White variety, stalk length is 25 cm and vase life of 9 days.
4.	Violet Cushion	Local Pink × AST-2	Developed through pedigree method, a floriferous variety bearing pompon type flowers of violet colour having 4 to 5 rows of ray florets with tubular disc florets, produces flower in 130 days, flower diameter is 4.5 cm and weight is 2 g each, bears 70 flowers/plant with stalk length of 20 cm and vase life of 8 days.

Mutation

In India, so far not much information on mutation breeding is available. Viable seeds of cvs. Poornima, Violet Cushion, Kamini and Local Pink were exposed to 10, 20 and 30 krad doses of gamma rays. Lower doses showed stimulatory effects and higher doses showed inhibitory effects with respect to seed germination (Janakiram *et al.*, 2001). In Poland, Wosinska (1982) irradiated seeds of China aster cultivars at 3-12 krad. In the M_1, variability was noted in leaf colour and size and in shape of lingual flowers. Changes induced by irradiation on plant height, shoot, leaf and flower colour and shape of lingual flowers were more frequent in the M_2 than the M_1 generation. Changes in M_1 generation were most pronounced after 9 krad dose and in the M_2 after 6-9 krad doses.

In another study, a radio mutant of China aster with a new type of head and ray florets is observed. All florets in the head are bisexual and tubular although dimorphic (varied in size and colour). The position of ray florets in the head is taken by bisexual open tubular florets which are often crowned with irregular teeth (lobes). The disc florets are short and parchment-like. A comparison of both types of florets is given in terms of number and size of stamens, vitality of pollen, floret fertility and number of teeth in the corolla (Wosinska, 2001).

Polyploidy

Applications of colchicine were performed on the swollen seeds of China aster cvs. Gerda and Princess Armida and on the apex of seedlings between cotyledon leaves after their opening. The first method of polyploidy induction was soaking of swollen achenes in a solution of colchicine at 0.01, 0.03, 0.05, 0.1, 0.3% and operating times were 6, 12 and 18 h. The second method was dropping colchicine solution (between 07.00 and 08.00 h) between cotyledon leaves after their opening (concentrations were 0.1, 0.3, 0.5, 1.0, 1.5% and operating times were 2, 3 and 5 days). Emergence rate was considerably suppressed after using low colchicine concentrations and after a short time of incidence. Longer operating time of colchicine action and lower colchicine concentration suppressed emergence rate more than higher concentration and shorter operating time. The control plants emerged 7-10 days after sowing while variants with higher colchicine concentration and longer operating time of incidence emerged after 21 days. Poor root regeneration was observed in deformed seedlings. High number of deformed seedlings was observed at the higher concentrations on colchicine solution within a relatively short time of colchicine action. The hypocotyledons of deformed seedlings were truncated and plumped while cotyledon leaves were small and plumped. Soaking of swollen achenes suppressed emergence rate and had high seedling dieback. After colchicine application on the apex, root damage was minimal and survival rate was relatively high. The highest dieback was found in the 5 days variant of colchicine application on the apex cv. Gerda had lower emergence rate, higher number of deformed seedlings and higher seedling mortality compared to cv. Princess Armida (Hanzelka and Kobza, 2001).

Polyploid leaves were thicker and more rounded with more entire leaf blade than diploids. Leaf index was more in diploid plants than polyploidy plants. Polyploid plant number was evaluated by flow-cytometric analysis. The number of polyploids i.e. tetraploidy and mixoploidy (part tetraploid and part diploid)) was determined in the C_0 population of *C. chinensis*. The most successful variants in cv. Princess Armida followed application of 0.3% colchicine concentration for 6 and 18 h on the swollen seeds. Five-day variants and 3-days with 1.5% on the apex had the highest number of polyploids. In var. Gerda, the most successful combination was 0.1% colchicine and 6 h of treatment and 0.05% and 18 h. Colchicine application on the apex resulted in maximum polyploidy plants in plants treated over 5 days with 1.5 and 1.0% colchicine (Hanzelka and Kobza, 2001a).

Resistance breeding

Pioneer work in breeding for wilt resistance was done by Jones and Riker. Through continuous pureline selection, wilt resistant lines in all types except pompons were developed. The cultivars Heart of France, Crego, Ostrich, Feather, Stardust, Roment, Kirkwell, American Branching (tall), Powderpull (bouquet type) and Laplator are reported to be resistance to this disease.

China aster also affected by root-knot nematodes, hence, in this direction trial was conducted at IIHR, Bengaluru. IIHR-2 was categorized as highly resistant pureline. AST-5, IIHR-17, IIHR-21 and cv. Shashank were found resistant, whereas, cv. Poornima was moderately resistant and Kamini, Violet Cushion and AST-16 were susceptible (Janakiram *et al.*, 2001).

BIOTECHNOLOGY

Tissue culture

Pillai *et al.* (1990) observed that leaf protoplasts of *Callistephus chinensis* entered sustained division only following the preconditioning for 24 h of peeled leaf tissues on agar solidified MS medium. Protoplasts were also isolated from all suspensions of *C. chinensis* and divided in MS or KM media, however, only the leaf mesophyll protoplasts produced callus, which developed shoots. In other study, protoplasts derived from pre-conditioned leaves of *C. chinensis* were planted in a range of media. On modified MS media, protoplasts of *C. chinensis* began dividing and gave callus resulting in plantlet regeneration (Pillai *et al.*, 1991).

Molecular breeding

Very little work so far has been done on improvement of China aster through biotechnological approaches. DNA was analysed from 98 genotypes of China asters (*Callistephus chinensis* Nees.) with the method of Dellaporta. With the method molecular genetic RAPD was made analysis of genetic variability. The 120 (10 mers) primers was replicated, used for screening and 26 primers with certificate level of polymorphism was choosed, which present 22% of primers with reproductivity of polymorphism. On the basis of realised screening of China asters genotypes and selected primers with polymorphism by RAPD reproduction was processed binary matrix and created dendrogram for genetical relationship. The average of similarity reached value of 0,7197, minimal value of similarity was determine between cv. Bukett Reinweiss and Pompon Hellviolet and reached value 0,5263; maximal value of similarity was 0,9255 and was reached between cv. Matsumoto Salmon and Matsumoto Light Blue (Neas and Kobza, 2004).

REFERENCES

Anonymous, 1977. Annual Report, IIHR, Bengaluru, pp. 76-77.

Anonymous, 1981. Annual Report, IIHR, Bengaluru, pp. 42-43.

Anonymous, 1982. Annual Report, IIHR, Bengaluru, pp. 45-46.

Anonymous, 1983. Annual Report, IIHR, Bengaluru, pp. 66-67.

Anonymous. 1950. The Wealth of India, CSIR, New Delhi, Vol. II: C, 427 p.

Ashwath, C. and Parthasarathy, V.A. 1993. Heritability and correlation studies in China aster (*Callistephus chinensis* Nees). *Indian J. Hort.*, **50**(1): 89-92.

Bailey, L.H. 1963. The Standard Cyclopedia of Horticulture, Mac Millan Co., New York, Vol. I, A-E, 1200 p.

Cockshull, K.E. 1985. *Callistephus chinensis*. *In*: A.H. Halevy (ed.), CRC Handbook of Flowering. CRC Press Inc., Florida, Vol. I: A, pp. 112-114.

Cockshull, K.E. and Hughes, A.P. 1969. Growth and dry weight distribution in *Callistephus chinensis* as influenced by lighting treatment. *Ann. Bot.*, **33**: 367-379.

Hanzelka, P. and Kobza, F. 2001. Genome induced mutation in *Callistephus chinensis* Nees. I. Effect of colchicine application on the early plant development. *Zahradnietvi Hortil. Sci.*, **28**(1): 15-20.

Hanzelka, P. and Kobza, F. 2001a. Genome induced mutation in *Callistephus chinensis* Nees. II. Results of colchicine application-efficiency of polyploidization.. *Zahradnietvi Hortil. Sci.,* **28**(1): 21-27.

Hou, JianWei, Zhao, ChunLi and Hou, TieMei. 2009. Selection and breeding of China aster Jinongda Blue. *J. Jilin Agril. Univ*., **31**(2): 168-169.

Huziwara, Y. 1954. Karyotype analysis in *Bellis, Callistephus* and *Solidago*. *Kromorsomo*, **21**: 773-776.

Janakiram, T., Patil, M.T. and Bhattacharjee, S.K. 2001. China aster (*Callistephus chinensis*). AICRP on Floriculture, Tech. Bull., ICAR, New Delhi.

Kumar, S., Shirol, A.M., Patil, B.R., Reddy B.S. and Kulkarni, B.S. 2004. Combining ability studies in China aster (*Callistephus chinensis* L. Nees). *Abst*: National Symposium on Recent Trends and Future Strategies in Ornamental Horticulture. ISOH, Dharwad, Karnataka. p. 20.

Kumar, S. Shirol, A.M., Reddy, B.S., Kulkarni and Mulge, R. 2004a. Heterosis studies in China aster. *J. Orna. Hort.,* **7**(3-4): 18-21.

Neas, T. and Kobza, F. 2004. Application of molecular method (RAPD) and determination of genetic variability in China asters (*Callistephus chinensis* Nees). *Acta Horticulturae et Regiotecturae*, **7**: 26-28.

Negi, S.S. and Raghava, S.P.S. 1990. Genetics of flower colour in China aster (*Callistephus chinensis* (L.) Nees). *Euphytica,* **48**(2): 117-122.

Negi, S.S. and Raghava, S.P.S., Sharma, T.V.R.S. and Srinivasan, V.R. 1983. Studies on variability and correlation in China aster (*Callistephus chinensis* Nees). *Indian J. Hort*., **40**: 102-106.

Novotna, I. 1989. The floral biology of *China aster* (*Callistephus chinensis* Nees) from the genetic and breeding aspects. Sbornik UVTIZ Zahradnictvi, **16**(1): 55-64.

Patil, S.S.D. and Rane, D.A. 1994. Combining ability studies in China aster. *J. Maharastra Agri. Uni.*, **19**(3): 418-420.

Patil, S.S.D. and Rane, D.A. 1994a. Heterosis studies in China aster. *J. Maharastra Agri. Uni*., **19**(3): 372-374.

Patil, V.S. and Ravi Kumar, H. 2003. Evaluation of China aster (*Callistephus chinensis* (L.) Nees) genotypes under transitional zone of north Karnataka. *Abst*: National Symposium on Recent Advances in Indian Floriculture. ISOH, Dharwad, Karnataka, p.84.

Pillai, V., Davey, M.R. and Power, J.B. 1990. Plant regeneration from mesophyll protoplasts of *Centaurea cyanus, Senecio hybridus* and *Callistephus chinensis*. *Plant Cell Reports*, **9**(7): 402-405.

Pillai, V., Davey, M.R. and Power, J.B. 1991. Protoplast isolation, culture and regeneration from some ornamental compositae species. *Physiologia Plantarum*, **82**(1): A 12.

Post, K. 1956. Florist Crop Production and Marketing. Orange Judd., New York. pp. 361-368.

Raghava, S.P.S. 1984. Genetical investigations of China aster [(*Callistephus chinensis* (L.) Nees]. Ph.D. Thesis, UAS, Bengaluru.

Raghava, S.P.S. and Negi, S.S. 1992. Inheritance of growth habit in China aster (*Callistephus chinensis* L. Nees). *Indian J. Hort.*, **49**(3): 281-283.

Raghava, S.P.S. and Negi, S.S. 1993. Combining ability in China aster (*Callistephus chinensis* L. Nees). *Indian J. Hort.*, **50**(2): 180-186.

Raghava, S.P.S. and Negi, S.S. 2001. Inheritance of flower type and doubleness in China aster. *J. Orna. Hort.,* **4**(1): 7-12.

Raghava, S.P.S., Negi, S.A. and Ramchandra, P.R. 1988. Heterobeltisis in China aster (*Callistephus chinensis* L. Nees). *Indian J. Hort.*, **45**: 336-343.

Ravi Kumar, H. and Patil, V.S. 2003. Genetic variability and character association studies in China aster (*Callistephus chinensis*) genotypes. *J. Orna. Hort.*, **6**(3): 222-228.

Singh, A.K. 2006. Flower Crops: Cultivation and Management. Pub. New India Publishing Agency, New Delhi. pp. 61-71.

Warren Auman, C. 1980. Minor cut crops, *In*: R.A. Larson (ed.), Introduction to Floriculture, Academic Press, New York. pp. 195-196.

Weigelt, K.H. 1984. China aster: tendencies and problems of selection. *Gb. and Gw.*, **45**: 1064- 1066.

Wit, F. 1937. Contributions to the genetics of China aster. *Genetica*, **19**(1-3):1-104.

Wosinska, A. 1982. Induction of variability and mutation in China aster (*Callistephus chinensis* L. Nees) with 60 Co gamma rays. *Acta Agrobotanica*, **35**(2): 285-301.

Wosinska, A. 2001. New form of China aster (*Callistephus chinensis* (L.) Nees). Structure and biology of ray florets with open tubular corollas (type RO). *Acta Agrobolanica*, **54**(2): 123-140.

Chapter – 20

Chrysanthemum

Chrysanthemum (*Dendranthema* spp.) belongs to family Compositae (Asteraceae). It was named by Carolus Linnaeus from the Greek prefix "*chrys*" means golden and "*anthemon*" means flower. *Chrysanthemum morifolium* is the modern florists chrysanthemum, which evolved as a result of endless variation and hybridization from *C. sinensis, C. indicum* and *C. ornatum* (Bailey, 1963). This group is also distinguished as *C. hortorum*. The name of *C. morifolium* Ramat has been changed to *Dendranthema grandiflora* Tzvelev (Anderson, 1987, Heywood and Humphries, 1977). Chrysanthemums are mainly herbaceous perennial and bear flowers in capitula. *Chrysanthemum perennates* in nature by means of stolons and is readily propagated through basal stem cuttings. It ranks second next to roses in importance among the cut flower crops in the world.

ORIGIN AND DISTRIBUTION

Chrysanthemum is originated in northern hemisphere (Europe and Asia). However, some workers reported that China is the native place of chrysanthemum (Carter, 1980). Most of the species in the lineage of present day cultivars are from China.

Chrysanthemum was first cultivated in China as a flowering herb and is described in writing as early as 15th century B.C. As herb, it was believed to have the power of life. In 18th century, they appeared in Japan as the crest and official seal of the Emperor. Japan also has a national chrysanthemum day, which is called the Festival of Happiness. In the western world, they are introduced in England, Japan, France and U.S. have developed a wide range of colours, shapes and sizes of flowers. Since the chrysanthemum was first introduced in the U.S. during colonial times, its popularity has grown such that mums now reign as undisputed 'Queen of the Fall Flowers'. The Netherlands is the foremost impórter of chrysanthemum cut flowers and is the centers for re-export to other parts of the world. In India, chrysanthemum is successfully grown under protected as well as open conditions in different states.

SPECIES

Genus *Dendranthema* comprises about 300 species of herbs and under shrubs. The important species which are used in improvement and breeding purpose are given in Table 1. The number of chrysanthemum cultivars is incredibly large with more than 15000 listed in Japan only. The National Chrysanthemum Society of Britain has listed over 60000 cultivars. In India, the number of chrysanthemum is around 550. A large number of new and novel varieties have been developed through hybridization, selection and mutagenesis. Research efforts are being made to develop standard cut flowers types, evolving high yielding small spray type varieties for decoration, garland and *Veni* and to get flowers all the year around with tolerance against water logging, pest and diseases (Singh, 2006).

Table 1: Important species of genus *Dendranthema.*

S. No.	Species	Characteristics
1.	*Dendranthema boreale*	Involved in the evolution of florists chrysanthemum.
2.	*D. carinatum*	Winter season annual with tricolour bloom (yellow, cream and white).
3.	*D. coronarium*	Winter season annual with double coloured bloom (yellow and white).
4.	*D. coccineum*	Perennial of white, pink and red colour and anemone shaped flowering plant.
5.	*D. cinerarifolium*	Used for making insecticide pyrethrum.
6.	*D. frutescens*	Shrub, grown as pot plant, double coloured bloom (white and yellow).
7.	*D. indicum*	Involved in the evolution of florist chrysanthemum.
8.	*D. japonicum*	Used as ornamental plant.
9.	*D. maximum*	Perennial grown for cut flowers of white and yellow colour.
10.	*D. morifolium*	Hybrid species, perennial, known as florists chrysanthemum.
11.	*D. ornatum*	Involved in the evolution of florists chrysanthemum bearing flowers of white centered yellow colour flowers.
12.	*D. rubellum*	It is exceptionally sturdy and used in breeding for hardiness.
13.	*D. satsumense*	Involved in evolution of present day cultivars.
14.	*D. sibiricum*	One of the parents of Korean hybrids.
15.	*D. sinense*	Involved in evolution of present day cultivars. The ray florets are white in colour.

FLOWER MORPHOLOGY AND BIOLOGY

Individual florets of chrysanthemum are very small but collectively they form a capitulum (head) that reflects showy inflorescence which is actinomorphic. The capitulum is surrounded by small green bracts within which the rays and disc florets are arranged marginally and in the center, respectively. The ray florets are female and flat corolla is fixed in a tube. The shape of corolla may be different and on the basis of various shape of corolla of ray florets, chrysanthemums have been divided into various classes. Disc florets are hermaphrodite and corolla consists of five distinct teeth. The proportion of ray and disc florets will vary with classes. The ray florets are more in double inflorescence while the disc florets are more in single inflorescence. This proportion in also affected by temperature (Cathey, 1955 and Runger, 1980) and day length (Post, 1943, Langton, 1977 and Runger, 1980). The number of ray florets is increased by higher daily light integrals and high temperature (Cockshull, 1972 and Hughes and Cockshull, 1972). In contrast to this, the number of disc florets is increased by shorter day length. The florets open centripetally and flowering starts from the outer most whorls. The flowers are protandrous (Kuklina, 2003).

Mode of pollination

Chrysanthemum is a cross-pollinated crop. Many insects are attracted due to distinct fragrance of the capitulum. However, large flowered cultivars are visited rarely by insect pollinators. This is because of enormous length of the ray florets act as mechanical barrier in pollination (Fryxell, 1957). Some insects like *Apis cerena indica, Apis florea* and *Musca domestica* were reported as frequent visitors in *Chrysanthemum indicum* field (Parthiban *et al.*, 1994). The maximum pollination occurred at about 11.00 am, however effective pollination period was recorded between 7.00 am and 1.00 pm.

Pollen viability

The pollen viability can be rapidly determined by shriveling of pistils after one day of pollination and the shriveled pistils showed a good seed set after 8 week later (De Jong and Kho, 1982). This system is correlated with dry papillate stigmas, trinucleate pollen and the incompatibility reaction at the stigmatic surface (Richards, 1986).

The flowers of chrysanthemum are of different types. They differ in size, form and arrangement of disc and ray florets.

To decide the breeding objectives, breeders must have knowledge about its classification. The flowers have been classified by National Chrysanthemum Society of America as given below.

DIVISION A (CLASSES 1-8)

Visible portions never tubular, ray florets are flattened to concave or convex.

Section 1 (Classes 1-3)

Disc more clear by outline, composed of many disc florets.

Class 1. Single: Single row of ray florets at right angles to the stem. Disc found in bright colour is flat to slightly rounded.

Class 2. Semi Double: More than one row of ray florets at right angles to stem, ray florets curve downward at the tips.

Class 3. Anemone: Ray florets are variable may be broad, flattened or reflexing type pointed at tip. Disc florets are tubular shaped and very prominent.

Section 2 (Classes 4-8)

Disc is not visible, disc florets are absent.

Class 4. Pompon: Blooms are globular, ray florets broad, incurved, smooth and firm.

Class 5. Incurve: Blooms are globular, ray florets narrow to broad which incurve in regular or in irregular manner.

Class 6. Reflexing incurve: Ray florets broad, bloom globular, less compact than incurve, some florets incurved and some may be reflexed. The lower florets are reflexed that gives a skirted effect.

Class 7. Decorative: Blooms more flattened than globular, ray florets vary from short to long, narrow to broad.

Class 8. Reflex: Blooms reflexed, globular with clear disc, ray florets narrow to broad.

DIVISION B (CLASSES 9-11)

Ray florets are tubular, thin and straight hooked at the distal end.

Class 9. Spoon: Disc is more visible, ray florets regular and tubular, distal portion flattened, open and spoon like.

Class 10. Quill: Disc is not visible; blooms are double, rays florets tubular, straight but not coiled, may be open or closed at the tip.

Class 11. Spider: Disk is not visible; ray florets are tubular and long. Tip may be closed or open and coiled.

DIVISION C (CLASSES 12-13)

Disc may or may not be visible, ray florets are flattened or tubular.

Class 12. Lacinated: Ray florets may be lacinated or feathered at tips.

Class 13. Brush or thistle like: Ray florets are arranged in brush form parallel to the stem.

GENETICS

The basic chromosome number of *Chrysanthemum morifolium* Ramat has been confirmed to be 9 by various workers (Dowrick, 1952, 1953, Miyazaki *et al*., 1982 and Nazeer and Khoshoo, 1982). The findings to these workers also showed a ploidy level of chrysanthemum which ranged from 2n=18, 36, 54, 72, 90 and 198 together with odd combinations like 19, 26 and 37. Most of chrysanthemum varieties are hexaploid with an average chromosome number of 54 (Dowrick, 1953 and Langton, 1989). Various Chinese varieties are also hexaploid and aneuploid (Li Maoxue *et al*., 1983). According to this study, there are three group of early and some large flowered (54±1) and many large flowered varieties (52-57 and 58-71). Liu *et al*. (2010) reported that doubling the chromosome of *C. nankingense* could overcome the distant hybridization barrier with the cultivated chrysanthemum.

Self incompatibility

Selfing is generally not possible and self incompatibility in chrysanthemum enhances the cross pollination. The species has a strong self incompatibility system, as do all members of the Asteraceae family (Richards, 1986). Fryxell (1957) and Drewlow *et al*. (1973) reported sporophytic type of self incompatibility in chrysanthemum, which is governed by a single locus with two alleles. The diploid chrysanthemum species like *Chrysanthemum carinatum*, *C. coronarium* and *C. segetum* are also reported to be self incompatible (Anderson *et al*., 1988).

Ronald and Ascher (1975), Zagarski *et al*. (1983), Anderson *et al*. (1992) and Anderson and Ascher (1996) found pseudo-self incompatibility in chrysanthemum. Zagarski *et al*. (1983) found that 3 genes might be involved in self incompatibility. One inbred sibling 77-AM3-3 showed high level of pseudo-self incompatibility in which high seed set occurred when self in comparison to crossing.

Inheritance of flower doubleness

The inheritance of flower doubleness and floret corolla shape was studied in 70 F_1 populations from 16 parents. Singleness is partially dominant over doubleness. The overall segregation for doubleness in crosses of the type double × single was 29% single, 58% semi double and 12% double. The percentages varied with the parents. The anemone flower type is inherited as if it were a single. Crosses between plants with ligulate (split lengthwise) and tubular corollas produced ligulate, tubular and intermediate types in various ratios depending on the parents used. Flower doubleness and corolla shape are not linked (De Jong and Drennan, 1984).

Inheritance of colour

In a study of colour inheritance, Jordan and Reimann-Philipp (1983) investigated cell-sap anthocyanin and chromoplast carotene in petal cells. The presence of one gene 'A' ensured anthocyanin production, while gene 'I' inhibited carotene production. Yellow flowers were formed in the absence of both 'A' and 'I', white flowers in the absence of 'A' alone. The combination of 'A'+'I' gave pink, carmine and bluish-red flowers whereas 'A' without 'I' resulted in bronze and brownish-red flowers. On the basis of segregation of flower colour, they also provided the tetrasomic and hexasomic inheritance.

In another study of colour inheritance, it was shown that red ray petals are not conditioned by a homozygous genotype. The orange colour is governed by a single dominant gene which causes mosaic like distribution of red vacuole pigmentation over yellow plastid pigmentation (Teynor *et al.*, 1989).

Heritability

Choudhary *et al.* (2003) observed high heritability for leaf number, number of primary branches, stem thickness, plant height, number of flowers/plant and number of ray florets. The number of flowers/plant and number of petals/flower are governed by additive gene action. These characters are amenable to selection for improvement (Panse, 1957, De Jong, 1984, Chattopadhyay *et al.*, 1992).

Genetic analyses showed a broad sense heritability of 70% for days to flowering and flower number and a highly significant GCA effect for days to flowering. The progenies and parents that flowered early at one temperature were generally early at all temperatures. It appears that in the selection of parents, special attention should be paid to earliness. If rapid flowering selections are sought for cultivation at lower temperatures at least one but preferably both parents should be early. Parents may be selected at the indicated low temperature on the basis of clonal performance which is a good guide to the breeding value (De Jong, 1984).

BREEDING OBJECTIVES

The purpose of breeding to evolve the high quality varieties for cut flower, loose flower and pot culture. Attempts were made to breed new varieties through planned crosses or open pollinated seedling selections. Various research institutions are involved in this directions *viz.*, NBRI (Lucknow), IIHR (Bengaluru), PAU (Ludhiana), BCKV (Kalyani) and TNAU (Coimbatore). Following are the centres for breeding and germplasm collection. In chrysanthemum, standard and spray types have been grown for various purposes.

Following are the objectives for breeding chrysanthemum.

A. Cut flowers

i. Long straight stem
ii. Tough florets
iii. Uniform bloom
iv. Attractive colour
v. Resistant to diseases and insects
vi. Healthy leaves
vii. Earliness
viii. Strong stems to withstand bloom weight
ix. Long vase life

B. Pot variety

i. Compact and dwarf plants
ii. More number of branches
iii. Uniform spreading of branches
iv. Uniform blooming
v. Good colour retention quality
vi. Healthy leaves

C. Garland variety

i. Diameter of bloom (about 5 cm)
ii. Fluffy blooms
iii. Invisible Disc
iv. Good quality of recovery from pressure
v. High yielding
vi. Specially yellow and white colour of flowers
vii. Good colour retention

viii. Long blooming Season

ix. Pollen less bloom

x. Development of varieties tolerant to water logged conditions

xi. Development of varieties suitable for out of season blooming

xii. New colours and forms through genetic engineering

ICAR identified various Coordinated Centres for breeding and germplasm collection as given below.

Centre for collection, maintenance and assessment of germplasm

- Acharya N.G. Ranga Agricultural University, Hyderabad
- Birsa Agricultural University, Kanke, Ranchi
- Dr. Y.S. Parmar University of Horticulture and Forestry, Solan
- Maharana Pratap University of Agriculture and Technology, Udaipur
- Regional Plant Resource Centre, Bhubaneswar
- Punjab Agricultural University, Ludhiana
- Sher-e-Kashmir University of Agricultural Sciences and Technology, Srinagar
- Tamil Nadu Agricultural University, Coimbatore
- G.B. Pant University of Agriculture and Technology, Pantnagar

Centre for breeding new cultivars suitable as cut flower export and domestic market

- Indian Institute of Horticultural Research, Bengaluru
- Indian Agricultural Research Institute, New Delhi

BREEDING METHODS

Introduction

A lot of varieties and species have been introduced from several countries to India. This method is very efficient in creating a wider gene pool as well in promoting new varieties evolved from their crossing. Some important introductions of chrysanthemum from other countries are given in Table 2.

Table 2: Some of the important introductions of chrysanthemum in India

S. No.	Country	Varieties and species
1.	USA	Casa Grande, Mountaineer, Nob Hill, Snow Ball, Potomac, Peacock, Royal Pomp, Yellow Glow, *Dendranthema cinerariefolium*, *D. corymbosum*, *D. maximum*, *D. parthenium* and *D. segetum*.
2.	Hungary	*D. corymbosum, D. fructescens, D. caucasium, D. segetum, D. coccineum, D. vulgare, C. maximum, D. leucanthemum, D. cinerariefolium, D. serotinum.*
3.	UK	*D. coronarium, D. parthenium D. segetum.* Alfred Wilson, Alfred Simpson, Balcombe, Perfection, Beatrice May, May Shoesmith, Maurice White, Levianthan, Pink Cloud, Princess Anne, and Woolman Century.
4.	Australia	J.S. Lloyad, Louisa Pockette, Willium Turner, T.W. Pockette.
5.	France	Gloria Deo, S.L. Andre Reffaud and Sancho.
6.	Japan	Ajina Purple, Kenroku Kangiku, Kiku Biori, Taiho Tozan, Tokyo, Shin Mei Getsu, Senkyo Emaki, Otome Zakura and Nanako.
7.	The New Zealand	Gusman Red, Icicles, Jane Sharpe, Orange Fair Lady and Nancy Ferneaux.

Besides the introduction of planting material, pollen can be introduced because of relatively less quarantine problems (Parlevliet *et al.*, 1979).

Selection

Selections from the open pollinated seedlings exhibited greater possibilities in evolution. Therefore, various workers took initiatives to collect seeds from open pollinated plants and evaluated for different traits. It resulted in release of several outstanding varieties which have performed very well in the industry. Improved varieties evolved through selection are given in Table 3.

Table 3: Improved varieties evolved through selection.

S. No.	Variety	Institute	Parentage	Characteristics
1.	Rakhee (1980)	IIHR	Lord Doonex	Plants bushy with shiny foliage, flower diameter 6 cm, anemone type, yellow flowers with red strips.
2.	Appu (1982)	NBRI	Original Dwarf	No pinch no stake type, double, dwarf, early blooming, purple coloured flowers.
3.	Apsara (1977)	NBRI	Nanako	Pompon type, rosy flush on white flower, most suitable for cut flower.
4.	Apurva Singar	NBRI	Double Korean	No pinch no stake type, dwarf, flowers double and mauve coloured. Suitable for pot culture.
5.	Arun Kiran (1983)	NBRI	Original Dwarf	No pinch no stake type, super dwarf, flowers of red colour.
6.	Arun Singar (1982)	NBRI	Original Dwarf	No pinch no stake type, dwarf with variegated leaves, flowers red, striped early blooming.

(Contd.)

(*Contd. Table 3*)

S. No.	Variety	Institute	Parentage	Characteristics
7.	Birbal Sahni (1976)	NBRI	Nanako	Snow white flowers with long vase life, Pompon type, suitable for cut flower purpose.
8.	Diana (1999)	NBRI	-	Dwarf plants with red coloured, anemone type flowers, profuse flowering.
9.	Guldasta (1986)	NBRI	-	No pinch no stake type, anemone type flowers of bronze colour, suitable for pot culture.
10.	Hemant Singar (1981)	NBRI	-	No pinch no stake type, Korean type, dwarf and early blooming, small flowers of purple colour.
11.	Haldi Ghati (1988)	NBRI	Original Dwarf	No pinch type, dwarf, Korean, out of season blooming with sulphur yellow colour flower.
12.	Apurva (1987)	NBRI	-	No pinch no stake type, very early blooming, small pink flowers, suitable for pot culture.
13.	Jayanti (1979)	NBRI	Nanako	Yellow coloured flowers of long vase life, suitable for cut flower, pot culture and garland making.
14.	Jubilee (1980)	NBRI	Nanako	Pompon type flowers of orange red colour, suitable for pot culture and cut flower.
15.	Kundan (1980)	NBRI	Jayanti	Pompon typed small flowers of sulphur yellow colour suitable for cut flowers, garland and pot culture.
16.	Maghi (1989)	NBRI	-	Pompon typed flowers of pink colour suitable for cut flowers and pot culture.
17.	Mother Teresa (1997)	NBRI	-	Anemone type, No pinch no stake type, dwarf, bushy and white colour flowers.
18.	Peet Singar (1981)	NBRI	-	No pinch no stake type, cineraria type, early blooming, small and sulphur coloured flowers, suitable for pot culture.
19.	Shanti (2000)	NBRI	-	White coloured, small, decorative type with long stem, profuse flowering, ideal for cut flower and garland making.
20.	Sharad Singar (1977)	NBRI	Original Dwarf	Out of seasons blooming type, No inch no stake type, flowers yellow with bronze stripes.
21.	Suhag Singar (1981)	NBRI	-	No pinch no stake type, of double Korean with coral red coloured flowers.
22.	Chandrakant (1994)	IIHR	-	Decorative type of white colour flowers, floriferous, ideal for loose flowers.
23.	Pankaj (1994)	IIHR	-	Decorative type with pink colour, suitable for cut and loose flowers.

(*Contd.*)

(*Contd. Table 3*)

S. No.	Variety	Institute	Parentage	Characteristics
24.	CO-1 (1985)	TNAU	-	Early flowering, medium sized flowers, canary yellow coloured flowers, with long flower duration.
25.	CO-2. (1985)	TNAU	-	Rhodamine purple coloured flowers with vigours flowering, high yielder.
26.	MDU-1 (1985)	TNAU	-	Early flowering, large flowers of sulphur yellow colour.
27.	IIHR–Sel-5 (1992)	IIHR	-	Pink blooms, high yielder, suitable for loose flowers.
28.	IIHR–Sel-6 (1992)	IIHR	-	White blooms, high yielder, suitable for loose flowers.
29.	Sadbhavana	NBRI	-	Double Korean type, dwarf, bushy, compact round shaped, profuse blooming.
30.	Kargil 99	NBRI	-	Small flowered, spoon shaped, purplish mauve florets with yellow disc, profuse flowering
31.	Y2K	NBRI	-	Small flowered, white anemone, mini type.
32.	Mini Queen	NBRI	Little Darling	Dark terracotta coloured, pompon type, round shape, profuse blooming.
33.	Rangoli	NBRI	Diana	Dark red, double Korean type, dwarf, profuse blooming.
34.	NBRI Little Pink (2008)	NBRI	Little Darling	Purple colour flowers, decorative type.
35.	NBRI Little Kusum (2008)	NBRI	Little Darling	Bronze flower, pompon type.
36.	NBRI Little Hemant (2008)	NBRI	Little Darling	Yellow coloured, pompon type, cut flower.
37.	NBRI Little Orange (2008)	NBRI	Little Darling	Yellow coloured, pompon type, cut flower tubular type.
38.	NBRI Himanshu (2009)	NBRI	-	Anemone type with central star shape disc floret, creamish white, ideal for pot culture, bedding and monochromatic landscaping.
39.	Pusa Aditya (2012)	IARI	Jaya	Spray type star shaped semi-double flowers, resemble to gazania flowers, flowers yellow, at periphery with orange red in center. Florets spatulate, suitable as a cut flower and as a potted plant.
40.	Pusa Chitraksha (2012)	IARI	Lal Pari	Spray type single flowers, deep magenta, florets are spatulate, silvery white ray floret tube, disc florets yellow, floriferous, suitable as a potted plant and for garden display.
41.	Pusa Sona (2012)	IARI	Sadbhawana	Extremely dwarf, spray type produces single yellow flowers, early flowering, suited for pot mums.

Some selections have been made for summer season production of chrysanthemum flowers and cvs. Kwibuin, Cheonsu, Baekkwang, Aengdu and Oweolocheong were selected as the most profitable for summer production (Choi *et al.*, 1993). Selection for adaptation to low light intensity has shown to be very successful (De Jong, 1986). Six chlorophyll variegated varieties (DWS-2, DWS-12, DWS-15, B-16, B-17 and OO-2) have been developed through selections. Variegated leaves of these cultivars provide an advantage to be attractive through out the year.

Hybridization

Due to complexity in *Chrysanthemum morifolium* Ramat and self incompatibility in chrysanthemums, few inbreds have been developed. But heterozygosity of hybrids can easily be sustained by vegetative means of propagation. For hybridization in chrysanthemum, male and female plants of desirable characters are selected. Anthers of disc florets and long ray florets (to expose stigma) are clipped before anthesis. The ray and disc florets are covered with butter paper bags to avoid natural pollination. Pollen grains are collected and dusted upon the stigma. All the crosses are properly labeled. Seeds are collected after complete drying of capitulum followed by storing them in dry place. In a study of chrysanthemum breeding, cvs. Bornholm and Axilia have been found to be the best material for breeding large flowered varieties and Klondike, Daymark, Palida and Cassa for breeding small flowered ones. Some developed hybrids, their parentage and characteristics are given in Table 4.

Table 4: Varieties evolved through hybridization.

S. No.	Hybrid	Released from	Parentage	Chief characteristics
1.	Indira (1980)	IIHR	Lord Doonex × (Flirt × Valentine)	Double Korean type flowers, flowers yellow with 6 cm diameter.
2.	Red Gold (1980)	IIHR	Flirt × Valentine	Double Korean, very floriferous Flowers golden yellow with 6 cm diameter.
3.	Chandrika (1994)	IIHR	Angela × G.P.I.	Decorative type, white flower, high yielder and for cut and loose flowers and for floral decoration.
4.	Kirti (1994)	IIHR	Angela × G.P.I.	White coloured, double Korean type of flowers, very floriferous and high yielder and for bedding and potting.
5.	Nilima	IIHR	Flirt × Valentine	Decorative type, purple coloured flowers, floriferous, ideal for cut and loose flowers.
6.	Ravikiran (1993)	IIHR	Flirt × Valentine	Gray-red to gray-orange coloured flowers of Stellate type, suitable as cut flower.
7.	Red Gold (1980)	IIHR	Flirt × Valentine	Grayish orange colour fades to golden yellow; flowers double Korean type,

(*Contd.*)

(*Contd. Table 4*)

S. No.	Hybrid	Released from	Parentage	Chief characteristics
				good for cut flower, bedding, potting and wreaths.
8.	Yellow Star (1994)	IIHR	Flirt × Valentine-22	Decorative typed flowers of yellow colour, high yielder and for cut flower and loose flower.
9.	Punjab Gold (1999)	PAU	Flirt × Gul-e-Sahir	Coppery red changes to golden yellow colour at full bloom and for pot culture, resistant to *Septoria* leaf spot.
10.	Wain's Pink Lady (1994)	-	Pink Arola × Charm	Unusual rose coloured eye with over all pink colours of flowers, upright to good branching and spreading habit.
11.	Jannel (1993)	-	Green Pea × Keystone	Attractive cream white flowers of Pompon type with green centre, less susceptible to leaf miner, used for cut flowers.
12.	NBRI Indiana (2002)	NBRI	Little Darling × Nanako	Small yellow pompon type, suitable for mini and pot culture.
13.	NBRI Kusum (2002)	NBRI	Haldighati × Sharad Kanti	Small, yellow disc with bright yellow florets, single Korean type, profuse blooming.
14.	NBRI Little Darling (2002)	NBRI	White Charm × Jubilee	Small, terracotta pompon type, profuse blooming, yellow–orange floret.
15.	NBRI Mini Jessie (2002)	NBRI	Cameo × Jessie	Small, cineraria type, attractive for mini culture, red purle florets.
16.	NBRI Kaul (2010)	NBRI	Lal Pari × Vijay	Floriferous and early blooming, single Korean type, purple coloured flower with yellow disc florets.
17.	NBRI Khoshoo (2010)	NBRI	Lal Pari × Vijay	Grayed red coloured flower head, Single Korean types, ideal for pot and cut flower, early bloomer.
18.	Arka Swarna	IIHR	Nanako × CO-1	Pompon yellow flowers, suitable for cut flowers and loose flower purpose.
19.	Arka Ravi	IIHR	Mundial × Hybrid 87-17-1	Flowers semi double, attractive peach, used for cut flower, vase life 12 days.
20.	Arka Ganga	IIHR	Flirt × Red Gold	Flower white with pink tinge, vase life 11 days, suitable for loose as well as cut flower.
21.	Anmol (2006)	PAU	Panchoo × Rage	Bright yellow anemone type, suitable for pot.
22.	Royal Purple (2006)	PAU	Bindiya × Rage	Purplish pink anemone, suitable for pot.
23.	Yellow Delight (2006)	PAU	White Bouquet × Gul-e-Sahir	Bright yellow pompon type, suitable for cut flower.
24.	Winter Queen (2006)	PAU	White Bouquet × Flirt	Pink spoon type, suitable for garden.
25.	Autumn Joy (2006)	PAU	White Bouquet × Flirt	Pink flowers, floriferous, suitable for garden display.

Earlier some other varieties of chrysanthemum have been developed through hybridization are Pride of Jamshedpur (Rajah × Alfred Simpson), Jamshedji (Alfred Simpson × Rajah) and Steel City Beauty (White Cloud × Temptation) in 1976 and Red Carpet (Mrs. W. A. Reid × Kikubiori) in 1978.

Interspecific hybridization

Huang *et al.* (2011) made interspecific hybridization between *Chrysanthemum morifolium* 'Hongxinju' (female), *C. morifolium* 'Xinbaiju' (female) and *C. indicum* (male) *C. nankingense* (male), and observed that bud pollination improves heterozygosity of interspecific hybridization of *C. morifolium* 'Hongxinju' and 'Xinbaiju'. Hybrids could be aneuploid chromosomes and show the significant segregation of character. Cheng *et al.* (2010) employed ovary rescue to create six interspecific hybrids from the cross between *Dendranthema morifolium* (Ramat.) Kitamura 'rm20-12' (2n=54) and its wild diploid relative *D. nankingense* (Nakai) Tzvel. (2n=18). The cold tolerance of five hybrids was significantly superior to that of their *D. morifolium* parent. Interspecific hybridization clearly provides an effective means of cultivar improvement in chrysanthemum.

Mutation

Mutation breeding is very successful in chrysanthemum improvement. A lot of work in chrysanthemum mutation has been done at NBRI, Lucknow. Apart from induced mutation, natural mutations also provide variability. However, natural or spontaneous mutations are very rare in nature but around 30% of the chrysanthemum varieties originated as bud sports (Wasscher, 1956). Several varieties of chrysanthemum have been developed in India, originated from spontaneous mutation. The most notable varieties are given in Table 5.

Table 5: Notable bud sports of chrysanthemum.

S. No.	Bud sport	Parentage
1.	Kasturba Gandhi (white)	Mahatma Gandhi (mauve)
2.	Sonar Bangla (yellow)	Snow Ball (white)
3.	White Cloud (white)	Pink Cloud (pink)
4.	Sharda (yellow)	Sharad Shobha (white)
5.	Queen of Tamluk (yellow)	Casa Grandi (white)
6.	R. Venkatraman (yellow)	S. S. Arnold (white)
7.	William Turner (white)	Pink Turner (pink)
8.	J.S. Lloyd (yellow)	William Turner (white)
9.	White Ball	Pride of Madford
10.	NBRI-Pushpangadan	A-22

The rate of mutation can be increased manifolds by mutation using physical and chemical mutagens. Among various physical mutagens, fast neutrons give a remarkable increase in mutation frequency, whereas X-rays at lower dose of 1500 R also provides marked mutations. Electrons are not so effective in creating variability (Breortjes, 1966). Gamma rays (Broertjes, 1966, Dowrick and El-Boyoumi, 1966, Gupta and Jugran, 1978) and chemical mutagens (Nayar *et al*., 1971, Drewlow and Widmer, 1971 and Balakrishnan *et al*., 2003) have also been reported to induce mutations. Various workers used a wide range of doses of gamma irradiation from 0.5 kR to 25 kR. The varieties of chrysanthemum have been classified on the basis of radio sensitivity to gamma rays. Varieties with LD_{50} dose of 1.5 kR have been termed as highly sensitive and with more than 2.5 kR as mildly sensitive, whereas rest of the varieties lying between 1.5 and 2.5 kR are sensitive. Some workers have also found variations in plant parts and flowers with high doses of gamma rays as up to 25 kR (Datta and Banerji, 1993). However for var. CO-1 and CO-2, LD_{50} value has been observed between 1.0 and 1.5 kR of gamma rays and from 10 to 20 mM of EMS. A number of varieties have been developed by induced mutations, which are as presented in Table 6.

Table 6: Developed mutants with their parents and doses of mutagens.

S. No.	Mutant	Released from	Parentage	Dose of mutagen	Chief characteristics
1.	Agnishikha (1987)	NBRI	D-5 and 2 krad	1.5(γ-ray)	Decorative type, Magnolia purple cultivar.
2.	Alankar (1982)	NBRI	D-5	1.5 krad (γ-ray)	Decorative type, Spanish orange coloured flower.
3.	Anamika (1979)	NBRI	E-13	1.5 krad (γ-ray)	Pompon type.
4.	Aruna (1982)	NBRI	Ashankit	(γ-ray)	Exhibition type, Semi-quilled, fringed ray florets with silvery reverse flower type.
5.	Asha (1974)	NBRI	Hope	1.5 krad (γ-ray)	Large flowered, Reflexed type, creamish coloured mutant.
6.	Ashankit (1974)	NBRI	Undaunted	1.5 krad (γ-ray)	Large flowered, Exhibition type bright mauve colour mutant.
7.	Basant (1979)	NBRI	Paul	1.5 krad (γ-ray)	Small flowered, Pompon type, yellow coloured mutant.
8.	Basanti (1974)	NBRI	E-13	1.5 krad	Small flowered, Pompon type, yellow colour mutant.
9.	Batik (1994)	NBRI	Flirt	2.0 krad(γ-ray)	Small flowered, double Korean type, strip colour mutant.
10.	Cosmonaut (1984)	NBRI	Nimrod	1.5 krad (γ-ray)	Small flowered, white, Anemone type cultivar.
11.	Gairik (1974)	NBRI	Belur Math	1 krad (γ-ray)	Large flowered, Incurved, lighter salmon colour mutant.

(*Contd.*)

(*Contd. Table 6*)

S. No.	Mutant	Released from	Parentage	Dose of mutagen	Chief characteristics
12.	Hemanti (1979)	NBRI	Megami	1.5 krad(γ-ray)	Small flowered, dwarf, Decorative, early blooming, Chinese yellow coloured cultivar.
13.	Himani (1974)	NBRI	E-13	2.0 krad (γ-ray)	Small flowered, Pompon type, white coloured flower.
14.	Jhalar (1975)	NBRI	Undaunted	1.5 krad (γ-ray)	Large flower, Exhibition type, mauve, flat and fringed ray floret.
15.	Jugnu (1991)	NBRI	Lalima	1.542.0 krad (γ-ray)	Small flowered, Double Korean type bicoloured (barium yellow colour at the tip of each floret).
16.	Kanak (1974)	NBRI	Undaunted	1.5 (γ-ray) krad	Large flowered, Exhibition type, dark brown flower coloured mutant.
17.	Kansya (1974)	NBRI	Roseday	1.5 krad(γ-ray)	Extra large, bronze coloured, reflexing ray floret.
18.	Kapish (1974)	NBRI	E-13	2.0 krad (γ-ray)	Small flowered, Pompon type, brownish coloured.
19.	Kumkum (1987)	NBRI	M-71	2 & 2.5 krad (γ-ray)	Garnet brown coloured.
20.	Kunchita (1974)	NBRI	Undaunted	1.5 krad (γ-ray)	Large flowered, Exhibition type, mauve coloured, incurved ray floret type.
21.	Lohita (1974)	NBRI	E-13	2.0 krad (γ-ray)	Small flowered Pompon type, dark reddish coloured.
22.	Man Bhawan (1982)	NBRI	Flirt	-	Small flowered, Double Korean type, bicoloured (yellow and red).
23.	Navneet (1987)	NBRI	Kalyani Mauve	2 krad (γ-ray)	Small flowered, Decorative type creamish white coloured.
24.	Navneet Yellow (1993)	NBRI	Navneet	1.5 krad(γ-ray)	Small flowered, yellow coloured mutant.
25.	Nirbhaya (1974)	NBRI	Undaunted	1.5 krad(γ-ray)	Large flowered, Exhibition type, semi-quilled fringed ray floret, light mauve coloured.
26.	Nirbhik (1974)	NBRI	Undaunted	1.5 krad(γ-ray)	Large flowered, Exhibition type, light mauve colour, almost flat fringed ray floret.
27.	Pingal (1974)	NBRI	Pink Casket	1.5 krad(γ-ray)	Large flowered, Spoon type, terracotta coloured.
28.	Pitika (1974)	NBRI	Kansya	1.5 krad(γ-ray)	Extra large, Reflexing ray floret, yellow coloured.

(*Contd.*)

(*Contd. Table 6*)

S. No.	Mutant	Released from	Parentage	Dose of mutagen	Chief characteristics
29.	Pitamber (1977)	NBRI	Otome Zakra	-	Large flowered, Pompon type, yellow coloured.
30.	Purnima (1977)	NBRI	Otome Zakra	-	Large flowered, Pompon type, white coloured.
31.	Rohit (1979)	NBRI	Kingsford Smith	1.5 krad (γ-ray)	Large flowered, Reflexed type, rhodonite red coloured.
32.	Shabnam (1987)	NBRI	D-5	1.5 krad (γ-ray)	Decorative type, small appendage like structure at the tip of each flower.
33.	Shafali (1974)	NBRI	Undaunted	2.0 krad (γ-ray)	Large flowered, Exhibition type, incurred ray floret, light reddish coloured.
34.	Sharad Har (1992)	NBRI	Sharad Mala	1.5 krad (γ-ray)	Small flowered, Decorative type, yellow coloured.
35.	Sheela (1985)	NBRI	Himani	2 and 2.5 krad (γ-ray)	Small flowered, Pompon type, canary yellow coloured.
36.	Sheveta (1974)	NBRI	Fish Tail	1.5 krad (γ-ray)	Large flowered, Incurving, white coloured.
37.	Surekha Yellow (1992)	NBRI	Seedling selection of Surekha	1.5 krad (γ-ray)	Yellow coloured.
38.	Sonali (1991)	NBRI	Ratna	2 krad (γ-ray)	Small flowered, Pompon type, yellow coloured.
39.	Subarna (1991)	NBRI	Flirt	2 krad (γ-ray)	Small flowered, Double Korean type, yellow coloured.
40.	Tamra (1974)	NBRI	Goldie	2 krad(γ-ray)	Large flowered, attractive, coppery red on yellow background.
41.	Taruni (1979)	NBRI	Kingsford Smith	2.5 krad (γ-ray)	Large flowered, Reflexed, azalea pink coloured.
42.	Navneet Yellow (1993)	NBRI	Navneet	1.5 krad(γ-ray)	Small flowered, decorative type, yellow colour.
43.	Nirbhaya (1966)	NBRI	Undaunted	1.5 krad(γ-ray)	Large flowered, semi quilled fringed ray floret, light mauve colour.
44.	Tulika (1985)	NBRI	M 24	1.5 krad (γ-ray)	Large flowered, purple coloured, Paint brush type flower shape.
45.	Yellow Gold	IIHR	Flirt	-	Small flowered, Double Korean type, Yellow colour with brick-red strips.

(*Contd.*)

(*Contd. Table 6*)

S. No.	Mutant	Released from	Parentage	Dose of mutagen	Chief characteristics
46.	Colchi Bahar (1985)	NBRI	Sharad Bahar	Colchicine	Small flowered, early blooming, terracotta red coloured.
47.	Pusa Anmol (2012)	IARI	Ajay	Gamma irradiation	Brilliant yellowish pink flowers, incurving type, easily propagated through vegetative means.
48.	Pusa Centenary (2012)	IARI	Thai Chen	Queen	Gamma irradiation brilliant yellow flowers, incurving type.
49.	Pusa Arunodya (2012)	IARI	Thai Chen	Queen	Gamma irradiation Pink ray florets, semi double flowers, large size, incurving ray florets, suitable for cut flower, pot culture.
50.	Pusa Kesari (2012)	IARI	Thai Chen	Queen	Gamma irradiation Red ray florets, semi double flowers, big size, incurving ray florets, suitable as a cut flower and pot culture.

The frequency of mutation is very high in pink cultivars (Jank, 1957). Variability in number of chromosome has also been observed as a result of gamma irradiation (Yamakawa and Ichikawa, 1970, Drewlow and Widmer, 1971, Datta and Banerji, 1995 and Furuya, 1998). The increase number of ray florets and disc florets has been reported with irradiation of gamma rays (Narayana Gowda, 2003). Broertjes and De Jong (1984) developed pollen free and male sterile mutants by irradiation of X-rays. Pollen sterility was also observed by treating chrysanthemum plants with gamma rays and colchicine (Datta and Datta, 1998). Low temperature tolerant mutant have also been obtained (Broertjes *et al*., 1983). A very long duration flowering mutant Hanrong No.1 from Fenmaju variety has been selected after exposing cultivar to 3.0 kR gamma radiation (Hu *et al*., 1989). Yamaguchi (2010) reported that the frequency of colour mutation increased when the total rate of radiation increased from 15 to 30 Gy concentration in chrysanthemum (*Chrysanthemum morifolium* cv. Taihei). Thus, the mutation frequency was mainly dependent on the total radiation use.

Furuya (1998) obtained one of the mutants after treating petals of cv. Magaretsutomamu with gamma rays. The mutant was able to produce flowers under long day condition. Somatic colour mutants have been observed by treating rooted cuttings with gamma radiation in cv. Ellen Van Langen with four yellow colour, Snow Ball with one deep yellow and one light yellow and one each, Gulmohar and Shyama, of brick red colour (Dilta and Sharma, 2003).

The mutation can be maintained by vegetative propagation in chrysanthemum but still along with low frequency of mutation, chimera formation remains as a problem in inducing mutations (Stewart and Derman, 1970). Mutations through *in vitro* culturing can solve the above said problem upto a great extent (Broertjes *et al*., 1976 and Nikaido and Onozawa, 1990). A combination of induced mutation and *in vitro* techniques has been applied to certain chrysanthemum (De Jong and Custer, 1986a,b Huttema *et al*., 1991 and Preit *et al.,* 1991). These methods also increase the mutation rate 10 times higher than the conventional chronic cutting method with non-chimeric mutants (Nagatomi *et al*., 2000). Misra *et al.* (2003) obtained mutants of changed colour and shape of florets by using gamma irradiation (0.05 kR) under *in vitro* condition. Some of the most promising mutant varieties are Sonali, Navneet, Purnima, Kumkum, Shabnam, Batik and Himani.

Nagatomi and Degi (2008) reported that the chronic culture method yielded the widest colour spectrum in chrysanthemum, while the acute culture method resulted in a relatively low mutation rate and a limited flower colour spectrum. Tanaka *et al.* (2010) found novel mutants and their responsible genes, such as UV-B resistant, serrated petals and sepals, anthocyaninless, etc. in chrysanthemum when induced by ion beams. It also indicated that the characteristics of ion beams for mutation induction are high mutation frequency and broad mutation spectrum and therefore, efficient induction of novel mutants.

Verma *et al.* (2012) worked out under *in vitro* condition to isolate red coloured mutant from chimeric ray florets of chrysanthemum induced by gamma-ray. They found that the isolated mutant produced flowers that was true-to-type to the original red mutant. When buds of orange flower chrysanthemum cv. Cheng Yun were treated with electron beam irradiation and cultured *in vitro*, it was found that the optimal medium for callus induction was with 1.0 mg NAA and 2.0 mg BAP/litre. Whereas, the dose of 40-60 Gy was optimal for inducing mutations. The mutants were true to type flowers (Sun *et al*., 2007). Datta and Chakrabarty (2008) developed a protocol that has been successfully used for the isolation of chimeric mutant tissues developed through sports, but also to develop a series of new flower colour/shape mutants through induced mutagenesis. Thus, it helps in the management of such chimeric tissues through direct shoot regeneration from chrysanthemum florets.

BIOTECHNOLOGY

Tissue culture

Barakat *et al.* (2010) carried out an *in vitro* experiment using the ray floret explants to determine the capability for plant regeneration on four media protocols. Study indicated that protocol A and B appear to be the best protocols for plant regeneration and it could be successfully utilized for further *in vitro* mutagenesis investigations.

IAA found best for regeneration of chrysanthemum plantlets through shoot tip culture followed by NAA and IBA, respectively (Waseem *et al.*, 2007). When nodal explants were cultured on ½ X MS medium containing different combinations of growth hormones, a maroonish embryogenic callus was formed on MS containing BAP. The callus exhibited various embryonic developmental stages and gave rise to normal seedlings. The plants ultimately flowered exhibiting superior quality and early maturity compared to stock plants (Ilahi *et al.*, 2007).

Somaclonal variation

There are several techniques to produce variants which are novel in colour as well as in morphology. Somaclonal variants are the result of chimaerism gene mutation and chromosomal changes. A number of white genotype plants were obtained from lilac coloured ray florets (Khalid *et al.*, 1989). The generation time (0.75 to 1.5 years) in florist's chrysanthemum impedes the rate of progress for sexual breeding programmes. Modifications to the crossing environment and embryo rescue were evaluated to minimize the generation cycle (Anderson *et al.*, 1990). Embryo rescued plants showed high per cent seed set, per cent germination and obtained average progeny generation time of approximately 100 days, thus allowing for three generations per year.

Molecular breeding

In a comparison of *Agrobacterium* and biolistic-mediated transformation methods, both techniques are found to be reliable for producing transferred plants (Yepes *et al.*, 1999). Various molecular markers have been used to identify the varieties and their ancestrals. In a study of rapid detection of genetic variability, Wolff and Peters–Van (1993) found that genetic variability among related chrysanthemum species was too high and because of the high level of polymorphism and clonal stability, RAPD fragments are useful for cultivar identification. Huang *et al.* (2000) also studied the complexity of chrysanthemum and found RAPDs as powerful tool to detect different molecular markers in hybrid populations of chrysanthemum cultivars. To identify the genetic relatedness between genotypes and species, phylogenic relationships have been investigated with molecular markers (Trigiano *et al.*, 1998 and Sehrawat *et al.*, 2002).

Genetic engineering

In further improvement programme, several efforts have been made in developing transgenic plants. Various workers have investigated the transformation system in chrysanthemum. In a series of vector mediated transformation, results have shown that *Agrobacterium* mediated transformation of chrysanthemum is very cultivar specific, both in susceptibility to infection as well as the regeneration transgenic plant (Deroles *et al.*, 1997). Many cultivars have been transformed using *A. tumefaciens* (Ledger *et al.*, 1991, Robinson and Firoozabady, 1993, Deroles *et al.*, 1997, Boase *et al.*, 1998 a, b and Kim *et al.*, 1998).

An efficient chrysanthemum transformation procedure was established by Dolgov *et al.* (1997). Dolgov *et al.* (1995) transformed two varieties Bornholm and White Harricon with vector constructions carrying the *Bacillus thuringiensis* delta endotoxin gene (this confers resistance to *Tetranynchus urticae* in transgenic tobacco plants). Transformation by a binary vector system based on armed strain A281 and plasmid pBTPu1, with the delta endotoxin gene under the control of the CaMV 35s promoter, resulted in the production of such 2 lines which were not affected by *T. urticae*. The transformation via *A. tumefaciens* inoculation of leaf explants transformed with vectors containing a cDNA of the chrysanthemum chalcone syntheses gene (chs) in either sense or antisense orientation has been reported (Courtney *et al.*, 1994). Some regenerated plants produced violet or white coloured flowers, whereas the parent was pink. Kim *et al.* (1998) found the tissue formation in cv. Fashion Yellow as 38.1% and in cv. Golden Glory it was 98.6% when transformed with vectors PCMAs CP121-123 and pTOK 233, respectively.

Development of transgenic

The first transgenic chrysanthemum plants were reported by Lemieux (1989). Micro projectile bombardment has been successfully used to transform chrysanthemum (Yepes *et al.*, 1995). He used four cultivars *viz.*, Tara, Gridon, Blush and Dark Bronze Charm for transformation. The explants were bombarded with the plasmid (npt II, nucleocapsid gene from tomato spotted wilt virus) carried on 1 μM tungsten particles. The method provided high number of stable transgenics.

Sherman *et al.* (1998) produced transgenic plants from the cultivar Polaris carrying sense and antisense genes of the nucleocapsid protein from tomato spotted wilt virus. Both sense and antisense strategies produced totally virus resistant plants. Transgenic plants have also been generated carrying genes for *Botrytis* resistance (Takatsu *et al.*, 1999). They allowed the transfer of rice chitinase gene (RCC2) by *Agrobacterium* mediated transformation system and found a higher production of RCC2 protein in resistant plants. The rol C gene sequence under 35 S promoter on plasmid pPcvooz in *Agrobacterium* stain GV 3101 was transferred into chrysanthemum variety White Snowdon and the transformed lines showed drastic changes in plant size, branching, flower and petal forms (Mitiochkina and Dolgov, 2000).

BREEDING FOR VARIOUS TRAITS

Disease resistance

White rust (*Puccinia horiana*) is a very infectious leaf disease of chrysanthemums. There are several resistant cultivars available which have not showed any breakdown of resistance. This is best achieved by complete resistance. Besides complete resistance, incomplete and partial resistance (where a form of necrosis inhibits sporulation) has been obtained. The cultivars of the resistance categories and susceptible varieties were intercrossed to study the inheritance (De Jong and Rademaker, 1986). Complete resistance is governed by single dominant gene. The cv. Refour was identified as incompletely resistant. The monogenic complete resistance is based on hypersensitivity, which is more stable while in partial resistance pesticides will still be required to achieve complete absence of symptoms.

A variety Punjab Gold hybrid between Flirt and Gel-e-Sahir, released in 1999 is resistant against septoria leaf spot and good for pot culture.

Insect resistance

The leaf minor (*Liriomyza trifolii*) attack ian chrysanthemum makes the plant unsaleable. During a screening process, *C. pacificum* and the cultivars Penny Lone and White Spider were found to be avoided by leaf minor, while Pink Pompon was found susceptible (De Jong and Van de Vries, 1987).

In a screening of chrysanthemum germplasm for resistance to two spotted spider mite *tetranychus urticae* and bud borers (*Helicoverma armigera*), Angel Bell, Arka Swarna, Chandrika, Nilima, Snow Ball and Collection No. 9 were found to be resistant (less than 10% damage) to bud bores, whereas Angel Bell was the least mite infested collection. These varieties can further be utilized in breeding programme through back cross method.

Transgenic plants of chrysanthemum with Cryl C gene showed resistance against beet armyworm (Schuler *et al.*, 1998).

Resistance to frost and physiological disorder

Variety Peach Centerpiece is a bud sport of Centerpiece which is frost resistant and has high winter hardiness. Frost resistant and has high winter hardiness F_1 hybrid Sesquicentennial Sun has been developed from a cross of Bandit × Allune (Anderson *et al.*, 2001).

The lines derived from Japanese cv. Chiyozakura (mainly pink singles) have been reported to be resistant against physiological disorder of premature budding combined with moderate speed of cropping under low temperature (Langton, 1984).

Development of no pinch no stake cultivars

Pinching and staking are important operations in chrysanthemums. These two operations are labour and cost intensive. Therefore, several varieties evolved which do not require pinching and staking. These cultivars are given below.

Haldighati: Developed by M.A. Kher in 1988. It is a out of season blooming Korean dwarf variety. It needs no pinching. Bears double flowers of sulphur yellow colour.

Hemant Singar: Developed by M.A. Kher in 1981. It is also a no pinch no stake type Korean dwarf variety as miniature. It is early blooming and bears small, purple coloured single flowers.

Arun Singar: Developed by M.A. Kher in 1983. It is no pinch no stake type variety with dwarf stature and variegated leaves. Flowers are red, striped, having early bloom.

Sadbhavana: Developed by S. K. Dutta in 2000. It is a small flowering variety belonging to Double Korean type mini chrysanthemum group. A no pinching and no staking type variety with dwarf, compact and profuse flowering nature.

Other no pinch cultivars are Rangoli, Suhag Singar, Bindiya, Mother Teresa and Dina.

REFERENCES

Anderson, N.O. 1987. Reclassification of the genus *chrysanthemum* L., *Hort. Sci*., **22**(2): 313.

Anderson, N.O. and Ascher, P.D. 1996. Inheritance of pseudo-self compatibility in self incompatible garden and greenhouse chrysanthemum, *Dendranthema grandiflora* Tzvelv. *Euphytica*, 87 (2): 153-164.

Anderson, N.O., Ascher, P.D. and Widmer, R.E. 1992. Inbreeding depression in garden and glasshouse chrysanthemum: germination and survivorship. *Euphytica*, **62**(3): 155-169.

Anderson, N.O., Ascher, P.D., Gesick, E., Walvatne, B., Eash, N., Fritz, V., Hebel, J., Poppe, S., Wagner, R. and Wildung, D. 2001. Garden chrysanthemums 'Peach Centerpiece' and 'Sesquicentennial Sun'. *HortScience*, **36**(7): 1349-1351.

Anderson, N.O., Ascher, P.D., Widmer, R.E. and Luby, J.J. 1990. Rapid generation cycling of chrysanthemum using laboratory seed development and embryo rescue techniques. *J. Am. Soc. Hortic. Sci*., **115**(2): 329-336.

Anderson, N.O., Lied, B.E., Ascher, P.D., Widmer, R.E. and Desborough, S.L. 1988. Evaluation self incompatibility in chrysanthemum: the influence of ovule number. *Sexual Plant Reproduction*, **1**(3): 173-181.

Bailey, L.H. 1963. The Standard Cyclopaedia of Horticulture. The Macmillan Company, New York.

Balakrishnan, S., Pappiah, C.M. and Anbu, S. 2003. Studies on induced mutation in chrysanthemum. *Abst*: National Symposium on Recent Advances in Indian Floriculture. ISOH, Kerala Agriculture University, 72 p.

Barakat, M.N., Fattah, R.S.A., Badr, M. and El-Torky, M.G. 2010. *In vitro* culture and plant regeneration derived from ray florets of *Chrysanthemum morifolium. African J. Biotec*., **9**(8): 1151-1158.

Barigidad, H., Patil, A.A. and Nalawadi, U.G. 1992. Variabilities studies in chrysanthemum. *Prog. Hort*., **24** (1-2): 55-59.

Boase, M.R., Butler, R.C. and Borst, N.K. 1998a. Chrysanthemum cultivar *Agrobacterium* interactions revealed by GUS expression time course experiments. *Sci. Hort*., **77**: 89-107.

Bosae, M.R., Bradley, J.M. and Berst, N.K. 1998b. Genetics transformation mediated by *Agrobacterium tumefaciens* of florists chrysanthemum (*Dendranthema* × *grandiflorum*) cultivar Peach Margaret. *In vitro Cell. Dev. Biol.-Plant,* **34**: 46-51.

Broertjes, C. 1996. Mutation breeding of chrysanthemums. *Euphytica*, **15**: 156-62.

Broertjes, C. and De Jong, J. 1984. Radiation-induced male-sterility in daisy-types of *Chrysanthemum morifolium* Ram. *Euphytica*, **33**(2): 433-434.

Broertjes, C., Koene, P. and Pronk, T. 1983. Radiation-induced low-temperature tolerant cultivars of chrysanthemum morifolium Ram Mutation breeding, X-rays. *Euphytica*, **32** (1): 97-101.

Broertjes, C., Roest, S. and Bokelmann, G.S. 1976. Mutation breeding of *Chrysanthemum morifolium* Ram. using *in vivo* and *in vitro* adventitious bud techniques. *Euphytica*, **25**: 11-19.

Carter, G.D. 1980. Potted Chrysanthemums. *In*: R.A. Larson (eds.). Introduction to Floriculture. Academic Press Inc., New York.

Cathey, H.M. 1955. Chrysanthemum temperatures study. D. effect of temperature shifts on the spray formation and flowering time of *Chrysanthemum morifolium. J. Amer. Soc. Hortic. Sci.,* **66:** 386-391.

Chattopadhyay, T.K., Biswas, M.R., Malla, D.K.S. and Jain, S.S. 1992. Studies on genetic variability, heritability and correlation in chrysanthemum. *Experimental Genetics*, **7**(1-2): 11-14.

Cheng, Xi, Chen, SuMei, Chen, Fadi, Fang, WeiMin, Deng, YanMing and She, LinFang. 2010. Interspecific hybrids between *Dendranthema morifolium* (Ramat.) Kitamura and D. nankingense (Nakai) Tzvel. achieved using ovary rescue and their cold tolerance characteristics. *Euphytica*, **172**(1): 101-108.

Choi, K.S., Nam, C.W., Kim, W.B., Ryu, S.Y., Yoo, D.L. and Suh, J.T. 1993. Selection of chrysanthemum varieties for summer season production in alpine area. *J. Agril. Sci.*, **35**(2): 436-441.

Choudhary, Talukdar, M., Mahanta, S., Sharma, B. and Das, S. 2003. Extent of genetic variation for growth and floral characters in chrysanthemum cultivars under Assam condition. *J. Orna. Hort.*, **6**(3): 207-211.

Cockshull, K.E. 1972. Photoperiodic control of flowering in the chrysanthemum. *In*: A.R. Rees, K.E. Cockshull, D.W. Hand and R.G. Hurd (*eds.*). Crop processes in controlled Environments. Academic Press, London. pp 235-250.

Courtney, G., Napoli, N., Lemieux, C., Morgan, A., Firoozabady, E. and Robinson, K. 1994. Modification of flower colour in florist's chrysanthemum: production of a white flowering variety through molecular genetics. *Bio Technology*, **12**: 268-271.

Datta, K. and Datta, S.K. 1998. Palynological interpretation of gamma ray and colchicine induced mutation in chrysanthemum cultivars. *Israel J. Plant Sci.*, **46**(3): 199-207.

Datta, S.K. and Banerji, B.K. 1993. Gamma ray induced somatic mutation in Chrysanthemum cv. Kalyani Mauve. *J. Nuclear Agri. Bio.*, **22**(1): 19-27.

Datta, S.K. and Banerji, B.K. 1995. Improvement of garden chrysanthemum through induced mutation. *Flora and Fauna,* **1**(1): 1-4.

Datta, S.K. and Chakrabarty, D. 2008. Management of chimera and *in vitro* mutagenesis for development of new flower colour/shape and chlorophyll variegated mutants in chrysanthemum. Induced plant mutations in the genomics era. *Proce*: International Joint FAO/IAEA Symposium, 303-305.

De Jong, J. 1984. Genetic analysis in *Chrysanthemum morifolium*. I. Flowering time and flower number at low and optimum temperature. *Euphytica*, **33**: 455-463

De Jong, J. 1986. Adaptation of *chrysanthemum morifolium* to low light levels. *Scientia Horticulturae*, **28**: 263-270.

De Jong, J. and Custer, J.B.M. 1986a. Induced changes in growth and flowering of chrysanthemum after irradiation and *in vitro* culture of pedicels and petal epidermis. *Euphytica*, **35**: 137-148.

De Jong, J. and Custers, J.B.M. 1986b. The effect of explant source, *in vitro* regeneration and irradiation on variation in yield induced in *Chrysanthemum morifolium. Proce*: Genetic manipulation in plant breeding: International Symposium organized by EUCARPIA, Berlin, pp. 607-609.

De Jong, J. and Drennan, D.L. 1984. Genetic analysis in *Chrysanthemum morifolium*. II. Flower doubleness and ray floret corolla splitting. *Euphytica*, **33**: 465-470.

De Jong, J. and Kho, Y.O. 1982. The shriveling of pollinated pistils as an aid to rapid determination of chrysanthemum pollen viability. *Euphytica*, **31**: 519-521.

De Jong, J. and Rademaker, W. 1986. The reaction of chrysanthemum cultivars to *Puccinia horiana*, and the inheritance of resistance. *Euphytica,* **35**: 945-954.

De Jong, J. and Van de Vries, M. 1987. Components of resistance to *Liriomyza trifolii* in *Chrysanthemum morifolium* and *C. pacificum. Euphytica*, **36**: 719-724.

Deroles, S.C., Boase, M.R. and Konczak, I. 1997. Transformation protocols for ornamental plants. *In*: R.L. Geneve, J.E. Preece, and S.A. Merkle (eds.), Biotechnology of Ornamental Plants, CAB International, Wallingford. pp 87-119.

Dilta, B.S. and Sharma, Y.D. 2003. Genetic improvement of *Dendranthema grandiflora* Tzueleu by gamma rays. *Abst*: National Symposium on Recent Advances in Indian Floriculture. ISOH, Kerala Agricultural University, 77 p.

Dolgov, S.V., Mityshkina, T.U., Rukavtsova, E.B., Buryanov, Y.I. and Vainstein, A. 1995. Production of transgenic plants of *Chrysanthemum morifolium* Ramat with the gene of *B. thuringiensis* delta endotoxin. *Acta Hort.,* **420**: 46-47.

Dologov, S.V., Mitiouchkina, T.Y. and Skryabin, K.G. 1997. Agrobacterial transformation of *Chrysanthemum. Acta Hort.*, **447**: 329-334.

Dowrick, G.J. 1952. The chromosomes of chrysanthemum I. *Heredity*, **6**: 365-376.

Dowrick, G.J. 1953. The chromosomes of chrysanthemum II. *Heredity*, **7**: 59-72.

Dowrick, G.J. and El-Bayoumi, A.S. 1966. The origin of new forms of the garden chrysanthemum. *Euphytica*, **15**: 32-38.

Drewlow L.W. and Widmer, R.E. 1971. Effect of radioactive phosphorus and apical bud removal on the incidence of apparent mutation in *Chrysanthemum morifolium* Ramat. *J. Amer. Soc. Hort. Sci.*, **96** (5): 633-6.

Drewlow, L.W., Ascher, P.D. and Widmer, R.E. 1973. Genetic studies of self incompatibility in the garden chrysanthemum (*C. morifolium* Ramat). *Theo. Appld. Genet.*, **43**: 1-5.

Fryxell, P.A. 1957. Mode of reproduction in higher plants. *Bot. Rev.*, **23** (3): 135-233.

Furuya, H. 1998. Mutation line characteristics of spray type chrysanthemum irradiated with gamma ray. *Bulletin of the Hiroshima Prefectural Agriculture Research Center*, **66**: 33-40.

Gupta, M.N. and Jugran, H.M. 1978. Mutation breeding of chrysanthemum. II. Detection of gamma ray induced somatic mutations in vM2. *J. Nuclear Agri. Bio.*, **7** (2): 50-54.

Heywood, V.H. and Humphries, C.J. 1977. Anthmideal systematic review. *In*: The Biology and Chemistry of Compositae. London Academic.

Hu, Y., Yin, D.C. and Yu, Q.C. 1989. Selection and breeding of a new chrysanthemum mutant cv. Hanrong No. 1. *Jiangsu Agricultural Sciences*, **6**: 31-32.

Huang, H.C., Tsai, C.C. and Sheu, C.H. 2000. Genetic analysis of chrysanthemum hybrids based on RAPD molecular markers. *Bot. Bull. Acad. Sin.*, **41**: 257-262.

Huang, Ying., Wang, KangCai., Zhang, Yuan. and Chen, Yun. 2011. Interspecific hybridization and F_1 hybrids identification of *Chrysanthemum morifolium. Zhongcaoyao = Chinese Traditional and Herbal Drugs*, **42**(4): 783-787.

Hughes, A.P. and Cockshull, K.E. 1972. Further effects of light intensity, carbon dioxide concentration and day temperature on the growth of *Chrysanthemums morifolium* cv. Bright Golden Anne in controlled environments. *Ann. Bot.*, **36**: 533-550.

Huttema, J.B.M., Prell, W. and Jong, J. De. 1991. Methods for selection of low temperature tolerant mutants of *Chrysanthemum morifollium* Ramat. using irradiated cell suspension cultures. III. *Plant Breed.*, **107**: 135-140.

Ilahi, I., Jabeen, M. and Sadaf, S.N. 2007. Rapid clonal propagation of chrysanthemum through embroyogenic callus formation. *Pakistan J. Bot.*, **39**(6): 1945-1952.

Jank, H. 1957. Experiment induction of mutations by X-rays in *Chrysanthemum indicum. Der Zuchter.*, **27**: 223-31.

Jordan, C. and Reimann-Philipp, R. 1983. Investigation into the type and degree of polyploidy in *Chrysanthemum morifolium* Ramat by genetical analysis of two flower colour characters. *Zeitschrift-fur-P flanzenzu chtung*, **91**(2): 111-122.

Khalid, N., Davey, M.R. and Power, J.B. 1989. An assessment of somaclonal variation in *Chrysanthemum morifollium*: the generation of plants of potential commercial value. *Scientia Horticulturae*, **38**(3-4): 287-294.

Kim, J.Y., Park, S.J., Um, B.Y., Pak, C.H., Chung, Y.S. and Shin, J.S. 1998. Transformation of chrysanthemum by *Agrobacterium tumefaciens* with three different types of vectors. *J. Koran Soc. Hortil. Sci.*, **39**(3): 360-366.

Kuklina, E.A. 2003. Flower development of greenhouse chrysanthemum. *Acta Biologica Cracoviensia.*, **45**: 1.

Langton, F.A. 1977. The response of early flowering chrysanthemums to day length. *Sci. Hortic.*, **7**: 277-289.

Langton, F.A. 1984. New GCRI lines will benefit the industry. *Grower*, **101**(23): 21-23.

Langton, F.A.1989. Inheritance in *Chrysanthemum morifolium* Ramat. *Heredity*, **62**: 419-423.

Ledger, S.E., Deroles, S.C. and Given, N.K. 1991. Regeneration and *Agrobacterium* mediated transformation of chrysanthemum. *Plant Cell Report*, **10**(4): 195-199.

Lemieux, C. 1989. Transformation of chrysanthemum cultivars with *Agrobacterium tumefaciens*. *Proce*: Hort. Biotechnology Symp., Davis. p. 49.

Li, Maoxue, Zhang, Xiaofang and Chen, Junyo. 1983. Cytogenetical studies on some Chinese wild *Dendranthema* species and chrysan*themum* cultivars. *Acta Horticulturae*, **10** (3): 199-206.

Liu, SiYu, Zhang, Fei, Chen, SuMei and Chen, FaDi .2010. Interspecific hybridization between the tetraploid *Chrysanthemum nankingense* and *Chrysanthemum grandiflorum* 'Zhongshanzixing' and the genetic performance of their F_1 hybrids. *Scientia Agricultura Sinica*, **43**(12): 2500-2507.

Misra, P., Datta, S.K. and Chakrabarty, D. 2003. Mutation in flower colour and shape of *Chrysanthemum morifolium* induced by gamma radiation. *Biologia Plantarum*, **47**(1): 153-156.

Misra, R.L. and Pragya. 2003. Genetic improvement in Ornamentals. *In*: P.K. Rajeevan, P.K. Valsalakumari and R.L. Misra (eds.), One Hundred Research Papers in Floriculture, KAU. pp. 242-254.

Mitiochkina, T.Y. and Dolgov, S.V. 2000. Modification of *Chrysanthemum* plant and flower architecture by Rol C gene from *Agrobacterium rhizogenes* introduction. *Acta Hort.*, **508**: 163-169.

Miyazaki, S., Tashiro, Y., Kanazawa, K. and Oshima T. 1982. On the flower characteristics and chromosome numbers of Higo chrysanthemums. *Bull. Fac. Agric. Saga. Univ*., **52**: 1-11.

Nagatomi, S. Degi, K. 2008. Mutation breeding of chrysanthemum by gamma field irradiation and *in vitro* culture. Induced plant mutations in the genomics era. *Proce*: International Joint FAO/IAEA Symposium, pp. 258-261.

Nagatomi, S., Miyahira, E. and Degi, K. 2000. Induction of flower mutation comparing with chronic and acute gamma irradiation using tissue culture techniques in *Chrysanthemum morifolium* Ramat. *Acta Hort.*, **508**: 69-73.

Narayana Gowda, J.V. 2003. Improvement in chrysanthemum. *Abst*: National Symposium on Recent Advances in Indian Floriculture. ISOH, Kerala Agricultural Universty, 83 p.

Nayar, K., Rao, M.G. and Goud, J.V. 1971. Chemically induced variation in annual chrysanthemum. *Mysore J. Agric. Sci.,* **5**(1): 116-18.

Nazeer, M.A. and Khoshoo, T.N. 1982. Cytological evolution of garden chrysanthemum. *Curr. Sci*., **51**(12): 583-585.

Nikaido, T. and Onozawa, Y. 1990. Establishment of non-chimeric flower colour mutation through *in vitro* culture of florets from a sport in chrysanthemum with special reference to the genetic background of the mutation line obtained. Scientific Reports of the Faculty of Agriculture Ibaraki University, p. 63-69.

Panse, V.G. 1957. Genetics of quantitative characters in relation to plant breeding. *Indian J. Genet.*, **17**: 318-328.

Parlevliet, J.E., Brewer, J.G. and Ottaro, W.G.M. 1979. Collecting Pyrethrum, *Chrysanthemum cinerariaefolium* Vis, in Yugoslavia for Kenya. *Proce*: Conf. Broadening the Genetic Base of Crops, A.C. Zaren and A. M. Van Harten (eds.). Wageningen, Netherlands *Pudoc*., pp. 91-96.

Parthiban, M., Baskaran, S. and Mohandoss, A. 1994. Insect pollination of the *Chrysanthemum indicum* Linn. *Environment* and *Ecology*, **12**(1): 9-13.

Post, K. 1943. The effect of an interval of long days in the short day treatment on the flowering of chrysanthemums. *J. Am. Soc. Hort. Sci.,* **43**: 311-315.

Preit, W. Huttema, J.B.M. and De Jong, J. 1991. Methods for selection of low temperature tolerant mutants of *Chrysanthemum morifolium* Ramat. using irradiated cell suspension cultures. II. *Plant Breed.*, **107**: 131-134.

Richards, A.J. 1986. Plant Breeding Systems. George Allen and Unwind, London.

Robinson, K.E.P. and Firoozabady, E. 1993. Transformation of floriculture crops. *Scientia Hort.*, **55**: 83-99.

Ronald, W.G. and Ascher, P.D. 1975. Self compatibility in garden Chrysanthemum occurrence, inheritance and breeding potential. *Theo. Appl. Genet.*, **46**(1): 45-54.

Runger, W. 1980. Einfluss der temperature wah rend Kurztag and Langtag perioden nach der bluhinduktion auf die Blutenentwicklung von chrysanthemen. *Gartenbauwissenschaft*, **45**: 264-269.

Schuler, T.H., Poppy G.M., Kerry, B.R. and Denholm, I. 1998. Insect resistant ransgenic plants. *Trends Biotech.*, **16**: 168-175.

Sehrawat, S.K., Boora, K.S., Kumar, R. and Dahiya, D.S. 2002. Randomly amplified polymorphic DNA (RAPD) markers for identification and genetic relatedness analysis of chrysanthemum. *Abst*: National Symposium on Indian Floriculture in the New Millennium, ISHO, Begaluru, 64 p.

Sherman, J.M., Moyer, J.W. and Daub, M.E. 1998. Tomato spotted wilt virus resistance in chrysanthemum expressing the viral nucleocapsid gene. *Plant Disease,* **82**: 407-414.

Singh, A.K. 2006. Flower Crops: Cultivation and Management. Pub. New India Publishing Agency, New Delhi. pp. 73-91.

Sirohi, P.S. and Behara, T.K. 2000. Genetic variability in chrysanthemum. *J. Orna. Hort.*, **3**(1): 34-36.

Stewart, R.N. and Derman, H. 1970. Somatic genetic analysis of the apical layers of chimeral sports in Chrysanthemum by experimental production of adventitious shoots. *Amer. J. Bot.*, **57**: 1061-1071.

Sun, M., Li, P. and Zhang, Q.X. 2007. Flower colour and florescence mutants obtained using electron beam irradiation of chrysanthemum buds. *Acta Horticulturae*, **760**(2): 667-672.

Takatsu, Y., Nishizawa, Y., Hibi, T. and Akutsu, K. 1999. Transgenic chrysanthemum [*Dendranthema grandiflorum* (Ramat.) Kitamura] expressing a rice chitinase gene shows enhanced resistance to gray mold (*Botrytis cinerea*). *Scientia Horticulturae*, **82**(1-2): 113-123.

Tanaka, A., Shikazono, N. and Hase, Y. 2010. Studies on biological effects of ion beams on lethality, molecular nature of mutation, mutation rate, and spectrum of mutation phenotype for mutation breeding in higher plants. *J. Radi. Res.*, **51**(3): 223-233.

Teynor, T.M., Ascher, P.D., Widmer, R.E. and Luby, J.J. 1989. Inheritance of flower colour in *Dendranthema grandiflora* Tzvelev. (*Chrysanthemum morifolium* Ramat.) using cultivars and inbreds. II. Vacuole pigmentation. *Euphytica*, **42**: 297-305.

Trigiano, R.N., Scott, M.C. and Caetano Anolles, G. 1998. Genetic signatures from amplification profiles characterize DNA mutation in somatic and radiation induced sports of chrysanthemum. *J. Am. Soc. Hort. Sci.*, **123**: 642-646.

Verma, A.K., Prasad, K.V., Singh, S.K. and Kumar, S. 2012. *In vitro* isolation of red coloured mutant from chimeric ray florets of chrysanthemum induced by gamma-ray. *Indian J. Horti.*, **69**(4): 562-567.

Waseem, K., Khan, M.K., Jaskani, J. and Khan, M.S. 2007. Impact of different auxins on the regeneration of chrysanthemum (*Dendranthema morifolium*) through *in vitro* shoot tip culture. *Pakistan J. Res.*, **20**: 1-2.

Wasscher, J. 1956. The importance of sports in some florist's flowers. *Euphytica*, **5**: 163-170.

Wolff, K., Peters-van Rijn, J. 1993. Rapid detection of genetic variability in chrysanthemum (*Dendranthema grandiflora* Tzveleo) using random primers. *Heredity*, **71**: 335-341.

Yamaguchi, H. 2010. Mutation breeding with ion beams and gamma rays. *Gamma Field Symposia*, **49**: 33-37.

Yamakawa, K. and lchikawa, S. 1970. Radiation induced mutants of chrysanthemum and their somatic chromosome number. Tech. News Inst. *Rad. Breeding*, **6**: 2.

Yepes, L.M., Mittak, V. and Slightom, J.L. 1999. *Agrobacterium tumefaciens* versus biolistic-mediated transformation of the chrysanthemum cvs. Polaris and Golden Polaris with nucleocapsid protein genes of three tospovirus species. *Acta Hort.*, **482**: 209-218.

Yepes, L.M., Mittak, V., Pang, S.Z., Gonsalves, C., Slightom, J.L. and Gonsalves, D. 1995. Biolistic transformation of chrysanthemum with the nucleocapsid genes of tomato spotted wilt virus. *Plant Cell Rep.*, **14**: 694-698.

Zagarski, J.S., Ascher, P.D. and Widmer, R.E. 1983. Multigenic self incompatibility in hexaploid chrysanthemum. *Euphytica*, **32**(1): 1-7.

[illegible] and [illegible] potential. [illegible]

[illegible] W. 198[illegible]. [illegible] 43: 20[illegible].

[illegible] 16: [illegible]

[illegible]

Sherman, J. M., Moyer, J. W. and Daub, M. E. 1998. [illegible] expressing the viral nucleocapsid gene. [illegible] 82: 40[illegible].

Singh, A. K. [illegible] New Delhi, [illegible]

[illegible]

[illegible]

[illegible] 2007. [illegible]

[illegible]

[illegible]

[illegible]

[illegible]

[illegible]

[illegible]

[illegible]

[illegible]

[illegible]

[illegible]

[illegible]

Chapter – 21

Rose

Modern rose (*Rosa* × *hybrida*) belongs to family Rosaceae. Rose is derived from Greek word "Rhedon" means fragrance. From the ancient time its beauty and attractiveness drag the attention of breeders to create new colours and types. Present cut roses are complex hybrids developed through interspecific hybridization, polyploidy with male and female sterility and these heterozygous genotypes are propagated vegetatively. Rose occupies the first rank among the cut flowers in international market and in India it occupied first position in cut flower export trade.

ORIGIN AND DISTRIBUTION

Roses have been in cultivation since the origins of civilization (Wylie, 1955). The evidence of rose cultivation is found about 5000 years ago in ancient civilization of China, western Asia and northern Africa (Shepherd, 1954). About 120 known wild species, all indigenous to temperate regions of northern hemisphere have been reported to be growing. Eleven species are reported to be growing wild in India. Among the rose species used for perfume extraction purpose *Rosa alba* originated perhaps in Crimea in Russia, *R. borboniana* originated in Isle of Reumon in 1817, *R. centifolia* is native to eastern Caucasus and is under cultivation since 1596. Original home of *R. damascena* is unknown and it was introduced to Europe from Damascus in Asia Minor during the sixteenth century. *R. moschata* is native to Middle East and Himalayas and is found at altitudes ranging from 1000 to 4700 meters extending from Europe. The Persian Musk, one of the forms of *R. moschata* is very common in India. It was introduced into England in 1599. *Rosa gallica*, known as French rose is native to central and southeast Europe and is in cultivation prior to 1542.

Significant information about roses in antiquity can be seen in the writings of the Greek historian Herodotus (490-420 B.C.), the Greek philosopher Theophrastus (372-287 B.C.) and the Roman naturalist Pliny (23-79 A.D.).

Originally all roses grew wild. About 120 known wild species of roses, all indigenous to the temperate regions of the northern hemisphere, have been reported. The species growing wild in India are *Rosa brunonii* (Himalayan Musk rose), *R. eglanteria, R. gigantea* (syn. *R. odorata* var. *gigantean*), *R. involucrate* (syn. *R. clinophylla*), *R. leschenaultiana* syn. *R. sempervirens, R. longicuspis, R. macrophylla, R. moschata* (Musk rose), *R. rubiginosa* (Sweet Briar, Eglantine rose), *R. walpoleana, R. sericea* (Ladakh rose). Till the nineteenth century only four species, *R. gallica* (Red rose), *R. canina* (Dog rose), *R. moschata* (Musk rose) and *R. phoenicia* (Phoenician rose) played a role in the development of cultivars grown at that time.

The French rose (*R. gallica*) is an ancestor of Moss roses, the Centifolias (Cabbage rose) and Damasks (*R. damascena*). In course of time, a variant of *R. damascena* with tendency to bloom in summer has produced autumn Damask (*R. damascena bifera* and *R. damascena semperflorens*).

The pink China rose (*R. indica*) was brought to Britain in 1790 by Sir Joesph Banks and planted in the garden of Mr. Parsons and was known as Parson's Pink China rose which had perpetual flowering quality. At about the same time, a red-flowering dwarf China rose was introduced by Mr. Gilbert Slater to Britain and it came to be known as Slater's Crimson China rose and had perpetual flowering habit. Crimson China rose is considered the father of modern roses. A derivative from the Crimson China rose and *R. gigantea* was the Park's Yellow Tea China rose. The Tea roses were so named because they had the flavour of tea. The Bourbon rose is a natural cross between Parson's Pink China rose and Autumn Damask rose in the island of Bourbon. One or two hybrids closely allied to *R. chinensis* were introduced to Europe during early 19th century due to recurred flowering habit and gave rose to Tea rose. The Tea roses crossed with hybrid perpetual, produced Hybrid Tea class.

A rose, growing in Japan, produced many flowers (*R. multiflora*) and was later knowns *R. polyantha.* These roses were crossed with the Tea roses and gave rise to low-growing, permanently flowering forms which came to be known as Polyantha rose.

In India, several species are found growing wild in Himalayan ranges. The cultivation of roses was perhaps taken up during Mughal period. The Edouard rose was introduced in 1840 during the British rule. The two species *viz.*, Damask and Edouard roses are bluffly scented and being extensively cultivated in Aligarh, Kannauj, Ghazipur, Ballia and Jaunpur in U.P., at Haldighatti in Rajasthan near Chandigarh in Punjab and in Tamil Nadu. Modern roses for cut flowers are grown most parts of India (Singh, 2006).

SPECIES

According to Rehder (1940), the genus *Rosa* comprises of approximately 120 species and is divided into four subgenera, Eurosa, Hulthermia, Platyrhodon and Hesperhodes. Out of these four, Eurosa is represented by 115 species divided into 10 sections. However, Maia and Venard (1976) reported 200 rose species under *Rosa* genus.

Modern garden roses are originated from only 11 species *viz*., *R. canina*, *R. chinensis*, *R. foetida*, *R. gallica*, *R. gigantean*, *R. moschata*, *R. multiflora*, *R. phoenicea*, *R. rugosa*, *R. rubra* and *R. wichuraiana* (Hurst, 1941 and Wylie, 1954). Following 3 species are also reported as ancestors of modern rose, but they are hybrids.

R. alba : *R. canina* × *R. gallica*
R. centifolia : (*R. rubra* × *R. moschata*) × *R. alba*
R. damascena : *R. gallica* × *R. moschata* (for the 'Summer' Damask group)
R. gallica × *R. phoenicia* (for the 'Autumn' Damask group)

Modern roses are classified into twenty classes. According to their parentage and characteristics their brief description is given in Table 1.

Table 1: Classes of modern roses.

S. No.	Class	Parentage	Characters
1.	Hybrid Perpetual	*R. gallica*, *R. centifolia* and *R. chinensis*	Very hardy, remontants in nature (flowers more than a season).
2.	Tea roses	*R. chinensis* × *R. gigantean*	Distinct aroma of tea leaves, free flowering.
3.	Hybrid Tea	Hybrid Perpetual × Tea roses	Bearing large, highly centered flower and long stems.
4.	Floribunda	Hybrid Tea × Polyantha	Clusters of large flowers, perpetual flowering habit.
5.	Grandiflora	Hybrid Tea × Floribunda	Clusters of relatively large flowers.
6.	Polyanthus	*R. multiflora*, *R. wichuraiana* and *R. indica major*	Enormous clusters of small flowers which bloom for several months.
7.	China rose	*R. chinensis*	Bears red to nearly white flowers in small clusters.
8.	Damask rose	*R. gallica* × *R. moschata* (Summer damask) *R. gallica* ×	Enormous clusters of very fragrant, pale pink to red, double petalled flowers.
9.	Bourbon rose	China rose × Damask rose	Large scented flowers have wide and silky petals.
10.	Cabbage rose	*Rosa centifolia*	Bear large, globular, solitary, very fragrant pink flowers.
11.	Moss rose	Bud sport of *R. centifolia*	Little bristles found on seed pod and sepals, sweetly scented.
12.	French rose	*Rosa gallica*	Fragrant flowers and having medicinal properties.

(*Contd.*)

(*Contd. Table 1*)

S. No.	Class	Parentage	Characters
13.	Albas roses	*R. corymbiferera* × *R. gallica*	Very hardy, bear white or pinkish flowers with fragrance.
14.	Musk roses	*R. moschata*	Having musky fragrance, good shrub roses.
15.	Noisette roses	China rose × Musk rose	Compact habit with inflorescence grouped in clusters, constant double, corollas and recurrent flowering.
16.	Rugosas	*R. rugosa*	Highly thorny nature, make good hedges, bear large red or white flowers.
17.	Austrian briars	*R. foetida*	Yellow coloured flowers.
18.	Multiflora ramblers	*R. multiflora*	Suitable for warm climate, bear small white flower, good for dense hedge, less thorny.
19.	Wichuraiana	*R. wichuraiana* ramblers	Hardy, bear white flowers, good for screening, mildew resistant foliage.
20.	Pernitiana	Hybrid Tea × *R. foetida*	Yellow coloured flowers.

FLOWER MORPHOLOGY AND BIOLOGY

Rose bears terminal flowers in general. In Hybrid Tea one or few terminal flowers and in Polyantha, clusters of flowers in the inflorescence are found. Flowers are hermaphrodite with inferior ovary. In most species, the floral structure is based on 5 sepals and 5 petals, however, the flowers are multipetalled. The combined number of petals and stamens are sometimes assumed to be constant (Honeywell, 1957). Roses are naturally cross pollinated; however selfing can easily be done by bagging.

Maximum stigma receptivity occurs one to two days after anthesis, depending upon the variety. The varieties of Hybrid Tea and Floribunda (except cv. Queen Elizabeth), have same receptivity of stigma in winter and spring. However, the variety Queen Elizabeth (Floribunda) shows earlier receptivity than Jadis (Hybrid Tea). Both groups, show receptivity before anthesis. Thus, the flower is protogynous type. Stigmatic exudates in female flower release maximum at pH 5 (Gudin and Arene, 1991). Pollen germination is improved at 23-30°C with 60-65% relative humidity (Gudin *et al.*, 1991). A putrescine treatment of the stigmata increases pollen germination and female fertility (Gudin and Arene, 1992). Self incompatibility is observed in roses (Lata, 1971). Male sterility is also one of the important tools for developing hybrids. Banerjee (1969) noticed that the physiological factors were being responsible for sterility in Polyantha roses cvs. Chattilon, Shahare and Shastry. Successful seed set is found in spring season (February-March). An optimal seed post maturation treatment involved stratification in darkness for 1 month at 23°C and then 2 months at 4°C (Gudin *et al.*, 1990).

GENETICS

The basic chromosome number in roses is 7. Several important species of Asian origin are diploid (2n=14), whereas modern roses are tetraploid (2n=28). Along with normal chromosome, B chromosomes also occur in roses (Jones, 1975). One or two fragments of B chromosome result into high pollen sterility (Lata, 1982). Ploidy level of important *Rosa* accessions assessed by flow cytometry (Jacob *et al.*, 1996) as given in Table 2.

Table 2: Important species and their ploidy level.

S. No.	Species	Ploidy level
1.	*R. gigantea, R. indica, R. moschata, R. multiflora, R. odorata, R. roxburghii, R. rugosa, R. rubra, R. wichuraiana*	2
2.	*R. centifolia major, R. chinensis, R. foetida persiana, R. simperflorens*	3
3.	*R. damascena, R. gallica*	4
4.	*R. canina*	5
5.	*R. alba*	6
6.	*R. spinosissima*	8

The ploidy level of *R. chinensis* is not clearly mentioned and it may be diploid, triploid or tetraploid. Various plant characters as well as growth habit and vigour of the plant are governed by different genes. The inheritance of these characters is given in Table 3.

Table 3: Inheritance of important characters.

S. No.	Characters	Inheritance
1.	Glossy leaves	Single dominant gene
2.	Climbing habit	Single dominant gene
3.	Double flowers	Single dominant gene
4.	Mildew resistance	Single dominant gene
5.	More duration of flowering	Single recessive gene
6.	Vigour	Multigenic
7.	Fragrance	Multigenic
8.	Strength of neck	Multigenic
9.	Width of leaf	Multigenic
10.	Length of stem	Multigenic
11.	Shape of bud	Multigenic
12.	Male sterility	Single dominant gene
13.	Dwarfness	Single dominant gene
14.	Moss character	Single dominant gene
15.	Thornlessness	Single recessive gene

Anthocyanin pigment study in 1200 progeny of 47 families showed that all pigments like cyanin, kaempferal, quercitin, peonin and pelargonin have quantitative inheritance (Marshall *et al.*, 1983). Each of the pigment in roses is controlled by additive gene action (De Vries *et al.*, 1974 and 1980). Magenta, pink or tyrian rose to orange red colour show dominance. Pink colour has dominance effect over dark red, orange-yellow, yellow, white and scarlet. Deep yellow colour is found to be recessive to light yellow and white is recessive to cream and light yellow. The greenish white colour of cv. Message is dominant over the pure white of cv. Virgo. Genetic variance for winter yield was found to be mainly additive and genotype with high GCA values was identified for use in further breeding (De Vries *et al.*, 1980). There are several characters of roses that show high heritability with high genetic advance such as number of branches, flower stems, flower buds petals and flower weight, this indicates that these all characters are governed by additive gene action which may further be improved by selection (Lal *et al.*, 1982). High heritability and genetic advance for thorn density, plant height, plant spread and petal number have been observed (Palai *et al.*, 2003).

BREEDING OBJECTIVES

i. Greater vigour
ii. Hardiness
iii. Novel colour
iv. Exclusive shape and size (long pointed or urn shaped bud)
v. Post harvest longevity
vi. Thornlessness of scion/rootstock
vii. Increased stem length
viii. Glossy or semi glossy, hard leathery foliage
ix. Strong neck
x. Fragrance
xi. Absence of blue fading reaction
xii. Slow opening of flower
xiii. A high centered open flower with 25-35 large petals
xiv. Cut flowers for export
xv. Cut flower for domestic use
xvi. Garden display

xvii. Tolerance to pest (for thrips and mites)

xviii. Resistance to diseases (for black spot and powdery mildew)

BREEDING METHODS

Introduction

Several species and commercial varieties of Hybrid Tea, Floribunda, Polyantha, Miniature and Climbing roses have been introduced from several countries through National Bureau of Plant Genetic Resources (NBPGR), New Delhi (Table 4).

Table 4: Important varieties/species of roses introduced from various countries.

S. No.	Country	Varieties/species
1.	USA	Torchy, Red Cushion, Marie Antonetta, Pilgrin, Elcid, Contoups, Cadenza (clg.), China Doll, Bigsplash (clg), Touch of Venus, Slerra, Dawan, Yellow Cushion, San Antoria, San Diago, First Prize, Aquarius (GR), Comlenche (GR), Angelface (A), Gene Boerner (F_1), Spotless Gold, Spotless Yellow, Spotless Pink, Orange Silk, Blessings, Sir Lancelot, Old Time (HT), Bridal Pink (F_1), Red Gold (F_1), First Federal Gold (HT), Golden Choice (HT), Brasilia(HT), Cathedral, Fleming Piece.
2.	The Netherlands	*R. canina* Brogs, *R. inermis*, *R. superba*, *R. multiflora*, *R. laxa*, *R. rubiginosa*, *R. heinsoln*, *R. smitstckelloze*, *R. schmidut.*
3.	U.K.	*R. canina, R. gallica, R. foetida, R. spinosissima, R. acicularis, R. rugosa, R. virginiana, R. mayesii, R. sericea*, Cabbage, Clg. Iceberg, Clg. Sterling Silver, Crimson China, City of Belfast, Lady Seton, Merlin, Hylo, Summer Holiday, Indian Chief, Beauty Secret, Dwarf King, Orange Silk, Blessings, Sir Lanceolet, Abundance, Summer Holiday, Indian Chief, Elizabeth Harkness, Casamba, Anne Mane, Tip Top, Abundance, Sweet Thoughts, California, Common China, Crimson China, Autumn Sunlight, Swan Lake.
4.	Germany	Ofterbl, Kletterose, Proud Land, Old Time , Tea Hybrid, Proud Land, Bridal Pink, City of Belfast, Bengali, First Federal Gold, Kadinal, Brasilia, Park Rose, Bayreuth Stranchrose, *R. trigonopetala*, Erotika, Stadt Kiel Stranchrose, Westerland Stranchrose, Lola Montez.

Selection

As to the selection process in rose, mainly negative selection is to be applied to their seedlings. Approximately 99% of the plants in each seedling population are discarded during some stage of selection. This type of strict selection in millions of new seedlings in each year must have lead to severe genetic erosion. The condition becomes worst when every breeder has same selection criteria for desirable phenotype. Each selection of productive plants is the key factor for higher yield of flowers. The selection time can be reduced by growing seedlings as De Vries (1976) found that the flower yield of seedlings on their own roots were similar to grafted seedlings. De Vries *et al.* (1978) evolved a method for selection of seedlings at low irradiance in order to select for low blind shoot occurrence.

Several new *in vitro* and *in vivo* selection techniques can help in identifying sources of resistance in rose germplasm. Wisniewska-Grzeszkiewicz and Wojdyla (1996) classified 120 rose cultivars according to their resistance level for stem canker, botrytis blight on petals and black spot disease of leaves. Voisin *et al*. (1996) also studied the host resistance in 32 rootstocks of rose to a nematode (*Meloidogyne hapla*).

Selection for higher vase life depends on the pH, conductivity and molarity of the petal cell sap and this bind of screening which is more precise and reliable than vase life test, can be applied early in the selection process (Gudin and Mouchotte, 1996).

The rootstock compatibility varies and is an important criterion for selection of rootstock. Zieslin *et al*. (1987) described an *in vitro* method using callus that may predict compatibility levels more quickly than traditional grafting test. *Rosa indica* var *odorata* rootstock was selected for better bud-take (80-100%) as well as high multiplication ability under north Indian conditions (Swarup and Malik, 1974). At IIHR, Bengaluru, thornless rootstock was identified and selected it for close planting system (Mukhopadhyay *et al*., 1980). For south Indian conditions, *Rosa multiflora* rootstock performed better in bud-take and was identified at IIHR, Bengaluru (Chadha, 1986). The varieties developed through selection are displayed in Table 5.

Table 5: Important varieties developed through selection.

S. No.	Variety	Parent	Characters
I.	**Hybrid Teas**		
1.	Dr. Homi Bhabha	Virgo	Very good white coloured variety for cut flower.
2.	Kanakangi	Mme Charles Sauvage	Apricot yellow coloured.
3.	Poornima	Farnand Arks	Creamy yellow, extra long buds.
4.	Mechak	Samurai	Dark red.
5.	Divaswapna	Sonia	Silvery pink and white, highly fragrant, lovely buds on stiff stem.
6.	Apsara	Sonia	Salmon colour, vigorous, floriferous.
7.	Aravalli Princess	Sonia	High centered blooms which unfurl slowly.
8.	Careless Love	Sport Red Radiance	Medium sized double, fragrant flowers of red colour splashed with white.
9.	Flaming Sunset	Sport of Mc Gredy's Sunset	Deep orange inside and outside yellow colour.
10.	Angar	Sport of Montezuma	Light red, attractive buds.
11.	Ganga	Sabine	Pointed buds, pale yellow to gold flowers ideal for bedding.
12.	Lal Makhmal	Scarlet Night	Large, symmetrical velvety ebony-red blooms.
II.	**Floribunda**		
1.	Rose Sherbat	Gruss-an-Teplitz	Crimson, 0.033% oil content.
2.	Summer Snow	Sport of Climbing Summer Snow	Large, pure white flowers.
3.	Chitchor	Pink Parfait	Double, white with pink to dark pink edges.
4.	Shabnam	Baby Sylvia	White flowers produced.

(Contd.)

(*Contd. Table 5*)

S.No.	Variety	Parent	Characters
5.	Sadabahar	Frolic	Deep pink flowers of medium size, tolerant to diseases.
6.	Pusa Barahmasi	Sadabahar	Deep pink flowers in cluster, good for standards.
7.	Pusa Virangna	Jantar Mantar	Large, deep red flowers in a cluster of 6 to 12, very good for garden.
III.	**Climbers**		
1.	Clg Pixie	Sport of Pixie	Bluish pink to white, small, double flowers.
2.	Delhi Pink Pearl	Bud sport of Echo	Pink flowers in clusters thornless.
3.	Delhi White Pearl	Bud sport of Prosperity	White flowers in cluster.
IV.	**Polyantha**		
1.	Priti	Bud sport of Margo Koster	Ball like soft pink flowers.
2.	Rashmi	Sport Priti	Globular flowers of deep blackish crimson.
3.	Swati	Winifred Coulter	Deep carmine pink buds, white flowers with dark pink edges, very floriferous.
V.	**Miniature**		
1.	Delhi Starlet	Goudvlinder	Small deep yellow buds.

Varieties developed for making rose water and extraction of rose oil

India is traditional producer of rose water and rose oil. Several local collections of oil bearing strains have been done by some organization to enhance the oil content and improve the oil quality. In this line CIMAP, Lucknow developed two varieties i.e. Noorjahan and Ranisahiba (*Rosa damascena*). Similarly, IHBT, Palampur (H.P.) also evolved two varieties Jwala and Himroj. Cultivar Jwala is suitable for subtropical regions of north plains, mid hills with mild temperate climate upto 1200 m altitude. The plants are short, compact and bushy with low tendency of blind shoots. Its main flush of flowers comes in March to April for about 30 to 40 days and the second flush in September-October. Whereas, cv. Himroz is suitable for temperate region with mild temperate to cold climate i.e. 1200 to 2500 m above mean sea level (Singh, 2006). The major components of rose water volatiles obtained from the bud, half bloom and full bloom stages of cultivar Ranisahiba were phenyl ethyl alcohol (66.2-79.0%), geraniol (3.3-6.6%) and citronellol (1.8-5.5%). The rose water volatiles of cultivar Noorjahan possessed phenyl ethyl alcohol (80.7%) as a major component at full bloom stage. The essential oil of cultivar Noorjahan was also obtained. The major components of these oils were citronellol (15.9-35.3%), geraniol (8.3-30.2%), nerol (4.0-9.6%), nonadecane (4.5-16.0%), heneicosane (2.6-7.9%) and linalool (0.7-2.8%) (Verma *et al.*, 2011). Description of some oil varieties of roses is given in Table 6.

Table 6: Rose varieties evolved for rose water and oil.

S. No.	Variety	Institute developed	Chief characteristics
1.	Noorjahan	CIMAP, Lucknow	An improved strain of locally grown *R. damascena* at Hasayan block in Aligarh (U.P). Produces 2.05 ton/ha flowers. Oil content is 0.04%.
2.	Ranisahiba	CIMAP, Lucknow	Plants are hardy, produces numerous flowers in bunches, more flowering period. Flower yield 52 q/ha.
3.	Jwala	IHBT, Palampur	This is a hardy cultivar with clusters of compact bunches. Weight per flower varies from 2.2 g to 5.0 g with 25 to 55 petals. Oil content varies from 0.03% to 0.06%.
4.	Himroz	IHBT, Palampur	The bushes are tall, bearing thin buds on loose bunches in clusters. Each flower weighs 1.8 to 4.0 g and has 25 to 45 petals.

Hybridization

For hybridization in rose, the variety selected for female flower should be vigorous, a good seed producer or with high female fertility and similarly for male with high pollen fertility. Varieties like Sweet Afton, Charles Mallerin, Ann Elizabeth, Esperanza, Arthur Bell, Anna Wheatcraoft, Pink Parfait, Charleston, First Prize, etc. are reported for production of good hips and seeds, while varieties Delhi Princess, Charles Mallerin and First Prize, etc. are highly male fertile.

The most appropriate time of crossing is November to April in plains and April to August on hills. The rainy or foggy days may be avoided to less number of viable pollen availability. Emasculation and collection of anthers must be done one day prior to flower opening. In both the cases dehiscence should not be occurred. Emasculation must be done in the afternoon. First of all, the petals are removed and then the anthers are taken out gently. Immediately after emasculation, flower is covered with butter paper bags having holes for ventilation. Similarly, the male flowers must be covered to avoid foreign pollen. The best pollination time is at 12 pm and this may be due to high temperature, where pollen tube growth is encouraged. In one study, 22° C was the most favourable temperature for crossing and subsequent growing of seed-bearing plants (De Vries and Dubois, 1987). Pollen can be stored at low humidity for 24 hours at 1°C. Number of pollinations also affects the seed set and it was found that two pollinations between 1 and 4 days after emasculation improved seed yield. Ripen hips are collected at least 4 months later. Seed germination depends on the thickness of endocarp (Khalatkar, 1992).

There are number of rose species which have specific characters, as given below and are being used to incorporate those characters through crossing. A Chinese originated, Australian var. Lee has ability to flower continuously for two years at a stretch. This character can also be incorporated into varieties. Important species used in breeding for specific purpose are tabulated below (Table 7).

Table 7: Important species used in breeding for specific purpose.

S. No.	Species	Specific characters for breeding
1.	*R. indica* var. *odorata*	Resistance against alkaline soils
2.	*R. rugosa*	Black spot resistance, winter hardiness
3.	*R. wichuraiana*	Black spot resistance
4.	*R. chinensis minima*	Dwarfness, ever blooming habit
5.	*R. foetida*	Yellow colour
6.	*R. clinophylla*	Warm climate suitability
7.	*R. multiflora*	Resistance against acid soils
8.	*R. blanda, R. arvensis*	Thornlessness
9.	*R. spinosissima*	Winter hardiness
10.	*R. macrantha*	Winter hardiness
11.	*R. laxa*	Winter hardiness

For winter hardiness and continuous flowering *Rosa rugosa* has been involved in crossing and 12 obtained cultivars meet the requirements (Rieksta *et al.*, 2003). In breeding of single flowered, yellow rose (*Rosa foetida*) with repeat flowering with white coloured rose, repeat flowering, yellow coloured flower was obtained (Dubois, 1978). A new cultivar Suryeo was developed for a standard-type cut flower with white-pink petals having 64 petals per flower that showed 12 days vase life. This cultivar showed high resistance against powdery mildew (Kim *et al.*, 2011).

Work has been done on disease tolerance in roses. Varieties have been developed which were tolerant to powdery mildew, leaf spot, etc. (Tejaswini *et al.*, 2007). A hybrid tea rose cultivar Zui Hongyan was selected from the offspring seedling population of Jumbo Meidiland × First Blush. Cv. Zui Hongyan has bright pink colour and double flowers with 13.9 cm average diameter of its flower. It has a delightful fragrance. This cultivar has certain resistance to black spot disease (Li *et al.*, 2013).

A floribunda rose cultivar Beijinghong was selected from the offspring seedling population of cvs. La Sevillana × Purple Mediland. Cv. Beijinghong has bright orange-red colour and semi-double flowers with 7.1 cm average diameter of its flower. The blossom amount in one plant is 285 flowers per year. This new cultivar has excellent resistance to black spot disease and high growth vigour (Kong *et al.*, 2013). The varieties developed through hybridization are given in Table 8.

Table 8: Important cultivars developed through hybridization.

S.No.	Variety	Parentage	Chief characteristics
I. Hybrid Tea			
1.	American Heritage	Queen Elizabeth × Yellow Perfection	Long pointed buds of ivory and salmon blend, double flowers with high centered bloom.
2.	Apollo	High Time × Imperial Gold	Slender, long pointed buds on long thornless stem, yellow flowers.
3.	Avon	Nocturne × Chrysler Imperial	High centered bloom with red colour, very fragrant.
4.	Charles Mallerin	(Rome Glory × Congo) × Tassin	Velvety blackish crimson, large double and fragrant flowers.
5.	Christian Dior	(Independence × Happiness) × (Peace × Happiness)	Velvety red, double, large, highly fragrant flowers.
6.	Crimson Glory	Catharine Kordes × W. E. Chaplin	Long pointed buds, large, double, very fragrant flowers with deep velvety crimson red.
7.	Happiness	(Rome Glory × Tassin) × [Charles P. Kilham × (Charles P. Kilham × Capucine Champard)]	Long pointed buds, brilliant red coloured flowers, large, double and highly centered bloom.
8.	June Bride	(Mme Butterfly × New Dawn) × Crimson Glory	Buds pointed, large, double highly centered bloom of creamy white colour.
9.	King's Ransom	Golden Masterpiece × Lydia	Long slender buds of golden yellow colour, high centered fragrant bloom.
10.	Montezuma	Fandango × Floradora	Large doubled, highly centered orange colour bloom.
11.	Nigrette	Chateau de Clos Vougeot × Lord casttereagh	Blackish maroon flowers with large, double, highly centered bloom.
12.	Oklahoma	Chrysler Imperial × Charles Mallerin	Long pointed buds, large, double flowers, highly centered, dark red blooms.
13.	Papa Meilland	Chrysler Imperial × Charles Mallerin	Long pointed buds, large, double, flowers of dark velvety crimson colour.
14.	Super Star	(Seedling × Peace) × (Seedling × Alpine Glow)	Coral orange coloured flower of double, large bloom.
15.	Zui Hongyan	Jumbo Meidiland × First Blush	Bright pink colour, double flowers. Flower diameter 13.9 cm. Delightful fragrance. Resistance to black spot disease.
II. Floribunda			
1.	All Gold	Goldilocks × Ellinor Le Grice	Large, semi-double, bright butter, yellow flowers, free bloomer.
2.	Arabian Nights	Spartan × Bea	Large, double, fragrant, salmon-orange flowers.
3.	Arthur Bell	Clare Grammerstorf × Piceadily	Large, semi-double, yellow to cream yellow flowers, very fragrant.

(Contd.)

(*Contd. Table 8*)

S. No.	Variety	Parentage	Chief characteristics
4.	Else Poulsen	Orleans Rose × Red Star	Single, medium sized, bright rose pink flowers.
5.	Flamenco	Cinnabar × Spartan	Large, double flower of salmon colour.
6.	Iceberg	Robin Hood × Virgo	Long pointed buds, fragrant, white coloured flowers.
7.	Independence	Baby Chateau × Crimson Glory	Large, double, fragrant, pure scarlet flowers.
8.	Queen Elizabeth	Charlotte Armstrong × Floradora	Pointed buds, large sized and double carmine rose to pink flowers.
9.	Beijinghong	La Sevillana × Purple Mediland	Bright orange-red colour and semi-double flowers. Flower diameter 7.1 cm. Excellent resistance to black spot disease and high growth vigour. Yield 285 flower stem/year.
III. Miniatures			
1.	Amy's Delight	Little Darling × Little Chief	Ovoid bud, single borne, compact plant, pink flowers.
2.	Antique Rose	Baccara × Little Chief	Pointed buds open to high centered rose-pink coloured flowers.
3.	Dusty Rose	Vanderbilt × Cecile Brunner	Pointed buds, high centered double, reddish purple bloom, upright growth, disease resistant.
4.	Gypsy Jewel	Little Darling × Little Buckaroo	Pointed buds, double, deep rose pink and long lasting blooms.
5.	Twinkle Twinkle	Contempo × Sheri Anne	Pointed buds, blooms white with apricot shade on edges.
Indian Varieties			
I. Hybrids			
1.	Anurag	Sweet Afton × Gulzar	Shapely blooms high centered, tyrian-rose colours.
2.	Arjun	Blith Spirit × Montezuma	Porcelain-rose colour, large bloom.
3.	Bhim	Charles Mallerin × Delhi Princess	Large flowers, bright red coloured, winter hardy.
4.	Jawahar	Sweet Afton × Delhi Princess	Large, high centered, highly fragrant, greenish white bud.
5.	Mrinalini	Pink Parfait × Christian Dior	Long pointed buds, large, phlox-pink flowers.
6.	Raktagandha	Christian Dior × Carrousel	Large, pointed buds, high centered flowers, red colour.
II. Floribunda			
1.	Chandrama	White Bouquet × Virgo	Large, double flowers borne in clusters of 4-6, moonlight white coloured.
2.	Mohini	Sea Pearl × Skola	Unusual chocolate brown colour with yellow shade at the base of petals.
3.	Neelambari	Blue Moon × Africa Star	Pointed buds with magenta-purple colour.
4.	Sindoor	Sea Pearl × Suryodaya	Long pointed buds, double.

(*Contd.*)

(*Contd. Table 8*)

S. No.	Variety	Parentage	Chief characteristics
III. Miniature			
1.	Chardrika	Katharina × Cindrella	Greenish white to white flower.
IV. Climber			
1.	Akash Pradip	Climbing Crimson Glory × Sugandha	Pillar rose of 4 meter height with red flowers.
2.	Pusa Komal	Pink Parfait × Suchitra	Very attractive pink coloured double flower with mild fragrance. Moderately tolerant to thrips and powdery mildew and black spot diseases.
3.	Pusa Shatabadi	Jadis × Century Two	Pink coloured flower bud very big in size about 6-8 cm. Moderately tolerant to powdery mildew and leaf spot diseases.
4.	Pusa Manhar	Pink Parfait × Alinka	Creamy white flower with pink shade and red coloured edges.
5.	Pusa Ranjana	Pink Parfait × Iceberg	Dark pink flower borne in big cluster, profuse flowering.
V. Hybrid Tea			
1.	Dr. Bharat Ram	Sweet Afton × Ganga	High centered pointed buds of peach pink colour long stem.
2.	Pusa Bahadur	Cara Mia × Century Two	High centered buds of red colour with high vase life.
3.	Pusa Garima	Cara Mia × Century Two	High centred, pointed buds of deep pink flowers on long stem.
4.	Pusa Gaurav	Pink Parfait × Arjun	Large sized flowers of deep pink to red colour on very long & sturdy stems.
5.	Pusa Priya	Jantar Mantar × Queen Elizabeth	Large buds, slow opening light pink flowers on long stems.
6.	Mother Teresa	Sweet Afton × First Prize	Long pointed soft pink buds fragrant flowers on long stem
7.	Pusa Pitamber	Jantar Mantra × Banjaran	Golden yellow flowers produced in a cluster of 15-20 flowers.
8.	Pusa Ajay	Pink Parfait × Queen Elizabeth	Attractive dark pink double flower. Moderately tolerant to powdery mildew and black spot diseases. Blooms are mildly fragrant and suitable for cut flower.
9.	Pusa Arun	Queen Elizabeth × Jantar Mantar	Pointed flower buds of dazzling dark red colour. Mildly fragrant, suitable for cut flower. Tolerant to red scale and powdery mildew.

Mutation

In roses, the variations obtained through mutation can be retained and multiplied vegetatively. There has been considerable work done on induced mutations in roses. In gamma irradiation LD_{50} dose for white and mauve coloured varieties is lower than for the yellow, red or pink coloured cultivars. The dark colour varieties are more prone to mutation and among Floribunda varieties, Pink Parfait is the most suitable for the induction of mutation (Lata, 1980). Mainly the irradiation of heterozygous rose cultivar cuttings alters the flower colour, through loss of dominant genes controlling colour and manifestation of recessive alleles. Although mutations (irradiation) reduce the essential oil content in flowers (Lata and Gupta, 1971). The effect of irradiance also causes low prickle density.

The mutants of Raktagandha, Raja Surendra Singh of Nalagarh, Jantar Mantar, Neelambri and Sukumari produce less number of prickles (Valsamma, 1989, Datta, 1990 and Kumar, 1990). There are number of varieties, which have been evolved through mutations (Table 8). The effective dose of gamma rays for the induction of somatic mutations in rose appeared to be in the region of 30-50 Gy. This depends on the cultivars (Singh *et al.*, 2003). The cultivars treated with 5-6 kr showed reduction in shoot elongation, rooting percentage and survival. In a series of mutagens, ion beam irradiation causes effective mutation in roses. Yamaguchi *et al.* (2003) found 9 mutants of 56 irradiated buds with helium ion beam, whereas 12 mutants of 33 irradiated buds with carbon ion. Koh and Ahn (2008) found a new rose variety from rooted cuttings of a rose 'Kardinal' irradiated with 70 Gy of gamma-rays from ^{60}Co source which showed a standard flower with beautiful shape. Koh (2011) irradiated rooted cuttings of 'Kardinal' at 70 Gy gamma-ray dose. Four mutant lines obtained from 'Kardinal' with red petal (Red group, 44A, 45B) include KA1 with light pink petal (Red group, 55B-55D), KA2 with pink petal (Red group, 63A-63B), KA3 with deep pink (Red purple, N57A-N57C), and KA4 with orange red (Red group, 43A-43B). Diameters of each flower in four mutant lines were different from 'Kardinal' and petal number per flower was also variable. It was found that the suitable dose for irradiation was 70-90 Gy dose for 'Spidella' and 90-110 Gy dose for 'Cabernet' for the induction of various mutants considering plant survival rate, shoot growth and mutant occurrence rate. The mutants obtained from 'Spidella' had white, ivory, pinky ivory, light pink and deep pink petal colours. The mutants obtained from 'Cabernet' had pink, deep pink, purple red (magenta), orange red and purple petal colours (Koh *et al.*, 2010). Kahrizi *et al.* (2011) found the most appropriate treatment dose for cultivars Apollo and Maroussia can be in between 60 to 70 Gy. This range provided sufficient survival rates and plant growth, basic requirements for performing a successful mutant selection.

Now a days *in vitro* techniques in combination with induced mutations, are used to speed-up plant breeding programmes. One of the important step when inducing mutations is to determine the radio-sensitivity (i.e., the lethal dose for 50% of samples; LD_{50}) of the species. An experiment was conducted to determine the LD_{50} of rose (*Rosa hybrida*), in which nodal sections of five rose genotypes (Apollo, Maroussia, Dolce Vita, Black Baccara, and Beauty by Ogre) and on leaf explants of two rose genotypes (Apollo and Maroussia) were used. Results revealed that as the dose of irradiation increased, growth parameters were significantly affected and leaf explants were more sensitive than nodal sections (Kahrizi *et al.*, 2013). Cut rose cultivar Candy

Avalanche is selected from mutation breeding of modern rose cultivar Avalanche. The flower is white with a broad and strong pink laciness and its flower is 10-13 cm large. The growth period of single cut flower in greenhouse is 38-42 days. The length of flowering branch is 90-110 cm. The yield of cut flower is 280-300 stems/m²/year (Fu *et al.*, 2013).

Up to now, induced mutations were mainly used to induce colour variants of commercialized cultivars (Table 9).

Table 9: Important varieties developed through mutation.

S. No.	Variety	Parent	Character
I. Hybrid Teas			
1.	Abhisarika	Kiss of Fire (Gamma rays)	Apricot colour, which changes into deep pink on age.
2.	Tata Centenary	Pigalle	Deep purple-mauve splashed with pale yellow strips on inside the petals and light yellow on outer side.
3.	Pusa Christina	Christian Dior	Pink coloured flowers.
4.	Madhosh	Gulzar, EMS 0.25% for 8 hours	Deep magenta red with mauve coloured streaks on petals (Kaicker and Swarup (1978).
5.	Light Pink Prize	First Prize	Light pink flowers.
6.	Stripped Christian Dior	Christian Dior (gamma rays)	Pink and white stripped on red base.
7.	Gippeumi	Kardinal	Bright orange-red with beautiful shape.
8.	Honami	Red Sandra	Standard flower of red purple petals with good shape.
9.	Candy Avalanche	Avalanche	Flower white with a broad and strong pink laciness. Stem length 90-110 cm. Yield of cut flower 280-300 stem/m^{-2}/every year.
II. Floribunda			
1.	Angara	Montezuma	Dark reddish flower.
2.	Saroda	Queen Elizabeth (3 kr gamma rays)	Almost white flower.
3.	Tangerine Contempo	Contempo	Tangerine orange flower.
4.	Yellow Contempo	Contempo	Yellow.
5.	Pink Contempo	Contempo	Pink flower.
6.	Curio	Imperator	Muth flowered curious.
7.	Twinkle	Imperator	Stripped (pink & red).
8.	Sharda	Queen Elizabeth	Pink carmine flower.
9.	Sukumari	America's Junior Miss (3 kr gamma rays)	Almost white flower.
III. Climbing			
1.	Clg Cinderella	Cinderella	White flowers with pale to pink shade in cluster.
2.	Clg Blue Moon	Blue Moon	High centered blooms of silver lilac, very fragrant.
3.	Clg. Caribia	Caribia	Pointed buds of yellow blend good recurling.

Polyploidy

Most of the modern rose varieties are tetraploids. The research demonstrated that in crossing diploid and tetraploids, triploid arise with relative ease. Triploids are usually female sterile but a number of them are fertile. It is notable that among recurrent flowering cultivars, vigour of the plant increases with the ploidy level. However, this has long been awarded that some wild species, mainly diploid, have a potential source of high resistance to disease like *R. rugosa*, *R. wichuraiana,* etc. especially to black spot (Palmer *et al*., 1966 and Saunders, 1970). Amphidiploids are used for the development of black spot resistant rose germplasm (Byrne *et al*., 1996). A natural fertile amphidiploid *R. kordesii*, cross of diploids *R. rugosa* and *R. wichuraiana* also showed disease resistance (Wulff, 1951). The amphidiploids have also been utilized for increased winter hardiness (Svejda, 1979). The doubling of chromosome numbers of diploid species can be made by colchicine treatment of seedling shoots (Semeniuk and Arisumi, 1968 and Basye, 1990) or *in vitro* culturing in a liquid medium (Roberts *et al*., 1990). Several triploid Hybrid Tea roses have been found good in performance. These cultivars are Penelope, Felcia, Fransesca, Nur Mahal and Vanity. Cv. Eva is one of the important selection as triploid from the cross of diploid Robin Hood and tetraploids J.G. Thornton. In IARI, one trisomic Mohini has been obtained on crossing Sea Pearl (4n) × Shola (2n), whereas its sister seedling Prema is a normal triploid. The extra chromosome in Mohini is responsible for the chocolate colour (Kaicker, 1995). Haploidization in roses has been successfully attempted by Meynet *et al*. (1994) in tetraploid cv. Sweet Promise through parthenogenesis induced by using irradiated pollen and *in vitro* culture of immature embryos. The resulting diploids are able to produce flowers and pollen. This could lead to a better understanding of basic rose genetics.

BIOTECHNOLOGY

Tissue Culture

Gene transfer to embryogenic callus derived from different types of explants is currently one of the most popular methods of improvement of rose cultivars, especially for novel characteristics such as blue colour. For every successful genetic engineering programme an efficient tissue culture procedure covering high frequency embryogenic callus induction and plant regeneration is pre-requisite. Thus, an efficient method for embryogenic callus induction from growing nodal stem segments on MS medium followed by a selection of appropriate young leaflets had been worked out by Vishlaghi *et al*. (2011). Where MS medium complemented with 2 mg/L 2,4-D and 0.5 mg/L NAA as well as medium B5 complemented with 0.5 mg/L 2,4-D and 2.5 mg/L NAA plus 0.5 mg/L kinetin were used for callus induction. High frequency of shoot development from mature somatic embryo and subsequent plant regeneration were obtained on a medium containing 3 mg/L of BAP. Micropropagation is a tool that allows solving the problems of spreading diseases due to cutting or grafting. In a study, different culture media were tested in order to establish, multiply and root *Rosa* × *hybrida* cv. Vegas *in vitro* via axillary buds. The best results for micropropagation were obtained with buds containing 5 to 7 primordia developed in MS culture media supplemented with 0.5 mg benzyladenine + 0.25 mg. GA_3 + 0.01 mg IAA/litre, multiplied in MS media supplemented with 1.0 mg benzyladenine + 0.1 mg GA_3/litre and rooted in 1/2MS without growth regulators (Drefahl *et al.*, 2007). Previati *et al.* (2008) obtained virus-free shoot culture lines by the treatment of single shoots at 37 degrees C for 2 weeks, followed by shoot-tip excision and culturing in gelled BA-containing MS medium in an ancient hybrid tea rose, the 'Rosa Sant'Antonio di Padova' and a selection of Rosa City FlorReg, the 'Domenica'. Nak-Udom *et al.* (2009) found *in vitro* flowering on *Rosa hybrida* cv. Perfume Delight cultured on MS medium containing 3 mg/l BA and 0.003 mg/l NAA. In recent years a lot of work had been done through *in vitro* regeneration in various rose species like *Rosa hybrida* cv. Babylon, *Rosa indica* var. odorata, *Rosa canina*, *Rosa damascena*, *Rosa hybrida* cv. Heirloom, *Rosa hybrida* cv. Al-Taif (Darsini and Anitha, 2008; Dixit and Kumar, 2009, Vantu, 2011, Noodezh *et al.*, 2012, Kanchanapoom *et al.*, 2010 and Attia *et al.*, 2012).

Embryogenesis

A lot of work has been done for *in vitro* studies. Hill (1967) was the first to observe morphogenesis of shoot primordia on rose calli. Since then many workers reported successful regeneration through organogenesis or somatic embryogenesis. Regenerants unlike those from cuttings or from direct *in vitro* regeneration (without callus formation), often provide a source of somaclonal variation. These variations are more often common from calli developed on zygotic embryos (70%) and less (21%) from calluses developed on vegetative explants (Arene *et al.*, 1993). Somaclonal variations might be a source for new cultivars, if it is combined with efficient *in vitro* selection pressure. Embryo rescue opens the new directions for wide hybridization. This technique has been used in roses (Gudin, 1994 and Marchant *et al.*, 1994). Embryo can either be rescued very early in ovule (heart shaped stage) or isolated at later stages. An original interspecific hybrid *R. rugosa* × *R. foetida* has been successfully obtained through embryo rescue (Gudin and Mouchotte, 1996).

Interspecific hybridization has been conducted by Mokadem *et al.* (1998). They produce parthenogenetic *R. hybrida* and regenerate them through embryo rescue method. They further crossed diploid *R. hybrida* with 6 diploid species *viz., R. gigantea, R. multiflora, R. wichuraiana, R. rugosa, R. laevigata* and *R. bracteata*, produce enlarged achenes with viable embryos. Fertile seeds from immature embryos have also obtained in early stage after incompatible crosses (Rieksta *et al.*, 2003).

The production of morphologically true to type plants has been performed through somatic embryogenesis (Noriega and Sondahl, 1991). In this method filament explants from young flower buds were cultured and produced plants were identical in morphology and colour to the donor plant. Work done in somatic embryogenesis and adventitious regeneration has been reviewed by Roberts *et al*. (1994). Cv. Sweet Yellow was found to be a very promising genotype in terms of getting embryos from embryogenic callus within 1-2 weeks when its *in vitro* root and leaf explants were induced with somatic embryogenesis (Lee *et al*., 2010).

Protoplast fusion

Protoplast fusion is another option for wide hybridization, though it is more complicated than embryo rescue. Matthews *et al.* (1991) demonstrated first time a protoplast to plant system in roses. Putative somatic intergeneric hybrid plants have been produced by crossing *R. hybrida* cv. Frensham and black-berry or cherry (Motley *et al*., 1996). Protoplasts of the diploid hybrid, *R. persica* Michx × *R. xanthina* have been self fused and tetraploid plants were obtained.

Protoplast fusion of tetraploid rose cultivars Pariser Charme or Heckenzauber and specific diploid genotypes of *R. multiflora*, *R. wichuraiana* or *R. roxburghii* have been performed to introduce resistance against black spot disease. Somatic hybridization with *R. roxburghii* is more important as sexual compatibility with cultivated roses is very limited (Schum *et al*., 2001).

Molecular breeding

The information about the genetic distance between species may be useful in roses, which have been bred only within a limited genepool. This information can also be used to resynthesized the hybrids species. Pair wise distances inferred from fragment data and dendrogram computations with UPGMA clustering are used for inference of genetic relationship between species (Ben-Meir and Vainstein, 1994, Debener *et al*., 1996 and Jan *et al*., 1998). A general molecular marker maps have been constructed for *Rosa* by using RAPD and AFLP markers (Debener and Mattiesch, 1999). Along with the molecular markers they also located genes for double flowers and pink flower colour on the rose maps. Mapping of a major dominant gene for black spot resistance was done with the help of markers, linked to gene (Malek and Debener, 2000). Several sports and seedlings of rose varieties were identified with the help of molecular markers (Debener *et al*., 2000).

Cytological studies are now-a-days replaced by genomic analysis using molecular markers. Among various markers, RFLP marker was used for the first time in USA (Rajapakse *et al*., 1992 and Hubbard *et al*., 1992). Then, in Spain, isozyme and RAPD markers were investigated (Torres *et al*., 1993). Later on RAPD markers have been successfully used in roses for characterization of present genetic diversity (Cubero *et al*., 1996, Moreno *et al*., 1996, Reynders-Aloisi and Bollereau, 1996 and Ballard *et al*., 1996). RAPD markers were also used to study the genetics of disease resistance genes and for marker assisted selection (Debener and Mattiesch, 1999). DNA fingerprinting was adopted as a method for extremely precise characterization (Vainstein and Ben-Meir, 1994 and Vainstein *et al*., 1995). Application of AFLP marker in roses is underway in France and the Netherlands. Apart from characterization of new cultivars; molecular markers are complementary to the morphological description of newly-introduced cultivars, which will help to breeders to defend their rights. Senapati *et al.* (2012) used random amplified polymorphic DNA (RAPD) and inter simple sequence repeats (ISSR) molecular marker techniques to validate the genetic fidelity in three varieties of *Rosa hybrida* (cvs. First Red, Cri Cri and Pusa Gaurav). The result indicated the clonal fidelity of the tissue culture raised *R. hybrida* plantlets and found as the safest mode for multiplication of true to type plants without any somaclonal variation.

A lot of work is going on to obtain blue colour roses from a decade. Flower colour is mainly determined by anthocyanins. *Rosa hybrida* lacks violet to blue flower varieties due to the absence of delphinidin-based anthocyanins, usually the major constituents of violet and blue flowers, because roses do not possess flavonoid 32, 52 -hydoxylase (F32 52 H), a key enzyme for delphinidin biosynthesis. Other factors such as the presence of co-pigments and the vacuolar pH also affect flower colour. In this line appreciable contribution has been done by Katsumoto *et al.* (2007). They analyzed the flavonoid composition of hundreds of rose cultivars and measured the pH of their petal juice in order to select hosts of genetic transformation that would be suitable for the exclusive accumulation of delphinidin and the resulting colour change towards blue. Expression of the viola F32 52 H gene in some of the selected cultivars resulted in the accumulation of a high percentage of delphinidin (up to 95%) and a novel bluish flower colour. For more exclusive and dominant accumulation of delphinidin irrespective of the hosts. They down-regulated the endogenous dihydroflavonol 4-reductase (DFR) gene and overexpressed the *Iris* × *hollandica* DFR gene in addition to the viola F3050H gene in a rose cultivar. The resultant roses exclusively accumulated delphinidin in the petals, and the flowers had blue hues not achieved by hybridization breeding. Moreover, the ability for exclusive accumulation of delphinidin was inherited by the next generations.

Genetic engineering

In rose the method of regeneration strongly depends on cultivars and the efficiency of regeneration is very low. The first successful transformation has been conducted in several clones of *R. canina* by using *Agrobacterium rhizogenes* (Vander Mark *et al*., 1990). For changing the colour of rose, Firoozabady *et al.* (1994) co-cultivated embryogenic callus of *R. hybrida* cv. Royalty with *A. tumefaciens* and *A. rhizogenes* and obtained transgenic plants, where the introduction of chalcone synthase gene reduced the colour intensity of original colour. Some other workers also incorporated chalcone synthase antisense gene in rose and this reduced the intensity of flower colour instead of total suppression of the chalcone synthase synthesis.

Some morphological variants have also been noticed when incorporated a gene modifying morphology (Souq *et al.*, 1996). The rooting abilities have been improved transferring *rol* gene in the rootstock 'Moneyway' (Vander Salm *et al.*, 1996). Marchant *et al.* (1998) have produced kitinase expressing transformants of *R. hybrida* with increased tolerance to the black spot disease. The target in rose improvement is to develop blue colour. A private biotechnology company is engaged in incorporating genes for blue colour in rose. Even after transformation and regeneration of transformed plants (containing delfinidin gene and expression system from petunia), some factors such as co-pigmentation, vacuolar pH, (Buxton and Darbishire, 1929, Lawrence, 1932 and Scott-Moncrieff, 1936) and gene silencing (Stam *et al.*, 1997) may affect the result. The plant genetic manipulation group at Nottingham University is now using high velocity bombardment techniques for the insertion of genes expressing defense related hydrolyses enzymes chitinase and β-1 gluconase. These two enzymes have activity against fungi in roses (Walker, 1996).

REFERENCES

Arene, L., Pellegrirco, C. and Gudin, S. 1993. A comparison of the somaclonal variation level of *Rosa hybrida* L. cv. Meirutral plants regenerated from callus or direct induction from different vegetative and embryonic tissues. *Euphytica*, **71**: 83-90.

Attia, A.O., Dessoky, E.S. and El-Tarras, A.E. 2012. *In vitro* propagation of *Rosa hybrida* L. cv. Al-Taif Rose plant. *African J. Biotech.*, **11**(48): 10888-10893.

Ballard, R., Rajapakse, S., Abbott. A. and Byrne, D.H. 1996. DNA markers in rose and their use for cultivar identification and genome mapping. *Acta Hort.*, **424**: 265-268.

Banerjee, S.K. 1969. Seed sterility in chattilon rose. *Sci. Cult.*, **35**: 268-270.

Basye, R. 1990. An amphidiploid of *Rosa banksiae* and *R. laevigata* induced by colchicines. *Am. Rose Annu.*, **75**: 82-88.

Ben-Meir, H. and Vainstein, A. 1994. Assessment of genetic relatedness in roses by DNA fingerprint analysis, *Sci. Hortic.*, **58**: 158-164.

Buxton, B.H. and Darbishire, F.V. 1929. On the behaviour of anthocyanins in varying hydrogen ion concentrations. *Proc. John lnnes Hort.* Gnst., **21**: 71-103.

Byrne, D.H., Black, W., Ma, Y. and Pemberion, H.B. 1996. The use of amphidiploidy in the development of black spot resistant rose germplasm. *Acta Hort.*, **424**: 269-272.

Chadha, K.K. 1986. Research on Ornamental Crops of IIHR. *In*: K.L. Chadha and B. Choudhury (eds.) Ornamental Horticulture in India. Publication and Information Division, ICAR, New Delhi. pp.1-6.

Cubero, J.I., Millan, T., Osuna, F., Torres, A.M. and Cobos, S. 1996. Varietal identification in rose by using isozyme and RAPD markers. *Acta Hort.*, **424**: 261-264.

Dabener T., Bartels, C. and Mattiesch, L. 1996. RAPD analysis of genetic variation between a group of rose cultivars and selected wild rose species. *Mol. Breed.*, **2**: 321-327.

Darsini, D.T.P. and Anitha, J. 2008. *In vitro* callus induction of *Rosa hybrida* L. cv. Babylon. *Biosciences, Biotechnology Research Asia*, **5**(2): 857-860.

Datta, S.K. 1990. The role of induced mutation on rose breeding. *American Rose Magazine*, **30** (15): 17-18.

De Vries, D.P. 1976. Selection of Hybrid Tea roses seedlings for cut flower yield. *Euphytica*, **25** (2): 361-65.

De Vries, D.P. and Dubois, L.A.M. 1987. Factors affecting fruit and seed set in the hybrid tea-rose 'Sonia'. *Acta Horticulturae*, **226**: 223-230.

De Vries, D.P. and Dubois, L.A.M. 1987. The effect of temperature on fruit set, seed set and seed germination in Soniax Hadley hybrid tea-rose crosses. *Euphytica*, **36** (1): 117-120.

De Vries, D.P., Dubois, L.A. and Smeets, L. 1978. Hybrid Tea roses under controlled light conditions. 3 Flower and blind shoot production in the irradiances in a growth room. *Neth. J. Agr. Sci.*, **26**: 399-404.

De Vries, D.P., Garretsen, F., Dubois, L.A.M. and Van Keulen, H.A. 1980. Breeding research on rose pigments. II Combining ability analysis of variance of four flavonoids in F_1 population. *Euphytica,* **29**: 115-120.

De Vries, D.P., Van Keulen, H.A. and De Bruyn, J.W. 1974. Breeding research on rose pigments. The occurrence of flavonoids and carotenoids in rose petals. *Euphytica,* **23**: 447-457.

Debener, T. and Mattiesch, L. 1999. Construction of a genetic linkage map of roses using RAPD and AFLP markers. *Theor. Apple. Genet.,* **99**: 891-899.

Debener, T., Janakiram, T. and Mattiesch, L. 2000. Sports and seedlings of rose varieties analyzed with molecular markers. *Plant Breeding,* **119**: 71-74.

Dixit, S.K. and Kumar, S. 2009. *In vitro* clonal propagation of rose (*Rosa indica* var. odorata). *Advances in Plant Sciences*, **22**(1): 15-16.

Drefahl, A. Quoirin, M.G. and Cuquel, F.L. 2007. Micropropagation of *Rosa × hybrida* cv. 'Vegas' via axillary buds. *Acta Horticulturae*, **751**: 407-411.

Dubois, A.M. 1978. Genetic study of yellow roses produces valuable selection. *Vakblad Voor de Bloemisteii*, **33**: 51-52.

Firoozabady, E., Moy Y., Courtney Gutterson, N. and Robinson, R. 1994. Regeneration of transgenic roses (*Rosa hybrida*) from embryogenic tissue. *Bio Technology*, **2**: 609-613.

Fu, C., Yang, C., Nic, P. and Wang, Q.G. 2013. A new Rosa cultivar Candy Avalanche. *Acta Horticulturae Sinica*, **40**(2): 403-404.

Gudin, S. 1994. Embryo rescue in *Rosa hybrida* L. *Euphytica*, **72**: 205-212.

Gudin, S. and Arene, L. 1991. Influence of the pH of the stigmatic exudates on male-female interaction in *Rosa hybrida L. Sex Plant Report*, **4**: 110-112.

Gudin, S. and Mouchotte, J. 1996. Integrated research in rose improvement. A breeder's experience. *Acta Hort.*, **424**: 285-292.

Gudin, S., and Arene, L. 1992. Putrescine increases effective pollination period in roses. *Hort Technology*, **2**: 211-213.

Gudin, S., Arene, L. and Pellegrino, C. 1991. Influence of temperature and hygrometry on rose pollen germination. *Adv. Hort. Sci.,* **5**: 96-98.

Gudin, S., Arene, L., Chavagnat, A. and Bulard, C. 1990. Influence of endocarp thickness on rose achieve germination: Genetic and environmental factors. *HortScience*, **25**: 786-788.

Hill, G. P. 1967. Morphogenesis of shoot primordia in cultured stem tissue of a garden rose. *Nature*, **216**: 596-597.

Honeywell, E.R. 1957. Roses. *Purdue Univ. Agric. Ext. Serv. Circ.*, **427**: 24-25.

Hubbard, M., Kell, J., Rajapakse, S., Abbott, A. and Ballard, R. 1992. Restriction fragment length polymorphism in rose and their use for cultivar identification, *HortScience*, **27**: 172-173.

Hurst, C.C. 1941. Notes on the origin and evolution of our garden roses. *J. Royal Hort. Soc.* **66**: 73-82.

Jacob, Y., Teyssier, C., Aloisi, S. and Brown, S.C. 1996. Use of flow cytometry for the rapid determination of ploidy level in the genus *Rosa. Acta. Horticulturae,* **424**: 273-278.

Jan, C.H., Byrne, D.H. Manhart, J. and Wilson, H. 1998. Rose germplasm analysis with RAPD markers, *HortScience*, **34**: 341-345.

Jones, R.N. 1975. B chromosome systems in flowering plants and animal species. *Int. Rev. Cytol.*, **40**: 1-18.

Kahrizi, Z.A., Kermani, M.J., Amiri, M., Vedadi, S. and Hosseini, Z. 2013. *In vitro* radio-sensitivity of different genotypes and explants of rose (*Rosa hybrida*). *J. Hortl. Sci. Biotech.*, **88**(1): 47-52.

Kahrizi, Z.A., Kermani, M.J., Amiri, M.E. and Vedadi, S. 2011. Identifying the correct dose of gamma-rays for *in vitro* mutation of rose cultivars. *Acta Horticulturae*, **923:** 121-127.

Kaicker, U.S. 1995. Rose breeding *In*: Advances in Horticulture vol. 12 Ornamental Plants, K.L. Chadha and S.K. Bhattacharjee (*eds*). Malhotra Publishing House, New Delhi. pp. 234-349.

Kaicker, U.S. and Swarup, V. 1978. Induced mutations in rose cv. 'Gulzar' and effect of chemical and physical mutagens on plant growth. *Acta Agronomie*, **27**: 43-48.

Kanchanapoom, K. Sakpeth, P. and Kanchanapoom, K. 2010. *In vitro* flowering of shoots regenerated from cultured nodal explants of *Rosa hybrida* cv. Heirloom. *Science Asia*, **36**(2): 161-164.

Katsumoto, Y., Mizutani, M.F., Fukui, Y., Brugliera, F., Holton, T.A., Karan, M., Nakamura, N., Sakakibara, K.Y., Togami, J., Pigeaire, A., Tao, G.Q., Nehra, N.S., Lu, C.Y., Dyson, B.K., Tsuda, S., Ashikari, T., Kusumi, T., Mason, J.G. and Tanaka, Y. 2007. Engineering of the rose flavonoid biosynthetic pathway successfully generated blue-hued flowers accumulating delphinidin. *Plant Cell Physiol.*, **48**(11): 1589-1600.

Khalatkar, A.S. 1992. Seed coat of rose. *Indian Rose Annual*, **10**: 106-107.

Kim, JuHyoung., Kim, SiDong., Kim, SeungDeok., Kim, TaeJung., Rho, ChangWoo., Min, KyeongBeom. and Paek, KeeYoeup. 2011. A new rose cultivar 'Suryeo' with white-pink flower colour. *Korean J. Hortil. Sci. Tech.*, **29**(2): 160-163.

Koh, GabCheon and Ahn, KyuBin. 2008. Breeding of rose 'Gippeumi' with orange-red flower through gamma irradiation. *Korean J. Hortil. Sci Tech.*, **26** (4): 445-448.

Koh, GabCheon, Ahn, KyuBin, Gi, GwangYeon, Na, TaekSang and Kang, SiYoung. 2008. Breeding of rose 'Honami' with red-purple flower through gamma irradiation. *Korean J. Hortil. Sci Tech.*, **26**(3): 302-305.

Koh, GabCheon, Kim, MinZa and Kang, SiYong. 2010. Induction of petal colour mutants through gamma ray irradiation in rooted cuttings of rose. *Korean J. Hortil. Sci. Tech.*, **28**(5): 796-801.

Koh, GabCheon. 2011. Characteristics comparison of mutants induced through gamma irradiation in 'Kardinal' rose. *Korean J. Hortil. Sci Tech.*, **29**(5): 456-460.

Kong, C., Chai, F., Cao, L., Yu, H.Q., You, J., Yang, C.Q., Yi, G.F. and Gu, S.G. 2011. A new floribunda rose cultivar Beijinghong. *Acta Horticulturae Sinica*, **38**(7): 1419-1420.

Kumar, S. 1990. Morphological and cytological studies in some induced mutants of rose cv. Folklore. M.Sc. Thesis, IARI, New Delhi.

Lal, S.D., Seth, J.N., Yadav, J.P. and Danu, N.S. 1982. Genetic variability and correlation in rose: Phenotypic variability, heritability and genetic advance. *Progressive Hort.*, **14**: 234-236.

Lata, P. 1971. Hybridization in modern roses. *Current Sci.*, **140**: 4-6.

Lata, P. 1980. Effect of ionizing radiation on roses: induction of somatic mutation. *Environmental and Experimental Botany*, **20** (4): 325-333.

Lata, P. 1982. Inheritance of B chromosomes in garden roses. *Genetica*, **58**: 51-54.

Lata, P. and Gupta, M.M. 1971. Mutation breeding of garden roses. Effect of gamma rays on essential oil content in some scented roses. *Parfumerie and Kosmetik*, **52**: 267-270.

Lawrence, W.J.C. 1932. Interaction of flavones and anthocyanins. *Nature*, **129**: 834-835.

Lee, SuYoung, Han, BongHee and Kim, YoungSoon. 2010. Somatic embryogenesis and shoot development in *Rosa hybrida* L. *Acta Horticulturae*, **870**: 219-226.

Li, H., Wu, C.Y., Wang, X.Y., Luo, J.J., Bai, B.B., Yu, H.Q., You, J. 2013. A new hybrid tea rose cultivar Zui Hongyan. *Acta Horticulturae Sinica*, **40**(3): 605-606.

Maia, N. and Venard, P. 1976. Cytaxonomyie do genre *Rosa* et originie des rosiers cultives. *In*: Travaux sur rosiers de serre. FNPHP, Antibes. pp 7-20.

Malek, B.V. and Debener, T. 2000. Identification of molecular markers linked to RDR1 a gene conferring resistance to black spot in roses. *Theor. Appl. Genet.*, **101**: 977-983.

Marchant, R., Power, J.B., Dazey, M.R. and Chartier-Hollis, J. 1994. Embryo rescue, for the production of F_1 hybrids in English rose. *Euphytica*, **74**: 187-193.

Marchant, R., Dazey, M.R., Lucas, J.A., Lamb, C.J., Dixson, R.A. and Power, J.B. 1998. Expression of kitinase transgen in rose (*Rosa hybrida*) reduce development of black spot disease (*Diplocarponrn rosae* volf.). *Molecular* Breeding, **4**: 187-164.

Marshall, H.H, Campbell, C.G. and Collicult, L.M. 1983. Breeding for anthocyanin colours in Rose. *Euphytica*, **32**: 205-216.

Matthews, D., Motley, J., Horan, I. and Roberts, A.V. 1991. A protoplast to plant system in roses. *Plant Cell, Tissue Org Cult.*, **24**: 173-180.

Meynet, J., Barrade, R., Duclos, A. and Siadous, R. 1994. Dihaploid plants of roses (*Rosa × hybrida*) obtained by parthenogenesis induced using irradiated pollen and *in vitro* culture of immature seeds. *Agronomie*, **2**: 169-175.

Mokadem, H. El., Meynet, J., Jacob, Y. and Gudin, S. 1998. Utilization of parthenogenetic diploid plants of *Rosa hybrida*. l. In interspecific hybridizations. *In*: XIX International Symposium on Improvement of Ornamental Plants. *Acta Hort.*, **508:** 185-190.

Moreno, M.T., Torres, A.M., Millan, T., Armada, J. and Cubero, J.I. 1996. Use of molecular markers in taxonomic studies of rose sp. *Acta Hort.,* **424**: 293-296.

Mottley, J., Yokoya, K., Matthews, D., Squirrell, J. and Wentworth, J.E. 1996. Protoplast fusion and its potential role in genetic improvement of roses. *Acta Hort.*, **424**: 389-392.

Mukhopadhyay, A., Banker, G.J. and Shukla, K.S. 1980. A thornless rootstock of roses. *Indian Hort.*, **25**(1): 1-2.

Nak-Udom, N., Kanchanapoom, K. and Kanchanapoom, K. 2009. Micropropagation from cultured nodal explants of rose (*Rosa hybrida* L. cv. Perfume Delight). *Songklanakarin J. Sci. Tech.*, **31**(6): 583-586.

Noodezh, H. M., Moieni, A. and Baghizadeh, A. 2012. *In vitro* propagation of the Damask rose (*Rosa damascena* Mill.). *In vitro Cellular & Developmental Biology – Plant*, **48**(5): 530-538.

Noriega, C. and Sondahl, M.R. 1991. Somatic embryo genesis in Hybrid Tea roses. *Bio- Technology*, **9**(10): 991-993.

Palai, S.K., Mishra, M., Bhuyan, S. and Mishra, H.N. 2003. Genetic variability in Hybrid Tea roses. *J. Orna. Hort.*, **6**(1): 29-33.

Palmer, J.G., Semeniuk, P. and Stewart, R.N. 1966. Roses and black spot. Pathogenicity to excised leaflets of *Diplocarpon rosae* from seven geographic locations. *Phytopathology*, **56**: 1277-1282.

Previati, A. Benelli, C. Re, F. da Ozudogru, A. Lambardi, M. 2008. Micropropagation and *in vitro* conservation of virus-free rose germplasm. *Propagation of Ornamental Plants*, **8**(2): 93-98.

Rajapakse, S., Hubbard, M., Kelly, J.W., Abbott, A.G. and Ballard, R.E. 1992. Identification of rose cultivars by restriction fragment length polymorphism. *Sci. Hort.*, **52**: 237-245.

Rehder, A. 1940. Manual of Cultivated Trees and Shrubs. Macmillon, New York. pp 426-451.

Reynders-Aioisi, S. and Bollereau, P. 1996. Characterization of genetic diversity in genus *Rosa* by randomly amplified polymorphic DNA. *Acta. Hort.*, **424**: 253-259.

Rieksta, D., Jakobsone, G., Rihtere, A., Porkmann, G., Hauser, B. and Michaelis, S. 2003. Classical breeding of *Rosa rugosa* and *in vitro* cultivation of immature embryo as an expanded resources for selection. *Acta Horticulturae*, **612**: 35-38.

Roberts, A.V., Yokoya, K., Walker, S. and Mottley, J. 1994. Somatic embryogenesis in *Rosa* spp. *In*: Somatic embryogenesis in woody plants. Kluwer Academic Pub., The Netherlands. pp. 277-289.

Roberts, A.V., Lloyd, D. and Short, K.C. 1990. *In vitro* procedures for the induction of tetraploidy in a diploid rose. *Euphytica*, **49**: 33-38.

Saunders, P.J.W. 1970. The resistance of some cultivars and species of *Rosa diplocarpon rosae*. Wolf causing black spot disease. *Rose Annu.*, 118-128.

Schum, A., Hofmann, K. and Felten, R. 2001. Fundamentals for integration on somatic hybridization in rose breeding. *Acta Horticultureae*, **552**: 29-34.

Scott -Moncrieff, R. 1936. A biochemical survey of some mendalian factors for flower colour. *J. Genetics*, **32**: 117-169.

Semeniuk, P. and Arisumi, T. 1968. Colchicine-induced tetraploid and cytochimeral roses. *Bot. Gaz.*, **129**: 190-193.

Senapati, S.K., Aparajita, S. and Rout, G. R. 2012. An assessment of genetic fidelity of *in vitro* grown plantlets of rose (*Rosa hybrida*) through molecular markers. *African J. Biotech.*, **11**(100): 16532-16538.

Shepherd, R.E. 1954. History of the Rose. Macmillan, New York.

Singh, A.K. 2006. Flower Crops: Cultivation and Management. Pub. New India Publishing Agency, New Delhi. pp. 297-336.

Singh, A.P., Prasad, K.V. and Chaudhary, M.L. 2003. Panorama of Rose Research, AICRPF, IARI, New Delhi. 181 p.

Souq, F., Coutos-Thevenot, P., Yean, H., Delbard, G., Maziere, Y., Barbe, J.P. and Boulay, M. 1996. Genitic transformation of roses, two examples: one on morphogenesis, the other on anthocyanin biosynthetic pathway. *Acta Hort.*, **424**: 381-388.

Stam, M., Mol, N.M. and Kooter, J.M. 1997. The silence of genes in transgenic plants. *Ann. Bot.*, **79**: 3-12.

Svejda, F. 1979. Inheritance of winter hardiness in roses. *Euphytica*, **28**: 309-314.

Swarup, V. and Malik, R.S. 1974. Studies on the performance of rose varieties on different rootstock. *Indian J. Hort.*, **31**: 268.

Tejaswini, P., Kumar, N., Misra, R.L., Sujatha, K., Sangama, N.R., Jhansi Rani and Pushkar, S. 2007. Rose. All Indian Coordinated Research Project on Floriculture, Division of Floriculture and Landscaping, IARI, New Delhi. pp. 56-57.

Torres, A.M., Millan, T. and Cubero, J.T. 1993. Identifying rose cultivars using random amplified polymorphic DNA markers. *HortScience*, **28**: 333-334.

Vainstein, A. and Ben-Meir, H. 1994. DNA finger printing analysis of roses. *J. Am. Soc. Hort. Sci.*, **119**: 1099-1103.

Vainstein, A., Ben-Meir, H.Zuker, A., Watad, A.A., Scovel, G., Ahroni, A. and Ovadis, M. 1995. Molecular markers and genetic transformation in the breeding of ornamentals. *Acta Hort.*, **420**: 65-67.

Valsamama, M. 1989. Mutation studies in garden roses. Ph.D. Thesis, IARI, New Delhi.

Vander Mark, F., Pijnacker Hordijk, J.P., Varega, G.A.I., de Vries, D.P. and Dons, J.M. 1990. *In vivo* transformation of clonal *Rosa canina* rootstocks with *Agrobacterium rhizogenes*. *J. Genet. Breed.*, **44**: 263-268.

Vander Salm, T.P.M., Vander Toorn, C.J.G., Hanisch ten cate, C.H., Van der Kricken, W.M. and Dons, J.J.M. 1996. The effects of exogenous auxin and *rol* genes on root formation in *Rosa hybrida* L. Moneyway. *Plant Growth Regulation*, **19**: 123-131.

Vantu, S. 2011. *In vitro* multiplication of *Rosa canina* L. *Biologie Vegetala*, **57**(1): 19-22.

Verma, R.S., Padalia, R.C., Chauhan, A., Singh, A., Yadav, A.K. 2011. Volatile constituents of essential oil and rose water of damask rose (*Rosa damascena* Mill.) cultivars from North Indian hills. *Nat. Prod. Res.*, **25**(17): 1577-84.

Vishlaghi, N., Javaran, M.J. and Moeini, A. 2011. Regeneration of miniature rose (*Rosa hybrida* cv Tanbakeda) through somatic embryogenesis. *Iranian J. Hortil. Sci.*, **41**(4): 347-357.

Voisin, R., Minot, J.C., Esmenjaud, D., Jacob, Y., Pelloli, G. and Aloisi, S. 1996. Host suitability of root stock to the root-knot nematode *Meloidogyne hapla* using a high inoculum pressure test. *Acta Hort.*, **424**: 237-240.

Walker, S. 1996. Discussion of the consume acceptance of genetic transformation in roses. *Acta Hort.*, **424**: 289-391.

Wisniewska-Grzeszkiewicz, H. and Wojdyla, A.T. 1996. Evolution of rose cultivars to fungal diseases susceptibility. *Acta Hort.*, **424**: 223-236.

Wright, P.H. 1978. Unexploited genes in rose species. *The American Rose*. **3**: 23.

Wulff, H.D. 1951. *Rosa kordesii*, eine neue amphidiploide rose. Z*uchter*, **21**: 123-132.

Wylie, A.P. 1954. The history of garden roses. *J. Royal Hort. Soc.*, **79**: 551-572.

Wylie, A.P. 1955. The history of garden roses, Part 2. *J. Royal Hort. Soc.*, **80**: 8-24.

Yamaguchi, H., Nagatomi, S., Morishita, J., Degi, K., Tanaka, A., Shikazono, N., and Hase, Y. 2003. Mutation induction with ion beam irradiation in rose. *Technical News Institute of Radiation Breeding*, **66**: 2.

Zieslin, N., Gavish, H. and Ziv, M. 1987. Growth interactions between calli and explants of rose plants *in vitro*. *Plant Sci.,* **49**: 57-62.

Chapter – 22

Tuberose

Tuberose (*Polianthes tuberosa* L.) belonging to family Amaryllidaceae is half hardy, perennial, bulbous plant. Bulbs are made of scales and leaf bases and stem remains concealed within scales. The generic name *Polianthes* is derived from Greek word "*polios*" meaning shining or white and "*anthos*" meaning flower, in allusion to the blooms of the common tuberose and species *tuberosa*, the plant being tuberous in nature. Tuberose is commercially cultivated in tropical and sub tropical areas for cut and loose flower trade and also for the extraction of it's highly valued natural flower oil.

ORIGIN AND DISTRIBUTION

Tuberose is a native of Mexico (Bailey, 1976) from where it spread to different parts of world during 16th century. The name tuberose is derived from tuberosa, this plant being the tuberous hyacinth as distinguished from the bulbous hyacinth. The name therefore, is tuber-ose, not tuberose. By the middle of the 18th century the tuberose was still referred for as *Hyacinthus indicus tuberosus*. Originally Linnaeus had used the name Polianthes floribus alternis in his Hortus Cliffortianus. Later it was grouped under the genus *Polianthes* and the species *tuberosa*.

Tuberose is cultivated on large scale in China, Egypt, France, Italy, Israel, South Africa, New Zealand, Spain, Sri Lanka, North Carolina, USA and in many tropical and subtropical countries.

In India tuberose is commonly known as Gulchari and Gulshabbo in Hindi, Rajanigandha in Bengali, Sukandaraji and Nelasampengi in Telegu, Nilasampangi in Tamil and as Sugandharaja in Kannada.

In India its commercial cultivations is confined to Bagnen, Midnapur, Krishnanagar, Ranaghat, Kolaghat and Pariskura in West Bengal, Guwahati and Jorhat in Assam, Pune, Thane, Sangli, Nasik and Ahmednagar in Maharastra, Udaipur, Ajmer and Jaipur in Rajasthan, Navsari and Valsad in Gujrat, Ambala, Gurgaon and Faridabad in Haryana, Ludhiana and Patiala in Punjab, Devanahalli, Tumkur, Kolar, Belgaum and Mysore in Karnataka, East Godavari, Guntur, Chitoor and Krishna district in Andhra Pradesh and Coimbatore and Madurai in Tamil Nadu. Growers of Uttar Pradesh, Delhi and Orissa are also concentrating in tuberose cultivation after evolvement of improved varieties namely Shringar, Suvasini, Prajwal and Vaibhav by Indian breeders (Singh, 2006).

SPECIES

Genus *Polianthes* contains a dozen of species (Rose, 1905). Bailey (1976) also reported 12 species but presently 14 species are existing in the world. Among them *Polianthes tuberosa* is commercially popular. The haploid number of chromosomes in *Polianthes tuberosa* is 30, of which 5 are large and 25 small. In the diploid set, two pairs of chromosomes are associated with the organization of nucleus. *P. tuberosa* is an erect herb, 60-120 cm high with stout and short bulbs. Leaves are six to nine in number, 30-45 cm long, about 1.3 cm wide, liner, bright green, bearing numerous dots near the base. Flower sessile, funnel shaped, about 5 cm long, waxy white, perianth segment 6, tube bent slightly near the base, filaments attached on upper part of corolla, fragrant. The flowers are born in about 30 alternate pairs on long terminal blooming from base upwards, some forms are double.

Several varieties are evolved from *P. tuberosa*. Other species are *P. blissii*, *P. durangensis, P. gracilis*, *P. graminifolia*, *P. germiniflora*, *P. howardii, P. longiflora*, *P. montana*, *P. nelsonii*, *P. platyphylla*, *P. palustris*, *P. pringlei* and *P. sessiliflora*.

FLOWER MORPHOLOGY AND BIOLOGY

Most of the species of *Polianthes* have white flower but *P. germiniflora* Rosa is red in colour (Bailey, 1976). The flowers have a funnel shaped perianth and are fragrant, waxy white, about 25 mm long, single or double, borne in a spike. Stamens are 6, anthers dorsifixed in middle; ovary 3 locular, ovules numerous and fruit is capsule.

Pollination, pollen viability and germination

Tuberose is a cross pollinated crop and usually pollinated by insect. Pollen viability in tuberose varies from genotype to genotype.

A study was carried out on the pollen viability and pollen germination in tuberose. Among the hybrids, the highest percentage of pollen viability was recorded from Suvasini (88.83%), whereas in the parents, the highest percentage of pollen viability was recorded from Double (87.72%). Among the open-pollinated cultivars developed from Variegated, the percentage of pollen viability was comparatively higher in IIHR-5 (89.90%) and IIHR-6 (89.37%), while it was 84.04% in Variegated. The percentage of germinated pollen grain in the 9 genotypes varied from 45.92% (Single) to 64.95% (IIHR-6). The highest percentage of germination by open pollination was recorded in Variegated and Shringar by 76 and 72%, respectively (Seetharamu *et al.*, 2000). They further reported that crosses involving Single type × Double type tuberose, the fruit set and fruit retention was maximum on fourth day after anthesis.

In an *in vitro* pollen germination study, among the single types, maximum percentage of pollen germination was recorded in IIHR-6 (62.95% and the minimum in cv. Single (45.92%). The fruit set by open- pollination was highest in IIHR-6 (76.0%) whereas it was lowest (18.0%) in IIHR-1(Seetharamu, 1993).

Fruit set and fruit retention

Shen *et al*. (1987) observed fruit setting when fertilization was done 2-3 days after anthesis. The seed germination percentage was maximum in cross pollinated seeds compared to open pollinated seeds.

Another study on tuberose seed and fruit set was carried out (Seetharamu, 1993). In single type to tuberose cv. Single and IIHR-6 were found to be better as seed parents. Among the double types, IIHR-2 and IIHR-3 were found to better as pollen parents. The crosses between IIHR-6 × IIHR-3 gave better fruit set (31.71%) followed by IIHR-1× cv. Double (35.29%). The open-pollinated seed germination was maximum in cv. Variegated (76.0%) followed by cv. Single (64%). In seeds obtained from artificial cross-pollination, the seed germination was maximum in crosses between IIHR-1 × cv. Double (90.0%) followed by cv. Single × IIHR-2 (88.8%).

GENETICS

Joshi and Pantulu (1941) have conducted an extensive cytological study on *P. tuberosa* and observed that the haploid number of chromosomes is 30, among these 5 are large and 25 are small. In a cytological study, Ayyanger (1963) found that multiple perianth variety was autotetraploid having 120 chromosomes of which 20 chromosomes were long and 100 were short. Schiva and Lanteri (1983) did cytological studies of Single (fertile) and Double (infertile) cultivars and reported chromosome number 60 and 50, respectively. Laxmi *et al.* (1984) detected different types of chromosomal aberrations in mutants. Different types of configuration were noticed in meiotic studies.

Self-incompatibility

Tuberose (*Polianthes tuberosa*) has the characters of dichogamy and self-incompatibility. Crosses between single and double varieties produce fruits and seeds when the female parent is 2–3 days after anthesis. Reciprocal crosses between the single and double varieties of tuberose grown in Taiwan produced many single and few double plants in the progenies. Twelve seedlings with improved characteristics were selected from these progenies. Work is now in progress to test stability of these characteristics to produce more prolific and higher quality tuberose varieties. Selection from hybrid seedlings is a very promising breeding method for improvement of cut flower production in tuberose (Shen *et al.*, 1987).

Pollination studies on tuberose revealed self-incompatibility in single with no fruit set and 63.78% fruit set on cross pollinating with variegated cultivar. This variegated cultivar is a both male and female fertile variety with sufficient pollen production and recorded a seed set of 12.13 and 28.84% when self- and cross-pollinated, respectively. But seeds obtained after cross pollination with the variegated variety as a female parent recorded lower seed viability and rate of seed germination, suggesting lower fertility and seedling vigour (Uma and Gowda, 2000).

Seed setting behaviour, self and cross compatibility and percentage of seed germination of tuberose hybrids Shringar, IIHR-2, Suvasini, IIHR-4, IIHR-5, IIHR-6, and other cultivars namely Single, Double and Variegated was observed (Seetharamu *et al.*, 2000). The seeds obtained from artificial crossing recorded maximum percentage of germination in crosses between Shringar × Double (90%) followed by Single × IIHR-2 (88.88%). Self-incompatibility was also observed in tuberose by Uma, 1990).

Genetic variability, heritability and genetic advance

Most of the varieties now used in cultivation are known as single or double and genetic base of commercial cultivars is relatively narrow. However, in a genetic study high heritability along with high genetic advance was observed for 100 flowers weight, number of leaves/plant and loose flower yield/m^2 in single types. In double types, high values were recorded for number of leaves/plant, spike length and 100 flowers weight. These results indicated the presence of additive gene effect for most of the traits and their amicability for direct selection in both types of tuberose (Radhakrishna *et al.*, 2003).

Genetic variability, heritabilitv and genetic advance of different yield contributing characters were studied in nine tuberose varieties viz., Local Single, Local Double, Shringar, Suvasini, Variegated, Vaibhav, Prajwal, Hyderabad Single and Hyderabad Double. Moderate degree of variability was observed for all the characters under study. Very high estimates of heritability coupled with moderate to high genetic advance as per cent of mean was noticed for 100 floret weight, spike weight, weight of bulbs/plant, length of flower stalk, plant height and rachis length (Gaurav *et al.*, 2005).

BREEDING OBJECTIVES

i. Good plant growth

ii. Better spike quality specially strength, good rachis length, symmetrical arrangement and facing of florets to all direction

iii. Coloured flower

iv. More number of florets

v. High fragrant variety

vi. More number of opened florets in vase

vii. More vase life

viii. High concrete recovery

ix. Varieties for higher yield

x. Varieties with enhanced keeping quality

xi. Resistance to tuberose mild mosaic virus and nematodes

In India many research institutions are engaged in breeding of tuberose to improve at various levels through introduction, hybridization, mutation and biotechnology. Following are the Coordinated Centres for germplasm collection and breeding:

1. Indian Institute of Horticultural Research, Bangaluru
2. Mahatama Phule Krishi Vidyapeeth, Pune
3. Bidhan Chandra Krishi Vishwavidyalaya, Kalyani
4. Acharya N.G. Ranga Agricultural University, Hyderabad
5. Assam Agricultural University, Kahikuchi
6. Indian Agricultural Research Institute, New Delhi
7. Punjab Agricultural University, Ludhiana
8. G.B. Pant University of Agriculture and Techonology, Pantnagar
9. Maharana Pratap University of Agriculture and Technology, Udaipur

BREEDING METHODS

Introduction and Selection

No literature is available about introduction of tuberose in India but it is a prediction that it was probably introduced by Europeans, while it was introduced in England by the end of sixteenth century. In an evaluation and selection study, Irulappan *et al.* (1980) studied 11 different single and double types of tuberose. Variety Calcutta Single produced maximum flowers (456 flowers/clump). Cv. Mexican Single ranked second (447 flowers/clump). Banker and Mukhopadhyay (1980) evaluated 4 cultivars for morphological and yield attributes. Among cvs. Single, Double, Semi Double and Variegated, cv. Single found to be best. In India, several local selections of tuberose cultivars were made according to their suitability and better performance in a particular agro climatic condition. These cultivars are described by Singh (2003) as given below.

Hyderabad Single: This cultivar has white colour single flowers. It is collected and maintained at ANGRAU, Hyderabad.

Hyderabad Double: It produces double florets. Flower colour is white. This cultivar is also maintained at ANGRAU, Hyderabad.

Variegated: This cultivar is collected from Gangtok (Sikkim) by NBRI, Lucknow. Its florets are almost similar to cv. Mexican Single, except their tinge have pinkish colour. Its spikes are tallest among the popular cultivars.

Calcutta Single: It is a single type variety. It is collected by Dr. B. Biswas and Dr. A.K. Pal at BCKV, Kalyani.

Calcutta Double: It has white colour florets tinged with pinkish red before opening. Florets are double petalled. This variety is maintained at BCKV, Kalyani.

An improved variety of tuberose 'Phule Rajni' was developed by MPKV, Pune and released in 2005. Besides above named varieties, few local selections were made at different places and were found superior in performance at respective regions and depicted in Table 1.

Table 1: Selection of tuberose varieties for cultivation by various organizations.

S. No.	Varieties	Organization
1.	Kahikuchi Single	HRS (AAU), Kahikuchi
2.	Kahikuchi Double	HRS (AAU), Kahikuchi
3.	Pune Single	MPKV, Pune
4.	Pune Double	MPKV, Pune
5.	Saurashtra Single	NAU, Navsari
6.	Saurashtra Double	NAU, Navsari
7.	Mexican Single	An introduction
8.	Pearl Double	An introduction

Hybridization

Tuberose has the characters of dichogamy and self incompatibility, but some improved varieties have been evolved in India through hybridization. Double petalled varieties generally do not set fruit and seed. Shen *et al.* (1987) did hybridization between

Single and Double cultivars. Reciprocal crosses between Single and Double cultivars produced many single and few double plants in the progenies. They selected 12 seedlings with improved characteristics.

In another study, hybrid progenies from 6 tuberose accessions characterized by high concrete content (7-IIHR-PT-1, 5-IIHR-PT-1 and 8-IIHR-PT-1 as female parents) and flowers yield (IIHR-PT-5, IIHR-PT-2 and 2-IIHR-PT-2 as male parents) were evaluated. One of the offspring of 8-IIHR-PT-1 × IIHR–PT-5 produced light pink floral buds, 5-IIHR-PT-1 × 2-IIHR–PT-2 gave light green flower buds, whereas 7-IIHR-PT × IIHR-PT-2 had yellow flower buds (Srivastava and Shridhar, 2002). In India, 4 varieties *viz*. Shringar (Single), Prajwal (Single), Suvasini (Double) and Vaibhav (Double) were developed by IIHR, Bengaluru (Srinivas, 2003). Recently, double type cultivars are gaining popularity for their attractive spike, hence, some more double type hybrids of tuberose *viz*., IIHR-2 (Mexican Single × Pearl Double), IIHR-4 (Mexican Single × Pearl Double) and IIHR-5 (Mexican Single × Pearl Double) were developed (Radhakrishna *et al*., 2003a). A hybrid variety Phule Rajani (GK-T-D-2) was developed under All India Co-ordinated Floriculture Improvement Project, Ganeshkhind (Pune). It was a cross between Local Single × Shringar. The hybrid has greenish flower bud, about 73.12 cm spike, 35.46 cm rachis length, good yielder (7.79 lakh spikes/ha), more vase life with better bulb multiplication ability (Katwate *et al*., 2006). Some varieties developed through hybridization are given in Table 2.

Table 2: Promising varieties developed through hybridization in India.

S. No.	Variety	Parentage	Characteristics
1.	Shringar (IIHR)	Mexican Single × Pearl Double	Single type florets on strong, medium spike length, florets pinkish, gives 36% higher yield than cv. Mexican Single, yield 150 q/ha loose flower. Produces seeds abundantly.
2.	Suvasini (IIHR)	Mexican Single × Pearl Double	Double petalled, bear's bold, big and white florets on long spike. Florets open regularly, 25% more spike yield than cv. Pearl Double. Does not set seeds.
3.	Prajwal (IIHR)	Shringar × Mexican Single	Single type florets on tall stiff spikes, buds pinkish, while open florets white, gives 20% more loose flower yield than cv. Shringar. Produces seeds abundantly.
4.	Vaibhav (IIHR)	Mexican Single × IIHR-2	Double petalled, buds greenish, florets white, medium spike, 50% higher spike yield than cv. Suvasini.
5.	Phule Rajani (MPKV)	Local Single × Shringar	Greenish flower bud. Spike length 73.12 cm. Rachis length 35.46 cm. More vase life with better bulb multiplication ability. Good yielder (7.79 lakh spikes/ha).
6.	Arka Nirantra		This hybrid has higher spike yield, early, white single flower and prolonged flowering period.

Breeding for development of coloured hybrids

Breeding with objective to develop coloured hybrids was done by Huang *et al.* (2001). Crossing and back crossing were done by using *Polianthes tuberosa* Single and Double and *P. howardii* which produced pink, reddish purple, purple orange and yellow flowers. They selected 9 bred lines of which 1 hybrid contained only carotenoids, 4 hybrids had only anthocyanins and other 4 had both carotenoids and anthocyanins in their petals.

In a study on the flower pigment constitutions of nine tuberose hybrids with pink, reddish purple, purple, orange and yellow flowers obtained from the crossings and back-crossings using *Polianthes tuberosa* 'Single' and 'Double' and *Polianthes howardii* were analysed (Huang *et al.*, 2001). Among the parents and hybrids, two white flowered cultivars of *P. tuberosa* had neither carotenoids nor anthocyanins, one hybrid had only carotenoids, four hybrids had only anthocyanins and an additional other four hybrids along with *P. howardii* had both carotenoids and anthocyanins in their petals. The main anthocyanidin in the petals of anthocyanin containing flowers is cyanidin with which some hybrids also contained delphinidin. Inducing of the anthocyanins and carotenoids from *P. howardii* into *P. tuberosa* can contribute to the extension of the diversity of flower colours in further tuberose breeding.

Breeding of coloured tuberose and cultural experiments was carried out in Taiwan (Huang *et al.*, 2002). Hybrid tuberoses of different colours were bred by crossing and backcrossing using *Polianthes tuberosa* 'Single' and 'Double' and *Polianthes howardii* and selected more than 50 hybrids. In addition, their pigment constituents were analyzed. Among the selected nine bred lines, one hybrid contained only carotenoids, four hybrids had only anthocyanins and the other four had both carotenoids and anthocyanins in their petals. The main anthocyanidin in the anthocyanin containing petals was cyanidin with some hybrids also containing delphinidin. Variation of pigment contents contributed the diversity of flower colours in hybrid tuberoses. One of the lines, '77A05', of which the main pigment was anthocyanin showed reddish-purple flowers when cultivated at 20ºC under natural light, whereas the colour was white at 30ºC. It was almost white at 25ºC with 45% shading of natural light, but pale with 25% shading. Further testing to find out colour content, cultivation of '77A05' in open fields was carried out at four different altitudes, 25, 75, 500 and 1,200 m in Taiwan. The anthocyanin content at 500 and 1,200 m was approximately two and three times higher than that at 25 m, respectively. The site selection for cultivation of coloured tuberose when the primary pigment is anthocyanin must be carefully considered.

Interspecific hybridization

Interspecific hybridization work is first reported by Howard (1977 and 1978). These hybrids were *P.* × *blissii* and *P.* × *bundrantii* (*P. tuberosa* × *P. howardii*). Some hybrids developed by Bundrant (1985) involved *P.* × *blissii* and *P.* × *bundrantii* which called Rainbow Tuberose and *P. tuberosa* × *P.* (*Manfreda*) *maculosa,* which is intermediate.

Shen *et al.* (2003) made crosses and back crosses among P. tuberosa 'Single', P. tuberosa 'Double', *P.* × *howardii* and *P.* × *blissii* were made. Several hybrids showing pink, reddish-purple, purple, orange and yellow flower colours were selected. However, the long spikes of flowers in these coloured hybrids were only suitable for cut flowers. To broaden the utilization of tuberose, further breeding programmes were conducted. Four hybrids showing a dwarf plant type were suitable for use as pot and/or bedding plants. The pedigrees and flowering performance of these selected dwarf tuberose hybrids described in detail.

Mutation

Mutation breeding mainly through gamma irradiation was followed in tuberose. In general, leaf chimaeras or variegation was observed by several workers. Irradiation of *P. tuberosa* bulbs with 0.5 krad gave rise to 1 mutant with larger flower size but higher dose (1.5 krad and above) caused abnormalities and found lethal (Patil *et al.*, 1975). Similar findings were also observed by Krasacchai (1992) and Estrada-Basaldua *et al.* (2011) at higher doses. Gamma rays were applied at 5, 10, 15, 20, 25 and 30 Gy at 7.12 Gy /minute. Doses of 10 Gy or above growth rate and no survival was recorded in 25 or 30 Gy irradiated bulbs. He observed only leaf chimaeras but no flower colour mutant was found. In a trial with 0.5-5 krad of X or gamma rays, 71-852 rad of fast neutrons or 0.5-2 krad of gamma rays combined with 284 rad of fast neutrons, it was observed that double flower type was more sensitive to all 3 radiation and yielded only 2 mutant, whereas single type produced 11 mutants with leaf colour variations. The fast neutron irradiation resulted in most mutant (Abraham and Desai, 1976). Datta (1993) reported that gamma radiation in single and double tuberose caused morphological abnormalities in leaves and chromosomal aberrations during root tip mitosis. Variegation in leaves was recorded after 2 krad treatment. Krishnan *et al.* (2003) treated bulbs of cvs. Single, Double, Shringar and Suvasini with gamma rays (5.0 to 25.0 Gy) and isolated dwarf mutants, high tiller mutant, non flowering mutants and compact inflorescence mutants and these characters were retained in VM_2 generation. They further observed high heritability and genetic gain for number of flowers/spike, spike length, flower diameter leaf length and leaf width in VM_2 generation. Shen *et al.* (2001) found that mutant plants showed a great deal of variations in flower colour, flower shape and floret number per flower spike when soaked in 1.0 and 1.5 mM sodium azide for 60, 90 and 120 minutes. They further suggested that use of sodium azide makes it possible to apply vegetative propagated crops for mutation breeding. Two gamma induced mutant var. Rajat Rekha and Swarna Rekha were developed at NBRI, Lucknow (Gupta *et al.*, 1984). Description of these varieties is given in Table 3.

Table 3: Mutant varieties of tuberose developed from NBRI, Lucknow.

S. No.	Vatiety	Type	Charecteristics
1.	Rajat Rekha	Induced	Single flowered type with silvery white streak along the middle of the leaf blade. Concrete content is 0.089%.
2.	Swarna Rekha	Induced	Double flowered type with golden yellow streak along the margins of leaf. Concrete content is 0.062%.

BIOTECHNOLOGY

Tissue culture

Different type of explants were used such as outer and inner scales of bulbs (Naz *et al.*, 2012), bulb segments (Rajasekaran *et al.*, 2000), bulb scale (Jyothi *et al.*, 2008). Lin *et al.* (2005) worked out to induce adventitious root formation of *P. tuberosa* cultivars Single and Double by using modified MS medium supplemented with IBA and NAA. They reported that the survival rate of *in vitro* rooted plantlets was higher than *in vitro* plantlets without roots.

A study was carried out to establish aseptic cultures and to standardize *in vitro* propagation protocol with petal segment and immature flower bud as explants. The explants pre-treated with carbendazim 0.1%+mancozeb 0.1% and 8-HQC (200 mg l^{-1}) for 2 1/2 h and surface sterilized with $HgCl_2$ (0.1%)+NaOCl (1.0%) for 8 min., respectively were found to be the best in getting aseptic culture. The best medium for shoot multiplication was MS medium supplemented with 6.0 mg/l BAP+0.5 mg/l NAA+0.7 mg/l 2,4-D+0.5 mg/l TDZ for both petals (4.0) and immature flower bud (4.33) enhanced shoot multiplication. Irrespective of explants, maximum roots were observed on half-strength MS medium+1.0 mg/l IBA. The *in vitro* regenerated micro-plantlets were successfully acclimatized in glass jar with polypropylene cap (Kadam *et al.*, 2010).

A protocol developed to propagate four genotypes of tuberose *viz.*, Phule Rajni, Shringar, Prajwal and Mexican Single through direct *in vitro* organogenesis from stem disc explants. Among the four cultivars Mexican Single was found significantly superior followed by Prajwal, Phule Rajni and Shringar for the most of the attributes. Regenerated plantlets were established successfully in the field after hardening. Phenotypically normal plants were regenerated from the stem disc explants (Gajbhiye *et al.*, 2011).

Somaclonal variation

Considerable somaclonal variation has been observed in floral traits *viz.*, inflorescence length, number of flower buds, flower size and colour when embryogenic callus differentiated into embryoids and plantlets after transfer medium (Gi and Tsay, 1989).

Somatic embryogenesis

The effects of plant growth regulators and culture conditions on the callus induction, somatic embryogenesis and regeneration of *P. tuberosa* were investigated *in vitro*. Addition of auxin to the MS basal medium increased callus formation from young petal explants of *P. tuberosa* 'Double'. More than 93% callus formation was achieved by culturing young petals on the basal medium containing 1, 2 and 4 mg 2,4-D/litre. Increasing the concentration of kinetin decreased callus formation. However, the combination of 2,4-D and kinetin in the basal medium induced the formation of yellowish, fine and smooth embryogenic calluses. Somatic embryos and secondary somatic embryos were obtained by culturing the embryogenic calluses on basal medium containing 0.01 and 1.0 mg thidiazuron (TDZ)/litre under low light intensity at 4 micro mol m^{-2} s^{-1}. Plant regeneration at a rate of 14% was obtained by subculturing somatic embryos on basal medium supplemented with 0.01 mg TDZ/litre for 4 weeks. Embryogenic calluses were derived from immature seed explants of *P. tuberosa* 'Chia-Nong Bright Jewel' cultured on basal medium containing 4 mg 2,4-D/litre and 1 mg kinetin/litre. Somatic embryogenesis and plant regeneration were obtained by subculturing embryogenic calluses on modified MS medium containing 0.1 mg NAA/litre and 4.5 mg BA/litre under low light intensity of 4 micro mol m^{-2} s^{-1} (Wu *et al.*, 2003).

Mutagenesis

The flower spikes of tuberose (*Polianthes tuberosa*), emerging to 15-20 cm long, were harvested and then soaked in 1.0 and 1.5 mM sodium azide for 60, 90 and 120 minutes. Young florets, 0.2-0.5 cm in length, were excised as explants and cultured on a modified MS medium for 5 weeks. The survival rate of floret explants treated with 1.0, 1.5 mM and 2.0 mM sodium azide for 90 minutes were 48.2%, 46.8% and 8.6%, respectively. However, all floret explants treated with 1.0 and 1.5 mM sodium for 120 minutes showed browning and/or etiolated. Flower spikes of tuberose were soaked in 1.0 and 1.5 mM sodium azide for 90 minutes. Young florets were then removed as explants and cultured on medium containing 0.1 mg/l NAA and 6.8 mg/l BA for inducing adventitious buds. *In vitro* regenerated plantlets were transferred to field soil. M_1 plants were found below 10% mutation rate. Mutant plants showed a great deal of variations in flower colour, flower shape and floret number per flower spike. Tuberose mutation breeding by sodium azide presented in this study makes it possible to apply vegetative propagated crops for mutation breeding model (Shen *et al.*, 2001).

Molecular breeding

Fifteen RAPD and twenty ISSR primers were used to assess genetic diversity among ten tuberose varieties (Shahaji, 2010). Both RAPD and ISSR primers revealed 53% and 73% polymorphism which generated 59 and 95 polymorphic markers respectively.

Polymorphic Information Content and Resolving Power for RAPD was found to be ranging from 0.35 – 0.46; 0.8 –3.6 respectively and that of ISSR was 0.36 – 0.49; 0.91 – 4.55 respectively. The dendrogram generated by UPGMA methodology using Jaccards co-efficient as distance matrix for RAPD and ISSR grouped the varieties into two major clusters and combined RAPD-ISSR cluster analysis grouped the 10 cultivars into 3 major clusters, which have been further grouped into sub-clusters based on their genetic relatedness/variation. The results of spatial arrangement of these 10 cultivars by Principal Component Analysis (PCA) using NTSys pc software were in congruent with earlier dendrogram analysis. Mantel's test indicated very good correlation with r = 0.86 for the combination of ISSR and RAPD-ISSR. To facilitate identification of tuberose cultivars, developed a cultivar identification diagram (CID) in which seven ISSR loci can differentiate all ten cultivars taken in the study. Molecular fingerprints for five IIHR released cultivars using 57 polymorphic loci generated by 11 ISSR primers. The sizes of these loci were ranging from 2.2kb to 252bp. These fingerprints can be used as standard reference source for quick cultivar identification.

Misra *et al.* (2010) used DNA marker technology for characterizing tuberose genotypes. Twenty tuberose genotypes comprising of both single- and double-petal types collected from different parts of India were selected for analysis. The DNA extraction and RAPD conditions were standardized. For RAPD analysis 20 mg DNA template, 2.5 mM $MgCl_2$ and 1U *Taq*DNA polymerase was found effective. Out of 80 random decamer primers tested, 17 were selected based on high level of polymorphism. On the basis of primer resolving power and marker index, RAPD primers OPC-13 and OPD-12 were identified as efficient primers for diversity analysis of tuberose. A total of 157 RAPD bands were generated by the 17 random decamer primers. The selected primers proved effective for DNA profiling in addition to diversity analysis. Genotypes Guwahati Double and Swarna Rekha showed good morphological similarity revealing high similarity coefficient (0.90) suggesting them to be very closely related. The RAPD analysis also confirmed their relatedness as they grouped in the same cluster. The suitability of this technique for genotyping and diversity analysis was also established. Genotypes, Vaibhav and Pune Single were found to have least pair-wise similarity although they had a greater morphological similarity with each other.

Twenty tuberose genotypes comprising of both single- and double-petal types collected from different parts of India were selected for analysis. The DNA extraction and RAPD conditions were standardized. For RAPD analysis 20 ng DNA template, 2.5 mM $MgCl_2$ and 1U *Taq*DNA polymerase was found effective. Out of 80 random decamer primers tested, 17 were selected based on high level of polymorphism. On the basis of primer resolving power and marker index, RAPD primers OPC-13 and OPD12 were identified as efficient primers for diversity analysis of tuberose. A total of 157 RAPD bands were generated by the 17 random decamer primers. The selected primers proved effective for DNA profiling in addition to diversity analysis. Genotypes Guwahati

Double and Swarna Rekha showed good morphological similarity revealing high similarity coefficient (0.90) suggesting them to be very closely related. The RAPD analysis also confirmed their relatedness as they grouped in the same cluster. The suitability of this technique for genotyping and diversity analysis was also established. Genotypes, Vaibhav and Pune Single were found to have least pair-wise similarity although they had a greater morphological similarity with each other (Sarkar *et al.*, 2010).

Genetic diversity evaluation among seven *Polianthes tuberosa* genotypes comprising of both single and double-petal types, of which collected from different regions of Iran, was determined using RAPD technique. The extraction of total genomic DNA was carried out using MBST kit. A total of 14 RAPD primers produced 102 bands, 60 (58.82%) of which were polymorphic and 42 (41.17%) monomorphic. Molecular analysis of seven populations based on the UPGMA method assigned the genotypes in two main groups. In this grouping, P.tub5 as double petal genotype in Dezful region showed the most difference to other groups especially to P.tub1 as single petal type in Varamin region, P.tub7 as double petal type in Tehran region was placed in separate sub group and the rest of accessions have not shown any difference. However, RAPD analysis was found to be a valuable diagnostic tool to evaluate genetic diversity of *P. tuberosa* genotypes in Iran (Ahmad Majd *et al.*, 2013).

SUMMARY

Tuberose is a commercial flower of India but due to less genetic variability and self incompatibility few varieties have been evolved. Some selections have been made on the basis of performance of single or double types in a particular agroclimatic condition. Two mutant varieties i.e. Rajat Rekha and Swarna Rekha are ornamentally important due to less flower production. In India, species *P. tuberosa* is commercially exploited whereas in other countries, different other species *P. howardii* and *P. blissii* were used in improvement programme. Our genetic pool is weak, which needs to be strengthened by introduction of important species. During 1993-94, two varieties namely Shringar and Suvasini were released and found with higher yield than their parents. Later during 2001-2002, two more cvs. Prajwal and Vaibhav were evolved from IIHR, Bengaluru and claimed that these are higher yielder than existing single and double varieties. Recently cv. Arka Nirantra was also developed by IIHR, Bengaluru. But there is great scope for improvement of tuberose and some high yielding varieties for loose and cut flower as well as for extraction of concrete can be explored by adopting classical and molecular breeding approaches.

REFERENCES

Abraham, V. and Desai, B.M. 1976. Radiation induced mutants in tuberose. *Indian J. Gen. Pl. Breeding*, **36**(3): 328-331.

Ahmad, M., Raheleh, A., Sayeh, J.M., Fahimeh, S. and Fariba, S. 2013. Genetic divergence analysis among seven populations of tuberose (*Polianthes tuberosa* L.) by RAPD- PCR in Iran. *Ann. Biological Res*, **4**(6): 65-69.

Ayyanger, R.K. 1963. Cytological basis of sterility in *Polianthes tuberosa* Linn. *Proce*: 50th Indian Sci. Cong. Part III, Section VI (Bot.), pp. 444-445.

Bailey, L.H. 1976. Hortus Third: a Concise Directory of Plants Cultivated in the United States and Canada, Macmillan Pub., New York, 893p.

Banker, G.J. and Mukhopadhyay, A. 1980. Varietal trial in tuberose (*Polianthes tuberosa* L.). *South Indian Hort.*, **28**(4): 150-151.

Bundrant, L.A. 1985. *Polianthes tuberosa* and its hybrids. *Herbertia*, **41**: 55-60.

Datta, S.K. 1993. Possibilities of directive mutation in ornamentals. *Proc*: SABRAO 7th International Cong. and WSAA Symposium. Academia Sinica, Nankang, Taiwan, **35**: 313-316.

Estrada-Basaldua, J.A., Pedraza-Santos, M.E., Cruz-Torres, E., de la Martinez-Palacios, A., Saenz-Romero, C. and Morales-Garcia, J.L. 2011. Effect of ^{60}Co gamma rays in tuberose (*Polianthes tuberosa* L.). *Revista Mexicana de Ciencias Agricolas*, **3**: 445-458.

Gajbhiye, S.S. Tripathi, M.K. Shankar, M.V. Singh, M. Baghel, B.S. Tiwari, S. 2011. Direct shoot organogenesis from cultured stem disc explants of tuberose (*Polianthes tuberosa* Linn.). *Interl. J. Agril. Tech.*, **7**(3): 695-709.

Gaurav, S.B., Katwate, S.M., Singh, B.R., Kahade, D.S., Dhane, A.V. and Sabale, R.N. 2005. Quantitative genetic studies in tuberose (*Polianthes tuberosa* L.). *J. Orna. Horti.*, **8**(2): 125-127.

Gi, H.S. and Tsay, H.S. 1989. Anther and somaclonal variation of tuberose (*Polianthes tuberosa* L.). *J. Agril. Res. China*, **38**(3): 346-352.

Gupta, M.N., Laxmi, V. and Dutta, S.K. 1984. Cytomorphological studies on control and gamma induced mutants of tuberose. *J. Nuclear Agri. Bio.*, **13**(4): 140-141.

Howard, T.M. 1977. *Polianthes × blissii* Worsley (Hybrid between *Polianthes germiniflora × P. tuberosa). Plant Life*, **33**: 82-84.

Howard, T.M. 1978. *Polianthes* hybrids. *Plant Life*, **34**: 126-129.

Huang, K.L., Miyajima, I., Okubo, H., Shen, T.M. and Huang, T.S. 2002. Breeding of coloured tuberose (*Polianthes*) and cultural experiments in Taiwan, *Acta Horticulturae*, **570**: 367-371.

Huang, K.L., Miyajima, I., Okubo, H., Shen, T.M. and Huang, T.S. 2001. Flower colours and pigments in hybrid tuberose (*Polianthes*). *Scientia Horticulturae*, **88**(3): 235-241.

Irulappan, I., Doraipandian, A. and Muthuswamy, S. 1980, Varietal evaluation in tuberose (*Polianthes tuberosa* Linn.). *Abst*: Nat. Sem. Production Tech. Commercial Flower Crops, TNAU, Coimbatore, pp. 69-70.

Joshi, A.C. and Pantulu, J.V. 1941. A morphological and cytological study of *Polianthes tuberosa* Linn. *J. Indian Bot. Soc.*, **20**: 37-71.

Jyothi, R., Singh, A.K. and Singh, K.P. 2008. *In vitro* propagation studies in tuberose (*Polianthes tuberosa* Linn.). *J. Orna. Hort.*, **11**(3): 196-201.

Kadam, G.B., Singh, K.P., Singh, A.K. and Jyothi, R. 2010. *In vitro* regeneration of tuberose through petals and immature flower buds. *Indian J. Hort.*, **67**(1): 76-79.

Katwate, S.M., Gaurav, S.B., Patil, M.T., Singh, B.R. and Chaudhary, S.M. 2006. Phule Rajani: A unique hybrid of tuberose. *Abst.*. Nat. Symp. on Ornamental Bulbous Crop, Indian Society of Ornamental Horticulture, SVBPUA&T, Meerut, p. 150.

Krasacchai, A. 1992. Effect of gamma radiation on tuberose (*Polianthes tuberosa*). *Kasetsart J. Natural Sci.*, **26**(1): 6-11.

Krishnan, A.G., Geetha, C.K., Rajeevan, P.K., Valsalakumari, P.K. and Saifudeen, N. 2003. Induced mutation in tuberose (*Polianthes tuberosa* Linn.) by gamma rays. *Proc*: Nat. Symp. on Recent Advances in Indian Flori., Kerala Agril. Uni., Trichur, pp. 255-261.

Laxmi, V., Gupta, M.N. and Datta, S.K. 1984. Cytomorphological studies on control and gamma induced mutants in tuberose. *J. Nuclear Agric Bio*., **13**: 140-141.

Lin, TzunSheang and Shen, TsaiMu. 2005. Rooting of *in vitro* plantlets in *Polianthes tuberosa* L. *J. Chinese Soc. Hortil. Sci.*, **51**(1): 51-62.

Misra, J.S., Bhat, R.L., Singh, K.V.A. and Singh, S.K. 2010. Genetic diversity analysis in tuberose (*Pollianthes tuberosa*) genotypes through Randomly Amplified Polymorphic DNA. *Indian J. Genet. Plant Breeding*, **70**(2):182-88.

Naz, Shagufta., Aslam, Farah., Ilyas, Saiqa., Shahzadi, Kiran. and Tariq, Amina. 2012. *In vitro* propagation of tuberose (*Polianthes tuberosa*). *J. Medicinal Plants res.*, **6**(24): 4107-4112.

Patil, A.V., Kale, P.N. and Kaulgud, S.N. 1975. Some observations of gamma irradiations on growth and flowering of tuberose bulbs. *Current Sci*., **44**(11): 406-407.

Radhakrishna, K.N., Janakiram, T. and Srinivas, M. 2003. Genetic studies in tuberose (*Polianthes tuberosa* L.). *Abst*: Nat. Symp. on Recent Advances in Indian Flori., Kerala Agril. Uni., Trichur, p. 86.

Radhakrishna, K.N., Srinivas., M. and Janakiram, T. 2003a. Performance of new promising genotypes of tuberose. *Abst*: Nat. Symp. on Recent Advances in Indian Flori., Kerala Agril. Uni., Trichur, p. 86.

Rajasekaran, V., Haripriya, K. and Arumugam, Shakila. 2000. *In vitro* propagation of tuberose (*Polianthes tuberosa* L.). Spices and Aromatic Plants: challenges and opportunities in the new century. *In*: Centennial Conference on Spices and Aromatic Plants, Calicut, Kerala, India.

Rose, J.N. 1905. Studies of Mexican and Central American plants. No. 3C ont. U.S. Nat.Herbarium.

Sarkar, Jayoti, Misra, R.L., Bhat, K.V., Singh, Anita, Singh, S.K. 2010. Genetic diversity analysis in tuberose (*Pollianthes tuberosa*) genotypes through Randomly Amplified Polymorphic DNA. *Indian J. Gene. Pl. Breed.*, **70**(2): 182-188.

Schiva, T. and Lanteri, S. 1983. Cytological observations on two cultivars of *Polianthes tuberosa. Annali dell' Instituto Sperimentale per la Floricultura,* **14**: 1-18.

Seetharamu, G.K. 1993. Hybridization studies in tuberose (*Polianthes tuberosa*). M.Sc. (Ag.) Thesis, UAS, Bangaluru.

Seetharamu, G.K., Bhat, R.N. and Rajanna, K.M. 2000. Studies on pollen viability, pollen germination and seed germination in tuberose hybrid and cultivars. *South Indian Hort*., **48**: 78-82.

Shen, T.M., Huang, K.K. and Huang, R.S. 1987. Study of tuberose hybridization. *Acta Horticulturae*, **205**: 71-74.

Shen, T.M., Huang, K.L. and Chang, P.S. 2001. Preliminary study on mutagenic effect of sodium azide on *Polianthes tuberosa*. *J. Chinese Soc. Hortil. Sci.*, **47**(2): 129-135.

Shen, T.M., Huang, K.L., Shen, R.S. and Du B.S. 2003. Breeding of dwarf tuberose (*Polianthes tuberosa* L.). *Acta Horticulturae,* **624**: 73-76.

Singh, A.K. 2006. Flower Crops: Cultivation and Management. Pub. New India Publishing Agency, New Delhi. pp. 358-370.

Singh. K.P. 2003. Tuberose. *In*: P.K. Rajeevan, K.P. Singh, P.K. Valsalakumari and C.K. Geetha (ed.). Bulbous Ornamentals, ISOH, New Delhi, India, pp. 73-85.

Srinivas, M. 2003. New tuberose hybrid Vaibhav. *Abst*: Nat. Symp. on Recent Advance in Indian Flori., 12-14 Nov. 2003, Kerala Agril. Uni., Trichur, p.82.

Srivastva, H.C. and Shridhar, C.J. 2002. Tuberose (*Polianthes tuberosa* L.) hybrids for high concrete yield and coloured floral buds. *Plant Archives,* **2**(2): 333-335.

Uma, S. 1990. Studies on causes of non seed setting in some ornamental bulbous crops. Ph.D.Thesis, UAS, Bengaluru.

Uma, S. and Gowda, J.V.N. 2000. Self-incompatibility studies in *Polianthes tuberosa L. Research on Crops*, **1**(3): 418-420.

Van Schaik, C.E., Van der Toorn, C., De Jeu, M.J., Raemakers, C.J.J.M. and Visser, R.G.F. 2000. Towards genetic transformation in the monocot *Alstroemeria* L. *Euphytica*, **115**: 17-26.

Wu, C.F., Hsu, S.T. and Shen, T.M. 2003. Studies on somatic embryogenesis and regeneration of *Polianthes tuberosa* L. *J. Chinese Soc. Hortil. Sci.*, **49**(2): 183-197.

Section IV

Cut and Pot Flower Crops

Chapter – 23

Cyclamen

Cyclamen (*Cyclamen persicum* Mill.), belonging to family Primulaceae, is cultivated throughout the Temperate Zone. Many people have enjoyed the charming plants as potted and/or garden plants. The cyclamen is one of the most famous and important commercial ornamental plants in many countries. It is commonly grown for its flowers, both outdoors and indoors in pots. It is used on a terrace, in flower boxes, in hanging baskets or in beddings. Several species, particularly *Cyclamen hederifolium*, are hardy and can be grown outdoors in mild climates such as Northwest Europe and the Pacific Northwest of North America. The cyclamen commonly sold by florists is *C. persicum*, which is frost-tender. Selected cyclamen cultivars can have white, bright pink, red or purple flowers. The use as cut flower is not important part of cyclamen production (Grey-Wilson, 2002), but not rare in Europe, especially in Germany. Some cultivars for cut flower production were bred in Europe, but cut flower cyclamen production is not popular in USA and Japan. Its popularity as pot plant led to many studies on the group, including cytology (Bennett and Grimshaw, 1991), hybridization (Gielly *et al*., 2001 and Grey-Wilson, 2002) and phenology (Debussche *et al*., 2004).

ORIGIN AND DISTRIBUTION

Cyclamen originated from extreme eastern Mediterranean region and naturally grows in countries such as Palestine, Syria, Israel, Greece, Turkey, Tunisia or even Mediterranean islands: Rhodes, Cyprus and Sicily (Grey-Wilson, 2002). Cyclamen was introduced in Europe at the end 16th century, beginning of the 17th century, passing first through France. Somewhat neglected in the following century, the cyclamen became fashionable again in France in the 1900s. But since 1860, the first variants of the botanical or "wild" cyclamen appeared in England, then in Germany, with much bigger flowers and with varying colour shades.

According to a report by Doorenbos (1950), *C. persicum* had already appeared in France in 1665, and the Persian cyclamen was firmly established in the Netherlands in 1739. Thus, *C. persicum* was introduced from the Eastern-Mediterranean region into Western Europe at least in the 17th century, but the exact year of the first introduction is not known. Cyclamen began to be a popular commercial plant in the nineteenth century. Doorenbos (1950) indicated three reasons for the development of cyclamen in the 19th century as follows: 1. Improvement of cultivation methods, 2. Improvement of economic circumstances (the period of great prosperity) and 3. Success of the breeding. The cultivation period was reduced from 3-4 years to 15-18 months by the latter part of the century (Grey-Wilson, 2002). The reduction of cultivation period was established by eliminating the resting period in summer, and this is regarded as the most important factor that made cyclamen popular (Doorenbos, 1950).

Since the early 1800s most cyclamen breeding has focused primarily on the florist cyclamen, *Cyclamen persicum*. Breeding has greatly increased the overall plant size and broadened the flower colour and form. Breeders also removed the summer dormancy period of the florist cyclamen, which greatly decreased the production time but made the plant less adaptable to outdoor conditions. Most of the work with florist cyclamens occurred in England, The Netherlands, Germany, and Japan. Modern *Cyclamen persicum* hybrids have many interesting traits including large size, ruffled petals, double flowers, picotee flowers, and a wide range of colours including the breakthrough hues, true red and pale yellow. The breeding of other cyclamen species pales in comparison to *Cyclamen persicum*. There are many nurseries and collectors who make hand crosses and evaluate seedlings, but there are few inbred breeding lines and breakthroughs in size, form, or flower colour. Only a handful of interspecies crosses have ever succeeded. Modern breeders have started using techniques such as colchicine-induced polyploidy and embryo rescue to make crosses that were previously impossible.

Breeders are focusing on creating hybrids of cold-tolerant and cold-sensitive species to create winter-hardy hybrids that have the traits of the frost sensitive species. Many desirable characteristics, especially new flower colours and disease resistance, are not available in *C. persicum*. Therefore, new breeding techniques such as interspecific hybridization and gene transformation should be the key to make cyclamen breeding programme rapidly progress, and more research will be desired.

SPECIES

Grey-Wilson (2002) reported that the genus *Cyclamen* consists of 22 species, but presently there are 26 known species reported. These are *C. abchasicum, C. africanum, C. alpinum* (*C. trochopteranthum*), *C. atkinsii* (*C. coum* × *C. vernum*), *C. balearicum, C. cilicium, C. colchicum, C. confusum, C. coum, C. creticum, C. cyprium, C. elegans,*

C. graecum, C. hederifolium, C. intaminatum, C. libanoticum, C. mirabile, C. parviflorum, C. persicum, C. pseudibericum, C. purpurascens, C. repandum, C. rohlfsianum, C. rhodium, C. somalanse and *C. vernum.*

Subgenus classification in the *Cyclamen* genus was made by Anderberg (1994) and subsequently by Grey-Wilson (2002), but the classifications did not coincide in some points. A molecular phylogenetic study was done by Anderberg *et al.* (2000) of the genus *Cyclamen* by DNA sequence. They suggested four monophyletic subgroups from the sequence data and morphological data in the genus. This study may be helpful in future to clarify relation among species.

Taskin *et al.* (2012) focused on study of genetic diversity of natural *Cyclamen alpinum* (formerly *C. trochopteranthum*). Habitat fragmentation, environmental degradation, and overharvesting of tubers have exerted pressure on native populations of this valuable ornamental species. Although the entire *Cyclamen* genus is in the Convention on International Trade in Endangered Species of Wild Fauna and Flora-Appendix II (CITES II), no species has yet been red-listed. Estimating the level and distribution of genetic variation in populations of rare and endemic species is important for conserving genetic diversity within a species in the context of well-developed conservation strategies. In this study, random amplified polymorphic DNA (RAPD) analysis was employed to assess the genetic diversity within and among 6 natural *C. alpinum* populations in the south and southwest of Turkey. A total of 190 loci were determined by using 15 polymorphic primers. Total genetic variation (H_T) was 0.27+or-0.02. A high proportion of this variation, 0.16+or-0.01 (59.26%), was due to within-population genetic variation (H_S). The genetic differentiation coefficient (G_{ST}) was 0.41, and the level of gene flow (*Nm*) within a generation among the 6 populations studied was 0.73. As a result of these findings, they proposed *in situ* combined with *ex situ* conservation of all *C. alpinum* populations. In addition, their results support prior recommendations to add *C. alpinum* to the International Union for Conservation of Nature and Natural Resources (IUCN) Red List under the critically endangered (CR) category.

FLOWER MORPHOLOGY AND BIOLOGY

In the autumn when most flowering plants are entering winter dormancy, cyclamen plants are just starting their season of growth and bloom. The cool wet seasons are when cyclamen flowers and foliage start to grow from the tuber. Growing through the fall, Winter, and into spring they enter a brief dormant period in the dry heat of summer when they release their seeds. All the *Cyclamen* species grow in this basic pattern with variations. Some will bloom in late summer and fall, others in winter, or spring. Growing several species of *Cyclamen* would make it possible to have blooming cyclamen in every season of the year.

Each flower is produced on a stem coming from a growing point on the tuber. Flowers have 5 petals, bent outwards or up, sometimes twisted, and connected at the base into a cup, and five sepals behind the cup. Petal shape varies depending on species, and sometimes within the same species. Cyclamen is a day-neutral plant that flowers irrespective of day length. However, Heo *et al.* (2003) reported that flowering can be controlled by manipulating the day length. Flowering time may be any month of the year, depending on the species. *Cyclamen hederifolium* and *Cyclamen purpurascens* bloom in summer and autumn, *Cyclamen persicum* and *Cyclamen coum* bloom in winter, and *Cyclamen repandum* blooms in spring. Initiation of flowering is associated with the development of 10-13 leaves on the plant (Stephens and Widmer, 1976; Sundberg, 1981a and Sundberg, 1981b and Widmer and Lyons, 1985). Temperature during early development has been found significant in determining leaf number at flowering (Karlsson and Werner, 2001). The hermaphroditic flowers are solitary and pendant at the tip of an erect leafless pedicel emerging from the floral trunk. They do not produce nectar. The corolla is divided into five reflexed lobes and the five sepals are fused at the base. Anthers are positioned inside the corolla and attached to the base of the corolla by a very short filament.

Mode of pollination

Although entomophilous type of pollination is seen in cyclamen, there is a rarity of buzz pollination which is performed by Anthophora bee only (Affre and Thompson, 1997, Affre *et al.*, 1995 and Tzachor *et al.*, 2006).

Pollination of *C. persicum* was studied in two wild populations in Israel. Buzz-pollination proved to be extremely rare, and performed by a large *Anthophora* bee only. The most frequent pollinators were various unspecialized species of thrips (Thysanoptera) and hoverflies (Syrphidae). In the Winter-flowering populations the commonest visitor was a small primitive moth, *Micropteris elegans* (Micropterigidae, Lepidoptera). These moths feed on pollen, copulate and oviposit within the flowers. From the rarity of buzz-pollination it is concluded that the genus *Cyclamen* co-evolved with large bees capable of buzz-pollination, but lost its original pollinators for unknown historical reasons. The vacant niche was then open to various unspecialized pollen consumers such as thrips, hoverflies and small solitary bees. While these insects are not specific to *C. persicum* and seem to play a minor role only, the moth strictly relies upon *Cyclamen* and seems to be the most efficient pollinator Schwartz-Tzachor *et al.* (2006).

Anther dehiscence and pollination

Anther dehiscence is introrse and pollen is mainly liberated through an apical pore. The superior ovary contains numerous ovules. Anther dehiscence and pollen viability are pre-requisites for successful fertilization. Schwartz-Tzachor *et al.* (2008) found that, in cyclamen pollen is viable up to 16 days. After fertilization the corolla falls and the erect pedicel coils from the apex downwards bringing the capsule to the substrate surface where the brownish seeds are liberated and dispersed by ants (Affre *et al.*, 1995).

Old flowers of *Cyclamen* generate few or no seeds. To understand the pollination problems of *Cyclamen* Reinhardt *et al.* (2007) investigated the general anatomy of the stigma and the style of *Cyclamen persicum* by scanning electron microscopy at different stages of floral maturity. This investigation confirms that there is a hollow style. Against data commonly found in the literature, they presented evidence of pollen germination and tube growth that show the stigma is not outside the style but inside it. Furthermore the maturation process of the style during the flowering time indicates a mechanism by which the stigma becomes shut off through closure at the terminal aperture of the style. At 3 to 5 days after anthesis there was the beginning closure of the style which was nearly completed at 21 days. The substance which leads to the closure is still unknown. The closure of the hollow style is a probable cause for failure of seed set in flowers not pollinated early in anthesis.

Seed set

It is natural that high-quality plants can produce a larger number of high quality seeds. Cultivation temperature also influences seed production. Although higher temperatures (more than 20° C) hastened the maturation of seeds, the number of seeds and seed set percentage were reduced (Takamura *et al.*, 1996). Number of seeds produced per plant varies with genotype, age, size of flower and pollination time. Ewald and Schwenkel (1999) reported that the highest seed yield was obtained, when the flowers were pollinated directly at the beginning of anthesis.

Fruit

The flower stem coils or bends when the fruit begins to form. The stems of *Cyclamen hederifolium* and *Cyclamen coum* coil starting at the end, *Cyclamen persicum* arches downwards but does not curl, *Cyclamen rohlfsianum* coils starting near the tuber, and *Cyclamen graecum* coils both directions, starting at the middle. The fruit is a round pod that opens by several flaps or teeth at maturity and contains numerous sticky seeds, brown at maturity. Natural seed dispersal is by ants (myrmecochory), which eat the sticky covering and then discard the seeds.

GENETICS

Wild forms of *C. persicum* are consistently diploid with a chromosome number of 2n=48. However, tetraploid (2n=4x=96) or aneuploid (2n=90, 92, 94 and 95) plants have been produced through the selection of natural mutants of *C. persicum* wild form and their varietal crossings.

In a study haploids, diploids and tetraploids of *Cyclamen persicum* (2n=x=24, 2x=48 and 4x=96), *C. graecum* (2n=84) and their interspecific hybrids (2n=45, 3x=66, 4x=90) were analyzed cytologically. The results indicate that *C. graecum* is an autotetraploid, the hybrid (2n=4x=90) is an amphidiploid, the microspore-derived plant (2n=45) is a polyhaploid and the hybrid (2n=3x=66) is a sesquidiploid. Based on studies of meiosis, the ploidy levels and the origin of *C. graecum* and *C. persicum* are discussed in comparison with several species in the genus *Cyclamen* (Ishizaka, 2003).

Inheritance of flower colour

The mode of inheritance of flower colour in reciprocal crosses between diploid cyanic and yellow-flowered *Cyclamen persicum* cultivars was investigated. No yellow-flowered F_1 progeny was obtained in the crosses. All the F_1 progenies were cyanic. In the F_2 progenies, yellow-flowered plants were found. The segregation ratio of flower colours in the F_2 progenies fitted the expected Mendelian ratio, showing the yellow-flowered phenotype was controlled by a recessive gene (Takamura *et al*., 2000).

Inheritance of anthocyanin in the petals in the crosses between diploid *Cyclamen persicum* cultivars (Cochineal, Golden Boy, Largo, Mirabelle, Strauss and Wase Murasaki) and *C. purpurascens* was investigated. In every cross combination, including the crosses by using acyanic cyclamen cultivars without the eye, all the F_1 progenies had a purple or reddish-purple eye in the petals, showing the eye expression was a dominant characteristic. All the F_1 progenies contained anthocyanins in the petals, whereas some progenies in the crosses by using a yellow-flowered cultivar contained very small amount of anthocyanin. The main flower pigment of *C. purpurascens* was malvidin 3,5-diglucoside. All the F_1 progenies also contained 3,5-diglucoside type anthocyanins, such as malvidin 3,5-diglucoside, peonidin 3,5-diglucoside and cyanidin 3,5-diglucoside, in the petals as the main pigments even in the crosses by using acyanic cultivars or cyanic cultivars containing malvidin 3-glucoside or peonidin 3-neohesperidoside in the petals as the main pigment. These results suggest that the expression of 5-glucosyltransferase in anthocyanin synthesis in the petals was dominant even in the crosses between cyclamen cultivars and *C. purpurascens* (Takamura *et al*., 2005).

In another study inheritance of anthocyanins in petals in interspecific crosses between diploid cyclamen (*C. persicum*) and *C. hederifolium* was investigated by Takamura and Aizawa (2007). The anthocyanin synthesis in the petal and expression of 5-glucosyltransferase during anthocyanin synthesis were dominant characteristics in the interspecific crosses. These findings should be useful for the genetic improvement of cyclamen using the interspecific crosses.

To identify the theoretic foundation for maintaining and improving high quality varieties and breeding new F_1 hybrids, flower colour segregation in the offspring in both self-pollination and cross- pollination of varieties of different colours was investigated. The heredity of *Cyclamen persicum* colour follows both quantitative and qualitative hereditary regularity. All genes can be classified as two groups of the throat-gene and coronal-gene occupying different spots on the chromosome. The coloured gene is the dominant gene, while the white gene is the recessive one. The gene groups and their quantity determine the colour of *Cyclamen persicum.* And sometimes the throat- gene and the coronal-gene take a recombination, forming a new type of flower (Lin *et al.*, 2006).

Song and Oh (2010) conducted experiment to characterize the inheritance of petal colour, flower-eye colour, flower shape, leaf variegation pattern, and leaf-end shape in *Cyclamen persicum*. The segregation of F_2 and backcross populations from crosses between inbred lines was tested for Mendelian inheritance mode. From four different combinations of crosses, it was found that the petal colour was controlled by a single incomplete dominant gene. Whereas, other characters controlled by single complete dominant genes which is given in Table 1.

Table 1: Genetics of leaf, flower colour and shape characters in cyclamen.

S. No.	Character	Number of genes	Type of gene action
1.	Petal colour	Monogenic	Incomplete dominant
2.	Flower-eye colour	Monogenic	Complete dominant
3.	Flower shape	Monogenic	Complete dominant
4.	Leaf variegation pattern	Monogenic	Complete dominant
5.	Leaf-end shape	Monogenic	Complete dominant

Heritability and genetic advance

Heritability and genetic advance are important selection parameters. Heritability estimates along with genetic advance are normally more helpful in predicting the gain under selection than heritability estimates alone. However, it is not necessary that a character showing high heritability will also exhibit high genetic advance (Johnson *et al.*, 1955). Variability studies in cyclamen revealed that heritability estimates were high for plant height, days to opening of first flower, days to full flowering, vase life, duration from bud opening to petal fall, display life and leaf area (Sharma *et al.*, 1998). Ewald and Schwenkel, (1999) reported that there was high variability for number of seeds per plant, 1000 seed weight and average period of ripening of pods.

BREEDING OBJECTIVES

i. Attractive plants

ii. Robustness

iii. Modified flower structure (poly petalous flowers) and floriferousness

iv. Intensity of flower colour

v. Novel flower coloured varieties

vi. Cultivars suitable for convenient cultivation

vii. Resistance to pests and diseases

viii. Resistance to heat

BREEDING METHODS

A large number of cyclamen cultivars have been bred until now. The breeding of *C. persicum* cultivars began seriously in the mid 19th century through selection and back-crossing, and resulting in flowers several times larger than the wild species. Much of this work took place in England and the Netherlands and many named forms were produced. In the 1960's a breeding programme at Wye College, University of London, crossed existing large flowered plants with wild forms, producing smaller plants with more refined, scented flowers and attractively marked foliage. These became known as the 'Wye College Hybrids'. Today, *Cyclamen persicum* cultivars are mass produced as pot plants particularly in the Netherlands and Germany, using F_1 hybrid seed strains.

In cyclamen, flowering time varied greatly species to species. Not only that, it varied at same place even planted at same time of different species. Some species produced attractive flower but not fragrant, but there are also strong fragrant species i.e. *C. purpurascens*. There are some barriers to produce interspecific hybrids because of cross-incompatibility (Legro, 1959). Interspecific hybrids from the crosses between commercial cyclamen (*C. persicum*) and some wild species were, however, produced through ovule (or ovary) culture (Ishizaka and Uematsu, 1990, 1992, 1995a, Ishizaka, 1996, Ewald, 1996 and Sibusawa and Ogawa, 1997). Several efforts have been made on interspecific hybridization and described well below in this chapter.

Introduction

The success of any breeding programme depends on the availability of germplasm, and introduction is the easiest way to enrich germ collection by the breeder. Cyclamen is the most popular ornamental crop in USA and several European countries, but it is very strange to India. In a first, a new variety of flower namely, Cyclamen, botanically *Cyclamen persicum*, has been introduced in the Government Botanical Garden (GBG), Ooty from Europe. Ideal Osena Pink and EC 118224 were introduced at Shimla (Rana and Singh, 1995). The cyclamen flowers of different colours have been planted in over 10,000 pots and displayed inside a glasshouse located at the upper level of the GBG, Ooty.

Some varieties and species of *Cyclamen* were introduced in India and maintained at NBPGR, Shimla and Mandsaur, M.P. (Table 2).

Table 2: Cyclamen varieties introduced from Germany and UK.

S. No.	Country	Introduced variety/species
1.	Germany	*C. persicum* cv. Engett Dianka, Engett Anneli, Engett Steffi, Engett Elbeland, Engett Rosalina, Engett Elbeporie, Engett Amelia, Engett Elbflarenz, Engett Elbrubin, Engett Elbroschen, Engett Elbfener, Engett Lauretta, Engett Rosoniarie, Engett Drigitta, Engett Kati, Engett Gotha, Engett Erfuri, Engett Jena.
2.	UK	*C. hederifolium*

Selection

Selection procedure is of two types, i.e. pure line selection (self pollinated crops) and mass selection (cross pollinated crops). Since cyclamen is a cross pollinated crop, mass selection is used for improvement and maintenance of varieties. This method is effective in improving the qualitative characters like flower colour. In the mid 18th century English and Dutch breeders used mass selection followed by back cross to obtain bigger flowers, which eventually led to the current mass production of cyclamen in different colours, double flowers, frilly petals and picotee forms.

The feasibility of pure lines for *Cyclamen persicum* with 30 inbred seedling lines showing different growth and flowering characteristics was investigated. The inbred seedling lines were obtained from selfing crosses 5 (S_5) or 4 (S_4) generation using several *Cyclamen persicum* cultivars. Eleven lines including 'Cy-99-20^5' flowered earlier than 330 days after sowing, however seven lines including 'Cy-99-17^5' were flowered later than 340 days after sowing. The number of leaves of 27 lines including 'Cy-99-6^5' was more than 30. The petal colour of four lines including 'Cy-99-6^5' was white, seven lines including 'Cy-99-43^5' were red, four lines including 'Cy-99-20^5' were violet, seven lines including 'Cy-99-17^5' were pink, four lines including 'Cy-99-28^5' were purple, and four lines including 'Cy-99-27^5' were salmon. Flower diameter ranged from 3.5 cm to 11.7 cm, 27 lines including 'Cy-99-6^5' were larger than 6.0 cm. The number of flowers of all lines was more than 9, eight lines including 'Cy-99-20^5' were more than 20. All the lines of various growth and flowering characteristics would be very promising to use as breeding materials for F_1 hybrids of *Cyclamen persicum* (Song, 2008).

Hybridization

Emasculation and pollination: Emasculation method comprises of cutting off petals of a selected bud or tearing off the petals by hand, peeling off all stamens by using a pair of hybridizing tweezers when the calyxes are exposed, and bagging the bud. The modern emasculation method comprises the following steps of: gently pinching the base of the flower stalk with the thumb, forefinger and middle finger of one hand when the large-flowered bud grows to 2.5-3 cm, the middle-flowered bud grows to

2-2.5 cm, the small-flowered bud grows to 1-1.5 cm, the petals of the bud are tightly packed and the colour just emerges, pinching the base of the bud with the same three fingers of the other hand, keeping the hand pinching the base of the flower stalk still, and rotating while quickly dragging the hand pinching the base of the bud outwards to peel off the petals and the stamens in good condition. The hand pollination is carried on with the help of a very fine brush.

Combining ability: An experiment was conducted to determine combining ability and correlation among plant height, plant width, leaf length, leaf number, floret height, floret diameter, number of florets and peduncle length in 10 F_1 crosses made from the partial five-parent diallel cross in *Cyclamen persicum*. The plant height showed high positive correlations with number of florets and peduncle length. The mean squares of general combining ability (GCA) and specific combining ability (SCA) were highly significant for all the investigated characteristics. Variance component values of SCA were greater than those of GCA for all parameters, implying preponderance of non-additive gene actions for these characters. Relatively high GCA effects were obtained from the line of D for plant height and plant width, and the lines of A and D for floret diameter, peduncle length, and number of florets. Crosses of A × B, A × C, A × D, B × E and D × E exhibited high SCA effects on number of florets. The broad sense heritability was generally high compared with narrow sense heritability. The number of florets showed particularly high effects in narrow sense heritability (Song, 2007).

Zheng *et al.* (2008) reported that more than 180 cultivars of *Cyclamen persicum* and F_1 been introduced in China from overseas since 2000 for study of domestication, selection, and industrialization cultivation. Individuals with excellent properties and good homogeneity were selected, and purified for several generations.

In a study based on character supplementary, male and female parents were selected for crossing and it was done. Observation on characters of their descendants demonstrated that 18 hybridized combinations were suitable for breeding, among which, 8 excellent hybridized combinations had better characters than their parents. Technical schedule was proposed for *C. persicum* industrialization production (Zheng *et al.*, 2008).

Varieties

There are two natural varieties and several named forms, which differ in flowering time and predominant petal colour.

i. *C. persicum* var. *persicum* (winter- and spring-flowering — all of range)
ia *C. persicum* var. *persicum* f. *persicum* (white to pale pink)
ib *C. persicum* var. *persicum* f. *albidum* (pure white)
ic *C. persicum* var. *persicum* f. *roseum* (rose-pink)
id *C. persicum* var. *persicum* f. *puniceum* (red to carmine)
ii *C. persicum* var. *autumnale* (autumn-flowering)

Hybrids

Cyclamen do not hybridize in the wild. In cultivation, they cross only with some difficulty because most species have different chromosome numbers. There are only 7 registered hybrids which are as follows.

i. Cyclamen × drydeniae (*Cyclamen coum* × *Cyclamen alpinum*). They generally have twisted petals and are fertile.

ii. Cyclamen × hildebrandii (*Cyclamen africanum* × *Cyclamen hederifolium*) is the first registered hybrid, created in 1955. The resulting plants resemble *C. hederifolium* on steroids, with a more muted leaf pattern.

iii. Cyclamen × meiklei (*Cyclamen creticum* × *Cyclamen repandum*).

iv. Cyclamen × saundersiae (*Cyclamen balearicum* × *Cyclamen rhodium* ssp. *peloponnesiacum*) has variable silver spotted and marbled leaves, and usually pink flowers.

v. Cyclamen × schwarzii (*Cyclamen libanoticum* × *Cyclamen pseudibericum*)

vi. Cyclamen × wellensiekii (*Cyclamen cyprium* × *Cyclamen libanoticum*) is named after Professor Dr. Ir. S.J. Wellensiek of the Agricultural University of Wageningen, The Netherlands, who first produced the hybrid in 1969. It flowers over a long period during winter and sets fertile seed. The flowering period is greatly extended and lasts as long as 6 months.

vii. Cyclamen × whiteae (*Cyclamen graecum* × *Cyclamen hederifolium*). The flowers of this cross are unusually stubby.

Interspecific hybridization

Interspecific hybridization is an effective method in several flowering plants to evolve a variety of various desirable traits. In different wild *Cyclamen* species have desirable characteristics i.e. cold or heat tolerance, disease resistance, and strong and sweet fragrance and different flowering time from that in commercially important cultivars of *C. persicum*. Hence, interspecific hybridization between *C. persicum* and other *Cyclamen* species should be useful for cyclamen breeding.

Ishizaka (1996) made interspecific crosses using *C. persicum* diploid cv. Pure White (2n = 48) and *C. persicum* tetraploid cv. Salmon Scarlet (2n = 96) as female parents and wild forms of *C. graecum* as male parents. The hybrids (2n = 66) derived from *C. persicum* diploid × C. *graecum* failed to yield viable seeds by self-pollination, although they showed some pollen fertility. In contrast, the hybrids (2n = 90) derived from *C. persicum* tetraploid × *C. graecum* showed high pollen fertility and yielded viable seeds by self-pollination. Furthermore, they were resistant to disease caused by *Fusarium oxysporum* f. sp. *cyclaminis*, *Erwinia herbicola* pv. *cyclamenae* and *Pseudomonas marginalis* pv. *marginalis*.

The effects of genotypes on the production of interspecific hybrids in cyclamen cultivars (*C. persicum*) × *C. purpurascens* crosses were investigated to establish efficient breeding protocols in the crosses. In the tetraploid cyclamen cultivars × *C. purpurascens* crosses, few fruits and no interspecific hybrids were obtained. Although more than half of cross combinations in the diploid cyclamen cultivars × *C. purpurascens* crosses produced interspecific hybrids, efficiency of the hybrid production depended on the genotypes of seed parents in the crosses. The results of the present study indicated that the ploidy level and genotypes of seed parents influenced hybrid production in the cyclamen cultivars × *C. purpurascens* crosses. The results also suggest that the selection of desirable genotypes as seed parents in the crosses may be effective for the efficient production of the interspecific hybrids (Takamura *et al*., 2002).

Mutation

Genetic variation is necessary in any plant breeding programme for crop improvement. Induced mutations are highly effective to enhance natural genetic resources and have successfully assisted in developing improved and new cultivars in cyclamen. Breider (1959) for the first time irradiated the young tubers of cyclamen and developed the mutant 'rotundifolia'. Some of the selected traits of the mutant plants are flower colour, flower morphology, plant architecture, compact growth, flower type, and variegated leaves. Among mutagens, gamma rays have been commonly used effectively for mutation induction. Recently, heavy-ion beam (HIB) has attracted increasing interest in floriculture for mutation induction. Increasing evidence suggests that ion beams efficiently induce mutations. Mutation-assisted breeding (MAB) together with biotechnology can contribute greatly for genetic improvement of cyclamen plants.

It is a generally accepted idea that ionizing radiation induces mutations randomly throughout the genome. The ability to efficiently induce mutations for a particular phenotypic trait, i.e., to control of the direction of mutations, would be a major breakthrough. However, no effective methods have been established to control the direction of mutations in practical plant breeding. Although the underlying mechanism was not clear, they hypothesized that the genes involved in flower colour synthesis were relatively mutable where those genes were highly expressed.

The advantage of mutation breeding lies in its ability to improve one or a few desirable traits without altering the remaining characteristics. Description of some mutants developed in cyclamen is given in Table 3.

Table 3: Mutants in cyclamen (Hase *et al.*, 2012)

S. No.	Mutant	Parent	Character
1.	ion3	Uruwashi-no-kaori', UR	Salmon red colour
2.	ion246	Uruwashi-no-kaori', UR	Salmon red colour
3.	ion257	Golden boy'× *C. purpurascens*	Yellow colour
4.	gra50	*C. graecum*	White colour
5.	KOpw	Kokou-nokaori', KO	White colour
6.	KMrp	Kaori-no-mai', KM	Red purple
7.	KMMv3	Kaori-no-mai', KM	Red purple

BIOTECHNOLOGY

Tissue culture

Cross-incompatibility is the major genetic barrier for conventional interspecific hybridization between *C. persicum* cultivars and other species (Legro, 1959). However, this trait conferred by embryo abortion after fertilization was overcome by ovule culture and interspecific hybrids have been produced in the following combinations: *C. persicum* cultivars (diploid and/or tetraploid) × *C. repandum*; × *C. hederifolium*; × *C. purpurascens*; × *C. libanoticum*; × *C. rohlfsianum* (Ishizaka and Uematsu, 1990, 1992, 1995 and Kobayashi and Ogawa, 1994). Furthermore, the sterility of the hybrids caused by meiotic abnormalities in the hybrids, such as lack of sufficient chromosome pairing and formation of micronuclei (Ishizaka, 1994; Ishizaka, 1997 and Ewald *et al.*, 2000) also leads to failure of interspecific hybridization. Sterility can be overcome through amphidiploidization by exposing the hybrid ovules to colchicine *in vitro* and by ovule culture (Ishizaka and Uematsu, 1994 and 1996).

A rapid micropropagation system for *Cyclamen persicum* was developed by the leaves of *Cyclamen persicum* as explants cultured on MS medium supplemented with different concentrations of 6-benzyladenine (6-BA) (2.5, 3.0, 3.5 and 4.0 mg/litre) in combination with different concentrations of 1-naphthaleneacetic acid (NAA) (0.3, 0.4, 0.5 and 0.6 mg/litre) or 2,4-dichlorophenoxyacetic acid (2,4-D) (0.3, 0.4, 0.5 and 0.6 mg/litre). The optimum induction medium was MS medium supplemented with 3.0 mg 6-BA/litre and 0.5 mg NAA/litre and the callus induction rate reached 80%. The solid MS medium supplemented with 3.5 mg 6-BA/litre and 0.5 mg NAA/litre was suitable for the multiplication of shoots and proliferation could reach more than 10 times. The optimum rooting medium consisted of MS supplemented by 3.0 mg NAA/litre and 100% plantlets with roots were obtained (Liu and Chen, 2012).

Somatic embryogenesis

Somatic embryogenesis in *Cyclamen persicum* was first reported in 1984 and has potential applications for propagation and breeding of this economically important ornamental crop. This *in vitro* regeneration system can be used for vegetative propagation of parental lines of F_1 hybrids and elite plants, production of artificial seeds, *Agrobacterium tumefaciens*-mediated genetic transformation, long-term cryo-preservation, protoplast to plant regeneration and somatic hybridization. Somatic embryogenesis was shown to be a powerful propagation system for some *C. persicum* genotypes, but commercial application in large scale so far is hindered by several limitations, i.e., asynchronous development, malformations or secondary somatic embryogenesis. However, recent molecular approaches by transcriptomic and proteomic analyses were undertaken in order to better understand and control this *in vitro* regeneration system and to overcome these problems (Winkelmann *et al*., 2011).

On somatic embryogenesis a lot of works has been done. Findings of some workers have been given here.

In *Cyclamen*, a protocol for somatic embryogenesis, applicable to a large number of cultivars, was established to produce artificial seeds. The use of 2,4-D to induce embryogenic competence is essential in this particular case (Laura *et al*., 2003).

Embryogenesis in seven *Cyclamen* species (*C. cilicium, C. coum, C. graecum, C. hederifolium, C. persicum, C. purpurascens* and *C. rohlfsianum*) was investigated using petioles and leaves of mature plants as explants. Somatic embryos were induced in *C. cilicium, C. coum, C. graecum* and *C. rohlfsianum*; unipolar tuber-like bodies were formed in *C. purpurascens* through culture shaking; and adventitious roots were produced in *C. persicum*. Somatic embryogenesis and organogenesis were not observed from the calluses of *C. hederifolium*. In *C. graecum* and *C. coum*, somatic embryos were induced in liquid media, and the embryos of *C. graecum* developed like microtubers. In *C. purpurascens*, unipolar tuber-like bodies were formed from compact calluses, but did not grow into plantlets. Induction of somatic embryogenesis was successful in 4 *Cyclamen* species. Results of this study demonstrate that micropropagation through somatic embryogenesis is possible in some *Cyclamen* species (Furukawa *et al*., 2001).

The effect of initial explants type (leaf and petiole) on callus growth and somatic embryo were studied. Somatic embryos were cultured on modified MS medium. Also embryogenic callus was induced on several media containing 2,4-dichlorophenoxyacetic acid (2,4-D) and kinetin (Kin). The greatest amount of callus was observed in two explants on media containing 1 mgL^{-1} 2,4-D+0.5 mgL^{-1} Kin. The optimal responses of somatic embryogenesis were recorded on media enriched with 4 mgL^{-1} 2,4-D, 0.1 mgL^{-1} Kin and leaf explants. These results were obtained from adult plants (Jalali *et al*., 2010).

Winkelmann *et al*. (2010) studied regeneration system for the first time with 30 genotypes which are valued as cut flowers. The aim of this study was to analyse the efficiency of the micropropagation through somatic embryogenesis, to investigate the genetic fidelity and performance of the regenerants and to estimate production costs of young plants. The genotype significantly influenced the frequency of callus induction that ranged from 8 to 80% and also the frequency of calluses with differentiation of somatic embryos that ranged from 0 to 92%. Only three of the 30 genotypes were not able to regenerate somatic embryos, while plants were obtained from the remaining 27 genotypes. In total, 2,783 regenerants were transferred to the greenhouse, 2,003 (72%) of which could be acclimatized. Some genotypes were extremely uniform in terms of flower colour and flower shapes, whereas in others somaclonal variation was detected.

Winkelmann *et al.* (2011) further comparing somatic embryos to their zygotic counterparts regarding their proteomes. Protein separation by two dimensional isoelectric focusing - sodium do-decyl sulfate polyacrylamide gel electrophoresis led to a resolution of about 1000 protein spots per gel, of which the first 253 were identified by mass spectrometry. Most were found to be involved in glycolysis/gluconeogenesis and stress response pathways. A proteome reference map of zygotic embryos will be publicly released soon and may serve as a basis for further investigations and improvements of somatic embryogenesis.

In a study, a high frequency cyclic secondary somatic embryogenesis system was established for *Cyclamen persicum* (You *et al*., 2011). Moreover, the influence of primary somatic embryos (PSEs) at different stages of development, number of passages, abscisic acid (ABA), and sucrose concentrations on secondary somatic embryogenesis were investigated. Primary somatic embryos, at different stages of development, were incubated on an induction medium consisting of half-strength MS, 2 mg/l 2,4-D, and 0.2 mg/l 6-benzyladenine (BA). These initial PSE cultures along with calli were transferred to a plant growth regulator (PGR)-free medium to promote secondary somatic embryo (SSE) development. Embryogenic calli were induced at a frequency >90% from all PSE cultures. Both morphology and size of PSEs influenced embryogenic competence. Large-sized globular embryos (GEs) yielded the highest number of SSEs followed by small- and large-sized torpedo-stage embryos (TEs). Passages of induced calli obtained from large-sized GEs had no effects on frequency of secondary embryogenesis; however, they significantly influenced SSE production. Calli from the first passage exhibited the highest competence for secondary embryogenesis, 100% frequency of SSEs, producing 3,306 and 1,296 SSEs per gram fresh weight callus and embryo, respectively. Incubating embryogenic calli derived from large-sized primary GEs in the first passage on a medium containing 30 g/l sucrose, and supplemented with either 0.5 or 1.0 mg/l abscisic acid (ABA) increased the frequency of TEs, consequently enhancing normal embryo development of SSEs. Transfer of such embryogenic calli to a PGR-free medium containing 60 g/l sucrose was optimal for SSE induction, producing 3,204 SSEs per gram callus and 8% TEs.

Interspecific somatic hybrids

By applying polyethylene glycol (PEG)-mediated protoplast fusion, the first somatic hybrids were obtained between *Cyclamen persicum* ($2n=2x=48$) and *C. coum* ($2n=2x=30$)-two species that cannot be combined by cross breeding. Heterofusion was detected by double fluorescent staining with fluorescein diacetate and scopoletin. The highest heterofusion frequencies (of about 5%) resulted from a protocol using a protoplast density of 1×10^6/mL and 40% PEG. The DNA content of *C. coum* was estimated for the first time by propidium iodide staining to be 14.7 pg/2C and was 4.6 times higher than that of *C. persicum*. Among 200 *in vitro* plantlets regenerated from fusion experiments, most resembled the *C. coum* parent, whereas only 5 plants showed typical *C. persicum* phenotypes and 46 had a deviating morphology. By flow cytometry, six putative somatic hybrids were identified. A species-specific DNA marker was developed based on the sequence of the 5.8S gene in the ribosomal nuclear DNA and its flanking internal transcribed spacers ITS1 and ITS2. The hybrid status of only one plant could be verified by the species-specific DNA marker as well as sequencing of the amplification product. RAPD markers turned out to be less informative and applicable for hybrid identification, as no clear additivity of the parental marker bands was observed. Chromosome counting in root tips of four hybrids revealed the presence of the 30 *C. coum* chromosomes and 2-41 additional ones indicating elimination of *C. persicum* chromosomes (Prange *et al.*, 2012). Interspecific hybrids between *C. persicum* and other *Cyclamen* species are given in Table 4.

Table 4: Interspecific hybrids between *C. persicum* and other species obtained by embryo rescue *in vitro*.

S. No.	Female parent	Male parent	Reference
1.	*C. persicum*	*C. hederifolium*	Ishizaka and Uematsu (1990)
2.	*C. persicum*	*C. purpurascens*	Ishizaka and Uematsu (1992)
3.	*C. persicum*	*C. africanum*	Ishizaka and Uematsu (1995a), Ewald (1996)
4.	*C. persicum*	*C. hederifolium*	Ishizaka (1996)
5.	*C. persicum*	*C. purpurascens*	Sibusawa and Ogawa (1997)
6.	*C. persicum*	*C. repandum*	Sibusawa and Ogawa (1997)
7.	*C. persicum*	*C. africanum*	Takamura *et al.* (2002)

Mutagenesis

Research findings focusing on the development of a logical and efficient mutation breeding by taking advantage of both ion beam irradiation and genomic information to obtain new plants with characteristics that could not be obtained from original cultivars, cross breeding or gene recombination discussed. Flower colour mutations on selected plants, such as glittering carnation, red creeping petunia and blue-purple and crimson fragrant cyclamen (*Cyclamen persicum* x *C. purpurascens*) were also discussed (Tanaka, 2012).

Ion beams have been used as a mutagen to improve the efficiency of plant mutation breeding. Mutation breeding is sometimes perceived as a random process. Hase *et al.* (2012) described recent progress in developing a more efficiency mutagenesis technique using ion beam irradiation combined with sucrose pretreatment or subsequent re-irradiation. To shorten the time required for breeding new cultivars of cyclamen, they identified anthocyanin biosynthesis genes and examined the effectiveness of PCR screening of irradiated deletion-mutant candidates at early growth stages. This research is a step toward more efficient and controlled mutation breeding using ion beams.

Nakayama *et al.* (2012) analyzed flower colour mutants of cyclamen (*Cyclamen* spp.) obtained by ion-beam irradiation with an idea that a comprehensive analysis of anthocyanin and its biosynthetically related compounds, such as flavonols and cinnamic acid derivatives, is necessary in order to understand flower colour expression mechanism. They discussed mechanisms for flower colour mutation and deduce the following ideas: anthocyanin and its biosynthetically related compounds are cooperatively and compensatively regulated; multiple factors are often concerned in the expression of the same colour phenotypes; and changes in chemical structure of a pigment induces new properties that generate novel phenotypes.

Polyploidization

Polyploidization is an important breeding technique in cyclamen. Hybrid sterility can be overcome through amphidiploidization by chromosome doubling. The chromosome number can effectively be doubled using colchicine. In cyclamen mitotic and meiotic polyploidization are possible and reported by several workers (Ishizaka and Uematsu, 1994, 1995b, Takamura and Miyajima, 1994, 1996a, 1996c and Takamura *et al.*, 1998). Jensen (1974) has reported several ways of using colchicine for chromosome doubling. To overcome from sterility of interspecific hybrids, Ishizaka and Uematsu (1994 and 1995b) induced amphidiploids of interspecific hybrids. Findings indicated that these plants are amphidiploids (2n = 82, AABB) between *C. persicum* and *C. purpurascens*. Three and 2 viable seeds were derived by the conventional crosses of diploid *C. persicum* × the amphidiploid and the amphidiploid × *C. purpurascens*, respectively. Flowering plants that developed from the seeds of diploid *C. persicum* × the amphidiploid were barely fertile and had 65 somatic chromosomes (2n = 65, AAB), whereas those that developed from the seeds of the amphidiploid × *C. purpurascens* were barely fertile and had 58 somatic chromosomes (2n = 58, ABB). The somatic chromosomes indicated that these plants are probably sesquidiploids between the amphidiploid and either *C. persicum* or *C. purpurascens*. In another experiment treatment through *in vitro* organogenesis was given to induce tetraploid yellow flowered cyclamen from the diploids by Takamura and Miyajima (1996c) and Takamura *et al.* (1998). It was interesting that the induced tetraploid yellow-flowered plants had larger and deeper yellow petals than the diploid relatives. It is indicating that polyploidization would be one of the methods for improving flower quality. Meiotic polyploidization by using unreduced pollen was also used for breeding of tetraploid yellow flowered cyclamen (Takamura and Miyajima, 1994). Hence, in future improved varieties of cyclamen would be evolved through polyploidization.

Haploid production

Establishment of pure lines in cyclamen is difficult, hence haploid production is useful for breeding programmes in cyclamen. Dihaploid must be homozygous, and these plants should be suitable for producing uniform plants as pure lines or parents of F_1 cultivars. Ishizaka and Uematsu (1993) have produced pollen derived haploids in cyclamen through anther culture. Haploid production technique has a great possibility to improve to improve the cyclamen.

Molecular breeding

Flower colour and plant form are important traits that influence consumer selection of ornamental plant cultivars. Molecular breeding using recombinant DNA techniques offers opportunities to produce ornamental cultivars displaying novel flower colours and form beyond the range possible with traditional breeding. Molecular breeding technologies provide opportunities to metabolically engineer the well-characterized flavonoid biosynthetic pathway for altered anthocyanin profile and hence the colour of the flower.

RAPD analysis was performed on embryogenic *Cyclamen* (cv. Halios) callus to study, in this propagation system, the variability related to the different 2,4-D concentrations and exposure times. The primary callus was induced on medium containing 2,4-D (2 mg litre^{-1}) and 2iP (0.8 mg litre^{-1}) from immature ovaries of one mother plant. From each ovary, one cell line was established. The embryogenic callus was then transferred in presence of increasing 2,4-D concentrations. After several subcultures, half of the calluses were transferred onto hormone-free medium to detect the embryogenic potential related to the 2,4-D levels. The second half was used for DNA analysis. Twenty-five random primers produced 161 reproducible bands. Six primers produced band patterns that differed among the samples. The four cell lines used in the experiments were divided into two groups showing different phenotypic and molecular behaviour. In one cell line, a different RAPD pattern was observed in the callus grown at the higher 2,4-D concentration. RAPD analysis appeared to be a useful tool for the early screening of genetic variation during callus culture and allowed precocious selection of the most stable embryogenic cell lines (Laura *et al.*, 2003).

Genetic engineering

Aida *et al.* (1999) documented on the transformation of cyclamen and they obtained transgenic plants by an *Agrobacterium tumefaciens*–mediated transformation. The gene transformations were made through organogenesis from explants of the etiolated petiole sections of aseptic cyclamen seedlings. The success should be a great step for progressing cyclamen breeding.

Several factors that affected genetic transformation of cyclamen cultivar 'Scarlet 1011' were examined and mediated by *Agrobacterium tumefaciens*. The suitable conditions of transformation contained 50 mg Kan/l, 500 mg Cef/l, 2 d pre-culture, 0.4-0.6 OD_{600}, 5-8 min infection, and 3 d co-culture. Thus, a simple and efficient genetic transformation system was developed (Zhou *et al*., 2009).

Boase *et al.* (2010) reported previously on a genetic transformation system for cyclamen. In a study they aimed to change pigment profiles and flower colours in cyclamen through the suppression of flavonoid 3',5'-hydroxylase, an enzyme in the flavonoid pathway that plays a determining role in the colour of anthocyanin pigments. In the study a full-length cDNA putatively identified as a F3'5'H (CpF3'5'H) was isolated from cyclamen flower tissue. Amino acid and phylogeny analyses indicated the CpF3'5'H encodes a F3'5'H enzyme. Two cultivars of minicyclamen were transformed *via Agrobacterium tumefaciens* with an antisense CpF3'5'H construct. Flowers of the transgenic lines showed modified colour and this correlated positively with the loss of endogenous F3'5'H transcript. Changes in observed colour were confirmed by colourimeter measurements, with an overall loss in intensity of colour (C) in the transgenic lines and a shift in hue from purple to red/pink in one cultivar. HPLC analysis showed that delphinidin derived pigment levels were reduced in transgenic lines relative to control lines while the percentage of cyanidin derived pigments increased. Total anthocyanin concentration was reduced up to 80% in some transgenic lines and a smaller increase in flavonol concentration was recorded. Differences were also seen in the ratio of flavonol types that accumulated. They claimed that this is the first report of genetic modification of the anthocyanin pathway in the commercially important species cyclamen. The effects of suppressing a key enzyme, F3'5'H, were wide ranging, extending from anthocyanins to other branches of the flavonoid pathway. The results illustrate the complexity involved in modifying a biosynthetic pathway with multiple branch points to different end products and provides important information for future flower colour modification experiments in cyclamen.

Isolation and characterization of the fragrant cyclamen *O-methyltransferase* involved in flower colouration was done by Akita *et al.* (2011) in cultivars KM and KMrp. They isolated and compared *OMT* genes expressed in KM and KMrp petals. RT-PCR analysis revealed that *CkmOMT2* was expressed in the petals of KM but not in KMrp. Three additional *CkmOMT*s with identical sequences were expressed in petals of both KM and KMrp. Genomic PCR analysis revealed that *CkmOMT2* was not amplified from the KMrp genome, indicating that ion-beam irradiation caused a loss of the entire *CkmOMT2* region in KMrp. *In vitro* enzyme assay demonstrated that CkmOMT2 catalyzes the 3' or 3',5' *O*-methylation of the B-ring of anthocyanin substrates. These results suggest that CkmOMT2 is functional for anthocyanin methylation, and defective expression of *CkmOMT2* is responsible for changes in anthocyanin composition and flower colouration in KMrp.

In order to identify the anthocyanin-related GST in cyclamen, degenerate PCR was performed using total RNA from immature young petals. Four candidates of GSTs (CkmGST1 to CkmGST4) were isolated. Phylogenetic analysis indicated that CkmGST3 was closely related to PhAN9, an anthocyanin-related GST of petunia, and this clade was clustered with other known anthocyanin-related GSTs. Expression analysis at different developmental stages of petals revealed that CkmGST3 was strongly expressed in paler pigmented petals than in fully pigmented petals, in contrast to the constitutive expression of the other three candidates during petal development. This expression pattern of CkmGST3 was correlated with those of other anthocyanin biosynthetic genes such as CkmF3'5'H and CkmDFR2. Molecular complementation of Arabidopsis tt19, a knockout mutant of an anthocyanin-related GST gene, demonstrated that CkmGST3 could complement the anthocyanin-less phenotype of tt19. Transgenic plants that expressed the other three CkmGSTs did not show anthocyanin accumulation. These results indicate CkmGST3 functions in anthocyanin accumulation in cyclamen (Kitamura *et al.*, 2012).

Transgenic Long-term expression of the *gusA* reporter gene in transgenic cyclamen

A protocol to produce transgenic cyclamen from etiolated hypocotyls of *Cyclamen persicum* has been developed. Precultured etiolated hypocotyls segments of cultivar Sierra Rose were infected with two disarmed *Agrobacterium tumefaciens* strains, LBA4404 or EHA105 containing the binary vectors pMOG410 or pART 27-10. Both binary vectors contain in their T-DNA region a nos-*npt*II gene and a 35S-*gus*A-intron. Transformed cells were selected with 50 mg l^{-1} kanamycin and shoots were regenerated *via* adventitious organogenesis in the presence of 1 mg l^{-1} 6-benzylaminopurine (BAP) and 1 mg l^{-1} indole-3-acetic acid (IAA). One hundred and twelve transgenic shoots produced from three experiments were rooted in the presence of 100 mg l^{-1} kanamycin and 0.1 mg l^{-1} alpha -naphthaleneacetic acid (NAA). Fifty-two of these plantlets (47%) were positive for beta -glucuronidase (GUS) enzyme activity, when leaf laminae were assayed histochemically. An enzyme-linked immunosorbent assay (ELISA) for the *npt*II protein confirmed expression of the selectable marker gene and validated the kanamycin rooting assay. Northern blot analyses of leaf laminae and petals of seven pMOG410 transformants and of leaf laminae of five pART27-10 transformants showed *gus*A RNA transcripts only in the transformants. Southern blot analysis of left border integration patterns of seven pMOG410 transformants confirmed the independent nature of the transformants. Petals, scapes, leaf laminae, petioles and corms from 14 independent transformants growing in a containment greenhouse were still GUS positive 33 months after explants were infected. Some seedlings from these transformants were GUS positive showing transgene transmission to sexual offspring and meiotic stability. These results indicate stable, long-term expression of a non-selected transgene can be obtained in transgenic cyclamen (Boase *et al*., 2002).

Production of yellow flower cyclamen

The yellow flower cyclamen was produced by *Agrobacterium tumefaciens*-mediated transformation with chalcone reductase gene. Kanamycin at 100 mg/litre or hygromycin at 10 mg/litre was suitable for selection after genetic transformation. Regeneration ability differed among the cultivars. Rondo-Pink, Piccolo and Npm showed good regeneration after transformation. Two transformants were obtained from Npm, and they showed pale yellow colour on petals from bud to early blooming stages (Mizukami *et al*., 2004).

Transgenic for abiotic stress (resistance to high temperature)

Chen *et al*. (2008) carried out a study to transform MnSOD gene to the Cyclamen plants, and to evaluate its effect on the endurance to high temperature stress and environmental stress. Leaves as acceptor, MnSOD gene was incorporated into cyclamen through *Agrobacterium tumefaciens*-mediated transformation. Resistant cyclamen plants were confirmed by molecular biological identification; SOD activity, MDA content, membrane permeability and protein content of high temperature stress at 36°C of transformed cyclamen and non-transformed cyclamen were measured. They found out that 35 resistant plants were regenerated, 3 of which were positive by GUS activity, PCR assay and Southern hybridization analysis. This result indicated that the MnSOD gene had integrated into cyclamen. The experimental data showed that the level of SOD activity in transformed cyclamen was higher than that in non-transformed plantlets; the membrane permeability and MDA content in transformed cyclamen rose more slowly than non-transformed plantlets; the protein content in transformed cyclamen was higher than that in non-transformed plantlets. Hence, it was concluded that the resistance of the transformed plantlets to high temperature stress was higher than non-transformed control.

Thermotolerant cyclamen

Reduced levels of trienoic fatty acids (TAs) in chloroplast membranes induce thermotolerance in several plant species, but the underlying mechanisms remain unclear. TA peroxidation in plant cell membranes generates cytotoxic, TA-derived compounds containing alpha, beta -unsaturated carbonyl groups. The relationship between low TA levels and the amounts of cytotoxic TA-derived compounds was examined using thermotolerant transgenic cyclamen (*Cyclamen persicum* Mill.) with low TA contents. Changes in the levels of the cytotoxic TA-derived acrolein (ACR), methyl vinyl ketone (MVK), (E)-2-hexenal, 4-hydroxy-2-nonenal, and malondialdehyde were analysed in the leaf tissues of wild-type (WT) and thermotolerant transgenic cyclamen under heat stress. Levels of ACR and MVK in the WT increased in parallel with the occurrence of heat-induced tissue damage, whereas no such changes were observed in the thermotolerant transgenic lines. Furthermore, exogenous ACR and MVK infiltrated into leaves to concentrations similar to those observed in heat-stressed WT leaves caused similar disease symptoms. These results suggest that thermotolerance in transgenic cyclamen depends on reduced production rates of ACR and MVK under heat stress, due to the low level of TAs in these plants (Kai *et al.*, 2012).

REFERENCES

Affre, L., Thompson, J.D. and Debussche, M. 1995. The reproductive biology of the Mediterranean endemic *Cyclamen balearicum* Willk. (Primulaceae). *Bot. J. Linn.Soc.*, **118**: 309-330.

Aida, R., Hirose, Y., Kishimoto, S. and Shibata, M. 1999. *Agrobacterium tumefaciens*-mediated transformation of *Cyclamen persicum* Mill. *Pl. Sci.*, **148**(1): 1-7.

Akita, Y., Kitamura, S., Hase, Y., Narumi, I., Ishizaka, H., Kondo, E., Kameari, N., Nakayama, M., Tanikawa, N., Morita, Y. and Tanaka, A. 2011. Isolation and characterization of the fragrant cyclamen O-methyltransferase involved in flower colouration. *Planta*, **234**(6): 1127-1136.

Anderberg, A.A. 1994. Phylogeny and subgeneric classification of *Cyclamen* L. (Primulaceae), *Kew Bulletin*, **49**: 455-467.

Anderberg, A.A., Trift, I. and Kallersjo, M. 2000. Phylogeny of *Cyclamen* L. (Primulaceae): Evidence from morphology and sequence data from the internal transcribed spacers of nuclear ribosomal DNA. *Plant Systematics Evolution*, **220**: 147-160.

Anderson, R.G. and Widmer, R.E. 1975. Improving vigour expression of cyclamen seed germination with surface disinfestations and gibberellin treatment. *J. Amer. Soc. Hort. Sci.*, **100**: 597-601.

Ando, T. and Murasaki, K. 1983. *In vitro* propagation of cyclamen by the use of etiolated petioles, *Tech. Bull. Fac. Hort. Chiba Univ.*, **32**: 1-5.

Bennett, S.T. and Grimshaw, J.M. 1991. Cytological studies in *Cyclamen* subg. *Cyclamen* (Primulaceae). *Plant Systematics Evolution*, **176**: 135-143.

Boase, M.R., Marshall, G.B., Peters, T.A. and Bendall, M.J. 2002. Long-term expression of the gusA reporter gene in transgenic cyclamen produced from etiolated hypocotyl explants. *Plant cell Tissue Organ Culture*, **70**(1): 27-39.

Boase, M.R., Lewis, D.H., Davies, K.M., Marshall, G.B., Patel, Deepa, Schwinn, K.E. and Deroles, S.C. 2010. Isolation and antisense suppression of flavonoid 3', 5'-hydroxylase modifies flower pigments and colour in cyclamen. *BMC Plant Biology*, **10**: 107-113.

Breider, H. 1959. Rontgeninduzierte mutationen bei *Vitis malus* and *Cyclamen*. *Bayer. Landwirtsch. Jahrb.*, **36**(4): 396-401.

Chen, L., Zhou, L.X., Ma, F.W., Liang, D. and Hua, Z.R. 2008. Transgenic cyclamen plants expressing MnSOD gene and its resistance to high temperature stress. *J. Northwest A. & F. Univ.*, **3**: 28.

Corbineau, F., Neveur, N. and Come, D. 1989. Seed germination and seedling development in *Cyclamen persicum*. *Annals Botany.* **63**: 87-96.

Debussche, M., Garnier, E. and Thompson, J.D. 2004. Exploring the causes of variation in phenology and morphology in Mediterranean geophytes: a genus- wide study of Cyclamen. *Bot. J. Linnean Soc.*, **145**: 469-484.

Doorenbos, J. 1950. The history of the Persian cyclamen, *Meded. Landbouwhogeschool, Wageningen*, **50**: 33-59.

Ewald, A. 1996. Interspecific hybridization between *Cyclamen persicum* Mill. and *C. purpurascens* Mill., *Plant Breeding*, **115**: 162-166.

Ewald, A. and Schwenkel, H.G. 1999. The influence of flower age and genotype on the seed set of *Cyclamen persicum* 'Sylvia'. *Gartenbauwissenschaft*, **64**(2): 49-53.

Ewald, A., Lepper, L., Lippold, R. and Schwenkel, H.G. 2000. Sexual reproduction of interspecific hybrids between *Cyclamen persicum* Mill. and *C. purpurascens* Mill., *Gartenbauwissenschaft*, **65**: 162-169.

Furukawa, K., Kakihara, F. and Kato, M. 2001. Somatic embryogenesis in seven wild Cyclamen species. *J. Society High Technology Agriculture*, **13**(4): 270-278.

Geier, T., Kohlenbach, H.W. and Reuther, G. 1990. Cyclamen. *In*: P.V. Ammirato, D.A. Evans, W.R. Sharp and Y.P.S. Bajaj (eds.), Handbook of Plant cell Culture. vol. 5, Ornamental Species, Macmillan Publishing Co., New York. pp. 352-374.

Gielly, L., Debussche, M. and Thompson, J.D. 2001. Geographic isolation and evolution of Mediterranean endemic Cyclamen: insights from chloroplast trnL (UAA) intron sequence variation. *Plant Systematics Evolution*, **230**: 75-88.

Grey-Wilson, C. 2002. Cyclamen: a Guide for Gardeners, Horticulturalists and Botanists: London: Batsford. Timber Press, Portland.

Hase,Y., Akita,Y., Kitamura, S., Narumi, I. and Tanaka, A. 2012. Development of an efficient mutagenesis technique using ion beams: toward more controlled mutation breeding. *Plant Biotechnology*, **29**(3): 193-200.

Hase,Yoshihiro, Akita, Yusuke, Kitamura, Satoshi, Narumi, Issay, Tanaka, Atsushi 2012. Development of an efûcient mutagenesis technique using ion beams: Toward more mutation breeding. *Plant Biotech*., **29:** 193-200.

Heo, Jeong, Lee, Chun, Murthy, H.N., Paek Kee, 2003. Influence of light quality and photoperiod on flowering of Cyclamen persicum Mill. cv. Dixie White. *Plant Growth Regulation*, **40**(1): 7-10.

Iakamura, I. and Aizawa, M. 2007. Inheritance of anthocyanins in petals in crosses between cyclamen cultivar and *Cyclamen hederifolium* Aiton. *Tech. Bulletin Faculty Agri., Kagawa Univ.*, **59**: 45-48.

Ishizaka, H. 1994. Chromosome association and fertility in the hybrid of *Cyclamen persicum* Mill. and *C. hederifolium* Aiton. and its amphidiploids. *Breed. Sci.,* **44**: 367-371.

Ishizaka, H. 1996. Interspecific hybrids of *Cyclamen persicum* and *C. graecum, Euphytica* **91**: 109-117.

Ishizaka, H. 1997. Interspecific hybridization using ovule culture and haploid production by anther in *Cyclamen. Spe. Bull. Saitama Hort. Exp. Sta.*, **5**: 1-95.

Ishizaka, H. 2003. Cytogenetic studies in *Cyclamen persicum*, *C. graecum* (Primulaceae) and their hybrids. *Plant Systematics Evolution*, **239**(1-2): 1-14.

Ishizaka, H. and Uematsu, J. 1990. Production of interspecific hybrids between *Cyclamen persicum* Mill. and *C. repandum* Sibth. Sm. through ovule culture. *Japan. J. Breed*. **40**(1): 60-61.

Jensen, C.J., 1974. Chromosome doubling techniques in haploids. *In*: K.J. Kasha (Ed). Haploids in Higher Plants. pp.153-190. Univ. Guelph, Guelph.

Ishizaka, H. and Uematsu, J. 1992. Production of interspecific hybrids of *Cyclamen persicum* Mill. and *C. hederifolium* Aiton. by ovule culture. *Japan. J. Breed*., **42**: 353-366.

Ishizaka, H. and Uematsu, J. 1993. Production of plants from pollen in *Cyclamen persicum* Mill. through anther culture. *Japan. J. Breed.*, **43**: 207-218.

Ishizaka, H. and Uematsu, J. 1994. Amphidiploids between *Cyclamen persicum* Mill. and *C. hederifolium* Aiton induced through colchicine treatment of ovules *in vitro* and plants, *Breed. Sci.,* **44**: 161-166.

Ishizaka, H. and Uematsu, J. 1995a. Interspecific hybrids of *Cyclamen persicum* Mill. and *C. purpurascens* Mill. produced by ovule culture. *Euphytica,* **82**: 31-37.

Ishizaka, H. and Uematsu, J. 1995b. Amphidiploids between *Cyclamen persicum* Mill. and *C. purpurascens* Mill. induced by treating ovules with colchicines *in vitro* and sesquidiploids between the amphidiploid and the parental species induced byconventional crosses, *Euphytica*, **86**: 211-218.

Jalali, N., Naderi, R., Babalar, M. and Mirmasoumi, M. 2010. Somatic embryogenesis in Cyclamen with two explants and combinations of plant growth regulators. *Horticulture, Environment and Biotechnology*, **51**(5): 445-448.

Johnson, H.W., Robinson, H.F. and Comstock, R.E. 1955. Estimation of genetic and environmental variability in soybean. *Agron. J.*, **47**: 477-483.

Kai, H., Hirashima, K., Matsuda, O., Ikegami, H., Winkelmann, T., Nakahara, T. and Iba, K. 2012. Thermotolerant cyclamen with reduced acrolein and methyl vinyl ketone. *J. Experimental Botany*, **63**(11): 4143-4150.

Karlsson, Meriam, G., Werner, Jeffrey, W. 2001. Temperature affects rate of leaf unfolding and flowering in cyclamen. *HortScience*, **36**: 292-294.

Kitamura, S. Akita, Y. Ishizaka, H. Narumi, I. and Tanaka, A. 2012. Molecular characterization of an anthocyanin-related glutathione S-transferase gene in cyclamen. *J. Plant Physiology*, **169**(6): 636-642.

Laura, M., Benedetti, L. de, Bruna, S., Burchi, G., Berio, T., Giovannini, A. and Ruffoni, B. 2003. *Cyclamen persicum* Mill.: somatic embryogenesis and RADP analysis of embryogenic callus. *Acta Horticulturae*, **625**: 137-141.

Legro, R.A.H. 1959. The cytological background of *Cyclamen* breeding. *Meded. Landbouwhogeschool, Wageningen,* **59**:1-51.

Lian, X. Z., Li, C., Ying, W. and Zhi Dong, Z. 2009. Established Agrobacterium-mediated genetic transformation system of cyclamen. *J. Jilin Agril. Univ.*, **31**(3): 258-261.

Lin, Yan, Yin, Shulian, Bi, Jun, Wang, Chunrong and Guo, Weizhen. 2006. Study on heredity rules of *Cyclamen persicum* colours. *Chinese Forestry Sci. Tech.*, **5**(3): 55-59.

Liu, H.M. and Chan, Q.Z.J. 2012. Study on the rapid micropropagation technology for *Cyclamen persicum* Mill. *Seed*, **31**(9):30-33.

Miyajima, I., Maehara, T., Kage, T. and Fujieda, K. 1991. Identification of the main agent causing yellow colour of yellow-flowered cyclamen mutant. *J. Japan Soc. Hort. Sci.*, **60**: 409-414.

Mizukami, Y., Fukuta, S. and Kanbe, M. 2004. Production of yellow flower cyclamen through *Agrobacterium tumefaciens* mediated transformation with chalcone reductase. *Res. Bullet. Aichioken Agril. Res. Center*, **36**: 59-63.

Nakayama, M., Tanikawa, N., Morita, Y. and Ban, Y. 2012. Comprehensive analyses of anthocyanin and related compounds to understand flower colour change in ion-beam mutants of cyclamen (*Cyclamen* spp.) and carnation (*Dianthus caryophyllus*). *Plant Biotech.*, **29**(3): 215-221.

Prange, A.N.S., Bartsch, M., Meiners, J., Serek, M. and Winkelmann, T. 2012. Interspecific somatic hybrids between *Cyclamen persicum* and *C. coum*, two sexually incompatible species. *Plant Cell Reports*, **31**(4): 723-735.

Rana, R.S. and Singh, B.P. 1995. Introduction of ornamental plants in India. *In*: K.L. Chadha and S.K. Bhattacharjee (eds.). Advances in Horticulture Vol. 12- Ornamental Plants. Malhotra Publishing House, New Delhi, India. pp. 47-86.

Reinhardt, S., Ewalda, A. and Hellwig, F. 2007. The Anatomy of the stigma and style from *Cyclamen persicum* (Mill.) cv. Pure White and its relation to pollination success. *Plant Biology*, **9**(1): 158-162.

Schwartz-Tzachor, R., Dafni, A., Potts, S. G. and Eisikowitch, D. 2006. An ancient pollinator of a contemporary plant (*Cyclamen persicum*): When pollination syndromes break down. *Flora-Morphology, Distribution Functional Ecology Plants*, **201**(5): 370-373.

Schwartz-Tzachor, R., Dafni, A., Potts, S.G. and Eisikowitch, D. 2008. Flower characteristics and breeding system of two phonological ecotypes of *Cyclamen persicum* Mill. (Myrsinaceae) in Israel. *Plant Syst. Evol.*, **274**: 127-134.

Sharma, Y.D., Sharma, Arvind and Sehgal, O.P. 1998. Variability studies in cyclamen. *J. Ornamental Horticulture*, **1**(2): 72-74.

Song, CheonYoung and Oh, DaeGeun 2010. Inheritance of several qualitative characters in *Cyclamen persicum. Korean Journal Breeding Science*, **42** (2): 188-194.

Song, CheonYoung. 2007. Combining ability and correlations for related to growth and flowering characteristics in F_1 hybrids by diallel cross of *Cyclamen persicum. Korean J. Hortil. Sci. Tech.*, **25** (1):67-73.

Song, CheonYoung. 2008. Selection of various pure lines with different growth and flowering characteristics in *Cyclamen persicum. Korean J. Hortil. Sci. Tech.*, **26**(1): 61-67.

Stephens, L.C., Widmer, R.E. 1976. Soil temperature effects on cyclamen flowering. *J. American Soc. Hort.*, **101**: 107-111.

Sundberg, M.D. 1981a. Apical events prior to floral evocation in *Cyclamen persicum* 'F1 Rosemunde' (Primulaceae). *Bot. Gaz.,* **142**: 27-35.

Sundberg, M.D. 1981b. The development of leaves and axillary flowers along the primary shoot axis of *Cyclamen persicum* 'F-1 Rosemunde' (Primulaceae). *Bot. Gaz.*, **142**: 214–221.

Takamura, T. and Miyajima, I. 1994. Creation of tetraploid yellow-flowered cyclamen by use of diploids having giant pollen grains. *Abst*: XXIVth International Horticultural Congress. p.125.

Takamura, T. and Miyajima, I. 1996a. Cross-compatibility and the ploidy of progenies in crosses between diploid and tetraploid cyclamen (*Cyclamen persicum* Mill.). *J. Japan Soc. Hort. Sci.*, **64**: 883-889.

Takamura, T. and Miyajima, I. 1996b. Embryo development in crosses between diploid and tetraploid cyclamen (*Cyclamen persicum* Mill.). *J. Japan Soc. Hort. Sci.*, **65**: 113-120.

Takamura, T. and Miyajima, I. 1996c. Colchicine induced tetraploids in yellow-flowered cyclamens and their characteristics. *Scientia Hortic.,* **65**: 305-312.

Takamura, T. and Miyajima, I. 2002. Origin of tetraploid progenies in 4x × 2x crosses of cyclamen (*Cyclamen persicum* Mill.). *J. Japan Soc. Hort. Sci.,* **68**: 55-61.

Takamura, T. and Yoshimura, N. 2001. Improvement of embryo rescue, and the production of pentaploids and hexaploids in 2x × 4x crosses of cyclamen. *Breeding research*, **3**(2): 65.

Takamura, T., Aizawa, M., Kim, S.Y., Nakayama, M. and Ishizaka, H. 2005. Inheritance of flower pigment in crosses between cyclamen cultivar and *Cyclamen purpurascens*. *Acta Horticulturae*, **673**(2):437-441.

Takamura, T., Matsumoto, Y., Yoshimura, N. and Tanaka, M. 2002. Effects of nitrogen source in medium on the production of interspecific hybrid between cyclamen and *C. africanum*. *J. Japan. Soc. Hort. Sci.*, **71**(1): 284.

Takamura, T., Sugimura, T., Tanaka, M. and Cadic, A. 2000. Inheritance of yellow-flowered characteristic in crosses between diploid cyanic and yellow-flowered cyclamen cultivars. *Proce*. 19th International Symposium on Improvement of Ornamental Plants. *Breeding Ornamentals in the Future*: *Goals, Genes, Tools, Angers,* France, **508**: 219-221.

Takamura, T., Sugimura, T., Tanaka, M. and Kage, T. 1998. Breeding of the tetraploid yellow flowered cyclamen with "eye". *Acta Hortic.*, **454**: 119-126.

Takamura, T., Tomihama, T. and Miyajima, I. 1995. Inheritance of yellow-flowered characteristic and yellow pigments in diploid cyclamen (*Cyclamen persicum* Mill.) cultivars. *Scientia Hortic.,* **64:** 55-63.

Takamura, T., Yamada, R. and Tanaka, M. 2002. Effects of genotypes of seed parents on the production of interspecific hybrids in *Cyclamen persicum* Mill. × *C. purpurascens* Mill. crosses. *Tech. Bulletin Faculty Agri., Kagawa Univ.*, **54:** 45-48.

Tanaka, A. 2012. Targeted mutation breeding of flower colour by taking advantage of ion-beam irradiation and genomic information. *Plant Biotechnology*, **29**(3): 191-227.

Taskin, B.G., Vardareli, N., Dogac, E., Mammadov, R. and Taskin, V. 2012. Genetic diversity of natural Cyclamen alpinum populations. *Turk. J. Biol.*, **36**: 413-422.

Wellensiek, S.J. 1952. The breeding of cyclamens. Report XIIIth International Horticultural Congress, 771-777.

Wellensiek, S.J. 1955. The genetics of diploid × tetraploid and reciprocal cyclamen crosses. *Züchhter,* **25**: 229-230.

Wellensiek, S.J. 1961. The breeding of diploid cultivars of *Cyclamen persicum. Euphytica,* **10**: 259-268.

Widmer, R.E. 1992. Cyclamen. *In*: R.A. Larson (eds.). Introduction to Floriculture, 2nd Edition. Academic Press, New York, pp. 385-407.

Widmer, R.E., Lyons, R.E., 1985. *Cyclamen persicum. In*: A.H. Halevy, (eds.). Handbook of Flowering, Vol. II. CRC Press, Boca Raton, FL, pp. 382-390.

Widmer, R.E., Stuart, M.C. and Lyons, R.E. 1991. Seven months 4" cyclamen-Widmer plan in G.V. Ball, *In*: Ball Red Book (eds.), 15th edition. Geo. J. Ball. Publishing, West Chicago. pp. 477-480.

Winkelmann, T., Ilczuk, A. and Wartenberg, S. 2010. Micropropagation through somatic embryogenesis of *Cyclamen persicum* Mill. genotype for cut flower production- feasibility study. *Propagation of Ornamental Plants*, **10**(4): 237-245.

Winkelmann, T., Rode, C. Bartsch, M., Prange, A., Heintz, D., Dorsselaer, A., and Van Braun, H.P. 2011. Towards a better understanding of somatic embryiogenesis in *Cyclamen persicum*. *Acta Horticulturae*, **923**: 15-23.

You, C.R., Fan, T.J. Gong X.Q. Bian, F.H., Liang, L.K. and Qu, F.N. 2011. A high-frequency cyclic secondary somatic embryogenesis system for *Cyclamen persicum* Mill. *Plant Cell, Tissue and Organ Culture*, **107**(2): 233-242.

Zheng, Y. P., Yu, J.Y., Zhang, M.X., Wang, C., Liang, Y.H. and Fan, W.F. 2008. Breeding by Crossing of *Cyclamen persicum* and technical schedule. *J. Zhejiang Forestry Sci. Tech.*, **28**(6): 63-66.

Chapter – 24

Freesia

Freesia is ornamental flowering plant belongs to family Iridaceae subfamily Ixioideae. Freesia is named after a physician and botanist from Germany named "Friedrich H.T. Freese" (1795–1876). Dr. Christian P. Ecklon, a collector of plants from South Africa, named the flower after his friend and student. Modern freesias are known as *Freesia × hybrida* and cultivated for cut flowers. Freesia is grown for two purposes i.e. cut flower and pot plant.

It is excellent cut flowers because of its appealing shapes and found suitable for various types of arrangements and bouquets. Wide range of colour increases its beauty. Hence, its flower demand increased dramatically in recent year in global market. Forced freesias are grown as pot plant. Colours in freesias consist of pink, red, orange, yellow, blue, lavender, cream and white. Some flowers are multicoloured and possess several variations of colour within the same bud. Because of their delectable fragrance, Freesias have been drafted as a vital ingredient in many fragrance oils. Freesia fragrance oils are utilized for many oil burners and vaporizers. It is used as an essential fragrance in potpourris. Freesia essence oil is high grade essence oil. This is because it is undiluted, alcohol free, and long lasting. The oil is used for aromatherapy to scent candles, soaps, massage oils, bath oils and perfumes.

ORIGIN AND DISTRIBUTION

Freesias are South African in origin. Most species were found in the Cape Province area in southern Africa and two others were found just north of the equator in a tropical climate and even in Sudan. *Freesia × hybrida* is the name of modern Freesia.

In 1866 *Freesia* was described as a distinct genus. *Freesia alba* was first described in 1878 by G.L. Meyer, and became well known in Europe, as *F. refracta* var. *alba*. Gumbleton recognized *Freesia alba* as a distinct species in 1896, but *Freesia alba* first appeared in the English nursery trade in 1878 and quickly spread to Europe and North America. It appears to have caused quite a sensation, and appeared in almost every horticultural publication of note in both Europe and America in the years following its introduction. Breeding began immediately after *F. alba* appeared on the market and continues to this day.

Most of the modern hybrids were first made in the 19th century from a few species. The genus contains some superbly ornamental and adaptable plants. These range from the lovely, scented *Freesia alba* and the brilliant red *Freesia grandiflora* and *Freesia laxa*, to the bizarre green-flowered *Freesia viridis*.

Today there are hundreds of hybrids and varieties in any imaginable colour. Most of these hybrids are derived from *F. alba, F. leichtlinii*, a rose pink form of *F. corymbosa* known then as *F. armstrongii*, and a deep yellow form of *F. corymbosa* known then as *F. aurea*.

SPECIES

Genus *Freesia* consists 16 species: *Freesia alba* (G.L. Mey.) Gumbl. (syn. *F. gentilis* N.E.Br., *F. herbertii* N.E.Br., *F.picta* N.E.Br.), *F. andersoniae* L. Bolus, *F. caryophyllacea* (Burm. F.) N.E.Br. (syn. *F. elimensis* L. Bolus, *F. parva* N.E.Br, *F. xanthospila* (DC.) Klatt, *F. corymbosa* (Burm.f.) N.E.Br. (syn. *F. armstorngii* W.Watson, *F. brevis* N.E.Br.), *F. fergusoniae* L. Bolus, *F. fucata* J.C. Manning and Goldblatt, *F. grandiflora* (Baker) Klatt, *F. laxa* (Thunb.) Goldblatt and J.C. Manning (syn. *F. cruenta* (Lindl.) Klatt), *F. leichtlinii* Klatt (syn. *F. middlemostii* F. Barker, *F. muirii* N.E.Br), *F. occidentalis* L. Bolus (syn. *F. framesii* L. Bolus), *F. refracta* (Jacq.) Klatt (syn. *F. hurlingii* L.Bolus), *F. sparrmannii* (Thunb.) N.E.Br., *Freesia marginata* J.C. Manning and Goldblatt, *F. speciosa* L. Bolus (syn. *F. flava* (E. Phillips and N.E.Br.), *F. verrucosa* (B. Vogel) Goldblatt and J.C. Manning (syn. *F. juncea* (Pourr.) Klatt), and *F. viridis* (Aiton) Goldblatt and J.C. Manning.

The genus was revised in 1935 by N.E. Brown and again in 1982 by Goldblatt. Goldblatt reduced the 19 species defined by Brown to 11, then the genus Anomatheca was sunk into Freesia and then Goldblatt and Manning discovered *Freesia fucata* in the Renosterveld in 2001. However, species of the former genus *Anomatheca* were now included in freesia, these are, *Anomatheca cruenta* Lindl. i.e., *Freesia laxa* subsp. *laxa*, *Anomatheca grandiflora* Baker i.e., *Freesia grandiflora*, *Anomatheca laxa* (Thunb.) Glodblatt i.e., *Freesia laxa*, *Anomatheca verrucosa* (B. Vogel) glodblatt i.e. *Freesia verrucosa*, *Anomatheca viridis* (Aiton) Glodblatt i.e. *Freesia viridis*, Anomatheca xanthospila (DC.) Ker Gawl.i.e. *Freesia caryophyllacea* and *Anomatheca juncea* (Pourr.) Ker Gawl. i.e. *Freesia verrucosa*. The plants usually called freesias are derived from crosses made in the 19th century between *F. refracta* and *F. leichtlinii*. Numerous cultivars have been bred from these species and the pink- and yellow-flowered forms of *F. corymbosa*. Modern tetraploid cultivars have flowers ranging from white to yellow, pink, red and blue mauve.

FLOWER MORPHOLOGY AND BIOLOGY

Freesia inflorescence is a cymose spike with sessile flowers borne along the upper part of the floral scape. The scape bends at a semi perpendicular angle at the base of the inflorescence. Each floret on the spike has membranous or dark green bracts, six tepals that are arranged in a broadly funnel-shaped tube, 6 stamen, and a bifid branched style. The flowering stalk may be branched. The leaves are narrow, with average lengths ranging from 10-30 cm long. The leaves are green and possess neither hairs nor extra-floral nectaries. Freesias also bear a single sparsely branched stem 10-40 cm tall. At the top of the stem, there is a one-sided spike hosting 5-12 flower buds. Freesias are zygomorphic, meaning that they grow along only the bottom side of the stem. The blossoms however are facing upwards because the stems take a sharp upward turn, just below the bud.

Temperature is the causal factor in flower bud initiation (FBI) and development in *F. × hybrida.* Hartsema (1962) described the apical changes associated with flower initiation and subsequent inflorescence development. It was well documented that the complete differentiation required 6 to 9 weeks at 12°C to 15°C, depending upon environmental conditions (Heide, 1965). Flower initiation will occur over a range of temperatures (5°C to 20°C) with 12°C to 15°C being the optimum (Heide, 1965 and Jensen and Bendixen, 1971), while temperature above 21°C are no inductive (Gilbertson-Ferriss *et al.*, 1981 and Mansour, 1968). The temperature during flowering and pollination affects the fertility of the flower and the ultimate seed yield of the plant. Temperature at 14°C during the growth period and 20°C at the time of pollination was found to be the optimum combination for a good seed yield in freesia. However, the effect of a lower night temperature was generally favourable, particularly so when measured by the number of seeds per pollinated flower (Kho and Baer, 1973).

However, photoperiod is less responsive towards flowering of *Freesia × hybrida.* Floral bud initiation may be stimulated by short days (Mansour, 1968) and development enhanced by long days (Gilbertson-Ferriss and Wilkins, 1978, Heide, 1965 and Mansour, 1968). It was observed that no flowering differences between colours under long day and short day condition on *F. × hybrida*. The responses of the various cultivars to the increase of light were not identical.

The pollen located on the flowers stamen is transported via external transportation, such as wind, bees, and other insects, to the stigma of another flower. This fertilizes the egg which is then encompassed in a non-fleshy fruit for protection.

GENETICS

The basic chromosome number in *Freesia* is x =11. Two of the total species have been counted and all are reported to be diploid, $2n$ =$2x$=22. The first count in the genus was made by Taylor (1926) who counted 2n=2x=22 in a cultivar which he referred to *F. refracta* 'Fischer'. Goldblatt (1971) reported the first published counts for wild populations for *F. refracta, F. alba* (=*F. muirii*), *F. caryophyllacea* (=*F. elimensis*) and *F.* cf. *speciosa*. In 1945, Lawrence reported the count of *F.* × *hybrida*, $2n$=$2x$=22, $2n$=$3x$=33, $2n$=$4x$=44 (Goldblatt, 1982). The karyotypic study of *Freesia refracta* revealed that there are six pairs of submetacentric chromosomes, four pairs of metacentric chromosomes, and one pair of telocentric chromosomes (Wang and Pu, 1988).

BREEDING OBJECTIVES

i. More number of florets/spike.
ii. Extra-large flower size.
iii. Long stem.
iv. Pure and clear colour.
v. Development of new and novel colours.
vi. Sweet fragrance.
vii. Longer vase life.
viii. Variety for pot.
ix. Resistant to biotic and abiotic stresses.

BREEDING METHODS

Introduction

A few improved cultivars freesia were introduced by NBPGR Regional Station, Shimla. These perfomed well in Shimla condition (Rana and Singh, 1995). These are given in Table 1.

Table 1: Freesia varieties introduced in India.

S. No.	Introduced from	Varieties
1.	Australia	EC 127716 (Mix)
2.	Denmark	Olympiad Jano and Olympiad Cloe

Hybridization

To develop a hybrid parent of desired traits to be selected. Parent plants to be raised in pots or field. After flowering female parent plant to be emasculated and should carefully be bagged. After one or two days later, the emasculated flower should be pollinated with pollen collected from male parent plant. Seeds may be collected after maturation of pods. Collected seed should be properly dried and sown in the pots carefully. After development of a plant variety with desired characteristics should be propagated by vegetative means i.e. corms. For fast and rapid multiplication, tissue culture via somatic embryogenesis can be employed.

Breeding work on freesia was begun in 19th century. Today there are hundreds of hybrids and varieties in any imaginable colour. Most of these hybrids are derived from *F. alba, F. leichtlinii*, a rose pink form of *F. corymbosa* known then as *F. armstrongii*, and a deep yellow form of *F. corymbosa* known then as *F. aurea*. There are many different modern hybrids that can commonly be found in garden centers all over the world. The hybrids include "Aurora", a yellow variety, "Pimpernel", a red flowering freesia, "Royal Blue" with blue to lavender flowers, "Talisman", an orange variety and "Alpine", which produces double white flowers. Conventional breeding and interspecific hybridization programmes introduced new commercial cultivars of Freesia Klatt, which is among the most sought after garden and container plant in the northern hemisphere (Fennell and Van Staden, 2003).

Cho *et al*. (2009) developed a variety Golden Amor by crossing cvs. Golden Crown and White Wing followed by seedling and line selection. The variety has vigorous growth and good production of cormlets. Another lemon double freesia variety Shiny Lemon was developed from crossing between the domestic breeding lines (Golden Crown and White Wings) and cv. Yvonne. This variety is found suitable for greenhouse conditions. It has double type flowers with multiple florets (Cho *et al*., 2010). A dark pink semi-double freesia (*Freesia hybrida*) cv. Happy Birthday was bred from a cross between 'Pink Jewel', F_1 progeny of 'Michelle' and 'Rossini' as a female and 'White Lace' as a male parent. It has multiple flowers, distinctive flower shape and upright leaves. It has also good production of cormlets and vigorous growth. This variety is suitable for the greenhouse conditions during winter season in Korea (Cho *et al*., 2011a).

An attempt was made by Cho *et al.* (2011b) to develop a light pink double freesia. They made cross of breeding a line 'Pink Jewel' derived Michelle × Rossini and cv. Purple Rain and evolved cv. Pink Ribbon. This variety has double type flower with light pink colour. Another cultivar Grace Girl was developed from a cross of cvs. Athene and Vienna. Cv. Grace Girl has double flowering florets with bright white colour. It has good production of cormlets, vigorous growth and low virus infection. It took approximately 135 days after planting for flowering of 'Grace Girl', and its vase life was 7.0 days (Cho *et al*., 2012). All these promising varieties have been developed by National Horticultural Research Institute, Rural Development Administration (RDA), South Korea. The details of these varieties are given in Table 2.

Table 2: Hybrid varieties of freesia developed by NHRI, South Korea.

S. No.	Hybrids	Parentage	Chief characteristics
1.	Golden Amor (2009)	Golden Crown × White Wing	Large, double type flower, bright golden colour, multiple florets, good flower shape, upright leaves and stalks. Average yield 6 cuttings/plant, resistance to *Fusarium* and virus.
2.	Shiny Lemon (2010)	(Golden Crown × White Wings) × Yvonne	Large double type flower, lemon colour, multiple florets, good flower shape, average yield 7.1 stems/plant under green house.
3.	Happy Birthday (2011)	Pink Jewel × White Lace	Semi-double flower, dark pink colour, multiple flowers, distinctive flower shape. Average yield 6.7 stems/plant under greenhouse.
4.	Pink Ribbon (2011)	Pink Jewel × Purple Rain	Double type flower, light pink, simultaneous flowering of many florets, good flower shape, average cut flower yield 7.7 stems/plant.
5.	Grace Girl (2012)	Athene × Vienna	Double flowering florets, bright white colour, low virus infection. Vase life 7 days. Average yield 8.3 stems/plant. Suitable for greenhouse.

Some other hybrids were also developed through hybridization but their parentages were not disclosed due to Plant Variety Protection Act. In Table 3 some hybrids along with their characteristics are discussed.

Table 3: Hybrids along with their chief characteristics.

S. No.	Hybrids	Parentage	Chief characteristics
1.	Varared	7130 AS2 × 6135 DS2	Very tall, red flowers, numerous buds, synchronous flowering, medium size.
2.	Vapano	86327-AP1 × 87320-AP1	Attractive pale yellow flowers, suitable for pots dwarf.
3.	Varablue	3208-AT-1 × 841-J	Attractive flowers, violet blue, early flowering, suitable as cut flower.
4.	Varatwisno	884135-AT3 × 72242H5	Attractive white flower, fast flowering, suitable for cut flower.
5.	Vaposma	86310-AP1 × 86304-AP1	Attractive pale yellow flowers, suitable for pots.
6.	Varayel	87316-AT1 × 87310AT4	Attractive flowers, early flowering, suitable for cut flower.

Breeding for fragrance

Studies showed that yellow flowered cultivars (Gold Flame, Rijnveld's Golden Yellow and Aladdin) emitted volatile compounds abundantly, while Rose Marie, Blue Heaven and Volcano showed hardly any emission when determination of volatile compounds were performed in 9 *Freesia* species (*F. alba*, *F. caryophyllacea*, *F.*

corymbosa, *F. elimensis*, *F. fergusoniae*, *F. leichtlinii*, *F. occidentalis*, *F. refracta* and *F. sparrmannii*) and 16 cultivars (Cherry, Elegance, Rapid Red, Sandra, Volcano, Aladdin, Agenta, Oberon, Rapid Yellow, Souvenir, Tonga, Evita, Gold Flame, Blue Heaven, Rose Marie and Rijinveld's Golden Yellow) by headspace adsorption/gas chromatography. Linalool, which accounted for at least one third of the total peak area was the dominant volatile compound common to the species and cultivars other than *F. caryophyllacea*, *F. refracta* and Evita. However, *F. fergusoniae* and *F. leichtlinii* and 5 cultivars, flowers had sweet fragrance comprised mainly of linalool and some other related monoterpenoids. However, *F. fergusoniae* and *F. leichtlinii* and 5 cultivars, flowers had sweet fragrance comprised mainly of linalool and some other related monoterpenoids. Whereas, Gold Flame, *F. sparrmannii* and *F. refracta* were outgroups having a characteristic scent (Wongchaochant *et al.*, 2005). Yang *et al.* (2007) collected aqua-space samples from the flowers of diploid *Freesia refracta*, tetraploid freesia cultivars and interspecific hybrids of tetraploid freesia cultivars were analyzed using gas chromatograph. About 75 different compounds were identified mainly terpenes, hydrocarbons, alcohols, fatty acid esters and aromatic class compounds. Among these, linalool was detected from all the sweet-scented flowers except for scentless white tetraploid *Freesia hybrida*.

Mutation

Two freesia varieties, yellow coloured and disease resistance were produced when ^{60}Co gamma rays were used. These varieties were used as parents and crosses were made in between them. 5 new varieties were developed having unique characters such as new colours, high flower stem, straight stem, long flower time, flower productive and better disease resistance (Lin *et al.*, 1993). A variety Shangnong Hongtaige of *Freesia* × *hybrida* evolved by radioactive breeding and selective breeding technique. It has deep red with golden yellow colour inside and have pleasant aroma. The number of flowers is 9-11 in main inflorescence. It is a double petal cultivar. Its average height of main stalk is 54.1 cm. The average diameter of the first flower on the base of main inflorescence is 4.95 cm (Tang and Qin, 2012).

Polyploidy

The first polyploid "Buttercup" was developed in 1910. Mohr (1958) stated that the majority of registered cultivars are polyploidy. Earlier research works confirmed the use of spontaneous tetraploid genotypes in contemporary Gladiolus and Freesia lineages. More recently, colchicines induced polyploidy has been incorporated into the breeding of these genera as a means of increasing flower size, producing original colour sport and strengthening peduncles (Kim and De Hertogh, 1997). During intervarietal crossing in freesias, an abnormal hypotriploid plant (2n=33-1) was obtained from crossing between tetraploid Golden Buttercup (2n=44) and diploid Super Giant Yellow (2n=22). In general, vigorous-growing aneuploid plants would not occur yet, although the normal polyploid levels are present and many tri or tetraploid strains are used as cultivars already (Saito, 1964).

BIOTECHNOLOGY

Through biotechnology it is possible to provide genetically identical flowers various pathways of *in vitro* morphogenesis can be induced from tissue cultures of *Freesia*. The young inflorescence, young leaf, and zygotic embryo are all suitable explants for induction of somatic embryos. The juvenile tissue and organs are more responsive to the inducing stimuli than the more mature tissue when used as explants. Wang *et al.* (1990, 1993, 1995 and 1996) reported that through *in vitro* from various parts of the plant *Freesia refracta* can be regenerated by three pathways through varying the exogenous hormonal conditions in culture media. Direct somatic embryogenesis, indirect somatic embryogenesis and organogenesis were the three ways. Kaminska *et al.* (2001) reported *Phytoplasma* disease through RFLP analysis of PCR, Freesia sneak virus had been identified through RT-PCR using ophio virus specific primers and by electron microscopy. The viral coat protein ORF was cloned and sequenced and the protein showed less than 50% identity with analogous sequences of three other ophio virus species. The viral infection was soil-transmitted to healthy freesia seedlings (Vaira *et al.*, 2005). Kumar *et al.* (2009) identified Freesia mosaic virus (FreMV) through ELISA, RT-PCR and pathogenicity tests, reported the occurance of disease in India for the first time. Similarly Kumar *et al.* (2009a) identified Bean yellow mosaic virus (BYMV) by ELISA and reverse transcription-polymerase chain reaction (RT-PCR). Complete coat protein (CP) gene and 3' UTR sequence of the isolate was determined. This study revealed the molecular characterization of BYMV infecting freesia for the first time in India.

Tissue culture

Explants of *Freesia hybrida* regenerated plantlets through somatic embryogenesis by two ways firstly from the epidermal cells and secondly via indirectly from an embryonic callus. Under direct embryogenesis, when the explants were cultured on MS medium supplemented with 11.43 micro M indole acetic acid (IAA) and 4.44 micro M 6-benzylaminopurine (6-BA), the induction rate was 84% for young inflorescence and 100% for rachillae. When sub cultured on the rooting MS medium containing 1.08 micro M alpha-naphthalene acetic acid (NAA), the rooting rate was close to 100%. However, in indirect embryogenesis, embryonic calluses were formed when the culture medium contained 22.22 micro M 6-BA and 4.52 micro M 2,4-dichlorophenoxy acetic acid (2,4-D), and the induction rate was 92.4% for young inflorescence and 100% for rachillae. The finding shows that tissue culture of flowering plants is a form of stress which can induce some heritable epigenetic variations and should be considered in future long-term genotype preservation programs of *Freesia hybrida* (Gao *et al.*, 2010).

Davies and Nichol (1971) carried out work on tissue culture and plant regeneration via organogenesis in *F. refracta*. However, the plant recovery from corms, stems, leaves flower buds and anthers under *in vitro* culture conditions in one freesia cultivar was described by Bajaj and Pierik (1974). Plants were regenerated by using buds and roots in several cultivars (Petru *et al.*, 1976). Similar work with freesia tissue culture was performed by various workers (Hussey, 1975, Mori *et al.*, 1975, Bach, 1984, Bajaj, 1989 and Wang, 1996). These findings showed that freesia is very amenable to *in vitro* culture and the organs and tissues of various parts of the plant have been demonstrated to be capable of regeneration and propagation.

Yang *et al.* (1999) reported that the callus of *Freesia* can be stored with the method of cryopreservation. The good candidates for cryopreservation were the calluses, which were subcultured for 15-20 days on medium. The better freezing procedure was to reduce the temperature from 0°C to –196°C directly. The better thawing condition for the stored calluses was at 4°C or in warm-water bath at 40°C. Qian *et al.* (2006) studied the influence of different plant growth regulators on the *in vitro* culture and plant regeneration of *Freesia refracta* cv. Shangnongjinghuanghou bulblets. The best medium for *F. refracta* callus and shoot induction was the MS medium supplemented with 2.0 mg benzyladenine and 0.2-0.5 mg NAA/litre and the best rooting medium was the MS medium supplemented with 0.2 mg NAA/litre and 0.25% activated carbon.

Somatic embryogenesis

Freesia can be regenerated *in vitro* to whole plants through two pathways of somatic embryogenesis, direct and indirect somatic embryogenesis (Wang *et al.*, 1990). In direct embryogenesis, somatic embryos can be induced directly from a single cell on the surface of explants of freesia (Wang *et al.*, 1994), whereas, in the indirect pathway, there is a stage of calluses, which initiated from the explants of the plant. Wang *et al.* (1996) found that the zygotic embryo, which exhibited a high potency of somatic embryo formation is another source of explants. Though, it is difficult to obtain sufficient embryo material due to the low seed setting rate of freesia.

Molecular breeding

Genome specific to pink flowered *F. corymbosa* and *F. corymbosa* random amplified polymorphic DNA (RAPD) marker was generated suggested that pinkish-tone perianth colour of Freesia cultivars originated from the pink-flowered *F. corymbosa*. Pair of primers were also designed for generating a 637 bp SCAR marker. It was also found that both RAPD and SCAR markers were detected in all three accessions of *F. corymbosa* but were absent in 32 accessions of eight *Freesia* spp. evaluated. The markers were detected in most cultivars having red, pink, blue, purple and violet perianths (Wongchaochant *et al.*, 2003).

ISSR (inter simple sequence repeat) markers were used to assessed the genetic diversity of 12 *Freesia refracta.* About 34 ISSR primers were used out of these 12 primers were selected to perform ISSR-PCR. The results demonstrated that ISSR is a useful tool to analyze the genetic diversity and genetic relationship among *F. refracta* accessions, which provides a scientific basis for cross breeding and new varieties protection of *F. refracta* (Wu *et al.*, 2009).

In assessing the two regeneration pathways in terms of genetic and epigenetic fidelity of the regenerants, two kinds of molecular markers i.e. amplified fragment length polymorphism (AFLP) and methylation-sensitive amplified polymorphism (MSAP) were employed. The AFLP analysis used 20 primer pairs that yielded 916 scorable bands among the donor plant and 11 regenerants from direct embryogenesis, of which 8 (0.87%) were polymorphic. The regenerants from indirect embryogenesis had 1075 clear bands of which 3 (0.27%) were polymorphic scorable bands from 18 primer pairs. Moreover, the variant band patterns included two types, that is, loss-of-original and gain-of-novel bands. MSAP analysis revealed that tissue culturing of the flower induced DNA cytosine methylation alterations in both CG and CNG levels and patterns compared with the donor plant. The variation rate was 1.1 and 1.3% for the direct and indirect embryogenesis pathways, respectively. The findings show that tissue culture of flowering plants is a form of stress which can induce some heritable epigenetic variations and should be considered in future long-term genotype preservation programs of *Freesia hybrida* (Gio *et al.*, 2010).

Genetic engineering

A cDNA clone (*Fh3GT1*) encoding UF3GT was isolated from *Freesia hybrida*, an enzyme that catalyzes the formation of the first stable anthocyanin in the biosynthesis of natural compounds. Isolation was helpful to modify plant flower colour by redirecting the anthocyanin biosynthesis (Sui *et al.*, 2011).

REFERENCES

Bach, A. 1984. The healthiness of *Freesia* × *hybrida* propagated *in vitro*. *In:* F.J. Novak, L. Havel and J. Dolezel (eds.). Plant Tissue and Cell Culture - Application to Crop Improvement. Czechoslovak Acad. Sci., (Prague), pp. 551-552.

Bajaj, Y.P.S. 1989. Freesia. *In*: P.V. Ammirato, D.A. Evans, W.R. Sharp and Y.P.S. Bajaj (eds.). Handbook of Plant Cell culture. Vol.5. Ornamental species, McGraw- Hill, New York, pp.413-428.

Bajaj, Y.P.S. and Pierik, R.L.M. 1974. Vegetative propagation of freesia through callus cultures. *Netherland J. Agril. Sci.*, **22**(3): 153-159.

Cho, H.R., Lim, J.H., Rhee, H.K., Kim, M.S., Choi, S.R. and Shin, H. 2009. A double freesia 'Golden Amor' with late flowering and high yielding for cut flower. *Korean J. Breed. Sci.*, **41**(3): 346-348.

Cho, H.R., Rhee, H.K., Lim, J.H., Kim, M.S. and Park, S.K. 2010. A lemon double multi-flowering freesia, Shiny Lemon with early flowering and high yield for cut flower. *Korean J. Breed. Sci.*, **42**(3): 248-251.

Cho, H.R., Rhee, H.K., Kim, M.S., Park, S.K., Shin, H., Joung, H.Y. and Lim, J.H. 2011a. A dark pink freesia 'Happy Birthday' with semi double multi flowering. *Korean J. Breed. Sci.*, **43**(3): 210-213.

Cho, H.R., Rhee, H.K., Kim, M.S., Park, S.K., Shin, H., Joung, H.Y. and Lim, J.H. 2011b. A light pink freesia 'Pink Ribbon' with double and simultaneous flowering of many florets. *Korean J. Breed. Sci.*, **43**(3): 214-217.

Cho, H.R., Rhee, J.H., Lim, J.H., Kim, M.S., Park, S.K., Shin, H., Choi, Y.J. and Joung, H.Y. 2012. Breeding of a new bright white double flowering freesia 'Grace Girl' with early flowering and high yielding for cut flower. *Korean J. Breed. Sci.*, **44**(4): 633-636.

Davies, D.R. and Nichol, M.A. 1971. *In vitro* propagation of freesia. Annu. Rep. John. Innes. Inst., 62: 45.

Fennell, C.W. and J. Van Staden. 2003. Biotechnology of southern African bulbs. *S. Afr. J. Bot.* **70**: 37-46.

Gao, Xiang, Yang, Dan, Cao, DongHui, Ao, Man, Sui Xin, Wang, QuiMei, Kimatu, J.N., Wang, Li. 2010. *In vitro* micropropagation of *Freesia hybrida* and the assessment of genetic and epigenetic stability in regenerated plantlets. *J. Pl. Growth Regu.*, **29**(3): 257-267.

Gilbertson – Ferriss, T.L. and Wilkins, H.F. 1978. Flower production of *Freesia hybrida* seedlings under night interruption lighting and shoot day influence. *J. Am. Soc. Hortil. Sci.*, **103**(5): 587-591.

Godblatt, P. 1982. Systematics of freesia (Iridaceae). *J. Southern Africa Bot.*, **48**: 39-92.

Goldblatt, P. 1971. Cytological and morphological studies in the southern African Iridaceae. *J. S. African Bot.*, **37**: 317-460.

Heide, O.M. 1965. Factors controlling flowering in seed raised freesia plant. *J. Hortil. Sci.*, **40**: 267-284.

Hartsema, A.M. 1962. Temperature treatment of freesia tubers, *Proc.* 16th Int. Hortic. Sci. Congr., Gembloux, Belgium, **5**: 298-304.

Hussy, G. 1975. Totipotency in tissue explants and callus of some members of the Liliaceae, Iridaceae and Amaryllidaceae. *J. Exp. Bot.*, **26**: 253-262.

Kaminska, M., Sliwa, H. and Startek, L. First report of phytoplasma infection in freesia plant. 2001. *Plant Disease*, **85**(3): 336.

Kho, Y.O. and Baer, J. 1973. The effect of temperature during growth and flowering on the seed set of tetraploid freesia. *Euphytica*, **22**(1): 35-38.

Kim, K.-W. and De Hertogh, A.A. 1997. Tissue culture of ornamental flowering bulbs (geophytes). *Amer. Soc. Hort. Sci.*, **18**: 87-169.

Kumar, Y., Hallan, V. and Zaidi, A.A. 2009. First finding of Freesia mosaic virus infecting freesia in India. *Plant Pathology*, **58**(2): 404.

Kumar, Y., Hallan, V. and Zaidi, A.A. 2009a. Identification and characterization of Bean yellow mosaic virus infecting freesia. *J. Pl. Boichem. Biotech.*, **18**(2): 253-255.

Mohr, O. 1958. Kromosomundersogelse hos *Freesia. Horticultura Odense,* **11**: 89-90.

Mori, Y., Hasegawa, A., Kano, K. 1975. Studies on the clonal propagation by meristem culture in freesia. *J. Japanese Soc. Hort. Sci.*, **44**: 294-302.

Petru, E., Jirsakova, E., Landa, Z. 1976. Clonal propagation of some freesia cultivars through tissue culture. *Biol. Pl.* (Prague) **18**: 304-306.

Qian, H.M., Zhang, H.L., Gao, Q., Yang, H. and Tang, D.Q. 2006. Study on *in vitro* culture of bulblets from *Freesia refracta* Shangnongjinghuanghou. *J. Shanghai Jiaotong Univ. Agril. Sci.*, **24**(5): 485-488.

Rana, R.S. and Singh, B.P. 1995. Introduction of ornamental plants in India. *In*: K.L. Chadha and S.K. Bhattacharjee (eds.). Advances in Horticulture Vol. 12- Ornamental Plants. Malhotra Publishing House, New Delhi, India. pp. 47-86.

Saito, K. 1964. Studies on the occurrence of polyploidy and its contribution to flower plants breeding. III. On the characteristics and biological significance of a hypotriploid plant obtained from intervarietal crossing in freesia. *Jap. J. Breed.*, **14**(3): 37-40.

Sui, X., Gao, X.A.M., Wang, Q.M., Yang, D., Wang, M.F. and Yang Wang L. 2011. cDNA cloning and characterization of UDP-glucose: anthocyanidin 3-O-glucosyltransferase in *Freesia hybrida*. *Plant Cell Report*, **30**(7): 1209-1218.

Tang, D.Q. and Qin, W.Y. 2012. A new cultivar of *Freesia* × *hybrida* Shangnong Hongtaige. *Acta Horticulturae Sinica*, **39**(10): 2097-2098.

Taylor, W.R. 1926. Chromosome morphology in Fritillaria, Asstroemeria, Silphium and other genera. *Am. J. Bot.*, **13**: 179-185.

Vaira, A.M., Torok, V., Accotto, G.P., Rapetti, S., Vecchaiati, M., Masenga, V., Vetten, H.J. and Milne, R.G. 2005. Freesia leaf necrosis: a new Ophio virus involved. *Proc*: Sixth symposium of the International Working Group on Plant Viruses with Fungal Vectors, Alma Mater Studiorum Universita di Bologna, Bologna, Italy. 44-46.

Wang, L., Huanh, B.Q., He, M.Y. and Hao, S. 1990. Somatic embryogenesis and its hormonal regulation in tissue cultures of *Freesia refracta*. *Ann. Bot.*, **65**: 271-276.

Wang, L., Huanh, B.Q., He, M.Y. and Hao, S. 1993. *In vitro* morphogenesis and its hormonal control in tissue cultures of *Freesia refracta*. *In*: C.B. You, Z.L. Chen and Y. Ding (eds.). Biotechnology in Agriculture, Kluwer Academic Publishers, Netherlands, pp. 379-382.

Wang, L., Huanh, B.Q., He, M.Y. and Hao, S. 1994. Origin of direct somatic embryos from cultured inflorescence axis segments of *Freesia refracta*. *International J. Pl. Sci.*, **155**(6): 672-676.

Wang, L., Huanh, B.Q., He, M.Y. and Hao, S. 1995. Somatic embryogenesis in *Freesia refracta*. *In*: Y.P.S. Bajaj (ed.). Biotechnology and Agriculture and Forestry, Vol.31, Somatic Embryogenesis and Synthetic Seed II, Springer- Verlag, Berlin Heidelberg, pp. 294-305.

Wang, L., Zou, M.Q. and Wang, X.G. 1996. Tissue culture of embryo and regeneration of plant in *Freesia refracta* Klatt. *Acta Horticulturae Sinica*, **23**(3): 281-284.

Wongchaochant, S., Doi, M., Inamoto, K. and Imanishi, H. 2003. Identification of genome specific RAPD market to *Freesia corymbosa* (Burm. F.) N.E. Br. Among Freesia cultivars (*F. hybrida* Hort.) to detect their lineage. *J. Japanese Soc. Hortil. Sci.*, **72**(6): 536-538.

Wongchaochant, S., Inamoto, K. and Doi, M. 2005. Analysis of flower scent of freesia species and cultivars. *Acta Horticulturae*, **673**(2): 595-601.

Wu, C.W., Zhou, L.Y., Wang, X.L., Song, H.S., Tang, D.Q. and Liu, Q.L. 2009. ISSR analysis of genetic diversity in *Freesia refracta* germplasm. *Bulletin Bot. Res.*, **29**(3): 357-361.

Yang, F., Gao, X., Xue, Y., Hui, Y., Chen, F.Q., Su, Q. and Wang, L. 2007. Volatile compounds in the flowers of freesia parental species and hybrids. *J. Integrative Pl. Bio.*, **49**(12): 1714-1718.

Yang, W., Liu, C.Y., Bu, X.L., Luan, H.C., Wang, L. 1999. Study on callus cryopreservation of *Freesia refracta* Klatt. *J. Northeast Normal University*, **4**: 70- 72.

Section V

Cut and Dry Flower Crops

Chapter – 25

Gypsophila

Gypsophila is herbacaeous annual or perennial flowering plant belongs to family Caryophyllaceae. It contains several ornamental species of which *Gypsophila paniculata* L. is the only species used as a cut flower (Zvi *et al.*, 2008) and it is applied as filler in flower arrangements (Jones, 1995 and Rehman, 2002). Gypsophila derived from "gypson" meaning chalk and "phylos" meaning lover, an etymology which alludes to its preference for calcareous soils (Vettory *et al.*, 2013). Mainly found in temperate or warm-temperate regions in the Northern Hemisphere, best developed in the Mediterranean region and Near East (Ataslar *et al.*, 2009 and Cronquist, 1981). It is a perennial plant often grown commercially as an annual crop. Root is rich source of triterpine saponins (Henry, 1993, Hostettmann, 1995 and Gevrenova *et al.*, 2010) which are used for commercially in medicines, detergents, adjuvents and cosmetics (Hostettmann, 1995, Tschesche and Wulff, 1973 and Gevrenova *et al.*, 2010). *G. paniculata* is one of the ten best selling cut flower species worldwide and thus becomes for the genetic improvement and selection of new varieties of commercial interest. Main regions of gypsophila production are Italy, Israel, Belgium, Germany and Holland while currently its cultivation is spreading across to other countries such as Ecuador, Egypt and Kenya due to lower labour costs. Infact, Gypsophila is among one of the most important varieties of cut flowers in Israel, where it is grown in more than 200 hectares of greenhouses and is almost entirely exported. For these reasons, the competition has increased dramatically, making the world market increasingly difficult for companies which feel the need to introduce new varieties that respond better to characteristics desired by consumers and florists.

ORIGIN AND DISTRIBUTION

Gypsophila is native to Caucasus and eastern Mediterranean region. The center of origin of the genus *Gypsophila* includes the Black Sea region, the Caucasus, north Iraq and north Iran (Barkoudah 1962). Whereas, the most common species *Gypsophila paniculata* occurs in east and central Europe, extending westward to central Austria, southward to Bulgaria, and northward to approximately 57' N in Russia (Barkoudah and Chater 1964).

G. paniculata was probably introduced into North America as a garden ornamental by early settlers. It was first introduced into the United States as a garden ornamental in the late 1800's. It is often cultivated in gardens and flower beds, and has escaped cultivation to pastures and rangelands (Rutledge and McLendon 1996 and Whitson *et al.*, 2000). It is widespread across the northern United States and Canada (Royer and Dickinson, 1999).

Macoun (1888) reported its first introduction into Manitoba in 1887. The first report of the weedy nature of the species in western Canada was noted in 1953 by members of the Saskatchewan Weed Survey (Darwent *et al.*, 1967). They reported a large infestation in a *Bromus inermis* hayfield in western Saskatchewan. The species has now been reported as an adventives species, or garden escape, from numerous locations across Canada and northern United States north of 40"N latitude (Gleason and Cronquist, 1963, Stevens, 1963, Moss, 1959 and Hitchcock *et al.*, 1964). *G. paniculata* has become an invasive species in parts of North America and is classified as a noxious weed in California, Montana, and Washington.

G. paniculata occurs in east and central Europe, extending westward to central Austria, southward to Bulgaria, and northward to approximately 57' N in Russia (Barkoudah and Chater, 1964). It occurs in Asiatic Russia, Mongolia and west China (Barkoudah 1962). The center of origin of the genus Gypsophila includes the Black Sea region, the Caucasus, north Iraq and north Iran (Barkoudah, 1962). Barkoudah (1962) reports it as an adventives species in west Europe, while in North America taxonomists (Fernald, 1950, Gleason and Cronquist, 1963 and Hitchcock *et al.*, 1964) have reported it as an adventive species between the latitudes of 40' and approximately 60'.

In Canada, *G. paniculata* occurs as a garden escape or adventive species in most provinces. Infestations have been observed as far north as Fort Smith, Northwest Territories. Largest infestations of the species occur in Saskatchewan, Alberta and British Columbia. The United States annually imports approximately 100 million gypsophila cut flower stems with an import value of about $20-24 million (Anon., 2010b). Over 87% of all fresh cut gypsophila entering the United States are grown in Ecuador with another 11% from Colombia and lesser amounts from Peru, Mexico, and Costa Rica. Conventional gypsophila is grown on large commercial farms in Ecuador and Colombia, with additional production in Mexico and Peru shipped by airfreight to the port of Miami, Florida.

Gypsophila is widely distributed in eastern Mediterranean region, Australia and New Zealand. The main areas of production are Italy, Israel, Belgium, Germany and The Netherlands.

SPECIES

There are about 120 species in the temperate zone of the Old World and in Australia (one species). In Russia there are about 70 species, growing mainly in the Caucasus and in Middle Asia. A wide spread species, *Gypsophila paniculata*, forms spherical shrubs (tum-bleweed). *Gypsophila paniculata* and *G. elegans* are cultivated as ornamental. Other species of *Gypsophila* are *G. acutifolia, G. altissima, G. arrostii, G. bicolor, G. bungeana, G. capituliflora, G. cephalotes, G. cerastioides, G. davurica, G. desertorum, G. fastigiata, G. huashanensis, G.licentiana, G. muralis, G. nana, G. oldhamiana, G. pacifica, G. paniculata, G. patrinii, G. perfoliata, G. petraea, G. pilosa, G. rokejeka, G. repens, G. scorzonerifolia, G. sericea, G. spinosa, G. tenuifolia, G. tschiliensi.* Description of *G. paniculata* is given below.

***Gypsophila paniculata*:** *Gypsophila paniculata* (baby's breath) is a species of flowering plant in the Caryophyllaceae family, native to central and eastern Europe. It is an herbaceous perennial growing to 1.2 m tall and wide, with mounds of branching stems covered in clouds of tiny white flowers in summer (hence the common name "baby's breath"). Its natural habitat is on the Steppes in dry, sandy and stony places, often on calcareous soils (*gypsophila* = "chalk-loving"). Specimens of this plant were first sent to Linnaeus from St Petersburg by the Swiss-Russian botanist Johann Amman.

Barkoudah (1962) has observed variations in the pubescence of stems, leaves, bracts and pedicels of *G. paniculata* populations in Hungary and has divided the species into two varieties based on this pubescence: *G. paniculata* var. *paniculata* Bark. has glabrous stems, leaves, bracts and pedicels whereas *G. paniculata* var. *adenopoda* Borbds ex Hallier-Koch has pubescence on the lower part of the stem, lower leaves and on the bracts and pedicels. Bailey (1958) reports considerable variation within the species as a garden plant. *G. paniculata* var. *compacta* Hort. is a dwarf garden form while the cultivars Double, Ehrle and Bristol Fairy have flowers that are double the size of those on unselected plants of the species. There are three varieties commercially important are described below.

***Gypsophila paniculata* 'Bristol Fairy':** 'Bristol Fairy' is an herbaceous perennial to 1.2 m in height, making a broad mound of slender stems with narrow, grey-green leaves and large, open panicles of pure white double flowers 12 mm in width.

***Gypsophila paniculata* 'Perfekta':** This cultivar is a diffusely branched, mound-forming plant that typically grows to 90-100 cm tall and features numerous, long-lasting, double white flowers (1.27 cm wide) in open, airy panicles on wiry, jointed stems in summer. Very similar to the popular *Gypsophilia paniculata* 'Bristol Fairy' except the flowers of 'Perfekta' are twice as large. Lance-shaped foliage is somewhat sparse.

***Gypsophila paniculata* 'Compacta Plena':** This cultivar is a compact, diffusely branched, mound-forming plant that only grows 38-45 cm tall, but otherwise is very similar to the popular *Gypsophilia paniculata* 'Bristol Fairy'. Features numerous, long-lasting, double white flowers (0.65 cm wide) in open, airy panicles on wiry, jointed stems in summer. Lance-shaped foliage is somewhat sparse. Provides excellent contrast and texture to the perennial border.

FLOWER MORPHOLOGY AND BIOLOGY

Gypsophila paniculata is a perennial herb that grows to 1.0 m tall with widely branching stems. The stems are smooth, linear leaves are opposite, hairless, with a prominent mid vein and 4 to 10 cm long and sharp-pointed. The plants have a deep, penetrating root system. Leaves are reduced in the upper stem, usually characterized by a rosette of basal leaves 5-8 cm long, gray-green, linear-lanceolate, from which the flower stems propagate and end in large panicles (Pignatti, 1982) and are mostly gone by the time the plant is in flower. The large and very dense inflorescences consist of numerous small round flowers, single or double, white or pink. Flowers are 1.5 to 3.5 cm wide, with a 5-lobed calyx. Calyx lobes are green and in some cases purple and the petals are typically white (Douglas *et al.*, 1988, Royer and Dickinson, 1999 and Whitson *et al.*, 2000).

The common name Baby's-breath arises from the particular structure of the inflorescence. Flower development was studied in cv. Bristol Fairy using a scanning electron microscope. It was found that plants which had been sufficiently chilled in the previous winter, flower bud initiation occurred when the stem reached approximately 18 cm in length (Doi *et al.*, 1990). Sepals and petals subsequently differentiated. In plants exposed to a limited amount of winter chilling, flower bud initiation of the terminal floret was delayed, and it did not occur in buds located from the middle to the basal nodes of the main stem. However, once the apical meristem had enlarged, the differentiation and development of the terminal floret was normal (Doi *et al.*, 1990).

Mode of pollination

Gypsophila paniculata is attractive to numerous species of pollinating bees and flies (Darwent and Coupland, 1966) and is considered to be predominately insect pollinated. The styles develop into long, spreading structures and tend to diverge outwards, while the filaments diverge in such a manner that the anthers are turned outwards. When in bloom the flowers have a pleasant, sweet odour. Milliron (1971) has observed floral visitations by bumble bees (*Bombus terricola occidentalis* Grne.).

At Beaverlodge, Alberta honey bees (*Apis mellifera* L.) have been observed to visit *G. paniculata* growing in a garden. However, since mature plants often have a multitude of flowers at various stages of development, it is likely that pollination can occur within the same plant.

Pollination

The structure of the flower suggests the possibility of cross pollination (Darwent and Coupland 1966). Hybridization of gypsophila in nature is facilitated by insect pollination and is only effectively achieved by Hymenoptera and Diptera (pollinating bees, flies). Pollen is not spread by wind. In general, production of viable pollen by cultivated gypsophila is much lower than that of wild or naturalized gypsophila. The gypsophila pollen is viable for approximately 12-24 hours. Gypsophila flowers are numerous, but relatively small, which would appear to diminish their relative attractiveness to pollinators. Seed is the only mechanism for natural reproduction with an average plant producing 13,700 seeds. Most seeds drop close to the plant; however some can be dispersed to remote locations by wind. Seeds show little or no dormancy (Anon., 2010a).

Seed production and dispersal

In North Dakota, Stevens (1957) reported that a single plant produced 13,700 seeds with a weight of 86 mg/100 seeds. The average weight of five samples of 100 seeds collected in Saskatchewan was 67 mg (Darwent and Coupland 1966). In Saskatchewan, wind was the most important agent of seed dispersal in the field (Darwent and Coupland 1966). During the latter part of the period of fruit development, the shoots became brittle and dry. Breakage occurred at the base and the plants were blown across the field. Most of the seeds in the capsules were loosely held and the majority dropped on the ground near the position of the parent plant. However, some capsules did not open completely, and seeds therein were carried for distances of nearly 1 km. The seeds exhibit little or no dormancy (Darwent and Coupland, 1966). The fruit is a small capsule containing 2 to 5 seeds (Douglas *et al*., 1988, Royer and Dickinson, 1999 and Whitson *et al*., 2000).

GENETICS

The basic chromosome number of *Gypsophila* is 17(n). Diploid number of the chromosomes (2n=34) is typical representative of this genus like *G. fastigiata* L., *G. muralis* L., *G. repens* L. (Góralski *et al*., 2009), *G. caricifolia* Boiss. (Ghaffari *et al*., 2005), *G. elegans* M. Bieb. (Gagnidze *et al*., 2006) and *G. paniculata* (Darwent, 1975). Although diploid number of the chromosomes (2n=34) is typical representative of this genus a relevant difference at ploidy level was found between the *G. paniculata* species, its variety (Flocon de Neige Doubles), the *G. libanotica* and the *G. pacifica* species.

Indeed, microscopic observations of metaphase cells of root tips of *G. paniculata*, *G. paniculata* Flocon de Neige Doubles and *G. libanotica* showed a chromosome number equivalent to half (2n=34) compared to that of *pacifica* (2n=68) (Troia *et al.*, 2012). The *G. pacifica* shows a lower height with shorter internodes and larger leaves in comparison with *G. paniculata* and *G. libanotica* which present much smaller and lanceolate leaves. This relevant difference in number of chromosomes of a *Gypsophila* species respect to the others was already found in *G. manginii* with a diploid number of 2n=66 (Kishi *et al.*, 1994). Major differences in chromosome number were found within the group of commercial hybrids of Gypsophila under study but also when compared with the wild type species.

BREEDING OBJECTIVES

i. Increase in flower number and size
ii. More compact inflorescences
iii. Free flowering cultivar
iv. Many petals per flower
v. Straight stem
vi. New colours of the petals
vii. Good flower colour retention
viii. Good vase life
ix. Cultivar for pot culture
x. Strong mechanical resistance to the manipulation of flower stems
xi. Resistance for biotic and abiotic stresses

BREEDING METHODS

Manual cross pollination is not allowed in this genus due to the very small size of pistils and their difficult handling and the lack of pollen production. Therefore, use of insect pollinators were followed but with limited success. The knowledge of the genetic basis of spontaneous species is necessary and may form the basis of the rational design of crossings and selection, from those relating to the reproductive system, the potential barriers of intra- and interspecies incompatibility and the chromosome number (Marino *et al.*, 2012, Peruzzi *et al.*, 2012 and Urdampilleta *et al.*, 2012). The flowers of commercial *G. paniculata* plants are sterile and do not produce seeds, therefore, breeding programmes are severely restricted (Shillo and Abraham, 1982 and Shillo, 1985). Very little variation exists among the gypsophila varieties used fresh and dried as fillers for bouquets. All these gypsophila genotypes have white flowers, with one exception: cv.

Flamingo, which has pale-pink flowers. Gypsophila is therefore an important target for classical breeding (Shillo, 1985).

Selection

Two new *Gypsophila paniculata* cultivars Yunxing 17 and Yunxing 23 were selected from *Gypsophila* Perfect and Fairy. The cultivar Yunxing 17 was much more uniform than cv. Perfect in flowering with a very high rate of inflorescence. Whereas, cv. Yunxing 23 had a strong sprouting ability and hardly went into rosette naturally and higher yield per plant than cv. Fairy (Zhao *et al*., 2009). Floriculture Resource Research Institute of Jeonbuk Agricultural Research and Extention Services developed a new cultivar Shine of *Gypsophila paniculata* for cut flowers. It was developed by selection of seedlings and lines in a seed-set bud mutant (Gyp99 line) derived from Bristol Fairy. The plant is upright type with many semi-double florets and emarginate petal tip and campanulate calyx shape. The new cultivar had a trait of simultaneous blooming of many florets and longer cut flower stalk, larger leaf length and calyx size (Cheong *et al*., 2010). Similarly, Zhao *et al*. (2011) selected a new *Gypsophila paniculata* cultivar Qianwanxing from abroad introduced cultivar Baiwanxing by clonal selection. It is fixed fine varieties of natural variation. Compared with the original cultivar, it has more petals (40-48 pieces), larger flowers (diameter 0.4-0.5 cm), thicker stems, broader leaves. The flowers are more tidy and it has less invalid flowering branches. A brief description has been given in Table 1 about the varieties developed through selection.

Table 1: Varieties developed through selection.

S. No.	Variety	Parentage	Chief characteristic
1.	Yunxing 17 (2009)	Perfect	More uniform, very high rate of inflorescence.
2.	Yunxing 23 (2009)	Fairy	Strong sprouting ability, higher yield.
3.	Shine (2010)	Bristol Fairy	Upright type, many semi-double florets, simultaneous blooming, longer cut flower stalk, excellent for cut flowers.
4.	Qianwanxing (2011)	Baiwanxing	More petals (40-48 pieces), larger flowers (diameter 0.4-0.5 cm), thicker stems, broader leaves.

Hybridization

A garden type hybrid (*G. bodgeri*) has been produced from a cross between a horticultural race of *G. paniculata* and *G. repens* L. var. *rosea* Hort. (Bailey, 1958).

Gypsophila paniculata is a potential cut flower worldwide and it is considered a "target" plant for breeding varieties. The study of basic chromosome number and the reproductive system (potential barriers to intraspecific and interspecific incompatibility) of the wild species and the commercial hybrids is necessary to set the job of cross breeding and selection variety. A significant difference in ploidy was found between the species *paniculata*, its variety Flocon duobles de Neige, the species *libanotica* (n=32) and *pacifica* (n=68). A major difference in the number of chromosomes was found among the commercial hybrids of *Gypsophila* observed; this difference was not only between themselves but also in comparison with the wild species. The flower structure and reproduction of commercial hybrids can be remarkably complex. Some hybrids as the "Perfecta" show pollen sacs empty, others as the "Mirabella" show pollen grains in anther. However, the pollen, when present, exhibits degenerated and immature appearance. The examined commercial hybrids showed a *gineceum* morphologically and anatomically conforming to that of wild species and it was potentially fertile. Therefore, the fertile pollen collected from the wild species could be used to fertilize the compatible eggs of commercial hybrids speeding the process of breeding varieties (Vettori *et al.*, 2011).

Through hybridization good number of varieties developed from Israel or The Netherlands. Name of parents are not disclosed because of several reasons. Some hybrid varieties of gypsophila are described in Table 2.

Table 2: Hybrids along with their parentage and chief characteristics.

S. No.	Cultivar	Parentage	Chief characteristics
1.	Dangy P39	P/3190 × OP 1	Pink flowers, globular-shaped, double, good post-production longevity, vase life 10-14 days, suitable as a cut flower.
2.	Esm Cirque	Line 114 × Line 56	Double, luminous white flowers, early, uniform and freely flowering, good longevity.
3.	Esm Chispa	Line 34 × (Unknown) *Gypsophila hybrida*	White flowers, compact, uniform, freely flowering habit, small semi-double, good vase life.
4.	Dangypfirm	Cv. 1087 × Cv. 16	Semi to full double flowers, flowering in cluster, prolonged vase life of 10-12 days.
5.	Dangypwhifa	Cv. 2088 × Cv.184	Bright white flower, globular shape, double flower, vase life of 10-14 days.
6.	Dangypink	96-184 × P2000	Pinkish white, semi double flower, globular shape, cut flower, vigorous growth habit.
7.	Dangypwhifa	Cv. 2088 × Cv. 184	Bright white flower, globular shape, double flower, 10-11mm diameter, conic arranged in clusters, vase life of 10-14 days.
8.	Fiorello	B-M-2 × B-PCC-2	Small flowers, vigorous growth habit, the flowers contain a large number of petals arranged in compound cymes.

A lot of hybrid varieties have been developed through hybridization but due to intellectual property right there crosses were not mentioned. Some of the varieties are given in Table 3. along with their characteristics.

Table 3: Varieties developed through hybridization.

S. No.	Variety	Chief characteristic
1.	Dangypurna	White semi double flowers, globular shape, conic inflorescence form.
2.	Danfestar	White flowers, perennial, small sized cut flower, compact, early flowering.
3.	Esmamerica (*G. hybrida*)	Luminous white, erect and straight flowering stems, freely flowering, good post-production longevity.
4.	Dangypday (*G. paniculata*)	White flower with a greenish centre, globular shaped, semi to full double flowers, symmetrical, suitable for cut flower.
5.	Esmasia (*G. hybrida*)	White flowers, erect strong, freely flowering, good post production longevity, resistant to *Botrytis* and *Alternaria*.
6.	Magic Gilboa	Pure white flower, double rows, suitable for cut flowers, good vase life.
7.	Dangyflash	White flower, upright uniform growth, yield 8-10 stems/plant, semi double flowers, early flowering, indoor and outdoor plant.
8.	Magic Tavor	White flower, sturdy stems, upright and conical growth habit, floriferous, early flowering.
9.	Danghappy	White flower, compact, good yield, floriferous habit.
10.	White Festival	Shiny white, semi double flower form, compact, full growth habit, floriferous.
11.	Magic Golan	Pure white, large flowers, long, strong and straight stem, good vase life.
12.	Magic Arbel	Large flower, white, excellent branching, floriferous, cut flower.
13.	Dangypmini (*G. paniculata*)	Luminous white, fully double flowers, fully branched flowering stem, head spray type, fast growing habit, indoor and outdoor cultivation.

Development of coloured and disease resistant varieties

New cultivar was developed from *Gypsophila paniculata* which was resistant to powdery mildew and pest injury. The cultivar named Dream Pink, developed for cut flowers by Jeollabuk-do Agricultural Research and Extension Services. It has morphological characteristics of the intermediate plant type, double florets with many petal and two pistils, emarginate petal tip, strong flower stalk, deep pink flower colour (Cheong *et al.*, 2011).

Another cultivar of *Gypsophila paniculata* 'Double Shine' with bright coloured flower and strong stem was developed cut flowers by Jeollabuk-do Agricultural Research and Extension Services. It has spreading plant type, double florets of many petals and two pistils, emarginate petal tip and cup shaped calyx. Cv. Double Shine has several traits such as strong flower stalk, bright coloured flower, narrow branching angle and more or less resistance to powdery mildew and pest injury. Furthermore, 'Double Shine' produced higher primary branches, and longer flower stalk length with internodal length and than a check variety Bristol Fairy during summer cultivation (Cheong *et al.*, 2012).

Mutation

The explants (bud microcuttings) were subjected to different radiations (15.705, 41.88 and 83.76 W/m^2), for a 16-h photoperiod and at 23+or-2°C to study the effect of luminous flux density on the multiplication on *Gypsophila paniculata*. The shoots being cultivated under low luminous flux density showed a brilliant green colour, while those developed under 41.88 and 83.76 W/m^2 were opaque green whereas, shoots under 83.76 and 41.88 W/m^2 presented a higher weight of fresh and dry matter (Radmann *et al.*, 2001). Yeowon a cultivar of *Gypsophila paniculata* was newly developed for cut flowers was selected via a bud mutant (Gyp99 line) from 'Bristol Fairy' in 1999. It produces more nodes and primary branches, thicker and stronger stems, shorter internode length, larger branching angle, and longer cut flower stalk (Cheong *et al.*, 2008). Brief about the variety has been given in Table 4.

Table 4: Mutant variety alongwith characteristics.

S. No.	Mutant	Parentage	Chief characteristics
1.	Yeowon (bud mutant)	Bristol Fairy	More primary branches and nodes, stems thicker and stronger, shorter internode length, larger branching angle and longer cut flower stalk. Produced normal flowers with 2 pistils and 7.8 stamens.

Tissue culture

The major hindrances in propagation of gypsophila are phytoplasma disease (Gera *et al.*, 2005) and low rooting frequency of vegetatively propagated cuttings, so the application of tissue culture techniques for gypsophila is very vital to overcome problems for breeding and propagation. Hence, the *in vitro* culture techniques can be the alternative for the continuous provision of plantlet stocks for large scale field cultivation (Barakat and El-Sammak, 2011). Gypsophila tissue culture has been used for micropropagation and viral elimination (Han *et al.*, 1991). Works of Kusey *et al.* (1980), Pauthe- Dayde *et al.* (1990) and Henry (1993) have dealt with multiplying shoot cultures, but stop short of producing a reliable adventitious shoot-regeneration method, an absolute prerequisite for establishing an efficient *in vitro* mutagenesis procedure.

Micropropogation through tissue culture was one of seedling production ways of gypsophila. Different plant parts were used so far under *in vitro* condition viz., shoot tip, leaf, meristem, woody plant, root etc. (Han *et al.*, 1991, Zamorano *et al.*, 1994, Castro *et al.*, 1996, Song *et al.*, 1996, Zuker *et al.*, 1997 and Gevrenova *et al.*, 2010). Study indicates that when fresh plantlets from stem tip cultures were treated by ABT and cultured in water rooted easily where rooting rate of the fresh plantlets was 90%. Plantlets cultured in water were 16% higher than in test tube and 50 higher than in solid media (Ling, 1997). Adventitious shoots were successfully regenerated from leaf explants of *Gypsophila paniculata*.

The highest efficiency of adventitious shoot regeneration was obtained with the oldest leaves originating from the youngest cutting analyzed on thidiazuron-containing medium, shoots regenerated on average from 67% of the leaves, with an average of seven shoots per explants (Zuker *et al.*, 1997). Lee and Bae (1999, 1999a) found the best condition for *in vitro* proliferation when different explants were used (shoot tip of *G. paniculata* cultivars Bristol Fairy and Red Sea and adventitious shoot regeneration from leaf segments from cv. Bristol Fairy respectively) were used. Results showed that formation and fresh weight of adventitious shoots were promoted by culture with 0.2 mg/litre BA (benzyladenine) and 0.2 mg/litre NAA (Bristol Fairy), and 0.2 mg/litre BA and 0.1 mg/litre NAA (Red Sea), and thidiazuron (TDZ) was remarkably effective for the regeneration of leaf segments compared with BA and kinetin. Similarly, it was also found most efficient cytokinin, with up to 100% of the explants forming shoots, at an average of up to 19 shoots per explant regenerated when adventitious shoot regeneration was developed for *Gypsophila paniculata*. using cv. Arbel (Ahroni, *et al.*, 1997). It was found that both the shoots and roots showed progressive growth when *Gypsophila paniculata* grown *in vitro* where, the root primordia appeared after 8 days in the rooting media (Bosa *et al.*, 2003).

The effects of two types of cytokinins [kinetin and CPPU (florchlorfenuron)] and cytokinin antagonists (2-chloro-4-cyclobutyl-amino-6-ethylamino-1,3,5-triazine (ACK 1) and N-(4-pyridyl)-O-(4-chlorophenyl) carbamate (ACK2)) were studied on the bud break, outgrowth of single nodes and ultra structure of the chloroplasts of *in vitro*-cultured *G. paniculata*. Results showed that both cytokinins stimulated bud opening, with CPPU resulting in higher number of open buds compared to kinetin. Both cytokinin antagonists reduced the number of sprouted and open buds (Kapchina-Toteva and Stoyanova, 2003).

When shoot tips of *Gypsophila paniculata* cv. Perfecta were excised and cultured in MS medium supplemented with NAA (0.0 and 0.3 mg/l) and BA (0.0, 0.5, 1.0, 2.0 and 4.0 mg/l). The highest rate of sucrose (60 g/l) with agar above 9.0 g/l produced normal plantlets without vitrification (Atta-Alla *et al.*, 2001) however, a new cultivar Kisyu-Kasumi 1 was developed from callus regeneration of cv. Perfecta during *in vitro* induction to increase the variability. The new cultivar produced early, large, white coloured double flowers (Yoshiki, 2003).

Shoot tip of gypsophila under *in vitro* condition were encapsulated in different concentrations of Na alginate dissolved either in MS-salts solution or sterilized distilled water (SDW) as solvent (Rady and Hanafy, 2004), similarly shoot tip and node cultures of *G. paniculata* cv. Bristol Fairy produced highest number of regenerated shoots and shoot length from half-strength MS medium supplemented with 0.3 mg benzyladenine/litre under *in vitro* condition. It was also found that callus mortality was higher in node culture than in shoot tip culture (Cheong *et al.*, 2004). The *in vitro* propagation of *G. paniculata* (cv. Golan) using nodal segments was studied, where, the multiplication of nodal segments was most pronounced in the MS medium containing 0.5 micro M kinetin and 0.3 micro M GA_3 (Quoirin *et al.*, 2008). Whereas, other plant parts (shoot tip and lateral buds) used as explants for regeneration of plantlets through four medium protocols then the explants gave high significant differences in callus induction. The bud explants gave higher values (16.66%) in callus induction compared to shoot tip explants (5.82%). The protocol A, which was MS supplemented with 0.5 mg/L BA and 0.5 mg/L NAA, was highly significantly superior to the other three medium protocols in (88.3%) shoot formation (Barakat and El- Sammak, 2011).

For induction of *in vitro* flowering, regenerated shoots could be induced to flower efficiently when cultured on MS medium containing 13.3 micro M BA and 50 g/l sucrose (Kanchanapoom *et al.*, 2011). On using different media for plant propagation where apical meristem was used as explants, multiple shoots were obtained on MS medium fortified with different concentrations of BAP along with various combinations of NAA and Kin. Among all the media tested the highest number of shoot emergence, multiple shoot formation, and maximum shoot length was obtained in medium supplemented with BAP 1mg/l after 4 days of inoculation (Rashid *et al.*, 2012).

In vitro interspecific hybrid production

A successful interspecific hybrid between *Gypsophila paniculata* cv. Red Sea ($2n = 34$) and *G. manginii* ($2n = 66$) was produced by ovule-embryo culture. Ovules were cultured on MS medium supplemented with 2.9 micro M GA_3, 3% sucrose and 0.8% agar. After 45 days of culture embryos were excised from the cultured ovules and cultured on the same media used for ovule culture. A torpedo-shaped embryos were regenerated into plantlets. The hybrid plant ($2n = 50$) had double, pale purplish-pink flowers. Hybridity was confirmed by morphological intermediacy, chromosome number and peroxidase isozyme patterns (Kishi *et al.*, 1994).

Intergeneric somatic hybrid

In gypsophila no much variation has been observed in colour and plant structure. Therefore, breeding of *Dianthus* spp. and *Gypsophila* spp. based on the intergeneric hybridization can be done to overcome sexual incompatibility barrier. Somatic hybridization by protoplast fusion offer an alternative method to conventional sexual hybridization for obtaining hybrids between distantly related species. Intergeneric transfer of desirable traits with respect to floral and marketable can be expected in *Dianthus* and *Gypsophila* by applying somatic hybridization technique. In this line, protoplasts isolated from cell suspension cultures of the two species of Caryophyllaceous family i.e. *Dianthus caryophyllus* and *Gypsophila paniculata* were fused using polyethylene glycol and cultured in MS medium containing 1 mg picloram/litre and 0.7 M sorbitol by Nakano and Mii (1993). Among the calluses produced, one showing green colouration and vigorous growth was identified as an intergeneric somatic hybrid based on isoenzyme analyses for esterase and glutamate dehydrogenase. Hybridity of this callus was also confirmed by Southern hybridization analysis using a rice rDNA probe. The callus regenerated roots after 2 months of culture on MS medium containing

1 mg picloram/litre, but shoot regeneration was not observed on any of the media tested.

Later on, intergeneric somatic hybrid plantlets between *Dianthus barbatus* and *Gypsophila paniculata* obtained by electrofusion (Nakano *et al.*, 1996). In this study they found out that hypocotyl-derived protoplasts of *Dianthus barbatus* that had been pretreated with iodoacetamide were fused electrically with cell suspension culture-derived protoplasts of *Gypsophila paniculata* that could divide to form callus but could not regenerate shoots under the culture conditions used in this study. Electrofusion-derived calli which produced shoots were selected as putative somatic hybrids and plantlets were subsequently regenerated from 2 of these selected calli. These plantlets, which *in vitro* produced flowers precociously, were identified as intergeneric somatic hybrids by nuclear ribosomal DNA analysis but normal plants have not been established.

Molecular breeding

RAPD technique could be successfully applied to the newly *Gypsophila* variants and can differentiate mutants from their parents. Regeneration of *Gypsophila paniculata*, using shoot tips explants were subjected to the random amplified polymorphic DNA (RAPD) analysis. Results indicated that very low variation at the DNA level occurred during *in vitro* culture of *Gypsophila* (Rady, 2006).

In a study, induce mutation in *Gypsophila paniculata* was performed through *in vitro* mutagenesis by treating the shoot tips and lateral buds with four doses of gamma irradiation (0.25, 0.5, 0.75 and 1 Gy). After that RAPD analysis were applied for the detection of genetic polymorphism among *Gypsophila* mutants and their parent. Analysis of RAPD recognized 105 different amplification products, where the genetic similarity among the ten genotypes ranged from 0.59 to 0.97 (Barakat and El- Sammak, 2011a). cDNA cloning was performed to analyzed the performance of CONSTANS (CO) homologs in gypsophila later, RNA was extracted from *G. paniculata* leaves. The sequencing results suggest that at least four *CO* homologs exist and are expressed in *G. paniculata*: *GpCOL1*, *GpCOL2*, *GpCOL3*, and *GpCOL4* (Kanayama *et al.*, 2008).

Genetic engineering

Gypsophila paniculata was genetically transformed by co-culturing cut stems with *Agrobacterium tumefaciens*. Occurrence of a crown gall during infection was essential for successful genetic transformation. After selection at 50 mg kanamycin/litre, a transformation rate of 89.8% was attained (Yu and Yang, 1997).

As a major contributor to the flower market, *Gypsophila paniculata* is an important target for the breeding of new varieties. However, gypsophila breeding is strongly hampered by the sterility of this species' genotypes and the lack of a genetic-transformation procedure for this genus. Zvi *et al.* (2008) described the establishment of a transformation procedure for gypsophila (*Gypsophila paniculata*) based on

Agrobacterium inoculation of highly regenerative stem segments. The transformation procedure employs stem explants derived from GA_3-pretreated mother plants and a two-step selection scheme. The GA_3 treatment was crucial for obtaining high gene-transfer frequencies (75-90% GUS-expressing explants out of total inoculated explants), as shown using three different gypsophila varieties. An overall transformation efficiency of five GUS-expressing shoots per 100 stem explants was demonstrated for cv. Arbel. The applicability of the transformation system to gypsophila was further reinforced by the generation of transgenic plants expressing *Agrobacterium rhizogenes rolC* driven by a CaMV 35S promoter. *rolC*-transgenic gypsophila plantlets exhibited extensive rooting and a bushy phenotype. While these traits are potentially beneficial for both propagators and growers, an extensive evaluation of transgenic plants under greenhouse conditions should be performed. As classical breeding of gypsophila has been held back by cultivar sterility, and since genetic-transformation protocols for gypsophila have not been reported to date, this work paves the way for advanced gypsophila breeding. Transgenic gypsophila plantlets exhibited extensive rooting and branching, traits that could be beneficial to the floriculture.

Danziger "Dan" Flower Farm, one of Israel's leading floriculture companies, is developing genetically engineered varieties of *Gypsophila paniculata* for the cut flower market. Danziger is developing varieties of *Gypsophila paniculata* that exhibit altered colour phenotypes through expression of a gene introduced using recombinant DNA techniques. Expression of the introduced gene results in a range of altered flower colours from dark purple and red to light pink. Danziger has confirmed the genetic stability of transformed lines and by PCR analysis for the presence of the genes. Danziger has selected new varieties for advancement and commercialization based on plant phenotype and agronomic performance, which includes the level and pattern of pigment expression, as well as the fulfilment of commercial criteria such as plant architecture, flower morphology, agronomic performance and economic yield.

REFERENCES

Ahroni, A, Zuker, A., Rozen, Y., Shejtman, H. and Vainstein, A. 1997. An efficient method for adventitious shoot regeneration from stem segment explants of gypsophila. *Pl. Cell. Tissue Organ Culture*, **49**(2): 101-106.

Anonymous. 2010a. Aggressive ornamentals. Baby's breath (*Gypsophila paniculata*), Government of British Columbia, Ministry of Agriculture, Food and Fisheries. Pest management. http://www.aaf.aov.bc.ca/crooorotlindex.htm.

Anonymous. 2010b. Global Agricultural Statistics, USDA FAS. http://www.fas.usda.aov /aats/default.ÿ

Ataþlar, E., Potoðlu Erkara, I. and Tokur, S. 2009. Pollen morphology of some *Gypsophila* L. (*Caryophyllaceae*) species and its taxonomic value. *Turk J. Bot.*, **33**: 335-351.

Atta-Alla, H.K., Moghazy, E.S., Waly, A.K. and Abo El- Soad, I.A. 2001. Micropropagation of *Gypsophila paniculata* var. Perfecta. *Ann. Agril. Sci.*, **39**(1): 629-644.

Bailey, L.H. 1958. Manual of Cultivated Plants. The MacMillan Company, New York. Il l6 pp.

Barakat, M.N. and El- Sammak, H. 2011. *In vitro* culture and plant regeneration from shoot tip and lateral bud explants of *Gypsophila paniculata* L. *J. Medicinal Pl. Res.*, **5**(15): 3351-3358.

Barakat, M.N. and El-Sammak, H. 2011a. *In vitro* mutagenesis, plant regeneration and characterization of mutants via RAPD analysis in Baby's breath *Gypsophila paniculata* L. *Australian J. Crop Sci.*, **5**(2): 214-222.

Barkoudah, Y.I. 1962. A revision of Gypsophila, Bolanthus, Ankyropetalum and Phryna. *Wentia* **9**: 1-203.

Barkoudah, Y.I. and Chater. A.O. 1964. Gypsophila L. *In*: T.G. Tutin, V.H. Heywood, N.A. Burges, D.H. Valentine, S.M. Walters and D.A. Webb. Flora Europaea. Univ. Press. Cambridge, England. pp. 181-184.

Bosa, N., Calvete, E.O., Suzin, M. And Bordignon, L. 2003. Growth evaluation of *Gypsophila paniculata* during *in vitro* rooting. *Horticultura Brasieira*, **21**(3): 510-513.

Castro, V.M., Dardel, H.C. and Verdugo, R.G. 1996. *In vitro* propagation of *Gypsophila paniculata* L. *Agricultura Tecnica* (Santiago), **56**(3): 224-228.

Cheong, D.C., Jeong, J.S., Chang, C.J., Lee, E.J., Choi, E.G. and Park, H.B. 2004. Shoot regeneration of callus from shoot tip and node culture of Gypsophila L. Bristol Fairy. *Korean J. Hortil. Sci. Tech.*, **22**(1): 100-106.

Cheong, D.C., Jeong, J.S., Lim, H.C., Song, Y.J., Chang, C.J., Choi, C.H. and Kang, M.H. 2008. A new cultivar *Gypsophila paniculata* Yeowon with hot tolerance for cut flower. *Korean J. Hortil. Sci. Tech.*, **26**(2): 168-171.

Cheong, D.C., Lim, H.C., Choi, C.H., Kim, K.C., Song, Y.J. and Kim, J.M. 2010. Simultaneous blooming 'Shine' a new cultivar of *Gypsophila paniculata. Korean J. Hortil. Sci. Tech.*, **28**(1): 159-161.

Cheong, D.C., Lim, H.C., Choi, C.H., Song, Y.J. and Kim, J.M. 2011. Anew cultivar of *Gypsophila paniculata* with powdery mildew resistance and double florets of pink colour Dream Pink. *Korean J. Hortil. Sci. Tech.*, **29**(5): 500-502.

Cheong, DongChun, Choi, ChangHak, Song, YoungJu, Lim, HoiChun, Kim, JeongMan, Lee, JinJe and Lee, JungSoo. 2012. A new cultivar of *Gypsophila paniculata* 'Double Shine' with bright coloured flower and strong stem. *Korean J. Hortil. Sci. Tech.*, **30**(1): 93-96.

Cronquist, A. 1981. An integrated system of classification of flowering plants. Columbia University Press, New York.

Darwent, A.L. 1975. The biology of Canadian weeds. 14. *Gypsophila paniculata* L. *Canadian J. Pl. Sci.*, **55**:1049-1058.

Darwent, A.L. and Coupland, R.T. 1966. Life history of *Gypsophila paniculata*. *Weeds* **14**: 313-318.

Darwent, A.L., Coupland, R.T. and Skoglund, N.A. 1967. The potential of *Gypsophila paniculata* as a weed in Saskatchewan. *Can. J. Plant Sci.* **47**: 125-134.

Doi, M., Takeda, Y. And Asahira, T. 1990. Flower bud initiation and development in *Gypsophila paniculata* L. *J. Japanese Soc. Hortil. Sci.*, **59**(3): 621-626.

Douglas, G.W., Straley, G.B., Meidinger, D. and Pojar, J. 1998. Illustrated flora of British Columbia. British Columbia, Ministry of Environment, Lands and Parks, Ministry of Forest, V.1. 436 pp.

Fernald, M.L. 1950. Gray's manual of botany. Eighth ed American Book Co., New York. 1632 pp.

Furusato, K. 1940. Polyploid plants produced by colchicine. *Bot and Zool.*, **8**: 1303-11.

Gagnidze, R., Gviniashvili, T. and Jinjolia, L. 2006. *Ggypsophila elegan*. *In*: Marol, K. (ed.) IAPT/IOPB chromosome data 2. Taxon **55**: 757.

Gera, A., Weintraub, P.G., Maslenin, L., Spiegel, S. and Zeidan, M. 2005. A new disease Causing stuting and shoot proliferation in *Gypsophila* is associated with phytoplasma. *Arab. J. biotec.*, **8** (1):155-168.

Ghaffari, S.M., Hejazi, A. and Pourahmad, A. 2005. New chromosome counts in nine endemic species from Iran. *Folia Geobotanica.* **40**: 435-440.

Gleason, H.A. and Cronquist, A. 1963. Manual of vascular plants of northeastern United States and Canada. D. Van Nostrand Co., Inc., Princeton, New Jersey. 810 pp.

Goralski, G., Lubczynska, P. and Joachimiack, A.J. 2009. Chromosome number database. http://www.binoz.uj.edu.pl:8080/chromosomes/.

Han, B.H., Paek, K.Y. and Choi, J.K. 1991. Micropropagation of *Gypsophila paniculata* using shoot tip culture *in vitro*. *J. Korean Soc. Hortil. Sci.*, **32**(3): 394-400.

Han, S.Y., Qi, L.W., Wei, Y.T., Yang, Y.L. and Zhang, Y.W. 1996. Tissue culture and *in vitro* propagation of *Gypsophila paniculata*. *Acta Horticulturae* Sinica, **23**(2): 175-178.

Henry, M. 1993. *Gypsophila paniculata* L. (Baby's Breath): *In vitro* culture and the production of gypsogenin saponins. *In*: Bajaj, Y.P.S. (ed.). Biotechnology in Agriculture and Forestry, vol. 21. berlin/Heidelberg: Springer- Verlag. pp. 187-206.

Hitchcock, C.L., Cronquist, A., Ownbey, M. and Thompson, J.W. 1964. Vascular plants of the Pacific Northwest. Part 2. Salicaeae to Saxifragaceae. Univ. Washington Press, Seattle, Washington. 597 p.

Hostettmann, K.M. 1995. A saponin. *In*: Phillipon, J.D., Ayres, D.C., Baxter, H. (eds.). Chemistry and Pharmacology of Natural Products. Cambridge, Cambridge University Press. pp. 326-327.

Jones, R.B. 1995. New ornamental crops and the market for floriculture products. *Acta Horticulture*, 397.

Kanayama, Y., Sato, A., Moriguchi, R and Kanahama, K. 2008. Molecular cloning and expression analysis of CONSTANS homologs from *Gypsophila paniculata*. *Acta Hort.*, **766**: 501-506.

Kanchanapoom, K., Jingjit, S. and Kanchanapoom, K. 2011. *In vitro* flowering of shoots regenerated from cultured nodal explants of Gypsophila paniculata L. *Notulae Botanicae, Horti Agrobotanici, Cluj Napoca*, **39**(1): 84-87.

Kaochina-Toteva, V. and Stoyanova, D. 2003. Effect of cytokinins and cytokinin antagonists on *in vitro* cultured *Gypsophila paniculata* L. *Biologia Planterum*, **46(3)**: 337-41.

Kishi, F., Kagami, Y., Shinohara, M., Hatano, S. and Tsurushima, H. 1994. Production of an interspecific hybrid in Gypsophila by ovule embryo culture. *Euphytica*, **74**(1-2): 85-90.

Kusey, W.E., Hammer, J.R. and Weiler, T.C. 1980. *In vitro* propagation of *G. paniculata* L. Bristol Fairy. *Hort. Sci.*, **15**(5): 600-601.

Lee, S.W. and Bae, J.J. 1999. Improvement of efficiency for multiplication of seedling by shoot cultured *in vitro* in *Gypsophila paniculata* L. *Korean J. Hortil. Sci. Tech.*, **17**(6): 768-769.

Lee, S.W. and Bae, J.J. 1999a. *In vitro* regeneration using leaf segment in *Gypsophila paniculata* L. Bristol Fairy. *Korean J. Hortil. Sci. Tech.*, **17**(6): 765-767.

Macoun, J. 1888. Catalogue of Canadian plants: IV Endogens. Dawson Bros., Montreal, Quebec. 428pp.

Marino, P., Geraci, A. and Schicchi, R. 2012. Notes on the Karyology, genetics and ecology of Genista sect. Voglera in Sicily. *Plant Biosystems*, **146(1)**: 324-329.

Moss, E.H. 1959. Flora of Alberta. Univ. Toronto Press, Toronto, Ontario. 546 pp.

Nakano, M. and Mii, M. 1993. Callus and root formation from an intergeneric somatic hybrid between *Dianthus caryophyllus* and *Gypsophila paniculata*. *Scientia Horticulturae*, **53**(1-2): 13-19.

Nakano, M., Hoshino, Y. and Mii, M. 1996. Intergeneric somatic hybrid plantlets between *Dianthus barbatus* and *Gypsophila paniculata* obtained by electrofusion. *Theo. App. Genet.*, **92**(2): 170-172.

Pauthe-Dayde, D., Rochd, M. and Henry, M. 1990. Triterpenoid saponins production in callus and multiple shoot cultures of *Gypsophila* spp. *Phytochemistry*, **29**(2): 483-487.

Peruzzi, L., Bedini, G. and Andreucci, A. 2012. Homoploid hybrid speciation in Doronicum L.(Asteraceae)? Morphological, karyological and molecular evidences. *Plant Biosystems*, **146**(4): 867-877.

Pignatti, S. 1982. Flora d'Italia. EdAgricole Bologna, Vol. I, p. 240.

Quoirin, M.G.G, Biasi, L.A., Rios, J.F. and Cuquel, F.L. 2008. Micropropagation of *Gypsophila paniculata* through culture of nodal segments. *Scientia Agraria*, **9**(1): 79-83.

Radmann, E.B., Braga, E.J.B., Karan, M.A.L., Posada, M.A.C. and Peters, J.A. 2001. Influence of the luminous flux density on the quality of micropropagated *Gypsophila paniculata* L. Plants. *Revista Brasileira de Agrociencia*, **7**(3): 171-175.

Rady, M.R. 2006. *In vitro* culture of *Gypsophila paniculata* L. and random amplified polymorphic DNA analysis of the propagated plants. *Biologia Plantarum*, **50** (4): 507-513.

Rady, M.R. and Hanafy, M.S. 2004. Synthetic seed technology for encapsulation and regrowth of *in vitro*-derived *Gypsophila paniculata* L. shoot-tips. *Arab J. Biotech*, **7**(2): 251-64.

Rashid, S., Ilyas, S., Naz, S., Aslam, F. and Ali, A. 2012. *In vitro* propagation of *Gypsophila paniculata* L. through plant tissue culture techniques. *Pakistan J. Sci.*, **64**: 6-10.

Rehman, A.U. 2002. Bioactive Natural Products. Elsevier Publications, **26**: 3-1347.

Royer, F. and Dickinson, R. 1999. Weeds of the Northern U.S. and Canada. The University of Alberta Press. 434 p.

Rutledge, C.R. and McLendon, T. 1996. An assessment of exotic plant species of Rocky Mountain National Park. Department of Rangeland Ecosystem Science, Colorado State University.

Shillo, R. 1985. *Gypsophila paniculata. In*: Halevy, A.H. (ed.). Handbook of Flowering, CRC Press, Inc, Boca Raton, Florida, **3**: 83-87.

Shillo, R. and H.H. Abraham. 1982. Interaction of photoperiod and temperature in flowering control of *Gypsophila paniculata* L. *Scientia Hort.,* **16**: 385- 393.

Song, R.H., Lu, J.P. and Zhon, X.G. 1996. Rapid Propagation of *Gypsophila paniculata* L. by tissue culture. *Acta Agricultura Shanghai,* **12**(3):71-73.

Stevens, O.A. 1957. Weights of seeds and numbers per plant. *Weeds*, **5**: 46-55.

Stevens, O.A. 1963. Handbook of North Dakota Plants. North Dakota Inst. Regional Stud., Fargo, North Dakota, 324pp.

Troia, A., Raimondo, F.M., Castellano, G. and Spadaro, V. 2012. Morphological and Karyological and taxonomic remarks on *Ferulago nodosa* (L.) Boiss. (Apiaceae). *Plant Biosystems*, **146**(1): 330-337.

Tschesche, R. and Wulff, G. 1973. Chemie und biologie der saponine. *Fortschr chem. Org naturst*, **3**: 461-606.

Urdampilleta, J.D., Ferrucci, M.S., Vanzela, A.L.L. 2012. Cytogenetic studies in South American species of *Serjania* (Sapindaceae: Paullinieae). *Plant Biosystems*, **146**(4): 835-846.

Vettori L., Schiff, S., Tani, C., Pasqualetto, P., and Bennici, A. 2013. Morphological and cytological observations of wild species and hybrids of Gypsophila. *Plant Biosystems - An Int. J. Dealing with all Aspects of Plt. Bio.: Official Journal of the Societa Botanica Italiana.* DOI: 10.1080/ 11263504.2013.824041.

Vettori, L., Schiff, S., Tani, C., Pasqualetto, P and Bennici, A. 2011. Studies on the breeding of wild species and hybrids of Gypsophila. *Colture Protette*, **40**(4): 72-77.

Whitson, T.D., Burrill, L.C., Dewey, S.A., Cudney, D.W., Nelson, B.E., Lee, R.D. and Parker, R. 2000. Weeds of the West. The Western Society of Weed Science in cooperation with the Western United States Land Grant Universities, Cooperative Extension Services. University of Wyoming. Laramie, Wyoming. 630 p.

Yoshiki, M. 2003. Studies on breeding of *Gypsophila paniculata* L. (2) breeding process and characteristics of a new cultivar 'Kisyu-kasumi 1'. *Wakayamaken norin suisan sogo gijutsu senta kenkyu hokoku.* **4**: 33-41.

Yu, P.T. and Yang, Y.P. 1997. Genetic transformation of *Gypsophila paniculata* by *A. tumefaciens. Acta Agriculturae Shanghai*, **13**(4): 17-20.

Zamorano, Mendoza, J.J. and Mejia Munoz, J.M. 1994. *In vitro* propagation of gypsophila (*Gypsophila paniculata* L.) cv. Perfecta. *Revista Chapingo, Serie Horticultura*, **1**(1): 67-71.

Zhao, P.F., Qu, Y.H., Wu, L.F., Yang, C.M., Znang Q. and Li, X. 2009. New *Gypsophila paniculata* cultivars Yunxing 17 and Yunxing 23. *Acta Horticulturae* Sinica, **36**(12): 1849-1850.

Zhao, P.F., Shan, Q.L., Wang, J.H., Yang, C.M., Qu, Y.H., Li, X., Wang, G.X., Wu, L.F. and Li, S.B. 2011. A new *Gypsophila paniculata* cultivar Quianwanxing. *Acta Horticulturae Sinica*, **38**(7): 1421-1422.

Zuker, A., Ahroni, A., Shejtman, H. and Vainstein, A. Adventitious shoot regeneration from leaf explants of *Gypsophila paniculata* L. *Plant Cell Report*, **16**(11): 775-778.

Zvi, M.M.B., Zuker, A., Ovadis, M., Shklarman, E., Ben-Meir, H., Zenvirt, S. and Vainstein, A. 2008. *Agrobacterium*-mediated transformation of gypsophila (*Gypsophila paniculata* L.). *Mol. Breeding*, **22**: 543–553.

Chapter – 26

Stock

Stock or Gillyflower (*Matthiola* spp.) belongs to family Brassicaceae. It is a hardy annual, biennial and perennial plant, flowering in terminal clusters. *Matthiola incana* is ornamentally important and grown for various purposes i.e. cut flower, fragrance, pot culture, etc. This species develops a woody base in areas where it grows as a biennial or short-lived perennial, hence the name of stock. Generic name "*Matthiola*" was given in the honour of "Dr. Pietro Andrea Gregoria Matthiole" (1501-1577), personal physician of Emperor Maximilian of Austria. "*Incana*" means "hairy or gray-white" in reference to plant leaves. It exists as both single and double forms and is valued for fragrant flowers that can be used as fresh or dried cut flower. They are excellent pot plants for the greenhouse in winter and spring and provide good cut flowers. Flowers are well arranged on a long column and colours are variable from white to rose, crimson, purple, yellow, mauve, pink, etc. The highly fragrant flowers are used as a garnish, especially with sweet deserts. Many old names were given to *Matthiola incana* including stocks, sea stocks, wallflowers and wall or gillyflowers.

ORIGIN AND DISTRIBUTION

Matthiola incana is native to the Iberian Peninsula, France, the Apennine Peninsula, the Balkan Peninsula, the British Isles, Turkey and Cyprus. According to Greuter *et al.* (1986), four species of *Matthiola* are known in Sicily and its surroundings: *M. fruticulosa* (L.) Maire, *M. incana* (L.) R. Br. & Ait., *M. tricuspidata* (L.) R. Br. and *M. sinuata* (L.) R. Br. Unlike its congeners, which only have a single taxon in this archipelago, *M. incana* consists of three infra-specific subdivisions (Lojacono-Pojero, 1888 and Greuter *et al.* 1986): ssp. *incana* (L.) R. Br., which is widespread in other insular and continental Mediterranean enclaves, and the narrow endemics ssp. pulchella (Conti) Greuter & Burdet (confined to the islet of Pantelleria) and ssp. *rupestris* (Raf.) Nyman, distributed in the surroundings of Palermo (Sicily) and in the Aeolids, the Egadian and Ustica, respectively (Raimondo *et al.* 1994).

Stock earlier known as *Leucoium album* or *Viola alba* was later renamed as Mattiola after Dr. Mattioli, an Italian physician and botanist. This was being grown by the ancient Greeks and was linked for its frangrance. In the wild state, stocks are found in the Mediterranean, Egypt, South Europe and in South Africa and two species, *Mattiola incana* and *M. sinuate* are among the rare British natives. Ruth *et al.* (2009) also reported that *Matthiola* is widespread in the Mediterranean and Irano-Turanian regions. Both single and double forms were grown in Europe from 10th century onwards and became popular garden flowers. Apart from its countries of origin, *Matthiola incana* can be found in Australia, Austria, Bolivia, Bulgaria, China, Colombia, Ecuador, Finland, France, Germany, Greece, Ireland, Israel, Japan, Kazakhstan, Mexico, The Netherlands, New Zealand, Peru, Romania, Sweden, United Kingdom and the United States.

SPECIES

Matthiola is a genus of 140 species. Important species of *Mattiola* are *M. incana*, *M. incana* var. *annua*, *M. bicornis*, *M. sinuate*, *M. odoratissima*, *M. tristis*, *M. fenestralis* and *M. tricuspidata*. Over the years, a huge number of stock varieties have been developed, mostly hybrids involving *M. incana*, *M. sinuata* and *M. odorata*, used extensively by the commercial floral industry because of its scent and long lasting blooms. *Matthiola incana* produces both single and double flowers and long grayish-green leaves. Important species of *Matthiola* are described here in detail:

***Matthiola incana*:** A branching stock, grown as annual (10-14 weeks) as well as biennial. It grows upto 45-90 cm in height. Flowers are fragrant, 2.5 cm across, in wide colour range from white, lilac, rose, lavender, apricot and crimson to canary-yellow. Pods are cylindrical without glands.

***Matthiola incana* var. *annua*:** It is commonly found from the Mediterranean costs to western Europe. Plant is erect and robust with a woody basal stalk. Leaves are narrow, lanceolate and obtuse. Flowers are dense on terminal racemes and 2.5 cm across. Pods are compressed and without glands.

***Matthiola bicornis*:** This species is known as Night Scented stocks. The long, slender siliqua has 2 pointed horns and hence it is called as *bicornis*. This species is native to Greece. Flowers are highly fragrant and open at night. It is an annual but sometimes perennial, growing upto 25-45 cm. Leaves are narrow, oblong, entire and irregularly lobed. The colour of leaves is grey green and pubescent. Flowers are small, pink, purple and with narrow petals. Pods are long, terete and cinereous.

***Matthiola sinuate*:** It is native to coastal region of western and southern Europe. Plant is erect and branched. Leaves are oblong, hairy and sinuate. Stigma compressed and velvety.

***Matthiola odoratissima*:** An evergreen, sub-shrubby plant growing upto 30-60 cm in height. Leaves are hairy and toothed. Flowers are creamish white fading to purplish brown and sweet scented in the evening. Pods are compressed, somewhat hoary.

***Matthiola tristis*:** It is a perennial, woody at the base and is native to southern Europe. Leaves are linear, entire or toothed and hairy. Flowers are almost creamish white.

***Matthiola fenestralis*:** It is a hardy sub-shrub with erect, simple stem. Leaves are crowded, obovate and revolute. Flowers are scarlet or pale purple in colour a little smaller than *Matthiola incana*. Pods are pubescent, without glands and broadest at the base.

***Matthiola tricuspidata*:** It is a hardy annual belonging to Mediterranean region. Leaves are obtuse, repandly dentate or slightly sinuate. Flowers are bright lilac, light coloured towards the base of each petal, flowering terminal, flexuose, many flowered racemes.

FLOWER MORPHOLOGY AND BIOLOGY

The different kinds vary with height from 30-90 cm and all have narrow, oblong or lance-shaped leaves, which are 3-8 cm in length and are usually glaucous. Extremely fragrant flowers, bloom in dense clusters (terminal racemes) in a large variety of colours including shades of pink, lavender, purple, white, yellow and red. Flowers bloom during winter, but bloom time is abbreviated considerably in beginning of summer climates. Long seed pods, called siliques, are set in summer.

The flower colours and anthocyanin constitution of eight cultivars of Vintage series bedding stock (*Matthiola incana*) surveyed to determine the relation between their flower colours and anthocyanin components. Thirteen anthocyanins were isolated from the flowers of these cultivars as major anthocyanins and their structures were identified by chemical and spectroscopic techniques. Among them, a novel anthocyanin, cyanidin 3-caffeoylsambubioside-5-malonyl-glucoside (pigment 1) was found in single and double flowers in cultivars 'Vintage Lavender and Vintage Burgundy. Furthermore, two anthocyanins, cyanidin 3-p-coumaroyl-sambubioside-5-malonyl-glucoside (pigment 2) and cyanidin 3-feruloyl-sambubioside-5-malonyl-glucoside (pigment 3), were also found in these cultivars for the first time in *Matthiola incana* flowers. Regarding the flower colour variation in these cultivars, the hue values (b*/a*) of these flower colours were roughly responsible for the numbers of hydroxycinnamic acid residues in anthocyanin molecules and also hydroxyl patterns of the B-ring in anthocyanidins. These flower colours were classified into eight groups, A-H, based on the hue values of their flowers, and were arranged as follows. In violet flowers (hue values b*/a*=-0.66 and -0.69, V 84A) of group A, cyanidin 3-dihydroxycinnamoyl-sambubioside-5-malonyl-glucosides were major anthocyanin pigments. In purple flowers (-0.43 and -0.45, P 75A) and red-purple flowers (-0.14 and -0.16, RP 74A) of groups B and D, pelargonidin 3-dihydroxycinnamoyl-sambubioside-5-malonyl-glucosides were major anthocyanin pigments. In red-purple flowers (-0.21 and -0.24, RP 72A) of group C, cyanidin 3-monohydroxycinnamoyl-sambubioside-5-malonyl-glucosides were major anthocyanin pigments. In red flowers (0.05 and 0.06, RP 66A) of group E, pelargonidin 3-monohydroxycinnamoyl-sambubioside-5-malonyl-glucosides were major anthocyanin pigments. In copper (0.23 and 0.16, R 54A) and peach (2.37 and 2.09, R38C) of groups F and G, pelargonidin 3-glucoside was a major anthocyanin pigment, and a small amount of pelargonidin 3-glucoside was present in yellow flowers of group H. From these results, the relation between flower colours and the bluing effects of acylated anthocyanins with hydroxycinnamic acids was discussed in flowers of *Matthiola incana* cultivars of Vintage series (Tatsuzawa *et al.*, 2012).

GENETICS

Chromosome number of *Matthiola incana* is 2n=14. Whereas, the haploid number of the chromosomes in *Matthiola* is 7. The number in the pure single strains is identical with that in the ever-sporting or double-throwing forms investigated (including the Sulphur White). Thus the genetical differences do not appear to be attributable to any observable difference in chromosome behaviour during pollen formation (Allen).

Inheritance of singleness and doubleness of flower

Matthiola produces constantly two types of singles—the "*no-d*-singles" which breed true, and produce pure singles only, and the ever-sporting or double-throwing singles (V-singles) which give both singles and doubles. The genetical facts are explicable on the supposition that at least two factors, *XY* are concerned in the production of singleness, doubleness being due to the absence of either or both factors. In the pure-breeding single the factors *X* and y appear to be linked—in the rf-single they are not. The probability is that in the d-single the pollen is all of one kind, and that it lacks the factor for singleness, but the ovules—from the evidence afforded by reciprocal crosses—are of two kinds, single-producing and double-producing. The Sulphur White variety which has also been investigated is peculiar in that it is ever-sporting as regards both plastid colour and doubleness. Doubleness, therefore, although apparently unrelated to sap colour or surface character is linked with the factor for plastid colour.

Slender trisomic is useful in producing seed to raise exceptionally high progeny proportions of double-flowered plants. Double matthiola flowers are completely sterile since they consist of masses of indefinitely replicated petals, doubles can be obtain by seed of heterozygous singles (Ss). Singleness is recessive character. The single line is kept permanently heterozygous for doubleness by pollen lethal, *l* gene, which sterilizes the single carring or *Sl* /*sL* plants and slightly reduces the viability of the *Sl* embryo sacs (Frost *et al.*,1959).

Inheritance of leaf morphology

The inheritance and linkage relationships of a leaf morphology gene of *Matthiola incana* had shown that the allele for sinuate leaf shape, c, was found to be recessive to the allele for normal entire leaf, C. The c allele was tightly linked to the recessive allele for double flowering, s. The recombination frequency between the two loci was close to zero. The mode of inheritance of the C gene was in accordance with the hypothesis that a pollen lethal gene is responsible for the constant 1:1 segregation ratio of double-flowered (male sterile) to single-flowered (fertile) plants in most *M. incana* breeding lines. The sinuate leaf allele seemed to reduce the frequency and delayed the flowering of double-flowered plants. From this the importance of the C gene as a double flowering marker in the cultivation and the breeding of *M. incana* is known (Ecker *et al.,* 1994).

Inheritance of plant height

Six cultivars of *Mathiola incana* belonging to series Giant Imperial viz., White, Yellow, Lavender, Violet, Crimson and Rose, were used in a cyclic crossing system with Giant Imperial White as tester parent. Fifty plants with double flowers were randomly taken from each plot and were analyzed concerning three plant characters *i.e.,* plant height, no. of inflorescences/plant, length of inflorescence. It is noticeable that the additive effects were the largest component of the total genetic effects while for plant height the dominance effects were slightly higher than the additive ones. For the above three studied characters additivity plays the major role in their inheritance with noticeable effects of dominance in case of height of plant. The choice of parents for crosses made for breeding purposes in gillyflower should take advantage of it (Ardelean *et al.,* 2009).

Inheritance of flowering

Six cultivars, White, Yellow, Lavender, Violet, Crimson and Rose of gillyflower (*M. incana* L.) belonging to series Giant Imperial, were used in a cyclic crossing system. All parental varieties, F_1 and F_2 hybrids as well as BCP1 and BCP2 backcross generations were grown on individual plots. Fifty plants with double flowers were randomly taken from each plot and were analyzed concerning several flower characters *i.e.,* number of flowers/inflorescence, diameter of flower, persistence of blooming and the main genetic effects involved in the inheritance of these characters were computed. It is noticeable that for the number of flowers/inflorescence and persistence of blooming, the additive effects were the largest component of the total genetic effects while for diameter of flower the dominance effects were slightly higher than the additive ones. There could be concluded that at least for these three studied characters additivity plays the major role in their inheritance with noticeable effects of dominance in case of persistence of flowering. The choice of parents for crosses made for breeding purposes should consider these characters (Cordea *et al.,* 2009).

Inheritance of flowering time

The inheritance of flowering time (FT) in a cross between early-flowering (P1) and late-flowering (P2) *Matthiola incana* genotypes, showed that the distribution of FT in F_1, F_2 and BC_1 generations indicated an additive genetic control with partial dominance of earliness alleles, particularly with regard to double-flowered plants. Single-flowered plants flowered earlier than double-flowered plants, averaged within both parental lines. However, the mean difference between singles and doubles was 16 days for P1 as compared with 6 days for P2. This flower-doubleness related delay in FT was found to be heritable, by analysis of F_3 family means (Ecker *et al.,* 1993).

Heritability

Heritability, both in wide (H) and narrow (h^2) sense, was computed for seven quantitative characters plant height, inflorescence number per plant, flower number per inflorescence, flower diameter, siliqua length, beginning of flowering and persistence of flowering in six cultivars of gillyflower (*M. incana*). There have been found quite a few quantitative traits *i.e.,* plant height, number of simple flowers per inflorescence, number of composite flowers per inflorescence, beginning of flowering, persistence of flowering, in which high values of H were accompanied by medium values of h^2. It is quite reasonable to expect a high efficiency of phenotypic selection for such traits since additive effects seem to play a major role in their inheritance (Pui *et al.,* 2007).

BREEDING OBJECTIVES

i. Compact plant growth
ii. Strong stem type
iii. New colour
iv. Desired form (compact, furlling)
v. Flower size
vi. Fragrant variety
vii. More vase life
viii. Resistance to diseases and insect pests

BREEDING METHODS

Hybridization

A total of 10 genotypes of 4 *Matthiola* spp. *M. tricuspidata*, *M. aspera*, *M. longipetala* and *M. incana* were evaluated for resistance to bacterial blight (*Xanthomonas campestris* pv. *incanae*). Accessions were highly resistant, while the breeding lines were severely infected. Interspecific F_1 hybrid plants resulting from the cross *M. incana* × *M. tricuspidata* also exhibited high resistance to the disease (Ecker *et al.*, 1995).

Heterosis: Six parental patterns (P_1-P_6) of cultivars of the *Matthiola incana* species and 10 direct and mutual hybrid combinations showed that regarding hybrid vigour, in F_1 descendants with simple flowers, positive heterosis was found in all or the majority of the studied combinations only in the case of the flower diameter, siliqua length and number of inflorescences per plant, when heterosis was calculated based on the parental mean. In descendants with plants presenting composite flowers, only the height of plants and the flower diameter had predominantly positive heterosis for both ways of calculation, the other characters having negative values. It was assumed that negative heterosis in this case, could be rather designated as the result of genetic transgressions (Pui and Ardelean, 2007).

Incomplete diallel crosses were performed involving stock (*Matthiola incana*) cultivars. The F_1 hybrids as well as their parents were grown and observations were performed on 6 quantitative characters viz., plant height, number of inflorescences per plant, number of flowers per inflorescence, diameter of individual flower, start of flowering and persistence of flowering. Additive effects of polygenes, emphasized by high GCA values in several characters, suggested that in stock for most of these characters there were groups of polygenes with positive effects and other groups with negative ones. For SCA, significant values were found for most of the characters, which could be considered as a preliminary proof that, in stock, there might be fair chances of developing commercial hybrids with an obvious phenotypic expression of the desired heterosis *i.e.,* negative or positive (Ardelean *et al.,* 2008).

Mutation

Sabaghi *et al*. (2013) evaluated the morphological diversity in the second generation of mutant stock var. Centum White. Based on cluster analyses, lines were grouped in three separated groups. The results of the stepwise regression analysis showed that the inflorescence length, plant height, number of true leaves below the flower and number of flowers in florescence remained in the model. The path coefficient analysis for the number of petals as the dependent variable for those four remained variables was performed. Results showed that mutation breeding effectively increased the morphological variation in the second generation of Stock plants var. Centum White.

Polyploidy

Euploids in genetics and breeding is examined with particular reference to a strategy for breeding double varieties of stock (Yabuya, 1991).

BIOTECHNOLOGY

Constant demands always exist for producing something new in the market of ornamental crops. Therefore, it is important to broaden the genetic variability through the suitable and available breeding strategies for each target ornamental plant. Plant biotechnologies such as tissue culture, embryo rescue, protoplast fusion and genetic transformation acts as the means for broadening the genetic variability and for creating novel useful germplasm to the breeders of ornamental crops. So far, interspecific hybrids have been produced in the different groups of plants with the aid of embryo rescue technique. Some of the products of these interspecific hybridizations have successfully been utilized to produce commercially available cultivars. Genetic transformation systems have successfully been established in various ornamental crops, including Matthiola. By utilizing these transformation systems, transgenic plants of useful traits, such as insect-resistant *Matthiola incana*, has already been produced (Mii, 2012).

Tissue culture

Callus induction and root formation on callus are possible when kinetin (Kn) and naphthalene acetic acid (NAA) are added to Murashige and Skoog's (MS) medium. Seeds of *Matthiola incana* were germinated on solid MS medium without plant growth regulators. Leaf micro-cuttings from four-week-old *in vitro* germinated seedlings were subcultured on solid MS medium containing Kn (0.0, 0.5 and 1.0 mg/L) and NAA (0.0, 0.5 and 1.0 mg/L). The callus was induced from these media after four weeks of culture, except for medium containing 0.5 and 1.0 mg/L NAA. The development of roots was observed from callus in MS medium containing suitable concentration of Kn and NAA. MS mediums containing 0.5 mg/L NAA (100%) and 0.5 mg/L Kn + 0.5 mg/L NAA (100%) were most effective in induction of callus on leaf micro-cuttings. The largest number (1.83) and the highest length (15.7 mm) of roots were obtained in MS medium supplemented with 0.5 mg/L Kn + 1.0 mg/L NAA. NAA did not stimulate callus induction and root formation when it was applied alone. Also, this hormone prevented root formation originated from callus with concentration of 1.0 mg/L along with 0.5 and 1.0 mg/L Kn in medium (Kaviani *et al.*, 2011).

Direct plant production from cultured explants is important to minimize somaclonal variation in regenerated plants. An efficient protocol for micropropagation of *Matthiola incana* using shoot tips is described. Seeds from mother plants were germinated on MS medium without growth regulators. Shoot tips from *in vitro* germinated seedlings were subcultured on solid MS medium supplemented with kinetin (KIN) (0, 0.5, 1 and 2 mg/L) and naphthalene acetic acid (NAA) (0, 0.5, 1 and 2 mg/L). Four-week-old *in vitro* plants, obtained from micro-cuttings, showed successful shooting and rooting. MS medium supplemented with 2 mg/L KIN without NAA resulted in the best shoot length (1.166 cm) and largest number of node (4.64). When the shoot tips were inoculated in the medium containing 2 mg/L NAA without KIN and medium containing the combination of 1 mg/L NAA+2 mg/L KIN, the best result was observed for root number (1.85) and root length (5.2 cm) (Kaviani *et al.*, 2011).

Gautam *et al.* (1983) reported that multiple shoot buds have differentiated from cotyledonary explants of *Matthiola incana* (Brassicaceae), cultured on Murashige and Skoog's (MS) basal medium supplemented with benzylaminopurine (BAP) (0.1–0.8 ppm). After subculturing on MS + naphthaleneacetic acid (NAA) (1 ppm), they produced roots. However, only a few shoots developed on hypocotyl explants reared on MS + kinetin (Kn) (0.1 ppm) or MS + adenine (Ad) (1 ppm). NAA (1 and 4 ppm) induced profuse rooting in both cotyledonary and hypocotyl explants, whereas 2,4-dichlorophenoxyacetic acid (2,4-D) (2 ppm) could mainly induce callusing with sporadic rooting only in cotyledonary explants.

In vitro single nodes of *Matthiola incana* were cultured in MS basal medium with different concentrations of kinetin to produce shoots and roots. Multiple shoots containing roots simultaneously obtained on MS medium only with supplementation of 0.5-2 mg/L kinetin. The best shoot length (11.72 mm) and the most number of nodes (4.64) were obtained when we used 2 mg/L of kinetin. The largest number (3.40) and the longest length of roots (54.0 mm) were achieved using 1 mg/L kinetin. The effect of kinetin was significant on the length of shoot and root and the number of node and root (Hesar *et al.*, 2011).

Intergeneric somatic hybridization

Lin *et al.* (2008) made an attempt for somatic hybridization between *Brassica napus* and *Matthiola incana*. Protoplast isolated from hypocotyls of *Matthiola incana* were fused by the PEG-High Ca_2^+–High pH method and the highest fusion frequency (17%) was obtained with a fusion solution containing 30% PEG and fusion was conducted at a density of 5.0 x 105 protoplasts/ml for 25 minutes. The fusion-protoplasts were then cultured by the "thin liquid layer" method in improved KM8p medium supplemented with 1.5 mg 2,4-D, 0.5 mg NAA and 0.5 ml benzyladenine, 0.3 mol sucrose and 0.1 mol glucose/litre used as osmotic regulators. The fusion-protoplasts started to divide after 3-6 days and the division frequency was 13.8% after being cultured for 7 days. The cell-masses and small calluses could be obtained after 35 days and the plating efficiency was 6.5%, then transferred to the proliferation MS medium supplemented with 0.1 mg 2, 4-D and 0.2 mg benzyladenine/litre. Calluses of 5 mm in diameter were transferred on MS differentiation medium supplemented with 0.5 mg benzyladenine, 0.1 mg NAA and 0.2 mg GA_3/litre, and whole plants were obtained after transferring 1-2 cm shoots to 1/2 MS medium supplemented with 0.5 mg IBA/ litre for 2 weeks. Cytology identifications shown that hybrid plants were obtained.

Production and analysis of intergeneric somatic hybrids was done between *B. oleracea* (2n=18) and *Matthiola incana* (2n=14). One hundred and eighteen plants were obtained from 96 calli. Only four plants (H1, H2, H3 and H4) showed an intermediate phenotype from the parents; among these, H1 and H3 arose from the same callus. Random amplified polymorphic DNA (RAPD), sequence-related amplified polymorphism (SRAP), and cytological analyses confirmed that H1 and H3 were hybrids. The nuclear DNA content of the regenerated plants was determined by flow cytometry. More than half of the plants contained a nuclear DNA content of 1.3 to 3.9 pg/cell, which was higher than the content of *B. oleracea* but lower than that of *M. incana*. H1 contained 4.89+or-0.02 pg of DNA per cell, while H3 nuclear DNA content was estimated at 4.87+or-0.06 pg/cell. Cytological study of the root-tip cells revealed that the majority of the H1 and H3 hybrid cells contained 28 chromosomes (Sheng *et al.*, 2008).

Molecular breeding

In various hybrids of *Matthiola incana* (2*n*=14) only 4 has shown intermediate phenotype only two of them, having 28 chromosomes, were found hybrids by Random amplified polymorphic DNA (RAPD), sequence-related amplified polymorphism (SRAP) and cytological analyses (Sheng *et al.*, 2008).

Genetic engineering

GL3 and MYC-146 genes from Arabidopsis were incorporated in white petal *Matthiola incana* by upon microprojectile bombardment resulted in anthocyanin development (Nicola *et al.*, 2003). *Matthiola incana* (Brassicaceae), used as a model system to study biochemical and genetical aspects of anthocyanin biosynthesis, several nearly isogenic coloured wild type lines and white-flowering mutant lines are available, each with a specific defect in the genes responsible for anthocyanin production (genes e, f, and g). For gene f supposed to code for chalcone synthase (CHS; EC 2.3.1.74), the key enzyme of the flavonoid/anthocyanin biosynthesis pathway belonging to the group of type III polyketide synthases (PKS), the wild type genomic sequence of *M. incana* line 04 was determined in comparison to the white-flowering CHS mutant line 18. The type of mutation in the chs gene was characterized as a single nucleotide substitution in a triplet AGG coding for an evolutionary conserved arginine into AGT coding for serine (R72S). Northern blots and RT-PCR demonstrated that the mutated gene is expressed in flower petals. Western blotting and enzyme assays with various starter molecules, revealed that the mutant protein had no detectable activity, indicating that the strictly conserved arginine residue is essential for the enzymatic reaction (Vera *et al.*, 2004).

SUMMARY

In spite of this rather extensive breeding work, little has been done so far to explain the genetic determinism of several quantitative characters of plant and flowers, which confer ornamental value to the gillyflowers. Double matthiola flowers are completely sterile since they consist of masses of indefinitely replicated petals, doubles can be obtained by seed of heterozygous singles (Ss). Singleness is recessive character. It is noticeable that the additive effects were the largest component of the total genetic effects for number of flowers/inflorescence, diameter of flower, persistence of blooming while, for plant height the dominance effects were slightly higher than the additive ones. Hence, plant height can be improved by heterosis breeding.

REFERENCES

Ardelean, M., Pui, D.A., Cordea, M. and Mihalescu, L. 2009. Genetic effects involved in the heritability of several quantitative characters of ornamental value in gilly flower (*Matthiola incana* L.). Plant Characters. *UASVM, Horticulture*, **66**(1): 13-16.

Ardelean, M., Sestras, R., Pui, D.A., Cordea, M. and Mitre, V. 2008. General (GCA) and Specific (SCA) combining ability in several intraspecific hybrids of gilly flower (*Matthiola incana* L.) and their value for breeding. *UASVM, Horticulture*, **65**(1): 44-47.

Chlebowski, B. 1985. Comparision of the effectiveness of five maintenance breeding methods for stock (*Matthiola incana annua fl. Pl.*). *Prace Instytutu Sadownictwa I Kwiaciarstwaw Skierniewcach B Rosliny Ozdobne*. **10**: 58-75.

Cordea, M., Ardelean, M., Pui, D.A. and Mihalescu, L. 2009. Genetic effects involved in the heritability of several quantitative characters of ornamental value in gillyflower (*Matthiola incana* L.). II. Flower Characters. *UASVM, Horticulture*, **66**(1): 40-43.

Ecker, R., Barzilay, A. and Osherenko, E. 1993. The inheritance of flowering time in garden stock (*Matthiola incana* R. Br.). *Euphytica*, **72**(1-2): 153-156.

Ecker, R., Barzilay, A. and Osherenko, E. 1994. Linkage relationships of genes for leaf morphology and double flowering in *Matthiola incana*. *Euphytica*, **74**(1-2): 133-136.

Ecker, R., Zutra, D., Barzilay, A., Osherenko, E. and Rav- David, D. 1995. Sources of resistance to bacterial blight of stock (*Matthiola incana* R. Br.). *Genetic Resources Crop Evolution*, **42**(4): 371-372.

Frost, H.B., Lesley, M.M. and Locke, W.F. 1969. Cytogenetics of a trisomic of Mattiola incana involving a ring chromosome and somatic instalibility of singleness (Versus doubleness) of Flowers and shapes of leaves. University of California Citrus Experiment Station, California, pp. 1-17.

Gautam, V.K., Mittal, A., Nanda, K. and Gupta, S.C. 1983. *In vitro* regeneration of plantlets from somatic explants *Matthiola incana*. *Plant Science Letters*, **29**(1): 25-32.

Greuter W., Burdet, H.M. and Long G. 1986. (eds.) Med-Checklist 3. Conservatoire et Jardin botaniques, Geneve.

Hesar, A.A., Kaviani, B., Tarang, A. and Zanjani, S.B. 2011. Effect of different concentrations of kinetin on regeneration of *Matthiola incana*. *Plant Omics*, **4**(5): 236-238.

Kaviani, B., Hesar, A.A. and Kharbian-Masouleh, A. 2011. *In vitro* propagation of *Matthiola incana* (Brassicaceae) an ornamental plant. *Plant Omics*, **4**(7): 435-440.

Kaviani, B., Hesar, A.A., Tarang, A., Zanjari, S.B., Hashemabadi, D. and Rezaei, M.A. 2011. Callus induction and root formation on the leaf micro-cuttings of *Matthiola incana* using Kn and NAA. *American-Eurasian J. Agril. Envir. Sci.*, **11**(3): 456-461.

Lin, Liang Bin., Ma, zhanQiang., Zhang, ChuanLi., Liv, YaTing., Mao, Xiao Qiang., Xia, Guo Yin. and Ma, ZhongQin. 2008. Studies on somatic hybridization between *Brassica napus* and *Matthiola incana* B. Br. *J. Yunnan Agil. Univ.*, **23**(3): 295-301.

Lojacono-Pojero, M. 1888. Matthiola R. Br. *In*: Flora Sicula 1. Stab. Tip. Virzý. Palermo, pp. 99-102.

Mii, M. 2012. Ornamental plant breeding through interspecific hybridization, somatic hybridization and genetic transformation. *Acta Horticulturae*, **953**: 43-54.

Nicola R.A., Walker, Amanda., Mooney, R., Mark, Gray and John, C. 2003. Two basic-helix-loop-helix genes (MYC 146 and GL3) from Arabidopsis can activate anthocyanin biosynthesis in a white-flowered *Matthiola incana* mutant. *Plant Mole. Bio.*, **52**(3): 679-688.

Preda, M. 1979. Floriculture (in Romanian), *Ed*. Ceres Bucuresti.

Pui, D.A. and Ardelean, M. 2007. Heterosis manifestation of several quantitative characters of ornamental interest in gilly flower (*Matthiola* sp.) breeding. *Contributii Botanice*, **42**: 99-104.

Pui, D.A., Ardelean, M., Cordea, M., Mitre, V. and Nicoara, A. 2007. Heritability of several quantitative characters of interest in gilly flower (*Matthiola incana*) breeding. *UASVM Horticulture,* **64**(1/2): 54-57.

Raimondo, F. M., Gianguzzi, L. and Ilardi, V. 1994. Inventario delle specie "a rischio" nella flora vascolare nativa della Sicilia. *Quad. Bot. Ambientale Appl.*, **3**: 65–132.

Ruth, J., Juli, C., Jorge, A., Reyes, B., Hossein, A., Olga, F., Julia, P., Rosa, F. and Águedo, M. 2009. The molecular phylogeny of Matthiola R. Br. (Brassicaceae) inferred from ITS sequences, with special emphasis on the Macaronesian endemics. *Mole. Phylogenetics Evolution*, **53**: 972–81.

Sabaghi, R., Arab, M., Lofti, M., Akbari, M. and Azizinia, S. 2013. Morphological characteristics evaluation of induced mutant lines of stock var. Cextrum White. *Annals Biol. Res.*, **4**(4): 152-57.

Selaru, E. 1998. Garden Flowers (in Romanian), *Ed*. Ceres Bucuresti.

Sheng, XiaoGuang., Liu, Fan., Zhu, YueLin., Zhao, Hong., Zhang, Li. and Chen, Bin. 2008. Production and analysis of intergeneric somatic hybrids between *Brassica oleracea* and *Matthiola incana*. *Plant Cell, Tissue and Organ Culture*, **92**(1): 55-62.

Singh, A.K. 2006. Flower Crops: Cultivation and Management. Pub. New India Publishing Agency, New Delhi. pp. 347-55.

Tatsuzawa, F., Saito, N., Toki, K., Shinoda, K. and Honda, T. 2012. Flower colours and their anthocyanins in *Matthiola incana* cultivars (Brassicaceae). *J. Japanese Soc. Hort. Sci.* **81**(1): 91-100.

Vera, H., Dressel Angela., Epping Brenhard., Lukacin Richard., Martens Stefan and Austin Michale. 2004. Characterization and structural features of a chalcone synthase mutation in a white-flowering line of *Matthiola incana* R. Br. (Brassicaceae). *Plant Mole. Bio.*, **55**(3): 455-465.

Yabuya, T. 1991. Aneuploidy of ornamental species. Genetics and Breeding of Ornamental Species. pp. 39-52.

Section VI

Loose Flower Crops

Chapter – 27

Crossandra

Crossandra belongs to family Acanthaceae. It is also known as Kanakambaram in Tamil or Kanakambara in Kannada and Fire cracker in English because of cracking sound during opening of seed pod. Crossandra is derived from Greek word "*krossoi*" meaning fringe and "*aner*" meaning male i.e. fringed anthers. Crossandra is an evergreen shrub and one of the important loose flower used for adornment. The flowers are commonly grown for flower adornment in the form of garlands, *venis* and *gajras*. Though it is not fragrant, flowers are very popular because of its attractive bright colour, light weight and good keeping quality. These are used for making garland, either alone or in combination with jasmine flowers. Using crossandra flowers in combination with jasmine is becoming increasingly popular in India, particularly in southern parts, because the jasmine flowers provide colour contrast and the desired fragrance. Crossandra varieties show a remarkable range of colours, varying from orange, pink, red, yellow and double coloured blue-type with white throat.

ORIGIN AND DISTRIBUTION

Crossandra is native to Indo-malaya region. They are all native to sub-tropoical regions including India, Southeast Asia, Africa, and Madagascar.

Crossandra Salisb is a genus of about 50 species which are shrubs or herbs from Tropical Africa, Madagascar and Arabia, where it is grown as a perennial shrub. *Crossandra undulaefolia* Salisb. (Syn *C. infundibuliformis* Nees) is a native of East Indies. However, it is sometimes considered to be native of south India and Ceylon. C. *guinensis* is native of east Africa while *C. nilotica* has its origin in east Africa and Mozambigue. *C. flava* is a native of tropical west Africa.

The crossandra plants are distributed in tropical as well as subtropical regions of the world such as south Asia, South America, South Africa and Madagascar. In India, it is commercially cultivated in southern states of Karnataka, Tamil Nadu and Andhra Pradesh (Singh, 2006).

SPECIES

Crossandra undulaefolia is an important species which is grown for commercial value across the world. The genus consists of around 50 species of tropical plants with only a few cultivated species such as *C. undulaeflia, C. mucronata, C. guineensis* and *C. sebaculis* are cultivated, whereas, *C. greenstockii* is a wild species. *C. nilotica* is another sp. grown in Africa. Other species are *Crossandra tridentate, Crossandra thomensis.* The following are few important species of *Crossandra*.

***C. guineensis*:** It is free flowering and about 30-60 cm in height. The bracts do not bear spines and are pale lilac in colour. It can be grown both in sunny situation as well as in partial shade.

***C. nilotica oliver*:** The plants are about 60 cm in height. Stem is semi-woody. Leaves are elliptic, dark green and glossy, borne in whorls of 4 and are about 5-7 cm in length. Brick red flowers are borne in dense spikes. Bracts are green, about 2 cm long, veined and hairy. Each flower consists of a long slender corolla tube at the end of which there are 5 petals like lobes. Lower 3 petals form a lip. This is a free flowering plant suited for partial shade.

***C. flava*:** It is an unbranched shrub about 15-20 cm tall with erect and green stem. Leaves are glabrous, opposite, obovate to lanceolate in shape and dark green in colour. Lower leaves are stalked while upper ones are sessile. Spike is 4 sided with spiny yellow green bracts. Flowers are bright yellow and the corolla tube is much exerted and jointed.

***C. undulaefolia* Salis (*C. infundibuliformis* Nees):** It is a short branching perennial shrub about 30-90 cm in height. Leaves are about 5-12 cm, long dark green, lustrous and pointed. Leaves are generally opposite, narrowly oval having more or less wavy margin. Flowers are showy, bright orange, salmon to scarlet in colour. They are borne in a large densely packed spike. The spikes are 7.5 to 12.5 cm long and 4 sided. Individual flowers consist of long slender tube about 2.5 cm long which terminates in a 5 lobed limb. Flowers have prominent spiny bract. The calyx is also 5 lobed and irregular and there are about 4 stamens. *C. undulaefolia* is also known as Fire Cracker plant due to the cracking sound of the splitting of the seed pod.

***C. pungens* Lindav:** It is dense sub shrub upto 60 cm tall. Leaves oblong, dull green, traced with pearly veins. Spikes are conjested, bracts broadly ovate, margins bristly to spiny and colour of flowers is orange. Cultivars Orange, Delhi, Lutea Yellow and Sebaculis Red belong to this species.

FLOWER MORPHOLOGY AND BIOLOGY

Inflorescence of crossandra is terminal, spicate, often on long peduncles, bracts large, ovate, usually pubescent, sometimes papery, bracteoles as long as calyx. Calyx sub irregular, often with stalked, glandular hairs, lobes 5, elliptic or lanceolate, with uppermost lobe broadest and often 2-toothed; tube shorter than lobes. Corolla is one lipped, five lobed, lobes are rotund; imbricate aestivation; tube is longer than lobes and are cylindric. Androecium consist of four fertile stamens, di-dynamous, arising in upper part of corolla tube, included, filaments linear, shorter than anthers; anthers are monothecous; theca sometimes ciliate, muticous; staminodes. Pollen is prolate, 3-colpate; surface is foveolate and discannular. Female parts called gynoecium consist of ellipsoid ovary, with 2 ovules in each locule, style is filamentous, included; two style branches equal, cylindriform. Capsule is oblong-elliptic, estipitate, with inelastic placental bases. Seeds are ovate, compressed, with hygroscopic scales over whole surface. Flowers are borne in terminal/auxillary dense sessile spike taking 18-25 days from initiation to full bloom stage. The pollen of one species, *C. flava* was found to be highly variable, with both ridged and unridged pollen, and some abnormally shaped sterile pollen grains. Crossandra comes to flowering stage 2-3 months after planting and continues to bear flowers throughout the year with a drop in production during the rainy season. The flowers open in sequence from the base of the spike. Two flowers which are diagonally opposite in the spike open at the same time. It takes about two days for complete opening of the flower.

GENETICS

Crossandra is highly self-pollinated crop and homozygous in nature. The diploid chromosome number of crossandra is 2n=40. Pandey *et al.* (1986) developed colchitetraploid with bigger, thicker and deeper green leaves, larger flower but reduced spike length. Meosis showed varying number of quadrivalents, bivalents and univalents.

Heritability and genetic advance

Genetic variation, heritability, and genetic advance for 12 quantitative characters (plant height, number of days to first flowering, number of branches and spikes per plant, spike length, number of flowers per spike, flower length and diameter, flower yield per plant, shelf life, 1000-seed weight, and seed germination) were estimated in 18 genotypes of crossandra grown in Usuppur, Tamil Nadu, India. The phenotypic coefficient of variation (PCV) was higher than the genetic coefficient of variation (GCV), indicating significant environmental effect on all characters. GCV and PCV were high for number of days to first flowering and number of spikes per plant. Seed germination, number of branches and spikes per plant, self life, flower yield per plant, and number of days to first flowering recorded high heritability values and high genetic advance. Spike length is an important character in crossandra, which along with flower weight and number of flowers decides the yield level of varieties. Long corolla tube is a desired varietal trait, which makes preparation of garlands easier as compared to short ones.

BREEDING OBJECTIVES

i. Good plant growth
ii. More number of branches
iii. Profuse flowering
iv. Good spike length
v. Long duration of flowering
vi. Good quality of recovery from pressure
vii. Light weight flowers
viii. More shelf life
ix. New flower colour
x. Adaptibility to various agroclimatic conditions
xi. Resistance Root-knot nematode (*Meloidogyne incognita*)
xii. Resistance wilt-causing fungus (*Fusarium solani*)

BREEDING METHODS

Most of the breeding work in crossandra has been done in Southern parts of India. The breeding work done at the Tamil Nadu Agricultural University (TNAU) resulted in a hybrid between two local types collected from Palni and Marudha malai hills. This hybrid produces larger flowers and has high degree of field tolerance to nematode fungus complex. The flowers are very attractive with yellowish orange petals. Another hybrid, evolved by crossing the orange-flowered cultivar and the Sebaculis Red, produces very attractive dark pink flowers.

Introduction

In India several morphologically different varieties have been identified but there is no introduced variety is reported, whereas breeders have introduced a number of colour variations including the Mona Wallhed, a Swedish cultivar, which has a lovely flower colour as given in Table 1.

Table 1: Introduced crossandra variety.

Variety	Country	Characters
Mona Wallhed	Sweden	Shiny black-green foliage and a very compact habit of growth. Leaves are lustrous and vivid green. Flowers are deep salmon to pink, irregular with prominent bract.

Selection

Crossandra is commercially grown in South India. There are several morphologically different germplam from which some varieties evolved by different breeding methods. A field study on characterization and evaluation of local germplasm of crossandra was conducted at ICAR Research Complex, Old Goa. Significant variation among seven accessions used for the study was noticed for various morphological and yield parameters. Based on the plant habit, accessions were classified into erect and semi erect. Plants were tall in CG-4, CG-2 and CG-7 whereas they were dwarf in stature in CG-6. Accession CG-4 was quite early in flowering while CG-2, CG-5 and CG-6 were found late in reaching flowering stage. Maximum spike length, corolla tube length and flower diameter were recorded in CG-2. Accession CG-5 produced the highest number of spikes/plant. The best performance in respect of number of flowers/spike and weight of 100 flowers and flower yield/plant was observed in CG-2 (Ramachandrudu and Thangam, 2010). Morphologically different genotypes of crossandra have been observed from Karnataka, Tamil Nadu and Kerala which can be selected in future for further improvement. Some varieties developed through selection are given in Table 2.

Table 2: Crossandra varieties evolved through selection.

S. No.	Variety	Characteristics
1.	Fortuna	It is a breakthrough in crossandra breeding. It has greatly improved root system, which makes the plant far more resistant to temperature fluctuations. It has more lavish flowering, glossier foliage, brighter coloured flowers and the extended life of the plant.
2.	Diane	A new and distinct cultivar of *Crossandra infundibuliformis*, having larger sized and more broadly shaped leaves, More robust and compact habit of growth and its flowers color is a dark, richer shade of salmon orange. Petals are more rounded and larger in size than those of the commonly cultivated variety.
3.	Soundarya	Selection from Delhi Kanakambaram.
4.	Mahatma	Selection from Delhi Kanakambaram.

Hybridization

Hybridization between the Local cultivars from Karnataka, Tamil Nadu, Andhra Pradesh, Pondicherry and improved cultivars of Lakshmi and Delhi Crossandra yielded many F_1 hybrids with varying morphological characters. Two F_1 hybrids IIHR 2004-09 and IIHR 2004-11 were selected among the large number of hybrids for their unique larger size, which is 67 and 69% higher than the Local cultivars, respectively. The Hybrid selections IIHR 2004-09 and IIHR 2004-11 are the result of crossing between two homozygous lines and the increase in size in these hybrids is due to heterotic effect. Later these two hybrids released as Arka Kanaka (Local Yellow × Delhi Crossandra and Arka Ambara (Local × Laxmi Crossandra). These recommended for commercial cultivation of loose flowers and potted plants. Since the flowers are novel in colour and size they may fetch fancy price in the market (Aswath *et al.*, 2009a).

Two promising hybrids were proposed for registration viz., IIHR 2005-1 which is a hybrid between Local × Arka Ambara. The plant and flower morphology of this hybrid resembles that of cv. Arka Ambara, however with a change in flower colour. The flower size and yield is superior to Arka Ambara and it has an unique flower colour (Orange Red group 32.A). The hybrid selection IIHR 2005-2 is a hybrid between *Crossandra nilotica* × Mangalore Local. The size and the yield of the flower are bigger than both the parents (Aswath *et al.*, 2009b). These two hybrids named as Arka Shreya and Arka Shravya, respectively. Some varieties developed from IIHR, Bengaluru through hybridization are described in Table 3.

Table 3: Crossandra varieties developed through hybridization.

S. No.	Hybrid	Parents	Remark
1	Arka Kanaka	Local Yellow × Delhi Crossandra	Crossandra hybrid with novel colour
2	Arka Ambara	Local × Laxmi Crossandra	Crossandra hybrid with bigger sized flowers.
3	Arka Shreya	Local × Arka Ambara	Its flower colour is reddish-orange and does not set seed.
4	Arka Shravya	*C. nilotica* × Mangalore Local	Flower colour is reddish-orange and does not set seed.

Mutation

Ornamental plants are ideal for the application of mutations induction techniques because many economically important traits for example flower characteristics or growth habit and are easily observed after the mutagenic treatment. Induced mutation in ornamental comprises traits such as altered flower characters, leaf characters, leaf characters, growth habit and physiological traits. Several varieties of crossandra have been developed through physical or chemical mutagens are given in Table 4.

Table 4: Mutant varieties of crossandra.

S. No.	Variety	Parent	Characteristics
1	Lakshmi	Delhi	Developed through gamma irradiation. Potential flower yield up to 75 kg/ha in a day, starting from second year of planting. Variety is same as parent in appearance but it is differentiated by pleasing orange colour large flowers.
2	Dr. APJ Abdul Kalam	Deep red mutant of Delhi	Developed by gamma irradiation. Bears deep red flowers which are first time larger than Delhi Crossandra flowers. Flower throughout the year and is resistant to major pests.
3	Maruvur Arasi	Red mutant of Delhi	Developed by gamma irradiation, bright deep red colour flowers. The flower stalks longer than normal, high yielding than Delhi Crossandra and each stalk bore over 75 flowers.
4	Kanakadhara	Delhi	Developed using sodium azide. Brilliant orange flowers of normal size and it had unique floral characteristics. It can yield 25 per cent more flowers than Delhi Crossandra.
5	Vijaya Kanakambaram	Local variety	Developed by treating seeds with colchicine. Large flowers light reddish orange colour. Yield 3 times more than the local varieties. Yield 75 kg flowers/ day/ha from the 90 days after planting.
6	Raj Kankambaram	Delhi	Its inflorescence is also long and it produces deep orange flowers. It has leaf variation.
7	Subasu	Delhi	It resembles the parent. It has long inflorescence and light orange, medium-sized flowers.
8	Neelambari	Local Yellow variety	It has dense green foliage. It produces normal sized flowers in long inflorescence.

Polyploidy

The allopolyploid has been synthesized in *C. infundibuliformis* with the use of colchicine. Gigas condition was observed in regard to the number of external morphological characters of the polyploid as compared to its diploid. The polypolid produced bigger sized and attractive flowers as compared to the normal plant (Cruz *et al.*, 1973). Polyploidy varieties of crossandra are described in Table 5.

Table 5: Polypoidy varieties of crossandra.

S. No.	Variety	Ploidy level	Remark
1	Delhi	Triploid (2n=40)	Produces more attractive flowers of bright deep orange colour.
2	Orange Crossandra	Tetraploid (2n=40)	Sets seeds profusely, breeds true and produces bright orange coloured flowers.
3	Lutea Yellow	Tetraploid (2n=40)	Flowers are orange yellow colour. It is used as front line of shrubbery and hanging baskets.
4	Sebaculis Red	Tetraploid (2n=40)	Hardy cultivar, which possesses high degree of tolerance to nematodes.

BIOTECHNOLOGY

Tissue culture

In India, the plant *Crossandra infundibuliformis* is considered an aphrodisiac. Recently, a promising triploid cultivar of crossandra has been released and there is a great demand from farmers for planting material. Unlike the earlier (wild) cultivars these cultivar can flower throughout the year, irrespective of the season. As they do not produce seeds, the only way of propagation is through cuttings. Propagation by conventional methods necessarily limits the rate of output and makes the end product expensive. Micropropagation is being used extensively for the rapid clonal propagation of many fruits, nuts and ornamental trees (Zimmerman, 1985). It also enables rapid clonal propagation and hastens the availability of planting material (Zimmerman, 1981). A preliminary study on crossandra was made by Bonke *et al.* (1983), but with limited success. However, none of the above studies optimized conditions for effective micropropagation of crossandra. For commercial micropropagation an efficient multiplication system is essential (Watad *et al.*, 1995). In order to meet the demand an attempt has been made to evolve mass multiplication through *in vitro* culture.

Multiple shoots were induced from axillary and apical buds of *Crossandra infundibuliformis* on MS medium containing BAP and kinetin individually. The maximum number of multiple shoots were obtained from MS medium containing BAP 1 mg^{-1} (Girija *et al.*, 1999).

Molecular breeding

ISSR primers were used to trace the genetic variation in 19 cultivars/lines in crossandra where these cultivars were grouped into two main clusters consisting of 12 and 7 cultivars. It was found that *C. nilotica* sp. were interspersed with Local, Lakshmi and Delhi. Crossandra is affected by fusarium wilt and the crop devastated by this wilt from 40 to 100% during rainy season. An attempt has been made by Aswath (2005) to provide information on the method of selection of new wilt resistant crossandra line through somaclonal induced variation. He studied the influence of fusaric acid, a non-specific toxin produced by *Fusarium*, on the germination of fire cracker (*Crossandra undulaefolia* [*C. infundibuliformis*]) seeds and on the induction, maintenance and regeneration of callus. The sensitivity to fusaric acid depended on the callus line. Regeneration ability of the callus decreased tremendously after the callus had been selected on the medium containing 0.5 mm fusaric acid. When the selected calluses were transferred to regeneration selection medium, their survival and regeneration were lower than their corresponding controls. These results indicated that the two steps of callus and regeneration selection were not practical. The results of this experiment may provide information on the methods of selection of new wilt resistant crossandra line through somoclonal-induced variation. Assessing genetic diversity of crossandra (*C. undualaefolia*) cultivars ISSR markers were used (Aswath *et al.*, 2009a).

REFERENCES

Aswath, C. 2005. *In vitro* selection for fusarium wilt resistance in Fire cracker (*Crossandra undulealia folia* Selisb). *J. Orna. Hort.*, **8**(4): 260-263.

Aswath, C., Rao, T. M., Ramachandran, N. 2009b. Promising crossandra F_1 hybrids IIHR 2005-1 and IIHR 2005-2. *J. Ornamental Hort.*, **12**(3): 211-212.

Aswath, C., Sane, A., Arthikirubha, A., Rao, T.M. and Ramchandran, N. 2009a. Assessing genetic diversity of crossandra (*C. undualaefolia*) cultivars using ISSR markers. *Abst*: National Conference on Floriculture for Livelihood and Profitability. ISOH, New Delhi. pp. 214.

Bonke, E., Dorfner, M. and Sonneborn, H.H. 1983. Micropropagation of *Crossandra infundibuliformis. Acta Horti.* **132**: 71-74.

Cruz, R.D., Jadhav, A.S. and Rane, M.S. 1973. Studies in polyploid *Crossandra infundibuliformis. Indian J. Hort.*, **30**: 551-52.

Girija, S. Ganapathi, A. and Vengadesan G. 1999. Micropropagation of *Crossandra infundibuliformis* (L.) Nees. *Scientia Horticulturae*, **82**: 331-337.

Pandey, R.M., Pal, M. and Bhatnagar, A. 1987. Colchitetraploidy in *Crossandra infundibuliformis* Nees. *New Bot.*, **14**: 73-77.

Ramachandrudu, K. and Thangam, M. 2010. Characterization and evaluation of local germplasm of crossandra (*Crossandra undulaefolia* Salisb.). *J. Orna. Hort.*, **13**(2): 138-141.

Sharmila, N. and Gomathi, N. 2011. Antibacterial, Antioxidant activity and Phytochemical studies of *Crossandra infundibuliformis* leaf extracts. *Internatl. J. Phytomedicine*, **3**: 151-156.

Singh, A.K. 2006. Flower Crops: Cultivation and Management. Pub. New India Publishing Agency, New Delhi. pp. 93-99.

Watad, A. A., Kochba, M., Nissim, A. and Goba, V. 1995. Improvement of Aconitum napellus micropropagation by liquid culture on floting membrane rafts. *Plant Cell Rep.*, **14**: 345-348.

Zimmerman, R.H. 1981. Micropropagation of fruit plants. *Acta Horti.*, **120**: 217-22.

Zimmerman, R.H., 1985. Application of tissue culture propagation to woody plants. *In*: Henke, R.R., Hughes, K.W., Constain, M.J., Hollander, A. (eds.). Tissue Culture in Forestry and Agriculture, Plenum, New York, pp. 165-77.

Chapter – 28

Jasmine

Jasmine (*Jasminum* spp.) belongs to the family Oleaceae, order Oleales are climbing, trailing or erect shrubby plants and there are both deciduous and evergreen species. The generic name *Jasminum* is derived from Arabic word "*Jassamine*", "*Jasmin*" and "*Jasminum*" (Bailey, 1963). Flowers are white, yellow or rarely reddish. Flowers are used for making garlands and *Veni*. Jamine oil extracted from the flower is highly valued in perfumery industry. France, The Netherlands, Italy, Belgium, Algeria, Turkey, Morocco and Tunisia are major jasmine oil producing countries (Singh, 2006).

ORIGIN AND DISTRIBUTION

Jasmine is native to tropical and subtropical regions; a large number of species are centred on the regions comprising India, China and Malaysia. Jasmines distributed in warm parts of Europe, Asia, Africa and Pacific region, nearly absent from America (Bailey, 1963).

Jasmines have adorned the gardens of central Asia, Afghanistan, Iran, Nepal and many other tropical and sub tropical countries. The distribution patterns of nearly 72 species in India, Malaysia and China provide a strong base to claim that India is one of the principal centers of origin of *Jasminum* species. Among these, about 40 species are reported to occur in India and mostly cultivated in South India (Bhattacharjee, 1980). Large number of species are however, centered to the region comprising of The Himalayas, China and Malaysia. Other countries where jasmine is cultivated are France, Italy, Morocco, Algeria, North Africa, Spain and Egypt. The cultivated jasmines fall into four species *viz., Jasminum sambac, Jasminum auriculatum, Jasminum grandiflorum,* and *Jasminum multiflorum* (Kakada).

In India, the three species *Jasminum sambac, Jasminum auriculatum* and *Jasminum grandiflorum* have been mentioned in the ancient Tamil Literature of 'Sangam' period (500 B.C. to 200 A.D.). Jasmines are commercially grown in Karnataka, Tamil Nadu and Andhra Pradesh. Other growing states are Madhya Pradesh, West Bengal, Uttar Pradesh and Bihar.

SPECIES

Although more than 200 species are known, 40 species have been identified in India and 20 are cultivated in South India (Bhattacharjee, 1980). Some species under the genus *Jasminum* are *J. abyssinicum* Hochst., *J. adenophyllum* Wall., *J. amplexicule, J. angulare* Vahl., *J. angustifolium, J. arborescens, J. atttenuatum, J. auriculatum, J. azoricum, J. beesianum, J. calophyllum, J. cinnamomifolium, J. coffeinum, J. crabibianum, J. cuspidatumkmkm, J. dichotomum* Vahl, *J. dispermum, J. favreri. J. floridum, J. fluminense, J. fruticans, J. gracillimum, J. didymum, J. grandiflorum* L., *J. humile* L., *J. lanceolarium* Roxb., *J. mesnyi* Hance, *J. multiflorum* Hance, *J. multipartitum* Hochst., *J. nervosum* Lour., *J. nitidum, J. nudiflorum, J. odoratissimum* L., *J. officinale* L., *J. parkeri* Dunn, *J. polyanthum* Franch., *J. primulinum, J. pubigerum, J. revolutum, J. rex, J. sambac* L., *J. smilacifolium, J. stephanense, J. sinense* Hemsl., *J. trinerve, J. urophyllum, J. wallichianum*, etc. The new species *J. cuneatum* Jongkind differs from *J. paucinervium* Benth., by its cuneate and not rounded or cordate leaf base and the absence of pit domatia. It differs from *J. kwangense* Liben from Central Africa by the more papery leaves without abundant glandular dots and pit domatia on their lower surface (Jongkind, 2010). Some commercially important species are described below (Singh, 2006).

***J. auriculatum*:** Plant is shrubby with shiny leaves and auricles. Flowers are white, sweet-scented borne in pubescent compound many flowered flax cymes. Corolla lobes are elliptical and fruit is black. Flowers are used for production of perfumes.

***J. grandiflorum* syn. *J. officinale* var. Grandiflorum:** Called as common or Spanish jasmine. It is woody bush with pinnate leaves bearing 3 to 5 leaflet of equal size. Bears large, white flowers with delightful fragrance borne in axillary or terminal cyme.

***J. sambac*:** It is also called Arabian or Tuscan jasmine. Evergreen twiner having simple cordate to oblong dark green leaves. Flowers are white, fragrant and small with three forked cymes. Essential oil is extracted from flowers to make perfume.

***J. officinale*:** Commonly termed as Poet's jasmine or White jasmine. It is a semi-evergreen or deciduous climber bearing stalkless leaflets. Flowers are white and fragrant, borne at branch terminal in loose clusters.

***J. multiflorum* (*J. pubescens*):** Commonly called as Fussy/Downy/Star Pinwheel jasmine. It is a strong woody vine bearing simple and opposite leaves. Flowers are sessile, fragrant, large, white and borne in terminal umbels, having 6 to 9 petals.

FLOWER MORPHOLOGY AND BIOLOGY

Flowers appear in clusters and sometimes solitary on the end of branchlets. Inflorescence is cymes which is bi or trichotomous. Flowers are bisexual, usually heterostylous and fragrant. Bracts are linear, simple and their shape is ovate. Sometimes it is petaloid. Calyx is bell shaped, campanulate, cupular, or funnelform, 4-16-lobed. Corolla is salver or funnel shaped and its colour is white, yellow, violet and sometimes reddish. It has 4-9 spreading lobes convoluted in the bud, much extending the calyx. Unlike most genera in the Oleaceae family, which have four corolla lobes petals, jasmines often have five or six lobes. Stamens are two included in corolla tube, filaments short, anthers are dorsifixed and introrse. Style is filiform, stigma capitate or 2-lobed. Ovary is bilobed. Ovules 1 or 2 in each locule. Fruit is berry rarely with separate capsules and each having 2 seeds. Seeds are without endosperm and radicle downward.

Jasminum is known as heterostylous, that is, some individuals bear flowers in which all the styles are long, while in other individuals they are all short. The anthers of the *J. sambac* are above the stigma, while those of *J. multiflorum* and *J. officinale* var. *grandiflorum* are much lower than the stigma. All these species do not produce any fruit. The individual bloom of *J. sambac* 'Grand Duke of Tuscany' has layers (3-7) of petals each layer with its own petiole. The anthers and the stigma degenerate making the plant sterile (Rosario *et al.*, 1993). In order to set seed, long styled flowers must receive pollen from those with a short style and *vice-versa*. Plants were incompatible within a morph and the polymorphism was characterized by the relative heights of the stigmas and anthers. In natural populations, viable seed production was higher in the short-styled morph, but in controlled crosses seed set was similar in both forms (Dommee *et al.*, 1992). Thompson and Dommee (1993) investigated the role of four sequential factors which may differentially influence fruit and seed set in two flower morphs as given below.

i. Differences in flower phenology
ii. A limitation of pollen transfer towards pin
iii. A differential capacity of the two morphs to act through female and male function
iv. Differential fruit abortion in the two morphs

Three natural populations of the distylous *J. fruticans* showed that mean fruit and seed production was significantly greater in short styled plants (thrums) than in long styled plants (pins). Supplementary pollination in a natural population significantly increased fruit set and reduced the difference in fruit set between the two morphs in relation to the controls, indicating a limitation on pollen transfer which was most severe in pin stigmas. Furthermore, individual pin plants with relative high percentage seed set as female parents gave poor seed set as male parents and *vice-versa*. Whereas fruit removal had no effect in seed number in thrum plants, a greater proportion of viable seeds produced on pin plants which were left to naturally mature their fruits than on pins which had fruits artificially removed, suggesting the occurrence of selective fruit abortion in pins but not in thrums. Phanachana *et al*. (1994), while experimenting with use of chemicals for control flowering of jasmine in winter reported that most and least number of flowers was produced by chemical biotica 0.3 cc/l and kinetin 100 ppm + GA_3 500 ppm, respectively. While, control gave longest and biggest flower and GA_3 gave shortest and smallest flower.

Guitian *et al.* (1998) studied various aspects of the reproductive biology of 2 populations of the distylous shrub *J. fruticans*. Two morphs (the L-morph with long-style/short-pistil, and the S-morph with short-style/long-pistil) occur. Morph frequency ratio in the study populations was 1:1. Experimental crosses revealed self-incompatibility, very weak within-morph compatibility and strong between-morph compatibility (typical of distylous species). Pollen grains were larger in S-morph plants, while both nectar reward and fruit set were higher in L-morph plants. No differences were detected in number of pollen grains/stamen or in pollinator spectrum. Gao *et al.* (2002) analyzed the changes in the endogenous plant growth regulator contents during development and flowering of jasmine (*J. sambac*) and revealed that the average flower bud weight increased in a reciprocal equation model during plant development and was simultaneous with the increase in zeatin (Z) and ABA contents, but the change in Z:ABA ratio showed a high-low-high trend. The average weight increase of the flower bud could be related to the change in Z:ABA ratio before flower opening. The Z:ABA ratio reached a peak 3 h after harvesting because of the decrease in ABA. The opening degree increased sharply after the high peak of Z:ABA ratio. IAA was not detected during the development and flowering of jasmine. In *J. grandiflora* the range of fertility are below 98% (Noor *et al.*, 2004).

Mode of pollination

Jasmine is a cross pollinated crop. Being fragrant in nature the flower attract insects to it. Both morphs of *J. angustifolium* anthese just before dusk and secrete 0-8 micro l nectar/flower. The nectar contains sucrose, glucose and fructose. Raju (1988) observed one bee species *Amegilla* sp., foraged (for pollen) on the flowers, 13 species of butterflies foraging for nectar and one of these, *Macroglossum gyrans*, was considered to be the main pollinator. One species of fly and one thrips species were also seen on the flowers. The most frequent visitors to flowers of jasmine being species *Anthophora*, *Melecta* and *Bombus* (Guitian *et al.,* 1998). Generally, the number of visits by Diptera and Rhopalocera order of insects was positively related to the number of open flowers in a patch. Short-tongued bees showed a positive relationship between visitation rate and the number of open flowers on the focal stem. The differences between insect types in patterns of variation in visitation rates in response to floral design and display may act to diversify selection on floral traits (Thompson, 2001).

In *J. auriculatum* selfing did not result in any fruit set. The fruit set under natural conditions varied from 33.1 to 71.8%. The seed germination ranged from 75 to 95%. Veluswamy (1981) failed to get any fruit set on selfing in *J. grandifloram* but open pollination yielded 18.0 to 52.1% fruit set. The seed germination was satisfactory ranging from 68.8 to 74.1%. Among the cultivated jasmines, only *J. sambac* cv. Soojimalli, *J. auriculatum*, *J. flexile* and *J. angustifolium* did set fruit.

GENETICS

The basic chromosome number n is 13. The chromosomal behaviour of *J. sambac*, *J. multiflorum* and *J. officinale* var. *grandiflorum* is generally normal. Thirteen bivalents can be observed at diakinesis. Differential disjunction and laggards can also be observed but there frequency is lower. However, the frequency of triads at meiosis is higher (Rosario *et al.*, 1993).

Cytological investigations in jasmines were made by Krishnaswamy and Raman (1948), Raman (1955), Khan *et al.* (1969) and Murthy and Khanna (1971). The genus with basic chromosome number of n=13 generally consists of diploid form with 2n=26. However, natural occurrence of forms with higher ploidy levels was also observed.

Chromosome number has been determined in jasmine from time to time by many workers as it is very important parameter for breeding programmes. Sax and Abbe (1932) reported 2n=26 in *J. fruticans* L. Taylor (1945) carried out extensive chromosomal investigation in 129 species of family Oleaceae and reported n=13 as a basic number of jasmine. Krishnaswamy and Raman (1948) carried out cytogenetical studies in Indian jasmine and confirmed chromosome numbers of four varieties of *J. sambac* and other species. Khan *et al.* (1970) reported new chromosomal form of superior ornamental value in jasmine. A detail of chromosome numbers of different species of jasmine has been shown in Table 1.

The studies carried out in *J. auriculatum* with seedling progenies raised from different clones have indicated the existence of a higher level of phenotypic variation (64 to 92%) for yield followed by length of primary laterals at flowering (43 to 82%). The genotypic variation for yield was relatively low and ranged from 24 to 60%, thereby suggesting role of environmental influence on yield. In contrast to this, the variations observed for corolla tube length, bud length, bud diameter and weight of flower buds were predominantly genetic. Heritability was found to be maximum for corolla tube length followed by flower bud diameter (90 to 98%) and length of flower bud (83 to 98%). The genetic advance was lower than heritability for corolla tube length and diameter of flower bud. For the other characters, the genetic advance was higher. Heritability estimates varied with species and also among diploids and tetraploids within the same species. Among the different yield parameters, weight of flower buds exhibits higher heritability in both diploids and tetraploids of *J. auriculatum* and in diploids of *J. grandiflorum* (Veluswamy, 1981).

Table 1: Chromosome numbers of different species of *Jasminum*.

S. No.	Name of species	Chromosome no. and ploidy level
1.	*J. fruticans* L.	2n=26 (diploid)
2.	*J. affinine* Blume	2n=26 (diploid)
3.	*J. floridum* Burge	2n=26 (diploid)
4.	*J. humile* L.	2n=26 (diploid)
5.	*J. officinale* L.	2n=26 (diploid)
6.	*J. wallichianum* Lindl.	2n=26 (diploid)
7.	*J. nudiflorum* Lindl.	2n=52 (tetraploid)
8.	*J. sambac* var. sujimillige	2n=26 (diploid)
9.	Iruvantige (Viruprakashi)	2n=26 (diploid)
10.	Elusuttumallige	2n=26 (diploid)
11.	Dundumallige	2n=39 (triploid)
12.	*J. auriculatum* Wild	n=13 (diploid)
13.	*J. malabaricum* Wild	n=13 (diploid)
14.	*J. primulinum*	2n=39 (triploid)
15.	*J. flexile*	n=13 (diploid)
16.	*J. rigidum* Wild	2n=26 (diploid)
17.	*J. angustifolium* (wild and cultivated)	2n=52 (tetraploid)
18.	*J. sambac* cv. Gundumalli	2n=39 (triploid)

The number of primary laterals and length of style were positively and significantly associated with flower yield in *J. auriculatum*, length of flower bud, internodal length and number of days taken for flowering had significant association with length of corolla tube as reported by Thangaraj *et al.* (1980). Veluswamy (1981) also studied the relationship between the weight of pruned shoots and yield of flower buds and jasmine concrete recovery. The nature of association of these charactersters were found to be contrasting in the two species, *J. auriculatum* and *J. grandiflorum*. The genotypic and phenotypic correlation between weight of pruned shoots and yield of flower buds and between jasmine concrete recovery and yield of flower were negative in *J. grandiflorum* but positive in *J. auriculatum*. In path analysis, the direct effect of the number of flower buds per shoot, number of pruned shoots and weight of flower buds on yield were negative in *J. grandiflorum* but positive in *J. auriculatum*. The direct effect of length of shoot and bud diameter on yield of flower were positive in *J. grandiflorum* whereas the effects were negative in *J. auriculatum* suggesting that different approaches are needed for crop improvement in these species.

Ploidy level

The ploidy level (2n= 2x, 3x, 4x) varies from species to species in *Jasminum*. *J. sambac* var. Gundumalli (2n=39) and *J. flexile* (2n=52) are the natural auto-polyploids. Polyploidy is a prominent process in plant evolution which has also been reported in *Jasminum* (Taylor, 1945). Polyploidy may be spontaneous in nature or artificially induced by colchicines. Triploidy in *J. grandiflorum* has been found to increase concrete content and thereby hold promise as useful avenue for improvement of this crop. Spontaneous triploid in (2n=39) *J. sambac* and in *J. autumnale* has been reported by Sharma and Sharma (1958), in *J. primulinus* Henst by Krishnaswamy and Raman (1948). Spontaneous tetraploidy (2n=52) in *J. flexile* Vahl. have been reported by Raman (1955).

BREEDING OBJECTIVES

i. New flower colours and shades
ii. High yield of bold floral buds with long corolla tube for fresh flower market
iii. Longer flowering duration over the year
iv. Long lasting fragrance
v. Higher concrete (jasmine extract) yield
vi. Flowers with short corolla tube for quality concrete production
vii. Resistance to biotic stress [*Rhizoctania solani*, *Coccus viridis* (green scale), *Hendescasis duplifascialis* (jasmine flower borer)]
viii. Resistance to abiotic stress

a. For fresh flower market: To evolve superior varieties giving high yield of floral buds, spread over many months of year, long corolla tube, bolder buds and better keeping quality.

b. For essential oil extraction: To develop varieties with high oil content, flowers with short corolla tube as corolla tube contributes mostly plant waxes, thus reducing the quality of the jasmine concrete.

BREEDING METHODS

In India majority of work related to breeding of jasmine is done at TNAU, Coimbatore and IIHR, Bengaluru. Large Round, Large Point, Short Point are the varieties of *J. auriculatum* and Adukkaumalli, Elusuttumallige, Iruvatchi, Mohra, Madanban, Rambanam and Virupakshi are the varieties of *J. sambac*, developed at TNAU, Coimbatore. One variety *i.e.* Surabhi has been developed at IIHR, Bengaluru.

Selection

The clones of *J. grandiflorum* from different states of India showed considerable variation in respect of flower production as well as yield of jasmine concrete. The mean flower yield of six clones varied from 4329 kg/ha to 10144 kg/ha, while that of jasmine concrete showed a range of 13.85 kg/ha to 29.42 kg/ha (Veluswamy,1981). This leads to release of cultivar CO-1 Pitchi (*J. grandiflorum*) giving of flower bud pink and attractive with pleasant fragrance suitable both for fresh flower trade and oil extraction. Similar work carried out in *J. auriculatum* to the identification of a high yielding, gall mite resistant clone Parimulli which was released by the TNAU in 1971 for commercial cultivation. The clones characterized by long flowering season of nine months and a mean flower bud yield of 10 t/ha (Muthuswami and Madhva Rao, 1972).

The selection, nevertheless, suffers from undesirable corolla length. Further, secondary selection has shown that the clones Long Round is more prolific in yield and possesses larger corolla tube as well as bud length and this was released as CO-1 Mullai which is capable of yielding 8.8 t/ha with good consumers preference. In an evaluation of 40 jasmine collections spanning ten species, *J. pubescence* and *J. auriculatum* were promising for flower yield, while *J. grandiflorum* outranked others for jasmine concrete recovery (More, 1980). Veluswamy (1981) found clones of *J. grandiflorum* superior to *J. auriculatum* in flower bud yield, while in jasmine concrete recovery, *J. auriculatum* was better than with 344 mg/100 g as against 288 mg/100 g in the former. Among diploid clones of *J. auriculatum*, the clone Short Point was best in flower bud yield, while Thimmapuram and Lucknow of *J. grandiflorum* were adjudged as best jasmine concrete yielders. Varieties of jasmine developed through selection and have been released from TNAU, Coimbatore and IIHR, Bengaluru are given in Table 2.

Table 2: Varieties developed through selection with their chief characteristics.

S. No.	Variety	Selection	Chief characteristics
1.	CO-1 Mullai (*J. auriculatum*)1980	Selected from open pollinated seedling	Bold bud with long corolla tube, flower yield (8.8 t/ha), 0.34% concrete recovery.
2.	CO-1 Pitchi (*J. grandifolium*) 1980	Clonal selection from Lucknow type	Flower buds pink and attractive, high flower buds yield (10t/ha), high concrete recovery (0.29%).
3.	Parimullai (*J. auriculatum*) 1972	Clonal selection from Medium Point variant possessing gall mite resistance	High yield of flower bud and high concrete recovery (0.29%).
4.	CO-2 Mullai (*J. auriculatum*) 1988	Clonal selection from progenies with desirable flower exhibits complete field tolerance to the phyllody disease and gall mite	Long corolla tube, longer bud. Length of corolla tube 1.70 cm while it is 1.5 cm in CO-1. Flower yield 11.19 t/ as against 8.8 t/ha in CO-1.
5.	Arka Arpan (*J. multiflorum*) 1998	Selected from *J. multiflorum* var. pubescence	Pinkish flower buds, mildly fragrant flower, yield 6 t/ha.
6.	Arka Aradhna (*J. sambac*) 2000	Selected from Single Mogra	Flower double whorled, used as fresh flower and concrete extraction, yield 8 t/ha.

IIHR-JA-13 is a promising selection of *J. auriculatum* was developed at IIHR, Bengaluru for higher concrete (jasmine extract) yield. The concrete yield was higher in IIHR-JA-13 (31.2 kg/ha) than the control cultivar Parimullai (18.8 kg/ha) (Srivastava and Kumar, 2004).

Hybridization

Though some attempts have been done in genetic improvement through hybridization in jasmine but it appears to be complicated due to non-fruitfulness and cross-incompatibility barriers. The problem is also backed by interference of insects due to fragrance.

Genetic improvement in jasmines through hybridization appears to be complicated. The most important barrier is non-fruitfulness in most species and cross combinations. Strong cross-incompatibility barrier among the *Jasminum* species like *J. auriculatum*, *J. grandiflorum, J. flexile, J. calophyllum* exist and also a strong self-incompatibility system exist in *J. auriculatum* and *J. grandiflorum*; success ranged from 0.5-5%. When an induced tetraploid of *J. auriculatum* was used as female and *J. grandiflorum, J. angustifolium, J. flexile* and *J. calophylum* as donors (male parent), seed failed to germinate. Under natural conditions, seeds set vary with variety in species.

Peak and active flowering occurs during night. Breeders normally face difficulty in emasculation and pollination as styles in some cases long and very delicate, so risk of damaging the stigmatic region is very high. The problems are further aggravated by greater impact of insect interference due to fragrance. It has been observed that very minute insects move up and down the corolla tube, thus possibly making the stigma dusted with pollen.

Normally jasmine fruit contains only one seed. Growth of seedling is slow. Keeping all the points in view jasmine breeder has to devise suitable breeding technique and method to control the minute insect before emasculation and after pollination, so as to get wide range of variability for different plant characters. Yet, TNAU has released one hybrid cultivar *viz.* CO-2 Mullai. It is a cross between a mutant (SI) and Long Point. It yields 11,198 kg of flower buds/ha, accounting for an increase of 26.88% over CO-1 Mullai. Several inter-varietal crosses of *J. grandiflorum* were attempted at IIHR, Bengaluru. Only one combination *viz.*, Selection × Pink yielded 6.8% seed set. In others seed set was 0 to 3.2%. Though seeds failed to germinate but successful fruit set was reported in crosses involving *J. auriculatum* with *J. grandiflorum* and *J. flexile; J. grandiflorum* with *J. auriculatum, J. flexile, J. calophyllum* and *J. sambac* with *J. grandiflorum.*

Correlations among 11 morpho-economic characters and their path analysis were studied using 16 genotype of jasmine. The weight of 100 flowers was significantly associated with the flower diameter, petal width, petal length, anther length, floral bud length and width, and 100 buds weight in positive direction. Path coefficient analysis revealed 100 flower bud weight and flower diameter to have high direct effect in positive direction. All the other characters affected the yield indirectly through these two characters, indicating that 100 floral bud weight and flower diameter could be taken as selection index (Srivastava and Deyaiah, 2006).

The Jasmine hybrid which is developed through hybridization and selected for their unique characteristics is given in Table 3.

Table 3: Hybrid of jasmine developed through hybridization.

S. No.	Variety	Parent	Chief characteristics
1.	CO-2 Mullai	Mutant S I × Long Point (LA1).	Flower bud yield 11,193 kg /ha

Intraspecific (Intervarietal or Interclonal) hybrids crosses: Intra-specific hybridization success ranged from 0.5 to 35% in *J. grandiflorum*. The crossed seed germination ranged from 50-100%. Cross-compatibility was evident in inter-clonal crosses. In *J. auriculatum*, success ranging from 0 to 27% was obtained in Short Round × Short Point and Parimullai × Short Point and induced tetraploid × diploid.

Polyploidy breeding

In general, triploids have been found to be more vigorous and higher yielders than diploids while tetraploids are robust, hardy and produce bolder flower.

Artificial induction of polyploidy has been feasible in *J. auriculatum* but not in *J. grandiflorum* (Veluswamy, 1981). Seed and seedling treatment with 1% colchicine recorded 11.4 to 23.1% ploidy in *J. auriculatum*.

Induction of polyploids was attempted in the diploid of *J. auriculatum* and *J. grandiflorum*. Seeds of Short Point clone of *J. auriciulatum* were treated with 0.5% colchicines for 12 to 24 hrs. One tetraploids seedling was obtained from Parimullai seeds. In general tetraploids were found to be poor yielder with low concrete recovery. Treatment of vegetative buds with 0.1 to 0.5% of colchicine did not yield in any useful polyploidy. In *J. grandiflorum* seed treatment with 0.5% colchicine resulted in six tetraploids which were, however not found to be economically useful. In *J. sambac*, tetraploids were more vigorous but produced less number of bold flowers per plant (Alikhan *et al.,* 1969).

Polyploidy has played important role in origin of new cultivars in ornamental and essential oil bearing plants. It may be spontaneous in nature or artificially induced by colchicine. Spontaneous triploid (2n=39) in *Jasmine sambac* has been reported by Sharma and Sharma (1958), in *J. grandiflorum* by Murthy and Khanna (1971), in *J. ilicifolium* and *J. nitidum* by Taylor (1945), in *J. primulinus* Henst by Krishnaswamy and Raman (1948) and in *J. autumnale* by Sharma and Sharma (1958).

Spontaneous tetraploidy (2n=52) have been reported by Dutt (1952) in *J. calophyllum* Wall by Raman (1955) in *J. flexile* Vahl. Triploidy in *J. grandiflorum* has been found to increase concrete content and thereby hold promise as useful avenue for improvement of this crop. Attempts to induce tetraploidy has made in jasmine. Sambandhmurthy and Abdul Khadar (1982) reported that induced tetraploidy in *J. grandiflorum* did not reveal superiority.

Mutation

In jasmine both chemical and physical mutagens have been tried in *J. auriculatum.* X ray treatment of active vegetative buds produced no mutagenic effects. Seed treatment with gamma rays beyond 10 Kr proved lethal. A mutant with longer corolla tube was obtained but it was susceptible to gall mite. Use of chemical mutagens showed that soaking in 0.1% EMS for 72 hours proved lethal. In *J. grandiflorum*, gamma rays of doses beyond 30 Kr proved to be lethal. Although 10 mutants were isolated with lower doses, none proved to be good for flower yield. Treatment of softwood cutting with gamma rays (1 Kr to 10 Kr) also met with the same fate suggesting the limited role of mutagenesis in jasmine improvement. However, Nambisan *et al*. (1980) obtained two promising mutants in the M_2 generation with 25 Kr gamma rays in *J. grandiflorum.* One was a dwarf mutant with short internodes and the other was found to be tolerant to leaf spot disease. But the yield of flowers and recovery of concrete were less than parent clone Thimmapuram.

Mekala *et al.* (2010) studied the effect of physical and chemical mutagens on floral characters of *J. sambac* cv. Gundumalli and isolation of important mutants. The semi hard wood cuttings of Gundumalli were treated with mutagens *viz.* gamma rays alone and in combination with ethyl methyl sulphonate (EMS) at different doses including control (untreated). In general, the treated population had manifested reduced expression than the control (untreated population) in most of the floral characters. Higher the dose of mutagens, there was reduction in the expressivity of the traits. The mean stalk length decreased with the increase in the irradiated population and it was minimum with 2.5Kr + 30mM EMS (0.89±0.021 cm). The mutagenic treatment 2.0 Kr followed by 1.5 Kr + 30 mM EMS manifested higher variability in terms of PCV and GCV recording the maximum flower range of (3–35) and (5–27). The floral characters like bud length and flower weight decreased with increase in the doses while the flower width did not follow a definite trend. Induced mutations produced higher amount of phenotypic and genotypic co-efficient of variation for number of flowers, flower weight indicating the scope for improvement of Gundumalli for yield and novelty by selection.

Chezhiyan *et al*. (1988) treated seed of *J. auriculatum* cv. CO-1 Mullai with 2.5-20 Kr doses of gamma rays or 10-30 mM ethyl methyl sulfonate (EMS) or with combination of both. Germination and establishment of seeds showed gradual decrease as the mutagenic treatment was increased. The LD_{50}s were 9.75 Kr gamma rays, 24 mM EMS and a combination of 25 mM+20 Kr for germination percentage and 16 Kr, 25 mM and 10 Mm + 13.5 Kr for establishment percentage. Terminal cuttings of *J. auriculatum* cv. Pink Pin and Pink Thrum, *J. sambac* cv. Gundumalli, *J. calophyllum* and *J. flexile* were irradiated with the gamma rays at 0.5 to 5.0 Kr to induce mutagenesis. The rooting percentage, number of roots/cutting and length and thickness of the roots decreased with irradiation intensity. LD_{50} intensities resulting in 50% rooting were 2.5, 0.5, 2.0, 1.0 and 2.5 Kr for the respective genotypes (Devaiah and Srivastava, 1989).

Christensen *et al*. (1997) investigated mutagenesis on the possibility to reduce or modify the emission of volatiles from *J. polyanthum* Pepita. The flower scent was dominated by benzyl acetate (57.8%), p-cresol (12.2%), and (E)-isoeugenol (9.7%), eugenol (3.5%), phenethyl acetate (2.1%) and (Z) 3-hexenyl butyrate (1.9%). The strong scent of papita reduces its production potentialities as pot plant. Kannan *et al*. (2002) observed that gamma rays at 1.5 and 2.0 Kr, EMS at 30 and 45 mM and 1.0 Kr + 15 mM and 1.5 Kr + 30 mM combined treatments gave higher mean values of the economic characters. The reproductive traits namely, length of corolla tube, length and width of flower bud and diameter of the open flower after mutagenesis showed, in general positive shift over control population in all doses except higher ones. Variety developed through induced mutation is given in Table 4.

Table 4: Variety developed through induced mutation.

S. No.	Variety	Parents	Chief characteristics
1.	CO-2 Pitchi (*J. grandiflorum*) (1991)	Induced mutant from CO-1 Pitchi	Bolder pink buds. Flower yield 3.64 kg/year/plant, flower bud yield is 11.68 t/ha, concrete recovery 0.30%.

BIOTECHNOLOGY

Tissue culture

Low seed setting rate in jasmine and varietal degeneration, low resistance and decline flower production in long term through cuttings has arisen the application of tissue culture and micropropagation to this crop. Bhattacharya and Bhattacharyya (1997) micropropagated *J. officinale* successfully, by culturing nodal segments. MS basal media with 3% sucrose and supplemented with 17.76 micro M benzyladenine + 0.53 micro M NAA was used for shoot proliferation. For elongation, medium containing BA and kinetin individually and in combination with NAA were used. Shoots 4 to 5 cm long were transferred to liquid rooting medium. Rooted plants grew normally under natural conditions. Leaf bit showed highest callus ability in MS medium supplemented with 1 mg IAA/litre and 10 mg BAP/litre within 23 days. Development of shoots were observed from leaf bit callus cultured on MS medium supplemented with 12 mg BAP/litre when compared to kinetin. Among the auxins tried, 3 mg NAA/litre proved to be the best for rhizogenesis (Rajasekaran *et al*., 2000). Layous (2004) found that kinetin (0.5 mg/l) on the MS medium gave the best result with regard to bud proliferation and multiplication. However, addition of NAA of different concentration was having inhibitory effect. Base and intermediate cuttings gave better response than terminal ones. Best time for bud proliferation in primary explants culture on MS medium supplemented with 0.5 mg/l kinetin was February and for bud multiplication February and March. IBA (1.5 mg/l), when used *in vitro* rooting medium gave 97% rooted plantlets, while *ex vitro* rooting of plantlets on agricultural media after soaking their bases in a solution containing IBA (20 mg/l) increased 99% rooting. In addition, the *ex vitro* rooting resulted in successful hardening of 94% of the plantlets and reduced time, costs and efforts required for rooting and hardening.

Cai Han *et al*. (2007) developed optimum media and condition for tissue culture and micropropagation of jasmine. The media favouring axillary bud germination, axillary bud reproduction and sugar-free rooting were MS combined with benzyladenine (2.0 mg/litre) and NAA (0.1 mg/litre); MS combined with benzyladenine (1.5 mg/litre), NAA (0.3 mg/litre) and GA_3 (0.5 mg/litre); and half-strength MS with NAA (1.2 mg/litre), respectively. The condition suitable for test-tube plantlet growth was light density at 5000-5500 lux, CO_2 concentration of 1500 mg/litre and three vent holes. The sugar free cultured plantlet grew strongly, where the survival rate reached 95%. Sun *et al*. (2009) studied the effects of different concentrations of 6-BA and IBA on induction of bud germination and proliferation of jasmine at different temperatures. Highest bud induction rate of 96.75% was obtained in the WPM medium supplemented with 2.0 mg/l 6-BA and 0.1 mg/l IBA. The best proliferation with a shoot proliferation coefficient of 5.31 was observed in the WPM medium supplemented with 1.0 mg/l 6-BA and 0.2 mg/l IBA at 35 p C. The optimal rooting culture mode was two-step rooting method, i.e. pre-culturing shoots on the rooting medium of 1/2 WPM+NAA 1.0 mg/l for 7 days, and then transferring pre-cultured shoots to 1/2WPM without supplement of any plant growth regulators. The rooting rate was up to 98.41%.

Among the different explants from *J. grandiflorum*, leaf bits were found ideal for callus induction, followed by petals and shoot tips. MS medium with 2,4-D at 1.25 ppm was the suitable medium for callus induction in leaf explants. Miller medium with NAA at 1.25 ppm and BAP at 1.25 ppm was the suitable medium for callus induction in petal explants collected from 18 days old buds of local varieties of *J. grandiflorum*. Among the different media combinations tried, MS medium supplemented with 0.2 ppm NAA and 0.3 ppm GA_3 is the best medium for callus induction in shoot apex explants of 10 mm size. Direct regeneration of plants was obtained in 10% of shoot apex explants cultured in MS medium supplemented with 0.2 ppm GA_3 and 3 ppm BAP in 60 days. A maximum of 95% embryogenicity is induced in leaf callus transferred to MS basal medium supplemented with 1.5% sucrose from MS medium with 2, 4-D at 1.0 ppm. The peroxidase assay indicated that embryogenic calli had six isoperoxidases while the non-embryogenic had only two. Biochemical analysis of the callus samples showed high phenylalanine ammonia lyase and phenol indicating the initiation of phenolic pathway, which branches out and ultimately leads to the synthesis of flavour compounds in callus (Gomathi *et al.*, 2005).

Bhattacharya and Bhattacharyya (2010) developed a protocol for *in vitro* propagation of *J. officinale* by shoot regeneration from existing as well as newly developed adventitious axillary buds *via* proper phytohormonal stimulation. To start with nodal segments as explants, March-April is the most ideal time of the year when planting material suitable for *in vitro* multiplication is abundantly available. Prior to inoculation of explants in the culture medium, special care is needed to reduce microbial contamination by spraying on selected spots of the donor plant with anti-microbial agents 24 h prior to collection; treatment with antiseptic solution after final cleaning and surface sterilization by treating explants with mercuric chloride. Inoculated explants are free from brown leaching from cut ends by two consecutive subcultures within 48 h in MS basal medium. Multiplication of shoots, average 4-5 at each node, takes place in MS medium containing 4.0 mg/l BAP, 0.1 mg/l NAA, and 40 g/l sucrose over a period of 8 weeks. For elongation of regenerated shoots, cultures are transferred to MS medium, supplemented with a single growth hormone, kinetin at 2.0 mg/l. Emergence and elongation of roots from shoot base is facilitated by placing on the notch of a filter paper bridge. The hardened *in vitro* propagated plants are able to grow normally in soil like other conventionally propagated *J. officinale*.

Somaclonal variation

Somaclonal variability often arises in tissue culture as a manifestation of epigenetic influence or changes in the genome of differentiating vegetative cells induced by tissue culture conditions (Larkin *et al.*, 1981 and Muller *et al.*, 1990). Joy and Raja (2010) reported that successful induction of regenerative calli from *J. auriculatum* leaf explants, cultured on Murashige and Skoog's (MS) medium supplemented with 2,4-D (auxin) and BAP (cytokines), kinetin and the extent of genetic variability in the plants regenerated from one of these calli as examined through RAPD analyses.

Molecular breeding

A study was made to find out the genetic diversity in 48 jasmine genotypes employing morphological and RAPD markers. The studies made on morphological quantitative characters revealed high variations within and between genotypes. The highest phenotypic coefficient of variation (123.8%) for number of whorls and lowest (28.7%) for flower diameter were observed. Among cultivated types *viz. Jasminum sambac and Jasminum multiflorum* showed significant variation within and between species for leaf length, bud width, number of whorls and petals, petal length and filament length. High variance was observed within the species in terms of length of leaf, corolla tube, petal and filament. Of the 48 genotypes studied, 66% of the genotypes were bushy in habit. Nearly 77% produced white/cream flowers and 79.2% exhibited seasonal flowering while, others flowered throughout the year. Among the genotypes, 52.1% had medium to very high aroma. The DNA fingerprints were generated using 54 primers of which, 26 primers gave good amplifications. Nineteen primers produced 100% polymorphism with an average of 90.6 per cent for 26 primers. The primers were very unique in identifying most of the genotypes. Clustering based on morphological and RAPD markers displayed 48 genotypes into two major clusters. This information enables easy selection of parents in crop improvement programme in jasmine (Champa, 2012).

To work out the genetic relationship among the twenty one accessions of *Jasminum* spp., RAPD was used. Forty decamer primers from OP-S and OP-V were used for this purpose out of which 14 primers, ten from OP-V and four from OP-S series produced 1173 reproducible, bright and clear bands which ranged from 250 bp to 10,000 bp. Overall 195 alleles were detected. The PIC value calculated for these primers ranged from 0.438 to 0.698 which indicated a good level of genetic diversity among the accessions as also revealed by generated similarity matrix. The similarity values among the genotypes ranged from 0.263 to 0.776. The resulting dendrogram divides the accessions into two distinct main clusters (A and B) at 0.40 similarity value. Cluster A has the most diverse jasmine accessions, while cluster B further divided the accessions into smaller groups. The results clearly indicated that RAPD analysis provide a good tool to detect and classify the genetic diversity of *Jasminum* spp. It is believed that these findings are helpful in further exploration, classification and improvement of *Jasminum* spp. (Mahmood *et al*., 2013).

REFERENCES

Bailey, L.H. 1963. The Standard Cyclopedia of Horticulture. The Macmillan Company, New York. pp. 1716-1719.

Bhattacharjee, S.K. 1980. Native jasmine of India. *Indian Perfumer*, **24**(3): 126-133.

Bhattacharya, S. and Bhattacharyya, S. 2010. *In-vitro* propagation of *Jasminum officinale* L.: a woody ornamental vine yielding aromatic oil from flowers. *Methods Mol. Biol.*, **589**: 117-26.

Bhattacharya, S. and Bhattacharyya, Sanghamitra. 1997. Rapid multiplication of *Jasminum officinale* L. by *in vitro* culture of nodal explants. *Plant Cell, Tissue Organ Cul.,* **51**(1): 57-60.

Cai, Han, Chen, XiaoQiang, Xiong, ZuoMing, Xie, LingLing and Zhao, LiangJun. 2007. Techniques of *in vitro* micropropagation and sugar-free rooting of jasmine (*Jasminum sambac*). *Jiangsu J. Agril. Sci.*, **23**(5): 464-468.

Champa, B.V. 2012. Molecular (RAPD) markers in jasmine species. Ph. D. Thesis, Department of Horticulture, University of Agricultural Sciences, Bangalore.

Chezhiyan, N., Ponnuswami, V., Khader, M.A. and Thambura, J.S. 1988. A note on the sensitivity of *Jasminum auriculatum* seeds for mutagenic treatments. *South Indian Horti.*, **36**(1): 92-93.

Christensen, L.P., Jakobsen, H.B., Kristiansen, K. and Moller, J. 1997. Volatiles emmited from flowers by gamma-radiated and nonradiated *Jasminum polyanthum* Franch. *in situ. J. Agril. Food Chemistry*, **45**(6): 2199-2203.

Devaiah, K.A. and Srivastava, M.C. 1989. Sensitiveness of jasmine cuttings to gamma radiation. *Herba Hungarica,* **28**: 11-14.

Dommee, B., Thompson, J.D. and Cristini, F. 1992. Distyly in *Jasminum fruticans* L.: hypothesis of optimal pollination based on variations in intrafloral ecology. *Bulletin de la Societe Botanique de France, Lettres Botaniques*, **139**(3): 223-234.

Dutt, M. 1952. Chromosome numbers in some ornamental jasmines. *Sci. Cult.*, **17**: 527-528.

Gao, LiPing, Xia, Tao, Zhang, YuQiong, Wang, LiMing and Li DaXiang. 2002. Studies on the endogenous hormone during the development and blooming of jasmine flower. *J. Tea Science*, **22**(2): 156-159.

Gomathi, K.S., Sambandamurthy, S., Sadasivam, S., Ramasamy, N.M. and Rajmohan, K. 2005. *In vitro* culture and biochemical analysis of *Jasminum* sp. *Proce*: Recent trends in horticultural biotechnology, ICAE National Symposium on Biotechnological Interventions for Improvement of Horticultural Crops: Issues and Strategies, Vellanikkara, Kerala, pp. 149-153.

Guitian, J., Guitian, P. and Medrano, M. 1998. Floral biology of the distylous Mediterranean shrub *Jasminum fruticans* (Oleaceae). *Nordic J. Bot.*, **18**(2): 195-201.

Jongkind, C.C.H. 2010. *Jasminum cuneatum* Jongkind (Oleaceae), a new species from Guinee, Liberia and Cote d'Ivoire. *Adansonia*, **32** (1): 141-143.

Joy, Priya and Raja, D.P. 2010. Genetic analysis of somaclonal variation among *Jasminum auriculatum* (Vohl.) and it's callus. *International J. Biological Tech.*, **1**(1): 94-98.

Kannan, M., Sathiyamurthy V.A. and Sankar, V. 2002. Mutagenic studies on *Jasminum sambac*. *Proce*: The National Symposium on Indian Floriculture in the New Millennium, Bengaluru. **5**: 209-211.

Khan, M.A., Muthuswamy, S. and Raman, K.R. 1969. New chromosomal forms of superior ornamental value in *Jasminum*. *S. Indian Hort.*, **17**: 79-85.

Krishnaswamy, N. and Raman, V.S. 1948. Cytogenetical studies of the Indian jasmines. Taxonomy and chromosome numbers of four varieties of *J. sambac* and other species. *J. Indian Bot. Soc.*, **27**: 77-83.

Larkin, P.J. and Scowcroft, W.R. 1981. Somaclonal variation: a novel source of variability from cell cultures for plant improvement. *Theor. Appl. Genet.*, **60**: 197-214.

Layous, L.N. 2004. Micropropagation of jasmine (*J. azoricum* L.) using tissue culture techniques. *Univ. Khartoum J. Agril. Sci.*, **12**(2): 293-308.

Mahmood, M.A., Hafiz, I.A., Abbasi, N.A. and Faheem, M. 2013. Detection of genetic diversity in *Jasminum* species through RAPD techniques. *International J. Agri. Biology*, **15**(3): 505-10.

Mekala, P., Kumar, N., Kannan, M. and Vijayakumar, G. 2010. Effect of physical and chemical mutagens on floral characters of *Jasminum sambac* cv. Gundumalli and isolation of important mutants. *J. Orna. Hort.*, **13**(3): 200-206.

More, T.A. 1980. Note on the effect of different growth retardants on flowering in *Jasminum grandiflorum* L. *Madras Agric. J.,* **72**(5): 293-295.

Muller, E., Brown, P.T.H., Hartke, S. and Lorz, H.T.I. 1990. DNA variation in tissue culture derived rice plants. *Theor. Appl. Genet.*, **80**: 673-679.

Murthy, A.S. and Khanna, K.R. 1971. A study of triploid *J .grandiflorum*. *Curr. Sci.*, **40**: 555-556.

Muthuswami, S. and Madhava Rao, V.N. 1972. Tamil Nadu's fragrant new Parimullai. *Indian Hort.,* **17** (3):28.

Nambisan, K.M.P., Krishnan, B.M., Veeraragavathatham, D. and Ramaswamy, N. 1980. Induced mutation in Jasmine (*Jasminum grandiflorum* L.): Leaf spot mutant and dwarf mutant. *Sci. Cult.,* **46**: 427-428.

Noor, M.J., Ahmad, M., Asghar, R., Kanwal, A. and Pervaiz, S. 2004. Palynological studies of cultivated plant species at university of arid agriculture, Rawalpindi, Pakistan. *Asian J. Plant Sci.*, **3**(4): 476-479.

Phanachana, Mikaewkunchon and Cho, Nitsiri Suisuwan. 1994. Use of some chemicals for induce flowering of jasmine in winter. Research abstracts of King Mongkut's Institute of Technology, Bangkok (Thailand). pp. 40.

Rajasekaran, V., Haripriya, K. and Arumugam S. 2000. Micropropagation studies in jasmine (*Jasminum sambac* Ait.). *In*: Centennial Conference on Spices and Aromatic Plants, Calicut, Kerala, India, pp. 47-49.

Raju, A.J.S. 1988. Pollination ecology of *Jasminum angustifolium* Vahl. (Oleaceae). Proceedings, Indian National Science Academy, *Biological Sci.,* **54**(2/3): 165-169.

Raman, V.S. 1955. Cytogenetics of the Indian jasmines II. The somatic chromosomes. *Cytologia,* **20**: 19-31.

Rosario, T.L., Jalandoni, L. and Siar, S.V. 1993. Cytology of 3 *Jasminum* species and one cultivar. *Philippine J. Crop Sci.,* **18**: 3.

Sambandhmurthy, S. and Khader, A. 1982. Essential oil in Jasmine. Proceedings of National Seminar on Medicinal and Aromatic Plants, pp. 183-185.

Sax, K. and Abbe, E.C. 1932. Chromosome numbers and the anatomy of the secondary xylem in the Oleaceae. *J. Arnold Arbor,* **13**: 37-48.

Sharma, A.K. and Sharma, A. 1958. Analysis of chromosome morphology and possible means of speciation in *Jasminum. Cytologia*, **23**: 172-185.

Singh, A.K. 2006. Flower Crops: Cultivation and Management. Pub. New India Publishing Agency, New Delhi. p. 463.

Srivastava, H.C. and Devaiah, K.A. 2006. Correlation and path analysis in jasmine. *Plant Breeding,* **100**(3): 235–237.

Srivastava, H.C. and Kumar, T.V. 2004. IIHR-JA-13 a promising selection of *Jasminum auriculatum* for high yield of concrete. *J. Medi. Aromatic Plant Sci.*, **26**(1): 2-4.

Sun, YanNi, Tang, FangPing, Fang, WeiMin and Tan, GuoHua. 2009. Rapid propagation *in vitro* of jasmine. *Acta Agriculturae Zhejiangensis*. **21**(4): 390-394.

Taylor, H. 1945. Cyto-taxonomy and phylogeny of the Oleaceae. *Brittonia*, **5**: 337-367.

Thangaraj, T., Muthukrishnan, C.R. and Muthuswamy, S. 1980. Correlation studies in open pollinated seedlings of *Jasminum auriculatum* Vahl. *Madras Agric. J.*, **67**(6): 391-393.

Thompson, J.D. 2001. How do visitation patterns vary among pollinators in relation to floral display and floral design in a generalist pollination system? *Oecologia*, **126**(3): 386-394.

Thompson, J.D. and Dommee, B. 1993. Sequential variation in the components of reproductive success in the distylous *J. fruticans* (Oleaceae). *Oecologia*, **94**(4): 480-487.

Veluswamy, P. 1981. Genetic studies in *J. auriculatum* Vahl. and *J. grandiflorum* L. M.Sc.(Agri.) thesis submitted to TNAU, Coimbatore.

Chapter – 29

Marigold

Marigold (*Tagetes* spp.) an annual flowering plant belongs to family Asteraceae, alternately known as Compositae. The word "marigold" means "Mary's Gold" and refers to the "Virgin Mary". However, word "*Tagetes*" named after the Etruscan God "Tages" who supposedly emerged from the earth as it was being ploughed and was imbued with the power of divination. Cultivated species include *Tagetes erecta* commonly referred as African marigold and *Tagetes patula* as French marigold. Marigold commonly propagated by seed and cutting is a leading loose flower of India. Selection of variety is favoured by a higher proportion of double inflorescences.

ORIGIN AND DISTRIBUTION

Origin of marigold is central and South America especially Mexico (Kaplan, 1960). Bailey (1951, 1963 and 1976) reported that marigold is a native to Arizona and Mexico to Argentina.

During early 16th century it spread to various parts of world from Mexico. African marigold (*Tagetes erecta*) first reached Spain and from there it become popular as rose of Indies in Southern Europe. Major work on improvement and cultural practices on marigold have been done in USA, Switzerland, France and Germany. In India, marigold was introduced by Portuguese and it became popular and spread quickly because of its wide adaptability to varying climatic and soil conditions. Marigold is well distributed in different states and union territories of India (Singh, 2006).

SPECIES

Earlier, Rydberg (1915) reported that genus *Tagetes* comprises about 33 species and Bailey (1951 and 1976) mentioned over 30 species, however, Turner and Nesom (1993) reported that *Tagetes* is made up of nearly 55 species distributed in the American continent, most of them located in Mexico. The two species *Tagetes erecta* and *Tagetes patula* are commercially important. Details of these species are given below.

***Tagetes erecta*:** *T. erecta* L. is commonly called African marigold. Plants of *T. erecta* are hardy annual, tall about 90 cm, erect and branched. Leaf is pinnate, divided and leaflets are lanceolate and serrated. It has single or double flowers and large sized globular heads. Flower colours are yellow and orange in various shades i.e. light yellow, golden yellow, canary yellow, bright yellow, cadmium orange, deep orange, golden orange, bright orange. The florets are quilled or 2-lipped.

***Tagetes patula*:** It is commonly called French marigold is comparatively shorter in height, about 30-40 cm tall, hardy and bushy plant. Leaves are dark green, stems are reddish in colour. Leaves are divided pinnately. Leaflets are linear lanceolate and serrated. Flowers are small, single or double. Flowers are borne on long peduncles. Flower colour are yellow and orange in various shades and crimson, mahogony red, bicolour like light yellow with maroon blotches, deep crimson edged yellow, gold and red colour. Many types and strains in African and French are available varying in plant height, growth habit, flower colour, shape and size.

Other important species are *T. tenuifolia* (*signata*), *T. lucida*, *T. lacera*, *T. minuta* and *T. lemmonii*. *T. minuta* is medicinally important and other species having ornamental value are used in improvement work to develop interspecific hybrids.

FLOWER MORPHOLOGY

In *Tagetes erecta*, peduncle of the solitary flower is swollen at the flower base and the teeth of the involucre are often elongate or dentate. The pappus is similar to that of *T. patula*. Flower heads of *Tagetes patula* are solitary and about 2 to 4 cm across. The involucre is thick and oblong with acute teeth and is dotted with glands. There are numerous rays and pappus is made up of 1 or 2 long-awned scales and 2 to 3 shorter blunt ones. The flower of *T. tenuifolia* is solitary and approximately 2.5 cm across and few ray flowers (approximately 5). The involucre has acute teeth with large glandular dots, while the pappus is made up of several scales, one or two of which are sharp-pointed or awned. The flower heads of *T. lucida* are approximately 1.5 cm across and occur in dense terminal clusters (Bailey, 1951 and 1976).

Flower development and anthesis

Flower bud development varies according to variety. Mohanty *et al*. (2002) tested 10 varieties of *T. erecta* and found that different varieties took 17 to 25 days for flower bud initiation after transplanting. Different cultivars took 13 to 24 days for bud maturity after initiation. From bud maturity to colour break and colour break to initiation of floret opening different cultivars took 2 to 5 days and 3 to 6 days, respectively, whereas full opening of flowers was noticed after 10 to 17 days of initiation of opening in individual flower heads. The apex of the flower bud gradually opened and the bracts as well as outer row of ray florets separated from each other. Then the florets started unfurling which started from outer most row of ray floret and proceeded towards the centre. In the study they observed that in cv. Coimbatore Local anthesis commenced between 5 to 6 am and in cv. Giant Orange between 10 to 11 am, whereas maximum intensity of floret opening (43-62%) was observed in most of the cultivars between 10 to 11 am.

Pollination and pollen viability

Marigold is a cross pollinated crop and visit of insects to marigold has been observed (Kaneko *et al.*, 1990). Various populations of marigold have been studied for pollen viability. Viability was 50 to 92% when inter population crossing was in one direction but in reciprocal direction pollen viability was 1 to 2% (Serrato-Cruz *et al.*, 1998). Dehiscence of anther took place by formation of a longitudinal slit which gradually appeared near the basal end of a pollen chamber. However, sometimes it also appeared in the centre and later moved towards the ends. In general, anther dehiscence started either with the commencement of floret opening or little later (Mohanty *et al.*, 2002). They further observed that dehiscence was earliest in cv. Early Yellow Single (8.45 am) and at last in cv. Coimbatore Local. Peak period of anther dehiscence was between 10-12 am and completion of dehiscence was varied between 1-4 pm.

Stigma receptivity and seed set

Studies on receptivity of stigma were observed by seed set method (Mohanty *et al.*, 2002). Very little or no seed set was observed when florets were pollinated on the day of opening. Seed germination ranged from 9 to 20% when pollination was done on third day of opening. However, maximum ranging from 70 to 80% when hand pollination was done on fifth day of anthesis indicating the maximum receptivity of stigma. It was also observed that the stigma retained its receptivity up to seventh day of floret opening when 8 to 15% of seed germination was noted. Serrato-Cruz *et al.* (1998) studied seed setting in various populations of marigold and found 53 to 93% seed set when pollination occurred in one direction, whereas in reciprocal direction seed set was 9 to 44%.

GENETICS

Studies conducted on the cytological status indicated that most *Tagetes* species are diploid with 2n= 2x= 24 chromosomes. *T. erecta*, *T. tenuifolia* (Towner, 1961 and Serrato-Cruz *et al.*, 2000), *T. jaliscensis* (Towner, 1962), *T. foetidissima* and 2 other species *Tagetes* sp1 and sp2 (Serrato-Cruz *et al*., 2000) have chromosome number 2n= 2x= 24 and *T. patula* is tetraploid, 2n= 2x= 48 (Towner, 1961 and Serrato-Cruz *et al*., 2000). Banerjee (1974) also reported that *T. erecta* is diploid while *T. patula* is tetraploid.

Inheritance of quantitative characters

The inheritance of quantitative characters was studied for days required to flowering, flower weight and number of flowers/plant in 7 varieties of *T. erecta*. Dominance and epistasis, particularly the dominance × dominance interaction, played a major role in the inheritance of days taken to flowering and of flower weight. The number of flowers/plant was governed mainly by additive gene effects. Transgressive segregation for all 3 characters observed in the F_2 generation (Singh and Swarup, 1973).

Inheritance of qualitative characters

In the study of qualitative character like flower doubleness in African marigold, Singh (1970) and Singh and Swarup (1973a) observed that doubleness is monogenically dominant over singleness and in F_2 segregation it was in 3:1 ratio. Flatness of florets are monogenically dominant over tubular floret and is designated as Tu and tu, respectively (Punnet, 1924). Punnet (1937) stated that flower colour is governed by two or more factors. Singh and Swarup (1973a) while working on the inheritance of a flower type and flower colour, observed that in a cross between carnation and cupid type, carnation type was dominant over cupid type flower and in F_2 generation segregation was observed in 9:7 ratio, indicating the presence of two complementary genes governing this character. In flower colour studies they found that orpiment and tangerine are orange colour but tangerine was deeper in shade. However, in a cross between yellow and tangerine coloured parents, in F_1 flower colour was tangerine, while in F_2 it segregated in the ratio of 12 tangerine : 3 yellow : 1 orpiment. It indicated that flower colour was governed by two dominant genes.

Heritability and genetic advance

In the variability studies several workers observed heritability and genetic advance for various characters in marigold. High heritability coupled with low genetic advance for 1000-seed weight and seedling length (Kennedy, 1997), the highest heritability value were recorded for days to flowering, flower diameter and shelf life of flowers followed by plant spread, primary branches/plant and seed yield (Nand Kishore and Raghava, 2001), plant canopy, 10-flower weight, plant height and stem diameter showed high heritability associated with high genetic advance (Verma *et al*., 2002), genetic advance was highest for seed yield whereas heritability was highest for flower weight (Patnaik *et al*., 2002). Sreekala *et al*. (2002) observed highly significant heritability for total xanthophylls and xanthophyll ester, whereas Gowrishankarayya and Hegde (2004) recorded high heritability and high genetic advance for days to flowering, flower yield and xanthophyll content. High heritability along with high genetic advance was also recorded for number of flowers/plant and plant height (Singh and Singh, 2004). Bhanu Pratap *et al.* (2002) observed that the heritability estimates were significant for first picking of the flowers and days to visibility of buds in French marigold. Singh *et al.* (2004) estimated genetic variability in 30 genotypes of *T. patula* for 14 characters. High estimates of heritability (broad sense) were observed for all the characters studied, whereas highest value of heritability was obtained for weight of flowers/plant followed by plant spread.

In a twenty-two genotypes of marigold (*Tagetes erecta* L.) high genotypic and phenotypic coefficient of variation was observed for flower yield per plant, fresh weight per flower and stalk length whereas lowest value for both GCV and PCV were obtained for number of seeds per head followed by days to flowering and harvest index. High heritability (>70%) were recorded for stalk length, stem diameter, flower diameter, fresh weight per flower, dry weight per flower, flower yield per plant and 1000 seed weight. High genetic advance as per cent of mean was observed for flower yield per plant followed by fresh weight per flower and stalk length revealing the importance of additive gene effects for these traits (Panwar *et al.*, 2013).

Male sterility

Khoshoo (1979) did intensive investigation for the male sterile lines and found a monogenic recessive male sterile 'apetalous' mutant located in one population of *T. erecta*. The hybrid between ms and mf lines segregated in a ratio of 3mf : 1ms. Similar apetalous male sterile lines have been detected at different centres by various workers (Anon., 1988 and Rao *et al.*, 1993).

Gupta *et al.* (1999) studied inheritance of a male sterile apetalous inflorescence in *T. erecta*. Male sterile apetalous lines, MS7 and MS8, were crossed with a male fertile petalous line (Selection-21). F_1 and F_2 (selfed F_1) plants were grown to investigate segregation of petalous and apetalous flowers. In F_2 generation, segregation of male fertile (petalous) and male sterile (apetalous) plants in a 3:1 ratio indicated that the apetalous sterile trait was governed by a single recessive gene. Male sterility in marigold is of two kinds, apetalous and double flowered. Apetalous sterility is preferred to full-double flowers because later type is prone to break down (Janakiram, 2004.)

BREEDING OBJECTIVES

For loose flower variety

i. More number of branches

ii. Doubleness of flower

iii. Profuse flowering

iv. Medium sized flowers
v. Good quality of recovery from pressure
vi. Good colour retention
vii. Long duration of flowering
viii. More shelf life
ix. Adaptibility to various agroclimatic conditions
x. Resistance to insect pests and diseases
xi. Uniform flower size

For pot variety

i. Dwarf and compact plant
ii. Profuse branching and uniform spreading
iii. Early flowering
iv. Novel or new colour
v. Good colour retention of flower
vi. Attractive flower colour, shape and size

BREEDING METHODS

Introduction

In India marigolds were introduced by Portuguese (Watts, 1893) and spread quickly because of the ease in cultivation and adaptability to varying soil and climate conditions. About 66 collections of *T. erecta* and *T. patula* have been introduced from different countries through NBPGR, New Delhi. Important collections in *T. patula* were Harmony from the Netherlands, Petite Orange, Petite Yellow and Golden Girl from Australia and Queen Sofia from USA. Among *T. erecta* important one was Falls WIR 630 from Russia. Exotic varieties of *T. erecta* were evaluated at NBRI, Lucknow by Sharga *et al*. (1972) and recommended 5 varieties Alaska, Doubloon, Fire Glow, Golden Jubilee and Yellow Fluffy for loose flower and cvs. Geraldine, Golden Climax Giant, Orange Fluffy, Orange Mums, Sovereign, Sun Giants, Super Chief Double and Yellow Climax for cut flower purpose. Marigold varieties introduced from different countries were evaluated at PAU, Ludhiana and Giant Double African Orange and Climax of *T. erecta* and cv. Rusty Red, Butter Scotch, Red Brocade of *T. patula* were recommended for loose flower production, whereas var. Valencia and Sussana were found promising for bedding purpose (Arora and Singh, 1980). Some varieties and hybrids of marigold introduced from the Netherlands, USA, Australia and Russia are given in Table 1.

Table 1: Marigold varieties and hybrids introduced from different countries.

S. No.	Country	Variety introduced
1.	The Netherlands	French marigold- Harmony.
2.	USA	African Marigold- Frills WIR 630, Queen Sofia.
3.	Russia	Frills, WIR 630.
4.	Australia	Petite Yellow, Petite Orange, Golden Girl, Hybrids- African Medium Orange F_1, hybrid Jubilee, Golden Jubilee.
5.	Thailand	EC 332176, EC 332185.

Selection

Marigold is a cross pollinated plant and commercially propagated by seed; therefore development of variety through selection with maintaining purity is not easy. However, 3 varieties evolved through selection. Cv. Pusa Arpita from IARI, New Delhi. Cvs. Hisar Beauty and Hisar Jaffri-2 of French marigold from CCS Haryana Agril. Uni., Hisar (Beniwal *et al.*, 2009). These varieties are described below.

Pusa Arpita: Selection from heterozygous material collected from rural area of Delhi. Produces medium sized, light orange flowers during mid-December to mid-February in northern plains of India. Flowers are compact with turmeric yellow colour of petal. The average yield is 18-20 t fresh flowers/ha.

Hisar Beauty: It released during 2008-09 from HAU, Hisar. Dwarf and compact plant, flower dark red, petals with yellow margin. The average yield is 40-50 flowers/plant. Suitable for exhibition, bedding and potting.

Hisar Jaffri-2: Released in 2008-09. Plant is dwarf and compact plant, orange flower. The average yield is 70-80 flowers/plant.

Hybridization

Being a cross pollinated crop, it is advisable to bag the female parent at bud stage. After 3-4 days of opening of outer ray florets, the disc florets must be removed. Pollens from desired plant are collected and pollinated on the stigma (ray florets). For good seed setting, pollination should be continuously done for 3-4 days as the ray florets developed. Successful pollination can be judged by shriveling of pistils and when the flowers become dry, the seeds are collected. Seed must be properly dried and stored.

Crossing between genetically diverse types is a commonly employed method to induce variation than existing cultivars in terms of novel type, more yield and better flower quality. Selection of new variety started from F_2 onwards till they become homozygous. Hybridization work in African marigold was first reported by Singh and Swarup (1971). They involved 7 parents for heterosis studies on a diallel set. The highest heterosis was recorded for number of flowers. The crosses Alaska × Hawaii,

Alaska × Cupid Orange Mum and Katrain Local × Cupid Orange Mum showed highest heterosis in flower size and flower weight, early flowering and number of flowers, respectively, whereas highest flower yield was recorded in the hybrid Hawaii × Cupid Orange Mum.

The combining ability of 8 maternal and 5 paternal lines was evaluated by pair crossing (Relichova and Petrova, 1988). The greatest heterosis effect occurred in the number of flowers/plant and flower diameter. In anther heterosis and combining ability studies, cvs. Orange Boy, Giant Double African Orange, Coimbatore Local and African Double Yellow had significant GCA effects for flower yield/plant, whereas Giant Double African Orange × Coimbatore Local and African Double Yellow × Orange Boy were the best specific combination for flower yield, having significant heterosis and SCA effects for this trait (Reddy *et al.* 1989). Out of 27 hybrids, 10 expressed significant and positive heterobeltiosis in respect of plant height which was observed by Kumar *et al.* (1989). A line × tester crossing in *T. erecta* was carried out using 3 male sterile lines, MS 7, MS 8 and MS 12 and 11 diverse pollinators as testers and highest heterosis was observed in flower yield during both winter and summer crops (Gupta *et al.*, 2001). In a combining ability analysis it was observed that high GCA and SCA values predominantly governed inheritance by additive and non-additive components. Cvs. Suttons Double Orange, Bhubaneswar Local Orange, Newtech Orange and Early Yellow Single are good combiners for yield and its component traits (Mohanty *et al.*, 2003). Mohanty *et al.* (2003a) also studied heterosis in marigold and observed that the hybrids which have high mean performance for different yield attributing characters did not exhibit high heterobeltiosis except some crosses because better or the best parent involved in these hybrids has high mean expression. The cross Suttons Double Orange × Newtech Orange was better than Suttons Double Orange × Royal Bengal in yield. Flowers of *T. erecta* is a major source of carotenoids, hence a study was conducted on combining ability and heterosis, aiming at genetic improvement by Sreekala and Raghava (2003). They reported that total content of carotenoids and lutien appears to be governed by dominance (or non-additive) gene action, while content of xanthophylls esters is governed by both additive and dominance (or non-additive) gene action. They also stated that heterosis can be exploited for total carotenoids and its commercially important fractions. Heterosis and combining ability effects have also been studied by Velmurugan *et al.* (2003) for xanthophyll content in *T. erecta*.

Male sterile and seed production

Significant achievements for production of F_1 hybrids and seeds by using male sterile lines have been reported by several workers. In production of F_1 hybrid seed, male sterile line and pollen parents are transplanted in alternate rows. The male sterile line gives rise to normal and sterile plants in a ratio of 50:50 and normal plants are rouged out. For pollination and seed production, only sterile plants are used. Raghava (1995) used apetalous male sterile line and developed several F_1 hybrids. One hybrid,

MS 8 × Pusa Narangi Gainda has been found to be very promising. A study on the development of F_1 hybrids of *T. erecta* was carried out at IARI, New Delhi using Line and Tester crossing programme. During winter, hybrids MS 7 × Selection 29 and MS 7 × Selection 56, while during summer hybrids MS 7 × Selection 22, MS 8 × Selection 21, and MS 8 × Selection 31 were among the high yielding hybrids. Hybrids MS 8 × Selection 7, MS 8 × Selection 8, and MS 8 × Selection 27, MS 8 × Selection 28, and MS 8 × Selection 56 performed very well over both, winter and summer seasons (Gupta *et al.*, 2002). Now several F_1 hybrids are available. First F_1 hybrid was 'Climax'. Some important African type F_1 hybrids are F_1 Gold Coin Series, F_1 Climax Series, Space Age Series, Galore Series, etc. Two varieties namely Pusa Narangi Gainda and Pusa Basanti Gainda of African marigold developed from IARI, New Delhi (Anon., 1998). The description of these varieties is given in Table 2.

Table 2: Varieties evolved through hybridization with their chief characteristics.

S. No.	Variety	Parentage	Chief characteristics
1.	Pusa Narangi Gainda	Cracker Jack × Golden Jubilee	Takes 125-136 days for flowering after sowing, plant 73.3 cm tall, vigorous, uniform, flowers orange colour, carnation type, double compact, 7.8 cm in size, disc florets present, yield 349 q/ha.
2.	Pusa Basanti Gainda	Golden Yellow × Sun Giant	Takes 135-145 days for flowering, flowering duration is 40-45 days, plant 58.8 cm tall, vigorous, uniform, flower sulphur yellow, carnation type, double, compact 6.9 cm in size, disc floret present but invisible, yield 58 flowers/plant.

Three male sterile lines of African marigold (*Tagetes erecta*), viz., MS 5, MS 7 and MS 8, were crossed with different promising lines of French marigold (*Tagetes patula*), viz, French maroon, French orange and French bicolour to obtain triploid interspecific hybrids. Among the hybrids developed, MS-5 C French maroon, proved to be very promising. It had intermediate growth habit and flower size. It was found most ideal for loose flower production (Anon., 2011).

Interspecific hybridization

The interspecific hybridization by crossing *T. erecta* × *T. tenuifolia* was done by Towner (1961) and observed that the natural and synthetic allotetraploids were intermediate between *T. erecta* and *T. tenuifolia* for nearly every character. *T. erecta-tenuifolia* showed a striking morphological similarity to *T. patula*. Interspecific crossing between *T. jaliscensis* × *T. erecta* was also studied by Towner (1962). Triploid hybrid was produced by interspecific crossing between *T. erecta* and *T. patula* which was relatively large double flowers (Jalil *et al.*, 1974). Hybrids of marigold are very popular as the interspecific hybrids (*T. erecta* × *T. patula)* have been produced in USA, which are intermediate in characters. Hybrids are early flowering, medium in height; plants are bushy and produce double flowers with delightful colour combination (Janakiram, 2004). Some promising interspecific hybrids were produced by various workers, among them important ones are Red and Gold hybrids, Nugget, Red Seven Star, etc.

Mutation

A peculiar and conspicuous bud-mutation observed simultaneously in French marigolds. Where since flower heads of completely different types appeared together on the same plant. The single French flower heads have a single cylindrical involucres with light bracts and six to nine well developed, flat, strapshaped petaloid ray flowers with golden brown velvet perianths (Iltis, 1945).

Geetha and Jayabalan (1992) studied the effectiveness and efficiency of mutagens and found it is higher in 2.5 kR gamma rays treatment. The chlorophyll deficient mutants observed in M_2 generation were albino, xantha, chlorina, viridis and chlorina viridis. Chlorophyll mutation frequency and spectrum were directly proportional to the doses of gamma rays. Mutants with different plant and flower characters were also isolated.

Two mutant plants displaying aberrant thiophene composition were identified after screening more than 300 M_2 generation plants, derived from ethyl methanesulfonate-treated seeds, using HPLC analysis of root extracts. Two M_3 generation mutant lines were subsequently obtained from these mutant plants. Both mutants, which may have originated from the same mutational event, contained high amounts of the C_{13} monothiophene 2-(but-3-en-1-ynyl)-5-(penta-1,3-diynyl)-thiophene and two C_{13} bithienyls compared to the wild-type. The mutant phenotype was also expressed in 21 *Agrobacterium rhizogenes* transformed root clones derived from both mutants. Feeding experiments with root cultures derived from one mutant and the wild-type indicated that the monothiophene accumulating in the mutant was the common precursor for all bithienyl thiophenes in wild-type and mutants. The above experiments also showed that one mutant was deficient in demethylation of the monothiophene (Jacobs *et al.*, 1995).

Singh *et al.* (2009) irradiated the seeds of African marigold cv. Pusa Narangi Gainda with different doses of gamma rays to induce mutation. Seeds were sown just after irradiation and 30-day old seedlings were transplanted into beds. Reduction in survival percentage, plant height, number of branches, leaf number, leaf size, plant-spread, stem diameter, increased foliage and floral abnormalities were observed upon irradiation and with increase in dose of gamma rays. LD_{50} was determined on survival basis. Leaf abnormality manifested itself as leathery texture of leaf, enhanced and irregular leaf thickness, asymmetric development of pinnate leaflets, reduction in pinnate number, chlorophyll variegation, pale and deep green leaves, narrow leaves and small leaves. Percentage of abnormal leaves and plants increased with increase in dose of gamma rays. Fasciation of stem was a common abnormality observed in all the treatments. Days to bud initiation, earliness in colour-appearance and days to full bloom were all significantly delayed upon exposure to gamma rays. Flower-head size, height and weight were highest at the lowest dose. Number of ray florets and size (length and width) decreased with increasing radiation dose. Floral abnormalities and percentage of plants with abnormal flower-heads increased with increasing dose of gamma irradiation. Floral abnormality included fasciation of flower-head and asymmetric development of ray florets. Stimulating effect of gamma irradiation was observed at 100 Gy where almost all the characters studied showed positive correlation, including growth and yield attributes. It is concluded that exposure to 100 Gy of gamma rays in African marigold cv. Pusa Narangi Gainda results in higher yield and marketable bloom.

Polyploidy

Colchicine was used successfully to produce fertile allotetraploids in *Tagetes* (Bolz, 1961). Mathew and Abraham (1980) induced polyploidy in *T. erecta* by using colchicine. They observed significant increase in size of capitulum and reduction in plant height. Kapelev and Robotyagov (1981) treated seeds of marigold for 12 hrs in a 0.10% colchicine solution. Plants raised from colchicine treatment resulted in exceeded flowering than original variety along with increased flower yield and essential oil content and similar results were recorded also in the second generation. Some polyploidy varieties are given below.

Zenith: The unique Afro-French hybrid (triploid) from Floranova with more branching, more flowers, more heat tolerance and more rain tolerance. With no seed set, Zenith requires no dead heading to maintain its phenomenal flower power all summer long.

Nugget Supreme: It is a triploid variety of yellow colour (hybrid between *Tagetes erecta* and *Tagetes patula*).

Jaguar: It is a tetraploid of French marigold (*Tagetes patula*).

BIOTECHNOLOGY

Tissue culture

For F_1 hybrid seed production, an efficient protocol for *in vitro* cloning male sterile plant was developed (Kumar *et al.*, 2004). A comparative field performance study of *in vitro* cultured and seed derived male sterile plants of two lines (MS 7 and MS 5) was also evaluated. Tissue cultured plants exhibited superiority over the seed derived plants.

Plant regeneration of *Tagetes patula* was achieved from anther explants *via* adventitious shoot differentiation from callus. The effects of genotype, temperature pretreatment, plant growth regulators, light regimes and sucrose concentration were studied. Eight of ten genotypes tested were successfully regenerated. The highest callus induction rate and regeneration frequency of line 21605 was obtained when inflorescence buds were stored at 4°C for 4 days, and anthers with microspores at the mid to late uninucleate stage were cultured on MS basal medium containing BA (2.2 micro M) and NAA (1.82 or 2.7 micro M). Frequencies of callus induction and shoot regeneration were 100 and 70.5%, respectively with the whole regeneration procedure completed in 40 days under light. This highly efficient, rapid regeneration system can be applied for both genetic transformation and doubled haploid plant induction (Qi *et al.*, 2011).

Embryogenesis

Embryogenic callus and somatic embryos were induced from cotyledonary explants of *T. erecta* (Bespalhok and Hattori, 1998). Cotyledons were first cultured on MS medium supplemented with 2.0 mg 2, 4-D and 0.2 mg kinetin/litre. After 5 weeks, calli were transferred to MS medium supplemented with 0.02 mg thidiazuron/litre where compact embryogenic callus developed. Friable embryogenic callus developed when the compact embryogenic callus was transferred to medium containing 2, 4-D and subcultured every 2 weeks. Friable embryogenic callus was maintained for 2 years. Somatic embryo development was obtained when friable embryogenic callus was transferred to MS medium supplemented with 3 mg ABA and 60 g sucrose/litre. In addition of 10-30 mM L-glutamine improved somatic embryo development.

Molecular breeding

Characterization of marigold (*Tagetes* spp.) genotypes was done by using SDS-PAGE and RAPD markers (Mor *et al.*, 2008). Nine genotypes of marigold, selected from two species, *Tagetes erecta* (3 genotypes) and *Tagetes patula* (6 genotypes), were characterized through electrophoresis of protein and using RAPD markers. The dendrogram based on protein electrophoresis grouped the nine genotypes in two clusters species-wise, whereas RAPD analysis showed clear-cut genotype and species difference, which confirms the reliability of RAPD markers over protein electrophoresis.

Various experimental protocols for RAPD analysis of genes related to fertility in *Tagetes erecta* were studied and an optimum procedure was developed for setting up a stable system. The CTAB and SDS method were compared to extract DNA from different tissues of *Tagetes erecta*. The quality of DNA extracted was evaluated by spectrophotometry and PCR. Results showed that total DNA extracted from young leaves by CTAB was suitable for the analysis of RAPD. By using this system, 100 sterile and fertile plants of W205, a two type male sterile line in *Tagetes erecta* were tested. Two hundred eighty eight bands were obtained using 64 RAPD primers, among which one band of 980 bp derived from the primer G-02(5'-GGCACTGAGG-3') was cosegregating with fertile plants. This RAPD marker named as G-02-980 could be used to identify male and fertile seedlings (Wang *et al.*, 2010).

Estimation of somaclonal variation in *Tagetes erecta* at callus stage using molecular markers (RAPD and ISSR) was done (Sunila Kumari *et al.*, 2012). Among both the systems, ISSR amplicons were found to be more polymorphic as compared to the RAPD products. Total 25 bands were generated from RAPD reaction of which 13 were found to be polymorphic. On the other hand, ISSR analysis showed 19 bands as polymorphic out of 25 bands generated. The Jaccard's similarity matrix was used to construct UPGMA based dendogram, total four major clusters were formed. The principal component analysis for 2D was also done, the eigen value showed a maximum of 52% variability distributed on the 1st dimension and 12% on the 2nd dimension. Though the callus was induced from a single ex-plant source it contained considerable amount of variability existing at molecular level. This suggested that variability was due to the treatments employed to induce callus.

Genetic diversity of *Tagetes erecta* and *Tagetes patula* was done by using PCR based RAPD and ISSR markers. Twenty three RAPD primers were used to generate polymorphism and mean per cent polymorphism was found as 57.72% ranging from 33.33% (OPF-2) to 83.33% (RAPD-07). Five most informative primers (OPA-02, OPA-04, OPO-10, RAPD-07 and RAPD-18) were identified on the basis of level of polymorphism detected by individual primer. Twelve ISSR primers were used to generate polymorphism and mean per cent polymorphism was 60.48% ranging from 40.00% (ISSR-02, ISSR-17) to 90.00% (ISSR-03). The five most informative primers (ISSR-03, ISSR-06, ISSR-12, ISSR-31 and ISSR-18) were identified on the basis of polymorphism. The dendrogram obtained from UPGMA cluster analysis of Jaccard's similarity values based on 23 RAPD and 12 ISSR markers grouped 15 genotypes into two clusters species wise (*Tagetes erecta* and *Tagetes patula*) which confirm the reliability of PCR based molecular markers. PCA analysis also confirmed the grouping of genotypes as they were present in same group as revealed by RAPD and ISSR markers (Namita *et al.*, 2013).

Genetic engineering

Hypocotyls, roots, leaf sections and shoot tips from *Tagetes erecta* plantlets were inoculated with *Agrobacterium tumefaciens*, harboring the binary vector pCAMBIA2301, containing the beta -glucuronidase gene. Histochemical GUS assays of infected tissues showed transient *gus* gene expression after 3 days (Godoy-Hernandez *et al.*, 2006).

Marigold (*Tagetes erecta*) leaf explants were bombarded with plasmid pBI426 in a low pressure particle inflow gun. Transient expression assays were performed with different pressures and distances. The best transient expression (GUS) level was obtained with explants induced in regeneration medium and bombarded at 10.5 cm and 551.6 kPa. Stable expression assays were performed and the callus tissue was selected on regeneration medium supplemented with 100 mg L^{-1} kanamycin. Stable genetic transformation of the resistant plantlets was demonstrated by PCR and Southern blot analysis; more than one transgene copy was observed in the transformed plantlet genomes. This method can be used for marigold gene transfer (Vanegas *et al*., 2006).

Kim *et al.* (2007) examined the possibility of transforming and obtaining transgenic marigold (*Tagetes erecta*) plants using the particle bombardment process. Conditions for optimal delivery of DNA to cotyledonous explants were established for three cultivars of marigold plant. The size of microcarriers, distance to target, and the pressure significantly affected transformation and regeneration efficiency. A total of approximately 1000 cotyledonous explants for each cultivar were bombarded with DNA-coated particles and placed on selection medium containing 50 mg litre^{-1} kanamycin. The regeneration rates before bombardment were between 17 and 38% in 1/2 MRM medium supplemented with 5 or 7 mg L^{-1} benzylaminopurine. However, the regeneration rate was reduced to 4% at most after bombardment.

REFERENCES

Anonymous. 1988. Annual Report, IARI, New Delhi. pp. 27-30.

Anonymous. 1998. New varieties of Flower crops. AICFIP Tech. Bull. 9. IARI, New Delhi, p. 35.

Anonymous. 2011. Report of the Quinquennial Review Team 2000-2008. Indian Council of Agricultural Research, New Delhi.

Arora, J.S. and Singh, J. 1980. Performance of marigold cultivars under north Indian conditions. *Abst*: National Sem. Prod. Tech. Comm. Flower Crops, TNAU, Coimbatore, p. 81-82.

Bailey, L.H. 1951. Manual of Cultivated Plants, Macmillan, New York.

Bailey, L.H. 1963. The Standard Cyclopedia of Horticulture, Vol. III, Macmillan Co. Ltd., New York, p. 3305.

Bailey, L.H. 1976. Staff of Hortorium, Hortus Third, A Concise Dictionary of Plants Cultivated in the United States and Canada, Macmillan, New York.

Banerjee, A.K. 1974. Chromosome studies in the cultivated *Tagetes* Linn. *Butt. Bot. Soc.*, **28**: 41-46.

Beniwal, B.S., Dahiya, B.S., Deswal, D.P. and Joon, M.S. 2009. New marigold varieties Hisar Beauty and Hisar Jaffri-2. *Abst.*: National Conference on Floriculture for Livelihood and Profitability, IARI, New Delhi. p. 38.

Bespalhok, F.J.C. and Hattori, K. 1998. Friable embryogenic callus and somatic embryo formation from cytoledon explants of African marigold (*Tagetes erecta* L.). *Plant Cell Reports*, **17**(11): 870-875.

Bhanu Pratap, Tiwari, G.N., Mishra, L.N. and Tripathi, V.K. 2002. Genetic variability in marigold. *Abst*: National Symposium on Indian Floriculture in New Millennium, Bangalore, p. 33.

Bolz, G. 1961. Genetisch-zochterische Untersuchungen bei *Tagetes*. III. Artkreuzungen, *Z. Planzenzuchtg*. **46**: 169-211.

Geetha, R.M. and Jayabalan, N. 1992. Induced chlorophyll and viable mutations in *Tagetes patula* L. *Acta Botanica Indica*. **20**(2): 312-14.

Godoy-Hernandez, G., Berzunza, E.A., Concha, L.C., Miranda-Ham, M. deL. 2006. *Agrobacterium*-mediated transient transformation of marigold (*Tagetes erecta*). *Plant Cell Tissue Organ Cult.*, **84**(3): 365-68.

Gowrishankarayya, V.M. and Hedge, L. 2004. Genetic variability studies in African marigold (*Tagetes erecta* L.). *Abst*: National Symposium on Recent Trends and Future Strategies in Ornamental Horticulture. ISHS, New Delhi and UAS, Karnataka, p. 14.

Gupta, Y.C., Raghava, S.P.S. and Misra, R.L. 1999. Inheritance of male sterile apetalous inflorescence in African marigold. *J. Orna. Hort.*, **2**(2): 65-66.

Gupta, Y.C., Raghava, S.P.S. and Misra, R.L. 2001. Heterobeltiosis in African marigold (*Tagetes erecta* L.). *Indian J. Genet. Pl. Breeding*, **61**(1): 65-68.

Gupta, Y.C., Raghava, S.P.S., Misra, R.L. and Pandey, H.N. 2002. Promising F_1 hybrids in African marigold. *Abst*: National Symposium on Indian Floriculture in New Millennium, Bangalore, p. 36.

Iltis, H. 1945. Bud mutations in the marigold. *Heredity*, **36**(10): 291-295.

Jacobs, J.J., Arroo, R.R., Koning, E.A., de Klunder, A.J., Croes, A.F., Wullems, G.J. 1995. Isolation and characterization of mutants of thiophene synthesis in *Tagetes erecta*. *Plant Physiol*., **107**(3):807-14.

Jalil, R., Khoshoo, T.N. and Pal, M. 1974. Origin, nature and limit of polyploidy in marigolds. *Curr. Sci*., **43**: 777-779.

Janakiram, T. 2004. Heterosis and hybrid seed production in ornamental crops. *In*: Souvenir of National Symposium on Harnessing Heterosis in Crop Plants, Varanasi, pp. 143-153.

Kaneko, J., Sugawara, N. and Kitamura, C. 1990. Relation between the number of flower visiting individuals and the number of male moths captured by sex pheromone traps in the silver Y moth (*Autographa gamma* L.). *Ann. Report Soc. Pl. Protection North Japan*, **41**: 160-163.

Kapelev, I.G. and Robotyagov, V.D. 1981. Increasing the yield of essential oil marigold by means of polyploidy. 4-i-s ezd-genetikov-i-seleksionerov-Ukrainy-Odessa-Tez-dok, **1**: 157-158.

Kaplan, L. 1960. Historical and ethnobotanical aspects of domestication in *Tagetes*. *Economic Botany*, **12**: 200-202.

Kennedy, R.R. 1997. Genetic variability for seed characters in marigold. *Madras Agric. J*., **84**(6): 328-330.

Khoshoo, T.N. 1979. Cytogenetics in relation to plant evolution and improvement. *In*: Khoshoo, T.N. and P.K. Nair (eds.). Progress in Plant Research. New Delhi, Today and Tomorrow Printers and Publishers, **2**: 1-74.

Kim, H., Hong, S., Mo, S. and Kim, D. 2007. Particle bombardment-mediated transformation and regeneration of marigold (*Tagetes erecta*). *Hort. Environ. Biotech*., **48**(3): 197-200.

Kumar, A., Singh, S.K., Sharma, S.K., Raghava, S.P.S. and Misra, R.L. 2004. Comparison of seed derived and micropropagated male sterile plants of *Tagetes erecta* L. for F_1 hybrid seed production. *J. Hort. Sci. Biotech*., **79**(2): 260-266.

Kumar, S., Shanmugavelu, K.G. and Irulappan, I. 1989. Hybrid vigour in marigold for economic characters. *South Indian Hort*., **38**(3): 173-174.

Mathew, P.M. and Abraham, M. 1980. Induced polyploidy in *Tagetes erecta* L. *Cytologia*, **45**: 803-807.

Mohanty, A., Mohanty, C.R. and Mohapatra, R.C. 2003. Combining ability for yield and its components in marigold. *J. Orna Hort*., **6**(1): 34-35.

Mohanty, A., Mohanty, C.R. and Mohapatra, R.C. 2003a. Heterosis studies in African marigold. *J. Orna. Hort*., **6**(1): 55-57.

Mohanty, C.R., Mohanty, A. and Mishra, N.K. 2002. Studies on floral biology of marigold under Bhubaneswar condition. *J. Orna. Hort*., **5**(1): 18-21.

Mor, V.S., Deswal, D.P., Mann, A., Dahiya, B.S. and Beniwal, B.S. 2008. Characterization of marigold (*Tagetes* spp.) genotypes using SDS-PAGE and RAPD markers. *Seed Sci. Tech*., **36**(3): 757-766.

Namita, N. Panwar, S., Sonah, H. Singh, K.P. and Sharma, T.R. 2013. Genetic diversity analysis of marigold (*Tagetes* sp.) genotypes using RAPD and ISSR markers. *Indian J. Agril. Sci*., **83**(5):484-90.

Nand Kishore and Raghava, S.P.S. 2001. Variability studies in African marigold. *J. Orna. Hort*., **4**(2): 124-125.

Panwar, S., Singh, K.P., Janakiram, T. and Namita. 2013. Genetic variability, heritability and genetic advance in African marigold (*Tagetes erecta* L.) genotypes. *Progressive Horticulture*, **45**(1): 135-140.

Patnaik, N., Mohanty, C.R. and Patnaik, N. 2002. Genetic variability, heritability and genetic advance in African marigold (*Tagetes erecta* L.). *Orissa J. Hort*., **30**(2): 35-38.

Punnet, R.C. 1924. Notes on the genetics of the African marigold (*Tagetes erecta*). *Studia Mendeliana*, pp. 187-191.

Punnet, R.C. 1937. Marigold breeding. *In*: Year book of Agriculture. U.S. Dept. Agriculture.

Qi, Y., Ye, Y. and Bao, M. 2011. Establishment of plant regeneration system from anther culture of *Tagetes patula*. *African J. Biotech.*, **10**(75): 17332-38.

Raghava, S.P.S. 1995. Genetic improvement of African marigolds. *Abst*: National Seminar on Ornamental Horticulture and Environment, Calcutta, p. 9.

Rao, T.M., Negi, S.S. and Raghava, S.P.S. 1993. Male Sterility in marigold. *Abst*: Jubilee Symposium Hort. Res. Changing Scenario, Bangalore, p. 91.

Reddy, E.N., Muthuswami, S., Irulappan, I. and Khader, M.A. 1989. Heterosis and combining ability for yield and yield components in African marigold (*Tagetes erecta* L.). *S.I. Hort.*, **36**(1-2): 51-61.

Relichova, J. and Petrova, I. 1998. Combining ability and heterosis in *Tagetes erecta* L. *Sbornik Uvtiz Genelika a Stenchteni*, **24**(1): 75-84.

Rydberg, P.A. 1915. Tagetes. *North American Flora*, **34**(2): 148-159.

Serrato-Cruz, M.A., Hernandez Rodriguez, M., Savidan, Y. and Barcenas Ortega, N.M. 2000. DNA content and ploidy level in *Tagetes* spp. using flow cytometry. *Agrociencia*, **34**: 729-734.

Serrato-Cruz, M.A., Miranda Colin, S., Garcia Velazquez, A. and Castillo Gonzalez, F. 1998. Reproductive isolation in Cempoalxochiti plants (*Tagetes* spp.). *Revista Fitotechnia Mexicana*, **21**(2): 127-128.

Sharga, A.N., Motilal, V.S. and Basario, K.K. 1972. Performance studies of some marigold varieties. *Prog. Hort.*, **4**(3-4): 71-76.

Singh, A.K. 2006. Flower Crops: Cultivation and Management. Pub. New India Publishing Agency, New Delhi. pp. 239-250.

Singh, B. 1970. Morphological and genetical studies in marigolds (*Tagetes erecta*, *T. patula* and *T. tenuifolia*). Ph.D. Thesis. Agra Univ., Agra.

Singh, B. and Swarup, V. 1971. Heterosis and combining ability in African marigold. *Indian J. Genet. Pl. Breeding*, **31**(3): 407-416.

Singh, B. and Swarup, V. 1973. Inheritance of flowering, flower weight and flower number in African marigold. *Indian J. Genet. Pl. Breeding*, **33**(2): 172-175.

Singh, B. and Swarup, V. 1973a. Inheritance of qualitative characters in African marigold. *Indian J. Genet.*, **33**: 373-377.

Singh, D. and Singh, A.K. 2004. Studies on genetic variability, heritability and genetic advance in *Tagetes* spp. *Abst*: National Symposium on Recent Trends and Future Strategies, ISHS, New Delhi and UAS, Karnataka, p. 14.

Singh, K.P., Saha, T.N. and Singh, A.P. 2004. Variability studies in French marigold. *Abst*: Improving Productivity, Quality, Post-harvest Management and Trade in Horticultural Crops. IHC, New Delhi. p. 62.

Singh, V.N., Banerji, B.K., Dwivedi, A.K. and Verma, A.K. 2009. Effect of gamma irradiation on African, marigold (*Tagetes erecta* L.) cv. Pusa Narangi Gainda. *J. Hortil. Sci.*, **4**(1): 36-40.

Sreekala, C. and Raghava, S.P.S. 2003. Exploitation of heterosis for carotenoid content in African marigold (*Tagetes erecta* L.) and its correlation with esterase polymorphism. *Theo. App. Gene.*, **106** (4): 771-776.

Sreekala, C., Raghava, S.P.S., Misra, R.L. and Voleti, S.R. 2002. Assessment of variability for carotenoids and yield components in African marigold. *J. Orna. Hort.*, **5**(2): 5-7.

Sunila Kumari, Panj, F.G., Gupta, P., Shah, R.R. 2012. Estimation of somaclonal variation in *Tagetes erecta* L. at callus stage using molecular markers (RAPD and ISSR). *Annals Agri. Bio. Res.*, **17**(1): 1-4.

Towner, J.W. 1961. Cytogenetics studies on the origin of *Tagetes patula* I. Meiosis and morphology of diploid and allotetraploid *T. erecta* × *T. tenuifolia*. *American J. Bot.*, **48**: 743-751.

Towner, J.W., 1962. Cytogenetics of *Tagetes jaliscensis* × *T. erecta*. *American J. Bot.*, **48**: 1064-1067.

Turner, B.L. and Nesom, G.L. 1993. Biogeography, diversity and endangered or threatened status of Mexican Asteraceae. *In*: Ramamoorty, T.P. Bye, R. Lot, A. and Fa, J. (eds.). Biological Diversity of Mexico: Origins and Distribution. New York, Oxford Univ. Press, pp. 290-299.

Vanegas, P.E., Valdez-Morales, M., Valverde, M.E., Cruz-Hernandez, A. and Paredes-Lopez, O. 2006. Particle bombardment, a method for gene transfer in marigold. *Plant Cell Tissue Organ Cult.*, **84**(3): 359-63.

Velmurugan, K., Vijayakumar, M. and Jawaharlal, M. 2003. Heterosis and combining ability effects for xanthophyll content of marigold (*Tagetes erecta* L.). *S.I. Hort.*, **51**(1/6): 237-240.

Verma, S.K., Singh, R.K. and Arya, R.R. 2002. Genotypic and phenotypic variability in marigold. *In*: R.L. Misra and Sanyat Misra (eds.). Floriculture Research Trend in India. ISOH, Bangalore. pp. 330-331.

Wang, Z., Yin, D., Wang, P., Zhao, J., Pan, B. and Zhang, H. 2010. RAPD markers developed for identifying male sterility in *Tagetes erecta*. *Acta Horticulturae*, **855**: 289-94

Watts, G. 1893. A Directory of the Economic Products of India. London, **4**(3): 403.

Section VII

Orchids

Chapter – 30

Cattleya

Cattleya belongs to the family Orchidaceae. It is highly important in floriculture and has been cultivated since the 19th century. The name "Cattleya" was given in the honour of "Sir William Cattley". Cattleyas have earned the reputation as the "Queen of Orchids" and are known to the public as the ultimate in floral corsages. While no longer the reigning queen of the orchid floral industry it is difficult not to be impressed by a well-flowered *Cattleya*. No longer limited to white and various shades of lavender and purple, high quality flowers occur in all colours except true blue and black and in a wide range of plant sizes. They are widely known for their large, showy flowers and were used extensively in hybridization for the cut-flower trade until the 1980s when pot plants became more popular.

ORIGIN AND DISTRIBUTION

The genus was named in 1824 by John Lindley after Sir William Cattley who received and was the first to bloom a specimen of *Cattleya labiata*. William Swainson had discovered the new plant in Pernambuco, Brazil, in 1817 and shipped to the Glasgow Botanic Gardens for identification. After the first importation from Mr. Swainson, no more plants similar to the original *Cattleya labiata* were discovered for many decades, although many varieties of this species were found. Each variety differed in some way from the plant of Mr. Cattley, either in colour, blooming season or some similar horticultural aspect. Brazil is the centre of distribution, where more than half the species are found (Pabst and Dungs, 1975 and Cruz *et al*., 2003). They are found naturally only in the western hemisphere, occurring along with Laelias in three major regions of Tropical America i.e. Mexico and Guatemala, northern part of South America from the western Cordilleras of Colombia to British Guiana and the maritime provinces of southern Brazil. Within these three large regions are found most of the species of *Cattleya* except for *C. dowiana* in Costa Rica and *C. skinneri* var. *autumnalis* from Panama. The plants occur at considerable elevation where the mean temperature is much lower than the mean temperature for the latitude at sea level.

SPECIES

Cattleya is a genus of 113 species of orchids from Costa Rica to tropical South America. Originally, the concept *Cattleya* included a requirement that the pollinarium contain only four pollinia. Starting some time after December, 2000, the Royal Horticultural Society (RHS) began re-organizing the generic boundaries with respect to hybrid registration. This has led to multiple senses of the term *Cattleya*, not only with respect to species, but also with respect to hybrids. Except for a half dozen species found in Mexico and Central America, all the bifoliate Cattleyas are natives of Brazil with an extension of range, in the case of several species, into neighbouring countries, for orchids do not respect national boundaries but geographical limits. Some important species are given in Table 1 and described.

Table 1: Important species of cattleya.

S. No.	Genus	Important species	Origin
1.	*Cattleya*	*Cattleya aurantiaea, C. bicolour, C. bowringiana, C. citrina, C. dowiana, C. intermedia, C. labiata, C. leopoldii, C. luteola, C. maxima, C. schilleriana, C. skinneri, C. violaceae, C. warscewiezii*	American Tropics

***Cattleya aclandiae*:** Origin is Brazil and introduced in 1839 by Lady Akland of Killerton near Exeter. It was named by Dr. Lindley while describing the species in the Botanical Register in 1840. A short peduncle bears one or two flowers from three to four inches across. Petals and sepals are similar fleshy, yellowish green transversely blotched and spotted with blackish purple. The rather fiddle-shaped lip is three-lobed, the lateral lobes small and curved toward the column, white tinged with rose, the mid lobe broadly kidney-shaped, wavy, bright rose-purple veined with deep purple.

***Cattleya amethystoglossa*:** Origin is Brazil. This species first appeared in the collection of Herr Reichenheim at Berlin and was described in 1856 in Bonplandia by Reichenbach as *Cattleya guttata var. prinzii* named to honour Herr Print who had sent the plant from Brazil. It appeared in England in the collection of Mr. F. Coventry of Shirley whose solitary plant went to Mr. Warner in 1860. Flowers are about 7-10 cm across, fleshy in an upright cluster containing five to twenty flowers. Petals are similar to sepals but somewhat broader, bright rose spotted with amethyst-purple, especially toward the margins. Lip is three-lobed, the lateral lobes folding over the column, the middle lobe spreading, notched in the front margin, deep amethyst-purple.

***Cattleya bicolour*:** Origin is Brazil introduced by Messrs Loddiges in 1836, described in the Botanical Register by Lindley in 1836. The inflorescence is nearly erect with two to five or more flowers. Flowers range from 8-10 cm across. The sepals and petals are fleshy with a distinct mid nerve, greenish brown to olive-brown spotted with purple, the petals somewhat wavy, the lateral sepals bowed inward. The lip is wedge-shaped without side lobes, curved downward with a central longitudinal depression or line, crimson-purple, occasionally margined with white. This species is unique in lacking the lateral lobes of the lip, a character usually inherited by its hybrid progeny, limiting its value in breeding.

***Cattleya dowiana*:** Origin is Costa Rica. Originally discovered by Warscewicz during year 1850. However, it was rediscovered during 1865 by Mr. Arce, a native naturalist, sending plants to Mr. G. Ure Skinner by way of a Captain Dow for whom it was named by Mr. Bateman in the Gardeners' Chronicle in 1866. The flowers are from 12-17 cm across usually two but up to seven on a cluster. Sepals and petals nankeen-yellow; middle lobe of lip broad and spreading with velvety texture, rich crimson-purple streaked with golden veins radiating from the center, three heavier golden veins passing longitudinally from the base to the apex.

***Cattleya gaskelliana*:** Origin is Venezuela, introduced to England during 1883 by Sanders. It was named for Mr. Holbrook Gaskell of Woolton. It is intermediate in character between *C. mossiae* and *C. lueddemanniana*. Flowers large, 15-17 cm, colour varying from medium amethyst purple to pure white. The lip is fairly large the tube the same colour as the petals and sepals, the lobe deep violet with pale border, the throat streaked with yellow and yellowish white.

***Cattleya granulosa*:** Origin is Brazil. Discovered in 1840 by Hartweg, its habitat was reported as Guatemala. The species was described by Lindley in the Botanical Register for 1842. The flower stem is stout, erect with five to eight flowers, each about 7-10 cm across. The sepals and petals are yellowish olive-green with scattered spots of red. Sepals oblong and obtuse and lateral sepals bowed inward. The petals are a little broader than the sepals, with the margin slightly waved. The lip is three-lobed, the lateral lobes erect, whitish outside and yellow inside, the middle lobe clawed with a fimbriate kidney-shaped blade, the claw yellow, the blade white, covered with numerous crimson-purple granulations.

***Cattleya intermedia*:** Origin is Brazil brought by captain Graham of the Royal Packet Service from Rio de Janeiro in 1824. It bloomed in the Botanic Garden at Glasgow in 1826 and was described by Hooker in the Botanical Magazine for 1828. Flower stems are stout, three to five or more flowers, each flower 10-12 cm across. Sepals and petals are arrow, the dorsal sepal strap-shaped, the lateral sepals and petals curved downward, pale rose to milk white, occasionally dotted with amethyst-purple. Lip is trilobbed, the side lobes rounded with smooth margins, overlapping around the column, the middle lobe amethyst-purple, spreading, with crisped and eroded margin.

***Cattleya labiata*:** It belongs to Brazil. The true *C. labiata* is a free-flowering species bearing from two to five flowers each from 12-15 cm across. The petals and lip are well balanced, the latter being more open than in *C. eldorado* but not as flat as in *C. dowiana*. Although quite variable in colour, the basic pattern consists of rose-coloured wavy petals, sepals the same colour, the ruffled lip rich crimson-purple bordered with lilac.

***Cattleya loddigesii*:** Origin is Brazil. Introduced to England from Rio de Janeiro by Messrs. Loddiges of Hackney under the name of *Epidendrum violaceum* early in the nineteenth century. It was placed into the new genus *Cattleya*, under the name *Cattleya loddigesii* by Lindley when he established the new genus in 1824. Flower stem bears from two to five flowers up to 10 cm across. Petals and sepals similar and nearly equal and the lateral sepals somewhat bowed, the petals slightly broader and waved along the margins, delicate rose-lilac. Lip is trilobed, the lateral lobes almost rectangular, erect, the front edge toothed, coloured as the petals and sepals on the outside, whitish inside, the middle lobe spreading, much crisped at the margin, pale amethyst-purple; the disk whitish to pale yellow.

***Cattleya lueddemanniana*:** Venezuela. Another species with a confused background, it has been grown under several names, including *C. speciossissima* and *C. dawsonii*. The type description, however, was by Reichenbach in Xenia Orchidacea in 1854, in compliment to the gardener of M. Pescatore of Paris. From three to four flowers are produced, up to 20cm across. Colour varying from purplish rose to pure white, the petals being nearly three times the width of the sepals. The lip is similar in colour to the petals and sepals except the front lobe which is rich amethyst-purple, with two pale yellow or whitish blotches at the throat.

***Cattleya luteola*:** Brazil and Peru. Described by Lindley in Gardeners' Chronicle in 1853. The flower stems are shorter than the leaves, bearing from two to five flowers. Each flower is about 5 cm across, pale lemon yellow with the margin of the front lobe whitish and the side lobes streaked with purple on the inner side. The petals are not much wider than the sepals, the lip tubular with a ruffled but not wide-spread front lobe.

***Cattleya maxima*:** Ecuador, Peru and Colombia. Originally collected in 1777 by the Spanish botanists, Ruiz and Pavon, who explored the Cinchona forests of Peru for their government, the specimens remained in their herbarium material until 1831 when it was partially described by Lindley in his genera and species of Orchidaceous plants. It reappeared in collections by Hartweg in 1842, flowering in the gardens of the Horticultural Society of London in 1844, but died out. In 1855 it was flowered again by Mr. W. G. The original form was deep lavender, the ruffled and spreading lip rose to lavender with a citron-yellow band running down the length of the tube, and a network of deep crimson veins covering the front lobe.

***Cattleya mossiae*:** Venezuela. First introduced by Mr. George Green who received it from La Guaira, Venezuela, in September, 1836, it flowered in the collection of Mrs. Moss of Otterpool and was named for her by Sir William J. Hooker who described it in the Botanical Magazine in 1839. A free-flowering, vigorous variety, it produces from three to four flowers on the spike, each flower from 15-20 cm or more in size. The petals broad, somewhat fluted on the margins, the sepals and petals ranging from deep lilac to white. The broad lip is flattened but has a frilled margin, is very variable in colour but basically is the same colour as the petals and sepals, with yellow in the throat and disc, the front lobe marked with broken streaks or blotched with violet, the margin being lilac.

***Cattleya percivaliana*:** Venezuela. Named for Mr. R. P. Percival, an English orchid grower, it was described by Reichenbach in Gardeners' Chronicle for 1882. Two to four flowers are produced on each stem, the flowers about 10-12 cm across. Petals are rosy lilac suffused with purple, the sepals frequently a little paler. The lip is smaller than in most labiata Cattleyas, quite variable in colouring and beautifully frilled. The front lobe of the lip is crimson-purple, shaded with maroon and spotted with yellow toward the throat.

***Cattleya tigrina* (*leopoldii*):** Brazil. This is the oldest name for what has long been known as *Cattleya guttata* var. *leopoldii*. A native of southern Brazil, found growing with *Cattleya* (*Laelia*) *purpurata* and *C. intermedia* in Santa Catarina and in other areas with *C. guttata* or *C. loddigesii*. The plants are large reaching as much as 120 cm tall and can produce 20 to even 30 flowers per inflorescence. While considered bifoliate, the pseudobulbs usually produce three leaves. The characteristic that most readily separates this species from *C. guttata* is its habit of flowering from green sheaths as opposed to dried sheaths in *C. guttata*.

***Cattleya violacea*:** British Guiana, Colombia, Venezuela and Brazil. First discovered by the traveler Humboldt, it was described in 1816 as *Cymbidium violaceum*. Subsequently introduced into cultivation in 1838 by Sir Robert Schomburgk it was named *Cattleya superba* in Lindley's Sertium Orchidaceum and became widely known under the latter name. In 1889, however, R. A. Rolfe in the Gardeners' Chronicle demonstrated that the two species were synonymous and the earlier specific name should stand as *Cattleya violacea*. The flower-stem bears three to five or more fleshy flowers, each from 10-12 cm across. The sepals and petals are spreading, bright rose-purple suffused with white, the sepals smooth and the petals broader, nearly rhomboidal and waved at the margins. The lip is three-lobed the side lobes acutely triangular and rolled over the column, the middle lobe transversely oblong, convex, eroded at the margin, deep crimson-purple with yellow disk with white blotches on each side, streaked with deep purple.

Cattleya walkeriana: Brazil. Discovered by the traveler Gardner around 1839, and was described by Hooker in the London Journal of Botany for 1843. *Cattleya bulbosa*, described by Lindley in Gardeners' Chronicle for 1847. A vegetative peculiarity distinguishes this species, for the flowers are not produced from the apex of the pseudobulb as in other Cattleyas but arise on separate short, slender shoots which spring directly from the rhizome at the base of the leaf-bearing growth. This character is found only in this species and in *Cattleya nobilior*. Flowers are large for the plant, 10-11 cm across ranging in colour from bright rose-purple to pale pink-lilac.

Guarianthe (Cattleya) aurantiaca: Mexico, Guatemala, El Salvador and Honduras. Originally described as *Epidendrum aurantiacum* by Lindley in the Botanical Register for 1838. The 2-3 cm long flowers are produced in an arching or slightly drooping raceme from the base of the leaves. The petals and sepals are similar, bright cinnabar-red in colour. The lip is obscurely three-lobed, the middle lobe somewhat acute, cinnabar-red with several dark streaks or veins. There are several forms of this species, some being very vigorous with numerous larger flowers.

Guarianthe (Cattleya) bowringiana: British Honduras, Guatemala. Introduced by Veitch in 1884 and described in the Gardeners' Chronicle for 1885, dedicated to Mr. J. C. Bowring of Forest Farm, an ardent English amateur orchid collector. The flower stem arises from a narrow, flattened sheath and bears from five to fifteen or more flowers which range in size from 5-7 cm across. Sepals and petals are bright rose-purple with somewhat deeper veining, the petals much broader than the sepals and somewhat wavy along the margins. The lip is narrowly tubular surrounding the column and externally similar in colour to the sepals and petals, the interior portion of the lip flaring outward without distinct lobes, deep purple with a central crescent of maroon surrounding the white to yellowish throat.

Guarianthe skinneri: Although the flowers of this bifoliate species from Central America are not as large as those of the labiate Cattleyas, this species is highly recommended to the hobbyist, particularly beginners, because of its very vigorous growth and free-flowering characteristics. The club-shaped stems are about a foot tall, bear two largish leaves and a cluster of five to ten fair-sized flowers, about four to five inches across. The petals and sepals range from pale to deep rose-purple, the lip a deeper purple with whitish disk and throat.

FLOWER MORPHOLOGY AND BIOLOGY

Plant epiphytic on trees or rocks thickened pseudo bulbous stems with one or two leaves. Leaves thick, leathery or fleshy. Inflorescence unbranched, arising from apex of pseudo bulb usually emerging from spathe-like sheath. The *Cattleya* flower is sometimes called a typical orchid flower simply because it is one of the most widely known types. The genus is divided into two sections, the first is distinguished by the club-shaped pseudo bulbs which bear a single leaf, while the second has long, spindle-shaped pseudo bulbs bearing two leaves. The flowers of the unifoliate group are larger, with broader flower segments, whereas, the flowers of the bifoliate group have narrower petals and sepals, smaller but thicker flowers.

The typical flower has three rather narrow sepals and three usually broader petals: two petals are similar to each other and the third is the quite different conspicuous lip, featuring various markings and specks and an often frilly margin. Lip attached to base of column, free above, usually three-lobed but rarely not; side lobes form tube enclosing column (except in *C. bicolour*); middle lobe spreading, sometimes distinct from side lobes, occasionally continuous with them. Column rather long, robust, half-rounded, without wings, slightly arching; anther at apex of column, incumbent, distinctly two-celled. Pollinia four in two pairs, waxy, somewhat flattened, parallel. The anther with its pollinia is found in a cavity near the apex of the column. Another highly specialized area the rostellum which excludes a viscid substance that adheres to the insects that enter seeking nectar lies below the anthers on the column. As the insect withdraws and trips the caudicle releasing the pollinia this substance then fixes the pollen masses to the insect's body until it rubs it off on another flowers. Moths wasps bees and flies usually pollinate orchid. A study shows that self pollination is possible in *Cattleya auranticum* where the barrier between the rostellum and the stigmatic surface break down and the pollinia drop onto the stigmatic surface. The stigmatic surface lies behind the anther on the lower side of the column usually in a saucer like depression area which is covered with a viscid substance to catch the pollina carried by insects. A solitary dehiscent anther is present on the top of column. Pollinia are 2-8 in numbers and anthers can be easily removed. The ovary is inferior and therefore lies below the column. It is usually three celled or one-celled by abortion and contains up to 1000000 ovules all of which may be fertilized by the pollinia from another flower of the same genus. The features of different species are given in Table 2.

Table 2: Distinguish features of different species.

Inflorescence	Species
I. Lateral lobes of lip small or sometimes absent. The column exposed or enveloped by the lateral lobes to a certain degree at the base of the lip.	
Basal	*Cattleya walkeriana, C. nobilior Cattleya*
Terminal	*aclandiae, C. × dolosa, C. bicolour, C. velutina*
II. Lateral lobes of lip large, enveloping the column entirely or for the greater part	
A. Midlobe of lip somewhat larger, more the greater part clawed.	*Cattleya granulosa, C. porphyroglossa,* or less long and the column entirely or for *C. schofieldiana, C. × brasiliensis, C. guttata, C. × patrocinii, C. tigrina, C. amethystoglossa, C. elongata, C. × victoriae-reginae, C. schilleriana, C. × Whitei.*
B. Midlobe of lip small, sessile or subsessile.	*Cattleya violacea, C. × brymeriana, C. loddigesii, C. harrisoniana, C. intermedia, C. forbesii, C. × isabella, C. dormaniana.*
C. Midlobe of lip large, more or less continuous with lateral lobes.	*Prosthechea citrina, Guarianthe aurantiaca, Gur. × guatemalensis, Gur. skinneri, Gur. Bowringiana*

Mode of pollination

Many orchids are pollinated by solitary Euglossid bees in the tropics and by social (hive-forming) bumblebees and wasps in temperate zones. Bees carry orchid pollinia on their thorax. Many bees can't see red (they are red-blind), so they are primarily attracted to orchid flowers that are yellow, blue-green, blue and violet. They also see ultraviolet (UV) light reflected by flowers that humans are unable to perceive. Many cattleyas are bee-pollinated, along with schomburgkias, encyclias, and some epidendrums. At higher elevations, birds take over from bees, but one of the major differences is that birds have no sense of smell, so bird-pollinated orchids are generally not fragrant. Also, birds and butterflies are attracted primarily by red, pink and orange hues. Flowers that are bird-pollinated are often tubular and have a heavy callus on the lip. Within the cattleya alliance, birds and butterflies are both found to pollinate epidendrums and birds are also attracted to laelias. For instance, humming birds are also seen pollinating the red *Comparettia falcata* and *Laelia milleri*. Fly-pollinated orchids are usually coloured dull-purple and offer a radial shape.

GENETICS

Chromosome numbers of several *C. bicolor* individuals have been determined, finding diploid chromosome numbers of both 2n = 40 and 2n = 80 (tetraploid). The diploid chromosome number of *C. trinae* has been determined as $2n = 40$ and the haploid chromosome number has been determined as $n = 20$. Similarly in case of *C. mossiae* the diploid chromosome number is 2n = 40 and haploid chromosome number has been determined as n = 20. Whereas, diploid chromosome number of *Cattleya nobilior* been established as 2n = 42 which contrasts with the value of 2n = 40 for *Cattleya walkeriana* (Felix and Guerra, 2010).

Inheritance of flower colour

The colour of cattleya flower is due to pigments which are present in the cell sap. There are two kinds of pigments, first is the anthoxanthones which are yellow and second is the anthocyanins which are lavender reddish or purple. The yellow colour of *Cattleya dowiana* is due to the presence of minute crystals of anthoxanthone in the cell sap.

Colour inheritance

It may be based on a single gene or multiple genes which can act either dominant or recessive. Additionally, there are intensifier or inhibitor genes. In cattleyas, yellow is usually recessive, hidden by genes for purple. In laelias, yellow is usually dominant and masks purple. Pigmentation patterns can be expressed differently in the floral and vegetative parts of plants and can be distributed in different manners, either solely in the epidermis or more uniformly throughout the plant tissue. Withner (1988) mentions that there are at least three classes of pigments in cattleya flowers.

Anthocyanins: Anthocyanins are flavonoids mainly responsible for blue-mauve-pink-lavender-purple-magenta-red tones. Anthocyanins are comprised of a colour compound (anthocyanidin) bonded to a sugar molecule. Common colour compounds include pelargonidin (scarlet-red), cyanidin (blue-red) and delphinidin (pure blue, but not found in any orchids). Anthocyanins can be modified through the addition or subtraction of hydroxyl groups or changes in pH of the cell sap. Acidic sap favours reds and whites and alkaline sap favours blues and yellows. Typically, sepal and petal tissue would contain two or three types of anthocyanins and lip (labellum) tissue would have several additional anthocyanins contributing to the typically richer colour. Anthocyanins may be co-pigmented with light yellow flavonoids.

Carotenoid derivatives: It forms yellow and orange colours in flowers. These plastid pigments are not sensitive to pH.

Chlorophyll: It produces different shades of green. Chlorophyll alone produces a dark jade green while chlorophyll in combination with carotenoids results in a clear, pale green. Chlorophyll, carotenoids and anthocyanins all together result in bronze and brown flowers. Muddy colours on the other hand result when anthocyanins are present in the surface layers, over chlorophyll in the deeper tissue, or carotenoids in the sepals and petals.

Genetic variation produces *alba* and semi *alba* forms which lack pigments, albescent or xanthotic forms, coerulea (blue or mauve), along with splash petalled and picoteed peloric varieties. Red flowers are a result of a combination of the yellow/orange carotenoids and redder anthocyanins. True red has no magenta component. It's important to remember that colour perception is highly dependent on the light source.

BREEDING OBJECTIVES

i. Improved growth habit

ii. Improved flower size

iii. Novel flower colour

iv. Form modification

v. Fragrant variety

vi. Disease and pest resistance

vii. Stress tolerance

BREEDING METHODS

Selection

Selection of cattleya was begun from tropical America, particularly from Brazil. Several attractive cattleya were selected from there. Cattleya varieties introduced in different countries and then improved varieties were selected from these introductions. Some varieties are developed through selection are given in Table 3.

Table 3: Cattleya varieties evolved through selection.

S. No.	Varieties	Characteristics
1.	*Cattleya mossiae* v. *coerulea* Alayon	A fine selection of a great Cattleya orchid species. Flowers lavender-pink, petals full and overlap the sepals, rounded appearance.
2.	Brassocattleya Cordelia	A selected clone from the cross between *Rhyncolaelia digbyana* and *Cattleya intermedia*. Colour and temperature tolerance, good size, showy, fimbriated labellum.

Hybridization

Cattleya orchids are among the most familiar orchids in the world. Enormous number of hybrid forms have been produced both by crossing different *Cattleya* species with each other and by crossing *Cattleya* species with species in related genera. The flowers of the largest modern hybrids can exceed 20 cm in diameter while the smallest hybrids have flowers of just 5-7 cm, often exceeding the size of the plant. For most of the second half of the 20th century cattleya was the standard orchid to which other orchids were compared by growers (Fitch, 2004 and Banks 2005). In the 1920s, Lewis Knudson, an American researcher, developed a sterile growing technique that dramatically reduced seedling mortality and represented an enormous efficiency improvement for commercial growers.

In Brazil, the species have fairly well-defined and restricted limits, the overall distribution being roughly similar to the unifoliate species and to the Laelias. The interpenetration of ranges has resulted in the formation of a number of natural hybrids, not only between two bifoliate species but also between bifoliate and unifoliate cattleyas. A few supposed natural hybrids between bifoliate cattleyas and species of *Laelia* are also known.

The breeding started with the hybrid of *Cattleya* Jungle Elf (*esalqueana* × *aclandiae*) made at and registered by Fennell's Orchid Jungle, Homestead, Florida, 1987. This grex produced star-shaped clear yellow flowers and a few had red spots. Its flowers were small (6.5 cm), the petals were narrow at 2 cm wide and 2 or 3 were produced per stem on small plants. The morphology of other yellow cattleya hybrids with red spots are discussed (Clarke, 2013). Pre requisites in cattleya hybrids are full form with very broad segments and a large highly coloured gold veined frilled labellum. Some interspecific hybrids are given in Table 4.

Table 4: Interspecific hybrids of cattleya.

S. No.	Hybrids	Parentage	Country
1.	*Cattleya* × *brasiliensi*	*C. bicolour* × *C. harrisoniana*	Brazil
2.	*Cattleya* × *brymeriana*	*C. violacea* × *C. wallisii*	N. Brazil
3.	*Cattleya* × *dayana*	*C. forbesii* × *C. guttata*	Brazil
4.	*Cattleya* × *dolosa*	*C. loddigesii* × *C. walkeriana*	Brazil
5.	*Cattleya* × *dukeana*	*C. bicolour* × *C. guttata*	S.E. Brazil
6.	*Cattleya* × *duveenii*	*C. guttata* × *C. harrisoniana*	S.E. Brazil
7.	*Cattleya* × *intricata*	*C. intermedia* × *C. leopoldii*	S. Brazil
8.	*Cattleya* × *imperator*	*C. granulata* × *C. labiata*	N.E. Brazil
9.	*Cattleya* × *hybrida*	*C. guttata* × *C. loddigesii*	S.E. Brazil
10.	*Cattleya* × *gravesiana*	*C. lueddemanniana* × *C. mossiae*	Venezuela
11.	*Cattleya* × *gransabanensis*	*C. jenmanii* × *C. lawrenceana*	Venezuela
12.	*Cattleya* × *resplendens*	*C. granulosa* × *C. schilleriana*	N.E. Brazil
13.	*Cattleya* × *picturata*	*C. guttata* × *C. intermedia*	S.E. Brazil
14.	*Cattleya* × *patrocinii*	*C. guttata* × *C. warneriana*	S.E. Brazil
15.	*Cattleya* × *moduloi*	*C. schofieldiana* × *C. warneri*	Brazil
16.	*Cattleya* × *mixta*	*C. guttata* × *C. schofieldiana*	Brazil
17.	*Cattleya* × *mesquitae*	*C. nobilior* × *C. walkeriana*	Brazil
18.	*Cattleya* × *measuresii*	*C. aclandiae* × *C. walkeriana*	E. Brazil
19.	*Cattleya* × *lucieniana*	*C. forbesii* × *C. granulose*	S.E. Brazil
20.	*Cattleya* × *kautskyi*	*C. harrisoniana* × *C. warneri*	S.E. Brazil
21.	*Cattleya* × *joaquiniana*	*C. bicolour* × *C. walkeriana*	Brazil
22.	*Cattleya* × *venosa*.	*C. forbesii* × *C. harrisoniana*	Brazil
23.	*Cattleya* × *isabella*	*C. forbesii* × *C. intermedia*	S.E. Brazil
24.	*Cattleya* × *victoria-regina*	*C. guttata* × *C. labiata*	N.E. Brazil
25.	*Cattleya* × *wilsoniana*	*C. bicolour* × *C. intermedia*	Brazil
26.	*Cattleya* × *undulata*	*C. elongata* × *C. schilleriana*	Brazil
27.	*Cattleya* × *tenuata*	*C. elongata* × *C. tenuis*	Brazil
28.	*Cattleya* × *scita*	*C. intermedia* × *C. tigrina*	S. Brazil
29.	*Cattleya* × *guatemalensis*	*C. aurantiaca* × *C. skinneri*	S.E. Mexico Central America
30.	*Cattleya* × *hardyana*	*C. dowiana* var. *aurea* × *C. warscewiczii*	Colombia

Intergeneric hybrid

Cattleyas have been hybridized both within the genus and with related genera for more than a century, but the last several decades have seen a remarkable increase in both the quantity and quality of the hybrids within the *Cattleya* alliance. Among the most popular are the Brassolaeliocattleya (Blc.) and Sophrolaeliocattleya (Slc.) hybrids. Some important genera related to Cattleya and involved in improvement of varieties through intergeneric hybridization are given below.

***Laelia* (L.):** Breeding with this genus refines the lip of the orchid, producing a more elongated closed cone that gracefully opens into the full lip of the blossom.

***Brassavola* (B.):** This cross is made in order to affect the fabulous fringed, feathered or ruffled lip of the bloom. It also expands the lip of the blossom and the most imposing cattleya hybrids almost always have this species in their ancestry.

***Sophronitis* (Soph.):** A tiny, flame-coloured orchid that introduces the most intense red colour to its descendants. Many crimson and scarlet cattleya hybrids betray sophronitis in their ancestry. *Sophronitis* is also used to miniaturize cattleya hybrids.

Potinara (Pot.): It is the combination of all three of the above with a *Cattleya*. Potinaras comes in all ranges of colours from light green to magenta. They are generally smaller than Blc's but larger than Slc's.

Yamadara (Yam.): The addition of *Epidendrum* appears to increase flower yield and some Yamadaras are intensely coloured. With the recent discovery that the first Yamadara was registered as Adamara, the nothogeneric epithet Yamadara has been extinguished and replaced with Adamara.

Cattleyas can be crossed with a large number of other allied genera, including *Cattleyopsis*, *Caularthron*, *Schomburgkia*, *Tetramicra*, etc. Hybridization can go all the way up to eight parent genera, such as *Brassavola* × *Broughtonia* × *Cattleya* × *Cattleyopsis* × *Caularthron* × *Epidendrum* × *Laelia* × *Sophronitis* in Gladysyeeara. Several hybrids were developed through it and are given in Table 5.

Table 5: Intergeneric hybrids of cattleya.

S. No.	Hybrids	Parentage
1.	Allenara	Cattleya × Diacrium × Epidendrum × Laelia
2.	Arizara	Cattleya × Domingoa × Epidendrum
3.	Bishopara	Broughtonia × Cattleya × Sophronitis
4.	Brassocattleya	Brassavola × Cattleya
5.	Brassolaeliocattleya	Brassavola × Cattleya × Laelia
6.	Brownara	Broughtonia × Cattleya × Diacrium
7.	Buiara	Broughtonia × Cattleya × Epidendrum × Laelia × Sophronitis
8.	Cattkeria	Barkeria × Cattleya
9.	Cattlassia	Brassia × Cattleya
10.	Cattleya	Cattleya × Cattleya
11.	Cattleytonia	Broughtonia × Cattleya
12.	Cattotes	Cattleya × Leptotes
13.	Clarkeara	Brassavola × Cattleya × Diacrium × Laelia × Sophronitis
14.	Cookara	Broughtonia × Cattleya × Diacrium × Laelia
15.	Dekensara	Brassavola × Cattleya × Schomburgkia
16.	Diacattleya	Cattleya × Diacrium
17.	Dialaeliocattleya	Cattleya × Diacrium × Laelia

(*Contd.*)

(*Contd. Table 5*)

S. No.	Hybrids	Parentage
18.	Epicatonia	Broughtonia × Cattleya × Epidendrumdendrum
19.	Epicattleya	Cattleya × Epidendrum
20.	Epilaeliocattleya	Cattleya × Epidendrum × Laelia
21.	Estelaara	Brassavola × Cattleya × Epidendrum × Tetramicra
22.	Fergusonara	Brassavola × Cattleya × Laelia ×Schomburgkia × Sophronitis
23.	Fialaara	Broughtonia × Cattleya × Laelia × Laeliopsis
24.	Fordyceara	Broughtonia × Cattleya × Laeliopsis × Tetramicra
25.	Fujiwarara	Brassavola × Cattleya × Laeliopsis
26.	Gladysyeeara	Glya Brassavola × Broughtonia × Cattleya × Cattleyopsis × Diacrium × Epidendrum × Laelia × Sophronitis
27.	Hasegawaara	Brassavola × Broughtonia × Cattleya × Laelia × Sophronitis
28.	Hattoriara	Brassavola × Broughtonia × Cattleya × Epidendrum × Laelia
29.	Hawkesara	Cattleya × Cattleyopsis × Epidendrum
30.	Hawkinsara	Broughtonia × Cattleya × Laelia × Sophronitis
31.	Herbertara	Cattleya × Laelia × Schomburgkia × Sophronitis
32.	Higashiara	Cattleya × Diacrium × Laelia × Sophronitis
33.	Hookerara	Brassavola × Cattleya × Diacrium
34.	Iacovielloara	Brassavola × Cattleya × Diacrium × Epidendrum × Laelia
35.	Iwanagara	Brassavola × Cattleya × Diacrium × Laelia
36.	Izumiara	Cattleya × Epidendrum × Laelia × Schomburgkia × Sophronitis
37.	Jewellara	Broughtonia × Cattleya × Epidendrum ×Laelia
38.	Johnyeeara	Brassavola × Cattleya × Epidendrum × Laelia × Schomburgkia × Sophronitis
39.	Kawamotoara	Brassavola × Cattleya × Domingoa × Epidendrum × Laelia
40.	Kirchara	Cattleya × Epidendrum × Laelia × Sophronitis
41.	Kraussara	Broughtonia × Cattleya × Diacrium × Laeliopsis
42.	Laeliocatonia	Broughtonia × Cattleya × Laelia
43.	Laeliocattkeria	Barkeria × Cattleya × Laelia
44.	Laeliocattleya	Cattleya × Laelia
45.	Laeliopleya	Cattleya × Laeliopsis
46.	Lyonara	Cattleya × Laelia × Sophronitis
47.	Mailamaiara	Cattleya × Diacrium × Laelia × Schomburgkia
48.	Matsudaara	Barkeria × Cattleya × Laelia × Sophronitis
49.	Maymoirara	Cattleya × Epidendrum × Laeliopsis
50.	Mizutara	Cattleya × Diacrium × Schomburgkia
51.	Mooreara	Brassavola × Broughtonia × Cattleya × Laelia × Schomburgkia × Sophronitis
52.	Northenara	Cattleya × Epidendrum × Laelia × Schomburgkia
53.	Opsiscattleya	Cattleya × Cattleyopsis
54.	Osmentara	Broughtonia × Cattleya × Laeliopsis
55.	Otaara	Brassavola × Broughtonia × Cattleya × Laelia
56.	Potinara	Brassavola × Cattleya × Laelia × Sophronitis
57.	Recchara	Brassavola × Cattleya × Laelia × Schomburgkia
58.	Rolfeara	Brassavola × Cattleya × Sophronitis
59.	Rothara	Brassavola × Cattleya × Epidendrum × Laelia × Sophronitis
60.	Sakabaara	Brassavola × Broughtonia × Cattleya ×Diacrium × Laelia

(*Contd.*)

(*Contd. Table 5*)

S. No.	Hybrids	Parentage
61.	Sallyyeeara	Brassavola × Broughtonia × Cattleya × Cattleyopsis × Diacrium × Epidendrum × Laelia × Schomburgkia × Sophronitis
62.	Schombocatonia	Broughtonia × Cattleya × Schomburgkia
63.	Schombocattleya	Cattleya × Schomburgkia
64.	Scullyara Scu	Cattleya × Epidendrum × Schomburgkia
65.	Sophrocattleya	Cattleya × Sophronitis
66.	Sophrolaeliocattleya	Cattleya × Laelia × Sophronitis
67.	Stacyara Stac	Cattleya × Epidendrum × Sophronitis
68.	Stellamizutaara	Brassavola × Broughtonia × Cattleya
69.	Susanperreiraara	Broughtonia × Cattleya × Tetramicra
70.	Symmonsara	Brassavola × Cattleya × Epidendrum × Schomburgkia
71.	Tetracattleya	Cattleya × Tetramicra
72.	Trisuloara	Barkeria × Brassavola × Cattleya × Epidendrum × Laelia × Sophronitis
73.	Tuckerara	Cattleya × Diacrium × Epidendrum
74.	Turnbowara	Barkeria × Broughtonia × Cattleya
75.	Vacherotara	Brassavola × Broughtonia × Cattleya × Epidendrum × Laelia × Sophronitis
76.	Vaughnara	Brassavola × Cattleya × Epidendrum
77.	Vejvarutara	Broughtonia × Cattleya × Cattleyopsis
78.	Westara	Brassavola × Broughtonia × Cattleya × Laelia × Schomburgkia
79.	Wilburchangara	Broughtonia × Cattleya × Epidendrum × Schomburgkia
80.	Yahiroara	Brassavola × Cattleya × Epidendrum × Laelia × Schomburgkia
81.	Yamadara	Brassavola × Cattleya × Epidendrum × Laelia
82.	Yeeara	Brassavola × Broughtonia × Cattleya × Epidendrum × Laelia

Mutation

Blue flower colour is rare in orchids and very rare in cattleyas growing in their natural range. While growing in the natural range, blue cattleya is the result of extremely infrequent mutations producing a plant that is typically called variety coerulea. Many *Cattleya* species have coerulea colour forms, though some species such as the yellow flowered *Cattleya dowiana* either doesn't have a coerulea form due to the yellow original colour or the coerulea form is extremely rare.

Table 6: Variety developed through mutation.

S. No.	Mutant	Parentage	Characteristics
1.	Maku'u Lemon Drop	Cattleya Chocolate Drop	Rare yellow, due to a genetic mutation. Flowers fragrant.

Polyploidy

The polyploidy has been an important factor in the development of improved commercial varieties and hybrids of the *Cattleya* orchids (Kamemoto, 1950). Although methods using colchicine for induction of polyploids are common for a range of plant species, few reports for orchids on the *in vitro* application of colchicine in culture medium are available. The drug has been used into culture medium in low concentrations during prolonged times, which reduces its phytotoxic effect and enhances the production of solid tetraploid plants (Vajrabhaya, 1983).

The effect of colchicine on the protocorm-like bodies of *Cattleya intermedia* L. (Orchidaceae) cultured *in vitro* was studied. Different concentrations of the drug (0.00, 0.05, 0.10 and 0.20%) as well as two times of culture (4 or 8 days) were tested to determine the best treatment for the induction of tetraploid plants. Of the 3 colchicine levels applied, 0.05 and 0.10% of drug treatments appeared to be effective on the production of mixoploids and tetraploids that could be further used in breeding programmes. The suitability of stomatal area and stomata density as criteria for the distinction between diploids and tetraploids was also tested.

BIOTECHNOLOGY

Tissue culture

Torres and Mogollon (2000) developed a protocol for micropropagating the endangered endemic Venezuelan orchid *Cattleya mossiae*. Clonal propagation was achieved in 4 phases: (I) Culture initiation, starting from caulinary apices of dormant vegetative buds from the latest mature growth flux on filter paper bridges, with Hang's modification (1984) of MS medium; (II) Shoot multiplication on medium containing 0.7% agar and 3.0 mg benzyladenine (BA)/litre, without coconut water. Shoot elongation and quality were reduced above 2800 lux. Over 3500 shoots/explant were produced in 5 subcultures using modified plant tissue culture techniques; (III) Shoots were rooted in half-strength liquid MS medium with 30 g sucrose/litre, without BA, on polypropylene membrane rafts; (IV) Plantlets were acclimatized under mist. Survival was highest in coconut fibre medium, tree fern fibre medium and a medium consisting of a mixture of both the above.

Ventura *et al.* (2002) investigated the morphogenic ability of shoot tips (ST; 24 explants), shoot axillary buds (SB; 19 explants) and pseudobulb axillary buds (PB; 10 explants) of *Laelia lobata* cv. Jenni × *L. lobata* cv. Alba to identify the best explants for this hybrid. All explants showed good morphogenic capability, particularly the shoot axillary buds, making the *in vitro* commercial propagation of this species possible. Fraguas *et al.* (2003) studied the optimum concentrations of Knudson culture medium and sucrose for the *in vitro* propagation of orchid plantlets obtained from the seeds of *Cattleya labiata* × *Laelia itambana*. Optimum growth of plantlets was observed in the Knudson culture in its original concentration (100%) supplemented with 20 g sucrose/litre.

Schneiders *et al.* (2012) determined an efficient medium for *in vitro* seed germination and initial seedling growth of *Cattleya forbesii* and *Cattleya harrisoniana*. Seeds of *Cattleya forbesii* were incubated in Murashige & Skoog (MS) medium (T1) and MS added with 2,5 g L^{-1} of activated charcoal (T2). The addition of activated charcoal also promoted an increase in the height of the *C. forbesii* seedlings after 180 days of *in vitro* cultivation. Use of MS medium added with 2.5 g L^{-1} of activated charcoal is significantly favourable for seed germination and initial growth for both species. Regeneration study of cattleya through different explants was carried by so many workers, a brief about the regeneration study is given in Table 7.

Table 7: Regeneration of cattleya through different explants.

S. No.	Species	Explant source	Reference
1.	*Cattleya*	Lateral bud	Reinert and Mohr, 1967
2.	*Cattleya*	Vegetative shoot	Scully, 1967
3.	*C. bowringiana* × *C. forbesii*	Leaf base	Champagnat *et al.*, 1970
4.	*Cattleya*	Meristem	Lindermann *et al.*, 1970
5.	*Cattleya*	Backbulb dormant bulb	Chuchill *et al.,* 1971
6.	*Cattleya* Dorothy Fried No.1	Apical/axillary bud	Kusumoto, 1979b
7.	*Cattleya*	Leaf tip	Vajrabhaya, 1978
8.	*Cattleya* Dorothy Fried No.1	Shoot	Kusumoto, 1979a
9.	*Cattleya labiata*	Callus	Stoltz, 1979
10.	*Cattleya*	Shoot	Huang, 1984
11.	*Cattleya* Georgiana × self	Plantlet	Adelberg *et al.*, 1997
12.	*Cattleya aurantica*	Shoot tip	Mauro *et al.*,1994
13.	*Cattleya leopoldii*	Embryo	Knapp *et al.*, 2000
14.	*Cattleya*	Fractionated PLB	Saiprasad *et al.*, 2002
15.	*Cattleya* Princess Margaret, Bonanazagaint	Protocorm-seedling	Useto, 1973

Somaclonal variation

A new and distinct variety of *Cattleya loddigesii* orchid plant, substantially as described and illustrated, resulting as a somoclonal variation from tissue culture of 'Harrisoniana', which is particularly distinguished from its parent and other members of the market and botanical class by having superior heavy flower texture and by having larger, flatter and more rounded flower segments which render longer lasting flowers.

Molecular breeding

A set of microsatellite markers for endangered *Cattleya labiata* and *Cattleya warneri* was developed, these were the first microsatellite loci for *Cattleya* that can be useful for studies in the conservation genetics of these species. Twenty-one new microsatellite primers pairs were characterized by cloning an enriched library and two libraries were obtained by cloning inter-simple sequence repeats. Thirteen polymorphic primers were found and the remaining eight were monomorphic for both species. These new microsatellite markers have the potential to be used as a tool for studying the genetic diversity of *C. labiata* and *C. warneri* and for future studies focusing mating systems and the detection of hybrids (Paulo *et al.*, 2013).

Neto *et al.* (2011) estimated the level of genetic diversity and interrelationships between wild and bred *Cattleya intermedia* with RAPD markers. A total of 65 polymorphic bands were used to generate a genetic distance matrix. No specific groupings were revealed by the cluster analysis as bred materials were not different from wild plants. The genetic differentiation (FST = 0.01626) was very low indicating a high gene flow in *C. intermedia* due to artificial crosses and a high differentiation between populations. The genetic variability available within this species is high enough to allow genetic progress in flower shape and size. Pinheiro *et al.* (2012) provided the first data on genetic diversity and structure of wild populations in the genus *Cattleya*, in particular *C. labiata*, using random amplified polymorphic DNA (RAPD) and intersimple sequence repeat (ISSR) markers. Data generated from 12 ISSR and 12 RAPD primers were used to determine genetic variability via a model-based Bayesian procedure (Structure) and molecular variance analysis. In addition, Shannon index, genetic diversity and Jaccard coefficients were also estimated. The marker data indicated that *C. labiata* has a high level of polymorphism and five reconstructed populations were identified by Structure. Other *Cattleya* species showed no relationship with any *C. labiata* sample.

Genetic engineering

Protocorm-like bodies (PLBs) of *Cattleya* orchid CM2450 cultured either under constant illumination with cool-white-fluorescent lamps or in the dark were cocultivated with *Agrobacterium tumefaciens* strain EHA105 carrying plasmid pSMAHdN627-ORSV harboring genes coding for *Odontoglossum ringspot* virus (ORSV) replicase and hygromycin phosphotransferase. The number of resistant PLBs generated using dark-cultured PLBs was higher than those cultured under constant illumination. Presence of acetosyringone (AS) in the pre-culture medium was also effective for the transformation. The highest frequency of transformation was obtained when dark-cultured PLBs were pre-cultured with 100m MAS for 3 days and inoculated with *Agrobacterium* liquid culture for 3 h. Transformation of hygromycin-resistant plantlets regenerated from different sites of inoculated PLBs was confirmed by Southern blot hybridization. Transcription of ORSV replicase gene in transgenic lines was successfully confirmed by Northern blot hybridization (Zhang *et al.*, 2010).

REFERENCES

Adelberg, J.W., Desamero, N., Hale, S.A. and Young, R.E. 1997. Long-term nutrient and water use during micropropagation of *Cattleya* orchid on liquid/membrane system. *Plant Cell, Tissue Organ Cult.*, **48**: 1-7.

Almeida, P.R.M., López-Roberts, M.C., Vigna, B.B.Z., Souza, A.P., Góes-Neto, A. and van den Berg, C. 2013. Microsatellite markers for the endangered orchids *Cattleya labiata* Lindl. and *C. warneri* T. Moore (Orchidaceae). *Cons. Genet. Res.*, **5**(3): 791-94.

Anonymous. 1993. *Cattleya loddigesii* var. Harrisoniana. Streeter's United States patent. US PP8541 P.

Banks, D.P. 2005. Orchid grower's companion: Cultivation, propagation, and varieties. Portland, Oregon (U.S.A.): Timber Press.

Chadwick, A.A. and Chadwick A.E. 2006. The Classic Cattleyas. Portland, Oregon (U.S.A.): Timber Press.

Champagnat, M., Morel, G. and Gambade, G. 1970. La multiplication vegetative des *Cattleya* a partir de jeunes feuiles cutlivees aseptiquement *in vitro. Annales des Sciences Naturelles Botanique et Biologie Vegetale Serie*, **11:** 97-114.

Churchill, M.E., Arditti, J. and Ball, E.A. 1971. Clonal propagation of orchids from leaf tips. *American Orchid Society Bulletin*, **42**: 109-113.

Clarke, F. 2013. Breeding yellow cattleyas with red spots: The current trends. *Orchids*, **82**(5): 274-81.

Cruz, D.T., Borba, E.L. and Van den Berg, C. 2003. O ge^nero Cattleya (Orchidaceae) no estado da Bahia, Brasil. *Sitientibus Ser Ci Bio.*, **3**: 26–34.

Felix, L.P. and Guerra, M. 2010. Variation in chromosome number and the basic number of subfamily Epidendroideae (Orchidaceae). *Botanical J. Linnean Soc.*, **163**: 250-51.

Fitch, C.M. 2004. The Best Orchid for Indoors. New York: Brooklyn Botanic Garden.

Fraguas, C.B., Villa, F., Souza, A.V., de Pasqual, M. and Dutra, L.F. 2003. Growing *in vitro* of orchid plantlets originated from the *Cattleya labiata* × *Laelia itambana* hybridization. *Revista Ceres.*, **50**(292): 719-26.

Huang, L.C. 1984. Alternative media and method for *Cattleya* propagation by tissue culture. *American Orchid Society Bulletin*, **53**: 167-170.

Kamemoto, H. 1950. Polyploidy in *Cattleyas*. *American Orchid Society Bulletin*, **1**(f): 366-73.

Knapp, J.E., Kausch, A.P. and Chandlee, J.M. 2000. Transformation of three generation of orchid using the *bar* gene as a selectable marker. *Plant Cell Reports*, **19**: 893-898.

Kusumota, M. 1979a. Effects of combinations of growth regulators and of organic supplemnts on the growth of *Cattleya* plantlets cultured *in vitro*. *J. Japanese Soc. Hortil. Sci.*, **47**: 492-501.

Kusumoto, M. 1979b. Effects of combinations of growth regulators and of organic supplements on the proliferation and organogenesis of *Cattleya* protocorm like bodies cultured *in vitro*. *J. Japanese Soc. Hortil. Sci.*, **47**: 502-510.

Lindemann, E.G.P., Gunckel, J.E. and Davidson, O.W. 1970. Meristem culture of *Cattleya*. *American Orchid Society Bulletin*, **39**: 100-27.

Mauro, M., Sabapathi, D. and Smith, R.A. 1994. Influence of benzylaminepurine and alpha-naphthaleneacetic acid on multiplication and biomass production of *Cattleya aurantiaca* shoot explants. *Lindleyana*, **9**:169-173.

Neto, M., Barbosa, N. and Luiz Gonzaga Esteves, V. 2011. Assessment of genetiç diversity in *Cattleya intermedia* Lindl. (Orchidaceae). *Brazilian Archives Biol. Tech.*, **54**(5): 939-46.

Pabst, G.F.J. and Dungs, F. 1975. *Orchidaceae Brasiliensis I*. Kurt Schmersow, Hildesheim.

Pinheiro, L.R., Rabbani, A.R.C., Silva, A.V.C., da Ledo, A., da, S., Pereira, K.L.G. and Diniz, L.E.C. 2012. Genetic diversity and population structure in the Brazilian *Cattleya labiata* (Orchidaceae) using RAPD and ISSR markers. *Plant Syst. Evol.*, **298**(10): 1815-25.

Reinert, R.A. and Mohr, H.C. 1967. Propagation of *Cattleya* by tissue culture of lateral bud meristems. *American Soc. Hortil. Sci.*, **91**: 664-671.

Saiprasad, G.V.S, Raghuveer, P., Khetrapal, S. and Chandra, R. 2002 Effects of various plant growth regulators on the production of protocorm like bodies in three orchid genera. *Indian J. Pl. Physio.*, **7**: 35-39.

Schneiders, D., Pescador, R., Booz, M.R. and Suzuki, R.M. 2012. Germination, growth and development *in vitro* of orchids of the genus *Cattleya*. *Revista Ceres*., **59**(2): 185-91.

Scully, R.M. 1967. Aspects of meristem culture in the *Cattleya* alliance. *American Orchid Society Bulletin*, **36**: 108.

Stoltz, L.P. 1979. Iron nutrition of *Cattleya* orchid grown *in vitro*. *J. American Soc. Hortil. Sci.*, **104**: 308-310.

Torres, J. and Mogollon, N. 2000. Clonal micropropagation of *Cattleya mossiae* Parker ex Hook. *Proce:* The Inter American Society for Tropical Horticulture. **42**: 87-92.

Uesato, K. 1973. Effects of different forms of nitrogen sources in the culture media on the growth of *Cattleya* young seedlings. *Scientific Rep. Agri.*, **20**: 1-12.

Vajrabhaya, M. 1978. Tissue culture of dormant buds from *Cattleya* backbulbs. *The Orchid Review,* **86**: 256-257.

Vajrabhaya, T. 1983. Variations in clonal propagation. *In:* J. Arditti (ed.) Orchid Biology: Reviews and Perspectives. Ithaca, New York: Comstock Publ. Assoc., pp. 176-01.

Ventura, G.M., Dias, J.M.M., Teixeira, S.L., Carvalho, V.S., Motoike, S.Y., Novais, R.F. and de Cecon, P.R. 2002. *In vitro* organogenesis from tip and axillary buds of adult plants of orchids from *Cattleya* group. *Revista Ceres*., **49**(286): 613-28.

Withner, C.L. 1988. The Cattleyas and their Relatives: The Cattleyas I. pp 147.

Zhang, L., Chin, D.P., Fukami, M., Ichikawa, H., Nakamura, I. and Mii, M. 2010. *Agrobacterium*-mediated genetic transformation of Cattleya with an Odontoglossum ringspot virus replicase gene sequence. *Plant Biotech.,* **27**: 421-26.

Chapter – 31
Cymbidium

Cymbidium is a genus in the orchid family Orchidaceae. It was first described by Olof Swartz in 1799. The name is derived from the Greek word "*kumbos*", meaning "hole", "cavity". It refers to the form of the base of the lip. The flowers are long-lasting, both on the plant and when cut large attractive and available in a wide range of colours. Smaller plants are now coming into vogue as decorative pot plants. These have smaller flowers but they are as desirable as the standard varieties and often more acceptable in modern homes. The popularity of miniature cymbidium is now spreading from Asia to the worldwide orchid community. These ancient flowers have been treasured in numerous societies for nearly two thousand years. Many orchid lovers are most charmed by its fragrance and form rather than its petite size.

ORIGIN AND DISTRIBUTION

Cymbidiums figure in the earliest historical records of human association with orchids from China and Japan. Cymbidium native of the monsoonal forest foothills of the Himalayas with the warmer growing Asiatic species from the tropical lowland monsoonal forests. The species come from the Himalayan regions of India and Nepal running through China and Japan, Indo-China down to Malaysia to Australia. This genus is distributed in tropical and subtropical Asia such as northern India, China, Japan, Malaysia, the Philippines, and Borneo and northern Australia. The larger flowered species from which the large flowered hybrids are derived grow at high altitudes. They have been cultivated for thousands of years especially in China. Cymbidiums became popular in Europe during the Victorian era. One feature that makes the plant so popular is the fact that it can survive during cold temperatures as low as 7° C. Originating in temperate and tropical parts of India, Nepal, China, Japan and down to Malaysia and Australia, cymbidiums are probably the most widely grown of all orchids because of their ease of cultivation and long-lasting flowers. From the 40-50 species, thousands of hybrids have been bred with flowers ranging in colour from white, yellow and orange to pink, red, brown and green, and varying in size from 5-10 cm across. There are several species existing in the Indian forest area. Several species are existing in the forest of Nagaland (Changkija *et al.*, 1992), some of them are given in Table 1.

Table 1: Distribution of *Cymbidium* species in the Indian forest area.

S. No.	Species	Distribution	Characteristics
1.	*Cymbidium aloifolium*	Intangki, Rangapahar, Tuli, Tsurang, Saring, Tezit, Namsa	FC: Yellow-purple, F.S.: 3 cm, FT: April May.
2.	*C. cochleare*	Japfu Range, Pfutsero, Hellipong, Chentang	FC: Dull brown flushed with green, F.S.: 4 to 5 cm long, F.T.: Oct- Nov.
3.	*C. devonianum*	Saramati, Hellipong	FC: Purplish-brown with green lines, FS: 2.5-3 cm, FT: May-June.
4.	*C. elegans*	Japfu Range, Phek, Pesao	FC: Pale yellow, FS: 4 cm, FT: Oct-Nov.
5.	*C. ensifolium*	Japfu Range, Plutsero, Hellipong, Longtok, Chentang	FC: Green with pink dotted lines, FS: 3 cm, FT: Nov.-Dec.
6.	*C. eburneum*	Viswema, Japfu Range, Noklak, Pangsa, Pesao	FC: Pure white in colour, FS: 7 to 8 cm, FT: Mar.-April.
7.	*C. iridioides*	Hellipong, Chentang, Saramati Sangsangyu, Japfu Range,	FC: Puriplish-red, FS: 7 to 8 cm, FT: Oct. Nov.
8.	*C. lancifolium*	Hellipong, Longtok, Chentang	FC: Pale green with red striper, FS: 4 to 5 cm, FT: June-July.
9.	*C. longifolium*	Japfu Range, Saramati, Hellipong	FC: Purplish, FS: 5 to 6 cm, FT: Sept.-Oct.
10.	*C. lowianum*	Chentang, Hellipong, Japfu Range	FC: Yellowish brown with reddish veins, FS: 12 cm, FT: April-May.
11.	*C. macrorhizon*	Meinkong, Phek, Kohima, Viswema	FC: White with pink spots, FT: June-July.
12.	*C. mastersii*	Viswema, Peren	FC: White with rose purple spot on the lip, FS: 5 to 6cm, FT: Oct.-Nov.
13.	*C. pendulum*	Saring, Tuli, Namsa, Naginimora	FC: Purplish-brown with yellow margin, FS: 3cm, FT: April-May.
14.	*C. tigrinum*	Japfu Range, Viswema, Pfutsero	FC: Yellowish to brownish-green in colour, FT: April.
15.	*C. tracyanum*	Chentang, Sangsangyu, Hellipong	FC: Greenish yellow with numerous red crimpson streaks and dots, FS: 12 cm, FT: Oct-Nov.

FC: Flower colour, FS: Flower size, FT: Flowering time.

SPECIES

The word lan is used for a wide variety of fragrant plants but referred to was almost certainly *Cymbidium ensifolium*. It is still cultivated in China and Japan, today, and displayed there in beautifully decorated containers whether or not it is in flower. This species was also one of the first cymbidiums to appear in Europe. Linnaeus named it in his Species Plantarum in 1753, but under the genus name *Epidendrum*. Live plants were first introduced by Dr. John Fothergill, on his return from a visit to China in 1778. They were sent to a relative in Yorkshire and flowered there quite soon afterwards, but they do not seem to have survived long or to have been propagated at that time. In Part IX published in 1893, he records that about 30 species of *Cymbidium* were then known, distributed from the Indo-Malayan region to tropical Australia and with outlying species in Japan and New Caledonia.

The most recent classification of the Orchidaceae which has been accorded wide acceptance among botanists is that of Robert Dressler (1981, and modified in 1986). He recognised six subfamilies in the Orchidaceae, with cymbidium included in the Epidendroideae. Within this subfamily there are many tribes, including the Cymbidieae which contains all the sympodial vandoid orchids with similar seeds and two pollinia. Dressler placed cymbidium in a small group of genera within the subtribe Cyrtopodiinae. All these plants have pseudobulbs of several internodes, articulated leaves, and usually a lateral inflorescence. The differences in structure of the pollinia and the whole pollinarium in cymbidium and related genera are small but very interesting. The most similar is the African genus *Ansellia*, but that genus has a particulate inflorescence borne at the apex, not the base, of the greatly elongated pseudobulb. The genus *Cymbidium* was established by Olof Swartz in 1799, based on *Epidendrum aloifolium* L. (*C. aloifolium* (L.) Sw.). At that time he included many species which have now been transferred elsewhere, but two of his species, *C. aloifolium* and *C. ensifolium*, are still recognised. Many botanists have described individual species of *Cymbidium* since then, or have removed some of them to other genera, notably the genus Cyperorchis Blume. DuPuy and Cribb review this information and then present their own proposal. They propose the division of cymbidium into three subgenera called Cymbidium, Cyperorchis and Jensoa. Each of these contains several sections, in some of which there is only one, rather distinctive species. They distinguish these subgenera according to the formation of the lip and the number of pollinia. In Cyperorchis the lip is fused with the column base while Jensoa with four pollinia has a free lip as does cymbidium with two pollinia

DuPuy and Cribb (1988) considered that the genus Cyperorchis, which was distinguished by the Dutch botanist Blume and others because all the flowers have the lip fused with the base of the column, should be reunited with cymbidium but maintained as a distinct subgenus within it. All the other species they classify in two further subgenera: Cymbidium, containing those where the anther bears two pollinia and Jensoa, whose species have four pollinia. Some important species of *Cymbidium* are given in Table 2.

Table 2: List of the some important species of cymbidium.

Genus	Number of species	Important species
Cymbidium	70	*C. aloifolium, C. devonianum, C. eburneum, C. ensifolium, C. insigne, C. giganteum, C. grandiflorum, C. longifolium, C. lowianum, C. madidum, C. mastersii, C. pendulum, C. simonsianum, C. tigrinum, C. tigrinum, C. tracyanum*

Asian cymbidiums or Chinese cymbidiums refer to mainly five species of cymbidiums orchids that are found throughout East Asia in areas of China, Korea, Japan, India, and in parts of Thailand and Vietnam. These species are usually grown for their variegated leaves. But plants are also grown for their fragrant flowers and peloric flower structure. Plants are usually grown in long and thin vase like pots. The five species are *Cymbidium sinense*, *Cymbidium ensifolium*, *Cymbidium kanran*, *Cymbidium goeringii* subsp. *gracillimum* and *Cymbidium faberi.*

Cymbidium iridioides: Previously known as *C. Giganteum.* It produces long arching spikes with upto 20 scented blooms of upto 10cm across. The flowers are coloured from ginger brown to red and the white labellum has a yellow centre outlined by red blotches. While somewhat similar to *C. tracyanum*, it is neither as showy nor as large and does not have the sinewy twisted petals and sepals of *C. tracyanum* never the less it is an attractive species especially when the plant grows large enough to carry several flower spikes.

***Cymbidium aloifolium*:** It is epiphyte on tree trunk in dense humid evergreen forest. Flowers 4-5 cm across, slightly fragrant, sepals and petals pale yellow to cream-yellow, with a broad central maroon-brown stripe, lip cream coloured. Flowering starts from April - May. Local distribution within Dibrugarh in district Jeypore, Jo-kai, Namdang, Telpani and Dihingmukh.

Cymbidium bicolour: It is epiphyte on tree trunk in dense humid evergreen forest Flowers 3-4 cm across. Sepals and petals yellow with a central purplish red longitudinal stripe, lip cream-yellow. Flowering from May-June. Local distribution within Dibrugarh in district Jeypore, Jo-kai and presently it is endangered.

***Cymbidium dayanum*:** It is epiphyte on tree trunk in dense humid evergreen forest. Flowers 4-5 cm across. Sepals and petals white with a central maroon stripe, lip maroon, tinged white at its base. Flowering from June -July. Local distribution within Dibrugarh in district Jeypore

Cymbidium elegans: It is a most distinctive species due to its yellow bell like flowers. The flower segments are long and thin but because they remain pointing forward the flowers retain their long bell like appearance. The plant is quite compact with narrow leaves and the flower spikes grow up and then arch over with the weight of the flowers giving a cascading effect. Specimen plants can look quite stunning with spikes of up to 30 closely packed flowers cascading out from the foliage.

Cymbidium erythraeum: It is also known as *C. longifolium* has similarities to *C. tracyanum*, though it produces smaller flowers to around 8 cm across. This species is particularly attractive in its own right and can make a fine specimen plant capable of carrying many spikes and flowers. Flower form is similar to *C. tracyanum* and flower colour is more predominantly green with degrees of red/brown on the petals and sepals. The labellum is white with a red midline and a few scattered red spots at the edges making a highlight against the darker petals and sepals.

Cymbidium tracyanum: It is easily recognized by its twisted petals and sepals and its distinctive spotted and striped patterning. Flowers are generally cream to greenish heavily overlaid with reddish brown to dark red and it's labellum is white spotted with purple/red. *C. tracyanum* is a large growing plant with long arching spikes carrying fifteen or so large flowers. Specimen sized plants with multiple spikes can look quite stunning. Beware of an old primary hybrid which is still quite common and is often incorrectly labelled as *C. tracyanum*. It flowers in August to September and has more pinkish coloured flowers. Although the flowers resemble *C. tracyanum* in shape it is not *C. tracyanum*.

Cymbidium lowianum: The striking red vee on *C. lowianum* labellum can still be clearly seen in the labellum's of many of today's hybrids. *C. lowianum* is a moderate sized plant which carries arching spikes of largish flowers of around fifteen to eighteen in number. There are three colour forms of *C. lowianum*, the type form having green petals and sepals often with a bronzy overlay, the labellum is white with a bold red vee. There is also an album form where the flowers are clean apple green and the white labellum has a yellow vee.

Cymbidium insigne: It comes in two forms, the type form produces pink blooms with white labellum's spotted purple/red, the other form is an album variant which produces pure white blooms with no purple pigment in the petals, sepals, or labellum. Flowers are large and more than twenty or so are carried on very tall and thin, but surprisingly strong spikes.

Cymbidium parishii: It is another moderately large growing plant with long arching flower spikes. Its flowers are of medium size, perhaps 75 to 80 mm across. Flower colour is white, with heavy dark red blotching on its white labellum's. There are several forms of *C. parishii* in circulation, though the colour is the same in all, flower form being the main difference between them. One particular clone named *C. parishii* var. Emma Menenger is described as being a tetraploid (4n) as opposed to the normal diploid (2n), it's generally fuller and more shapely flowers being attributed to the doubling of chromosomes in the tetraploid form.

***Cymbidium erythrostylum*:** It is a more compact growing plant than others. Its flower spikes are shorter though its flowers are quite large. *C. erythrostylum* has a very distinctive flower form with its forward pointing petals shielding a small labellum lined with red, it's sepals are large and full shaped and a bit unruly looking petals and sepals are pristine white. The overall effect is quite beautiful and due to its compact plant size an impressive specimen can be produced with many flower spikes in a 200 mm pot.

FLOWER MORPHOLOGY AND BIOLOGY

The cymbidium plant is epiphytic, lithophytic or terrestrial. Plants which normally grow on trees are sometimes encountered on rocks, and one species, *C. macrorhizon*, is a saprophyte which grows entirely beneath the soil surface except when the flower spike emerges. Most species have thick roots which are covered in a spongy white velamen and have only a thin core of vascular tissue. The erect stems are usually short and swollen to form a prominent pseudobulb which is often slightly flattened. Many species produce a new growth annually. In one section of the genus the pseudobulbs grow and flower for several years before a new shoot is produced, and in *C. mastersii*, *C. elongatuni* and *C. suave*, each shoot grows continuously for many years producing an elongated stem rather than a typical pseudobulb. The pseudobulbs remain encased by the leaf bases. The leaves may be thick, rigid and leathery, as in *C. aloifolium* and its allies, but are more often flexible with prominent ribs. The shade loving species, *C. lancifolium* and *C. devonianum* have broad, elliptic leaves, and many of the forest epiphytes have long, thin and relatively narrow leaves.

The inflorescence in cymbidium is unbranched and may be erect, arching or pendulous. Each mature pseudobulb usually produces one or two inflorescences from leaf axils near the base. In *C. eburneum* and its allies the inflorescences arise from the leaf axils near the apex of the pseudobulb, as they do in *C. suave* and *C. elongatum*. The inflorescences bear up to 50 flowers in *C. canaliculatum* and only one in *C. goeringii* and *C. eburneum*. Most species bear 10-20 flowers.

The raceme grows from the base of the most recent pseudobulb. Each flower can have a diameter of 5 to 10 cm, according to the species. They bloom during the winter, and each plant can have up to fifteen or more flowers. The fantastic range of colours for this genus include white, green, yellowish-green, cream, yellow, brown, pink, and red and black, but not blue. The flowers last about ten weeks. They have a waxy texture. The rounded sepals and petals have about the same dimensions. They show very diverse colour patterns, different for every species. They comprise a dorsal sepal, two lateral sepals, two free petals and a three lobed lip which is hinged at the base of the column. There is usually a callus of two distinct ridges along the upper surface of the lip. The anther contains two pollinia or four pollinia fused in two pairs.

Mode of pollination

The path of pollination in *Cymbidium pendulum* (Sw.) Roxb. had been traced. The honey bee identified as *Apis mellifera* was found to act as main pollinator and this bee is the only insects among others who succeeded in performing pollination because probably due to its structural compatibility with the plant species. Pollination by *Apis mellifera* bees was suggested to occur in a number of families but rare phenomenon in orchids and it is first time that the species was observed to pollinate the *C. pendulum* flowers. Bee moved around the flowers for some times, entered the flower and carried on pollinia along with on the back during its journey. It revisited the different flower and deposited its pollinia on to it and the act of pollination was accomplished. SEM study showed an intricate network on the back of bee thus clearly indicates its role in firm attachment to pollinia (Attri and Kant, 2011). The Chinese orchid, *C. serratum* is pollinated by wild mountain mouse, *Rattus fulvescens*. The flowers use both odour and colour as attractants and provide the lips as a food reward for the pollinators. The mice pollinate the flowers during their endeavour to eat the lips. Chan *et al*. (1994) reported that *C. finlaysonianum* is efficiently pollinated by Giant Indian Honey bee, *Apis dorsata.*

Sugahara *et al*. (2010) reported that foragers of the Japanese honeybee (*Apis cerana japonica*) were attracted by flowers of an oriental orchid (*C. floribundum*) and were observed to carry the pollinia on their scutella. After the removal of pollinia from the flowers, their labial colour changed from white to reddish brown. Both artificial removal of pollinia and ethrel treatment of the flowers also induced this labial colour change. Labia in colour-changed flowers showed a decreased reflectance of wavelengths less than 670 nm compared to control intact flower. It was resulted that the honeybee foragers preferentially visited flowers having white labia rather than reddish brown. Japanese honeybees discriminate between the floral phases of *C. floribundum* using colour vision. Sasaki *et al*. (1991) observed that drone honey bees (*Apis cerana japonica*) were attracted by the fragrance of *C. pumilum* flowers during their mating flights in April and May. Some clustered on the flower racemes others inserted their heads deep into the flowers causing pollinia to become attached to the scutellum. Workers and swarming colonies were also attracted to *C. pumilum,* but European honey bees (*A. mellifera*) were not.

GENETICS

The diploid chromosome number in all species of cymbidium was found to be 40. A few exceptions were the triploid or tetraploid named cultivars of a few species, e.g. *C. insigne* 'Bieri' (2n = 60), *C. floribundum* 'Geshohen' (2n = 80) and *C. floribundum* 'Yoshina' (2n = 60). Chromosome numbers have been determined for 38 of the approximately 70 known species. Few deviations from 2n number of 40 have been reported.

Self incompatibility

Chung and Chung (2003) investigated the breeding system of *Cremastra appendiculata* and *Cymbidium goeringii* populations in Oenaro Island, South Korea. The two orchid species showed low rate of fruit set in natural conditions. However, a great increase in fruit set was achieved by hand-pollinations, artificial self-pollination and geitonogamous pollination indicating that the two species are highly self-compatible. Fruit set for pollinator excluded flowers of *Cremastra appendiculata* and *Cymbidium goeringii* was 0%, indicating that the two species are neither autogamous nor agamospermous. The extremely low fruit set observed in the two species under natural conditions may be attributed to pollinator limitation.

Gene expression

The *Flowering Locus T* (*FT*) gene plays crucial roles in regulating the vegetative-to-reproductive phase transition. The *FT*-like gene of spring orchid (*Cymbidium goeringii* Rchb. f.), *CgFT*, was isolated and characterized. *CgFT* mRNA was detected in leaves, pseudobulb, and flowers. In flowers, Cg*FT* was expressed more in young flower buds than in mature flowers and was predominantly expressed in young ovary and sheath. Seasonal expression analysis in leaves of a three-year-old spring orchid showed that a large increase in transcription which started on June 20 for *CgFT*. The ectopic expression of *CgFT* in transgenic tobacco plants showed novel phenotypes by flowering earlier than wild-type plants. Further analysis of the flowering time-related genes indicated that the expression of *Leafy*, *Apetalla1*, *Fruitfull* and *Sepallata1* were significantly upregulated in 35S::*CgFT* transgenic tobacco plants. These results indicated that *CgFT* is a putative *FT* homolog in spring orchid that regulates flower transition similar to its homolog in *Arabidopsis*. This study provides the first information on the spring orchid floral gene to elucidate the regulation of the flowering transition in spring orchid (Xiang *et al.*, 2012)

BREEDING OBJECTIVES

i. More number of flowers per spike.

ii. Attractive colour.

iii. More vase life.

iv. Miniature cymbidiums.

v. Development of varieties resistant to biotic and abiotic stresses (diseases, particularly to those caused by viruses).

vi. Late blooming hybrids (demand of late blooming hybrids is high, *Cymbidium simonsianum* can be used as one parent hybrids).

BREEDING METHODS

Introduction

Demand of Indian cymbidium is high, but several varieties of cymbidium introduced in India from Austria and Australia for improvement and commercial production. These varieties are given in Table 3.

Table 3: List of some introduced varieties of cymbidium.

S. No.	Country	Varieties
1.	Austria	Cymbidium (mericlones) – cv. Narda Jennieer Girl M 12, M. 16 and Appleby Carolina M. 20, Astronaut Rajah M. 34, Harry B. Ireland Modestry M. 41, Cymbidium (mericlones). Miniature- Nancy Maxwell Shirley, Orchids Yanka Illa leprechaaun M. 31, Lady Bug Sentinal M. 32, Alice Williams Camden M. 33, Amesbury Franslatlery M. 51.
2.	Australia	Cymbidium (Seedling orchid) - cv. M. 106, C. 2044 and C. 2143.

Hybridization

Most attention was paid to the Indian species with larger flowers and the first hybrid which had been created in 1878 from *C. lowianum* and *C. eburneum*. It took nine years for the resulting seedlings to reach flowering size and the Royal Horticultural Society gave its first award to a cymbidium hybrid, *C. eburneolowianum*, in March 1889. A list is given of natural hybrids (Table 4), Interspecific hybrids (Table 5) and hybrids between Classic type and Chinese type (Table 6).

Table 4: List of natural hybrids of cymbidium.

S. No.	Hybrids	Parentage	Country
1.	*Cymbidium* × *ballianum*	*C. eburneum* × *C. mastersii*	Myanmar
2.	*Cymbidium* × *baoshanense*	*C. lowianum* × *C. tigrinum*	SC. Yunnan
3.	*Cymbidium* × *chiu-lih*	*C. lancifolium* × *C. ensifolium*	China
4.	*Cymbidium* × *nishiuchianum*	*C. goeringii* × *C. kanran*	Taiwan
5.	*Cymbidium* × *nishiuchianum*	*C. goeringii* subsp. *goeringii* var. *formosanum* × *C. kanran*	Taiwan
6.	*Cymbidium* × *florinda*	*C. erythrostylum* × C. *iridioides*. *Cyperorchis* × *florinda*	Vietnam
7.	*Cymbidium* × *gammieanum*	*C. elegans* × *C. erythraeum*. *Cyperorchis* × *gammieana*	Nepal to Sikkim
8.	*Cymbidium* × *glebelandensis*	*C. insigne* × *C. schroederi*	Vietnam
9.	*Cymbidium* × *jy-shiang*	*C. lancifolium* × *C. sinense*	China
10.	*Cymbidium* × *rosefieldense*	*C. hookerianum* × C. tracyanum. Cyperorchis × rosefieldensis	Vietnam
11.	*Cymbidium* × *woodlandense*	*C. mastersii* × *C. tracyanum*. *Cyperorchis* × *woodlandensis*	Myanmar

Table 5: List of interspecific hybrids of cymbidium.

S. No.	Hybrids	Parentage
1.	*Cymbidium* Alexanderi 'Westonbirt'	*C. eburneolowianum* × *C. insigne* 'westonbirt'
2.	*C.* Pumander	*C.* Louis Sander × *C. pumilum*
3.	*C.* Jean Brummitt	*C. devonianum* × *C. eburneum*
4.	*C.* Flirtation	*C. pumilum* × *C. Zebra*
5.	*C.* Putana	*C. pumilum* × *C.* Rutana
6.	*C.* Gloria 'Lella'	*C.* Mary Pinches 'Sunbeam' × *C.* Showgirl 'Julie christie')
7.	*C.* 'Green Honey'	*Cymbidium* 'A-one' × *Cymbidium* 'Oji'

Table 6: Hybrids between Classic type and Chinese type.

S. No.	Hybrids	Parentage
1.	Peter Pan	Ensifolium × Miretta
2.	Maureen Carter	Sinense × Sleeping Beauty
3.	Golden Elf	Ensifolium × Enid Haupt

Mutation

Doubling of chromosome in orchids was incorporated by using colchine. Submerging back bulbs of *Cymbidium coningsbyanum* Brockhurst in colchine solution twice within on interval of ten days has been reported to produce tetraploids. *Cymbidium* was mutated with UV radiation and found the effect was similar as with colchicine treatment (Lin and Deng, 1997). However, Rosario and Maningas (1998) observed late germination and no root formation in Spathoglottis 'Yellow' when embryos treated with 0.01% and 0.05% colchicine and gamma rays (500r, 800r, 1Kr and 2Kr).

Protocorms of spring orchid *Cymbidium* developed by tissue culture of aerial parts, seeds and stems, were treated with ultraviolet (UV) radiation and colchicine. After UV irradiation the cell nucleus became deformed. Frequencies of cell division increased with UV dose, causing phenotypic variation among regenerated plants. Similar phenotypic variation was observed in plants regenerated from material treated with colchicine (Lin and Deng, 1997).

Polyploidy

Good cymbidium diploids are certainly of great importance. Because of the fact that many of the most famous polyploids in cymbidiums have been brought about by much inbreeding. There are some poor growth characteristics that have carried along and it is often the free-growing habit of the diploid that influences the progeny into being good, free performers. Cymbidium diploids are also characterized by often having more flowers per spike than many of the more popular tetraploids. Together with this popularity in the tetraploid line has centered around the full-shaped white tetraploid.

Triploids normally are the resultant progeny from the mating of a tetraploid with a diploid. These comprise the great bulk of present day cymbidiums. They are distinguished by uniform good growing characteristics and freeness of performance. Their visual traits vary from the extremes of one parent to the other and combinations of both with their typical average lying about one-third from the tetraploid parent and two-thirds from the diploid. The idea is to find those few plants from a particular cross which exhibit the good features of both parents and in these exceptional cases of improvements. Kim *et al.* (1997) usually found that triploid cymbidiums are sterile and will not produce seed. There are, however, some exceptions which give rise to other categories of polyploidy. In another study, 82.3% survival of polyploids of *Cymbidium kanran* was noticed with 0.01% of colchicine for 1 week treatment. Number and length of stomata were not correlated with polyploidy, but width tended to increase as the number of chromosomes increased.

Yin *et al.* (2010) successfully induced tetraploid from *in vitro* plantlets of diploid cymbidium interspecific hybrids treated with different concentration colchicine. Rhizomes were treated with colchicine at concentration of 0.02%, 0.05%, 0.10% and 0.20% for 24, 48 and 72hr. The results indicated that the treatment of rhizomes in colchicine solution at 0.10% for 48h was the best and its induction rate was 36%. The chromosome number of induced plants was 2n=4x=80, while that of the control plants was 2n=2x=40. Compared with the normal diploid plants, tetraploids were characterized by thicker and stronger roots, stems and leaves. The colour of leaves and base of stems were darker and the plants shown slower growth rate. These new materials could be used in breeding of orchid. *Cymbidium* Fanfare 'St. Francis' an example of a tetraploid (4n) flower.

Ji *et al.* (2011) soaked the shoots of *Cymbidium hybridium* in colchicine solutions at different concentrations and treatment times to induce doubling of body cell chromosome. Ploidy levels were identified by morphological and cytological traits. The treated shoots of *C. hybridium* with 0.05% colchicine for 24 h had a higher inducing rate which was up to 28.2%. The external morphology, the number of chloroplast as well as the number and size of stomata were obviously different between the diploid and polyploid. The nucleus of polyploid cell was obviously bigger and the number of chromosome was increased.

Wang *et al.* (2010) investigated the effects of colchicine treatment on proliferation, differentiation and tetraploid induction of protocorm-like bodies (PLBs) of cymbidium Sunrise by using of tissue culture method. The colchicine treatment had obviously inhibition effects on the growth and differentiation of PLBs and increased the browning and death rate of PLBs; the higher the concentration of colchicine and the longer the treatment time was the stronger the inhibition effects were. The tetraploid plant with darker and harder leaves was more sturdy and compacted in appearance compared to diploid one. It is a practical way of creating polyploids in *Cymbidium hybridium* by the use of tissue culture combined with colchicine treatment method.

BIOTECHNOLOGY

Tissue Culture

During mass-propagating the cymbidium Dongyang using the meristem culture technology, a useful leaf mutant was selected from the protocom like bodies. The cymbidium variety by *in vitro* mutagenesis from Dongyang was named as Daegook in 1998. The mutant has white or yellow stripes along the margin of leaves and flowers. The plant length was shorter than Dongyang and the mean length and width of leaf in mutant was 40 cm and 1.6 cm, respectively. The mutant is expected to be a popular cymbidium variety among consumer as a ornamental orchid mutant for pot culture by its fine and unique stripes and growth character (Kang *et al.*, 2009).

The technique of embryo culture i.e., culturing immature embryos from the unripened fruits has been adopted and an attempt has been made to determine the suitable nutrient medium and plant growth regulator combinations for two *Cymbidium* species for rapid multiplication *in vitro*. Mitra medium supplemented with 1 mgl^{-1} of IBA, BAP, NAA or IBA+BAP were used to study the regeneration response. Seed germination in both the species was highest in basal medium supplemented with activated charcoal (AC). BAP supplemented medium with AC produced best response for seedling formation for *C. pendulum* while NAA enriched medium produced the same result in *C. giganteum*. The fully developed seedlings after hardening are transplanted to community pots for further acclimatization to the nursery conditions (Nongdam *et al.*, 2006).

Nongdam and Chongtham, (2011) attempted to preserve *Cymbidium aloifolium* (L.) important orchid by establishing an efficient *in vitro* regeneration protocol using seed culture. Two culture media viz., Mitra (M) and Murashige and Skoog (MS) supplemented with different plant growth regulators like Indole-3-butyric acid (IBA), 6-Benzyl Amino Purine (BAP) and 1-Nathyl Acetic Acid (NAA) singly or in combination were used for the present study. In 0.5 mg L^{-1} BAP supplemented media, seed germination percentage was very low and seedling formation was not observed. A similar response was observed when a combination of both IBA and BAP at 0.5 mg L^{-1} was used. Protocorm development was delayed and insignificant protocorms turned brown in colour with the lost of chlorophyll and finally died even after repeated subculturing. Presence of Activated Charcoal (AC) significantly improved seed germination, protocorm formation and shoot multiplication. Shooting was best observed in MS medium supplemented with 1 mg L^{-1} BAP and AC. Auxins like NAA and IBA proved to be beneficial for rapid root initiation. However, more superior rooting was observed when MS medium was enriched with 0.5 mg L^{-1} of NAA. Healthy seedlings with well developed roots were obtained which were then hardened and transplanted to community pots for further acclimatization to the nursery conditions. The regenerated plants can subsequently be used for commercial and conservation purposes. The detail regeneration study of cymbidium through different explants is given in Table 7.

Table 7: Regeneration of cymbidium through different explants.

S. No.	Plant	Explant source	References
1.	*Cymbidium* sp.	Shoot meristem	Morel,1970
2.	*Cymbidium* sp.	Leaf tip callus	Churchill *et al.,*1973
3.	*Cymbidium* sp.	Shoot tip/callus	Steward and Mapes, 1971
4.	*Cymbidium* sp.	Leaf tip/protocorm	Fonnesbech, 1972
5.	*Cymbidium* sp.	Dormant bud	Tran Thanh Van, 1974
6.	*Cymbidium* 16 hybrids	Shoot tip	Dalla Rosa and Laneri, 1977
7.	*Cymbidium* sp.	Shoot apex	Kusumoto, 1980
8.	*Cymbidium* sp.	Shoot tip	Gu *et al.*, 1987
9.	*Cymbidium* sp.	Apical meristem	Yang *et al.*, 1999
10.	*Cymbidium* sp.	Immature seed	Bannerjee and Mandal, 1999
11.	*Cymbidium aloifolium*	ShootRhizome, PLB thin cross section	Nayak *et al.*, 1997bNayak *et al.*, 1998aNayak *et al.*, 2002
12.	*Cymbidium ensifolium*	Axillary/flower bud	Jia *et al.*, 2000
13.	*Cymbidium ensifolium*	Apical/axillary bud	Wang *et al.*,1988
14.	*Cymbidium goeringii* Qiulan	Apical/axillary bud	Wang, 1988b
15.	*Cymbidium ensifolium* var. Misericors	Pseudobulb/rhizome/root	Chang and Chang, 1998
16.	*Cymbidium ensifolium* YuhHwa	Rhizome apex	Lu *et al.*, 2001
17.	*Cymbidium faberi*	Rhizome tip	Hasegawa *et al.,* 1985
18.	*Cymbidium faberi,C. forrestii, C. goergii, C. hakuran, C. insigne,*	Rhizome tip Shoot tip	Hasegawa *et al.*, 1985 Hasegawa *et al.*, 1987
19.	*Cymbidium forrestii* Garyung	Rhizome(from seed)	Paek and Yueng, 1991
20.	*Cymbidium giganteum*	Shoot tip	Corrie and Tandon, 1993
21.	*Cymbidium Goeringii* Reichenbach	Apical flower bud	Shimasaki and Uemoto, 1991
22.	*Cymbidium goeringii* Qiulan, *C. insigne*	Shoot	Ueda and Torikata, 1974
23.	*Cymbidium goeringii, C. kanran*	Shoot tip	Shimasaki and Uemoto, 1991
24.	*Cymbidiumin signe* Rolfe, *C. eburneum* Lindley	Shoot	Shimasaki *et al.*, 2003a
25.	*Cymbidium Karan* Makino	Rhizome Rhizome Rhizome Rhizome	Kokubu *et al.*, 1980 Lee, 1986Kim *et al.*, 1988 Shimasaki and Uemoto, 1990
26.	*Cymbidium langifolium*	Rhizome segment	Kim and Lee, 1992
27.	*Cymbidium langifolium*	protocorm	Chetia *et al.,*1998b
28.	*Cymbidium pendulum*		Vij *et al.*,1994
29.	*Cymbidium sinense* Wild	Rhizome (from seed)	Chang and Chang, 2000
30.	*Cymbidium* April Rose, Tsurugi Awabijin, Waltz SS	Shoot tip	Momose and Yoneda, 1989
31.	*Cymbidium*Lancelot Yagoto+Misono	PLB	Kusumoto, 1981a
32.	*Cymbidium* Lois Kelly Cherry, Gawain Empress	PLB	Tanaka, 1999

(*Contd.*)

(*Contd. Table 7*)

S. No.	Plant	Explant source	References
33.	*Cymbidium* Melody Fair' Marilyn Monroe'	PLB	Ogasawara *et al.*, 1995
34.	*Cymbidium* Reporsa	2-3 leaves	Kozai *et al.*, 1990
35.	*Cymbidium* Thanksgiving cv. Nativity, C. Lucky Rainbow Lapine Dancer	PLB outer tissue	Fujii *et al.*, 1999 Begum *et al.*, 1994
36.	*Cymbidium* twilight Moon 'Day Light	PLB	Huan *et al.*, 2004

In vitro flowering

In vitro flowering in orchids has attracted attention considerably because orchids has long juvenile phase (1-10 years). In orchids *in vitro* flowering may be influenced by ABA induced stress and complex addition of sucrose, banana homogenate, nutrient application and manipulating temperature, pH, etc. Work on this aspect is carried out by various workers is given in Table 8.

Table 8: *In vitro* flowering studies in cymbidium.

S. No.	Plant species	Medium composition and pH	References
1.	*Cymbidium* (Maya-Ran, Shun-Ran)	Solid	Wang, 1988a
2.	*C. ensifolium*	W/MSs + 1/2 BA + 0.1/0.2 NAA	Wang, 1988b
3.	*C. goeringii*	MSs + 10% CW	Wang ,1988
4.	*C. nipponicum*	Hy (pH 5.4)	Mizuno *et al.*,1991
5.	*C. niveomarginatum*	MS/Hy (pH 5.8)	Niimi *et al.*,1993

In vitro mutagenesis

Ma *et al.* (2008) obtained two kinds of leaf colour mutants with white and yellow stripes in regenerated seedlings of *C. hybridum*. The total mutation frequency was ~0.12%. The mutants were compared with the control by random amplified polymorphic DNA (RAPD) using 100 UBC arbitrary primers. The polymorphism between the yellow stripe mutant and the control was detected by using primer P17 or P31. The former mutant showed a polymorphic fragment with size of 995 or 1320 bp, respectively. The polymorphism between the white stripe mutant and the control was detected by using primer P84 or P86. A specific band with size of 1120 or 1315 bp, respectively, was found. The results of the RAPD analysis suggest that the genomic DNA of the 2 kinds of mutants has changed.

Somatic embryogenesis

A protocol for the induction of direct somatic embryogenesis from seed derived protocorms was developed in *Cymbidium bicolor* by Mahendran and Bai (2012). 90 day-old protocorms cultured on half strength MS medium alone or supplemented with auxin and cytokinin induced direct somatic embryogenesis. The best response was observed in protocorms cultured half strength MS medium supplemented with BAP at 1.0 mg/L and 2-4,D at 2.0 mg/L. Both epidermal and sub epidermal cells were involved in the formation of embryos. The pro-embryos developed into globular stage and subsequently developed into protocoms. Complete plantlets were formed after 60 days culture. The plantlets were acclimatized in plastic pots containing sterilized vermiculite. Survival rate was 90% when maintained in culture room condition (25 + or -2°C).

Embryogenic callus (EC) was initiated from inner tissue (IT) of protocorm-like bodies (PLB) of cymbidium cv. Nativity cultured on MS medium supplemented with NAA or 2, 4-D. EC was obtained within 15 days on a MS medium containing 2 mg NAA or 0.5 mg 2,4-D/litre. When EC were transferred to a hormone-free MS medium, protocorms were produced within one month and complete plants in another 3 months. In contrast, in the presence of BA or NAA EC produced light green, hairless, string-like structures in one month. These string-like structures grew 2-3 cm in length and died within 3 months. It produced no callus in hormone-free MS medium or in a medium to which BA and NAA were added. After 5 days of culture, with NAA or 2, 4-D, a small mass of EC differentiated from the parenchyma cells of the vascular bundle on the basal cut end of IT sections. These ECs produced proembryoid-like structures after 10 days of culture which became globular embryos 5 days later. The globular embryos differentiated into protocorms and plantlets after being transplanted to the hormone-free MS medium. These findings demonstrate that cymbidium plantlets can be obtained from IT of PLB through somatic embryogenesis by manipulating the culture conditions (Begum *et al*., 1994).

Silva *et al*. (2007) studied on the capacity to induce protocorm-like bodies (PLBs) or somatic embryogenesis in ornamental hybrid cymbidium are still rare. Often *in vitro* regeneration studies claim no variation based on morphological traits alone without cytogenetic confirmation. The induction of PLBs and embryogenic callus (EC) by different plant growth regulators (singly-applied, or in combination) were tested. Resulting organs (shoots, adventitious roots, EC, PLBs) were tested for genetic variability and somaclonal variation using flow cytometry and RAPD analysis.

Molecular breeding

Obara and Kako (1998) examined the DNA from 36 cymbidium cultivars using PCR to determine the efficiency of RAPD markers in identifying cultivars and determining levels of genetic variability. A total of 132 RAPD markers, 78% of which were polymorphic produced from 15 10-mer arbitrary primers. All cultivars were distinguishable when a number of primers were considered. Blue Smoke Green Meadow was distinguished from all others based only on the lack of the OPA5-370 fragment. Genetic distances among the cultivars were estimated based on the amount of band sharing and ranged from 0.08 to 0.50, with an average of 0.29. Cluster analysis of genetic distance estimates grouped siblings and parents with offspring, thereby agreeing with known parentage information and corroborating isoenzyme data obtained from a separate study.

Gan (2006) evaluated the genetic diversity among 20 cymbidium accessions by using RAPD markers and PCR-RFLP analyses of organellar DNAs. In RAPD analysis, the amplified products of 36 primers (72%) were polymorphic. The RAPD-based genetic similarity (GS) among 20 accessions ranged from 0.503 to 0.765, with a mean of 0.598. Based on the genetic similarity matrix from RAPD markers, the genetic relationship among 20 accessions were estimated by UPGMA clustering analysis. All 20 accessions could be distinguished by RAPD markers. In cpDNA PCR-RFLP analyses, 6 out of 7 markers (87.5%) could produce one or more than one distinct bands by direct electrophoresis in 2% agarose gel. After the amplified products were digested by 7 restriction enzymes, a total of 53 bands were detected in 19 cpDNA PCR-RFLP marker/enzyme combinations, among which 37 bands (69.8%) were polymorphic. Li *et al.* (2007) assessed the random amplified polymorphic DNA (RAPD) marker to detect the genetic relationships among 48 hybrid cymbidium cultivars from Japan, Korea, China, and USA and 2 species of native *Cymbidium*. All cultivars were distinguishable when a number of primers were considered. Genetic similarities among the cultivars and species were estimated based on the amount of band sharing ranging from 0.364-0.817 with an average of 0.581. According to the data, a dendrogram of genetic relationship, which was constructed using the UPGMA method, showed that all the tested cultivars and native species were classified into five cluster groups with the similarity coefficient of 0.592. It revealed that the genetic relationships among tested accessions were to some extent related with their origin, flower colour, branch type, and genealogy. Similarly, DNA from thirty-eight *Cymbidium ensifolium* cultivars was examined using polymerase chain reaction (PCR) to determine the efficiency of randomly amplified polymorphic DNA (RAPD) markers in genetic diversity and genetic relationship. Unweighted pair-group method with arithmetic averaging (UPGMA) of genetic distances estimates grouped 'suxin' cultivars and 'caixin' cultivars together with each other, and 'suxin' cultivars diverged from 'caixin' cultivars, thereby agreeing with known traditional classification information. Four cultivars Minxi Yuchen, Yuchen Dagong, Yuchen and Yinbian Yuchen could be distinguished from all the rest based only on the S153-650 fragment, and cvs. Shiliu Luohan and Yuchen based only on the S38-1200 fragment. Cultivar Ma2 er Siji could be differentiated from all the rest by lack of the S38-800 fragment (Hu *et al.*, 2008).

Petolescu and Lazar (2009) determined *in vitro* variability using randomly amplified polymorphic DNA (RAPD) markers. DNA from eleven *in vitro* obtained plantlets of cymbidium was examined using polymerase chain reaction (PCR) to determine the efficiency of randomly amplified polymorphic DNA (RAPD) markers in identifying of somaclonal variability. Concerning RAPD analysis, template DNA produced clear PCR profiles. As a result of UV light screening of agarose gels, it has been observed slight differences between the genetic fingerprints of the studied variety induced by amplification using RAPD primer.

Shi *et al*. (2010) assessed the random amplified polymorphic DNA marker (RAPD) to detect the genetic relationships among 18 *Cymbidium hybridum* cultivars from 2 companies. The results showed that primers were screened from 50 random decamer primers and a total of 92 DNA bands were amplified, 90 of which (97.8%) were polymorphic. The average number of polymorphic DNA bands amplified by each primer was 6. Genetic similarities among the cultivars ranged from 0.1975 to 0.6704 with an average value of 0.4761, and the genetic distance among those ranged from 0.3296 to 0.8025 with an average value of 0.5239. A dendrogram of genetic relationship which was constructed using UPGMA method showed that all the tested cultivars were classified into 4 cluster groups with the genetic distance coefficient of 0.5239. It revealed that the genetic relationships among tested cultivars were related with flower colour and branch type to some extent.

Microsatellite primers were developed for the important and popular ornamental spring orchid, *Cymbidium goeringii* to investigate its genetic diversity and population genetic structure. Using the Fast Isolation by AFLP of Sequences Containing (FIASCO) repeats protocol 15 primer sets were identified in two wild populations. The number of alleles per locus ranged from 2 to 5, with a mean of 3.3. The observed and expected heterozygosities varied from 0.167 to 0.917 and from 0.159 to 0.822, respectively. All these primers successfully amplified in the congener *C. tortisepalum*, and 12 primers were found useful in *C. faberi* and *C. sinense*. They concluded that these markers will facilitate further studies on the population genetics of *Cymbidium goeringii* and other congeneric species (Huang *et al*., 2011).

Genetic engineering

Transgenic cymbidium orchids through particle bombardment have been produced by Yang *et al*. (1999). The technique consisted vector pK N200 (npt II/GUS-INT) using the PDS 1000/He device and the transformation efficiency is of the 15%. The technique is being used to alter flowering time and morphology as well as to develop viral resistance. Liquid culture was used to generate a large number of protocorm-like bodies (PLBs) to be bombarded and to promote continued development of the bombarded meristematic tissue. Shoot regeneration was then induced from the kanamycin-resistant PLBs and transgenic plantlets were produced. Both the kanamycin-resistant PLBs and regenerated shoots expressed the GUS-INT gene. The presence of the introduced gene in the transformed orchid plants was confirmed by PCR analysis, sequencing and Southern blot analysis of the PCR product. The recovered transgenic plants were established in soil and acclimatized in the greenhouse.

A transformation protocol based on a particle-inflow-gun was successfully used to transfer multiple transgenes into 'Jungfrau Dos Pueblos', a commercial cultivar of cymbidium orchid. Protocorm-like bodies (PLBs) were bombarded with three transgenes, a dihydroflavonol reductase (*DFR*) antisense construct, the green fluorescent protein reporter gene (*GFP*) and the neomycin phosphotransferase II (*NPTII*) selectable marker, all of which are under the control of the CaMV-*35S* promoter. It was found that antisense *DFR* (as*DFR*) construct was expressed in floral tissue but no obvious change in anthocyanin biosynthesis was observed initially. Induced anthocyanin accumulation in the flower lip was less in the as *DFR* flowers than in the non-transgenic flowers, possibly due to reduced DFR activity which is consistent with the action of this transgene (Boase, 2012).

Flower colour

Cymbidium hybrid 'Jung Frau dos Pueblos' (JDP) is a commercial white flowered orchid cultivar. Albert *et al.* (2010) investigated whether anthocyanin pigment production could be restored to JDP petals by introducing known anthocyanin-regulating transcription factors through biolistic transformation. Simultaneous transformation of white JDP petals with the maize anthocyanin regulators *Leaf colour* and *Colourless1* restored anthocyanin biosynthesis, producing a characteristic biolistic blast pattern of scarlet anthocyanin-producing cells. The effectiveness of different MYB/bHLH combinations to induce anthocyanin pigmentation in JDP petals was assessed, and they demonstrated that the previously uncharacterized R2R3 MYB transcription factor *AaMYB1* from *Anthurium andraeanum* is capable of activating anthocyanin production in cymbidium petals.

REFERENCES

Albert, N.W., Arathoon, S., Collette, V.E., Schwinn, K.E., Jameson, P.E., Lewis, D.H., Zhang, H. and Davies, K.M. 2010. Activation of anthocyanin synthesis in cymbidium orchids: variability between known regulators. *Plant Cell Tissue Organ Cult.*, **100**(3): 355-60.

Attri, L.K. and Kant, R. 2011.Orchid Pollination: An Observation on Pollination-Pollinator Interaction in *Cymbidium pendulum* (Sw.) Roxb. *Current Botany*, **2**(7): 05-08.

Bannerjee, B. and Mandal, A.B. 1999. *In vitro* germination of immature cymbidium seeds for rapid propagation of plantlets in inlands. *Cell and Chromosome Research,* **21**: 1-5.

Begum, A.A., Tamaki, M., Tahara, M. and Kako, S. 1994. Somatic embryogenesis in cymbidium through *in vitro* culture of inner tissue of protocorm-like bodies. *J. Japanese Soc. Hortil. Sci.*, **63**(2): 419-27.

Boase, M.R., Albert, N.W., Zhang, H., Arathoon, S., Wang, L. and Lewis, D.H. 2012. Long-term stable expression of multiple transgenes under control of the same promoter in cymbidium orchid. *Acta Horticulturae,* **937**: 597-04.

Chan, C.L., Lamb, A., Shim, P.S. and Wood, J.J. 1994. Orchids of Borneo Vol. I. Introduction and selection of species. The Sabah Society Kota Kinabalu, in association with Royal Botanic Gardens, Kew. pp. 420.

Chang, C. and Chang, W.C. 1998. Plant regeneration from callus culture of *Cymbidium ensifolium* var. *misericors*. *Plant Cell Reports,* **17**: 251-255.

Changkija, S., Kumar, Y. and Gurung, P.B. 1992. Orchids of Nagaland. Forest Department, Govt. of Nagaland.

Chung, M. and Chung, M. 2003. The breeding systems of *Cremastra appendiculata* and *Cymbidium goeringii*: high levels of annual fruit failure in two self-compatible orchids. *Annales Botanici Fennici.,* **40**(2): 81-85.

Churchill, M.E., Ball, E.A. and Arditti, J. 1973. Tissue culture of orchids.I. Methods for leaf tips. *New Phytologist,* **72**: 161-166.

Corrie, S. and Tandon, P. 1993. Propagation of *Cymbidium gigantium* Wall. Through high frenquency conversion of encapsulated protocorms under *in vitro* and *in vivo* conditions. *Indian J. Experimental Bio.,* **31**: 61-64.

Dalla, R.M. and Laneri, U. 1977. Modification of nutrient solutions for germination *in vitro* of some cultivated orchids and for the vegetative propagation of cymbidium cultivars. *American Orchid Soci. Hortil. Sci.,* **46**: 813-820.

Fonnesbech, B.1972. Growth hormones and propagation of cymbidium *in vitro*. *Physiolegia Planterum,* **27**: 310-312.

Fujii, K., Kawano, M. and Kako, S. 1999. effects of benzyladenine and á-naphthaleneacetic acid on the formation of protocorm like bodies (PLBs) from explants of outer tissue of cymbidium PLBs cultured *in vitro*. *J. Japanese Soc. Hortil. Sci.,* **68**: 35-40.

Gan, N., Tan, X., Chen, Q., Wei, Y. and Zheng, Y. 2006. Genetic diversity in cymbidium based on RAPD markers and PCR-RFLP analyses of organellar DNAs. *Acta Horticulturae Sinica,* **33**(2): 349-55.

Gu, Z., Arditti J. and Nyman, L.P. 1987. The effect of benzyladenine, 2, 4-dichloro-phenoxyacetic acid and indoleacetic acid on shoot tip cultures of cymbidium. *Lindleyana,* **2**: 88-90.

Hasegawa, A. 1987. Studies on the propagation of oriental cymbidium. Memoires of the Faculty of Agriculture, Kagawa University, **50**: 1-108.

Hasegawa, A., Ohashi, H. and Goi, M. 1985. Effects of BA, rhizome treatment, mechanical treatment and liquid shaking culture on the shoot formation in *Cymbidium faberi* Rolfe. *Acta Horticulturae,* **166**: 25-40.

Hu, W., Huang, R., Pan, X., Li, J. and Sun, D. 2008. RAPD analysis of thirty-eight *Cymbidium ensifolium* cultivars. *Acta Horticulturae Sinica,* **35**(2): 289-94.

Huan, L.V.T., Takamura, T. and Tanaka, M. 2004. Callus formation and plant regeneration from callus through somatic embryo structures in cymbidium orchid. *Pl. Sci.,* **166**: 1443-1449.

Huang, J., Zeng, C., Li, H. and Yang, J. 2011. Isolation and characterization of 15 microsatellite markers from the spring orchid (*Cymbidium goeringii*) (Orchidaceae). *American J. Bot.,* **98**(4): 76-77.

Ji, B., Chen, D., Zhang, C., Min, D., Huang, W. and Wang, Y. 2011. High efficient polyploid induction of *Cymbidium hybridium*. *Bulletin of Botanical Research.* **31**(5): 558-62.

Jia, Y.J., Cao, Y.L., Wang, S., Tang, L., Xu, Y. and Chen, F. 200. Studies on the culture of internode of flower branch of China orchid. *J. Sichuan University* (*Natural Science Edition*), **37**: 94-97.

Kang, K., Park, K., Mo, S., Kim, D. and Kang, S. 2009. A new cymbidium orchid variety Daegook bred by *in vitro* mutagenesis. *Korean J. Breeding Sci.,* **41**(4): 510-14.

Kim, M.,Won, J., Song, C., Eun, J. and Lee, D. 1997. Polyploidy induction of *Cymbidium kanran* by treatment of colchicine *in vitro*. *RDA J. Hortil. Sci.,* **39**(1): 73-76.

Kokubu, T., Kaieda, Y., Higashi, Y., Kitano, T. and Fukamizo, Y. 1980. Organogenesis in sterile culture of oriental cymbidium, *Cymbidium kanran* Makino. Memoirs of Faculty of Agriculture, Kagoshima University, **16**: 53-64.

Kozai, T., Oki, H. and Fujiwara, K. 1990. Photosynthetic characteristics of cymbidium plantlets *in vitro*. *Plant Cell Tissue Organ Cult.,* **22**: 205-211.

Kusumoto, M. 1980. Interform variations of the proliferation, organogenesis and effects of growth of the growth regulating substances on cymbidium protocorms cultured *in vitro. J. Japanese Soc. Hortil. Sci.,* **48**: 510-518.

Kusumoto, M. 1981. Studies on the tissue culture of orchids-I. effects of additional fertilization of culture solution and growth regulating substances on the growth of cymbidium plantlets cultured *in vitro.* Bulletin of the College of Agriculture and Veterinary Medicine, Nihon University, **38**: 108-117.

Lee, J.S., Shim, K.K., Yoo, M.S., Lee, J.S. and Kim, Y.J. 1986. Studies on rhizome growth and organogenesis of *Cymbidium kanran* cultured *in vitro. J. Korean Soc. Hortil. Sci.,* **27**: 174-180.

Li, D., Ye, Q. and Zhu, G. 2007. Analysis of the germplasm resources and genetic relationships among hybrid cymbidium cultivars and native species with RAPD markers. *Agril. Sci. China,* **6**(8): 922-29.

Lin, F. and Deng, G. 1997. Studies on the inductive mutation of spring orchid. *J. Hunan Agril. Univ.,* **23**(4): 336-40.

Lu, I.I., Sutter, E. and Burger, D. 2001. Relationship between benzyladenine uptake, endogenous free IAA level and peroxidase activities during upright shoot induction in *Cymbidium ensifolium* cv. Yuh Hwa rhizomes *in vitro. Plant Growth Regulation,* **35**: 161-170.

Ma, H., Yao, Y., Xin, Y., Chen, W., Tong, J. and Cui, H. 2008. Gain and DNA identification of leaf colour mutant derived from tissue culture of *Cymbidium hybridum. Acta Agriculturae Zhejiangensis,* **20**(3): 149-53.

Mahendran, G. and Bai, V.N. 2012. Direct somatic embryogenesis and plant regeneration from seed derived protocorms of *Cymbidium bicolor* Lindl. *Scientia Horticulturae.* **135**: 40-44.

Mizuno, N., Hiyama, I. and Higuchi, H. 1991. Aseptic culture *in vitro* flowering and *in vitro* fruiting of a mycoparasitic orchid *Cymbidium nipponicum. Proce:* Nagoyo International Orchid Show, pp: 141.

Momose, H. and Yoneda, H. 1989. Studies on the histochemistry of an orchid I. Sugar metabolism on PLB propagation of cymbidium. Bulletin of the College of Agriculture and Veterinary Medicine, Nihon University, **43**: 63-68.

Morel, G. 1970. Neues ouf dem gebiet der meristem-forshungi. *Die orchidae,* **20**: 433-443.

Nayak, N.R., Chand, P.K. Rath, S.P. and Patnaik, S. 1998. Influence of some plant growth regulators on the growth and organogenesis of *Cymbidium aloifolium* (L.) Sw. seed-derived rhizomes *in vitro. In vitro Cellular Developmental Bio. Pl.,* **34**: 184-188.

Nayak, N.R., Rath, S.P. and Patnaik, S. 1997. *In vitro* propagation of three epiphytic orchids, *Cymbidium aloifolium* (L.) Sw., *Dendrobium aphyllum* and *Dendrobium moschatum* (Buch-Ham) Sw. through thiodizuron-induced high frequency shoot proliferation. *Scientia Horticulturae,* **71**: 243-250.

Nayak, N.R., Sahoo, S., Patnaik, S. and Rath, S.P. 2002. Establishment of thin cross section (TCS) culture method for rapid propagation of *Cymbidium aloifolium* (L.) Sw. and *Dendrobium nobile* Lindl. (Orchidaceae). *Scientia Horticulturae,* **84**: 107-116.

Niimi, Y., Tanaka, C. and Hayata, Y. 1993. Proliferation and organogenesis of cymbidium using rhizome. *Proce:* Nagoyo International Orchid Showroom, pp: 72-79.

Nongdam, P. and Chongtham, N. 2011. *In vitro* rapid propagation of *Cymbidium aloifolium* (L.) Sw.: a medicinally important orchid via seed culture. *J. Bio. Sci.,* **11**(3): 254-60.

Nongdam, P., Nirmala, C. and Tewari, R. 2006. *In vitro* multiplication of *Cymbidium* orchids via embryo culture. *Plant Cell Biotech. Molecular Biology,* **7**(3/4): 145-50.

Obara-Okeyo, P. and Kako, S. 1998. Genetic diversity and identification of cymbidium cultivars as measured by random amplified polymorphic DNA (RAPD) markers. *Euphytica,* **99**(2): 95-101.

Ogasawara, N., Haneishi, S., Suzuki, C. and Takagi, H. 1995. The growth and CO2 exchange of cymbidium protocorm like bodies related to sugar in the culture medium. *Acta Horticulturae,* **393**: 85-90.

Paek, K.Y. nad Yeung, E.C 1991. The effect of 1-naphthaleneacetic acid and N^6-benzyladenine on the growth of *Cymbidium forestii* rhizomes *in vitro. Plant Cell Tissue Organ Cult.,* **24**: 65-74.

Petolescu, C. and Lazar, A. 2009. *In vitro* variability of cymbidium sp. based on random amplified polymorphic DNA (RAPD) markers. *J. Hort. Forestry Biotech.,* **13**: 475-77.

Sasaki, M., Ono, M., Asada, S. and Yoshida, T. 1991. Oriental orchid (*Cymbidium pumilum*) attracts drones of the Japanese honeybee (*Apis cerana japonica*) as pollinators. *Experientia*, **47**: 1229-31.

Shi, L., Wang, X. and Lu, J. 2010. The genetic relationships of *Cymbidium hybridum* germplasm resources by RAPD molecular markers. *Acta Agriculturae Zhejiangensis*, **22**(4): 414-19.

Shimasaki, K. and Uemoto, S. 1991. Rhizome induction and plantlets regeneration of *Cymbidium goeringii* from flower buds cultures *in vitro*. *Plant Cell, Tissue Organ Cult.*, **25**: 49-52.

Shimasaki, K., Shiraga, T. and Fukomoto, Y. 2003. effect of methyl jasmonte on organogenesis in shoot cultures of epiphytic and terrestrial *Cymbidium* species. *Environmental Control in Biology,* **41**: 179-182.

Silva, J.A.T., Norikane, A. and Tanaka, M. 2007. *Cymbidium*: successful *in vitro* growth and subsequent acclimatization. *Acta Horticulturae*, **748**: 207-13.

Stewart, F.C. and Mapes, M.O. 1971. Morphogenesis in aseptic cell cultures of cymbidium. *Botanical Gazette*, **132**: 65-70.

Sugahara, M., Minamoto, T., Fuchikawa, T., Michinomae, M. and Shimizu, I. 2010. *Apis cerana japonica* discriminates between floral colour phases of the oriental orchid, *Cymbidium floribundum*. *Zoological Sci.,* **27**(12): 901-06.

Tanaka, M. 1991. Disposable film culture vessels. *In*: Bajaj YPS (Ed.) Biotechnology in Agriculture and Forestry (Vol. 17). High Tech and Micropropagation I, Springer, Berlin, pp. 212-228.

Tran Thanh Van. 1974. Methods of acceleration of growth and flowering in a few species of orchids. *American Orchid Soci. Bulletin,* **43**: 699-707.

Ueda, H. and Torikata, H. 1974. Organogenesis in meristem culture of cymbidiums VII. Study on the extract from mycorrhizomes of *Cymbidium goeringii* Reichb. f. (*C. virens* Lindl). *J. Japanese Soc. Hortil. Sci.,* **43**: 281-285.

Wang, G.T and Xiong, N.H. 1988. Tissue culture of cymbidium: plant and flower induction *in vitro*. *Lindleyana,* **3**: 184-189.

Wang, M., Zeng, R., Xie, L., Li, Y., Zeng, F., Du, B. and Zhang, Z. 2010. *In vitro* induction and its identification of tetraploid *Cymbidium hybridium*. *Acta Botanica Boreali-Occidentalia Sinica*, **30**(1): 56-62.

Wang, X. 1988. Studies on the growth and development of cymbidium in tissue culture. *J. Shandong Agril. University,* **19**: 19-24.

Wang, X. 1988a. Tissue culture of cymbidium: Plant and flower induction *in vitro*. *Lindleyana,* **3**: 184-189.

Xiang, L., Li, X., Qin, D., Guo, F., Wu, C., Miao, L. and Sun, C. 2012. Functional analysis of flowering locust orthologs from spring orchid (*Cymbidium goeringii* Rchb. f.) that regulates the vegetative to reproductive transition. *Plant Physiol. Biochem.*, **58**: 98-105.

Yang, J., Lee, H.J., Shin, D.H., Oh, S.K., Seon, J.H., Paek, K.Y. and Han, K.H. 1999. Genetic transformation of cymbidium orchid by particle bombardment. *Plant Cell Reports*. **18**(12): 978-84.

Yin, C., Zhang, Y., Zhang, J., Chen, Y. and Wang, G. 2010. Tetraploid induction by colchicine and identification in cymbidium interspecific hybrids. *J. Nuclear Agril. Sci.* **24**(3): 518-521.

Chapter – 32

Dendrobium

Dendrobium is a huge genus of orchids. It was established by Olof Swartz in 1799. The name is from the Greek word "*Dendron*" means "tree" and "*bios*" means "life"; it means "one who lives on trees", or essentially, "epiphyte". The truly spectacular genus *Dendrobium* contains the largest diversity of horticulturally interesting specimens. More than 1,000 natural species make dendrobium the second-largest orchid genus. Dendrobiums are among the most commonly encountered orchids in the retail trade. Flowers are very long-lasting both on the plant and once cut, looking fresh for up to three weeks or more. Faded flowers on the bottom of the stem can be removed to encourage upper buds to open. Dendrobiums are sturdy orchids that can add a focal point to any room.

ORIGIN AND DISTRIBUTION

Dendrobiums are originally from Southeast Asia. The geographic distribution is from India through China, Japan, Malaysia, Philippines and the islands of the South Pacific with the largest diversity in New Guinea. *Dendrobium* species are either epiphytic or occasionally lithophytic. They have adapted to a wide variety of habitats from the high altitudes in the Himalayan mountains to lowland tropical forests and even to the dry climate of the Australian desert.

Several species of *Dendrobium* are originated from India. These species are well distributed in the Himalayan region of North-East India. About 37 species of *Dendrobium* are existing the forest of Nagaland (Changkija *et al*., 1992). These species are described in Table 1.

Table 1: *Dendrobium* species distributed in Nagaland (India).

S. No.	Species	Distribution	Characteristics
1.	*Dendrobium acinacifore*	Medziphema, Tuli, Tsurang Valley, Mon, Tezit, Saring, Aonokpho	FC: Pale-yellow, FS: 8 cm, FT: June - July.
2.	*D. anceps*	Pangsa, Tobou, Peren, Jaluke	FC: Greenish-yellow, FS: 1cm, FT: April-May.
3.	*D. aphyllum*	Dzuja, Jaluke, Sanis, Doyang	FC: Creamy or rosy white, FS: 3-4 cm, FT: April-May.
4.	*D. aurum*	Chare, Changtongya, Sanis, Akuluto	FC: Pale yellow streaked with reddish purple in the lip, FS: 5-6 cm, FT: April-May.
5.	*D. bensoniae*	Kohima, Wazeho, Khepri, Samator	FC: Pure white, FS: 6 cm, FT: May-June.
6.	*D. bicameratm*	Doyang, Jaluke, Chare	FC: Yellow with dull red stripes, FS: 1cm, FT: Aug.-Sept.
7.	*D. candidum*	Saramati, Japfu Range, Phek	FC: Pure white, FT: April-May.
8.	*D.chrysanthm*	Wokha, Mokokchung, Longkhum, Changki	FC: Orange-yellow, FS: 4-5 cm, FT: Sept.-Oct.
9.	*D. chrysotoxm*	Jaluki, Medziphena, Dzuja, Changki, Changtongya, Longleng, Akuluto	FC: Deep yellow, FS: 5 cm, FT: April-May.
10.	*D. crepidatum*	Peren, Phek	FS: 4cm, FT: March-May.
11.	*D. densiflorum*	Longkhim, Noklak, Longsa, Changki, Longleng, Longwa, Changtongya, Peren, Doyang	FC: Yellow, FT: April-May. Inflorescence axillary, hanging with dense cluster flowers.
12.	*D. denudans*	Pesao, Changlangshu	FC: Pale-green in colour, FS: 1.8 cm, FT: Sept.-Oct.
13.	*D. devonianum*	Viswema, Peren, Pfutsero, Noklak, Tuensang, Changki	FC: White with purple tipped petals, FS: 4-5 cm, FT: May-June.
14.	*D. eriaeflorum*	Viswema, Phek	FC: Greenish-yellow, FS: 6-10cm, FT: Sept.-Oct.
15.	*D. falconeri*	Japfu Range, Dzukou, Longtok, Pangsa	FC: White flushed with pale-rose and tipped with amethyst purple, FS: 10 cm, FT: May-June.
16.	*D. farmeri*	Longsa, Akuluto, Chesore	FC: Pink-White with yellow lip, FS: 5 cm, FT: April-May.
17.	*D. fimbriatum*	Dzuja, Tseminyu, Peren, Mokokchung, Longkhim, Wokha	FC: bright yellow, FS: 7.5 cm, FT: April-May.
18.	*D. formosum*	Viswema, Japfu Range, Peren	FC: White, FS: 3-5 cm, FT: May-June.
19.	*D. gibsonii*	Mokokchung, Chare, Changki, Tobou, Dzuja	FC: Dull yellow, FS: 4 cm, FT: July-Aug.
20.	*D.heterocarpm*	Longkhim, Tuensang, JapfuRange, Viswema	FC: White, FS: 5 cm, FT: April.
21.	*D. hookerianm*	Dzuja, Jaluki, Longjang, Doyang, Mon, Tezit	FC: Yellow, FS: 7-10 cm, FT: April-May.
22.	*D. infundibulm*	Hellipong,Tuensang, Kohima, Mokokchung	FC: Pure white with large blotch of golden-yellow on the funnel-shaped throat of the lip, FS: 10 cm, FT: April-May.

(Contd.)

(*Contd. Table 3*)

S. No.	Species	Distribution	Characteristics
23.	*D. jenkinsii*	Saramati Range, Phek, Changlangsu	FC: Orange-yellow, FS: 1-2, 2.5-3 cm, FT: April-May.
24.	*D. lindleyi*	Mokokchung, Tuensang, Wokha, Kohima, Zunheboto	FC: Pale-yellow, FS: 3 cm, FT: Feb March.
25.	*D. longicornu*	Japfu Range, Chentang, Hellipong, Saramati, Pesao	FC: White, FS: 3-4 cm, FT: Oct-Nov.
26.	*D. moschatum*	Kiphire, Longkhim, Longkhum, Wokha	FC: Orange-yellow, FS: 7cm, FT: May June.
27.	*D. nobile*	Peren, Phek, Kiphire, Noklak	FC: Purplish white, FS: 6 cm, FT: April-May.
28.	*D. ochreatum*	Viswema, Kohima, Tuensang, Wokha, Phek, Mokokchung,	FC: Golden yellow, FS: 4-5 cm, FT: April-May.
29.	*D. porphyrochilum*	Chentang, Noklak, Samator	FC: Pale-Green with red lines, FS: 0.8 mm, FT: Sept.-Oct.
30.	*D. primulinum*	Dzuja, Peren, Phek, Changki, Longjang, Chare, Akuluto	FC: White or rose-pink with deeper coloured nerves, FT: Mar.-April.
31.	*D. pulchellum*	Pangsa, Saramati, Viswema, Changlangshu, Peren	FC: Orange, FS: 1 cm, FT: April-May.
32.	*D. stuposum*	Noklak, Pangsa, Meluri, Wazeho	FC: White, FS: 1 cm, FT: June-July.
33.	*D. terminale*	Jaluke, Peren, Phek	FC: White or pale pink, FT: Sept.-Oct.
34.	*D. thyrsiflorum*	Longkhim, Tobou, Chare, Longsa, Longkhum, Changki	FC: Pure white to light yellow, FS: 5 cm, FT: April-May.
35.	*D. transparens*	Viswema, Phek, Peen, Samator	FC: Pinkish-white, FS: 3-4 cm, FT: April-May.
36.	*D. wardianum*	Japfu Range, Peren, Pfutsero	FC: White with purple tips and 2 maroon blotches on the lip, FT: April-May.
37.	*D. williamsonii*	Peren, Pangsa, Saramati, Noklak	FC: Pale-green, FS: 4 cm, FT: March-April.

FC: Flower colour, FS: Flower size, FT: Flowering time.

SPECIES

The *Dendrobium* genus of orchid is a tropical flower that contains over 1600 species. In 1981, Briegar reclassified all terete-leaved *Dendrobiums* from Australia and New Guinea into a new genus, *Dockrillia*. The Winika orchid from New Zealand was formerly *D. cunninghamii*, but has now been moved into a monotypic genus *Winika*. In 1989, Clements upgraded the *D. speciosum* complex into individual species, similarly, the *D. bigibbum* complex has recently been split up. Numerous species originated in Australia, New Guinea, the Philippines, Thailand, India, China and Indonesia, among other countries. Horticulturally important species are clustered into taxonomic sections including *Phalaenanthe, Spatulata* (*Ceratobium*), *Latourea, Formosae* (*Nigrohirsutae*) *Dendrobium* (*Eugenanthe*) and *Callista* (Schelpe and Stewart, 1990). Species comprising the warm group (night temperatures above 16° C) are *D. phalaenopsis, D. gouldii, D.*

bigibbum, D. antennatum, and *D. discolor*, and the cool group (nights of 10° C) include *D. lindleyi (D. aggregatum), D. parishii, D. pierardii, D. densiflorum, D. chrysotoxum and D. anosmum.* Following are some important species of *Dendrobium.* Some important species of *Dendrobium* are given in Table 2.

Table 2: List of some important species of *Dendrobium.*

Genus	No. of species	Important species	Origin
Dendrobium	1600	*D. aggregatum, D. amplcim, D. bigibbum, D. bellatulum, D. bensoniae, D. canaliculatum, D. chrysanthum, D. phalaenopsis, D. nobile*	Varies from hot sea level areas to cool mountain climates

***Dendrobium acinaciforme*:** Stem upto 30 cm long, compressed, narrow at base. Leaves ensiform, acute, fleshy, sessile whereas, flowers about 8 cm across, pale-yellow, many on the long leafless extremities of the stem. Its flowering time is from June-July.

***Dendrobium acinaciforme*:** Epiphyte on tree trunk in dense humid evergreen. Flowering: June -August. It is distributed within Dibrugarh district: Jeypore (India). Its present ecological status is common.

***Dendrobium aduncum*:** Flowers 1.2-2 cm across, pale purple; sepals and petals pale pink; lip white; column white. Habitat: Epiphyte on a small tree trunk in dense humid evergreen forest. Its flowering time is May. Its local distribution within Dibrugarh district: Jeypore, Jo-kai (India). Its present ecological status is endangered.

***Dendrobium aphyllum*:** Epiphyte in mixed deciduous forest or open forest. Flowers, pale rose, 4.5-5 cm across; lip yellow. Flowering from April to May. It is locally distributed within Dibrugarh district: Jeypore, Jo-kai, Namdang, Telpani and Dihingmukh (India) and its present ecological status is common.

***Dendrobium anceps*:** Stem upto 40 cm long, compressed; leaves lanceolate, acute, fleshy; flower about 1 cm across, greenish-yellow, solitary, axillary. Its flowering time is from April-May.

***Dendrobium bensoniae*:** Pseudobulb stout, teate, pale yellow to purplish, 30-75 cm long; leaves 5-8 cm × 12-15 mm, linear, apex notched; flower 1-3 from the nodes of leafless. Pseudobulbs, Pure white, upto 6 cm across; blotch with two yellow disc having two maroon eye spots at the lip. Its flowering time is May-June.

***Dendrobium bigibbum*:** Pseudobulb; 40-122 cm long. Leaves; 3-12 per growth. The oblong-lanceolate leaves are 8-15 cm long. They are flushed with red or purple, and are leathery, rigid, and evergreen for 2 years. The apical 30% of the stems are leafy. The inflorescences are usually arching or horizontal, but they may be pendent. Flowers are nicely spaced along the upper half of the raceme. Flowers are 8-20 per inflorescence, occasionally less. The flowers are 3-5 cm across. Sepals and petals usually curve backward. The showy flowers have a heavy texture, broad overlapping sepals, and narrower petals. Flowers are normally violet, but colours include white, pale to deep lilac, magenta, and purple. All colours are bright and rich.

***Dendrobium candidum*:** Pseudobulb upto 30 cm long; leaves linear-lanceolate, tip unequally biloped; flowers 2-3 on short peduncle from leafless pseudobulbs, pure white in colour. Flowering time is from April- May.

***Dendrobium chrysanthum*:** Pseudobulbs upto 1 cm tall, leafy, many noded; leaves lanceolate, acuminate; flowers 4-5 cm across, in fascicles of 2-4 at the nodes, orange-yellow in colour. Flowering time is Sept.- Oct.

***Dendrobium chrysotoxum*:** Pseudobulb is upto 35 cm long, grooved, shinning yellow; leaves 3-7 in number, oblong- lanceolate, tip bilobed; inflorescence upto 15 flowered, hanging; flowers 5 cm across deep-yellow in colours. Its flowering time is from April- May.

***Dendrobium cumulatum*:** Habitat: Epiphyte on tree trunk in forest along a small stream. Flowers white with purple; lip white suffused with yellow. Flowering: March-May. Local distribution within Dibrugarh district: Jeypore (India). Its present ecological status is endangered.

***Dendrobium densiflorum*:** This is a densely flowered *Dendrobium* and also known as Pineapple orchid because of the inflorescence resembling pineapple. This is one of the most spectacular orchid species among the Dendrobiums Pseudobulbs is upto 40 cm, long erect, 4 angled, gradually tapering towards the base, sheathed; leaves 3-5 in number, clustered near the apex, lanceolate, acute; inflorescence axillary, hanging with a dense cluster of yellow flower. Flowering time is April-May.

***Dendrobium fimbriatum*:** Habitat: Epiphyte in mixed deciduous forest, also in humid evergreen forest. Flowers 4-5 cm across, sepals and petals golden yellow; lip golden yellow, with or without maroon blotch. Flowering: March-May. Local distribution within Dibrugarh district: Jeypore (India). Its present ecological status is vulnerable.

***Dendrobium herbaceum*:** The white coloured flowers of this sympodial orchid usually bloom in the months of January-May. This bushy leafless epiphyte with its inflorescence having innumerable flowers sitting high on trees being the highlight helping in identification of this species. The typical size of the flower is less than a centimeter in width and when observed closely the greenish lip or labellum is seen. Usually the flowering lasts several weeks. Normally once the flowering season is complete and rains set in, the same plant can be seen full of new leaves and shoots. Typical habitat to find this species is mid to lower elevations of the Southern Western Ghats. It is also observed in other parts of India as well.

***Dendrobium lituiflorum*:** Habitat: Epiphytic on tree trunks in open forests. Flowers 4-5 cm across, pale purple inner surface of lip with a deep purple spot surrounded by a white circle. Flowering: April-May. Local distribution within Dibrugarh district: Jeypore, Jo-kai, Namdang, Telpani and Dihingmukh (India). Present ecological status: Common.

***Dendrobium moschatum*:** Habitat: Epiphytic on tree trunks in open forests. Flower orange yellow, 5-7 cm across, deep yellow disk with a pale purplish brown blotch on either side. Flowering: April - June. Local distribution within Dibrugarh district: Jeypore, Jo-kai, Namdang, Telpani and Dihingmukh (India). Present ecological status: Common.

***Dendrobium nobile*:** Pseudobulb is 60-90 cm long. The stems are swollen at the apex and taper to a narrower base. They are clustered on a short connecting rhizome. The canes are often yellowish, somewhat zigzag, round in cross-section, and become furrowed with age. The nodes are usually thickened and flattened. Leaves are 6-7 in number, distichous, 7-10 cm long, oblong to strap-shaped, softly leathery, and deciduous after 2 years. Many inflorescences emerge simultaneously from the upper nodes of both leafy and older leafless canes. Flowers are 1-4 per inflorescence. The blossoms, which are 6-10 cm across, have a waxy and heavy texture. The oval sepals and much wider, wavy-margined petals are normally white with rose tips. The downy lip, which is tubular at the base, is cream-white with rose at the apex and has deep crimson or crimson-purple markings in the throat. It is occasionally pure white.

***Dendrobium phalaenopsis*:** Pseudobulb 40-122 cm long. Leaves are 3-5, rarely as many as 12. The oblong-lanceolate leaves are 8-15 cm long. They are leathery, rigid, and flushed with red or purple. The inflorescences are usually arching or horizontal, but they may be pendent. Flowers are nicely spaced along the upper half of the raceme. Flowers are 3-20 per inflorescence. The showy flowers are 3.5-7.0 cm across with broad overlapping sepals and petals and a heavy texture. Flowers are normally deep-lilac, but they may be white, pale lilac, magenta, or purple. All colours are bright and rich.

***Dendrobium stuposum*:** Habitat: Epiphytic on tree trunks in open and mountain forests Flowers small, white with a dull yellow spot on the lip. Flowering: June-August. Local distribution within Dibrugarh district: Jeypore (India). Present ecological status: Endangered.

***Dendrobium sulcatum*:** Habitat: Epiphytic on tree trunks in dense forests. Flowers golden yellow, 2.5-3.5 cm across; lip with red-purple markings at base. Flowering: April-May. Local distribution within Dibrugarh district: Jeypore (India). Present ecological status: Critically Endangered.

***Dendrobium terminale*:** Habitat: Epiphytic on tree trunks at forest. Flowers 1-1.2 cm long, white flushed pink. Flowering: April-June. Local distribution within Dibrugarh district: Jeypore (India). Present ecological status: Endangered.

Dendrobium transparens: Habitat: Epiphytic on tree trunks at forest. Flowers 4-5 cm across; sepals and petals white linged with purple at tips, lip white with large purple patch on the disc. Flowering: April – May. Local distribution within Dibrugarh district: Jeypore, Jo-kai, Namdang, Telpani and Dihingmukh (India). Its present ecological status is vulnerable.

Dendrobium thyrsiflorum: It is one of the most spectacular of all orchids. The large bunches of flowers look like festive Chinese lanterns with each individual flower about 5 cm across and pleasingly fragrant. The sepals and petals are white, the lip canary-yellow. The plant originates from South Asia, from the Himalayas to Thailand.

FLOWER MORPHOLOGY AND BIOLOGY

This genus of sympodial orchids develop pseudobulbs, which vary in length from under a centimetre (e.g. *Dendrobium leucocyanum*) to several metres long (e.g. *Dendrobium discolor*), resembling canes. A few grow into long reed like stems. Leaf bases form sheaths that completely envelope the stem. In the section *Formosae* (e.g. *Dendrobium infundibulum*), the sheaths and undersides of leaves are covered with fine short black hairs. Other species (e.g. *Dendrobium senile*), are covered with fine white hairs.

Dendrobiums have cane-like stems with racemose (clustered) flowers. Sepal bright, consists of 3 strands of lancet-shaped, tapered or rounded and sizes vary. Sepal middle sepallum called the dorsal or eyelids back. While the two sides called sepal or petal sepal lateral side. Sepal base spurs united to form a triangle. Petal of three strands with third petal below is part of a united and form the lips of flowers and petal shaped generally more rounded and bigger and smoother texture than the sepal. Lip (labellum) is the development of the third petal. In some species the size of the lips can be enlarged and more brightly coloured. Lips are split into 3 and the bottom one with the spur interest. Pollen numbered 4, arranged in small rostellum and 2nd round. Large size range, small, even very subtle. Polinia pale yellow to bright yellow, which appears on the monument bagain. Pistil normally located next to the monument is covered by the male genitalia. Fruits would lie in the base of the flower.

Xu *et al*. (2006) proposed the ABCDE model as a general model to explain flower development in diverse plant groups. To investigate the regulatory mechanisms underlying orchid flower development, they isolated candidates for A, B, C, D and E function genes from *Dendrobium crumenatum*. These include AP2-, PI/GLO-, AP3/DEF-, AG- and SEP-like genes. The expression profiles of these genes exhibited different patterns from their Arabidopsis orthologs in floral patterning. Functional studies showed that DcOPI and DcOAG1 could replace the functions of PI and AG in Arabidopsis, respectively. By using chimeric repressor silencing technology, DcOAP3A was found to be another putative B function gene. Yeast two-hybrid analysis demonstrated that DcOAP3A/B and DcOPI could form heterodimers. These heterodimers could further interact with DcOSEP to form higher protein complexes, similar to their orthologs in eudicots. There is partial conservation in the B and C function genes between Arabidopsis and orchid. However, gene duplication might have led to the divergence in gene expression and regulation, possibly followed by functional divergence, resulting in the unique floral ontogeny in orchids.

Mode of polliantion

Members of the Orchidaceae are well-known for their complex flowers. A particular type of floral morphology, colour pattern, fragrance or reward, all result in the attraction of a specific pollinator, and thus, the tendency for autogamy is reduced (Darwin, 1890, Pijl and Dodson, 1960 and Dressler, 1993). *Dendrobium* can be naturally pollinated by insects as are many other orchids (Arditti, 1992). A few species such as *D. antennatum* show cleistogomy and can self-pollinate in closed buds (Kamemoto *et al.*, 1999). In breeding with most *Dendrobium* species, no emasculation is required for cross-pollination as the plants cannot physically self-pollinate. In pollinator-free environments and under stressful growing conditions, the facultative process of self-pollination occurs at the end of anthesis (Bonatti *et al.*, 2006). Self-pollination mechanisms have been confirmed for two species of *Dendrobium* Sw. (Catling, 1990). In *D. chryseum* Rolfe and *D. mastersianum* F. Muell. and Kraenzl, entire pollinia, owing to the close proximity of anther and stigma were observed to fall or slide down onto the stigma thus resulting in self-pollination.

Anthesis

In dendrobium varieties Candy Stripe, Pink Tips and Emma White, flower opening started from 7:30 am to 11:30 am, whereas the flowers of Sabine, Sonia-28, Pramott-11, New Pink and Hieng Beauty started opening around 8:00 am to 8:30 and lasted upto 11:30 am. In general, anthesis in dendrobium varieties occurs during the day time with a peak between 9:00 am and 11:00 am (Sobhana and Rajeevan, 2004). Varghese (1995) also reported a peak time of flower opening between 9:00 am and 10:00 am in dendrobium varieties. It took 1-2 days to the complete opening of each flower.

Stigma receptivity

Pollen of *Dendrobium phalaenopsis* retained its viability for 4-6 months, whereas those of *D. undulatum* remained viable for more than 12 months at 7.2°C. Stigma receptivity continues throughout the day after anthesis, but the total of stigma receptivity varied in different orchids. Stigma remains receptive throughout the day after anthesis in *D. amoenum*, *Spathoglottis plicata* and *Aerides odoratum* (Devi and Deka, 1992). They also observed that the stigma receptivity was higher upto five day after anthesis in *D. amoenum*. Although the flowers retained freshness for a longer period. Stigma receptivity was observed maximum for New Pink (3rd to 10th day after anthesis) followed by Pink Tips (2nd to 8th day after anthesis) of *Dendrobium* (Sobhana and Rajeevan,

2004). In other genera such as *Spathoglottis*, *Aerides* and *Phaius*, stigma will remain receptive for three, four and eleven days, respectively.

Seed maturity

After fertilization, the petals become faded and the upper part of the column start to thicken and this ultimately develop into capsule. Orchid seeds are minute like fine powder. The number of seeds vary from 1,300 to 40,00,000 per capsule. Seeds are without endosperm and lose their viability very fast. Under natural conditions a small part of them germinate on humus, if all the other conditions are favourable. The capsule dehisces at maturity, scattering the seeds by wind.

GENETICS

The basic and 2n chromosomal number of *Dendrobium* is 19 and 38, respectively but some species shows 2n = 40, 76 and 114. Bunnag and Sipoh (2010) compared chromosome number, anatomy and expression of antisense ACC oxidase gene between non transgenic and transgenic *Dendrobium draconis* Rchb.f. obtained by Homkrajae (2005). It was found that the frequency of hygromycin-resistant *D. draconis* was 80% while 20% were escape plants. One hundred percent of the hygromycin-resistant plants were viable when further cultured on the same medium for eight months. GUS assay used to verify transient gene expression in transgenic *D. draconis* showed positive results. An anatomical study in leaves, stems and roots of transgenic plants showed positive blue colouration in all tissues while those of non transgenic plants showed negative results. PCR analysis confirmed the integration of 35S promoter, NOS terminator and antisense ACC oxidase gene in *D. draconis*. A cytological study revealed that there was no change in chromosome number (2n=38). Furthermore, both transgenic and nontransgenic orchids exhibited no significant difference in chlorophyll content.

Gene sequences

In 1993 about 64 genes from seven orchid genera have been published (Mudalige and Kuehnle, 2004). Over two-thirds (67%) of orchid genes were discovered between 1998 and 2001. Some countries active in the research include Singapore, Taiwan, Korea, Germany, and the United States. Only a few of sequenced genes are directly applicable to orchid production and improvement. These genes concern floral transition (2000, 2000a), flower colour (Mudalige and Kuehnle, 2002), and senescence (Yang *et al*., 1996).

Reproductive barriers

Interspecific crossing barriers are largely absent for many *Dendrobium* hybrid combination. The *Phalaenanthe*, *Ceratobium*, and *Eleutheroglossum* combinations are the most compatible, as represented by 50 to 100% viable progeny obtained (Kamemoto and Wilfret 1980). The diploid hybrid progeny resulting from intersectional *Dendrobium* crosses do have poor chromosome pairing in meiosis and reduced fertility. However, some functional unreduced gametes are produced during microsporogenesis such that offspring can be produced when the hybrids are used in further crosses. More normal meiosis and fertility are restored upon doubling of the chromosome number in the hybrids. Such amphidiploids become valuable for breeding programs. Hybrid endosperm-embryo incompatibility appears absent in *Dendrobium* due to insignificant endosperm formation (Kamemoto *et al.*, 1999a). Inbreeding depression, expressed as loss in vigour and pollen degeneration is observed with *D. phalaenopsis* and *D. bigibbum*, but not with other species such as *D. antennatum* (reviewed in Kamemoto *et al.*, 1999a). Moreover, several cycles of inbreeding and selection using an amphidiploid *D.* Jaquelyn Thomas proved useful to increase flower size and lighten colour (Bobisud and Kamemoto, 1982) and produced valuable stud plants when used subsequently in outcrosses to restore vigour.

Self incompatibility

In the genus *Dendrobium*, 72% of species were found to be self-incompatible (Johansen, 1990). This involved production of ethylene by the stigma resulting in premature senescence and abscission of the flower during pollen tube growth. A similar situation was observed in the Oncidiinae; 73% of species studied were found to be fully or partially self-incompatible (Charanasri and Kamemoto, 1977), again resulting in abscission. In contrast, pollen germination was found to be low and pollen tube growth slow. Self-incompatibility has been shown to result in extremely low levels of fruit set (Neiland and Wilcock, 1998). In field studies on *Dendrobium monophyllum*, Bartareau (1995) found that only 6.9% of the 71.5% of flowers that initiated capsule formation produced matured capsules.

Huda and Wilcock (2012) studied the timing and pattern of floral senescence and occurrence of fruit formation of self, geitonogamous and cross-pollination in 14 epiphytic and two terrestrial orchid species to determine their breeding system and assess the occurrence of floral abscission following pollinaria removal. Both pollination and pollinaria removal caused rapid floral senescence and the pattern and timing of the floral changes were the same in all treatments. Six *Dendrobium* species and *Pelatantheria insectifera* were self-incompatible (SI) and eight other species including one terrestrial species were self-compatible (SC). Capsules produced from outcrossing in four SC species, *Phalaenopsis cornu-cervi*, *Eria pubescens*, *Cleisostoma appendiculatum* and *Arundina graminifolia*, were larger and heavier than those produced after selfing. Reductions in flower life span following pollinaria removal were positively correlated with flower size and longevity of unpollinated flowers but not with position in the inflorescence or nature of the breeding system. Rapid flower senescence following pollinaria removal reported here suggests that it may be widespread in tropical species. The significant association of the response with size of flowers and inflorescences among the species studied suggests that the cost of flower maintenance outweighs the benefit of remaining open for female function after pollinaria have been removed. Both SC and SI species were found among tropical orchids, but variation in capsule size following self and cross-pollination indicates that there may be a reduction in seed production following selfing, even in SC species and that fruit formation alone should not be taken as reliable evidence of full self-compatibility.

The pollinia from five cultivars of dendrobium (Miss Teen, Karen, Sakura, Willie and Pompadour) were placed on the stigma of open flowers of dendrobium cv. Miss Teen (or Kenny). The ethylene production and post-pollination development of cross-pollinated flowers were monitored. The pollinated flowers showed two different development patterns depending upon pollination compatibility. The first group, which was pollinated with incompatible pollen from flowers of 'Karen' and 'Miss Teen' did not exhibit post-pollination symptoms and ovary growth. The second group which was pollinated with compatible pollen from flowers of 'Pompadour', 'Sakura' and 'Willie' showed premature perianth senescence and induced ovary growth within a day. Compatible pollination of 'Miss Teen' flowers resulted in an immediate burst in the production of ethylene and a rapid ovary growth followed by an increase in pollen tube length. In contrast, when 'Miss Teen' flowers were self-pollinated or pollinated with 'Karen' pollinia, these changes did not occur. In the pollinia of cvs. Pompadour, Sakura and Willie there were higher levels of auxin and 1-aminocyclopropane-1-carboxylic acicd (ACC) than in those of 'Miss Teen' and 'Karen'. Compatible pollination also resulted in higher ACC synthase and ACC oxidase activities within 12 h after pollination than after pollination with incompatible pollen. There was a low level of ACC oxidase activity in non-pollinated flowers and a higher ACC oxidaseactivity in the column plus pedicel than petal plus sepal and lip (Luangsuwalai *et al.,* 2013).

Heritability and genetic advance

Moniruzzaman *et al.* (2012) estimated genetic variability, genetic parameters and correlation coefficient among different flower components of different *Dendrobium* spp. highly significant variations were observed among the genotypes for all the characters. High broad sense heritability with high genetic advance was observed in spike length (94.00% and 98.29), rachis length (95.34% and 93.85), flower durability (94.00% and 89.00) and floret number (70.50% and 65.47) indicating additive gene action, suggesting the possibility of improvement of these traits through selection. Other characters exhibited moderate broad sense heritability with low genetic advance. The number of florets per spike recorded a highly significant association with flower durability, rachis length and spike length. Durability of flower had significant positive correlation with rachis length, flower size and spike length at both genotypic and phenotypic level. Highly significant positive correlation was also found between rachis length and spike length. However, the correlation study revealed that selection of parents should be done based on characters such as floret number, rachis length and flower durability which would be useful in a breeding program. The path analysis revealed that rachis length, durability of flower and number of flowers per spike had contributed maximum direct effects on flower yield, indicating the importance of these three characters as selection criteria for native *Dendrobium* orchid.

BREEDING OBJECTIVES

i. Colour improvement and correct blending of the colours in sepals, petals and lip.
ii. Development of round full form of sepals and petals with a minimum fenestration and twists.
iii. Long spikes with higher number of flowers per spike.
iv. Florets alignment must be such that all flowers look appealing during display.
v. Extension of blooming period.
vi. Higher vase life.
vii. Variety suitable as pot plants.
viii. Variety resistant to biotic and abiotic stresses (diseases, particularly to those caused by viruses).
ix. Development of fragrant variety.

BREEDING METHODS

Hybridization

Some species are in great demand by orchid lovers. This has resulted in numerous varieties and hybrids, such as the Noble Dendrobium (*D. nobile*) breeds, which have greatly extended the range of colours of the original plant from the Himalayas.

A regular peloria appeared in a *Dendrobium* hybrid registered by D'Bush Nursery of Australia as *D.* D'Bush Pansy in 1988. Unlike the typical lip of *Dendrobium*, the lip of *D.* D'Bush Pansy is similar to the 2 lateral sepals, and the flat petals and sepals resemble those of the pansy; hence the name D'Bush Pansy. A few seedlings of *D.* D'Bush Pansy were obtained from Australia through Bangkok Flowers, Inc. in 1989. *Dendrobium* D'Bush Pansy crossed to several *Dendrobium* plants with normal-lips produced offspring with a normal-lip, indicating that pansy-lip is recessive to normal-lip. Three F_1 offspring backcrossed to the pansy-lip parent produced 1:1 ratios, and one F_2 progeny segregated into 3 normal-lip: 1 pansy-lip. Thus, pansy-lip is controlled by a single recessive gene pair (Amore and Kamemoto, 1997). Several interspecific hybrids have been developed; some of them are given in Table 3.

Table 3: Varieties developed through interspecific hybridization.

S. No.	Interspecific hybrids	Parentage
1.	*Dendrobium* Neo-Hawaii (Registered by Walter Carter in 1945)	*D. phalaenopsis* × *D. grantii*
2.	*D.* Jaquelyn Thomas (Registered by Rolla K. Thomas in 1949).	*D. phalaenopsis* × *D. gouldii*
3.	*D.* Wiganiae	*D. nobile* × *D. signatum*
4.	*D.* ainsworthii	*D. aureum* × *D. nobile*
5.	*D.* cybele Oak wood	*D. nobile* 'Burfordiense' × *D. findlayanum*
6.	Snow King	*Dendrobium* Lovely Virgin 'Angel' × *Dendrobium* Cassiope 'Seto'

Breeding for flower colour

Anthocyanidins were identified in 28 *Dendrobium* species and hybrids selected for analysis based on colour and suitability in cut flower breeding. Flowers designated pink, red, maroon, orange, bronze, and brown in the trade were placed in RHS colour groups red-purple, purple-violet, violet on yellow, greyed-purple on yellow or yellow-orange, and brown. This colour range contained anthocyanins based on cyanidin, with peonidin occurring as a minor pigment. The colours of three blue genotypes, *D. gouldii* K280-6, *D. biggibum* 'blue', and D. Kultana 'blue', were light violet to purple by RHS standards and contained anthocyanins based on cyanidin. Peach-coloured flowers were classified as red or red-purple and included pelargonidin glycosides. Anthocyanin concentrations ranged from 0.13 to 0.18 moles/g FW in light lavender and peach, and up to 3.66 moles/g FW in brown. Combined cellular and vacuolar pH ranged narrowly from 4.67 to 5.09 among white, peach, lavender, and brown lines. Predominant copigments were flavonol glycosides based on kaempferol, quercetin, myricetin, and methylated derivatives. Flavonol aglycones and glycosylation sites differed little among two colour forms of D. gouldii and two D. Jaquelyn Thomas hybrids. Accumulation of quercetin, myricetin and cyanidin indicated flavonoid 3' and 3',5' hydroxylation activities in several Dendrobium. Additional accumulation of isorhamnetin, syringetin, and peonidin indicated active flavonoid 3'- and 3', 5'- O-methyltransferase enzymes (Kuehnle *et al.*, 1997).

Anthocyanin pigments comprised of cyanidin glycosides are the predominant pigments extracted from lavender and purple *Dendrobium*. These pigments are produced in the phenylpropanoid biosynthesis pathway. Key enzymes that make pigmented products are chalcone synthase (CHS) at the top of the pathway, followed by dihydroflavonol 4-reductase (DFR) and the flavonoid hydroxylases (F-3'H and F-3'5'H) that turn orange to lavender/purple and/or to blue by a simple chemical change. Genetic studies conclude that two complementary gene pairs, *C* and *R*, control expression of purple via duplicative recessive epistasis (Kamemoto and Amore, 1990). Genotypes *ccRR* and *CCrr* will comprise white such that selfing produces white progeny and crossing with a member of the other genotype produces colour. Thus, breeding for alba by crossing two white parents often results in non-white progeny due to complementary gene action. A dominant gene *P* is responsible for lip colour, or the semi-alba trait of white petals and sepals with a coloured labellum, based on crosses between semi-alba *D. dicuphum* and white *D. affine* or white *D. phalaenopsis* var. *compactum* 'Mauna Kea' (Kamemoto and Amore 1990). Semi-alba (*P_*) is dominant to alba (*pp*).

Breeding for fragrance

Many *Dendrobium* species have fairly strong, pleasant scents. Of 140 species evaluated, 40% produced scents ranging from floral to fruity to herbaceous (Kaiser, 1993). The mechanism of scent production has also been investigated (Arditti, 1992). However, little is known about fragrance genetics. Despite several commercial hybrids being scented, notably *D.* Jaquelyn Thomas hybrids, fragrance has not been marketed per se for *Dendrobium*, either as a cut flower or a potted plant. Nevertheless, fragrance is a desirable attribute that may play a more important role in future breeding. One parent of possible interest for breeding fragrant potted plants is *D. antennatum* (synonym *D. d'albertisii*), with its long lasting scented and attractive flowers (Kaiser 1993).

Breeding for miniaturization

Several species useful for producing compact hybrids with short pseudobulbs are shown in Table 4. Some members of the section *Formosae* (*Nigrohirsutae*) are also relatively short but bear short inflorescences of only two to five flowers (Kuehnle, 2007). Such varieties are suitable as potted plant.

Table 4: *Dendrobium* species and hybrids with short pseudobulbs for potted plant breeding.

S. No.	Species	Section	Pseudobulb height	Colour
1.	*D. carronii*	*Eleutheroglossum*	5 cm	Pink-purplish red
2.	*D. canaliculatum*	*Eleutheroglossum*	5 – 12 cm	White-yellow-brown
3.	*D. phalaenopsis* var. *Compactum*	*Phalaenanthe*	5 – 8 cm	Light purple
4.	*D. bigibbum* var. *Compactum*	*Phalaenanthe*	5 – 8 cm	Mauve
5.	*D.* 'Mini Gem'*	—	5 – 12 cm	Light lavender

**D. carronii* × *D. bigibbum* var. *compactum;* the amphidiploid is used in breeding.

Mutation

In vitro mutagenesis is one of the simplest ways to produce a wide range of new flower forms, types, colours and shades. Nirmala and Elangbam (2007) attempted *in vitro* mutation induction using the highly potent mutagen ethyl methane sulphonate (EMS) in a floriculturally important orchid *Dendrobium chrysotoxum* Lind. The mutagen affected survival rate brings about changes in growth habit, phenotype and induced chlorophyll mutations. The survival rate and chlorophyll mutation rate depended on the dose of the mutagen. The higher doses were lethal. Five chlorophyll variants - albina, chlorina, xantha, viridis and maculata were detected. Phenotypic variants with altered morphology such as stunted growth, prominent nodes and long leaves were also obtained. Since chlorophyll variation is a reliable and most sensitive index of gene mutation, EMS appears to be a potent mutagen for inducing mutation in *Dendrobium*.

Polyploidy

Dendrobiums that are polyploidy with more than two sets of chromosomes have emerged spontaneously just as in other groups of orchids- cattleyas, cymbidiums, paphiopedilums, phalaenopsises and vanda. Flowers from polyploids generally are fuller and of greater substance than are flowers from diploids, although they may be borne on slightly shorter racemes with fewer flowers per raceme. The first account of polyploidy in *Dendrobium* was by Ito and Mutsuura (1957) who reported six triploids and six tetraploids or near tetraploids mostly belonging to the Eugenanthe section. They pointed out the desirable horticultural characteristics associated with polyploidy. The first documentation of polyploidy in *Dendrobium* was by Ito and Mutsuura (1957). For potted plant breeding, using section *Phalaenanthe* is of interest due to the presence of overlapping, firm sepals and petals in polyploids. A liquid culture phase is generally used for induction of chromosome doubling. Protocorm-like bodies can be treated with autoclaved 0.1% (w/v) colchicine in orchid seed germination medium for 5, 7, and 10 days at 100 rpm under continuous light before transfer onto solid transflasking medium (Sanguthai *et al.*, 1973). Orchid cytogenetics or the study of orchid chromosomes using cytological techniques, was well underway by the 1940s and 1950s. This permitted documenting the value of polyploidy for show flowers, cut flowers and potted plants and enabled development of genome breeding as a strategy for commercial cultivar production (Kamemoto, 1987). Polyploids may also arise spontaneously during tissue culture or during early development of the embryo. In any case, polyploids among plantlets in flask can be recognized by vigorous growth and enhanced substance of leaves. They can be ascertained by counting chromosomes in root tips or pollen mother cells (Tanaka and Kamemoto, 1984) or by determination of nuclear DNA content by flow cytometry (Jones *et al.*, 1998).

Kosaki (1958) reported that two outstanding *D. phalaenopsis* plant were tetraploid, five hybrids were triplod and three hybrids, D. Lady Hamilton, D. Cummins and D. Ruby King were tetraploid. Kamemoto *et al.* (1961) recorded five additional triploids and three tetraploids. Vajrabhaya and Randolph (1960) found that of 25 commercial hybrids, 14 were diplod, 8 triploid or near triplod, two tetraploid and one near pentaploids. Subsequent investigations have uncovered several other polyploids in *Dendrobium* (Kamemoto *et al.*, 1972 and Tanaka and Kamemoto 1984). Polyploids are generally exhibit improved characteristics over their diploid counterparts. Crosses between two diploid dendrobiums have produced triploids and pentaploids (Vajrabhaya and Randolph 1960, Kosaki and Kamemoto 1961, Dorn and Kamemoto, 1962 and Kamemoto and Tara, 1968).

BIOTECHNOLOGY

Tissue culture

Kuehnle (1996) described tissue culture and genetic strategies developed to eliminate cymbidium mosaic potexvirus (CyMV) from *Dendrobium* cultivars and breeding plants. The possibility of virus transmission by seed from infected plants is negated by results showing no spread of virus from infected pod and pollen parents to over 7000 seedlings of seed-propagated cultivars. The possibility of virus transmission by pollinia from an infected pollen-source plant to an uninfected mother plant during pollination is considered remote. While pollinia from infected plants harboured the virus, mechanical inoculation of *Dendrobium* leaves with large numbers of macerated pollinia produced only infrequent systemic transmission of CyMV. Chemotherapy using the antiviral compounds dithiouracil or ribavirin was suitable for the recovery of CyMV-negative tissue and subsequently normal plants developed from protocorm-like bodies that were originally virus-positive. A genetic alternative to chemotherapy, androgenesis coupled with chromosome doubling by colchicine treatment, is reported as effective for ridding amphidiploid pollen-source plants of virus. The feasibility of introducing virus resistance genes into *Dendrobium* by genetic engineering is discussed and the genome of *Dendrobium* is shown to be of a size suitable for techniques used in molecular breeding.

Wang (2002) studied the shoot tip culture and propagation of two species of *Dendrobium* showed that MS medium supplemented with 0.5 mg BA/litre+0.5 mg NAA/litre + 1.0 mg CH (casein hydrolysate)/litre recorded the best induction from small buds. MS medium supplemented with 2.0 mg BA/litre + 0.5 mg NAA/litre + 1.0 mg CH/litre had the best increment effect on protocorm-like bodies. Half-strength MS medium supplemented with 0.3 mg IBA/litre+0.5 g activated carbon/litre promoted root growth. The detail regeneration study of dendrobium through different explants is given in Table 5.

Table 5: Regeneration of dendrobium through different explants.

S. No.	Plant species	Explants source	References
1.	*Dendrobium* sp.	Seedling	McConnell and Tomomitsu, 1983
2.	*D. densiflorum*	Node	Vij *et al.*, 2001
3.	*D. moniliforme*	ProtocormFloral bud	Lim *et al*., 1993Kishi and Takagi, 1997a
4.	*D. nobile*	Protocorm-seedling	Uesato, 1974
5.	*D. phalaenopsis*	Flower stalkStem node Seedlimg tipApical/axilary budShoot tip	Intuwong *et al.,* 1972Mosich *et al.*, 1974Gandawijaja, 1980Lim-Ho, 1981Sagawa and Kunisaki, 1982
6.	*D. phalaenopsis, D. antennatum*	Axillary bud	Kukulezanka and Wojciechowska, 1983
7.	*D. phalaenopsis* Banyan Pink	Germinated seed	Men *et al*., 2003
8.	*D. aduncum, D. loddigesii, D. transparens*	Stem node	Yam and Weatherhead, 1990
9.	*D. aphyllum, D. moschatum*	shoot	Nayak *et al*., 1997
10.	*D. candidum*	Seed/protocorm	Wang *et al.*, 1990
11.	*D.chrysanthum*	Pseudobub segment	Vij and Pathak, 1989
12.	*D. crumenatum*	Seedling leaf	Manorama *et al*., 1986
13.	*D. fimbriatum* var. *oculatum*	Shoot tip	Roy and Banerjee, 2003
14.	*D. formosum*	leaf	Nasiruddin *et al.,* 2003
15.	*D. linawianum*	Lateral bud	Nalawade *et al.,* 2003
16.	*D. macrostachyum*	node	Pyati *et al.*, 2002
17.	*D. moniliforme*	Floral bud	Lim *et al.*,1993Kishi and Takagi, 1997
18.	*D. moschatum*	Shoot stem disc	Kanjila *et al.*,1999
19.	*D. nobile*	PLB thin cross section	Nayak *et al*., 2002
20.	*D. wardianum*	Shoot apex	Sharma and Tendon, 1991
21.	*Dendrobium* Alice Spalding(*D. tokai* × *D. Undulatum*)	Leaf base, Axillary bud Leaf tip/baseApical/ axilllary bud	Fu, 1979, Irawati, 1978, Lay, 1978
22.	*Dendrobium* jaquelyn Thomas White	Shoot tip	Soediono, 1983
23.	*Dendrobium* Madame Pompadour	Apical meristem	Mujib and Jana, 1994
24.	*Dendrobium* Madame Thong-In	PLB thin section	Yu *et al*., 2001
25.	*Dendrobium* Miss Hawaii	Scape node	Nuraini and Shaib, 1992
26.	*Dendrobium* Multico White	Apical meristem	Hew and Mah, 1989
27.	*Dendrobium* Ng Eng Cheow	Shoot tip	Singh, 1976
28.	*Dendrobium* Sonia (*D. caesar* × *D. tomie* Drake)	Shoot tip/fractionated PLB	Saiprasad and Polisetty, 2003
29.	*Dendrobium* Sonia 17	PLB	Tee *et al*., 2003
30.	*Dendrobium* White Fairy 1	Plantlets	Lim *et al.*,1992

Zhang and Fang (2005) cultured the *D. candidum* apical meristem to produce protocorms and then the protocorms were cultured into seedlings. The best medium to induce the protocorms was 1/2 MS+0.3 mg 6-BA (benzyladenine)+0.1 mg kinetin (KT)+0.1 g LH l^{-1}, and the optimum medium to induce roots was MS+0.05 mg NAA+0.1 mg KT+2.0 mg folic acid+10 mg La l^{-1}. Anatomical examination of the vegetative organs and protocorms of *D. candidum* showed that the roots were composed of multi-epidermis, cortex and vascular cylinder their cortex was well developed and their vascular cylinder contained primary pentarch xylem and pith at its central part. The stem consisted of epidermis, ground tissue and irregularly distributed vascular bundle. The leaves were isolateral whose upperside and underside contained sclerenchyma. The protocorms developed from the embryonic globular embryos formed by the parenchyma cells behind apical meristem through dedifferentiation.

Bunnag and Hongthongkham (2012) determined the optimal concentrations of colchicine and incubation time of polyploidy induction in the *Dendrobium* hybrid 'Miss Singapore' through the tissue culture. *Dendrobium* 'Miss Singapore' protocorms were treated with colchicine at the concentration of 0, 0.01, 0.05 and 0.1% amended in the MS medium for 1, 2, 3 and 4 days at 25+or-2° C under dark condition. The results showed that the percentage of surviving plantlets decreased when the colchicine concentration and incubation time increased. The optimal treatment was a colchicine concentration of 0.01% and incubation time of 2 days. Only tetraploids were obtained under such condition. The chromosome number was $2n=4x=76$ as compared to the diploid control plantlets $2n=2x=38$.

Pradhan *et al.* (2013) carried out study for plant regeneration from shoot tip explants of *D. densiflorum* by tissue culture technique. The shoot tip explants of this species obtained from *in vitro* grown seedlings were cultured on Murashige and Skoog (MS) media alone and MS medium supplemented with different combination and concentration of 6-benzylaminopurine (BAP) and napthalenacetic acid (NAA). The maximum number of healthy shoot was observed on MS+BAP (2 mg/l) + NAA (0.5 mg/l) (4 shoots/culture). Among these, the most effective condition for *in vitro* rooting was observed on MS+ IBA (1.5 mg/l). The *in vitro* propagated plantlets were transferred in 2:1:1 ratio of coco peat, litter and clay containing earthen pot for acclimatization. About 85% plantlets were successfully acclimatized in the greenhouse.

Breeding for resistance

Tissue culture and genetic strategies developed to eliminate cymbidium mosaic potexvirus (CyMV) from *Dendrobium* cultivars and breeding plants are described. The possibility of virus transmission by seed from infected plants is negated by results showing no spread of virus from infected pod and pollen parents to over 7000 seedlings of seed-propagated cultivars. The possibility of virus transmission by pollinia from an infected pollen-source plant to an uninfected mother plant during pollination is considered remote. While pollinia from infected plants harboured the virus, mechanical inoculation of *Dendrobium* leaves with large numbers of macerated pollinia produced only infrequent systemic transmission of CyMV. Chemotherapy using the antiviral compounds dithiouracil or ribavirin was suitable for the recovery of CyMV-negative tissue and subsequently normal plants developed from protocorm-like bodies that were originally virus-positive. A genetic alternative to chemotherapy, androgenesis coupled with chromosome doubling by colchicine treatment is reported as effective for ridding amphidiploid pollen-source plants of virus. The feasibility of introducing virus resistance genes into *Dendrobium* by genetic engineering is discussed and the genome of *Dendrobium* is shown to be of a size suitable for techniques used in molecular breeding (Kuehnle, 1996).

In vitro flowering

Generally vegetative juvenile phase in various orchids is varying from 1-10 years, therefore *in vitro* flowering in orchids has attracted attention considerably. *In vitro* flowering has been reported in various species of *Dendrobium* (Wang *et al*., 1993, Duan and Yazawa, 1994 and Chia *et al*., 1999). Work on this aspect is carried out by these workers is given in Table 6.

Table 6: *In vitro* flowering studies in dendrobium.

S. No.	Plant species	Medium composition &pH	Reference
1.	*Dendrobium candidum*	MS + NAA, BA, polamines	Wang *et al.*, 1990
2.	*D. madame*	Solid + BA	Chia *et al.*, 1999
3.	*D. monilliforme*	VW + 5BA	Duan and Yazawa, 1994
4.	*D. sonia*	Solid + BA	Goh, 1996
5.	*Dendrobium* (*striolatum* × *monophyllum*)	Solid	Chia *et al.*,1999

Somaclonal variation

Many outstanding *Dendrobium* cultivars and hybrids came from conventional methods and somaclonal variation resulting from tissue culture techniques. Besides recombinant traits from hybridization, orchid somaclonal variances that have arisen from micropropagation procedures provide many novel genetic variations (Vajrabhaya, 1977). Patch colouration on petals is a novel trait in orchids. It was first found in a *Dendrobium* Ekapol 'Red' cut-flower farm in Nakhon Pathom, Thailand. The mutant was named *D.* Ekapol 'King Dragon'. Patch colouration on petals is a novel trait for cultivar improvement of the genus. The trait was obtained from somaclonal variances of both cut flower and potted plant cultivars. Most of the variance is male-sterile. Hybrids between the variants and the wild-type cultivars yielded some progenies with patch petals. Genetic analysis on the inheritance of some flower traits of the variance was performed (Wannakrairoj and Rattamanee, 2012). Only a cross between both parents with patch petals could be produced from *Dendrobium* Anucha Flare 'Suriyon' × *D*. Ekapol 'Annie', the only male-fertile mutant. From the segregating populations, it was proposed that a dominant *M* allele controls the pinkish purple colouration and a recessive *m* allele controls the white coloration on sepals. For petal colour, a dominant *C* allele was confi rmed to control the pinkish purple and a recessive *c* allele controls the white. Importantly, dominant alleles in five loci were found to be essential for the petal patch coloration. In addition, it was postulated that the *El* allele controls the enlarge labellum while the *el* allele controls the normal labellum.

Asymbiotic seed germination and the conventional vegetative method have been commonly used by growers to propagate these plants. To overcome somaclonal variation, which is commonly exhibited by *Dendrobium* (Nobile group) when micropropagated from protocorm-like bodies, a protocol for propagating *Dendrobium* Second Love *in vitro* using axillary buds in the presence of thidiazuron was developed (Ferreira *et al*., 2006). Random amplified polymorphic DNA analysis was also carried out to check for possible genetic alterations in plants originating from six consecutive subcultures. The results revealed that the established protocol was efficient for the *in vitro* cloning of this orchid hybrid and the plants obtained from the six subcultures did not exhibit any type of polymorphism.

Among the varieties listed below (Table. 7) some of them were introduced into China from Japan, Thailand and Taiwan. Whereas, another varieties were produced by Zhejiang Senhe Seed Company Limited through somaclonal variation and other ways.

Table 7: Some of the varieties along with their varietal characters.

S. No.	Cultivar	Origin	Cultivar property
1	Senlan No. 1	Taiwan	White flower, pink flower corner, pink flower eye.
2	Senlan No. 2	Taiwan	Strong pink flower, yellow flower center, pink flower eye.
3	Senlan No. 3	Taiwan	White flower, pink flower corner, white flower center, pink flower eye.
4	Senlan No. 4	Taiwan	Light pink flower, green flower center.
5	Senlan No. 5	Taiwan	Pink flower, white flower center, pink flower eye.
6	Senlan No. 6	Taiwan	White flower, pink flower corner, yellow flower center, orange red flower eye.
7	Senlan No. 7	Taiwan	Light pink flower, white flower center, purple flower eye.
8	Senlan No. 8	Taiwan	Light pink flower, light yellow flower center.
9	Pittero Gold Grace	Japan	Light yellow flower, green flower center.
10	Senlan No. 9	Unknown	White flower, pink laciness, green flower center.
11	Senlan No.10	Unknown	Pink flower, yellow flower center.
12	Pink Dell Elegame	Japan	Pinkish white flower, yellow flower center.
13	Senlan No. 11	Unknown	White flower, pink laciness, yellow flower center.
14	Senlan No. 12	Unknown	Light yellow flower, faint pink flower corner.
15	Senlan No. 13	Unknown	Light yellow flower.

(Contd.)

(*Contd. Table 7*)

S. No.	Cultivar	Origin	Cultivar property
16	Senlan No. 14	Unknown	Pink flower, pink laciness, yellow flower center.
17	Senlan No. 15	Unknown	Light yellow flower, faint red flower center.
18	Senlan No. 16	Unknown	Light yellow flower, pink laciness.
19	Senlan No. 17	Unknown	Pinkish white flower, pink laciness, yellow flower center.
20	Sanya	Unknown	Pinkish white flower, green flower center.
21	China Doll	Unknown	White flower, faint pink flower corner.
22	SnowBoy Romance	Thailand	Light pink flower, green flower center.
23	Senlan No. 18	Unknown	Yellow flower, red flower center.
24	Hamana Lake Maki	Thailand	Purple flower, red flower center.
25	White Rabbit Sakura	Thailand	White flower, yellow flower center.
26	Sakura Hime	Thailand	Pinkish white flower, green flower center, green flower center, mini plant shape.
27	Santana Canary	Thailand	Yellow flower.
28	Lai's Pearl Queen	Thailand	White flower, comparatively small plant shape.
29	Snowflake Otome	Japan	Pink flower, white flower center.
30	Snowflake Red Star	Japan	Pink flower, purple flower center.

Somatic embryogenesis

Zhan *et al*. (2005) studied the histological processes of plant regeneration from the root tips of *D. officinale*. The segments produced clusters of somatic embryos from totipotent calluses or directly from root apical meristems. The results suggest that protocorm-like bodies of *D. officinale* can be considered a true somatic embryos that originated from a single cell.

Chung *et al*. (2007) developed a protocol for induction of direct somatic embryogenesis, secondary embryogenesis and plant regeneration of *Dendrobium* cv. Chiengmai Pink. Thidiazuron (TDZ) at 0.3, 1 and 3 mg dm^{-3} induced 5-25% of leaf tip segments of *in vitro* grown plants to directly from embryos after 60 days of culture, and 1 mg dm^{-3} TDZ was the best treatment. Somatic embryos mostly formed from leaf surfaces near cut ends, and occasionally found on leaf tips. Higher frequency of embryogenesis was obtained in light than in darkness.

Molecular breeding

Genetic diversity was examined within and among nine natural populations of *Dendrobium officinale* using inter-simple sequence repeat (ISSR) and random amplified polymorphic DNA (RAPD) for conservation. Both molecular markers revealed a high percentage (>89%) of polymorphic bands and ISSR markers detected more diversity than RAPD markers. Analysis of molecular variance (AMOVA) revealed that 78.84% (ISSR) and 78.88% (RAPD) of variability was partitioned among individuals within populations. From the conservation point of view populations GL, GS and GSD with higher genetic diversity should be protected firstly to maintain the species potential for evolutionary change and population YG with lower diversity but representing a novel evolutionary unit should also be paid more attention to during *D. officinale* conservation practice (Ding *et al.*, 2009).

Ma and Yin (2009) investigated the genetic variability of five populations from xishuangbanna using inter-simple sequence repeat (ISSR). A total of 117 unambiguous and repetitious bands were obtained from 114 individuals of sampled populations using twelve primers selected. The results showed that *Dendrobium fimbriatum* exhibited a relatively low genetic diversity at population level. At species level, the value of the average percentage of polymorphic bands (*PPB*) was 89.74%, expected heterozygosity (*H*) was 0.3227, Shannon's Information index (H_{sp}) was 0.4779; while at population level, the value of the average percentage of polymorphic bands (*PPB*) was 23.93%, expected heterozygosity (*H*) was 0.0871, Shannon's Information index (Ho) was 0.1290. Most of genetic variation partitioned among populations (74.79%). The coefficient of gene differentiation among populations (G_{st}) was 0.7443 showed that genetic differentiation among population based on Nei's genetic distance had a high degree. Pairwise genetic identity (*I*) among populations ranged from 0.5882 to 0.8331. No significant correlation was found between geographical and genetic distance. Based on the findings in present study, we proposed that conservation management of *Dendrobium fimbriatum* should include protection of nature habitats occupied by all five wild populations and preservation of germplasm in botanical gardens from multiple seed sources to accelerate the gene flow.

The genome organization of *Dendrobium* species remains largely unexplored. An F_1 family derived from the controlled interspecific hybridization between *Dendrobium moniliforme* and *D. officinale* were used to create genetic linkage maps of parental *Dendrobiums* using double pseudo-testcross strategy. A total of 422 markers, including 66 EST-SSR, 126 SRAP, 74 ISSR, and 156 RAPD markers showed expected Mendelian segregation ratio and were used for genetic map construction. Genetic linkage groups with Kosambi genetic distances were constructed using a LOD score >=4 and a recombination threshold of 0.4. The map of *D. moniliforme* was 1127.9 cM in total length with 165 marker loci distributed in 17 linkage groups; and the *D. officinale* map consisted of 19 genetic linkage groups with a total length of 1210.9 cM positioned by 169 marker loci. Thus, the genetic maps constructed in this study can serve as a reference linkage map of *Dendrobium* species for efficient studies of comparative genetics, genetic mapping of traits with economic interest and marker-assisted selection in horticultural breeding (Lu *et al.*, 2012).

Linkage mapping

Two preliminary genetic linkage maps were constructed using 90 F_1 progeny individuals derived from an interspecific cross between *D. nobile* and *D. moniliforme* (both, $2n = 38$), using random amplified polymorphic DNA (RAPD) and intersimple sequence repeat (ISSR). A total of 286 RAPD loci and 68 ISSR loci were identified and used for genetic linkage analysis. Maps were constructed by double pseudo-testcross mapping strategy using the software Mapmaker/EXP ver. 3.0 and Kosambi map distances were constructed using a LOD score 4 and a recombination threshold of 0.4. Both maps showed 76.91% and 73.59% genome coverage for *D. nobile* and *D. moniliforme*, respectively. These primary maps provide an important basis for genetic studies and further medicinal and horticultural traits mapping and marker-assisted selection in *Dendrobium* breeding programmes (Feng *et al.*, 2013)

Genetic engineering

Genetics transformation of *Dendrobium secundum* mediated by *Agrobacterium tumefaciens* strain LBA 4404 which harbored the plasmid pCAMBIA 1305.1 containing antisense ACC oxidase, beta -glucuronidase (GUS) and hygromycin resistance (*hptII*) genes. The efficiency of transformation was 40 min co-cultivation of protocorms with *A. tumefaciens* that has been induced with 200 micro M acetosyringone. Atichart *et al.* (2007) determined the effect of antibiotic on protocorm growth. It was found that hygromycin at 25 mg L^{-1} was effective for transformant selection. The maximum concentration of cefotaxime that protocorm could tolerate was 500 mg L^{-1}. Hygromycin resistant protocorms showed histochemical blue staining due to GUS activity was 60%. Successful transformation was confirmed by PCR analysis. Breeding of *Dendrobium* orchids using sexual hybridization is limited in its ability to manipulate certain important traits, such as virus disease resistance, due primarily to a lack of suitable genetic variability. Genes with potential utility for orchid improvement are becoming available through recent genetic engineering. To use such genes in a breeding program an effective gene transfer procedure is desirable for *Dendrobium*. Stable transformed *Dendrobium* orchids were recovered from protocorms of some Den. × Jaquelyn Thomas hybrid crosses treated by microprojectile bombardment. The plasmid DNA used in these treatments was the *Agrobacterium* binary vector p GA482GG/cpPRV-4 (Fitch *et al.*, 1990) carrying the *kan* gene for resistance to the antibiotic kanamycin. The plasmid also carried chimeric genes encoding the coat protein of papaya ringspot virus (*prv*) and the reporter protein beta -glucuronidase, both driven by the CaMV 35S promoter. Approximately 70 protocorms from each of four different crosses were bombarded three times using a procedure described previously (Klein *et al.*, 1987). Transformed plantlets were obtained by culture of tissue on hormone-free half-strength Murashige and Skoog medium containing first 50 mg l^{-1}, then 100 mg l^{-1}, kanamycin sulfate. These results demonstrate that microprojectile bombardment is suitable for the transfer of engineered genes, including potential virus resistance genes, into *Dendrobium* (Kuehnle and Sugii, 2008).

Chemical analysis shows that DFR are both F- 3'H and F-3'5' H are active in lavender *Dendrobium*, due to accumulation of myrcetin and syringetin as well as cyanidin (Kuehnle *et al.*, 1997). DFR is expressed predominantly in buds (Mudalige and Kuehnle, 2002) and results in accumulation of the pigment through all layers of the epidermis and into the mesophyll of the petal (Mudalige *et al.*, 1999).). Yellows and greens in *Dendrobium* are due to carotenoids and chlorophylls (Thammasiri et al. 1986). Triploidy contributes to improved yellow cultivars. These plants themselves are poor breeders and efforts must be made to produce tetraploid plants. One strategy was described in Kamemoto *et al.* (1999a) in which a triploid yellow (*D.* Mary Mak) was crossed to a diploid (*D. helix*). Several tetraploids were recovered among a population of mostly aneuploids for use in subsequent breeding.

REFERENCES

Amore, T.D., Kamemoto, H. and Lindleyana. 1997. Inheritance of pansy-lip in *Dendrobium*. **12**(1): 12-15.

Arditti, J. 1992. Fundamentals of orchid biology. John Wiley & Sons, New York.

Atichart, P., Bunnag, S. and Theerakulpisut, P. 2007. Agrobacterium-mediated transformation of *Dendrobium secundum* (Bl.) Lindl with antisense ACC oxidase. *Asian J. Plant Sciences*, **6**(7):1065-71.

Bartareau, T. 1995. Pollination limitation, costs of capsule production and the capsule-to-flower ratio in *Dendrobium monophyllum* F. Muell. (Orchidaceae). *Australian J. Ecology,* **20**: 257-65.

Bobisud C.A. and H. Kamemoto. 1982. Selection and inbreeding in amphidiploids *Dendrobium* (Orchidaceae). *J. Amer. Soc. Hort. Sci.*, **107**: 1024-27.

Bonatti, P.M., Sgarbi, E. and Del Prete, C. 2006. Gynostemium mi-cromorphology and pollination in Epipactis microphylla (Orchidaceae). *J. Pl. Res.*, **119**(5): 431–37.

Bunnag, S. and Sipoh, T. 2010. Genetics, anatomy and expression of antisense ACC oxidase gene in transgenic *Dendrobium draconis* Rchb.f. *AAB Bioflux*. **2**(3): 259-64.

Catling, P.M. 1990. Auto-pollination in the Orchidaceae. *In:* J. Arditti (ed.). Orchid biology: reviews and perspectives. Portland. Timber Press. pp. 121-58.

Changkija, S., Kumar, Y. and Gurung, P.B. 1992. Orchids of Nagaland. Forest Department, Govt. of Nagaland.

Charanasri, U. and Kamemoto, H. 1977. Self incompatibility in the Oncidium alliance. *Hawaii Orchid J.*, **6**: 12-15.

Chia, T.F., Arditti, J., Segeren, M.I. and Hew, C.S. 1999. Review: *In vitro* flowering of orchids. *Lindleyana,* **14**: 60-76.

Chung, H.H., Chen, J.T. and Chang, W.C. 2007. Plant regeneration through direct somatic embryogenesis from leaf explants of Dendrobium. *Biologia Plantarum*, **51**(2): 346-50.

Darwin, C. 1890. The various contrivances by which orchids are fertilized by insects. London: John Murray.

Devi, J. and Deka, P.C. 1992. Pollen viability, stigma receptivity and cross compatibility of some Indian orchids. *J. Orchid Soc.*, **6**(1-2): 79-84.

Ding, G., Li, X., Ding, X. and Qian, L. 2009. Genetic diversity across natural populations of *Dendrobium officinale*, the endangered medicinal herb endemic to China, revealed by ISSR and RAPD markers. *Russian J. Genetics*, **45**(3): 327-34.

Dorn, E. and Kamemoto, H. 1962. Chromosome transmission of Dendrobium phalaenopsis Lyons Light No. 1. *Amer. Orchid Soc. Bull.*, **32**(12): 997-1006.

Dressler, R.L. 1993. Phylogeny and classification of the orchid family. Dioscorides Press Portland.

Duan, J.X. and Yazawa, S. 1994. *In vitro* flowering of *Doriella*, *Phalaenopsis* and *Dendrobium. Proce*: Nagoyo International Orchid Show, pp: 87-96.

Ferreira, W.D., Kerbauy, G.B. and Costa, A.P.P. 2006. Micropropagation and genetic stability of a *Dendrobium* hybrid (*Orchidaceae*). *In Vitro Cellular and Developmental Biology - Plant*. **42**(6): 568-71.

Feng, S., Zhao, H., Lu, J., Liu, J., Shen, B. and Wang, H. 2013. Preliminary genetic linkage maps of Chinese herb *Dendrobium nobile* and *D. moniliforme*. *J. Genet.* **92**(2): 205-12.

Fitch, M.M.M., Manshardt, R.M., Gonsalves, D., Slightom, J.L. and Sanford, J.C. 1990. Stable transformation of papaya via microprojectile bombardment. *Plant Cell Reports*, **9**: 189-94.

Fu, F.M.L. 1979. Studies on the tissue culture of orchids. 2. Clonal propagation of *Aranda, Ascocenda* and *Cattleya* by lateral buds from flower stems. *The Orchid Review*, **87**: 343-346.

Gandawijaja, D. 1980. Effects of nitrate and ammonium on growth of tissue culture of *Dendrobium phalaenopsis* Fitzg. *Annals bogoriensis,* **7**: 63-69.

Goh, C.J. 1996. Production of flowering orchid seedlings and plantlets. *Malaysian Orchid Review,* **30**: 27-29.

Hew, C.S. and Mah, T.C. 1989. Sugar uptake and invertase activity in *Dendrobium* tissues. *New Phytologist,* **111**: 167-171.

Homkrajae, B. 2005 Transformation of Dendrobium draconis Rchb.f. and D. MissSingapore with antisense ACC oxidase. M.Sc. thesis. Khon Kaen University.

Huda, M.K. and Wilcock, C.C. 2012. Rapid floral senescence following male function and breeding systems of some tropical orchids. *Plant Biol* (*Stuttg*), **14**(2): 278-84.

Intuwong, O., Kunisaki, J.T. and Sagawa, Y. 1972. Vegetative propagation of flower stalk cuttings. *American Orchid Society Bulletin,* **41**: 13-18.

Irawati, 1978. The influence of sucrose on tissue cultures of *Oncidium* Goldiana, *Dendrobium* Alice Spalding and *Aranthera* Beatrice N.G. *Gardener's Bulletin Singapore,* **31**: 217-221.

Ito, I. and Mutsuura. S. 1957. Chromosome numbers in Dendrobium. *Japan Orchid Soc. Bull.*, **5**: 1-2.

Johansen, B. 1990. Incompatibility in Dendrobium (Orchidaceae). *Botanical J. Linnean Society,* **103**(2): 165-96.

Johnson, S.D. and Nilsson, L.A. 1999. Pollen carryover, geitonogamy and the evolution of deceptive pollination systems in orchids. *Ecology*, **80**(8): 2607-2619.

Jones W.E., Kuehnle A.R. and K. Arumuganathan. 1998. Nuclear DNA content of 26 orchid (Orchidaceae) genera with emphasis on *Dendrobium*. *Ann. Bot.*, **82**: 189-95.

Kamemoto H. 1987. Four decades of research on orchid cytogenetics and breeding. *Proce:* 12th World Orchid Conference. Tokyo, pp. 59-73.

Kamemoto, H. and G.J. Wilfret. 1980. Inter- and multi-sectional *Dendrobium* hybrids. *Proc*: 9th World Orchid Conf. Bangkok, pp 255-261.

Kamemoto, H. and T.D. Amore. 1990. Inheritance of semi-alba and alba in *Dendrobium*. *Proc*: 13th World Orchid Conference. 13th World Orchid Conference Proceedings Trust, Auckland, pp. 242-244.

Kamemoto, H., Amore T.D. and A.R. Kuehnle. 1999. Breeding *Dendrobium* Orchids in Hawaii. University of Hawaii Press, Honolulu, pp. 166.

Kamemoto, H. and Shindo, K. 1962. Genome relationships in interspecific and intergeneric hybrids of Renanthera. *Amer. Jour. Bot.*, **49**: 737-48.

Kamemoto, H. and Tara, M. 1968. Chromosome Inheritance in Reciprocal Crosses of Dendrobium-Phalaenopsis-M Lyons Light No 1 Dendrobium-Dalbertsii-M Dendrobium-Stratiotes-M. *American Society for Horticultural Science*, **92**: 665-71.

Kamemoto, H., Amore, T. D. and Kuehnle, A. R. 1999. Breeding *Dendrobium* orchids in Hawaii. Breeding *Dendrobium* orchids in Hawaii. Vol. IX pp-166.

Kamemoto, H., Meeyot, W. and Takeshita, M. 1972. Breeding behaviour of polyploid Dendrobium orchids. *J. Am. Sot. Hort. Sci.*, **97**: 3-5.

Kanjilal, B., Sarkar, D., Mitra, J. and Datta, K.B. 1999. Stem disc culture: Development of rapid mass propagation method for *Dendrobium moschatum* (Buch-Ham) Swartz: An endangered orchid. *Current Science,* **77**: 497-500.

Kishi, F. and Takagi, K. 1997. Efficient method of preservation and regeneration of orchid protocorm-like bodies. *Scientia Horticulturae,* **68**: 149-156.

Kishi, F., Hoshini, K, Takagi, K. 1997. Low frequency of somaclonal variation in floral bud culture-derived plants of *Vandofinetia* (Orchidaceae). *Plant Biotechnology,* **14**: 67-69.

Klein, T.M., Wolf, E.D. and Sanford, J.C. 1987. High velocity microproctiles for delivering nucleic acids into living cells. *Nature,* **327**: 70-73.

Komemoto, H., Tanaka, R. and Kosaki, K. 1961. Chromosome numbers of orchids in Hawaii. *Univ. Hawaii Agr. Exp. Sta. Bull.* **127**: 32.

Kosaki, K. 1958. Preliminary investigations on the cytogenetics of Dendrobium. *Proc:* Second World Orchid Conf., pp. 25-29.

Kosaki, K. and Kamemoto, H. 1961. Chromosomes of some Dendrobium species and hybrids. *Na Pua Okika Hawaii Nei.,* **11**: 75-86.

Kuehnle, A. R. 1996. Towards control and eradication of virus in *Dendrobium* by hybridization, tissue culture, and genetic engineering. *J. Orchid Soc. India,* **10**(1/2): 43-51.

Kuehnle, A.R. and Sugii, N. 1992. Transformation of *Dendrobium* orchid using particle bombardment of protocorms. *Plant Cell Rep*., **11**: 484–88.

Kukulczanka, K. and Wojciechowska, U. 1983. Propagation of two *Dendrobium* species by *in vitro* culture. *Acta Horticulturae,* **131**: 105-110.

Lay, F.M.M. 1978. Clonal propagation of *Aranda, Ascocenda, Cattleya* by leaf tissue culture. *Gardener's Bulletin Singapore,* **31**: 132-138.

Lim, H.C., Park, N.B., Choi, D.C., Jin, S.G., Park, K.H. and Choi, B.J. 1993. Studies on micropropagation *in vitro* and hardening culture in *Dendrobium moniliforme. RDA J. Agril. Sci. Biotech.,* **35**: 221-225.

Lim, L.Y., Hew, Y.C., Wong, S.C. and Hew, C.S. 1992. Effects of light intensity, sugar and CO_2 concentration on growth and mineral uptake of *Dendrobium* plantlets. *J. Hortil. Sci.,* **67**: 601-611.

Lim-Ho, C.L. 1981. Experimental findings of the tissue culture of orchid hybrids at the Singapore Botanical Gardens. *Garden's Bulletin,* **34**: 148-160.

Lu, J., Zhao, H., Suo, N., Wang, S., Shen, B., Wang, H. and Liu, J. 2012. Genetic linkage maps of *Dendrobium moniliforme* and *D. officinale* based on EST-SSR, SRAP, ISSR and RAPD markers. *Scientia Horticulturae*. **137**: 1-10.

Luangsuwalai, K., Paull, R.E. and Ketsa, S. 2013. Compatible and incompatible pollination and the senescence and ovary growth of Dendrobium flowers. *European J. Envirl. Sci.,* **3**(1): 35-42.

Ma, J. and Yin, S. 2009. Genetic diversity of *Dendrobium fimbriatum* (Orchidaceae), an endangered species, detected by inter-simple sequence repeat (ISSR). *Acta Botanica Yunnanica.* **31**(1): 35-41.

Manorama, P., Rao, A.N., Goh, C.J. and Loh, C.S. 1986. Leaf callus development in *Aranda* and *Dendrobium. Proce*: 5th ASEAN orchid Congress, pp: 102-109.

McConnell, J. and Tomomitsu, H. 1983. Kappa carrageenan as an agar substitue for transflasking *Dendrobium* seedlings. *American Orchid Society Bulletin,* **52**: 1174-1177.

Men, S., Ming, X., Liu, R., Wei, C. and Li, Y. 2003. *Agrobacterium*-mediated genetic transfer of a *Dendrobium* orchid. *Plant Cell, Tissue and Organ Culture,* **75**: 63-71.

Moniruzzaman, M., Zaman, M.A., Hossain, M.E., Bhuiyan, M.M.H. and Rahman, M.Z. 2012. Genetic variability and character association in some native orchid species (*Dendrobium* spp.). *The Agriculturists*. **10**(1): 1-9.

Mosich, S.K., Ball, E.A. and Arditti, J. 1974. Clonal propagation of orchids by means of node cultures. *American Orchid Society Bulletin,* **57**: 949-958.

Mudalige, R.G. and Kuehnle, A.R. 2002. Floral flavonoid genes in orchids. *In:* Plant, Animal and Microbe Genome Conferences. Final Abstracts Guide. Applied Biosystems, Foster City CA, pp. 59.

Mudalige R.G. and A.R. Kuehnle. 2004. Orchid biotechnology in production and improvement. *HortSci.*

Mujib, A. and Jana, B.K. 1994. Clonal propagation of *Dendrobium* Madame Pompadour through apical meristem culture. *Advances in Plant Science,* **7**: 340-346.

Nalawade, S.M., Sagare, A.P., Lee, C.Y., Kao, C.L. and Tsay, H.S. 2003. Studies on the tissue culture of Chinese medicinal plant resources in Taiwan and their sustainable utilization. *Botanical Bulletin of Academia Sinica,* **44**: 79-98.

Nasiruddin, K.M., Begum, R. and Yasmin, S. 2003. Protocorm-like bodies and plantlet regeneration from *Dendrobium formosum* leaf callus. *Asian J. Pl.Sci.,* **2**: 955-57.

Nayak, N.R., Rath, S.P. and Patnaik, P. 1997. *In vitro* propagation of three orchids, *Cymbidium aloifolium* (L.) Sw., *Dendrobium aphyllum* and *Dendrobium moschatum* (Buch-Ham) Sw. through thidiazuron-induced high frequency shoot proliferation. *Scientia Horticulturae*, **71**: 243-50.

Nayak, N.R., Sahoo, S., Patnaik, S. and Rath, S.P. 2002. Establishment of thin cross-section (TCS) culture method for rapid propagation of *Cymbidium aloifolium* (L.) Sw. and *Dendrobium nobile* Lindl. (Orchidaceae). *Scientia Horticulturae,* **94**: 107-116.

Neiland, M.R.M. and Wilcock, C.C. 1998. Fruit set, nectar reward, and rarity in the Orchidaceae. *American J. Bot.*, **85**: 1657-71.

Nirmala, C. and Elangbam, D. 2007. Effect of the mutagen ems on a commercially important orchid *Dendrobium chrysotoxum* Lind. *Phytomorphology*, **57**(1/2): 51-57.

Nuraini, I. and Shaib, M.J. 1992. Micropropagation of orchid using scape nodes as the explant material. *Acta Horticulturae,* **292**: 169-72.

Pijl, L. Dodson, C.H. 1966. Orchid flowers: their pollination and evolution. Coral Gables FL: University of Miami Press.

Pradhan, S., Paudel, Y.P. and Pant, B. 2013. Efficient regeneration of plants from shoot tip explants of *Dendrobium densiflorum* Lindl., a medicinal orchid. *African J. Biotech.,* **12**(12): 1378-83.

Pyati, A.N., Murthy, H.N., Hahn, E.J. and Paek, K.Y. 2002. *In vitro* propagation of *Dendrobium macrostachyum* Lindl. a threatened orchid. *Indian J. Experimental Bio.,* **40**: 620-623.

Roy, J. and Bannerjee, N. 2003 Induction of callus and plant regeneration from shoot-tip explants of *Dendrobium fimbriatum* Lindl. var. *oculatum* Hk. *f. Scientia Horticulturae,* **97**: 333-40.

Sagawa, Y. and Kunisaki, J.T. 1982. Clonal propagation of orchids by tissue culture. *Proce:* 5th Congress on Plant Tissue and Cell Culture, pp: 683-84.

Saiprasad, G.V.S. and Polisetty, R. 2003. Propagation of three orchid genera using encapsulated protocorm-like bodies. *In-vitro Cellular and Developmental Biology-Plant*. **3**: 42-48.

Saiprasad, G.V.S., Polisetty, R. and Raj, A. 2003. Effects of growth regulators on the production of PLBs and multiple shoots in orchid: *Dendrobium* Sonia: Assessment of role of methane and ethylene. *Phytomorphology,* **53**: 63-71.

Sanguthai O., Sanguthai S. and H. Kamemoto. 1973. Chromosome doubling of a dendrobium hybrid with colchicine in meristem culture. *Na Okika o Hawaii* **2**: 12-16.

Sharma, A. and Tandon, P. 1991. Nutritional studies on *in vitro* raised protocorms of endangered dendrobe; *Dendrobium wardianum* Warner. *Indian Botanical Contractor,* **8**: 75-78.

Singh, H. 1976. Meristem tissue culture of *Dendrobium* Ng Eng Cheow. *Garden's Bulletin,* **28**: 259-267.

Sobhana, A. and Rajeevan, P.K. 2004. Floral biology of a commercially important *Dendrobium*. *J. Orna. Hort.*, **7**(1): 115-117.

Soediono, N. 1983. Use of coconut water, NAA, 2,4-D and vitamins in shoot tips culture of *Dendrobium* cv Jaqueline Thomas White. *The Orchid Review*, **91**: 86-87.

Tanaka R. and H. Kamemoto. 1984. Chromosomes in orchids: counting and numbers. *In:* J. Arditti (ed.), Orchid Biology: Reviews and Perspectives, III. Cornell University Press, Ithaca, New York, pp. 323-410.

Tanaka, R. and Kamemoto, H. 1984. Chromosomes in orchids: counting and numbers. *In:* J. Arditti (ed.). Orchid Biology-Reviews and Perpestives, Vol. III. Cornell University Press, Ithaca, NewYork. pp. 324-10.

Tee, C.S., Marziah, M., Tan, C.S. and Abdullah, M.P. 2003. Evaluation of different promoters driving the GFP reporter gene and selected target tissues for particule bombardment of *Dendrobium* Sonia 17. *Plant Cell Reports,* **21**: 452-58.

Useta, K. 1974. Effects of different sources of nitrogen sources in the culture media on the growth of *Dendrobium nobile* seedlings. Scientific Reports of Agriculture, Ryukyu University, **21**: 73-81.

Vajrabhaya, T. 1977. Variations in clonal propagation. *In*: J. Arditti, (ed.). Orchid Biology. Reviews and Perspectives, I. Cornell University Press. London, UK., pp. 177–01.

Vajrabhaya, T. and Randolph, L.F. 1960. Chromosome studies in Dendrobium. *Amer. Orchid Soc. Bull.*, **29**: 507-17.

Varghese, S. 1995. Floral biology and Compatibility studies in Dendrobium. M.Sc. (Hort.) Thesis, Kerala Agricultural University, Thrissur, Kerala, India.

Vij, S.P. and Pathak, K.K. 1989. Micropropagation of *Dendrobium chrysanthum* Wall. through pseudobulb segments. *J. Orchid Soc. India,* **3**: 25-28.

Vij, S.P., Kaur, P. and Gupta, A. 2001. Synseeds and their utility in orchids: *Dendrobium densiflorum* Lindl. *Phytomorphology*, **51**: 159-165.

Wang, C. 2002. Study on tissue culture and quick propagation of *Dendrobium. J. Zhejiang Forestry Sci. Tech.,* **22**(2): 38-40.

Wang, G.Y., Xu, Z.H. Chia, T.P., Wong, P. and Chua, N.H. 1990. *In vitro* flowering of *Dendrobium candidum. In:* 13th World Orchid Conference, pp: 67.

Wannakrairoj, S. and Rattamanee, C. 2012. Inheritance of some Flower Traits in Patch Petal Dendrobium Orchids. *Kasetsart J. (Nat. Sci.)*., **46**: 515 – 21.

Xu, Y., Teo, L.L., Zhou, J., Kumar, P.P. and Yu, H. 2006. Floral Organ Identity Genes in the orchid *Dendrobium crumenatum. Plant J.* **46**(1): 54-68.

Yam, T.W. and Weatherhead, M.A. 1990. Nodal culture of several native orchids of Hong Kong. *Lindleyana,* **5**: 218-223.

Yang, X.H., Pua E.C. and C.J. Goh 1996. Isolation of a cDNA clone encoding 1- aminocyclopropane-1-carboxylate synthase from *Dendrobium crumenatum* PGR96-088 (Accession No. U64031). *Plant Physiol.*, **112**: 863.

Yu, H. and Goh, C.J. 2000. Differential gene expression during floral transition in an orchid hybrid *Dendrobium* Madame Thong-In. *Plant Cell Rep.*, **19**: 926-931.

Yu, H. and Goh, C.J. 2000a. Identification and characterization of three orchid MADS-box genes of the AP1/AGL9 subfamily during floral transition. *Plant Physiol.* **123**: 1325- 1336.

Yu, H., Yang, S.H. and Goh, J.H. 2001. *Agrobacterium*-mediated transformation of *Dendrobium* orchid with the 1 knox gene *DOH1. Plant Cell Reports,* **12**: 2143-2159.

Zhan, Z., Xu, C., Zhang, M. and Luo, Z. 2005. Plant regeneration through somatic embryogenesis on root-tip explants of Dendrobium officinale Kimuraet Miga. *J. Zhejiang University.* **31**(5): 579-80.

Zhang, Q. and Fang, Y. 2005. Tissue culture and *in vitro* seedling and protocorm-like body examination of *Dendrobium candidum. Acta Botanica Boreali-Occidentalia Sinica*. **25**(9): 1761-765.

Chapter – 33

Paphiopedilum

Paphiopedilum is a genus of the orchid subfamily Cypripedioideae and family Orchidaceae. It is derived from Paphos and ancient Greek word "*pedilon*" means "slipper". Paphiopedilums are often called "slipper orchids" because of their unique pouch. The pouch traps insects seeking nectar and to leave again they have to climb up past the staminode, behind which they collect or deposit pollinia. *Paphiopedilum* has a highly ornamental value being the treasure of Orchidaceae. They are easily grown as houseplants and their care is very similar to African Violets.

ORIGIN AND DISTRIBUTION

The genus name *Paphiopedilum* was established by Ernst Hugo Heinrich Pfitzer in 1886. Ironically, no paphiopedilum occurs on Cyprus- at least not as the genus is understood today. But it was long mixed up with its holarctic relative *Cypripedium*, which indeed grows in the Mediterranean region. *Paphiopedilum* was finally decided to be a valid taxon in 1959, but its use has become restricted to eastern Asian species. The genus is native to Indo-Malaysia (South China, Southeast Asia and the Pacific Islands) and India. *Paphiopedilum armeniacum* is found from China (west Yunnan province) to northern Myanmar. It occurs at 1,600 to 2,000 metres above sea level. It is found growing on limestone hills, in semi-shaded deciduous forest, usually on cliffs above rivers.

SPECIES

Over fifty species of *Paphiopedilum* have been discovered in the far east from the higher, cooler elevations of the Himalayas, across to China and Southeast Asia and down to the warmer, sea-level regions of Malaysia, Indonesia and the Philippines (Hawkes, 1965). Most species, like *Paphiopedilum callosum* are terrestrial in nature growing under the shaded and moist conditions of the forest floor (Fitch, 1982). Some species, such as *Paphiopedilum niveum* are lithophytic found in colonies on the coastal rock formations of the Malaysian islands (Fitch, 1982a). While a few of the larger strap-leaf *Paphiopedilum* species exposed to the higher light conditions found in their loftier habitats. *Paphiopedilum philippinense* var. *roebelenii* and *Paphiopedilum lowii* provide a strong contrast to such as *Paph. lowii,* are usually epiphytic *Paphiopedilum charlesworthii* previously described and serve to illustrate the extent of variation in both plant and flower habit within this genus. Unlike the single-flowered spikes of *Paphiopedilum charlesworthii,* which are usually around 15 cm tall, those of *Paph. philippinense* var. *roebelenii* and *Paph. lowii* can bear up to a half-dozen flowers and can measure 60 cm or more in length. Paphiopedilums with this kind of flowering habit are referred to as multifloral paphiopedilums (Batchelor). The plants themselves can be quite large particularly in the case of *Paphiopedilum lowii,* where individual leaves are often well over 30 cm long, though narrow.

The flowers of *Paphiopedilum philippinense* var. *roebelenii*, with their pendent, highly twisted petals measuring 13 cm long (Day, 1974) also contrast strikingly with those of *Paphiopedilum charlesworthii*. Some notable species are *P. bellatulum*, *P. callosum*, *P. concolor*, *P. fairrieanum*, *P. hirsutissimum*, *P. insigne*, *P. lawrenceanum*, *P. niveum*, *P. parishii*, *P. spicerianum*. *P. venustum* and *P. villosum*.

***P. hirsutissimum*:** Stem reduced, covered by leaves; leaves 6-12 in number, 30-40 cm long, linear-oblong, green rather fleshy; scape 1 flowered; flower about 7 cm across or more, greenish spotted with blackish-purple spots, hirsule. Flowering time: April-May.

***P. insigne*:** Stem absent; leaves radical, linear-ligulate, green; scape 1-flowered rarely 2-flowered; flowers about 12 cm across; dorsal sepal oval, apical part white irregularly brownish-purple dots; lateral sepals smaller, greenish; petals linear oblong, yellowish-green with purple veins; lip yellowish green with brownish shade. Flowering time: October-December.

FLOWER MORPHOLOGY AND BIOLOGY

Paphiopedilum species naturally occur among humus layers as terrestrials on the forest floor while a few are true epiphytes and some are lithophytes. These sympodial orchids lack pseudobulbs. Instead, they grow robust shoots each with several leaves some are hemicryptophytes. The leaves can be short and rounded or long and narrow and typically have a mottled pattern. When older shoots die, newer ones take over. Each new shoot only blooms once when it is fully grown, producing a raceme between the fleshy, succulent leaves. The roots are thick and fleshy. Members of this genus are considered highly collectible by orchid fanciers due to the curious and unusual form of their flowers. In the *Paphiopedilum* group, pollen is contained in two sticky pollen masses, one on each side of the sterile staminodium. This means every plant is unique. *Paphiopedilum armeniacum* has elongated rhizomes which grow up to 25 cm long, usually spreading through leaf litter or around rocks and forming large clumps. Plantlets occur up to 15 cm apart and each bears 5 to 7 leaves. The upper surface of the leaves is greenish-white with a dark green marbled pattern. The lower leaf surface is heavily spotted with irregular dark purple-red patches and the edges of the leaves are finely serrated. Various flower parts are given below.

1. **Sepals:** Although they may look like petals, they are actually the glorified remains of the flower bud. There are usually three of approximately equal sizes, but not always. The two lateral sepals are fused into one and the dorsal (top) sepal is greatly enlarged in the case of the paphiopedilum and others.
2. **Petals:** Orchids always have three petals. Two are normal and the third becomes a highly specialized structure called a lip.
3. **Lip or Labellum:** The lower petal of an orchid. Used by the flower to provide a landing platform (or a trap, in this case) for its pollinator.
4. **Column:** A finger-like structure that carries the orchid's reproductive organs—the stigmatic surface (female organs) and the pollinia (male organs) located under the anther cap (operculum).

Paphiopedilum delenatii is an endangered slipper orchid with delicate flowers that is native to Vietnam. An understanding of this plant species reproductive biology is required for its effective propagation and conservation. *P. delenatii* has a bisporic pattern of embryo sac development with a six-celled number in a mature embryo sac. Prior to pollination, the nucellar filaments are present within the ovary. The archesporial cell at the tip of the nucellar filament differentiates into a megasporocyte fifteen days after pollination. As the megasporocyte matures, the nucleus becomes located at the micropylar end of the cell, resulting in an uneven distribution of the cytoplasm with the starch granules becoming more abundant at the chalazal end. After the first meiotic division, a distinct callose layer separates the micropylar dyad from the chalazal one. Soon after, a callose wall completely encloses the functional dyad at the chalazal end. As the functional dyad increases its size, the callose remains in the common wall separating the two dyads as well as in the chalazal end wall of the functional dyad. One notable finding of this study is the presence of cuticular material in the lateral walls of the mature embryo sac. The deposition of the cuticular material may serve to strengthen the fairly large embryo sac and helps to conserve water for the developing embryo after fertilization (Lee and Yeung, 2012).

Mode of pollination

Rare calcicolous *Paphiopedilum thaianum* and *P. niveum* were studied (Banziger *et al.*, 2012). Flowers are rewardless kettle-traps the pouch-like lips keeping pollinators prisoner for many minutes. Since the flowers were unusual, being fragrant and conspicuously white they were compared with two typically malodorous and mottled/yellow species, *P. godefroyae* and *P. concolor*. Unexpectedly, melittophily was discovered in *P. thaianum* and *P. niveum*, the pollinators being the newly described halictid bee *Lasioglossum orchidodeceptum* and the meliponine bee *Tetragonula testaceitarsis*, respectively. Pollinators of *P. godefroyae* and *P. concolor* were, as expected, milesiine hoverflies, viz. *Eumerus nicobarensis* and *E. figurans*, respectively, confirming these two orchids as myiophilous like *Paphiopedilum bellatulum*. The five orchid species belong to the subgenus *Brachypetalum*, the second-most basal subgenus of *Paphiopedilum*. The most basal subgenus, *Parvisepalum*, is the closest relative of the mainly holarctic genus *Cypripedium*, in flower structure and in melittophily. Hence *P. thaianum* and *P. niveum* are proposed as the hitherto missing link between *Cypripedium* (via *Parvisepalum*) and the more derived *Paphiopedilum*, all myiophilous. The very similar *P. thaianum* and *P. niveum* were found to be reproductively isolated by enticing pollinators differing at family rank and by allowing pollen acquisition only by appropriately-sized bees, smaller in *P. thaianum*.

GENETICS

Cytological studies showed that probable number of chromosomes of Orchidaceae family is 20. *Paphiopedilum* is a horticulturally and ecologically important genus. About 80 species of lady's slipper orchids native to Southeast Asia. These plants have long been of interest regarding their chromosomal evolution, which involves a progressive aneuploid series based on either fission or fusion of centromeres. Chromosome number is positively correlated with genome size, so rearrangement processes must include either insertion or deletion of DNA segments. Lan and Albert (2011) conducted Fluorescence *in situ* hybridization (FISH) studies using 5S and 25S ribosomal DNA (rDNA) probes to survey for rearrangements, duplications and phylogenetically-correlated variation within *Paphiopedilum*. Results shows that 5S and 25S rDNA loci among *Paphiopedilum* species, representing all key phylogenetic lineages, exhibit a considerable diversity that correlates well with recognized evolutionary groups. 25S rDNA signals range from 2 (representing 1 locus) to 9, the latter representing hemizygosity. 5S loci display extensive structural variation and show from 2 specific signals to many, both major and minor and highly dispersed. The dispersed signals mainly occur at centromeric and subtelomeric positions, which are hotspots for chromosomal breakpoints. Phylogenetic analysis of cloned 5S rDNA non-transcribed spacer (5S-NTS) sequences showed evidence for both ancient and recent post-speciation duplication events, as well as interlocus and intralocus diversity. It was concluded that *Paphiopedilum* species display many chromosomal rearrangements like, duplications, translocations and inversions but only weak concerted evolutionary forces among highly duplicated 5S arrays which suggests that double-strand break repair processes are dynamic and ongoing. These results make the genus a model system for the study of complex chromosomal evolution in plants.

Sub-haploid pollen grain

Sub-haploid pollen grains with chromosome numbers ranging from n = 5-12 were seen in the cultivar 'Maulei' of *Paphiopedilum insigne* Pfitz. Many grains showed n = 5 and 6 (38%) while 3% showed a hyperploid number (n = 18, 20). It was also observed that pollen grains with the same chromosome number differed in karyotype. The high percentage (81%) of sub-haploid pollen grains may be due to the high degree of segregational irregularities during meiosis or to split spindle or failure of wall formation. It may also be due to changes in the environment of the species which may influence the metabolic activity responsible for the initiation and subsequent development of the cell wall and the process of chromosome duplication (Singh, 1985).

BREEDING OBJECTIVES

i. Novelty in colour.

ii. Small plant size and produce flowers in proportion to the plant's size.

iii. Resistance to diseases (root rot, viral infections).

iv. Resistance to abiotic stresses.

BREEDING METHODS

Selection

Selection from the collected orchid plants contributed much to the improvement of orchids. The yellow flowered forms of *Paphiopedilum insigne*, *P. insigne* var. *Sanderae* and var. *Sanderianum* which played significant role in the development of white and yellow flowers. They are both variant forms of the species and were observed for the first time among plants collected from wild habitats.

Hybridization

Lee *et al*. (2011) analysed the meiotic behaviour of interspecific hybrids by genomic *in situ* hybridization and cytologically evaluates the genomic relationships among parental species. Results shows that Genomic *in situ* hybridization was demonstrated as an efficient method to differentiate between *Paphiopedilum* genomes and to visualize the chromosome pairing affinities in interspecific F_1 hybrids, clarifying the phylogenetic distances among these species. Comparatively regular chromosome pairing observed in the hybrids of *P. delenatii* × *P. bellatulum*, *P. delenatii* × *P. rothschildianum* and *P.*

rothschildianum × *P. bellatulum* suggested high genomic affinities and close relationships between parents of each hybrid. In contrast, irregular chromosome associations, such as univalents, trivalents and quadrivalents occurred frequently in the hybrids derived from distant parents with divergent karyotypes, such as *P. delenatii* × *P. callosum*, *P. delenatii* × *P. glaucophyllum*, *P. rothschildianum* × *P. micranthum* and *P. rothschildianum* × *P. moquetteanum*. The existence of multivalents and autosynthesis demonstrated by genomic *in situ* hybridization indicates that some micro-rearrangements and other structural alterations may also play a part in differentiating *Paphiopedilum* species at chromosomal level, demonstrated as different chromosome pairing affinities in interspecific hybrids. Some hybrids developed through interspecific hybridization are given in Table 1.

Table 1: Hybrids developed through interspecific hybridization.

S. No.	Hybrids	Parental crosses
1.	*Paphiopedilum* Lecanum	*P. insigne* × *P. spicerianum*
2.	*P.* Arthurianum	*P. insigne* × *P. fairrieanum*
3.	*P.* Harrisianum	*P. villosum* × *P. barbatum*
4.	*P.* Aylingii	*P. niveum* × *P. ciliolare*
5.	*P.* Demura	*P. bellatulum* × *P.* Blendia
6.	*P.* Charles Richman	*P. bellatulum* × *P. barbatum*
7.	*P.* Lathamianum	*P. villosum* × *P. spicerianum*
8.	*P.* Winnifred Hollington	*P. niveum* × *P. callosum*
9.	*P.* Tsselatum	*P. barbatum* × *P. concolor*
10.	*P.* Callosum Wottonii	*P. bellatulum* × *P. callosum*
11.	*Paphiopedilum* × *wenshanense*	*P. bellatulum* × *P. concolor*, including *P.* × *conco-bellatulum*
12.	*Paphiopedilum* × *affine*	*P. appletonianum* × *P. villosum*
13.	*Paphiopedilum* × *areeanum*	*P. barbigerum* × *P. villosum* var. *Annamense*
14.	*Paphiopedilum* × *burbidgei*	*P. dayanum* × *P. javanicum* var. *Virens*
15.	*Paphiopedilum* × *dalatense*	*P. callosum* × *P. villosum var. Annamense*
16.	*Paphiopedilum* × *dixlerianum*	*P. callosum* × *P. wardii*
17.	*Paphiopedilum* × *expansum*	*P. hennisianum* × *P. philippinense*
18.	*Paphiopedilum* × *fanaticum*	*P. malipoense* × *P. micranthum*
19.	*Paphiopedilum* × *frankeanum*	*P. superbiens* × *P. tonsum*
20.	*Paphiopedilum* × *grussianum*	*P. dianthum* × *P. hirsutissimum* var. *Esquirolei*
21.	*Paphiopedilum* × *herrmannii*	*P. helenae* × *P. hirsutissimum* var. *Esquirolei*
22.	*Paphiopedilum* × *kimballianum*	*P. dayanum* × *P. rothschildianum*
23.	*Paphiopedilum* × *littleanum*	*P. dayanum* × *P. lawrenceanum*
24.	*Paphiopedilum* × *mattesii*	*P. barbatum* × *P. bullenianum*
25.	*Paphiopedilum* × *pereirae*	*P. exul* × *P. niveum*
26.	*Paphiopedilum* × *petchleungianum*	*P. dianthum* × *P. villosum*
27.	*Paphiopedilum* × *powellii*	*P. callosum* × *P. exul*
28.	*Paphiopedilum* × *pradhanii*	*P. fairrieanum* × *P. venustum*
29.	*Paphiopedilum* × *shipwayae*	*P. dayanum* × *P. hookerae*
30.	*Paphiopedilum* × *siamense*	*P. appletonianum* × *P. callosum*

Polyploidy

Numerous interspecific hybridizations have been made for improving horticultural qualities. However, the interspecific hybrids often present apparent sterility or low fertility, therefore cytological study is important to deepen our understanding for breeding works. In a study Lee and Chung (2010) reported the karyomorphology of 14 hybrids for the reference to breeding programs. The basic chromosome number of the genus *Paphiopedilum* is 2n=26. The parental species of *P.* Magic Lantern, *P.* Vanda M. Pearman, *P.* Woluwense and *P.* (Susan Booth × Saint Swithin) belong to the basic chromosome number groups, therefore the somatic chromosome number of each is 2n=26. In the sections Barbata and Cochlopetalum of subgenus *Paphiopedilum*, the species have diverse chromosome numbers, ranging from 2n=28 to 42 and 2n=30 to 37, and the progenies have a variety of chromosome numbers. In the present investigation, some hybrids are considered as the polyploidy.

BIOTECHNOLOGY

Tissue culture

Propagation of *Paphiopedilum* orchid genera has been hampered by the naturally slow growth rate of the plant, which renders it very difficult to be propagated through conventional methods. *In vitro* culture techniques have provided a useful alternative technology for propagating this recalcitrant species. The propagation of *P. rothschildianum* was achieved through the *in vitro* formation of secondary protocorm-like bodies (PLBs) from the primary PLB that developed from stem-derived callus. The PLBs were cultured on half-strength MS medium supplemented with different concentrations (1.0, 2.0, 3.0, and 4.0 micro M) of 6-benzyladenine (BA) and kinetin for the induction of secondary PLBs. The highest number of secondary PLBs formed was obtained on half-strength MS medium supplemented with 4.0 micro M kinetin, with an average of 4.1 PLBs per explant after 8 weeks of culture. The secondary PLBs continued to proliferate further and formed 9.5-12.1 new PLBs per secondary PLB after being subcultured onto half-strength plant growth regulator-free MS medium supplemented with 60 g/L banana homogenate (BH). These tertiary PLBs were subcultured onto media containing different organic additives, such as BH, coconut water, potato homogenate, and tomato homogenate for plantlet regeneration. Among the organic additives tested, the addition of 20% CW to half-strength MS medium resulted in the best average plantlet regeneration percentage from the PLBs, 67.9%, after 8 weeks of culture (Ng and Norihan, 2011).

Paphiopedilum species are recalcitrant in tissue culture and no explant from mature plants has been successfully mass propagated *in vitro*. Liao *et al.* (2011) aimed at inducing shoots and regenerating plants from the flowering plants of a sequentially flowering *Paphiopedilum* Deperle and a single floral *Paphiopedilum* Armeni White.

By using cross-sectioned flower buds (FBs), it was found that in both species only sections that contained the base tissue of FBs were able to produce shoots and plants. Microscopic observations revealed that the small bract at the FB base harbored a new miniature FB, which further harbored a primitive FB with dome-shaped meristem-like tissues that presumably led to the plant induction. The reiteration of this pattern resulted in a scorpioid cyme inflorescence architecture in the multifloral *Paphiopedilum* species and its failure to reiterate resulted in a single flower. The induction rates were 57-75%, and all plants survived in a greenhouse. This method is potentially applicable for the micropropagation and conservation of slipper orchids.

Paphiopedilum orchids are generally propagated through the division of axillary buds from mother plant, which limits commercial production due to its unproductive proliferation and time consuming. Therefore investigation was carried out to see the effect of *in vitro* cutting methods and medium composition on efficient shoot multiplication of *Paphiopedilum* Hsinying Rubyweb. Among three different *in vitro* stem cutting methods used, vertical cutting was able to produce more new shoots than horizontal and cross cutting when cultured on Hyponex based medium. After 12 weeks of culture, plantlets regenerated from vertical cutting were able to produce new healthy and well rooted shoots higher than without cutting on the same medium. Moreover, the newly-formed shoots which were divided into single plantlets and subcultured onto half-strength Murashige and Skoog (MS) medium without growth regulators could remain higher shoot multiplication than in other media. The micropropagation procedure developed in this study provides a simple means to *in vitro* propagate *Paphiopedilum* plantlets which are able to produce large numbers of uniform plantlets in a shorter time compared to the conventional propagation method (Udomdee *et al.*, 2012). The detail regeneration study of paphiopedilum through different explants is given in Table 3.

Table 2: Regeneration of paphiopedilum through different explants.

S. No.	Plant species	Explants source	References
1.	*Paphiopedilum*	Apical/axillary bud	Bubeck, 1973
2.	*P. insigne, P. villosum, P. fairieanum*	Apical bud/shoot tip	Stewart and Button, 1976
3.	*Paphiopedilum*	Shoot tip	Huang *et al.*, 2001
4.	*P. callosum*	Shoot/leaf tip	Huang, 1988
5.	*P. callosum* × *P. lawrenceanum*	Protocorm	Lin *et al.*, 2000
6.	*P. philippinense* (hybrid PH59, PH60)	Stem node, Leaf	Chen *et al.*, 2002, 2004

In vitro mutagenesis

Le *et al.* (2012) found that the basic medium containing half-strength of MS basal medium, full-strength of vitamin MS, 20 ml l^{-1} coconut water and 2 g l^{-1} activated charcoal supplemented with 170 mg l^{-1} NaH_2PO_4, 10 mg l^{-1} 2,4D and 0.1 mg l^{-1} TDZ was appropriate for callus formation of *Paphiopedilum delenatii* and *Paphiopedilum callosum*. The basic medium supplemented with NaH_2PO_4 was suitable for plantlet regeneration. No variant line was found for both orchid samples irradiated by gamma-rays, whereas 24 variant lines (which are divided among the two tested *Paphiopedilum*) were screened for samples irradiated by C^{6+} ion-beams (3 Gy). The genetic relationships among six generated variant lines and wild types were analyzed using RAPD techniques. Irradiation-induced variation is promising for developing new mutant varieties of *Paphiopedilum*.

Table 3: Some mutants developed through mutation with their characteristics.

S.No.	Mutants	Characteristics
1.	*Paphiopedilum* Barge	Produced a normal flower apart from two lips.
2.	*Paphiopedilum* Pinocchio	Double back, flower and no reproductive organs.
3.	*Paphiopedilum* Barge	Two dorsal sepals and two staminodes.
4.	*Paphiopedilum* Kubele	Showing a deformed flower.
5.	*Paphiopedilum* Moustache	Deformed bloom.

Molecular breeding

Chung *et al.* (2006) determined the genetic diversity and relationship between *Paphiopedilum* and *Phragmipedium* species and cultivars by randomly amplified polymorphic DNA (RAPD) analysis. Twenty-one species and 13 varieties of these two orchids genera were analysed and the similar values ranged from 0.629, between *P. kolosand* and *Paphiopedilum koloparkngii*, to 0.882, between *P. koloparkingii* and *Phragmipedium* 'Hanne Popow'. *Paphiopedilum* and *Phragmipedium* were successfully differentiated by RAPD and the results were in good agreement with morphological based classifications. Therefore, it was suggested that the phylogenetic information could be obtained using molecular markers to address the interspecies genetic relationships of *Paphiopedilum* and *Phragmipedium*.

Paphiopedilum armeniacum, *Paphiopedilum micranthum* and *Paphiopedilum delenatii* are endangered orchid species. These three Paphiopedilum species and their hybrids are difficult to distinguish morphologically. Sun *et al.* (2011) used rDNA-ITS (internal transcribed spacer) sequences to design species-specific SCAR (sequence characterized amplified regions) markers to distinguish *P. armeniacum*, *P. micranthum*, *P. delenatii* and their respective hybrids. The developed markers efficiently amplified 600 bp DNA product for *P. armeniacum* and its hybrids (SCAR-600armF/Pap-ITS2R), 300 bp product for *P. delenatii* and its hybrids (SCAR-300delF/Pap-ITS2R) and 700 bp product for *P. micranthum* and its hybrids (SCAR-700micF/Pap-ITS2R). The effectiveness of designed species-specific markers was also confirmed by using multiplex polymerase chain reaction amplification with a combination of developed three SCAR markers.

Chung and Choi (2012) investigated genetic variability and relationships among several hybrid cultivars and parental species of *Paphiopedilum* for the purposes of further breeding. Genetic information was obtained by means of phylogenetic analysis, using nuclear ribosomal DNA (rDNA) nucleotide sequence data. Nuclear rDNA was derived from internal transcribed spacer (ITS) regions and the 5.8S regions were amplified by polymerase chain reaction. The resulting nucleotide sequences were examined and compared for genetic variability, and the relationships were analyzed by pair-wise multiple alignment. ITS region analysis revealed the following lengths of the ITS1, ITS2, and 5.8S regions: 382-428, 207-215, and 170 bp, respectively. Sequence identity at the ITS site, including the 5.8S region, ranged from 75.9% between *P. emersonii* and *P. sanderianum*, to 99.4% between *Paphiopedilum* 'Hamana Wave' and *Paphiopedilum* 'Kolosand'. The phylogenetic tree revealed that hybrids derived from the subgenus *Brachypetalum* were clearly differentiated from those derived from the subgenus *Paphiopedilum*. Moreover, interspecific hybrids showed unique phylogenetic positioning among parental species, according to the ITS sequence.

REFERENCES

Banziger, H., Pumikong, S. and Srimuang, K. 2012. The missing link: bee pollination in wild lady slipper orchids *Paphiopedilum thaianum* and *P. niveum* (Orchidaceae) in Thailand. *Mitteilungen der Schweizerischen Entomologischen Gesellschaft*. **85**(1/2):1-26.

Bubeck, S.K. 1973. A study of *Paphiopedilum* meristem culture. Ph.D. thesis, Rutgers University, Michigan, USA.

Chen, T.Y. Chen, J.T. and Chang, W.C. 2004. Plant regeneration through direct shoot formation from leave culture of *Paphiopedilum* orchids. *Plant Cell, Tissue and Organ Culture,* **76**: 11-15.

Chen, T.Y., Chen, J.T. and Chang, W.C. 2002. Multiple shoot formation and plant regeneration from stem nodal explants of *Paphiopedilum* orchid. *In vitro Cellular and Developmental Biology- Plant*, **38**: 597-598.

Chung, S. and Choi, S. 2012. Genetic variability and relationships among interspecific hybrid cultivars and parental species of *Paphiopedilum* via ribosomal DNA sequence analysis. *Plant Systematics and Evolution,* **298**(10): 1897-07.

Chung, S.Y., Choi, S.H., Kim, M.J., Yoon, K.E., Lee, G.P., Lee, J.S. and Ryu, K.H. 2006. Genetic relationship and differentiation of Paphiopedilum and Phragmepedium based on RAPD analysis. *Scientia Horticulturae*, **109**: 153-159.

Day, J.C. 1974. *Paphiopedilum philippinense* var. *Roebelenii*-One of the Real Charmers. *Amer. Orchid Soc. Bull.* **43**(4): 288-89.

Fitch, C.M. 1982. Paphiopedilum callosum in Malaysia-Drive-Along Orchid Viewing. *Amer. Orchid Soc. Bull.* **51**(7): 713-16.

Fitch, C.M. 1982a. Paphiopedilum Paradise by the Sea. *Amer. Orchid Soc. Bull.* **51**(4): 377-80.

Hawkes, A.D. 1965. Encyclopaedia of Cultivated Orchids. Faber and Faber, Limited, London, England.

Huang, L.C. 1988. A procedure of asexual multiplication of *Paphiopedilum in vitro*. *American Orchid Society Bulletin,* **57**: 274-278.

Huang, L.C., Lin, C.J., Kuo, C.I., Huang, B.L. and Murashige, T. 2001. *Paphiopedilum* cloning *in vitro*. *Scienctia Horticulturae,* **191**: 111-121.

Lan, T. and Albert, V.A. 2011. Dynamic distribution patterns of ribosomal DNA and chromosomal evolution in *Paphiopedilum*, a lady's slipper orchid. *Plant Biology*, **11**: 126.

Le, Q., Luan, N., Huynh, P., Uyen, V.T. and Thu, H. 2012. *In vitro* mutation breeding of *Paphiopedilum* by ionization radiation. *Scientia Horticulturae*, **144**:1-9.

Lee, Y. and Yeung, E.C. 2012. Embryology of the lady's slipper orchid, *Paphiopedilum delenatii*: ovule development. *Botanical Studies*. **53**(1): 97-104.

Lee, Y.I. and Chung, M.C. 2010. Karyomorphological observation on some *Paphiopedilum* hybrids. *Acta Horticulturae*. **878**: 99-105.

Lee, Y., Chang, F. and Chung, M. 2011. Chromosome pairing affinities in interspecific hybrids reflect phylogenetic distances among lady's slipper orchids (*Paphiopedilum*). *Ann. Bot.*, **108**(1): 113-21.

Liao,Y., Tsai, Y., Sun, Y., Lin, R. and Wu, F.S. 2011. *In vitro* shoot induction and plant regeneration from flower buds in *Paphiopedilum* orchids. *In Vitro Cellular and Developmental Biology-Plant*. **47**(6): 702-09.

Lin, Y.N. Chang, C. and Chang, W.C. 2000. Plant regeneration from callus culture of *Paphiopedilum* hybrid. *Plant Cell, Tissue and Organ Culture,* **62**: 21-25.

Ng, C. and Norihan, M.S. 2011. *In vitro* propagation of *Paphiopedilum* orchid through formation of protocorm-like bodies. *Plant Cell, Tissue and Organ Culture*. **105**(2):193-202.

Singh, F. 1985. Sub-haploid pollen in *Paphiopedilum insigne* (Orchidaceae). *Genetica*. **67**: 57-61.

Stewart, J. and Button, J. 1976. Tissue culture studies in *Paphiopedilum. Proce:* 8th World Orchid Conference, pp: 372-378.

Sun, Y., Liao, Y., Hung, Y., Chang, J. and Sung, J. 2011. Development of ITS sequence based SCAR markers for discrimination of *Paphiopedilum armeniacum*, *Paphiopedilum micranthum*, *Paphiopedilum delenatii* and their hybrids. *Scientia Horticulturae*. **127**(3):405-10.

Udomdee, W., Wen, P., Chin, S. and Chen, F. 2012. Shoot multiplication of *Paphiopedilum* orchid through *in vitro* cutting methods. *African J. Biotech.*, **11**(76): 14077-82.

Chapter – 34

Vanda

Vanda is a genus in the orchid belongs to family Orchidaceae. The name Vanda is derived from the Sanskrit name for the species *Vanda tessellate.* The name Vandas will be used here to cover all vandaceous orchids including *Vanda, Ascocentum, Aerides, Renanthera, Rhynchostylis*, etc. and hybrids between them. Although there are some vandaceous orchids that originate in mountainous areas most are low-level plants that love warm temperatures. This has contributed much to the work of hybridists producing flowers for the cut flower market. It is one of the five most horticulturally important orchid genera because it has some of the most magnificent flowers to be found in the entire orchid family. It is highly prized in horticulture for its showy, fragrant, long lasting and intensely colourful flowers.

ORIGIN AND DISTRIBUTION

Vanda was discovered in 1613 by Alvin Semedo, who called it 'tiao hua' or 'air plant'. It is originated from the warm islands of the South Pacific. Places like Malaysia, Thailand, Borneo and the Philippines are natural home to these orchid species. These are mostly epiphytic but sometimes lithophytic or terrestrial orchids. These orchids were distributed in India, Himalaya, S.E. Asia, Indonesia, the Philippines, New Guinea, southern China and northern Australia. About 7 species of *Vanda* exist in the forest of Nagaland (Changkija *et al.*, 1992). These species are described in Table 1.

Table 1: Species of *Vanda* existing in the forest of Nagaland, India.

S. No.	Species	Distribution	Characteristics
1.	*Vanda alpina*	Japfu Range, Hellipong, Changlangshu	FC: Deep maroon colour, FS: 2 cm, FT: June-July.
2.	*V. bicolor*	Phek, Peren, Viswema	FC: Brownish purple, FS: 5 cm, FT: July-Sept.
3.	*V. coerulea*	Kohima, Wokha, Tuensang, Mokokchung, Zunhebhoto	FC: Whitish-blue or dark-blue, FS: 10 cm, FT: Oct.-Nov.
4.	*V. cristata*	Kiphire, Meluri, Chare	FC: Greenish yellow, FS: 4 cm, FT: March-April.
5.	*V. pumila*	Phek, Peren, Jaluke	FC: Cream coloured marked with red in the tip, FS: 2.5 cm, FT: May-June.
6.	*V. tessellata*	Intangki, Rangapahar, Jaluke	FC: Yellowish-green with blue tingue, FS: 3.5-5cm, FT: June-July.
7.	*V. testacea*	Doyang, Merapani, Magkolemba, Tuli	FC: Pale-puff coloured, FT: May-June.

FC: Flower colour, FS: Flower size, FT: Flowering time.

SPECIES

Almost all of the species in this genus are very large epiphytes that are found in disturbed areas in habitat and prefer very high light levels and the plants have large root systems. Some of the important species are *V. amesiana, V. batemanni, V. coerulea, V. coerulescens, V. dearci, V. denisoniana, V. hookeriana, V. insignis, V. lamellate, V. loatica, V. limbata, V. liouvillei, V. luzonica, V. marvilii, V. parviflora, V. roeblingiana, V. roxburghii, V. sanderiana, V. spathulata, V. suavis, V. sumatrana, V. teres* and *V. tricolor*. Some of these species have a monopodial vine-like growth habit and the plants can quickly become quite massive. Some species of *Vanda* are given in Table 2.

Table 2: List of some important species of *Vanda*.

S. No.	Species	Regions
1.	*Vanda alpina*	Himalaya to China - S. Yunnan
2.	*V. arbuthnotiana*	India
3.	*V. arcuata*	Indonesia – Sulawesi
4.	*V. barnesii*	North Luzon, Philippines
5.	*V. bensonii*	Assam to Thailand
6.	*V. bicolor*	Bhutan
7.	*V. bidupensis*	Vietnam
8.	*V. brunnea*	China - Yunnan to Indo-China
9.	*V.celebica*	Indonesia – Sulawesi
10.	*V. chlorosantha*	Bhutan
11.	*V. coerulea*	Assam to China - S. Yunnan
12.	*V. coerulescens*	Arunachal Pradesh to China - S. Yunnan
13.	*V. concolor*	S. China to Vietnam

(*Contd.*)

(*Contd. Table 2*)

S. No.	Species	Regions
14.	*V. cristata*	Himalaya to China - NW. Yunnan
15.	*V. dearei*	Borneo
16.	*V. denisoniana*	China - Yunnan to N. Indo-China
17.	*V. devoogtii*	Sulawesi
18.	*V. foetida*	S. Sumatra
19.	*V. furva*	Java, Maluku
20.	*V. fuscoviridis*	S. China to Vietnam
21.	*V. griffithii*	E. Himalaya
22.	*V. hastifera*	Borneo
	V. hastifera var. *gibbsiae*	N. Borneo
	V. hastifera var. *hastifera*	Borneo
23.	*V. helvola*	W. Malaysia to Philippines
24.	*V. hindsii*	Papuasia to N. Queensland
25.	*V. insignis*	Lesser Sunda Is
26.	*V. jainii*	Assam
27.	*V. javierae*	Philippines – Luzon
28.	*V. lamellata*, *V. lamellata* var. *boxallii* *V. lamellata* var. *calayana*, *V. lamellata* var. *flava*, *V. lamellata* var. *lamellata* *V. lamellata* var. *remediosae*	Taiwan to Philippines, North Borneo
29.	*V. leucostele*	N. & W. Sumatra
30.	*V. lilacina*	China - Yunnan to Indo-China
31.	*V. limbata*	Java, Lesser Sunda Is., Philippines- Mindanao
32.	*V. lindenii*	Maluku
33.	*V. liouvillei*	Assam to Indo-China
34.	*V. lombokensis*	Lesser Sunda Is.
35.	*V. luzonica*	Philippines – Luzon
36.	*V. merrillii*, *V. merrillii* var. *Rotorii*, *V. merrillii* var. *immaculata*	Philippines
37.	*V. petersiana*	Myanmar
38.	*V. pumila*	Nepal to Hainan and N. Sumatra
39.	*V. punctata*	Pen. Malaysia
40.	*V. roeblingiana*	Philippines – Luzon
41.	*V. scandens*	Borneo, Philippines – Mindanao
42.	*V. spathulata*	India -Kerala, Tamil Nadu, Sri Lanka
43.	*V. stangeana*	India - Arunachal Pradesh to Assam
44.	*V. subconcolor*, *V. subconcolor*	China - SW. Yunnan to Hainan
	var. *disticha*	Hainan
45.	*V. sumatrana*	Sumatra
46.	*V. tessellata*	Indian Subcontinent to Indo-China
47.	*V. testacea*	Indian Subcontinent to SC. China
48.	*V. thwaitesii*	South India, Sri Lanka
49.	*V. tricolor*	Laos, Java to Bali
50.	*V. ustii*	Philippines – Luzon
51.	*V. vipanii*	Myanmar
52.	*V. wightii*	South India

***Vanda alpine*:** Stem is 15-30 cm long with many leaves of about 8-10 × 1.5-2.8 cm, narrowly oblong, conduplicate, unequally bilobed at the apex, dilated towards the base. Inflorescence raceme and short with 1-3 nodding flowers partly open, 2 cm across, greenish in colour, lip striated with deep maroon colour. Flowering time is from June-July.

***Vanda bicolour*:** Stem is upto 90 cm strap-shaped and obtusely bilobed. Inflorescence stout and erect, with 4-6 flowers of about 5cm across, brownish purple. Flowering time is from July-Sept.

***Vanda coerulea*:** Stem upto 40 cm long erect stout and leafy. Leaves oblong unequally bilobed, keeled and fleshy. Inflorescence erect, lax-flowered racemes, flower upto 10 cm across whitish-blue or dark purple in colour. Flowering time is from Oct-Nov.

***Vanda cristata*:** Stem upto 20 cm long, erect, stout, leafy. Leaves oblong, apex 3 toothed, coriaceous, keeled, fleshy. Inflorescence is a few flowered raceme, flowers about 4 cm across, greenish-yellow in colour. Flowering time is from March-April.

***Vanda pumila*:** Stem 10-15 cm high, stout, covered with old sheaths. Leaves numerous, narrowly oblong, unequally bilobed at the apex, 7-10 × 2-3 cm, dilated towards the base. Inflorescence shorter than the leaves, 2-4 flowered; flowers 2.5 cm across, drooping, cream- coloured marked with red in the tip, fragrant. Flowering time: May-June.

***Vanda tessellate*:** Stem erect 30 cm long. Leaves several, distichous, 10-20 × 1-2 cm, linear, keeled, articulate at the base, unequally bilobed at the apex; inflorescence a raceme, axillary, 5-10 flowered, longer than the leaves, zigzag; flowers 3.5-5 cm across, yellowish-green with blue tingue, scented. Flowering time: June-July.

***Vanda testacea*:** Stem upto 30 cm long, erect leaves many keeled, unequally bi- or tri fid at apex, fleshy; flower pale-puff coloured. Flowering time: May-Jane.

FLOWER MORPHOLOGY AND BIOLOGY

Inflorescences erect to pendent, spicate, racemose, or paniculate, one to many flowered, basal, lateral, or terminal, flowers rarely secund, subumbellate, or distichously arranged. Flowers small to large, often quite showy, hermaphroditic or rarely monoecious and polymorphic, sessile or variously pedicellate, most often twisted through 180, occassionally not twisted or twisted through 360. Ovary inferior, 1-locular and placentation parietal or rarely 3-locular and placentation axile. Sepals usually free but sometimes variously adnate, median (dorsal) one often dissimilar to lateral, laterals sometimes adnate to a column foot to form a saccate, conic, or spurlike mentum. Petals free or rarely partly adnate to sepals, similar to sepals or not, often showy; lip entire, variously lobe or 2- or 3-partite, ornamented or not with calli, ridges, hair cushions, or crests, with or without a basal spur or nectary, margins entire to laciniate. Column short to long with or without a basal foot occasionally winged or with lobes or arms at apex or ventrally; anther 1 or rarely 2 or 3 terminal or ventral on column, caplike or opening by longitudinal slits, pollinia mealy, waxy or horny, sectile or not, 2, 4, 6, or 8, sessile or attached by stalks to 1 or 2 sticky viscidia; stigma 3-lobed, midlobe often modified to form a rostellum, other lobes either sunken on ventral surface ofcolumn behind santher or with 2 lobes porrect. Fruit a capsule, usually opening laterally by 3 or 6 slits. Seeds numerous, dustlike, lacking endosperm, rarely winged.

GENETICS

The basic and 2n chromosome number of vanda is 19 and 38, respectively. Meiotic chromosome behaviour of 11 groups of diploid intergeneric hybrids (2n= 38) of the *Vanda alliance* was investigated by Tanaka and Kamemoto, (1961). Vanda Miss Joaquim × *Luisia teretifolia* and *Ascocentrum curvifolium* × *V. lamellate* usually produced 19 bivalents at metaphase I. *Trichoglottis brachiata* × *V. sanderiana*, *V. tricolor* var. *purpurea* × *Phalaenopsis denevei*, *V. tatzeri* × *Aerides lawrenceae*, *Renanthera monachica* × V. luzonica, *Renanthera storiei* × *V. clara* shipman fisher and *Aerides lawrenceae* × *Saccolabium giganteum* formed an average of 2-7 bivalents, thereby indicating poor homology of parental genomes. The hybrids *Renanthera monachia* × *Phalaenopsis sanderina* and *Arachis hookeriana* × *V. suavis*, lacked chromosome pairing. On the basis of chromosome affinity at meiosis, a cytotaxonomic scheme was developed for several genera of *Vanda alliance*.

Inheritance of flower colour

The blue colour of *Vanda coerulea* petals is slowly produced during flower development but quickly disappears when the flowers are pollinated. To investigate the molecular basis of the phenomenon, Ratanasut *et al.* (2011) isolated a novel cytochrome P450 gene, *CYP76AB1*, from this plant by polymerase chain reaction (PCR) based on the conserved regions of flavonoid 3´, 5´-hydroxylase amino acid sequences. *CYP76AB1* transcripts were detectable at low level in the late phase of flower development when the petals were light blue, but became abundant in the subsequent opening flower stage as the petals turned blue. When the flowers were pollinated they turned white and the *CYP76AB1* transcripts returned to low level. This indicates that the expression of *CYP76AB1* may regulate petal colour of *V. coerulea* and be a target for developing permanently blue flowers.

BREEDING OBJECTIVES

i. More number of flowers.

ii. Precocious bearing.

iii. To develop free-flowering plants.

iv. Possess great vigour.

v. Develop plant type suitable for cultivation in temperature greenhouses or under lights.

vi. Resistance to diseases (cymbidium mosaic potexvirus).

vii. Resistance to abiotic stresses.

BREEDING METHODS

Introduction

Vanda plants are existing in different part of India particularly North-East India and also domesticated for cultivation and improvement. Apart from this some species and hybrids introduced in India from Thailand are given in Table 3.

Table 3: Important introductions in vanda orchid hybrids.

S. No.	Country	Hybrids
1.	Thailand	Ble. Mem Crispin Rosales, Le Dorset Gold, Orchid Hurst AM; Gift AV 164, *Vanda rothschildiana* AM × *V. kapielana*, Gift PN 100, *V.* Chussi W × *V.* Diane Organa, Gift PN 322, Suwat Sandcriane × *V. coerulea*, Gift Jo 227 *V. rothschildiana* AM × *V.* Mad Pranerma and Gift Jo 3 Aeridovanda Vieny Ping × Axda Mad Kenny.

Hybridization

Vanda orchid has become so popular since they are an excellent orchid species that lends itself to hybridization with several other orchid species. It has been used in so many successful hybridizations. Any mention of the *vanda* orchid is not complete without mentioning the *Vanda sanderiana* (also known as the *Euanthe sanderiana*). This is a truly beautiful orchid that bears bright pink and maroon orchid flowers that have spectacularly long-lasting characteristics. Another spectacular example of the *vanda* orchid hybridization is the *Vanda rothschildiana* × *Vanda coerulea*, which is possibly the bluest of all orchid flowers with heavy contrasts of white. The orchid flowers itself are saucer-sized, flat blooms with an extremely heavy substance and a crystalline texture that glistens in the light.

Vanda coerulea is one of the few botanical orchids which can produces varieties with blue flowers a property much appreciated for producing interspecific and intergeneric hybrids. The colour blue is rare among orchids and only *Thelymitra crinita*, a terrestrial species from Australia produce flowers that are truly blue among the orchids, the other being *Aganisia cyanea*, a lowland species from Northern South America that is difficult in cultivation but has metallic blue flowers. Both of these species much like vandas also have a bluish purple tint towards the inner petals of the flowers. *Vanda dearei* is one of the chief sources of yellow colour in *Vanda* hybrids. *Vanda* Miss Joaquim, a terrestrial Vanda variety is a natural hybrid with terete leaves.

Interspecific hybrids

Hybrids between different subspecies within a species are known as intra-specific hybrids. Hybrids between different species within the same genus are sometimes known as interspecific hybrids or crosses. Once the obstacle of raising orchid seedlings was overcome, intensive hybridisation in vandaceous and other orchids was carried out by local breeders with great success. Several early interspecific hybrids of Arachnis and intergeneric hybrids of Arachnis × Renanthera and Arachnis × Vanda were first flowered in the Gardens in the 1940s and 1950s and were later grown as cut-flower crops on commercial scale.Some interspecific hybrids of vanda are given in Table 4.

Table 4: List of interspecific hybrids of vanda.

Hybrids	Parentage	Regions
Vanda × *boumaniae*	*V. insignis* × *V. Limbata*	Lesser Sunda Is.
Vanda × *charlesworthii*	*V. bensonii* × *V. Coerulea*	Myanmar
Vanda × *confuse*	*V. coerulescens* × *V. Lilacina*	Myanmar
Vanda × Miss Joaquim	*V. hookeriana* × *V. Teres*	Singapore
Vanda Mimi Palmer	*Vanda* Tan Chay Yan × *Vanda tessellate*	-

Intergeneric hybrids

Hybrids between different genera are known as intergeneric hybrids. In the nomenclature of intergeneric hybrids in orchids, a compound generic name is given to any class of bigeneric hybrids to indicate its parental genera (Johnson, 1990). Thus the name Aranthera is used for hybrids of Arachnis × Renanthera, Aranda for Arachnis × Vanda and Ascocenda for Ascocentrum × Vanda. For hybrids of trigeneric and multigeneric crosses the name of the first breeder of the new hybrid or other names with the suffix 'ara' is used. Thus Mokara is given to hybrids of Arachnis × Vanda × Renanthera. Intergeneric hybridisation followed by progeny selection has been the main approach in vandaceous orchid breeding. In 1963 there were 32 bigeneric and 8 trigeneric classes of hybrids in vandaceous orchids (Kirsch, 1963). Intergeneric hybrids along with their parentage have been given in Table 5. Whereas, some hybrids along with their parentage and characteristics are given in Table 6.

Table 5: List of intergeneric hybrids of vanda.

S. No.	Hybrids	Parentage
I. Bigeneric		
1.	Aeridovanda	*Aerides × Vanda*
2.	Aranda	*Arachnis × Vanda*
3.	Ascocenda	*Ascocentrum × Vanda*
4.	Opsisanda	*Vanda × Vandopsis*
5.	Luisanda	*Luisia × Vanda*
6.	Renantanda	*Renanthera × Vanda*
7.	Rhynchovanda	*Rhynchostylis × Vanda*
8.	Sarcovanda	*Sarcochilus × Vanda*
9.	Trichovanda	*Trichoglottis × Vanda*
10.	Vandachnis	*Arachnis × Vandopsis*
11.	Vancampe	*Acampe × Vanda*
12.	Vandaenopsis	*Phalaenopsis × Vanda*
13.	Vandaeranthes	*Aeranthes × Vanda*
14.	Vandofinetia	*Neofinetia × Vanda*
15.	Vandoritis	*Doritis × Vanda*
16.	Vanglossum	*Ascoglossum × Vanda*
II. Tri-generic		
1.	Ascovandoritis	*Ascocentrum × Doritis × Vanda*
2.	Aeridovanisia	*Aerides × Luisia × Vanda*
3.	Burkillara	*Aerides × Arachnis × Vanda*
4.	Charlieara	*Rhynchostylis × Vanda × Vandopsis*
5.	Christieara	*Aerides × Ascocentrum × Vanda*
6.	Debruyneara	*Ascocentrum × Luisia × Vanda*
7.	Devereuxara	*Ascocentrum × Phalaenopsis × Vanda*
8.	Eastonara	*Ascocentrum × Gastrochilus × Vanda*
9.	Fujiora	*Ascocentrum × Trichoglottis × Vanda*
10.	Goffara	*Luisia × Rhynchostylis × Vanda*
11.	Hawaiiara	*Renanthera × Vanda × Vandopsis*
12.	Hagerara	*Doritis × Phalaenopsis × Vanda*
13.	Holttumara	*Arachnis × Renanthera × Vanda*
14.	Joannara	*Renanthera × Rhynchostylis × Vanda*
15.	Kagawara	*Ascocentrum × Renanthera × Vanda*
16.	Leeara	*Arachnis × Vanda × Vandopsis*
17.	Luivanetia	*Luisia × Neofinetia × Vanda*
18.	Maccoyara	*Aerides × Vanda × Vandopsis*
19.	Moirara	*Phalaenopsis × Renanthera × Vanda*
20.	Mokara	*Arachnis × Ascocentrum × Vanda*
21.	Nakamotoara	*Ascocentrum × Neofinetia × Vanda*
22.	Nobleara	*Aerides × Renanthera × Vanda*
23.	Pantapaara	*Ascoglossum × Renanthera × Vanda*
24.	Pereiraara	*Aerides × Rhynchostylis × Vanda*
25.	Phalaerianda	*Aerides × Phalaenopsis × Vanda*

(Contd.)

(*Contd. Table 5*)

S. No.	Hybrids	Parentage
26.	Raganara	*Renanthera* × *Trichoglottis* × *Vanda*
27.	Ramasamyara	*Arachnis* × *Rhynchostylis* × *Vanda*
28.	Renafinanda	*Neofinetia* × *Renanthera* × *Vanda*
29.	Renanda	*Arachnis* × *Renanthera* × *Vanda*
30.	Ridleyare	*Arachnis* × *Trichoglottis* × *Vanda*
31.	Trevorara	*Arachnis* × *Phalaenopsis* × *Vanda*
32.	Vascostylis	*Ascocentrum* × *Rhynchostylis* × *Vanda*
33.	Vandofinides	*Aerides* × *Neofinetia* × *Vanda*
34.	Wilkinsara	*Ascocentrum* × *Vanda* × *Vandopsis*
35.	Yapara	*Phalaenopsis* × *Rhynchostylis* × *Vanda*
36.	Yonezawaara	*Neofinetia* × *Rhynchostylis* × *Vanda*
III. Tetra-generic		
1.	Alphonsoara	*Arachnis* × *Ascocentrum* × *Vanda* × *Vandopsis*
2.	Andrewara	*Arachnis* × *Renanthera* × *Trichoglottis* × *Vanda*
3.	Bokchoonara	*Arachnis* × *Ascocentrum* × *Phalaenopsis* × *Vanda*
4.	Bovornara	*Arachnis* × *Ascocentrum* × *Rhynchostylis* × *Vanda*
5.	Darwinara	*Ascocentrum* × *Neofinetia* × *Rhynchostylis* × *Vanda*
6.	Himoriara	*Ascocentrum* × *Phalaenopsis* × *Rhynchostylis* × *Vanda*
7.	Isaoara	*Aerides* × *Ascocentrum* × *Phalaenopsis* × *Vanda*
8.	Knappara	*Ascocentrum* × *Rhynchostylis* × *Vanda* × *Vandopsis*
9.	Lewisara	*Aerides* × *Arachnis* × *Ascocentrum* × *Vanda*
10.	Micholitzara	*Aerides* × *Ascocentrum* × *Neofinetia* × *Vanda*
11.	Okaara	*Ascocentrum* × *Renanthera* × *Rhynchostylis* × *Vanda*
12.	Onoara	*Ascocentrum* × *Renanthera* × *Vanda* × *Vandopsis*
13.	Pageara	*Ascocentrum* × *Luisia* × *Rhynchostylis* × *Vanda*
14.	Pehara	*Aerides* × *Arachnis* × *Vanda* × *Vandopsis*
15.	Robinaria	*Aerides* × *Ascocentrum* × *Renanthera* × *Vanda*
16.	Ronnyara	*Aerides* × *Ascocentrum* × *Rhynchostylis* × *Vanda*
17.	Sanjumeara	*Aerides* × *Neofinetia* × *Rhynchostylis* × *Vanda*
18.	Shigeuraara	*Ascocentrum* × *Ascoglossum* × *Renanthera* × *Vanda*
19.	Stamariaara	*Ascocentrum* × *Phalaenopsis* × *Renanthera* × *Vanda*
20.	Teohara	*Arachnis* × *Renanthera* × *Vanda* × *Vandopsis*
21.	Vandewegheara	*Ascocentrum* × *Doritis* × *Phalaenopsis* × *Vanda*
22.	Yusofara	*Arachnis* × *Ascocentrum* × *Renanthera* × *Vanda*
23.	Knudsonara	*Ascocentrum* × *Neofinetia* × *Renanthera* × *Rhynchostylis* × *Vanda*
24.	Macekara	*Arachnis* × *Phalaenopsis* × *Renanthera* × *Vanda* × *Vandopsis*
25.	Paulara	*Ascocentrum* × *Doritis* × *Phalaenopsis* × *Renanthera* × *Vanda*
26.	Sutingara	*Arachnis* × *Ascocentrum* × *Phalaenopsis* × *Vanda* × *Vandopsis*

Table 6: Hybrids of vanda and their characteristics.

S. No.	Hybrid varieties	Parentage	Characteristics
1.	Vanda Amy	*Vanda tricuspidata* × *Vanda hookeriana*	Flower shape similar to *Vanda tricuspidata* but the hybrid lip is much bigger and rounder due to the influence of *Vanda hookeriana.*
2.	Vanda Miss Joaquim Agnes	*Vanda teres* × *Vanda hookeriana*	It is a very free flowering robust plant when given very high light.
3.	Vanda Miss Jaoquim Rose Marie	*Vanda teres* var Alba × *Vanda hookeriana*	The Singapore Botanical Garden also repeated this cross in 1945. This rare hybrid also has pretty much disappeared in cultivation and only exists in a handful at private collections.
4.	Vanda Cooperi White Wing	*Vanda Miss Joaquim* × *Vanda hookeriana* var. Alba	Special white flower.
5.	Vanda Poepoe Diana	*Vanda teres* var Alba × *Vanda Cooperi* var White Wing	This is a robust free flowering terete vanda with pure white flower.

Polyploidy

Occurrence of endopolyploidy in somatic tissues of the hybrid orchid vanda Miss Joaquim (*Vanda hookeriana* × *Vanda teres*) with respect to tissue type and developmental stage was investigated. Effects of naphthalene acetic acid (NAA) and gibberellic acid (GA_3) on endopolyploidy during embryo development were also studied. Multiploid cells were observed in leaves, roots and column but not in shoot apex, stem, perianth and pedicel. Furthermore, differential distribution of multiploid cells was found among different parts of leaves and roots. The degree of endopolyploidy in embryos increased with development. NAA was shown to induce endoreduplication in germinating embryos to a much larger extent than GA_3. The pattern of endopolyploidy was characteristic of tissue type and developmental stage (Lim *et al.*, 2003). Hawaiian breeders produced exciting tetraploids by repeated selfing of V. Miss Joaquim which is diploid (2n = 38). Few varieties that have been developed through polyploidy are given in Table 7.

Table 7: Polyploidy varieties along with their chief characteristics.

S. No.	Variety/ Cultivar	Ploidy level	Characteristics
1.	Vanda Miss Joaquim Douglas	Tetraploid form of Atherton	Size of the flower is twice the size of Atherton much deeper in colour.
2.	4×-Aranda × 2×-Vanda	Fertile amphidiploid	Arandas are not easily available although they can be obtained by colchicine treatment of diploid arandas.
3.	Mariana Lila	Triploid	Ar. Wendy Scott × *V. sanderiana*, 1990.

Amphidiploid of cv. Wendy Scott was obtained through spontaneous chromosome doubling during meristem culture (Lee and Kamemoto, 1986). Fertile clone showed that the anthers produced over 60% of normally reduced microspores, presumably with AV genomes and functional. It was crossed successfully to diploid vandas but only as the female parent. The production of new hybrids through 2x-Arachnis × 4x-Vanda combination of crosses is limited by the lack of suitable tetraploid vandas. The discovery of a fertile Vanda Dawn Nishimura, an amphidiploid from *V. Rothschildiana* × *V. Hilo* Blue, enabled breeders to produce several triploid arandas of AVV genomic constitutions directly (Lee *et al.*, 1991). The 4 x -Vanda hybrid was found to behave like a diploid cytologically and produce over 99% of n gametes containing two V genomes. It was used commonly as the pollen parent in crosses with diploid arachnis, giving rise to triploid progeny having AVV genomes. Several outstanding aranda cultivars have been bred this way (Eng *et al* ., 1984). Among them are cv. Noorah Alsagoff (*A. hookeriana* × V. Dawn Nishimura, 1975), cv. Ang Hee Seng (A . Maroon Maggie × V. Dawn Nishimura, 1975), cv. Neo Hoe Kiat (A. Maggie Oei × V. Dawn Nishimura, 1975), and cv. Etsuko Nishio (A . Ishbel × V. Dawn Nishimura, 1979).

Ploidy effects: Triploid arandas with AVV genomes are generally superior to their diploid (AV genomes) and tetraploid (AVVV genomes) counterparts in several horticultural characteristics. Their flowers are of better form and desirable size for display as cut-flowers.

The flower sizes of triploid arandas range from 8 to 9 cm in width, while those of diploid and tetraploid cultivars measure 6 to 7 cm and 11 to 12 cm respectively (Lee, 1991). Triploid aranda cultivars are also generally more vigorous and floriferous than diploid cultivars. Although tetraploid arandas produce even larger vegetative organs and flowers than triploids, they tend to suffer from slow growth rate and delayed development and lower yield of inflorescences. Thus commercial production of tetraploid arandas for cut-flowers has been phased out. The desirability of triploid arandas is also reflected in the numbers of award-winners as compared to the other two classes (Lee *et al.*, 1988).

Tetraploid Arandas: Several tetraploid aranda cultivars with one A and three V genomes were produced from crossing 4x-Vanda Dawn Nishimura with 2x-Aranda hybrids (Eng *et al.*, 1984). Among those of some commercial importance are cv. How Yee Peng (Ar. Tay Theng Suan × V. Dawn Nishimura, 1975), cv. Multico Deluxe (Ar. Christine × V. Dawn Nishimura, 1981) and cv. Wong Bee Yeok (Ar. Lily Chong × V. Dawn Nishimura, 1975). Their flowers are significantly larger than those of their triploid counterparts. They also appear to be more similar to that of vandas in various features, reflecting the diminished influence of Arachnis and the increased dominance of Vanda genomes (Lee, 1991).

Mutation

Mutation breeding in orchid vanda was reported in the RIOP Report 2000. In this research, the protocorms were irradiated with gamma rays with the dose levels of upto 100 Gy and then cultured in Vacin and Went media. The results found that mutations both in Dendrobium and Vanda were detected in the gamma irradiation treatment with the dose of 30 Gy. These mutations have been confirmed by RAPD analysis which indicated different genetic changes found in the mutants compared to the original varieties. Later, these mutants were being investigated further through *in-vitro* techniques to learn the stability of the mutations. Phenotypic performance of the mutants showed more rapid growth and more leaves if compared to the original varieties.

In vitro mutation was also reported in orchids by various workers (Sanguthai and Sagawa, 1973 and Sanguthai *et al.*, 1973). The optimum dose varies from approximately 10 Gy to 40 Gy x or γ rays but the dose varies with every gene and species. Doubling of chromosome in orchids was incorporated by using colchicine. Submerging back bulbs of *Cymbidium coningsbyanum* 'Brockhurst' in colchicine solution twice within on interval of ten days has been reported to produce tetraploids. Dendrobium (Sanguthai *et al.*, 1973) and Vanda (Sanguthai and Sagawa, 1973) also showed the successful application of the above technique and exhibited variability. The optitmum treatment depends on the species but will be around 0.05-0.1% colchicine applied for several days to *in vitro* material. The recommended gamma ray irradiation dose for vanda Genta Bandung orchid is 10 Gy-25 Gy, whereas LD_{50} dose for this orchid is 3,188 Gy (Suskandari *et al.*, 1999). Sterile mutant *V. kogawara* was resulted from mericloning of the normal hybrid *V. diana* × *V. kamperanon* (Rosario *et al.*, 1991).

Vanda Miss Joaquim Peloric is rare varietal occurance and probably the first ever reported for terete vandas. It is a peloric mutation from the variety Atherton. Pelorism mutation whereby either the lip is replaced by an extra petal or both petals are replace by 2 extra lips.

BIOTECHNOLOGY

Tissue culture

Efficient shoot regeneration of *Vanda coerulea* was achieved using thin shoot tip sections and thidiazuron. Protocorm-like bodies or proliferating shoot buds was observed when thin shoot tip sections were cultured on Vacin and Went's (1949) basal medium supplemented with 11.35 μM thidiazuron. The highest percentage of protocorm-like bodies (95%) survived and ultimately produced healthy shoots with 2-3 leaves when subjected to a 4 week thidiazuron treatment. A culture period longer than 8 weeks with thidiazuron resulted in the formation of fasciated or distorted shoots. Shoots produced roots when cultured on half strength VW basal medium supplemented with 11.42 μM

IAA. The well rooted shoots were transferred to pots containing charcoal chips, coconut husk and broken tiles (2:2:1) and a 98% survival rate was achieved. (Malabadi *et al.*, 2004).

Lang and Hang (2006) established the high-frequency somatic embryogenesis and plant regeneration from callus cultures of vanda (*Vanda coerulea*). Embryogenic calluses emerged from root tips, stem internodes and leaf segments cultured in 1/4 MS basal medium supplemented with TDZ [thidiazuron] (0.0, 0.3, 1.0 or 3.0 mg/litre) and 2,4-D (1, 3 or 10 mg/litre) for 4 weeks. Embryogenic callus was maintained by subculture in the same medium for callus induction and proliferated twice in one month. The survival rate of plantlets under *in vitro* conditions was 90% after 2 months. The high regeneration capacity may be due to the callus origin. The capacity for somatic embryogenesis significantly varied between cell lines of the same callus origin.

Orchid produces a huge number of minute seeds but the seeds cannot germinate easily in nature due to the lack of endosperm in the seeds is an incompatibility barrier that limits its propagation in nature. To develop *in vitro* culture techniques for quick propagation of *Vanda tessellate*, a commercially important orchid species. Shoot tips were used (Rahman *et al.*, 2009) as experimental materials. The isolated shoot tips were cultured in MS medium supplemented with different concentration and combinations of auxin and cytokinin. The combination of 1.5 mgl^{-1} NAA and 1.0 mgl^{-1} BAP was proved to be the best medium formulation for multiple shoot formation as well as maximum shoot elongation. Maximum root induction was obtained in MS agarified medium having 0.5 mgl^{-1} NAA and 1.0 mgl^{-1}IBA. The well rooted plantlets were hardened successfully in the potting mixture containing coconut husk, perlite, charcoal, brick pieces in the ratio of 2:1:1:1 and eventually established under natural condition.

Gantait *et al.* (2012) demonstrated the single bead alginate-encapsulation and conversion (complete plantlet regeneration) from protocorm-like bodies (PLBs) of *Aranda* Wan Chark Kuan 'Blue' × *Vanda coerulea* Grifft. ex. Lindl. (AV) (a monopodial orchid hybrid) for the first time. PLBs induced from leaf segments of AV were isolated from *in vitro* proliferating PLB clusters. Individual PLBs (4+or-1 mm diameter) were encapsulated in calcium alginate beads to manage mass propagation, short-term storage and germplasm sharing. The superior gel matrix for encapsulation was obtained using 3% sodium alginate and 75 mM calcium chloride ($CaCl_2.2H_2O$). Highest percentage of germination (98.1%) and conversion (96.2%) of encapsulated PLBs (capsules) was obtained on plant growth regulator-free half-strength MS medium. Successful storage of capsules until 180 days was achieved at 25° C under zero-irradiance with germination and conversion frequency of 76.9 and 70.2%, respectively. Plantlets regenerated from capsules were acclimatized successfully with 92% survival rate.

In Table 8 explant source were given through which vanda regeneration were studied by various workers.

Table 8: Regeneration of vanda through different explants.

S.No.	Plant species	Explants source	Reference
1.	*Vanda*	Apical/axillary bud Root/ leaf tip Flower stalk/bud	Khaw *et al*., 1978 Chaturvedi and Sharma, 1986 Valmayor *et al*., 1986
2.	*Vanda coerulea* (blue vanda)	Leaf base EmbryoShoot tip	Mitra *et al*., 1976; Seeni, 1988 Malabadi *et al*., 2004
3.	*Vanda cristata*	Leaf	Sharma and Vij, 1997
4.	*Vanda insignis* × *V. tessellata*	Shoot tip	Teo *et al*., 1973
5.	*Vanda spathulata*	Stem node leaf	Decruse *et al*., 2003 Decruse *et al*., 2003a
6.	*Vanda teres*	protocorm	Sinha and Roy, 2004
7.	*Vanda* (terete leaf) Miss Joaquim	Shoot tip/axillary bud/root	Goh, 1970
8.	*Vanda* (terete leaf) Miss Joaquim	Shoot tip	Kunisaki *et al*., 1972

Molecular breeding

Lim *et al*. (1999) determined the genetic closeness of various species of *Vanda* by using random amplified polymorphic DNA (RAPD). Strap-leaved *Vanda* species (including *Vanda sanderiana*) and *Ascocentrum miniatum* were more closely related to each other than the terete-leaved *Vanda* species studied. RAPD analysis supports the suggestion that terete-leaved *Vanda teres* and *Vanda hookeriana* be classified in the separate genus *Papilionanthe* and that *Vanda sanderiana* should remain in the genus *Vanda*.

The effect of thidiazuron (TDZ) on direct multiple shoot induction in axenic seedlings of a monopodial orchid hybrid *Aerides vandarum* × *Vanda stangeana*, using a dual phase culture medium was studied (Kishor and Devi, 2009). The culture system consisted of a basal agar solidified half-strength Murashige and Skoog medium overlaid by a liquid fraction of the same composition. Highest regeneration of multiple shoots (15.8 shoots per seedling) was obtained in the medium containing 2% sucrose (w/v) supplemented with 2 mgl"1 TDZ. Genetic stability of the regenerated shoots was assessed using random amplified polymorphic DNA (RAPD), inter simple sequence repeat (ISSR) and restriction fragment length polymorphism of the PCR amplified (PCR-RFLP) nrITS region, as well as those of the coding (*mat*K) and non-coding (*trn*L-F) regions of the cpDNA. Across the randomly selected mother plant and nine of its regenerated shoots, 2,680 bands were generated by 19 RAPD and 12 ISSR primers, exhibiting monomorphic banding profiles. Homogenous PCR-RFLP profiles were generated for nrITS using four restriction enzymes (REs), *mat*K using five REs and *trn*L-F using six REs. These molecular analyses showed no genomic alterations in all regenerated shoots obtained on medium containing 2 mgl^{-1} TDZ.

Microsatellite (SSR) markers were developed for vanda orchid from the enriched library using dinucleotide repeats [$(CA)_{15}$ and $(GA)_{15}$] and trinucleotide repeats [$(ACC)_{10}$ and $(CCT)_{10}$] as probes. Positive clones were selected using dot blot hybridization. The results showed that 82.45% of dinucleotide-enriched libraries but only 9.91% of trinucleotide-enriched libraries gave positive signals. Based on the nine loci of these microsatellite markers, the probability of identity (PI) of any two vanda and related orchid cultivars having the same genotype was approximately 1 in 1,000,000. Therefore, these markers could be used for identification of the vanda orchid samples studied. After analyzing the genetic relationships of 33 vanda and related orchid cultivars using NTS YS-pc 2.1 m, the result indicated that the vanda and related orchid cultivars could be divided into four groups. The first three groups were strap-leaved vanda, while the fourth one was terete vanda which was clearly clustered separately from the other vanda groups. This study showed the isolation efficiency of the enrichment procedure, the abundance of microsatellites in vanda orchids and their potential use for the individual identification of vanda and related orchid samples (Phuekvilai *et al.*, 2009).

Molecular genetic fingerprints of seven populations of *Vanda coerulea* comprising of thirty-two genotypes from Northeast India were developed using PCR based markers. Genetic variability in the wild genotypes of *V. coerulea* was analyzed using two different single primer amplification reactions (SPAR) methods, viz., random amplified polymorphic DNA (RAPD) and inter-simple sequence repeats (ISSR). A total of 32 genotypes were used to investigate the existing natural genetic diversity at intra-specific level. Two hundred and twenty six (226) amplification products were scored by RAPD and ISSR, both of which collectively showed 58.88% polymorphism with a mean intra-population genetic diversity (*Hpop*) of 0.119. The study suggests the efficiency of SPAR methods to estimate the genetic diversity of *V. coerulea* and can be seen as a starting point for future research on the population and evolutionary genetics of this species (Manners *et al.*, 2013).

Genetic engineering

Since particle bombardment is basically used irrespective of the species and genotypes, it was initially applied to orchids before the establishment of the protocol for *A. tumefaciens*-mediated transformation. The first successful orchid transformation using particle bombardment was reported in Vanda (Chia *et al.*, 1990).

Transgenic plants of vanda were successfully regenerated after co-cultivating protocorm-like bodies (PLB) with *Agrobacterium tumefaciens* strain EHA101 (pIG121Hm) that harbored genes for beta-glucuronidase (gus), hygromycin phos-photransferase (hpt) and neomycin phosphotransferase II (nptII). PLB of Tokyo

Blue maintained in liquid New Dogashima medium (NDM) supplemented with 0.1 mg l-1 NAA, 1 mg l-1 BA and 30 g l-1 maltose under dark condition were subjected to transformation experiments. The PLBs infected with *Agrobacterium* produced secondary PLB 4 weeks after transfer onto 3 g l-1 gellan gum-solidified NDM containing 10 mg l-1 meropenem and 10 mg l-1 hygromycin. Transformation efficiency was increased by prolonged period of inoculation (4 h) with *Agrobacterium* suspension culture and wounding treatment of PLBs by sonication (5 min) during inoculation. Transformation of the hygromycin-resistant plantlets regenerated from different PLBs was confirmed by histochemical GUS assay, PCR analysis and Southern hybridization. With this transformation system, approximately 17 independent transgenic plants were obtained from 1g PLBs (Shrestha *et al.*, 2010).

REFERENCES

Changkija, S., Kumar, Y. and Gurung, P.B. 1992. Orchids of Nagaland. Forest Department, Govt. of Nagaland.

Chaturvedi, H.K. and Sharma, A.K. 1986. Mericloning of orchids through culture of tips of leaves and roots. *In:* S.P. Vij. (ed.) Biology, Conservation and Culture of Orchids, Affiliated East West Press, New Delhi, pp: 473-78.

Chia, T.F., Chan, Y.S. and Chua, N.H. 1990. Large-scale screening of *cymbidium mosaic* and *odontoglossum mosaic* ringspot viruses in cultivated orchids by nucleic acid spot hybridization. *Proce:* 13th World Orchid Conference, 13th WOC Proceedings Trust, Auckland, New Zealand. pp. 284.

Decruse, S.W. Gangaprasad, A., Seeni, S. and Menon, V.S. 2003. Micropropagation and ecorestoration of *Vanda spathulata*, an exquisite orchid. *Plant Cell, Tissue and Organ Culture*, **72**: 199-02.

Decruse, S.W., Gangaprasad, A., Seeni, S. and Menon, V.S. 2003a. A protocol for shoot multiplication from foliar meristem of *Vanda spathulata* (L) Spreng. *Indian J. Exp. Bio.*, **41**: 924-27.

Eng, P.S., Loh, C.S., Lee, Y.H. and Hew, C.S. 1984. Origin of some polyploid Aranda hybrids derived from Vanda Dawn Nishimura. *Malaya Orchid Rev.*, **18**: 98-01.

Gantait, S., Bustam, S. and Sinniah, U.R. 2012. Alginate-encapsulation, short-term storage and plant regeneration from protocorm-like bodies of *Aranda* Wan Chark Kuan 'Blue' × *Vanda coerulea* Grifft. ex. Lindl. (Orchidaceae). *Plant Growth Regulation*, **68**(2): 303-11.

Goh, C.J. 1970. Tissue culture of Vanda Miss Joaquim. *J. Singapore National Aca. Sci.*, **2**: 31-33.

Johnson, H. 1990. What is a genus? An introduction to orchid classification. *Malayan Orchid Rev.*, **24**: 65-68.

Khaw, C.H., Ong, H.T. and Nair, H. 1978. Hormones in the nutrition of orchid tissues in mericloning. *Malaysian Orchid Reviews*, **13**: 60-65.

Kirsch, M.O. 1963. Intergeneric hybrids in vandaceous genera. *Proc*: 4th World Orchid Conference. Singapore. pp. 98-103.

Kishor, R. and Devi, H.S. 2009. Induction of multiple shoots in a monopodial orchid hybrid (*Aerides vandarum* Reichb.f × *Vanda stangeana* Reichb.f) using thidiazuron and analysis of their genetic stability. *Plant Cell, Tissue and Organ Culture*, **97**(2): 121-29.

Kunisaki, J.T., Kim, K.K. and Sagawa, Y. 1972. Shoot tip culture of vanda. *American Orchid Society Bulletin*, **41**: 435-439.

Lang, N. and Hang, N. 2006. Using biotechnological approaches for vanda orchid improvement. *Omonrice*, **14**: 140-43.

Lee, Y.H. and Kamemoto, H. 1986. Cytology and breeding behaviour of diploid and amphidiploid Aranda Wendy Scott (Orchidaceae). *Proc*: 5th Asian Orchid Congress. Ministry of National Development, Singapore.

Lee, Y.H., 1991. Genomic constitutions and flower characteristics of selected Aranda orchid cultivars. *Euphytica,* **54**: 251-54.

Lee, Y.H., How, Y.P. and Ede, J. 1991. Breeding polyploid Aranda orchids. *Malayan Orchid Rev.*, **25**: 82-87.

Lee, Y.H., Tham, F.Y. and Gwan, W.L. 1988. Origin and significance of polyploid Aranda and related hybrids. *Proc*: 1st International Symposium of Chromosome Engineering in Plants . China.

Lim, S., Teng, P.C., Lee, Y. and Goh, C. 1999. RAPD Analysis of some species in the genus *Vanda* (Orchidaceae). *Ann. Bot.*, **83**(2): 193-96.

Lim, W.L. and Loh, C.S. 2003. Endopolyploidy in *Vanda* Miss Joaquim (Orchidaceae). *New Phytologist*, **159**(1): 279–87.

Malabadi, R.B., Mulgund, G.S. and Nataraja, K. 2004. Efficient regeneration of *Vanda coerulea,* an enlarged using thidiazuron. *Plant Cell, Tissue and Organ Culture,* **76**: 289-293.

Manners, V., Kumaria, S. and Tandon, P. 2013. SPAR methods revealed high genetic diversity within populations and high gene flow of *Vanda coerulea* Griff ex Lindl (Blue Vanda), an endangered orchid species. *Gene*, **519**(1): 91-97.

Mitra, G.C., Prasad, R.N. and Roychowdhury, A. 1976. Inorganic salts and differentiation of protocorms in seed callus of an orchid and correlated changes in its free amino acid content. *Indian J. Exp. Bio.*, **14**: 350-51.

Phuekvilai, P., Pongtongkam, P. and Peyachoknagul, S. 2009. Development of microsatellite markers for vanda orchid. *Kasetsart J.*, **43**(3): 497-06.

Rahman, M.S., Hasan, M.F., Das, R., Hossain, M.S. and Rahman, M. 2009. *In vitro* micropropagation of orchid (*Vanda tessellata* L.) from shoot tip explant. *J. Bio-Sci.*, **17**: 139-44.

Ratanasut, K., Wongkhamprai, B. and Maknoi, S. 2011 Expression of a *CYP76AB1* correlates with the sequential white-blue-white colour transition of *Vanda coerulea* petals. *Biologia Plantarum,* **55**(2): 353-356.

Rosario, T.L., Francisco, R.C. and Siar, S.V. 1991. Morpho anatomical and cytological studies of some somaclonal variants of orchids. *Philipines J. Crop Sci.*, **16**: 52.

Sanguthai, O., Sanguthai, S. and Kamemoto, H. 1973. Chromosome doubling of a Dendrobium hybrid with colchicines in meristem culture. *Hawai Orchid J.*, **2**: 12-16.

Sanguthai, S. and Sagawa, Y. 1973. Induction of polyploidy in *Vanda* by colchicines treatment. *Hawaii Orchid J.*, **2**(2): 17-19.

Seeni, S. 1988. Micropropagation of blue vanda using leaf bases. *In*: S.P. Vij and S.P. Khullar (eds.) Current Research Trend in Indian Orchids, Orchid Society of India, Departmewnt of Botany, Punjab University, Chandigarh, India. pp-22.

Sharma, V. and Vij, S.P. 1997. Effect of $CuSO_4.5H_2O$ on *in vitro* regenerative capacity of foliar explants excised from mature *Vanda cristata* Lindl. plants. *Phytomorphology,* **47**: 203-208.

Shrestha, B.P.,Chin, D.P., Tokuhara, K. and Mii, M. 2010. *Agrobacterium*-mediated transformation of vanda using protocorm-like bodies. *AsPac J. Mol. Biol. Biotechnol.,* **18**(1): 225-28.

Sinha, P. and Roy, S.K. 2004. Regeneration of an indigenous orchid, *Vanda teres* (Roxb.) Lindl. through *in vitro* culture. *Plant Tissue Culture,* **14**: 55-61.

Suskandari, K., Rainawati, S. and Soertini, S. 1999. Induced mutation of gamma ray on Vanda Genta Bandung Orchid. *AGRIS*, **10**(1): 57-61.

Tanaka, R. and Kamemoto, H. 1961. Meiotic chromosome behaviour in some intergeneric hybrids of the vanda alliance. *Amer. J. Bot.*, **48**(7): 573-82.

Teo, C.K.H., Kunisaki, J.T. and Sagawa, Y. 1973. Clonal propagation of strap-leafed vanda by shoot-tip culture. *American Orchid Society Bulletin,* **42**: 402-405.

Valmayor, H.L. and Sagawa, Y. 1967. Ovule culture in some orchids. *American Orchid Society Bulletin*, **36**: 766-69.

Chapter – 35

Milestones in Transgenic Science

Year	Milestones
1951	: First amino acid was sequenced by F. Sanger and H. Tuppy.
1953	: J.D. Watson and F.H.C. Crick proposed that DNA is a double helix.
1958	: Discovery of DNA polymerase by Kornberg's group and DNA replication studies began.
1959	: Isolation of RNA polymerase.
1970	: Discovery of reverse transcriptase showed that RNA can act as the template for the transcription of DNA and opened the way for cloning using cDNA libraries.
1970	: General method was developed by M. Mandel and A. Higa for introducing DNA into cells by calcium dependent method.
1970	: G.H. Khorana synthesized an artificial gene from DNA nucleotides.
1971	: Restriction endonucleases for mapping DNA was first used by K. Danna and D. Nathans.
1972	: The recombinant DNA Era begins. J. Mertz and R.W. Davis used T_4 ligase to join DNA molecules.
1973	: A DNA fragment containing the Kannamyin-resistance gene was cut out of plasmid using EcoR 1 and ligated into the unique EcoR1 site of the plasmid pSCI01 that was tetracycline resistant. Doubly resistant clones were isolated.

1973 : Cohen and his co-workers invented cloning.

1975 : E.M. Southern detected specific DNA sequences in gels.

1976 : cDNA cloning of eukaryotic genes - the rabbit fl-globin gene.

1977 : A.M. Maxam, W. Gilbert and Sanger and others gave the first DNA sequencing method.

1983 : Ti plasmid used as vectors for transformation of plant cells by Herrera-Estrella and co-workers.

1983 : First plant species "*Nicotiana*" was transformed using *Agrobacterium tumefaciens* system by Baston and others.

1984 : Biolistic process was invented by E.D. Wolf, N.K. Allen and J.C. Sanford.

1984 : First transformation was performed in "Tobacco" using direct gene transfer technique by Paszkowski and others.

1985 : The polymerase chain reaction (PCR) technique was first used by R.K. Saiki, G. Mullies, F. Faloona and their co-workers to artificially multiply DNA sequences.

1987 : A novel orange colour was first introduced into "Petunia" by incorporating maize *DFR* gene into the genome of Petunia by Meyer, Heidmann, Forkmann and Sadler.

1988 : First transgenic plant produced a pharmaceutical.

1989 : The first transgenic "Chrysanthemum" plants were reported by Lemieux.

1990 : Suntory Limited tied up with Australian Company Florigene Ltd. beginning of joint project to create a Blue Rose using biotechnology.Vander Mark, Pijnacker Hordijk, Varega, de Vries and Dons conducted the first successful transformation in several clones of *Rosa canina* by using *Agrobacterium rhizogenes*.Chia, Chan and Chua reported the first successful orchid transformation using particle bombardment in "Vanda orchid".

1991 : Blue gene (Flavonoid 3'.5'-hydroxylase) successfully isolated from "Petunia".Calgene Pacific (CP) isolated the key genes responsible for the colours blue and red.

1992 : Blue gene was patented.Blue transgenic "Petunia" was reported.Kuehnle and Sugi were the first to report the production of transgenic "Dendrobium orchids" using microprojectile bombardment.

1993 : First transgenic plant in "Gerbera" was produced from the cultivar Terra Regina carrying the antisense chalcone gene (*gchs*) for the alteration of flower colour.

1995 : Robichon, Renou and Jalouzet produced transgenic "Pelargonium (*Pelargonium × hortorum*)" plants *via Argrobacterium tumefaciens* mediate transformation.

1996 : Florigene developed mauve-coloured Carnation, Florigene "Moondust" and it was the world's first genetically modified flower on sale.Chen and Kuehnle first reported the production of transgenic "Anthurium" plant.

1997 : Second genetically-modified Carnation was developed as Florigene "Moonshadow" with a richer and true purple colour.

1998 : Yellow gene (AS Gene) successfully isolated from Snapdragon (Antirrhinum).

2000 : Hoshino, Sato, Murata and Shinoda reported the genetic transformation of "Alstroemeria" *via* co-cultivation of embryonic suspension cells with *Argrobacterium tumefaciens.*

2002 : Yellow gene (4'CGT Gene) successfully isolated from "Snapdragon".

2003 : Vainstein and Zuker invented a method for transforming a "Gypsophila" plant with a nucleic acid of interest.

2004 : World's first "Blue Rose" developed.

2005 : Suntory Ltd. and Florigen Ltd. announced the production of a "Blue Rose" by introducing three transformation constructs simultaneously into roses.

2009 : The transgenic Rose variety "Applause" was commercially released in Japan.

2010 : Teruhiko Terakawa, Plant-Bio Team, Bioscience Research Division, Central Research Laboratories, Hokko Chemical Industry Co., Ltd. and Masaru Takagi and his coworkers, Research Institute of Genome-based Biofactory of the National Institute of Advanced Industrial Science and Technology have succeeded in developing "Polypetalous" variety of "Cyclamen" using a novel gene silencing technology (CRES-T method).

Section VIII

Colour Plates

Alstroemeria

Hybrids

Dusty Rose
(Chilean Hybrid)

Freedom
(*A. caryophyllae* × Chilean hybrid)

Mutant varieties

Harmony Stabroza

Yellow Tiger Stavero

Anthurium

Selection

Eternity

Jupiter

Eidibel

Hybrids

Kalapana
(Paradise Pink × Diamond Jubilee)

Tropic Ice
(Anuenue × UH515)

Tropic Mist
(Sea Pearl × Manoa Mist)

Interspecific hybrid

A. × Bubble Gum

A. × Green Queen

A. × Gemini

Gerbera

Hybrids

Arka Krishika

Yellow Queen

Purple Queen
(Selma × Marathon)

Pink Melody

Gladiolus

Selection

Pusa Kiran

Jyotasana

Urvashi

Bindia

New varieties

Pusa Srijana
(Berlew × Heady Wine)

Pusa Unnati
(Berlew × Heady Wine)

PusaVidushi
(Melody × Berlew)

Pusa Manmohak
(Mayur × Hunting)

Hybrids

Pusa Shubham (Lucky Shamrock × Green Lilac)

Gulal (Snow Princess × Viola)

Punjab Dawn (Suchitra × Melody)

Pusa Archana (Creamy Green × American Beauty)

Sagar (Melody × Wild Rose)

Shubhangini (Parent- Fidelio)

Heliconia

Selection

Pink Bush

Pearl

Cracker Jack

Hybrids

Golden Torch
(*H. psittacorum* × *H. spathocircinata*)

Yellow Parrot
(*H. psittacorum* × *H. spathocircinata*)

Cinnamon Twist
(*H. bihai* × *H. spathocircinata*)

Lilium

Hybrids

Red Velvet

Dunyazade

Stargazer

Polyploidy

Tiger babies (3x)

John Hancock (4x)

Captain Kidd (4x)

Resistant to disease

Royal Sunset

Mutant variety

Flying Circus

Lisianthus

Hybrids

Florida Blue
(GCREC-46 × GCREC-34)

Heat Tolerant

Florida Pink

Tulip

Resistant to fusarium wilt

Lucky Striker

Black Parrot

Aristocrat

Tetraploid

Judith Leyster

Darwin Hybrids

Pink Impression

Marit

Zantedeschia

Selection

Flamingo

Goldilocks

Captain Ventura

Child's Perfection

Hybrid

Hot Chocolate
(Cameo × Dark Purple)

Apricot Glow
(Z. elliottiana × Z. Rehmanii)

Mutant variety

Snow White

China Aster

Phule Ganesh White

Phule Ganesh Pink

Hybrids

Phule Ganesh Purple

Poornima (AST-29 × AST-3)

Kamini (AST-6 × AST-36)

Violet Cushion (Local Pink × AST-2)

Resistant to Root Knot Nematodes

Shashank (Local Pink × AST-2)

Chrysanthemum

Selection

Pusa Chitraksha

Pusa Aditya

Pusa Sona

Mini Queen

Hybrids

Khoshoo (Lal Pari × Vijay)

Kaul (Lal Pari × Vijay)

Mutant varieties

Pusa Kesari

Pusa Anmol

Pusa Centenary

Pushpangadan

Pusa Arunodaya

Vijay Kiran

Rose

Selection

Pusa Virangana

Hybridization

Pusa Arun
(Queen Elizabeth × Jantar Mantar)

Pusa Ajay
(Pink Parfait × Queen Elizabeth)

Raktagandha
Christian Dior × Carrousel)

Pusa Komal
(Pink Parfait × Suchitra)

Rose

Mutant Varieties

Pusa Mansij

Pusa Abhishek

Chitra

Resistant

Arka Parimala
(Moderately resistant to thrips and black spot)

Pusa Gaurav
(Tolerant to black spot and die back)

Tuberose

Hybrids

Suvasini
(Mexican Single ×
Pearl Double)

Shringar
(Mexican Single ×
Pearl Double)

Vaibhav
(Mexican Single × IIHR-2)

Fressia

Hybrids

Varablue
(3208-AT-1 ×
841-J)

Varared
(7130 AS2 ×
6135 DS2)

Varatwisno
(884135-AT3 × 72242H5)

Varayel
(87316-AT1 × 87310AT4)

Gypsophilla

Hybrids

Dangy White

Dangy Pink

Stock

Centum Lavender

Centum Rose

Crossandra

Selection

Fortuna

Polyploidy

Lutea Yellow (4x)

Hybrids

Arka Ambara
(Local × Laxmi Crossandra)

Arka Kanaka
(Local Yellow × Delhi Crossandra)

Jasmine

Hybrids

CO-1 Pitchi (*J. grandiflorum*)

CO-1 Mullai (*J. auriculatum*)

CO-2 Mullai (*J. auriculatum*

Marigold

Selection

Pusa Arpita

Hisar Beauty

Hybrids

Pusa Basanti Gainda
(Golden Yellow × Sun Giant)

Pusa Narangi Gainda
(Cracker Jack × Golden Jubilee)

Cattleya

Selection

Brassocattleya Cordelia

Hybrids

Cattleya × brasiliensi
(*C. bicolour* × *C. harrisoniana*)

Cattleya × brasiliensi
(*C. forbesii* × *C. granulose*)

Cymbidium

Hybrid

Cymbidium × *baoshanense*
(*C. lowianum* × *C. tigrinum*)

Dendrobium

Hybrids

Dendrobium Neo-Hawaii
(*D. phalaenopsis* × *D. grantii*)

D. Jaquelyn Thomas
(*D. phalaenopsis* × *D. gouldii*)

Paphiopedilum

Hybrids

Paphiopedilum Lecanum (*P. insigne* × *P. spicerianum*)

Paphiopedilum × affine (*P. appletonianum* × *P. villosum*)

Paphiopedilum × *pradhanii* (*P. fairrieanum* × *P. venustum*)

Vanda

Vanda × charlesworthii (*V. bensonii* × *V. coerulea*)

Vanda Mimi Palmer (*Vanda* Tan Chay Yan × *Vanda tessellate*)

Vandoritis (*Doritis* × *Vanda*)